Albrecht Unsöld
Bodo Baschek

The New Cosmos

Springer
Berlin
Heidelberg
New York
Barcelona
Hong Kong
London
Milan
Paris
Tokyo

Physics and Astronomy ONLINE LIBRARY

http://www.springer.de/phys/

Albrecht Unsöld
Bodo Baschek

The New Cosmos

An Introduction to Astronomy and Astrophysics

Translated by
William D. Brewer

**Fifth Edition
With 278 Figures
Including 20 Color Figures**

Springer

Professor Dr. *Albrecht Unsöld* †

Professor Dr. *Bodo Baschek*
Institut für Theoretische Astrophysik
Universität Heidelberg
Tiergartenstrasse 15
69121 Heidelberg, Germany

Translator:
Professor *William D. Brewer*, Ph. D.
Fachbereich Physik, Freie Universität Berlin, Arnimallee 14,
14195 Berlin, Germany

ISBN 3-540-67877-8   5th Edition
Springer-Verlag Berlin Heidelberg New York

ISBN 3-540-52593-9   4th Edition
Springer-Verlag Berlin Heidelberg New York

Library of Congress Cataloging-in-Publication Data.
Unsöld, Albrecht, 1905– [Neue Kosmos. English] The new cosmos : an introduction to astronomy and astrophysics. – 5th ed. / Albrecht Unsöld, Bodo Baschek ; translated by William D. Brewer. p. cm. Includes bibliographical references and index. ISBN 3540678778 (acid-free paper) 1. Astronomy. I. Baschek, B., 1935– II. Title QB43.3 .U5713 2001   520–dc21   2001034462

Cover: Optical image of the radio galaxy NGC 5128 = Centaurus A. Recorded by the Anglo-Australian Telescope. (© Anglo-Australian Observatory, photograph by David Malin)

Frontispiece: The galaxy group (NGC 6769-71) in Pavo with indications of gravitational interaction between the galaxies: faint haloes and connecting arcs, deforming of spiral arms and central regions. (By S. Laustsen with the 3.6 m-telescope of the European Southern Observatory)

Title of the German edition:
A. Unsöld, B. Baschek: *Der neue Kosmos*. Siebte Auflage
© Springer-Verlag Berlin Heidelberg
1967, 1974, 1981, 1988, 1991, 1999, 2002
ISBN 3-540-42177-7
Springer-Verlag Berlin Heidelberg New York

This work is subject to copyright. All rights are reserved, whether the whole or part of the material is concerned, specifically the rights of translation, reprinting, reuse of illustrations, recitation, broadcasting, reproduction on microfilm or in any other way, and storage in data banks. Duplication of this publication or parts of thereof is permitted only under the provisions of the German Copyright Law of September 9, 1965, in its current version, and permission for use must always be obtained from Springer-Verlag. Violations are liable for prosecution under the German Copyright Law.

Springer-Verlag Berlin Heidelberg
a member of BertelsmannSpringer Science+Business Media GmbH

http://www.springer.de

© Springer-Verlag New York Inc. 1969, 1977, 1983
© Springer-Verlag Berlin Heidelberg 1991, 2001
Printed in Germany

The use of general descriptive names, registered names, trademarks, etc. in this publication does not imply, even in the absence of a specific statement, that such names are exempt from the relevant protective laws and regulations and therefore free for general use.

Production Editor: Claus-Dieter Bachem, Heidelberg
Typesetting: LE-TeX, Jelonek, Schmidt & Vöckler GbR, Leipzig
Layout: Schreiber VIS, Seeheim
Cover Design: Erich Kirchner, Heidelberg
Printing and binding: Appl, Wemding

Printed on acid-free paper
SPIN: 10710924   55/3141/ba   5 4 3 2 1 0

Preface

Albrecht Unsöld's *Der Neue Kosmos*, in which he gave an overview of the whole field of astronomy that was intended to be accessible to all students and practitioners of the natural sciences, first appeared in 1967. The title was deliberately chosen with reference to Alexander von Humboldt's *Kosmos* and expressed the author's intention of "making our new understanding of the Universe clear" to a wide group of readers and of "allowing the basic ideas of the various areas of astronomical research with their factual and historical-humanistic connections to come into the foreground". After the third edition, new editions of the book were prepared with the collaboration of B. Baschek, who gradually took over responsibility for it. Prof. Dr. Dr. h.c. mult. Albrecht Unsöld died in 1995 at the age of 90; he was a pioneer and an Old Master of astrophysics. Although he announced his retirement from active research "officially" in 1988 in the Foreword to the 4th German edition of this book, he remained keenly interested in the further development of this introduction to astronomy and astrophysics and continued to contribute to its revisions through discussions and observations.

Now, ten years after the appearance of the 4th edition, a completely revised and updated version of the *New Cosmos*, translated from the 7th German edition, has been completed; it takes into account the wealth of new results from astronomical research which have appeared in recent years, ranging from our Solar System to the most distant galaxies. In this new edition, the organization of the material has been made clearer by introducing some changes in the order of presentation as well as by a finer subdivision of the topics covered.

Within our Solar System, space probes have investigated the Moon and Mars, the Jupiter system, some of the asteroids, and the solar wind from close up. New satellites have brought an enormous increase in observation power ranging from the gamma-ray region down to the radiofrequency spectral range; especially noteworthy is the Hubble Space Telescope with its incomparable angular resolution in the optical and near-ultraviolet, as well as the two large X-ray satellites, Chandra and XMM Newton. Furthermore, in the past decade, a new generation of large earthbound telescopes based on active and adaptive optics has come into use.

Among the discoveries and events of this period we mention the following: the numerous planetoids found outside the orbit of Neptune, the impact of a comet onto Jupiter, and the appearance of the bright comets Hyakutake and Hale-Bopp; advances in solar seismology, the resolution of very detailed structures in the regions of star formation and in planetary nebulae, the evidence for black holes in the centers of our Milky Way and of other galaxies; the observation of very distant supernova explosions, and the precise determination of the fluctuations in the 3 K cosmic background radiation, giving indications of a flat Universe with a nonzero cosmological constant; the view of most distant galaxies in the Hubble Deep Field, obtained with the Hubble Space Telescope; the new evidence for neutrino oscillations, the explanation of the origin of the gamma-ray bursts which had remained a riddle for decades; and finally the discoveries of numerous planets orbiting nearby stars.

My thanks go to all those readers of the earlier editions who have contributed to the improvement of this book through their suggestions, criticisms and detection of errors. Furthermore, I wish to thank my colleagues W.J. Duschl, D. Fiebig, B. Fuchs, H. Holweger, G. Klare, M. Scholz, A. Schwope, P. Ulmschneider, C. van de Bruck, R. Wehrse, and G. Weigelt for critical readings of various sections and for their comments and suggestions on numerous topics. In particular, I owe sincere thanks to Prof. Wolfgang J. Duschl for his aid in the choice of new illustrations and for their acquisition and electronic image processing.

It was a great asset for the preparation of this book that Prof. William D. Brewer was once again willing to undertake the translation. I wish to express my heartfelt thanks for his excellent job as well as for very agreeable and constructive collaboration. Prof. W. Beiglböck, Dr. H. Lotsch, and Mr. C.-D. Bachem of Springer-Verlag are due my sincere gratitude for their excellent cooperation, as always, in the completion of this edition.

Heidelberg, June 2001   *Bodo Baschek*

Contents

1. Introduction 1

I. Classical Astronomy and the Solar System 5

Humanity and the Stars: Observing and Thinking
Historical Introduction to Classical Astronomy 6

2. **Classical Astronomy**
   - 2.1 **Spatial Coordinates and Time; the Motions of the Sun, the Earth, and the Moon** 10
     - 2.1.1 The Celestial Sphere and Astronomical Coordinate Systems 10
     - 2.1.2 The Motions of the Earth. Seasons and the Zodiac 12
     - 2.1.3 Time: Days, Years, and the Calendar 15
     - 2.1.4 The Moon 17
     - 2.1.5 Eclipses of the Sun and the Moon 19
   - 2.2 **Orbital Motions and Distances in the Solar System** 20
     - 2.2.1 Planetary Motions and Orbital Elements 20
     - 2.2.2 Comets and Meteors 23
     - 2.2.3 Distance Determination, the Doppler Effect and Aberration of Light 25
   - 2.3 **Mechanics and Gravitational Theory** 27
     - 2.3.1 Newton's Laws of Motion 28
     - 2.3.2 The Conservation of Linear Momentum 29
     - 2.3.3 Conservation of Angular Momentum: the Area Theorem 29
     - 2.3.4 Conservation of Energy 30
     - 2.3.5 The Virial Theorem 31
     - 2.3.6 The Law of Gravitation. Gravitational Energy 31
   - 2.4 **Celestial Mechanics** 33
     - 2.4.1 Kepler's First and Second Laws: Planetary Orbits 33
     - 2.4.2 Kepler's Third Law: Determination of Masses 34
     - 2.4.3 Conservation of Energy and the Escape Velocity 35
     - 2.4.4 Rotation and the Moment of Inertia 35
     - 2.4.5 Precession 36
     - 2.4.6 The Tides 36
     - 2.4.7 The Ptolemaic and the Copernican Worldviews 37
   - 2.5 **Space Research** 38
     - 2.5.1 The Orbits of Artificial Satellites and Space Vehicles 39
     - 2.5.2 Astronomical Observations from Space 40
     - 2.5.3 The Exploration of the Moon 42
     - 2.5.4 Space Probe Missions in the Solar System 43

3. The Physical Structure of the Objects in the Solar System

- **3.1 Global Properties of the Planets and Their Satellites** 46
  - 3.1.1 Ways of Studying the Planets and Their Satellites 46
  - 3.1.2 The Global Energy Balance of the Planets 48
  - 3.1.3 Interior Structure and Stability 48
  - 3.1.4 The Structures of Planetary Atmospheres 50
- **3.2 The Earth, the Moon, and the Earthlike Planets** 51
  - 3.2.1 The Internal Structures of the Earthlike Planets 52
  - 3.2.2 Radioactive Dating. The Earth's History 53
  - 3.2.3 Magnetic Fields. Plate Tectonics 55
  - 3.2.4 The Lunar Surface 57
  - 3.2.5 The Surfaces of the Earthlike Planets 60
  - 3.2.6 The Atmospheres of the Earthlike Planets 66
- **3.3 Asteroids or Small Planets (Planetoids)** 69
  - 3.3.1 The Orbits of the Asteroids 70
  - 3.3.2 Properties of the Asteroids 70
- **3.4 The Major Planets** 72
  - 3.4.1 Jupiter 72
  - 3.4.2 Saturn 77
  - 3.4.3 Uranus 80
  - 3.4.4 Neptune 83
- **3.5 Pluto and the Transneptunian Small Planets** 84
  - 3.5.1 Pluto and Charon 85
  - 3.5.2 The Transneptunian Small Planets 86
- **3.6 Comets** 87
  - 3.6.1 Structure, Spectra, and Chemical Composition 87
  - 3.6.2 The Evolution of Comets 89
- **3.7 Meteors and Meteorites** 89
  - 3.7.1 Meteorites and Impact Craters 89
  - 3.7.2 Meteors in the Earth's Atmosphere 90
  - 3.7.3 Properties and Origins of Meteorites 91
- **3.8 Interplanetary Matter** 94

II. Radiation, Instruments, and Observational Techniques 97

The Development of Astronomical Observation Methods
Historical Introduction to Our Knowledge of the Electromagnetic Spectrum 98

4. Radiation and Matter

- **4.1 Electromagnetic Radiation** 101
- **4.2 The Theory of Special Relativity** 102
  - 4.2.1 The Lorentz Transformation. The Doppler Effect 103
  - 4.2.2 Relativistic Mechanics 103
- **4.3 The Theory of Radiation** 104
  - 4.3.1 Phenomenological Description of Radiation 104
  - 4.3.2 Emission and Absorption. The Radiation Transport Equation . 107
  - 4.3.3 Thermodynamic Equilibrium and Black-Body Radiation 109

		4.4	**Matter in Thermodynamic Equilibrium** 111
		4.4.1	Boltzmann Statistics 111
		4.4.2	The Velocity Distribution 112
		4.4.3	Thermal Excitation 112
		4.4.4	Thermal Ionization 112
		4.4.5	The Law of Mass Action 113
		4.5	**The Interaction of Radiation with Matter** 114
		4.5.1	Mean Free Paths 114
		4.5.2	Interaction Cross Section and Reaction Rate 114
		4.5.3	Collision and Radiation Processes: Kinetic Equations 115
		4.5.4	Elementary Atomic Processes 116
		4.5.5	Extinction and Emission Coefficients 119
		4.5.6	Energetic Photons and Particles 119
5.	**Astronomical and Astrophysical Instruments**	5.1	**Telescopes and Detectors for the Optical and the Ultraviolet Regions** 123
		5.1.1	Conventional Telescopes 123
		5.1.2	Resolving Power and Light Gathering Power. Optical Interferometers 128
		5.1.3	Adaptive and Active Optics. Large Telescopes 131
		5.1.4	Optical Detectors 135
		5.1.5	Spectrographs 138
		5.1.6	Space Telescopes 141
		5.2	**Telescopes and Detectors for Radiofrequencies and the Infrared** 143
		5.2.1	Radiotelescopes 143
		5.2.2	Receivers and Spectrometers for the Radiofrequency Region 149
		5.2.3	Observation Methods in the Infrared 150
		5.3	**Instruments for High-Energy Astronomy** 152
		5.3.1	Particle Detectors 152
		5.3.2	Telescopes for Cosmic Radiation 154
		5.3.3	Gamma-ray Telescopes 155
		5.4	**Instruments for X-rays and the Extreme Ultraviolet** 157
		5.4.1	Detectors and Spectrometers for the X-ray Region 158
		5.4.2	Telescopes and Satellites for the X-ray Region 158
		5.4.3	Telescopes for the Extreme Ultraviolet 160

III. The Sun and Stars: The Astrophysics of Individual Stars 161

Astronomy + Physics = Astrophysics
Historical Introduction 162

6.	**The Distances and Fundamental Properties of the Stars**	6.1	**The Sun** 167
		6.1.1	The Spectrum of the Photosphere. Center–Limb Variation ... 168
		6.1.2	The Energy Distribution 169
		6.1.3	Luminosity and Effective Temperature 169

	6.2	**Distances and Velocities of the Stars**	171
	6.2.1	Trigonometric Parallaxes	171
	6.2.2	Radial Velocities and Proper Motions	172
	6.2.3	Stream Parallaxes	172
	6.2.4	Star Positions and Catalogs	173
	6.3	**Magnitudes, Colors and Radii of the Stars**	174
	6.3.1	Apparent Magnitudes	174
	6.3.2	Color Index and Energy Distribution	175
	6.3.3	Absolute Magnitudes	178
	6.3.4	Bolometric Magnitudes and Luminosities	178
	6.3.5	Stellar Radii	179
	6.4	**Classification of the Stellar Spectra. The Hertzsprung–Russell Diagram**	180
	6.4.1	The Spectral Type	180
	6.4.2	The Hertzsprung–Russell Diagram. Luminosity Classes	181
	6.4.3	The MK Classification	183
	6.4.4	Two-Color Diagrams	185
	6.4.5	Rotation of the Stars	185
	6.5	**Binary Star Systems and the Masses of Stars**	186
	6.5.1	Visual Binary Stars	186
	6.5.2	Spectroscopic Binary Stars and Eclipsing Variables	187
	6.5.3	Periods and Rotation in Binary Systems	188
	6.5.4	The Stellar Masses	188
	6.5.5	Close Binary Star Systems	189
	6.5.6	Pulsars in Binary Star Systems	190
	6.5.7	Companions of Substellar Mass: Brown Dwarfs and Exoplanets	192
7. The Spectra and Atmospheres of Stars	7.1	**Spectra and Atoms**	195
	7.1.1	Basic Concepts of Atomic Spectroscopy	195
	7.1.2	Excitation and Ionization	198
	7.1.3	Line Absorption Coefficients	201
	7.1.4	Broadening of Spectral Lines	202
	7.1.5	Remarks on Molecular Spectroscopy	204
	7.2	**The Physics of Stellar Atmospheres**	205
	7.2.1	The Structure of Stellar Atmospheres	205
	7.2.2	Absorption Coefficients in Stellar Atmospheres	207
	7.2.3	Model Atmospheres. The Spectral Energy Distribution	209
	7.2.4	Radiation Transport in the Fraunhofer Lines	211
	7.2.5	The Curve of Growth	213
	7.2.6	Quantitative Analysis of Stellar Spectra	215
	7.2.7	Element Abundances in the Sun and in Other Stars	216
	7.3	**The Sun: Its Chromosphere and Corona. Flow Fields, Magnetic Fields and Activity**	219
	7.3.1	Granulation and Convection	219
	7.3.2	Magnetic Fields and Magnetohydrodynamics	220

	7.3.3	Sunspots and the Activity Cycle. Magnetic Flux Tubes	222
	7.3.4	The Chromosphere and the Corona	225
	7.3.5	Prominences	231
	7.3.6	Solar Eruptions or Flares	233
	7.3.7	The Solar Wind	236
	7.3.8	Oscillations: Helioseismology	240
	7.4	**Variable Stars.**	
		Flow Fields, Magnetic Fields and Activity of the Stars	243
	7.4.1	Pulsating Stars. R Coronae Borealis Stars	244
	7.4.2	Magnetic or Spectrum Variables.	
		Ap Stars and Metallic-Line Stars	247
	7.4.3	Activity, Chromospheres, and Coronas of Cool Stars	248
	7.4.4	Coronas, Stellar Winds and Variability of Hot Stars	251
	7.4.5	Cataclysmic Variables: Novae and Dwarf Novae	253
	7.4.6	X-ray Binary Systems: Accretion onto Neutron Stars	255
	7.4.7	Supernovae and Pulsars	259
	7.4.8	Stellar Gamma-ray Sources	266
	7.4.9	Gamma Bursters	267
8. The Structure	**8.1**	**The Fundamental Equations of Stellar Structure**	271
and Evolution of Stars	8.1.1	Hydrostatic Equilibrium and the Equation of State of Matter .	271
	8.1.2	Temperature Distribution and Energy Transport	272
	8.1.3	Energy Production Through Nuclear Reactions	273
	8.1.4	Gravitational Energy and Thermal Energy	277
	8.1.5	Stability of the Stars	279
	8.1.6	The System of Fundamental Equations	
		and Their General Results	279
	8.2	**Stellar Evolution**	281
	8.2.1	Main Sequence Stars: Central Hydrogen Burning	281
	8.2.2	The Internal Structure of the Sun. Solar Neutrinos	282
	8.2.3	From Hydrogen to Helium Burning	285
	8.2.4	Late Phases of Stellar Evolution	287
	8.2.5	Nucleosynthesis in Stars	291
	8.2.6	Close Binary Star Systems	294
	8.2.7	The Physics of Accretion Disks	297
	8.3	**The Final Stages of Stellar Evolution**	298
	8.3.1	Brown Dwarfs	299
	8.3.2	The Equation of State of Matter with Degenerate Electrons ..	299
	8.3.3	The Structure of the White Dwarfs	301
	8.3.4	Neutron Stars	301
	8.4	**Strong Gravitational Fields**	303
	8.4.1	The Theory of General Relativity	303
	8.4.2	Spherically Symmetric Fields in Vacuum	304
	8.4.3	Light Deflection and Gravitational Lenses	305
	8.4.4	Black Holes	308
	8.4.5	Gravity Waves	308

IV. Stellar Systems. Cosmology and Cosmogony 311
The Advance into the Universe
Historical Introduction to the Astronomy of the 20th Century 312

9. Star Clusters

- **9.1 Open Star Clusters and Stellar Associations** 318
  - 9.1.1 Open (Galactic) Star Clusters 318
  - 9.1.2 Stellar Associations 319
  - 9.1.3 Color–Magnitude Diagrams and the Age of Open Clusters ... 319
- **9.2 Globular Clusters** 322
  - 9.2.1 Globular Clusters in the Milky Way 322
  - 9.2.2 Metal Abundances and Two-Color Diagrams 323
  - 9.2.3 Color–Magnitude Diagrams and the Ages of the Globular Clusters 324
  - 9.2.4 Globular Clusters in Other Galaxies 326

10. Interstellar Matter and Star Formation

- **10.1 Interstellar Dust** 330
  - 10.1.1 Dark Clouds 330
  - 10.1.2 Interstellar Extinction and Reddening 330
  - 10.1.3 Polarization of Starlight by Interstellar Dust 332
  - 10.1.4 Properties of the Dust Grains 332
  - 10.1.5 Diffuse Interstellar Absorption Bands 334
- **10.2 Neutral Interstellar Gas** 335
  - 10.2.1 Atomic Interstellar Absorption Lines 335
  - 10.2.2 The 21 cm Line of Neutral Hydrogen. H I Clouds 337
  - 10.2.3 Interstellar Molecular Lines. Molecular Clouds 339
- **10.3 Ionized Gas: Luminous Gaseous Nebulae** 341
  - 10.3.1 H II Regions 342
  - 10.3.2 Planetary Nebulae 346
  - 10.3.3 Supernova Remnants 347
  - 10.3.4 Hot Interstellar Gas 354
- **10.4 High-Energy Components** 354
  - 10.4.1 Interstellar Magnetic Fields 355
  - 10.4.2 Cosmic Radiation 355
  - 10.4.3 Galactic Gamma Radiation 360
- **10.5 Early Evolution and Formation of the Stars** 363
  - 10.5.1 Pre-Main-Sequence Stars 364
  - 10.5.2 Regions of Star Formation 366
  - 10.5.3 Gravitational Instability and Fragmentation 368
  - 10.5.4 The Evolution of Protostars 369
  - 10.5.5 Matter Flows in the Vicinity of Protostars 371
  - 10.5.6 Stellar Statistics and the Star Formation Rate 373

11. The Structure and Dynamics of the Milky Way Galaxy

- **11.1 Stars and the Structure of the Milky Way** 377
  - 11.1.1 Galactic Coordinates 377
  - 11.1.2 Star Gauging 377

	11.1.3 Spatial Velocities of the Stars	378
	11.1.4 Star Clusters: Distance Determinations and the Structure of the Milky Way	380
11.2	**The Dynamics and Distribution of Matter**	381
	11.2.1 The Rotation of the Galactic Disk	382
	11.2.2 The Distribution of the Interstellar Matter	383
	11.2.3 The Galactic Orbits of the Stars. Local Mass Density	387
	11.2.4 The Mass Distribution in the Milky Way Galaxy	389
	11.2.5 The Dynamics of the Spiral Arms	390
	11.2.6 Stellar Populations and Element Abundances	392
11.3	**The Central Region of the Milky Way**	395
	11.3.1 The Galactic Bulge ($R \leq 3$ kpc)	395
	11.3.2 The Nuclear Region of the Galactic Bulge ($R \leq 300$ pc)	396
	11.3.3 The Circumnuclear Disk and the Minispiral ($R \leq 10$ pc)	398
	11.3.4 The Innermost Region ($R \leq 1$ pc) and Sgr A*	399

12. Galaxies and Clusters of Galaxies

12.1	**Normal Galaxies**	402
	12.1.1 Distance Determination	402
	12.1.2 Classification and Absolute Magnitudes	405
	12.1.3 The Luminosity Function	409
	12.1.4 Brightness Profiles and Diameters	410
	12.1.5 Dynamics and Masses	411
	12.1.6 Stellar Populations and Element Abundances	415
	12.1.7 The Distribution of Gas and Dust	418
12.2	**Infrared and Starburst Galaxies**	422
	12.2.1 Infrared Galaxies	423
	12.2.2 Starburst Activity	424
12.3	**Radio Galaxies, Quasars, and Activity in Galactic Nuclei**	425
	12.3.1 Synchrotron Radiation	425
	12.3.2 The Non-thermal Radiofrequency Emissions of Normal Galaxies	428
	12.3.3 Radio Galaxies	429
	12.3.4 Quasars (Quasistellar Objects)	434
	12.3.5 Seyfert Galaxies	441
	12.3.6 Activity in Galactic Nuclei	443
12.4	**Clusters and Superclusters of Galaxies**	446
	12.4.1 The Local Group	447
	12.4.2 The Classification and the Masses of the Galaxy Clusters	448
	12.4.3 The Gas in Galaxy Clusters	450
	12.4.4 Interacting Galaxies. The Evolution of Galaxy Clusters	452
	12.4.5 Superclusters of Galaxies	452
12.5	**The Formation and Evolution of the Galaxies**	455
	12.5.1 The Formation of the Galaxies and the Clusters of Galaxies	455
	12.5.2 The Intergalactic Medium and Lyman α Systems	458

		12.5.3 Interacting Galaxies	460
		12.5.4 Evolution of the Galaxies	461
		12.5.5 Galaxies in the Early Universe	466
13. Cosmology: the Cosmos as a Whole	**13.1**	**Models of the Universe**	468
		13.1.1 The Expanding Universe	468
		13.1.2 Newtonian Cosmology	469
		13.1.3 Relativistic Cosmology	471
		13.1.4 The Matter Cosmos	474
	13.2	**Radiation and Observations. Element Synthesis in the Universe**	475
		13.2.1 The Propagation of Radiation	475
		13.2.2 The Microwave Background Radiation	477
		13.2.3 The Radiation Cosmos	479
		13.2.4 Element Synthesis in the Universe	481
		13.2.5 Observed Parameters of the Present-Day Universe	482
		13.2.6 Olbers' Paradox	485
	13.3	**The Evolution of the Universe**	486
		13.3.1 The Planck Time	486
		13.3.2 Elementary Particles and Fundamental Interactions	487
		13.3.3 Cosmic Evolution According to the Standard Model	489
		13.3.4 The Inflationary Universe	491
		13.3.5 Other Cosmologies	492
14. The Cosmogony of the Solar System	**14.1**	**The Formation of the Sun and of the Solar System**	494
		14.1.1 A Survey of the Solar System	494
		14.1.2 The Protoplanetary Disk and the Formation of the Planets	496
		14.1.3 The Origin of the Meteorites	498
		14.1.4 The Formation of the Earth–Moon System	500
		14.1.5 Planetary Systems Around Other Stars	503
	14.2	**The Evolution of the Earth and of Life**	505
		14.2.1 The Development of the Atmosphere and of the Oceans	505
		14.2.2 Fundamentals of Molecular Biology	506
		14.2.3 Prebiotic Molecules	509
		14.2.4 The Development of Lifeforms	510
Appendix	**A.1**	**Units: the International and the Gaussian Unit Systems**	517
	A.2	**Names of the Constellations**	521

Selected Exercises 523
Literature and Sources of Data 531
Acknowledgements 537
Subject Index 543
Fundamental Physical Constants (Inside back cover)
Astronomical Constants and Units (Inside back cover)

1. Introduction

Astronomy, the study of the stars and other celestial objects, is one of the exact sciences. It deals with the quantitative investigation of the cosmos and the physical laws which govern it: with the motions, the structures, the formation, and the evolution of the various celestial bodies.

Astronomy is among the oldest of the sciences. The earliest human cultures made use of their knowledge of celestial phenomena and collected astronomical data in order to establish a calendar, measure time, and as an aid to navigation. This early astronomy was often closely interwoven with magical, mythological, religious, and philosophical ideas.

The study of the cosmos in the modern sense, however, dates back only to the ancient Greeks: the determination of distances on the Earth and of positions of the celestial bodies in the sky, together with knowledge of geometry, led to the first realistic estimates of the sizes and distances of the objects in outer space. The complex orbits of the Sun, the Moon, and the planets were described in a mathematical, kinematical picture, which allowed the calculation of the positions of the planets in advance. Greek astronomy attained its zenith, and experienced its swan song, in the impressive work of Ptolemy, about 150 a.D. The name of the science, *astronomy*, is quite appropriately derived from the Greek word "$\alpha\sigma\tau\eta\rho$" = star or "$\alpha\sigma\tau\rho\text{o}\nu$" = constellation or heavenly body.

At the beginning of the modern period, in the 16th and 17th centuries, the Copernican view of the universe became generally accepted. Celestial mechanics received its foundation in Newton's Theory of Gravitation in the 17th century and was completed mathematically in the period immediately following. Major progress in astronomical research was made in this period, on the one hand through the introduction of new concepts and theoretical approaches, and on the other through observations of new celestial phenomena. The latter were made possible by the development of new instruments. The invention of the telescope at the beginning of the 17th century led to a nearly unimaginable increase in the scope of astronomical knowledge. Later, new eras in astronomical research were opened up by the development of photography, of the spectrograph, the radio telescope, and of space travel, allowing observations to be made over the entire range of the electromagnetic spectrum.

In the 19th and particularly in the 20th centuries, physics assumed the decisive role in the elucidation of astronomical phenomena; astrophysics has steadily increased in importance over "classical astronomy". There is an extremely fruitful interaction between astrophysics/astronomy and physics: on the one hand, astronomy can be considered to be the physics of the cosmos, and there is hardly a discipline in physics which does not find application in modern astronomy; on the other hand, the cosmos with its often extreme states of matter offers the opportunity to study physical processes under conditions which are unattainable in the laboratory. Along with physics, and of course mathematics, applications of chemistry and the Earth and biological sciences are also of importance in astronomy.

Among the sciences, astronomy is unique in that no experiments can be carried out on the distant celestial objects; astronomers must content themselves with *observations*. "Diagnosis from a distance", and in particular the quantitative analysis of radiation from the cosmos over the widest possible spectral range, thus play a central role in astronomical research.

The rapid development of many branches of astronomy has continued up to the present time. With this revised edition of *The New Cosmos*, we have tried to keep pace with the rapid expansion of astronomical knowledge while maintaining our goal of providing a comprehensive – and comprehensible – introductory survey of the *whole* field of astronomy. We have placed emphasis on observations of the manifold objects and phenomena in the cosmos, as well as on the basic ideas which provide the foundation for the various fields within the discipline. We have combined description of the observations as directly as possible with the theoretical approaches to their elucidation. Particular results, as well as information from physics and the other natural sciences which are required for the understanding of astronomical phenomena, are, however, often simply stated without detailed explanations. The complete bibliography, together with a list of important reference works, journals, etc., is intended to help the reader to gain access to the more detailed and specialized literature.

We begin our study of the cosmos, its structure and its laws, "at home" by considering our *Solar System* in Part I, along with *classical astronomy*. This part, like the three following parts, starts with a historical summary which is intended to give the reader an overview of the subject. We first become acquainted with observations of the heavens and with the motions of the Earth, the Sun, and the Moon, and introduce celestial coordinates and sidereal time. The apparent motions of the planets and other objects are then explained in the framework of the Newtonian Theory of Gravitation. Before considering the planets and other objects in the Solar System in detail, we give a summary of the development of space research, which has contributed enormously to knowledge of our planetary system. Part I ends with a discussion of the individual planets, their moons, and other smaller bodies such as asteroids, comets and meteors.

Prior to taking up the topic of the Sun and other stars, it is appropriate to describe the basic principles of astronomical *observation methods*, and we do this in Part II. An impressive arsenal of telescopes and detectors is available to today's astronomer; with them, from the Earth or from space vehicles, he or she can investigate the radiation emitted by celestial bodies over the entire range of the electromagnetic spectrum, from the radio and microwave regions through the infrared, the visible, and the ultraviolet to the realm of highly energetic radiations, the X-rays and gamma rays. The use of computers provides an essential tool for the modern astronomer in these observations.

Part III is devoted to *stars*, which we first treat as individual objects. We give an overview of the different types of stars such as those of the main sequence, giants and supergiants, brown dwarfs, white dwarfs and neutron stars, as well as the great variety of variable stars (Cepheids, magnetic stars, novas, supernovas, pulsars, gamma sources ...) and of stellar activity, and become acquainted with their distances, magnitudes, colors, temperatures, luminosities, and masses. In this part, the *Sun* plays a particularly important role: on the one hand, as the nearest star, it offers us the possibility of making incomparably more detailed observations than of any other star; on the other, its properties are those of an "average" star, and their study thus yields important information about the physical state of stars in general.

The treatment of the *physics of individual stars* occupies an important place in Part III. Along with the theory of radiation, atomic spectroscopy in particular forms the basis for quantitative investigation of the radiation and the spectra of the Sun and other stars, and for the understanding of the physical-chemical structure of their outer layers, the stellar atmospheres. Understanding of the mechanism of energy release by thermonuclear reactions and by gravitation is of decisive importance for the study of stellar interiors, their structures and evolution.

We then discuss the development of the stars of the main sequence, which includes the phase of intensive stellar hydrogen burning, continuing to their final stages (white dwarf, neutron star or black hole). The formation of stars and their earliest development are treated in the following sections in connection with the interstellar material in our galaxy. At the end of Part III, we deal with strong gravitational fields, which we describe in the framework of Einstein's General Relativity theory; here, we concentrate in particular on black holes, gravitational lenses, and gravitational waves.

In Part IV, we take up *stellar systems* and the macroscopic structure of the universe. Making use of our knowledge of individual stars and their distances from the Earth, we first develop a picture of stellar clusters and stellar associations. We then discuss the interstellar matter which consists of tenuous gas and dust clouds, and treat star formation. Finally, we develop a picture of our own Milky Way galaxy, to which the Sun belongs together with about 100 million other stars. We treat the distribution and the motions of the stars and star clusters and of the interstellar matter. After making the acquaintance of methods for the determination of the enormous distances in intergalactic space, we turn to other galaxies, among which we find a variety of types: spiral and elliptical galaxies, infrared and starburst galaxies, radio galaxies, and the distant quasars. In the centers of many galaxies, we observe an "activity" involving the appearance of extremely large amounts of energy, whose origins are still a mystery.

Galaxies, as a rule, belong to larger systems, called galactic clusters. These are in turn ordered in clusters of galactic clusters, the superclusters, which finally form a "lattice" enclosing large areas of empty intergalactic space and defining the macroscopic structure of the Universe. Like individual stars, the galaxies and galactic clusters evolve with the passage of time. The mutual gravitational influence of the galaxies plays an important role in their development.

At the conclusion of Part IV, we consider the *Universe as a whole*, its content of matter, radiation, and energy, and its structure and evolution throughout the expansion which has taken place over the roughly $2 \cdot 10^9$ years from the "big bang" to the present time.

Finally, after pressing out to the far reaches of the cosmos, we return at the end of Part IV to our Solar System and take up the problems of the *formation and evolution of the Sun and the planets* as well as the existence of planetary systems around other stars. In this section, we give particular attention to the development of the Earth and of life on Earth.

I. Classical Astronomy and the Solar System

Humanity and the Stars: Observing and Thinking
Historical Introduction to Classical Astronomy

Unaffected by the evolution and the activities of mankind, the objects in the heavens have moved along their paths for millenia. The starry skies have thus always been a symbol of the "Other" – of Nature, of deities – the antithesis of the "Self" with its world of inner experience, striving and activity. The history of astronomy is at the same time one of the most exciting chapters in the history of human thought. Again and again, there has been an interplay between the appearance of new *concepts and ways of thinking* on the one hand and the discovery of new *phenomena* on the other, the latter often with the aid of newly-developed *observational instruments*.

We cannot treat here the great achievements of the ancient Middle Eastern peoples, the Sumerians, Babylonians, Assyrians, and the Egyptians; nor do we have the space to describe the astronomy of the the Far Eastern cultures in China, Japan, and India, which was highly developed by the standards of the time.

The concept of the *Universe* and its investigation in the modern sense dates back to the ancient Greeks, who were the first to dare to shake off the fetters of black magic and mythology and, aided by their enormously flexible language, to adopt forms of thinking which allowed them, bit by bit, to "comprehend" the phenomena of the cosmos.

How bold were the ideas of the pre-Socratic Greeks! Thales of Milet, about 600 B.C., had already clearly understood that the Earth is round, and that the Moon is illuminated by the Sun, and he predicted the Solar eclipse of the year 585 B.C. But is it not just as important that he attempted to reduce understanding of the entire universe to a *single* principle, that of "water"?

The little that we know of Pythagoras (in the middle of the 6th century B.C.) and of his school seems surprisingly modern. The spherical shapes of the Earth, the Sun, and the Moon, the Earth's rotation, and the revolution of at least the inner planets, Venus and Mercury, were already known to the Pythagorans.

After the collapse of the Greek states, *Alexandria* became the center of ancient science; there, the quantitative investigation of the heavens made rapid progress with the aid of systematic measurements. The numerical results are less important for us today than the happy realization that the great Greek astronomers made the bold leap of applying the laws of *geo*metry to the cosmos! Aristarchus of Samos, who lived in the first half of the 3rd century B.C., attempted to compare the *distances* of the Earth to the Sun and the Earth to the Moon with the *diameters* of the three bodies by making the assumption that when the Moon is in its first and third quarter, the triangle Sun-Moon-Earth makes a right angle at the Moon. In addition to carrying out these first quantitative estimates of dimensions *in* space, Aristarchus was the first to teach the *heliocentric system* and to recognize its important consequence that the distances to the fixed stars must be incomparably greater than that from the Earth to the Sun. How far he was ahead of his time with these discoveries can be seen from the fact that by the following generation, they had already been forgotten. Soon after Aristarchus' important achievements, Eratosthenes carried out the first measurement of a degree of arc on the Earth's surface, between Alexandria and Syene: he compared the difference in latitude between the two places with their distance along a much-traveled caravan route, and thereby determined the circumference and diameter of the Earth fairly precisely. However, the greatest observer of ancient times was Hipparchus (about 150 B.C.), whose *stellar catalog* was still nearly unsurpassed in accuracy in the 16th century A.D. Even though the means at his disposal naturally did not allow him to make significantly better determinations of the basic dimensions of the Solar System, he was able to make the important discovery of *precession*, i.e. the yearly shift of the equinoxes and thus the difference between the tropical and the sidereal years.

The theory of *planetary motion*, which we shall treat next, was necessarily limited in Greek astronomy to a problem in *geometry* and *kinematics*. Gradual improvements and extensions of observations on the one hand, and new mathematical approaches on the other, formed the basis for the attempts of Philolaus, Eudoxus, Heracleides, Appollonius, and others to describe the observed motions of the planets; their attempts employed the superposition of ever more complicated circular motions. Ancient astronomy and planetary theory attained its final development much later, in the work of Claudius Ptolemy, who wrote his 13-volume Handbook of Astronomy (Mathematics), $M\alpha\theta\eta\mu\alpha\tau\iota\chi\eta\varsigma\ \Sigma\upsilon\nu\tau\alpha\xi\epsilon\omega\varsigma$, in Alexandria about 150 B.C. His "Syntax" later acquired the adjective $\mu\epsilon\gamma\iota\sigma\tau\eta$, "greatest", from which the arabic title *Almagest* is derived. The Almagest is based to a large extent on the observations and research of Hipparchus, but Ptolemy also added much new material, particularly in the theory of planetary motion. At this point, we need only sketch the outlines of Ptolemy's geocentric system: the Earth rests at the midpoint of the Universe. The motions of the Sun and the Moon in the sky may be represented fairly simply by circular orbits. The planetary motions are described by Ptolemy using the *theory of epicycles*: each planet moves on a circle, the so-called epicycle, whose nonmaterial center moves around the Earth on a second circle, the deferent. We shall not delve further into the refinements of this system involving additional, in some cases eccentric circular orbits, etc. The intellectual posture of the Almagest clearly shows the influence of Aristotelian philosophy, or rather of *Aristotelianism*. Its modes of thought, originally the tools of vital research, had long since hardened into the dogmas of a rigid school; this was the principal reason for the remarkable historical durability of the Ptolemaic world-system.

We cannot go into detail here about how, following the decline of the academy in Alexandria, first the Nestorian Christians in Syria and later the Arabs in Bhagdad took over and continued the work of Ptolemy.

Translations and commentaries on the Almagest were the basic sources of the first Western textbook on astronomy, the *Tractatus de Sphaera* of Ioannes de Sacrobosco, a native of England who taught at the University of Paris until his death in the year 1256. The *Sphaera* was issued again and again and often commentated; it was still "the" text for teaching astronomy in Galileo's time, three centuries later.

The intellectual basis of the new thinking was provided in part by the conquest of Constantinople by the Turks in 1453: thereafter, numerous scientific works from antiquity were made accessible to the West by Byzantine scholars. For example, some very fragmentary texts concerning the heliocentric system of the ancients clearly made a strong impression on Copernicus. The result was a turning-away from the rigid doctrine of the Aristotelians in favor of the much more lively and flexible thinking of the schools of Pythagoras and Plato. The "Platonic" idea that the process of understanding the Universe consists of a progressive adaptation of our inner world of concepts and ways of thinking to the more and more precisely-studied outer world of phenomena has become the hallmark of modern research from Cusanus through Kepler to Niels Bohr. Finally, with the blossoming of a practical approach to life exemplified by the rise of crafts and trades, the question was no longer "What did Aristotle say?", but rather "How can you do this . . . ?".

In the 15th century, a completely new spirit in science and in life arose, at first in Italy and soon thereafter in the North as well. The sententious meditations of Cardinal Nicholas Cusanus (1401–1464) have only today begun to be properly appreciated. It is fascinating to see how his ideas about the infinity of the Universe and about quantitative scientific research arose from religious or theological considerations. Near the end of the century (1492), the discovery of America by Christopher Columbus added the classic expression "il mondo e poco" to the new spirit. A few years later, Nicolas Copernicus (1473–1543) founded the *heliocentric system*.

About 1510, Copernicus sent a letter to several noted astronomers of his time; it was rediscovered only in 1877, and was entitled "De Hypothesibus Motuum Caelestium A Se Constitutis Commentariolus". It foreshadowed the major part of the results which were later published in his major work, "De Revolutionibus Orbium Coelestium", which appeared in Nuremberg in 1543, the year of his death.

Copernicus held fast to the idea of the "perfection of circular motion" which had formed the basis for astronomical thought throughout antiquity and the Middle Ages; he never considered the possibility of another form of motion.

It was Johannes Kepler (1571–1630) who, starting from the phythagorian-platonic traditions, was able to break through to a more general point of view. Making use of the observations of Tycho Brahe (1546–1601), which were vastly more precise than any that had preceded them, he discovered his three *Laws of Planetary Motion*. Kepler derived his first two laws from an enormously tedious trigonometric calculation of the motions of Mars reported by Tycho in his "Astronomia Nova" (Prague, 1609). The third law is reported in his "Harmonices Mundi" (1619). We can only briefly mention Kepler's ground-breaking works on optics, his Keplerian telescope, his Rudolphinian Tables (1627), and numerous other achievements.

About the same time, the Italian Galileo Galilei (1564–1642) directed the *telescope* which he had built in 1609 to the heavens and discovered, in rapid succession: the "maria", the craters, and other mountain formations on the Moon; the numerous stars of the Pleides and the Hyads; the four largest moons of Jupiter and their free orbits around the planet; the first indication of the rings of Saturn; and sunspots. His "Galileis Sidereus Nuncius" (1610), in which he describes the discoveries with his telescope, the "Dialogo Delli Due Massimi Sistemi Del Mondo, Tolemaico, e Copernico" (1632), and the "Discorsi e Dimonstrazioni Matematiche Intorno a Due Nuove Scienze" (1638), which was written after his condemnation by the Inquisition and contained the beginnings of theoretical mechanics, are masterworks not only in the scientific sense but also as works of art. The observations with the telescope, Tycho Brahe's observation of the supernova of 1572 and that of 1604 by Kepler and Galileo, and finally the appearance of several comets required what was perhaps the most essential scientific insight of the time: that, in contrast to the opinion of the Aristotelians, there is *no fundamental difference* between cosmic and earthly matter and that *the same natural laws* hold in the realms of *astronomy* and of *terrestial physics* (this had already been recognized by the ancient Greeks in the case of the laws of geometry). This leap of thought, whose difficulty only becomes clear to us when we look back at Copernicus, gave impetus to the enormous upswing of scientific research at the beginning of the 17th century. W. Gilbert's investigations into electricity and magnetism, Otto v. Guericke's experiments with vacuum pumps and electrification machines, and much more, were stimulated by the revolution in the astronomical worldview.

We have no space here to pay tribute to the many observers and theoreticians who developed the new astronomy, among whom such important thinkers as J. Hevelius, C. Huygens, and E. Halley are particularly prominent.

An entirely new era of natural science began with Isaac Newton (1642–1727). His major work, "Philosophiae Naturalis Principia Mathematica" (1687), begins by placing theoretical mechanics on a firm basis using the *calculus of infinitesimals* ("fluxions"), which he developed for the purpose. Its connection with the *Law of Gravitation* explains Kepler's Laws and in one stroke provides the justification for the whole of terrestrial and *celestial mechanics*. In the area of optics, he invented the reflecting telescope and investigated the interference phenomena known as "Newton's Rings". Almost casually, he developed the basic approaches leading to numerous branches of theoretical physics.

Only the "Princeps Mathematicorum", Carl Friedrich Gauss (1777–1855), is of comparable importance; to him, astronomy owes the theory of *orbit calculation*, important contributions to *celestial mechanics* and advanced geodesics as well as the method of Least Squares. Never again has a mathematician shown such a combination of intuition in the choice of new areas of research and of facility in solving particular problems.

Again, this is not the place to pay tribute to the great theoreticians of celestial mechanics, from L. Euler to J.L. Lagrange and P.-S. Laplace to H. Poincaré; however, to finish this historical overview, we describe briefly the discovery of those planets which were not known in ancient times.

The planet *Uranus* was discovered quite unexpectedly in 1781 by W. Herschel. Kepler had already supposed that there should be a celestial body in the gap between Mars and Jupiter (Fig. 2.15); the first planetoid or asteroid, *Ceres*, was discovered in this region on 1.1.1801 by G. Piazzi, but in mid-February, it was "lost" when it passed near the Sun. By October of the same year, the 24-year-old C.F. Gauss had already calculated its orbit and ephemerides, so that F. Zach could find it again. Following this mathematical achievement, Gauss solved the general problem of determining the orbit of a planet or asteroid based on three complete observations. Today, several thousand asteroids are known, most of them between Mars and Jupiter (Sect. 3.3).

From perturbations of the orbit of Uranus, J.C. Adams and J.J. Leverrier concluded that there must be a planet with a still longer orbital period, and calculated its orbit and ephemerides. J.G. Galle then found *Neptune* near the predicted position in 1846.

Perturbations of the orbits of Uranus and Neptune led to the postulate that there was a transneptunian planet. The long search for it, in which P. Lowell (d. 1916) played a decisive role, was finally crowned with success: C. Tombaugh discovered *Pluto* in 1930 at the Lowell Observatory as a "faint star" of 15th magnitude.

Lengthy search programs for a "planet X" beyond the orbit of Pluto have remained unsuccessful; there are *no* indications for the existence of a further large planet. However, in 1992, D. Jewitt and J. Luu succeeded in discovering a *small* object outside Pluto's orbit, whose size is comparable with that of many of the asteroids. Soon thereafter, a number of "planets" were observed outside the orbits of Neptune and Pluto.

2. Classical Astronomy

Following the historical overview of classical astronomy from ancient times up through the founding of the heliocentric worldview and the discovery of the basic principles of celestial mechanics, we begin in Sect. 2.1 our treatment of astronomy with a description of the motions of the Sun, the Earth, and the Moon in terms of the coordinates on the celestial sphere and of astronomical determinations of time. In Sect. 2.2, we then give a summary of the motions of the other planets, the comets etc. and of the determination of distances within the Solar System. After a brief treatment of the basic principles of mechanics and gravitational theory (Sect. 2.3), we give some applications to celestial mechanics in Sect. 2.4. Finally, in Sect. 2.5, we treat the orbits of artificial satellites and space probes and summarize the most important space research missions within our Solar System.

2.1 Spatial Coordinates and Time; the Motions of the Sun, the Earth, and the Moon

As a beginning of our study of astronomy, in Sect. 2.1.1 we describe apparent motions on the celestial sphere and the coordinate system used to specify the positions of celestial objects. In Sect. 2.1.2, we treat the motions of the Earth, its rotation and its revolution around the Sun, which are reflected as apparent motions on the celestial sphere. Section 2.1.3 is devoted to the astronomical measurement of time. Following these preparatory topics, we gradually become familiar with the objects in our Solar System, beginning this process in Sect. 2.1.4 with our Moon, its motions and its phases. We then treat lunar and solar eclipses in Sect. 2.1.5.

2.1.1 The Celestial Sphere and Astronomical Coordinate Systems

Since antiquity, human imagination has combined the easily-recognized groups of stars into *constellations* (Fig. 2.1). In the northern sky, the Great Bear (or the Big Dipper) is readily seen. We can find the Pole Star (Polaris) by extending the line joining the two brightest stars of the Big Dipper until it is about five times longer. Continuing about the same distance past Polaris (which is the brightest star in the Little Dipper or Small Bear), we see the "W" of Cassiopeia. Using a sky globe or a star map, we can readily find the other constellations. In his "Uranometria Nova" (1603), J. Bayer named the stars in each constellation $\alpha, \beta, \gamma \ldots$, as a rule in the order of decreasing brightness. Besides these Greek letters, we also use the numbering system of the "Historia Coelestis Britannica" (1725), compiled by the first Astronomer Royal, J. Flamsteed. The Latin names of the constellations are usually abbreviated to 3 letters (see Appendix A.2).

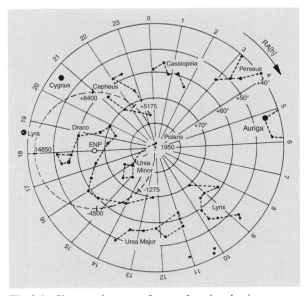

Fig. 2.1. Circumpolar stars from a location having a geographic latitude of $\varphi = +50°$ (about that of Frankfurt or Prague). The coordinate lines indicate the right ascension RA and the declination ($+40°$ to $+90°$). Precession: the celestial pole circles about the pole of the ecliptic ENP once every 25 700 years. The location of the celestial north pole is indicated for several past and future dates

Celestial Sphere. On the celestial sphere (in mathematical terms, the infinitely distant sphere on which the stars seem to be projected), we in addition define the following quantities (Fig. 2.2):

1. the *horizon* with the directions North, West, South, and East,
2. vertically above our position the *zenith*, directly under us the *nadir*,
3. the curve which passes through the zenith, the nadir, the celestial pole, and the north and south points is the *meridian*, and
4. the curve which is perpendicular to the meridian and the horizon, passing through the zenith and the east and west points, is the *principal vertical*.

In the coordinate system defined by these features, we denote the momentary position of a star by giving two angles (Fig. 2.2): (a) the *azimuth* is measured along the horizon in the direction SWNE, starting sometimes from the S- and sometimes from the N-point; (b) the *altitude* is 90° – the angle to the zenith.

The celestial sphere apears to rotate once each day around the *celestial axis* (which passes through the celestial North and South Poles). The *celestial equator* is perpendicular to this axis. The *position* of a star

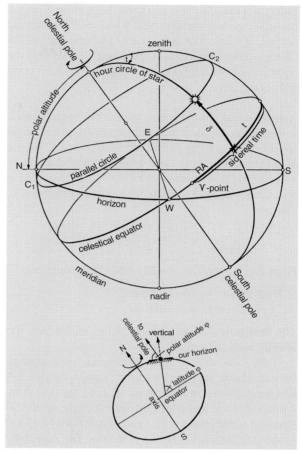

Fig. 2.3. Celestial coordinates: right ascension RA and declination δ. The hour angle t = sidereal time minus the right ascension RA. C_1 = lower culmination, C_2 = upper culmination. *Lower right:* the Earth (polar flattening exaggerated). Polar altitude = geographic latitude

(Fig. 2.3) at a given time on the celestial sphere, imagined to be infinitely distant, is also described by the *declination* δ, which is positive from the equator to the North Pole and negative from the equator to the South Pole, and by the *hour angle t*, which is measured from the meridian in the direction of the diurnal motion, i.e. towards W.

In the course of a day, a star therefore traces out a circle on the sphere; its plane is parallel to the plane of the celestial equator. On the meridian, the greatest height reached by a star is its *upper culmination*, and the least height is its *lower culmination*.

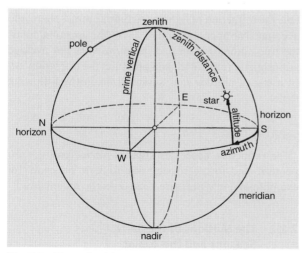

Fig. 2.2. The celestial sphere. The horizon with north, east, south, and west points. The (celestial) meridian passes through the north point, the (celestial) pole, zenith, south point, and the nadir. Coordinates: altitude and azimuth

Sidereal Time. We also mark the *Aries Point* ♈ on the celestial equator; we shall deal with it in the following section. It marks the point reached by the sun on the vernal equinox (March 21), on which the day and the night are equally long. The hour angle of the Aries point defines the *sidereal time* τ.

Astronomical Coordinates. We are now in a position to determine the coordinates of a celestial object on the sphere independently of the time of day: we call the arc of the equator from the Aries point to the hour-circle of a star the *right ascension* RA of that star. It is quoted in hours, minutes, and seconds. 24 h (hora) correspond to 360°, or

$$1\,\text{h} = 15°, \quad 1\,\text{min} = 15', \quad 1\,\text{s} = 15'',$$
$$1° = 4\,\text{min}, \quad 1' = 4\,\text{s}.$$

From Fig. 2.3, one can readily read off the relation:

$$\text{Hour angle } t = \text{sidereal time } \tau \qquad (2.1)$$
$$- \text{right ascension RA}.$$

The declination δ, our second stellar coordinate, has already been defined.

If we now wish to train a telescope on a particular star, planet, etc., we look up its right ascension RA and declination δ in a star catalog, read the time from a sidereal clock, and adjust the setting circles of the instrument to the angle hour t calculated from (2.1) and to the declination (+ north, − south). The especially precisely determined positions of the so-called *fundamental stars* (especially for determinations of the time, see Sect. 2.1.3) are to be found, along with those of the Sun, the Moon, the planets, etc. in the astronomical yearbooks or ephemerides; the most important of these is the Astronomical Almanac.

Astronomical Coordinates. The Copernican system attributes the apparent rotation of the celestial sphere to the fact that the Earth rotates about its axis once every 24 h of sidereal time. The horizon is defined by a plane tangent to the Earth at the location of the observer; more precisely, by an infinite water surface at the observer's altitude. The zenith or vertical is the direction of a plumb-bob perpendicular to this plane, i.e. the direction of the local acceleration of gravity (including the centrifugal acceleration caused by the Earth's rotation). The *polar altitude* (the altitude of the celestial pole above the horizon) is given from Fig. 2.3 by the *geographic latitude* φ (the angle between the vertical and the Earth's equatorial plane); it can be readily measured as the average of the altitudes of the Pole Star or a circumpolar star at the upper and lower culminations.

The *geographic longitude l* corresponds to the hour angle. If the hour angle of the same object is measured *simultaneously* at Greenwich (zero meridian, $l_G = 0°$) and, for example, in New York, the difference gives the geographic longitude of New York, l_G. The determination of the latitude requires only a simple angle measurement, while that of a longitude necessitates a precise time measurement at two places. In earlier times, the "time markers" were taken from the motions of the Moon or of one of the moons of Jupiter. The introduction of the "seaworthy" chronometer by John Harrison (ca. 1760–65) brought a great improvement, as did the later transmission of time signals by telegraph and still later by radio.

A few further facts: at a location having (northern) latitude φ, a star of declination δ reaches an altitude of $h_{\max} = 90° - |\varphi - \delta|$ at its upper culmination and $h_{\min} = -90° + |\varphi + \delta|$ at its lower culmination. Stars with $\delta > 90° - \varphi$ always remain above the horizon (circumpolar stars); those with $\delta < (90° - \varphi)$ never rise above the horizon.

Refraction. In measuring stellar altitudes h, we must take the refraction of light in the Earth's atmosphere into account. The apparent shift of a star (the apparent minus the true altitude) is termed the *refraction*. For average atmospheric tmperature and pressure, the refraction Δh of a star at altitude h is summarized in the following table:

$h =$	0°	5°	10°	20°	40°	60°	90°
$\Delta h =$	34′50″	9′45″	5′16″	2′37″	1′09″	33″	0″

The refraction decreases slightly for increasing temperature and for decreasing atmospheric pressure, for example in a low-pressure zone or in the mountains.

2.1.2 The Motions of the Earth. Seasons and the Zodiac

We now consider the *orbital motion* or *revolution* of the Earth around the Sun in the Copernican sense, and then the daily *rotation* of the Earth about its own axis, as well as the motions of the axis itself. We first place ourselves in the position of an observer in space. In Sect. 2.4, we shall derive Newton's theory of the motions of the Earth and the planets starting from his principles of mechanics and law of gravitation.

Ecliptic and Seasons. The apparent annual motion of the Sun in the sky was attributed by Copernicus to the revolution of the Earth around the Sun on a (nearly) circular orbit. The plane of the Earth's orbit intersects the celestial sphere as a great circle called the *ecliptic* (Fig. 2.4). This makes an angle of $23°27'$ with the celestial equator, the *obliquity of the ecliptic*. This means that the Earth's axis retains its direction in space relative to the fixed stars during its annual revolution around the Sun; it forms an angle of $90° - 23°27' = 66°33'$ with the Earth's orbital plane.

A brief summary will suffice to explain the *seasons* (Figs. 2.4, 5), starting with the Northern Hemisphere.

In the Northern Hemisphere, the Sun reaches its maximum altitude (midday altitude) at a geographical latitude φ on the 21st of June (the first day of Summer or Summer solstice), $h = 90° - |23°27' - \varphi|$. On the 22nd of December (Winter solstice), it has its lowest midday altitude, $h = 90° - \varphi - 23°27'$. It can reach the zenith at latitudes up to $\varphi = +23°27'$, the Tropic of Cancer. North of the Arctic Circle, $\varphi \geq 90° - 23°27' = 66°33'$, the Sun remains below the horizon around the Winter solstice; near the Summer solstice, the "midnight Sun" acts as a circumpolar star.

In the Southern Hemisphere, Summer corresponds to Winter in the Northern Hemisphere, the Tropic of Capricorn to the Tropic of Cancer, etc.

The *zodiac* is the term for a band in the sky on each side of the ecliptic. Since ancient times, it has been divided into 12 equal "signs of the zodiac" (Fig. 2.5).

It is often expedient for calculating the motions of the Earth and the planets to use a coordinate system oriented on the ecliptic and its poles. The (ecliptical) longitude is measured along the ecliptic starting from the Aries or ♈ point, like the right ascension in the direction of the annual motion of the Sun. The (ecliptical)

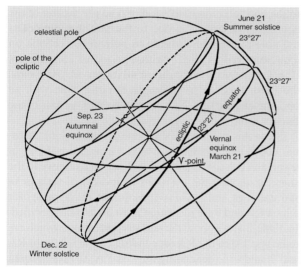

Fig. 2.4. Annual (apparent) motion of the Sun among the stars. The Ecliptic. The seasons

Start date	Name	Coordinates of the Sun		The Sun enters the constellation
		Right ascension RA [h]	Declination δ	
March 21 Spring	Vernal equinox[a]	0	0°	Aries ♈
June 21 Summer	Summer solstice	6	$+23°27'$	Cancer ♋
Sept. 23 Autumn	Autumnal equinox[a]	12	0°	Libra ♎
Dec. 22 Winter	Winter solstice	18	$-23°27'$	Capricorn ♑

[a] On these days, the day and night arcs of the Sun are equal and each correspond to 12 hours.

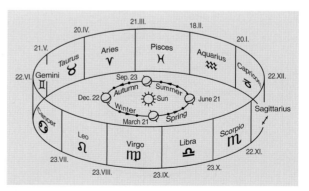

Fig. 2.5. The orbit of the Earth around the Sun. The seasons. The zodiac and the signs of the zodiac. The Earth is at perihelion (closest approach to the Sun) on the 2nd of January, and at aphelion (furthest distance from the Sun) on the 2nd of July

latitude is measured, analogously to the declination, in a direction perpendicular to the ecliptic. The *ecliptical coordinates* in the sky must naturally not be confused with the similarly-named geographical coordinates!

Kepler's Laws. The apparent annual motion of the Sun contains irregularities which were already known to ancient astronomers; they were recognized by Johannes Kepler as consequences of his first two laws of planetary motion, which we shall treat in more detail in Sect. 2.4.1:

Kepler's 1st Law: The planets move on elliptical orbits, with the Sun at one (common) focus.

Kepler's 2nd Law: The radius vector of a planet sweeps out equal areas in equal times.

Kepler's 3rd Law: The squares of the orbital periods of two planets are in the ratio of the cubes of their orbital semimajor axes.

The quantities needed for the geometric definition of the Earth's orbit or that of another planet around the Sun are shown in Fig. 2.6: we first note the semimajor axis a. The distance from a focus to the midpoint is denoted by $a \cdot e$, where the pure number e is called the eccentricity of the orbit. At the *perihelion*, the closest approach to the Sun, the distance of the Earth or other planet from the Sun is $r_{min} = a(1-e)$; at the *aphelion*, the point furthest from the Sun, $r_{max} = a(1+e)$. The diurnal motion of the Sun in the sky, i.e. the angle passed through by the radius vector of the Earth each day, is found from Kepler's 2nd law to obey $(r_{max}/r_{min})^2 = [(1+e)/(1-e)]^2$. The corresponding apparent diameters of the Sun's disk have the ratio $(1+e)/(1-e)$. Measurements yield an eccentricity for the Earth's orbit of $e = 0.0167$. At present, the Earth passes through its perihelion about January 2nd. The approximate coincidence of this date with the beginning of the year is purely accidental.

Precession. Hipparchus had already discovered that the Aries point (♈ point) is not fixed with respect to the celestial equator, but rather moves forward by about 50″ each year. This has led to the advancing of the ♈ point from the constellation Aries in ancient times to the constellation Pisces today. The *precession* of the equinoxes described is due to the fact that the celestial pole rotates about the fixed pole of the ecliptic on a circle of radius 23°27′ with a period of 25 700 years (Fig. 2.1); or, expressed differently: every 25 700 years, the Earth's axis of rotation traces out a cone with an opening angle of 23°27′, centered around the axis of the Earth's orbit.

Since this precession shifts the position of the celestial coordinate system in which we measure the right ascension RA (or α) and the declination δ relative to the stars, in quoting the positions of individual stars or in star catalogs, the *equinox* (position of the ♈ point) for which RA and δ are measured must always be specified. In Table 2.1, we show the various corrections which are to be applied to the RA (depending on the values of RA and δ) and to δ (depending only on RA) to take account of the precession during a 10-year interval. The stellar positions also change because of the stars' proper motions (Sect. 6.2.2). We shall come back to the question of the stellar positions and star catalogs in Sect. 6.2.4.

Nutation. Superimposed on the precession, which has a period of 25 700 years, is a superficially similar motion having a period of 19 years, the *nutation*. Finally, the axis of the Earth's rotation fluctuates relative to the body axis of the Earth itself by about ±0.2″; analysis of this motion shows it to have a periodic part, with the so-called Chandler period of 433 days, as well as an annual contribution and an irregular part. The resulting fluctuations in the polar altitude are continuously checked at a series of observation stations. We shall return to the explanation of the various motions of the Earth's axis in Sect. 2.4.

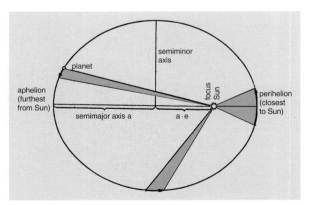

Fig. 2.6. The orbital ellipse of a planet. The semimajor axis is a, and the distance from the midpoint to a focus (the Sun) is $a \cdot e$, where e is the eccentricity. (The actual eccentricity of the orbits of the planets is much smaller than shown here)

2.1.3 Time: Days, Years, and the Calendar

Our daily life depends to a considerable extent on the position of the Sun. Thus, the

true solar time = hour angle of the Sun + 12 h

was first defined. This is the time indicated by a simple sundial; 12:00 h corresponds to the upper culmination of the Sun. However, due to the nonuniformity of the Earth's orbital motion around the Sun (Kepler's 2nd law) and to the inclination of the ecliptic, this time varies throughout the year. The *mean solar time* was therefore introduced; it is based on a fictitious "mean Sun", which passes through the equator uniformly in the same time required by the true Sun in its passage around the ecliptic. The hour angle of this fictitious Sun defines the mean solar time. The difference

true solar time − mean solar time
= equation of time

is thus composed of two contributions, which derive from the eccentricity of the Earth's orbit and from the inclination of the ecliptic, respectively. Their extrema are:

	Febr. 12	May 14	July 26	Nov. 4
Equation of time:	−14.3 min	+3.7 min	−6.4 min	+16.4 min

Mean solar time is different for each meridian. As a simplification for transportation and communication, it has been agreed upon to use the local time of a *particular* meridian within a suitably defined "time zone", for example Greenwich Mean Time (GMT) in England and Western Europe, Central European Time (CET) in Central Europe, Eastern Standard Time (EST) on the East Coast of the USA and Canada, etc.

For scientific purposes, for example astronomical or geophysical measurements at stations which may be spread around the globe, the same time is used *everywhere*, namely *universal time*:

world time or universal time (UT)
= mean solar time
at the Greenwich meridian .

This time is measured in 24-hour days, starting with 0:00 h at midnight. For example, 12:00 h corresponds to 07:00 h EST or 13:00 h CET.

For astronomical observations, the relation between mean solar time and sidereal time is required. The "mean Sun" moves relative to the ♈ point by 360° or 24 h from west to east in the course of a year (365 d). The mean solar day is thus 24 h/365 d or 3 min 56 s longer than a sidereal day. A sidereal clock gains about 2 h per month relative to a "normal" UT clock. To make

Table 2.1. Precession during a 10-year interval

a) $\Delta(RA)$ in minutes of time (+ increase, − decrease)

RA for northern objects [h]	6 / 5	7 / 4	8 / 3	9 / 2	10 / 1	11 / 0	12 / 23	13 / 22	14 / 21	15 / 20	16 / 19	17 / 18	
$|\delta|$ 80°	+1.77	+1.73	+1.60	+1.40	+1.14	+0.84	+0.51	+0.19	−0.12	−0.38	−0.58	−0.70	−0.75	80°
70°	1.12	1.10	1.04	0.94	0.82	0.67	0.51	0.35	+0.21	+0.08	−0.02	−0.08	−0.10	70°
60°	0.898	0.885	0.846	0.785	0.705	0.612	0.512	0.412	+0.319	+0.240	+0.178	+0.140	+0.126	60°
50°	0.778	0.768	0.742	0.700	0.645	0.581	0.512	0.444	+0.380	+0.324	+0.282	+0.256	+0.247	50°
40°	0.699	0.693	0.674	0.644	0.606	0.560	0.512	0.464	+0.419	+0.380	+0.350	+0.332	+0.335	40°
30°	0.641	0.636	0.624	0.603	0.576	0.546	0.512	0.479	+0.448	+0.421	+0.401	+0.388	+0.384	30°
20°	0.593	0.590	0.582	0.570	0.553	0.533	0.512	0.491	+0.472	+0.455	+0.442	+0.434	+0.431	20°
10°	0.552	0.550	0.546	0.540	0.532	0.522	0.512	0.502	+0.492	+0.484	+0.478	+0.476	+0.473	10°
0°	+0.512	+0.512	+0.512	+0.512	+0.512	+0.512	+0.512	+0.512	+0.512	+0.512	+0.512	+0.512	+0.512	0°
RA for southern objects [h]	18 / 17	19 / 16	20 / 15	21 / 14	22 / 13	23 / 12	0 / 11	1 / 10	2 / 9	3 / 8	4 / 7	5 / 6		

b) $\Delta\delta$ in minutes of arc (+ increase in δ, in the southern sky thus decreasing absolute values $|\delta|$!)

RA [h]	0 / 24	1 / 23	2 / 22	3 / 21	4 / 20	5 / 19	6 / 18	7 / 17	8 / 16	9 / 15	10 / 14	11 / 13	12
$\Delta\delta$	+3.34′	+3.23′	+2.89′	+2.36′	+1.67′	+0.86′	0.0′	−0.86′	−1.67′	−2.36′	−2.89′	−3.23′	−3.34′

this clearer, we list the *sidereal time* for several dates at 0:00 h local time (midnight). This is given by the hour angle of the ♈ point or the right ascension of stars which cross the meridian at midnight (and are thus favorable for observation):

0 h local time sidereal time or RA at the meridian	January 1 6.7 h	April 1 12.6 h	July 1 18.6 h	October 1 0.6 h

The Year. The suitable unit for long periods of time is the year. We define: one

$$\text{siderial year} = 365.25637 \text{ mean solar days}$$

as the time between two passages of the Sun through the same point in the sky (from sidus = "star"); it is thus the true orbital period of the Earth. One

$$\text{tropical year} = 365.24220 \text{ mean solar days}$$

is the time between two passages of the Sun through the Aries point ♈ or beginning of Spring (from $\tau\rho o\pi\varepsilon\tilde{\iota}\nu$ = "to turn"). Since this point moves $50.3''$ to the west each year, the tropical year is correspondingly shorter than the sidereal year.

Calendar. The seasons and the calendar are calculated relative to the tropical year. Because it is preferable for practical reasons that each year comprise an integral number of days, in everyday life we use the

$$\text{calendar year} = 365.2425 = 365 + \tfrac{1}{4} - \tfrac{3}{400} \text{ mean solar days}$$

It corresponds to the prescription for leap years given by the *Gregorian calendar*, introduced in 1582 by Pope Gregory XIII. Every 3 years with 365 days are followed by a leap year with 366 days, except for the whole centuries which are *not* divisible by 400. We cannot discuss here the earlier Julian calendar introduced by Julius Caesar in 45 B.C., nor other problems of chronology which are interesting from a cultural-historical point of view.

The Julian Day. To simplify chronological calculations dealing with long periods of time, and in particular for observations and ephemerides of variable stars, etc., it is desirable to avoid the variation in the lengths of years and months. Following a suggestion of J. Scaliger (1582), the *Julian days* are simply counted in an unbroken succession. Each Julian day begins at 12:00 h UT (mean Greenwich midday). The beginning of Julian day 0 was fixed at 12:00 h UT on January 1st in the year 4713 B.C. On January 1, 2000, the Julian day number 2 451 545 began at 12:00 h UT.

Independently of the detailed definition, we use the symbols yr for "year" and d for "day".

For the calculation of astronomical times over long periods, the unit

$$\text{Julian century} = 36\,525.0 \text{ d}$$

with $1 \text{ d} = 24 \cdot 60 \cdot 60 \text{ s} = 86\,400 \text{ s}$ has been defined.

The Measurement of Time. Astronomical time reckoning was long based on the assumed uniformity of the Earth's rotation. The basic physical principle underlying terrestrial time measurements was already recognized by Christian Huygens ("Horologium Oscillatorium", 1673): every clock consists of an oscillatory mechanism which is isolated as far as possible from its surroundings (a pendulum, clock movement, etc.) and held in motion by a driving mechanism (weight, spring, etc.) with the least possible feedback. The *pendulum chronometer*, steadily improved over the years, was for three centuries one of the most important instruments in every astronomical observatory. The *quartz clock*, considerably less susceptible to disturbances, consists of an oscillating piezoelectric quartz crystal, which is kept in motion by a loosely-coupled electrical oscillator circuit. The pinnacle of metrological precision, however, was attained in recent times by the *atomic clock*, which uses the oscillation frequency of cesium atoms (the isotope ^{133}Cs) in the vapor phase to measure time. The frequency corresponds to the transition between the two energetically lowest hyperfine levels. Other atomic clocks are based on oscillations in rubidium atoms or hydrogen masers. The extreme exactness of atomic clocks, which attain a relative accuracy of their oscillation frequencies of better than 10^{-14}, is the basis for various fundamental measurements and observations in physics and astronomy.

The comparison of astronomical observations with groups of quartz clocks and later with atomic clocks has shown that the rotational period of the Earth is not constant, but rather exhibits fluctuations of the order of a millisecond, in part with an annual period, in part aperiodic; these are related to changes in the mass distribution on the Earth.

The second was *defined* in 1967 as 1 s = the time required for 9 192 631 770 oscillations of the radiation from the transition between the lowest two hyperfine structure levels of the ground state of ^{133}Cs; it is a base unit of the international system of units (SI).

The atomic-clock measurements of time by standards institutes are combined into International Atomic Time (TAI). From it, Coordinated Universal Time (UTC), the basis for civil time, is derived as an approximation to universal time; it is corrected only for the variations in the Earth's axis and thus is not uniform. It is termed UT1. It is a measure of the true rotation of the Earth about its axis and differs from UTC by the correction

$$\Delta \text{UT} = \text{UT1} - \text{UTC} ,$$

which is promulgated regularly by the Bureau International des Poids et Mésures (BIPM) in Sèvres, near Paris (prior to 1987, it was called the Bureau International de l'Heure or BIH) and by other similar institutions.

In everyday life, for navigation, geophysics, etc., the measurement of time is still necessarily based on the Earth's rotation; therefore, UTC is "switched" by one whole second ahead or back whenever the magnitude of ΔUT appraoches 1 s.

Independently of the progress in the physical measurement of time it has been discovered that the motions of the planets and the Sun, or of the Earth and in particular the Moon, exhibit small, common variations over long periods of time relative to the ephemerides calculated according to the laws of Newtonian mechanics and gravitation theory. On the one hand, there is a secular (i.e. progressive) increase in the length of the day, which is caused by the braking effect of tidal friction (Sect. 2.4.6). Another contribution shows no such obvious origins. However, comparison of the variations for different objects forces us to attribute them to deviations of "astronomical time", which is based on the Earth's rotation, from "physical time", based on Newton's laws. Because of this empirical fact, it was decided in 1950 to base all astronomical ephemerides on a time scale derived from the basic laws of physics; the latter is termed *Ephemeris Time*, ET. The small corrections (Ephemeris Time minus Universal Time) are found for the most part by very accurate observations of the motion of the Moon. They can only be determined retrospectively; for most predictive purposes, they can be extrapolated with sufficient accuracy. In 1956, the ephemeris second was defined as the 31 556 925.9747th part of tropical year 1900.

Ten years later, it was decided to relate the ephemeris second to the atomic time unit. This, however, disturbs the internal consistency of the system of ET. On the recommendation of the International Astronomical Union, ephemeris time was replaced by *Terrestrial Time*, TT, which is based on the SI second.

2.1.4 The Moon

The *Moon* appears to us as a disk in the sky of mean diameter 31'; it is thus just the same apparent size as the Sun. Its distance from the Earth is small enough to be determined by triangulation from two widely-separated points on the ground (e.g. on the same meridian). Astronomers refer to the angle subtended by the equatorial radius of the Earth, seen from the Moon, as the *equatorial horizontal parallax* of the Moon. It has a mean value equal to 3422.6''. Since the Earth's radius is known to have the value 6378 km, one can calculate from these two numbers the average distance of the Moon from the center of the Earth:

$$r_\text{M} = 60.3 R_\text{E} = 384\,400 \text{ km}$$

and therefore the Moon's radius:

$$R_\text{M} = 0.272 R_\text{E} = 1738 \text{ km} .$$

We shall take up the physical structure of the Earth and the Moon in Sect. 3.2. First, we consider the Moon's orbit and its motions from the viewpoint of an observer.

The Moon orbits around the Earth, in the same direction as the Earth around the Sun, in one *sidereal month* = 27.32 d; that is, after that time it has returned to the same point in the heavens.

The origin of the *phases* of the Moon is illustrated in Fig. 2.7. Their period, the *synodic month* = 29.53 d ($1 \rightarrow 3$ in Fig. 2.8), is the time after which the Moon returns to exactly the same position relative to the Sun, and is longer than the sidereal month ($1 \rightarrow 2$ in Fig. 2.8). The Moon moves in an easterly direction relative to the

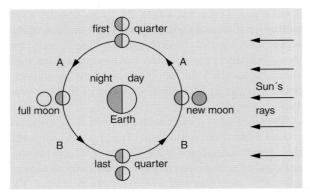

Fig. 2.7. The phases of the Moon; the Sun is at the right. The outer pictures indicate the Moon as seen from the Earth: A = waxing Moon, B = waning Moon

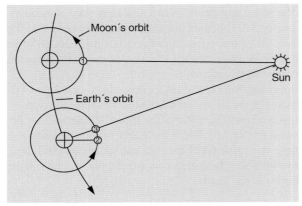

Fig. 2.8. The synodic month (1 → 3) is longer than the sidereal month (1 → 2), because the Earth moves onwards in its orbit in the meantime

Sun by $360°/29.53\,\mathrm{d} = 12.2°$ each day, and relative to the stars, by $360°/27.32\,\mathrm{d} = 13.2°$.

The difference between the sidereal and synodic daily motion of the Moon is equal to the daily motion of the Sun, i.e. $360°/365\,\mathrm{d} \approx 1°\,\mathrm{d}^{-1}$. This becomes immediately clear if we consider that the daily motion is nothing other than the *angular velocity* (2π/period) in astronomical units. We could just as well write

$$\frac{1}{\text{sidereal month}} - \frac{1}{\text{sidereal year}} = \frac{1}{\text{synodic month}}.$$

Orbit. More precisely, the orbit of the Moon around the Earth is an ellipse with eccentricity $e = 0.055$. The point in the orbit where the Moon is closest to the Earth (analogous to the perihelion in the Earth's orbit around the Sun) is called the *perigee*, and the most distant point is the *apogee*. The plane of the Moon's orbit is inclined relative to the Earth's orbit (the plane of the ecliptic) by an angle $i = 5.1°$. The Moon crosses over the ecliptic from the south to the north at ascending *nodes*, and passes "below" the ecliptic (for observers in the Northern Hemisphere) at descending nodes.

As a result of the perturbation (gravitational attraction) caused by the Sun and the planets, the Moon's orbit also includes the following motions:

1. the perigee rotates "directly" around the Earth in the plane of the Moon's orbit, i.e. in the same sense as the revolution of the Earth around the Sun, with a period of 8.85 yr.
2. the nodes of the Moon's orbit, or the *line of nodes*, in which the orbit of the Moon crosses the Earth's orbital plane, has a retrograde motion in the ecliptic, i.e. in the opposite sense from the Earth's revolution, with a period of 18.61 yr, the so-called *period of nutation*.

This "regression of the lunar nodes" furthermore causes a corresponding "nodding" of the Earth by a maximum of $9''$; this is the nutation of the Earth's axis mentioned previously.

The average time between two successive passages of the Moon through the same node is called the *draconitic month* = 27.2122 d. It is important for the prediction of eclipses (see Sect. 2.1.5).

If we were to observe the orbits of the Moon and the Earth around the Sun from a spaceship, we would see, in agreement with a simple calculation, that the Moon's orbit is always concave as seen from the Sun (Fig. 2.9).

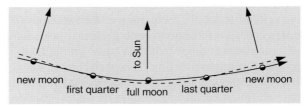

Fig. 2.9. The orbits of the Earth and the Moon around the Sun (— Earth's orbit, --- Moon's orbit)

Rotation. Let us now consider the rotation of the Moon and its other motions around its center of gravity. These can be determined very precisely by observing the motion of a sharply defined crater or similar feature on the lunar disk.

The fact that the Moon always shows us more or less the same face is because the lunar rotational period is equal to the orbital period, i.e. equal to a sidereal month. The two periods apparently were equalized by tidal interactions (Sect. 2.4.6) between the Moon and the Earth.

Librations. Careful observation shows, however, that the "face" of the Moon wobbles somewhat. The so-called *geometric* librations of the Moon have the following causes:

1. the equator and the orbital plane of the Moon form an angle of $\simeq 6.7°$; the latitudinal libration caused by this is equal to about $\pm 6.7°$.
2. the rotation of the Moon is uniform (following Newton's law of inertia), but its revolution, from Kepler's second law taking the eccentricity of its orbit into account, is not; this causes a longitudinal libration of about $\pm 7.6°$.
3. the equatorial radius of the Earth appears to subtend an angle of $57'$, from the Moon, the lunar horizontal parallax; the daily rotation of the Earth thus causes a diurnal libration.

Furthermore, there is the considerably smaller *physical* libration, which is due to the fact that the Moon is not quite spherical in shape and therefore performs small oscillations in the gravitational field (mainly that of the Earth).

All together, the librations have the effect that we can observe 59% of the Moon's surface from the Earth.

2.1.5 Eclipses of the Sun and the Moon

Having studied the motions of the Sun, the Earth, and the Moon, let us turn to the impressive spectacle of the lunar and solar eclipses!

A *lunar eclipse* occurs when the full Moon is covered by the shadow of the Earth. As with shadows on the Earth, we distinguish between the central part of the shadow, the umbra, and the surrounding half-shadow, the penumbra. If the Moon is completely in the umbra of the Earth, we speak of a total eclipse; if only a part of the Moon's disk is in the umbra, we have a partial eclipse. From the known geometrical facts we can calculate that a lunar eclipse can last for at most 3 h 40 min, while totality lasts for at most 1 h 40 min. Because the Sun's light is absorbed and scattered by the Earth's atmosphere more strongly at the blue end of the visible spectrum than at the red end, the outer edge of the penumbra on the Moon is not sharp and that of the umbra is also noticeably fuzzy. Furthermore, the penumbra and to a lesser extent the umbra seem to have a reddish-coppery color.

If the new Moon passes in front of the Sun, a *solar eclipse* occurs (Fig. 2.10). It can be partial or total. If the apparent diameter of the Moon is smaller than that of the Sun, we will observe only a ring-shaped eclipse when the Moon's shadow is centered on the Sun. In a partial eclipse, an observer on the Earth is in the penumbra of the Moon; in a total eclipse, the observer is in the umbra. In the case of a ring-shaped eclipse, the vertex of the Moon's shadow cone is between the Moon and the observer.

Total eclipses are particularly important for astrophysical observations of the outer layers of the Sun and the nearby interplanetary material; the bright sunlight is then completely blocked off outside the Earth's atmosphere.

Relative to the Sun, the Moon moves through an angle of $0.51''$ per second in the sky, in agreement with the

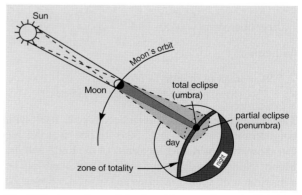

Fig. 2.10. An eclipse of the Sun (shown schematically). The Moon moves from W to E across the Sun's disk. In the umbra, a total eclipse is observed; in the penumbra, a partial eclipse

length of the synodic month. This corresponds to a distance of 370 km on the Sun. Observations of eclipses carried out with good time resolution therefore yield an angular resolution which is generally better than that of available telescopes.

Stellar occultations by the Moon, which, like solar eclipses, must be predicted individually for each location, are also very sharply defined in time, since the Moon has no atmosphere. They are important for checking the orbit of the Moon, for determining the fluctuations in the Earth's rotation. Since the Moon moves 0.55″ per second relative to the stars, measurements of these occultations with good time resolution may, in favorable cases, yield the angular diameters of the tiny stellar disks. Still more important for radio-astronomical observations with high angular resolution are occultations of radio-emitting objects by the Moon.

It was already known to the ancient eastern cultures that eclipses of the Sun and Moon (in the following, we shall refer for short simply to eclipses) succeed each other with a period of about 18 yr 11 d, the so-called *Saros cycle*. This cycle is based upon the fact that an eclipse can occur only when the Sun *and* the Moon are rather close to a *node* in the lunar orbit. The time which the Sun requires to return to a particular lunar node is, due to the regression of the nodes, slightly less than a tropical year, namely 346.62 d; this time is called an *eclipse year*. As we may readily verify, the Saros cycle corresponds to an integral number of eclipse years:

223 synodic months = 6585.32 d

and of

19 eclipse years = 6585.78 d

furthermore,

239 anomalous months = 6585.54 d
(from perigee to perigee, 27.555 d) .

Thus, an eclipse configuration indeed repeats itself with good accuracy after 18 yr 11.33 d. In one year, as one can show by considering the orbits of the Earth and the Moon, taking their diameters into account, there can be a maximum of 3 lunar eclipses and 5 solar eclipses. At a particular location, lunar eclipses, which can be seen from a whole hemisphere of the Earth, are relatively frequent; a total solar eclipse is, in contrast, very rare.

2.2 Orbital Motions and Distances in the Solar System

The planets known since ancient times (with their time-honored symbols), Mercury ☿, Venus ♀, Mars ♂, Jupiter ♃, and Saturn ♄, have fascinated people again and again; their motions in the sky often appeared erratic, but were found, step by step, to obey regular laws.

In the historical introduction to Part I, we briefly summarized the efforts made in ancient times to explain the motions of the planets. Here, we immediately adopt the *heliocentric* point of view, as developed by N. Copernicus in 1543. We shall furthermore drop the insistence on circular orbits which Copernicus had retained as a last vestige of Aristotelianism and make use of J. Kepler's elliptical orbits and his three laws of planetary motion (1609 and 1619). We thus place ourselves at the threshold of modern mathematical-physical thinking, which took on a clear form through the work of Galileo (1564–1642) and was consolidated into the beginnings of classical mechanics and gravitation theory in Newton's Principia (1687).

In Sect. 2.2.1, we describe the planets and their orbits and define the orbital elements necessary to fully specify their motions. In Sect. 2.2.2 we then summarize the orbits of the comets and meteors. Finally, in Sect. 2.2.3, we discuss the determination of the Earth-Sun distance, the fundamental "astronomical unit" (AU), as well as the Doppler effect which results from the motion of the Earth, and the aberration of light.

We shall defer the discussion of the physical strucure of the planets and their satellites, the comets and other objects in the Solar System to Chap. 3.

2.2.1 Planetary Motions and Orbital Elements

The origin of the *direct* (west–east) and of the *retrograde* (east–west) motions of the planets is explained in Fig. 2.11, taking Mars as an example.

Referring to Fig. 2.12, we first consider the motion around the Sun of an *inner* planet, e.g. Venus, as seen from our more slowly revolving Earth. The planet is closest to us at its *lower conjunction*. It then moves away from the Sun in the sky and, as Morning Star, reaches its greatest westerly *elongation* of 48°. At the *upper* conjunction, Venus is at its greatest distance from the Earth

2.2 Orbital Motions and Distances in the Solar System

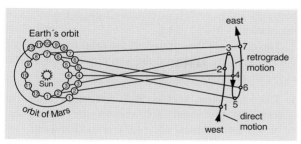

Fig. 2.11. Direct (west–east) and retrograde (east–west) motions of the planet Mars. The positions of the Earth and Mars on their orbits are numbered from month to month. At 4, Mars is in apposition to the Sun; it is overtaken here by the Earth and thereafter shows retrograde motion. At this time, it is closest to the Earth and most readily observed. The orbit of Mars is inclined relative to that of the Earth, i.e. the ecliptic, by 1.9°

and is closest to the Sun as seen in the sky. It then again moves away from the Sun and reaches its greatest easterly elongation of 48° as the Evening Star. The ratio of the orbital radii of Venus and the Earth is established by the maximum elongation of ±48° (for Mercury, ±28°). The *phases* of Venus, which are readily recognized in Fig. 2.12, and the corresponding changes in its apparent diameter (9.9″ to 64.5″) were immediately discovered by Galileo with his telescope; they prove that the Sun is also at the center of the true orbit of Venus. The planet reaches its maximum apparent brightness, as can be seen from Fig. 2.12, in the neighborhood of its greatest elongations. In the lower conjunction, Venus (and Mercury) can pass in *front* of the Sun. These transits of Venus were previously of interest for the determination of the distance to the Sun or of the solar parallax.

An *outer planet*, for example Mars (Fig. 2.13), is nearest to us at its *opposition*; it then has its culmination at midnight true local time, when it has its largest apparent diameter and is most favorable to observe. When it is near the Sun in the sky, it is said to be in *conjunction*. The outer planets do not go through the full cycle of *phases* from "full" to "new". The angle between the Earth and the Sun as seen from the planet is called the phase angle φ. The fraction of the planet's hemisphere which faces the Earth and is dark is thus $\varphi/180°$. The phase angle of an outer planet passes through a maximum at the *quadratures*, i.e. when the planet and the Sun form an angle of 90° in the sky. The largest phase angle of Mars is 47°, that of Jupiter is only 12°.

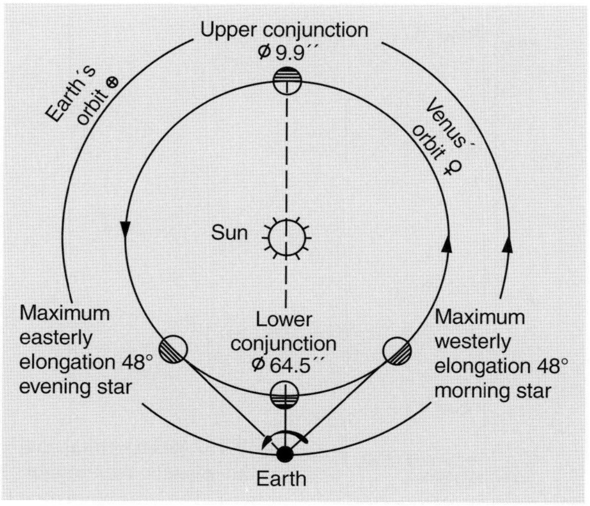

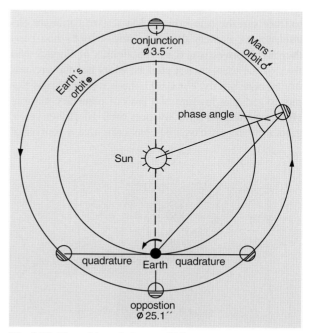

Fig. 2.12. The orbit and phases of Venus, an inner planet. The elongation of Venus in the sky cannot exceed ±48° (Mercury, ±28°). The phases are similar to those of the Moon. The maximum brightness occurs near the maximum elongation

Fig. 2.13. The orbit and phases of Mars, an outer planet. The maximum brightness and largest angular diameter of 25.1″ occur when it is in opposition

Planetary Period. The true time taken by a planet to revolve around the Sun is termed its *sidereal* period. The *synodic* period is the time it requires for a revolution in the sky relative to the Sun, i.e. the time between two successive, corresponding conjunctions. In analogy to the Moon, the following relations hold for the planets (subtraction of the angular velocities):

$$\frac{1}{\text{Synodic Period}} = \left| \frac{1}{\text{Sidereal Period}} - \frac{1}{\text{Sidereal Year (Earth)}} \right|. \quad (2.2)$$

For example, the sidereal period of Mars is found from the observed synodic period of 780 d and the length of the sidereal year, 365 d, to be 687 d.

Planetary Orbits. Kepler was the first to derive the true form of *Mars' orbit*, by combining pairs of observations of Mars which were taken at intervals equal to Mars' sidereal period, that is when the planet had returned to the same point on its orbit. Thus he could localize Mars from two points on the Earth's orbit separated by 687 d; since the latter was sufficiently well-known, he was able to trace out the true orbit of Mars. Two fortunate circumstances, namely that the conic sections had been thoroughly investigated by Apollonius of Pergae; and that, of the then-known planets, Mars has the greatest orbital eccentricity, $e = 0.093$, made it possible for Kepler to arrive at his first two laws of planetary motion. He discovered the third law only after 10 further years, guided by the unshakeable conviction that a "universal harmony" must somehow express itself in the orbits of the planets.

The complete description of the orbit of a planet or comet (Sect. 2.2.2) around the Sun requires the *orbital elements* defined in Fig. 2.14:

1. The semimajor axis a. It is measured either in terms of the semimajor axis of the Earth's orbit = 1 astronomical unit (AU), or in kilometers.

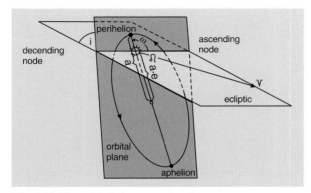

Fig. 2.14. The orbital elements of a planet or comet

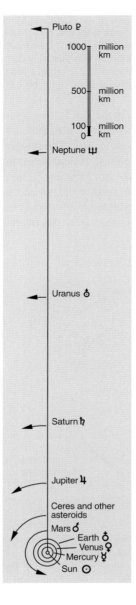

Fig. 2.15. Mean orbital radii of the planets (semimajor axes a). The arcs correspond to the average motion per year; in one year, Venus circles the Sun 1.62 times and Mercury 4.15 times

2. The eccentricity e [distance at perihelion $a(1-e)$; distance at aphelion $a(1+e)$].
3. The inclination i of the orbital plane relative to the ecliptic.
4. The length of the ascending node ☊ (angle from the Aries point ♈ to the ascending node).
5. The distance ω of the perihelion from the node (angle from the ascending node to the perihelion). The sum of the two angles, ☊ $+\,\omega$, where the first is measured in the ecliptic and the second in the orbital plane, is called the perihelion length, $\widetilde{\omega}$.
6. The period P (sidereal period, measured in tropical years) or the mean daily motion n (in degrees or arc seconds per day).
7. The epoch E or the time of the passage through the perihelion, T.

The orbital elements a and e determine the size and shape of the orbit (Fig. 2.6), i and ☊ fix the orbital plane, and ω determines the position of the orbit within the plane. The motion along the orbit depends upon P and T; the period P is, in fact, determined by the value of a, from Kepler's third law, apart from small corrections.

To conclude this section, in Table 2.2 we have listed the orbital elements of the planets which are of interest to us here (Fig. 2.15). In addition to the planets known in ancient times, we have included Ceres, the brightest of the many thousand planetoids or *asteroids* which orbit between Mars and Jupiter, as well as, of course, Uranus (discovered in 1781), Neptune (1846), and Pluto (1932).

Pluto's orbital elements (eccentricity $e = 0.25$, inclination $17.1°$), along with its diameter and mass (cf. Sect. 3.5.1) make it quite different from the other outer planets. Pluto spends some time within the orbit of Neptune, and its orbital period is in "resonance" (3:2) with that of Neptune. Pluto's last perihelion occurred in 1989.

We will consider the orbits of the many "small objects" which have been recently discovered (since 1992) outside the orbit of Neptune (and to some extent outside that of Pluto) in Sect. 3.5.2.

2.2.2 Comets and Meteors

Our planetary system also contains, besides the planets and their moons and the asteroids, *comets* and *meteors* ("shooting stars").

The comets are characterized by their extended, diffuse outer shell, the *coma*, which surrounds the bright core or nucleus. When a comet approaches the Sun, its coma develops a noticeable *tail* (see Fig. 3.27).

In ancient times and in the Middle Ages, the comets were relegated to the Earth's upper atmosphere, in accordance with the doctrine of the immutability of the heavenly regions. The first proof that this doctrine was incorrect was given by Tycho Brahe's precise observations of the comets of 1577 and 1585, from which he

Table 2.2. Some orbital elements of the planets (epoch 1990.0)

Name		Symbol	Sidereal period [yr]	Semimajor axis of the orbit [AU]	[10^6 km]	Eccentricity e	Inclination i to ecliptic	Mean orbital velocity v [km s^{-1}]
Inner Planets	Mercury	☿	0.241	0.387	57.9	0.206	7.0°	47.9
	Venus	♀	0.615	0.723	108.2	0.007	3.4°	35.0
	Earth	♁	1.000	1.000	149.6	0.017	—	29.8
Outer Planets	Mars	♂	1.881	1.524	227.9	0.093	1.8°	24.1
	Asteroids:	Ceres	4.601	2.766	413.5	0.077	10.6°	17.9
	Jupiter	♃	11.87	5.205	779	0.048	1.3°	13.1
	Saturn	♄	29.63	9.576	1432	0.055	2.5°	9.6
	Uranus	♅	84.67	19.28	2884	0.047	0.8°	6.8
	Neptune	♆	165.5	30.14	4509	0.010	1.8°	5.4
	Pluto	♇	251.9	39.88	5966	0.248	17.1°	4.7

derived parallaxes that showed that the comet of 1577, for example, must be at least six times more distant from the Earth than the Moon. Isaac Newton recognized that the comets move on elongated ellipses or parabolas around the Sun, with eccentricities equal to or a little smaller than 1. His contemporary E. Halley improved the method of determining their orbits and in 1705 was able to show that *Halley's comet* of 1682, which bears his name, must have a period of 76 yr. From Kepler's third law, the semimajor axis of its orbit is equal to $2 \cdot 76^{2/3} = 36$ AU, i.e. its aphelion lies somewhere outside the orbit of Neptune. Halley's orbit calculation also showed that the bright comet of 1682 was identical with those of 1531 and 1607, and he could predict its return in 1758. All together, 29 appearances of Halley's comet have been witnessed since the year 240 B.C. The most recent perihelion of Halley's comet took place in 1986.

In searching for comets, observations in the *far infrared*, which are sensitive to the thermal radiation emitted by the comet's dust (cf. Sect. 3.6.1), can be very useful. Thus, the first infrared observation satellite, IRAS, was able to discover six new comets in the single year 1983, while they were still extremely faint in the visible region.

Nomenclature. The modern (since 1995) method of naming comets is similar to that used for asteroids (Sect. 3.3.1): when a comet is discovered, a letter corresponding to the time interval (in half months) and a number for the order of the discovery within that interval are appended to the year of discovery; the name of the discoverer is often included, also, with a number added after the name to distinguish different comets discovered by the same person. Furthermore, the type of comet is indicated by a prefix: P/ stands for comets with short orbital periods, less than 200 yr, C/ for comets of long period or non-periodic comets, and D/ for comets which have ceased to exist. (A/ indicates an asteroid or planetoid). If a clearcut determination of the orbit has been made, a series number is put before the P/ or D/, as is also done in the case of asteroids.

For example, Halley's comet has been denoted (retroactively) as 1P/Halley or 1P/1682 Q1 Halley; the two bright comets of the 1990's are C/1995 O1 Hale-Bopp (Fig. 3.28) and C/1996 B2 Hyakutake.

Previously, comets were given a provisional name according to their year of discovery, with lower-case letters indicating the order of discovery in that year. Then, after determination of the orbit, they were denoted by the year of their perihelion followed by Roman numbering in the order of their passage through perihelion in that year; comets of short period were characterized by P/ before the year. Examples are 1P/Halley = 1835 III = 1910 II = 1986 III, C/1956 R1 Arend-Roland = Arend-Roland 1956 h = 1957 III, 29P/1927V1 Schwassmann-Wachmann 1 = 1925 II = 1974 II.

One of the reasons for introducing the new terminology was that comets are not always clearly distinguishable from asteroids in terms of their appearance and orbit. For example, the asteroid 2060 Chiron (semimajor orbital axis $a = 3.7$ AU) developed a coma many years after its initial discovery and thus became a comet (95 P/Chiron). Conversely, comets can lose their coma (through "sealing" of their surfaces by a thin layer of dust) and become inactive asteroids.

The Orbits and Periods of the Comets. The orbits of the comets fall into two groups:

a) Comets of *long orbital period* with periods of revolution between 10^2 and 10^6 years, and perhelia in the range of 1 AU (high probability of discovery): the inclinations i of their orbits are randomly distributed; direct and retrograde motions are about equally probable. The eccentricities e are slightly smaller than or nearly equal to 1, so that their orbits are long, thin ellipses or, as a limiting case, parabolas. Hyperbolic orbits, with $e > 1$, are only rarely produced by perturbations from the major planets.

C/1995 Hale-Bopp – the third-brightest comet which has ever been observed – had its perihelion in 1997. According to its calculated orbit, it last passed near the Sun 4210 years ago and will return in 2380. The changes in its period are caused by perturbations from the major planets.

Since the velocities v of these comets are quite small at large distances from the Sun, it is likely that they originate from a cloud which accompanies the Sun on its path in the Milky Way. It is estimated that this "Oort Cloud" (with a diameter of about 50 000 AU) contains some 10^{12} comets, whose total mass is however equal to only about 50 times the mass of the Earth.

b) Comets of *short orbital period*, with periods less than 200 years. They move for the most part in elliptical orbits with small inclinations i (mean inclination of this group $\bar{i} \simeq 20°$). Nearly half of these comets have their aphelia in the range 5...6 AU, i.e. in the neighborhood of Jupiter's orbit ($a_J = 5.2$ AU). The mean values of the orbital data correspond to $a \simeq 3.6$ AU and $e \simeq 0.56$. This "Jupiter family" of comets evidently arose through capture of longer-period comets by the planet Jupiter. Similar comet families associated with other planets have not been identified with certainty.

Encke's comet (2P/1786B1 Encke) has the shortest known orbital period of 3.3 yr. The comets 29P/Schwassmann-Wachmann ($a = 6.1$ AU, $e = 0.11$) and 39P/1943 G1 Oterma ($a = 4.0$ AU, $e = 0.14$) move on nearly circular orbits.

Comets decay in time by breaking up and by vaporization of cometary matter, so that the swarm of short-period comets must be constantly replenished by capture of new objects. Model calculations for the short-period comets with their small orbital inclinations indicate a flattened, ring-shaped reservoir outside the orbit of Neptune at a distance of 50 to 500 AU from the Sun, the *Kuiper ring*, containing 10^8 to 10^{12} comets.

We can thus observe only a vanishingly small fraction both of the long-period and of the short-period comets.

The Breakup of Comets. Not only the orbits of the comets are unstable with respect to gravitational perturbations; the comets themselves are also not stable. Since they consist of weakly bound matter (cf. Sect. 3.6.2), they can be strongly influenced by a near passage through the gravitational field of Jupiter or the Sun. In fact, several comets have been observed to break up: the comet 3D/1772 E1 Biela broke into two parts in 1846, which were both last observed in 1852 and have since disappeared. Comet 16P/1889 N1 Brooks 2 was discovered after a near passage by Jupiter, when it already consisted of (at least) two separate comets. Only the larger of these fragments has "survived" until the present as a returning comet with a period of 6.9 yr.

An unusual object was discovered in 1993 by C. and G. Shoemaker und D. Levy: a comet which had already split up into a chain of many fragments which were arrayed along its orbit. Each fragment was surrounded by its own dust-filled coma. This comet D/1993 F2 Shoemaker-Levy 9 orbited Jupiter(!) on a strongly elliptical orbit, approaching to within about 0.3 AU and having a period of around 2 yr. In 1992, it approached the big planet so closely that it was torn apart; one orbital period later, in July 1994, about 20 fragments crashed one after another within 6 days at velocities near 60 km s^{-1} onto the surface of Jupiter. We shall describe this spectacular event later in connection with the structure of Jupiter's atmosphere (Sect. 3.4.1).

Meteors. The showers of "falling stars" or meteors which on certain days of the year appear to emanate from a particular point in the sky (their "radiant", similar to a vanishing point in perspective drawings) are, as indicated e.g. by their periodicity, simply the debris of comets, whose orbits crossed or nearly crossed that of the Earth. In some cases, the cometary matter seems to be fairly well concentrated along the orbit, so that especially lively meteor showers are observed with the corresponding period: an example is the famous case of the Leonids (radiant $\alpha \simeq 152°$, $\delta \simeq +22°$), which were observed by Alexander v. Humboldt from Venezuela in 1799, and can be attributed to the comet 55P/1865Y1 Tempel-Tuttle, with a 33 year period. In addition, there are sporadic meteors which show no recognizable periodicity.

The fact that "falling stars" are really small objects from space, which enter the Earth's atmosphere and are heated to incandescence by it, was first demonstrated in 1798 by two students in Göttingen, Brandes and Benzenberg. They made observations of meteors from two sufficiently distant points and calculated the altitude of their tracks. Earlier, E.F.F. Chladni had shown that *meteorites* are just meteoric material (from larger meteors) which has reached the Earth's surface.

Comets are not the only source for the meteors and meteorites. They can also be debris from larger objects originally in the asteroid belt (Sect. 3.3).

Comets or meteors whose orbits were originally hyperbolic, i.e. objects which have entered the Solar System from outer space, have not been observed either among the comets or among the meteors.

2.2.3 Distance Determination, the Doppler Effect and Aberration of Light

We need now to turn to the important question of how the *distance* from the Earth to the Sun (or, more precisely, the semimajor axis of the Earth's orbit, which we have defined as the astronomical unit, 1 AU) can be measured in established units such as kilometers. Astronomers prefer to refer to the *solar parallax* $\pi_\odot$, i.e. the angle which the equatorial radius of the Earth, $R_E = 6378$ km (known from geodetic measurements) subtends when seen from the center of the Sun. The solar parallax is too small for direct measurement, as can be done in the case of the lunar parallax. Therefore, the first step is to determine the distance to a planet or asteroid whose orbit brings it sufficiently near to the Earth, by making observations from several observatories in both the Northern and Southern Hemispheres. In former times, Mars at opposition or Venus at its lower conjunction were used; more recently, extensive series of observations of the oppositions of the asteroid Eros, which is more favorable for such determinations, have been carried out. These astronomical methods of distance determination have recently acquired serious competetion from radar techniques, which allow the precise determination not only of the distance to the Moon, but also that to Venus and Mars by direct measurement of the round-trip travel time of reflected radiofrequency signals, together with the velocity of light c which is known from terrestrial experiments. Combining the individual measurements allows the calculation of the radius of the Earth's orbit using Kepler's 3rd law; the details involve difficult celestial-mechanical calculations.

Instead of the radius of the Earth's orbit, the *orbital velocity* of the Earth can be determined in [km s^{-1}] by using the *Doppler effect* (Fig. 2.16), which was derived in 1842 by C. Doppler in connection with considerations of the orbital motions of binary stars using the wave theory of light:

When a radiation (light) source moves relative to an observer with a radial velocity v (the velocity component in the direction of the line of sight between source and observer), the wave length λ_0 of the radiation (or its frequency $\nu_0 = c/\lambda_0$) appears to be shifted by $\Delta\lambda = \lambda - \lambda_0$ or $\Delta\nu = \nu - \nu_0$, where:

$$\frac{\Delta\lambda}{\lambda_0} = -\frac{\Delta\nu}{\nu_0} = \frac{v}{c} \quad (v \ll c) . \qquad (2.3)$$

A source whose relative motion is *away* from the observer (by definition a positive velocity) produces an *increase* in the wavelength λ, i.e. a *red shift* of the spectral lines and a reduction in their frequency ν, and *vice versa*. (One speaks of a "red shift" even for example

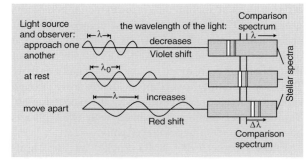

Fig. 2.16. The Doppler effect, $\Delta\lambda/\lambda_0 = v/c$

in the radiofrequency range, when a shift occurs *away* from red, to still larger wavelengths.)

In practice, either the radial velocity of a fixed star relative to the Earth is followed over most of a year, using the Doppler shift in its spectral lines, or else the relative velocity between the Earth and e.g. Venus is determined from the frequency shift of reflected radar signals (reflection from a moving mirror yields twice the frequency shift quoted above).

A similar consideration formed the basis of the historically important first *measurement of the velocity of light* by O. Römer in 1675: he determined the *frequencies of revolution* ν of the larger moons of Jupiter from their transits behind the planet's disk. When the Earth is moving away from Jupiter, these frequencies appear to be reduced, due to the finite propagation velocity c of light; when it is moving towards the planet, the frequencies are apparently increased. Starting from the contemporary value of the solar parallax, Römer obtained a relatively good numerical value for the velocity of light. The fact that he anticipated Doppler's principle (2.3) by nearly two hundred years before its first spectroscopic application is seldom mentioned in texts on astronomy and physics.

The first terrestrial determination of the velocity of light was carried out by A. Fizeau in 1849 using a rotating chopper in a light beam.

Another effect which is due to the finite velocity of light is the *aberration* of light (Fig. 2.17), which was discovered by J. Bradley in 1728 while he was attempting to measure the parallax of fixed stars. When a star is observed which would be perpendicular to the Earth's orbit for an observer at rest, the Earth's (and thus the observer's) orbital motion makes it necessary to incline

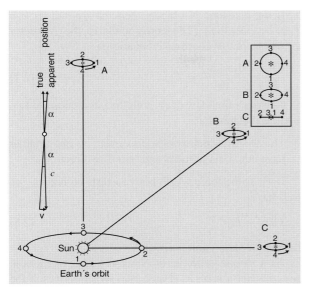

Fig. 2.17. The aberration of light. The light from the stars seems to be deflected in the direction of the velocity vector of the Earth (*left*) by an angle v/c, where v is the component of the Earth's velocity perpendicular to the light propagation direction and c is the velocity of light. A star at the celestial pole thus describes a circle of radius equal to the ratio of Earth's velocity/velocity of light: $\alpha = 20.49''$; in the eclipte, it moves on a line of maximum extension $\pm\alpha$; and in between, it describes an ellipse (*indicated at the upper right*). (∗) shows the true position of the star; an observer looking to the right sees the star at intervals of 1/4 year in the positions *1–2–3–4*

the telescope by a small angle in the direction of the Earth's orbital velocity v in order to see the star; this aberration angle is v/c. As a result, a star at the celestial pole appears to describe a small circle in the course of a year (Fig. 2.17, upper right); in the ecliptic it seems to move back and forth on a straight line, and in between, it moves on an ellipse. The usual analogy is that of an astronomer who is walking rapidly through rain which is coming straight down, and must incline his umbrella forwards in order to stay dry.

This demonstration, like our elementary derivation of the Doppler effect, is imperfect; it neglects the principle of the constancy of the velocity of light in all frames of motion, independently of the motion of the source, which was demonstrated by the experiment of A.A. Michelson and E.W. Morley in 1887. A consistent explanation of all the effects of order v/c in that experiment, and especially of those of order $(v/c)^2$, is given

by Einstein's special relativity theory (1905). We treat the relativistic Doppler effect in Sect. 4.2.1.

Finally, we give a summary of the numerical values of the quantities discussed and their various relationships:

Equatorial radius of the Earth	$R_E = 6.378 \cdot 10^6$ m
Solar parallax (equatorial horizontal parallax)	$\pi_\odot = 8.794''$
Astronomical Unit [AU] = semimajor axis of the Earth's orbit	$A = R_E/\pi_\odot = 1.496 \cdot 10^{11}$ m
Velocity of light	$c = 2.9979 \cdot 10^8$ m s^{-1}
Light propagation time for 1 AU	$\tau_A = A/c = 499.0$ s
Average orbital velocity of the Earth	$\bar{v} = 2.98 \cdot 10^4$ m s^{-1}
Rotation of the Earth: Angular velocity	$\omega_E = 7.29 \cdot 10^{-5}$ s^{-1}
Velocity at the equator	$v_{\rm rot} = \omega_E R_E = 465$ m s^{-1}
Aberration constant (Epoch 2000)	$\chi = \bar{v}/c = 9.94 \cdot 10^{-5}$ $\alpha = 20.496''$

2.3 Mechanics and Gravitational Theory

Following the tedious and even dangerous beginnings made by Galileo Galilei and Johannes Kepler, Isaac Newton in his *Principia* (1687) gave the first complete treatment of the mechanics of terrestrial and extraterrestrial systems. Combining it with his law of gravitation, in the same work he derived Kepler's laws and many other observed regularities in the motions of objects in the Solar System. It is not surprising that the further development of celestial mechanics remained an important field for the great mathematicians and astronomers for nearly two hundred years.

We begin our treatment here by stating the most important concepts and laws of *mechanics* and the *theory of gravitation* for later reference: Newton's laws of motion (Sect. 2.3.1), conservation of momentum (Sect. 2.3.2), conservation of angular momentum (Sect. 2.3.3), the law of conservation of energy (Sect. 2.3.4) and the virial theorem (Sect. 2.3.5), and finally Newton's Law of Universal Gravitation (Sect. 2.3.6). Its application to celestial mechanics is then treated in the next section, Sect. 2.4.

2.3.1 Newton's Laws of Motion

We formulate Newton's three *basic laws of mechanics* in modern language:

1. A body remains in a state of rest or moves with constant velocity in a straight line as long as it is not subject to an external force (law of inertia).

We denote *velocities* in direction and magnitude by a vector (arrow), v; in a similar way, we denote the vector quantity *force* by F: the vector quantities are indicated in print by using boldface characters. The magnitude of a vector quantity, e.g. of the velocity – the "length of the arrow", so to speak – is an ordinary number (scalar quantity) and is denoted by absolute value signs $|v|$ or simply by the corresponding nonboldface character, v.

For the addition and subtraction of vector quantities, we use the vector parallelogram: the two vectors are represented by their components in a rectilinear coordinate system x, y, z, i.e. by their projections on the corresponding axes. Thus for example $v = \{v_x, v_y, v_z\}$.

If a moving body has the mass m, the vector mass times velocity

$$p = mv \qquad (2.4)$$

is defined as its momentum. This important concept allows the formulation of the *law of momentum*:

2. The rate of change with time of the momentum of a body is proportional to the magnitude of the external force which acts on it, and is in the direction of that force.

Mathematically formulated, we write for *one* body (where t is the time):

$$\frac{dp}{dt} = \frac{d(mv)}{dt} = F, \qquad (2.5)$$

Law 1 is clearly just a special case for $F = 0$ of law 2. We can interpret the velocity v as the rate of change of the position vector r with components $\{x, y, z\}$, and we thus write $v = dr/dt$ and, for $m = $ constant, also

$$m \frac{d^2 r}{dt^2} = F. \qquad (2.6)$$

This formulation (force = mass × acceleration) is however valid only for constant masses, while (2.5) remains generally valid within special relativity theory, where the mass depends on the velocity (Sect. 4.2.2).

If we consider N objects, denoted by indices $k = 1, 2, 3 \ldots N$, then (2.5) corresponds to the N vector equations or $3N$ coordinate equations:

$$\frac{dp_k}{dt} = \frac{d(m_k v_k)}{dt} = F_k. \qquad (2.7)$$

Newton's final law deals with the interactions of two bodies; it states that:

3. The forces which two bodies exert on one another have equal magnitudes and opposite directions.

If F_{ik} is the force which body i exerts on body k, we thus have:

$$F_{ik} = -F_{ki}, \qquad (2.8)$$

the law of *action* and *reaction*.

As a simple example of Newton's laws of motion, we consider a mass m (Fig. 2.18a) which moves at the end of a string of length r in a horizontal circle with the constant velocity $v = |v|$. Its *angular velocity* is then

$$\omega = \frac{d\varphi}{dt} = \frac{2\pi}{P} = \frac{v}{r} \qquad (2.9)$$

(angle φ in arc measure, units = radians; P is the period of the rotational motion.) Conversely, $v = \omega \cdot r$.

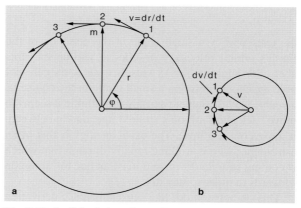

Fig. 2.18a,b. Calculation of centrifugal force. (**a**) A point mass m on a circular orbit, with the position vector r at the times $1, 2, 3, \ldots$ and the velocity vector $v = dr/dt$ tangential to the orbit. The magnitude of v is $v = r \cdot d\varphi/dt = \omega r$. (**b**) Hodograph. The velocity vector v at the times $1, 2, 3, \ldots$ The acceleration vector dv/dt points in the direction of the tangent to the hodograph and is therefore parallel to $-r$. The magnitude of the acceleration is $v \cdot d\varphi/dt = v^2/r = \omega^2 r$

If we trace successive velocity vectors v using the same starting point, thus drawing a socalled hodograph (Fig. 2.18b), we can immediately see that the acceleration $|dv/dt|$ is equal to $(v/r)v$ and points towards the center of the circular path. We thus obtain the law of *centrifugal force* which was derived by C. Huygens even before Newton:

$$F = mv^2/r = m\omega^2 r . \tag{2.10}$$

Newton's 3rd law states that the string pulls on its anchor at the center with the *same* force that pulls the object towards the center of its circular orbit.

When the details of the internal structure of a sufficiently small object of mass m are not important in mechanics, we speak of a *point mass*. In the theory of planetary motion, for example, we can consider the Earth to be a point mass.

From Newton's three laws for the motion of individual point masses, we go on to the equations of motion for a *system* of point masses. From them, we derive the three conservation laws of mechanics, which we shall often use later in this book.

2.3.2 The Conservation of Linear Momentum

In a system of N point masses m_k, which we denumerate by the indices i or k ($i, k = 1, 2, 3 \ldots N$), we distinguish between internal forces F_{ik}, which for example mass i exerts on mass k, and external forces $F_k^{(e)}$, which are exerted on the mass k from "outside". This point mass obeys the equation of motion (2.7):

$$\frac{dp_k}{dt} = F_k^{(e)} + \sum_i F_{ik} . \tag{2.11}$$

Here and in the following sections, all summation signs $\sum$ imply a summation from $k = 1$ to N over all values of k. Defining the total linear momentum of the system by

$$P = \sum_k p_k \tag{2.12}$$

an the total external force acting on the system by

$$F = \sum_k F_k^{(e)} , \tag{2.13}$$

we find, summing (2.11) over all values of k and using (2.8), the *equation of motion* for the system, analogous to that for a single point mass:

$$\frac{dP}{dt} = F . \tag{2.14}$$

If no external forces act on the system ($F = 0$), then the law of *conservation of total linear momentum* applies:

$$P = \sum_k p_k = \text{const} . \tag{2.15}$$

The content of equations (2.14) and (2.15) can be perhaps more intuitively clarified if we define the position vector R of the *center of gravity* of our system of overall mass $\mathcal{M} = \sum m_k$:

$$\mathcal{M} R = \sum_k m_k r_k . \tag{2.16}$$

With this definition, (2.14) is converted into the equation of motion of the center of gravity:

$$\mathcal{M} \frac{d^2 R}{dt^2} = F \tag{2.17}$$

in analogy to that of a single point mass. We can see from this last equation that in the case of no net external force, $F = 0$, the center of gravity must exhibit a straight-line inertial motion with a constant velocity, $dR/dt = $ constant (in agreement with Newton's 1st law).

2.3.3 Conservation of Angular Momentum: the Area Theorem

We first consider a point mass m (Fig. 2.19), which is free to rotate about a fixed point 0 on a lever arm r. A force F acts on m. This force tends to rotate the mass around an axis through 0 and perpendicular to the plane containing r and F; only the tangential component of the force $|F| \sin \alpha$ is effective in producing such a rotation, where α is the angle between r and F. The quantity "lever arm r times effective force component $|F| \sin \alpha$", drawn as a vector perpendicular to the plane containing r and F, is, in mathematical terms, the vector product $r \times F$, and in physical terms, it is the moment of the force F around 0, also known as the *torque* $M = r \times F$.

The torque is the equivalent of the force for rotational motion; similarly, we can define a quantity

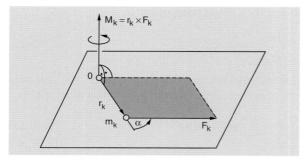

Fig. 2.19. The torque $M_k = r_k \times F_k$. The absolute magnitude of M_k is $|r_k| \cdot |F_k| \sin \alpha$, i.e. equal to the area of the parallelogram spanned by r_k and F_k. The order of the factors is defined in such a way that a right-handed screw which is screwed in the direction from r_k to F_k would move parallel to M_k

corresponding to the linear momentum $p = mv$, the *angular momentum*:

$$L = r \times p = r \times mv. \tag{2.18}$$

We now multiply Newton's equation of motion, (2.14), vectorially from the left by r:

$$r \times \frac{dp}{dt} = r \times F = M, \tag{2.19}$$

and then take the time derivative of the angular momentum

$$\frac{dL}{dt} = \frac{dr}{dt} \times p + r \times \frac{dp}{dt}. \tag{2.20}$$

Here, the first term on the right is zero, since $dr/dt = v$, $p = mv$, and the vector product of the two parallel vectors v and p is zero. We thus obtain the law of conservation of angular momentum, initially for a point mass k:

$$\frac{dL_k}{dt} = M_k. \tag{2.21}$$

If we consider a *system* of N point masses m_k, we define the total torque of all internal and external forces relative to the fixed point 0 as

$$M = \sum_k r_k \times F_k$$

$$= \sum_k r_k \times \left(F_k^{(e)} + \sum_i F_{ik} \right) \tag{2.22}$$

and the total angular momentum is defined by

$$L = \sum_k r_k \times p_k = \sum_k r_k \times m_k v_k. \tag{2.23}$$

The equation of motion then becomes

$$\frac{dL}{dt} = M, \tag{2.24}$$

i.e. the time derivative of the angular momentum vector is equal to the combined torque due to all the forces.

If only *central forces* act in our system, such as gravitational forces, for example, which act along the line connecting two point masses, then the contribution to M from internal forces vanishes and the right side of (2.24) contains only the torque due to external forces, $M^{(e)}$.

If, furthermore, no external forces are present or if the resultant torque is zero, then $dL/dt = 0$ and the important theorem of *conservation of angular momentum* holds for the system of masses:

$$L = \sum_k r_k \times m_k v_k = \text{const}. \tag{2.25}$$

2.3.4 Conservation of Energy

If a point mass m moves under the influence of a force F along a differential path element dr, the latter making an angle α with the force F, then the net *work*:

$$dA = |F| |dr| \cos \alpha = F \cdot dr \tag{2.26}$$

is performed. The scalar product of the two vectors is denoted by a raised dot "·". If we calculate the work performed on passing along a finite section of an orbit, $1 \to 2$, using the Newtonian equation of motion (2.5) and $v = dr/dt$, we find:

$$\int_1^2 F \cdot dr = \int_1^2 \frac{d(mv)}{dt} \cdot dr = \left. \frac{1}{2} mv^2 \right|_1^2. \tag{2.27}$$

The scalar quantity which appears on the right, or its sum over several point masses, is called the *kinetic energy*:

$$E_{\text{kin}} = \frac{1}{2} mv^2 = \frac{p^2}{2m}. \tag{2.28}$$

In the following, we shall limit our discussion to *conservative* forces, among which in particular is the gravitational force (Sect. 2.3.6). Such forces may be derived from a *Potential* $\Phi(r)$ (which may, for example, depend only on position):

$$F = -m \operatorname{grad} \Phi = -m \left\{ \frac{\partial \Phi}{\partial x}, \frac{\partial \Phi}{\partial y}, \frac{\partial \Phi}{\partial z} \right\}. \quad (2.29)$$

Here, the gradient of Φ is a vector, whose components are given in (2.29) for a cartesian coordinate system. The zero point of the potential can be arbitrarily chosen; it then follows from (2.29) that $F \cdot dr$ is an exact differential and the work performed by the forces, $\int F \cdot dr$, is independent of the actual orbits of the point masses and depends only on the initial and final positions of m.

If we now introduce the *potential energy*

$$E_{\text{pot}} = m \Phi \quad (2.30)$$

then from (2.27), the important theorem of *conservation of energy* holds:

$$E = E_{\text{kin}} + E_{\text{pot}} = \text{const}, \quad (2.31)$$

i.e. the sum E of kinetic energy E_{kin} and potential energy E_{pot} is constant.

In a *system* of point masses m_k with conservative forces, (2.31) holds when E_{kin} and E_{pot} are interpreted as the sum over the contributions from all the point masses k. In a system with gravitational forces – as in many other cases – E_{kin} depends only on the velocities and E_{pot} only on the positions of the point masses.

2.3.5 The Virial Theorem

An important basis for the understanding of many problems, not only in celestial mechanics but also in the structure and evolution of stars and stellar systems, is the virial theorem of R. Clausius (1870). In an isolated system of point masses, we consider the time variation of the quantity $\sum p_k \cdot r_k$. By differentiating, we find

$$\frac{d}{dt} \sum_k p_k \cdot r_k = \sum_k \frac{dp_k}{dt} \cdot r_k + \sum_k p_k \cdot v_k, \quad (2.32)$$

or – cf. (2.27) – using the equation of motion (2.7) and the kinetic energy defined above (2.28),

$$\frac{d}{dt} \sum_k p_k \cdot r_k = \sum_k F_k \cdot r_k + 2 E_{\text{kin}}. \quad (2.33)$$

If we now average this expression over a sufficiently long time, the mean value of the left side is equal to zero (as long as r_k and p_k are finite for all the point masses, and therefore $\sum p_k \cdot r_k$ remains bounded), and we obtain the *virial theorem*:

$$\overline{E_{\text{kin}}} = -\frac{1}{2} \overline{\sum_k F_k \cdot r_k}. \quad (2.34)$$

In order to evaluate the *virial* $\overline{\sum_k F_k \cdot r_k}$, we must know the forces F_k. We initially consider only a *single* particle of mass m under the influence of a central (conservative) force $\propto r^n$ with the corresponding potential $\Phi \propto r^{n+1}$ (2.29), where $r = (x^2 + y^2 + z^2)^{1/2}$ is the distance from the origin of the coordinate system. The virial is then defined with $r(d\Phi/dr) = (n+1)\Phi$ and (2.30) by:

$$\overline{\sum_k F_k \cdot r_k} = -m \cdot r \frac{d\Phi}{dr} = -(n+1) \overline{E_{\text{pot}}}, \quad (2.35)$$

i.e. $(-n-1)$ multiplied by the time average of the potential energy. In the important case that the point masses are held together by *gravitational forces* (Sect. 2.3.6), $n = -2$, so that the total energy E distributes itself in such a way between the kinetic and the potential energies that the *time averaged* energies obey

$$\overline{E_{\text{kin}}} = -\frac{1}{2} \overline{E_{\text{pot}}} \quad (2.36)$$

and, with (2.31),

$$\overline{E_{\text{kin}}} = -E \quad (2.37)$$

holds.

For a *system* of gravitationally interacting point masses, E_{pot} consists of individual contributions from r_{ik}^{n+1}, where $r_{ik} = |r_i - r_k|$ is the relative distance of two point masses i and k. As can be verified by calculation, for such a system Eqns. (2.36) and (2.37) likewise hold.

2.3.6 The Law of Gravitation. Gravitational Energy

In order to obtain a theory of celestial motions, Newton had to add his law of universal gravitation (ca. 1665) to his basic laws of mechanics:

Two point masses m_i and m_k at a distance r attract each other along the line joining them with a force of magnitude

$$F(r) = G \frac{m_i m_k}{r^2}. \quad (2.38)$$

By carrying out an integration, which we shall not repeat here, Newton first demonstrated that exactly the same law (2.38) holds for the attraction of two *spherical* mass distributions of finite size (e.g. the Sun, a planet, etc.) as for the corresponding point masses. He then verified the law of gravitation by assuming that free fall near the Earth's surface (Galileo) and the lunar orbit are both dominated by the gravitational attraction of the Earth.

The *acceleration* (force/mass) for a free fall can be determined in experiments with falling objects, or, more precisely, with a pendulum. Its exact value depends on the location on the Earth, and is reduced by the centrifugal acceleration of the Earth's rotation (Sect. 3.2.1). Its standard value for an "average" Earth, after correction for the Earth's rotation, is equal to:

$$g_E = 9.81 \text{ m s}^{-2}. \quad (2.39)$$

On the other hand, the Moon moves on its circular orbit of radius $r_M = 60.3 R_E = 3.84 \cdot 10^8$ m with a velocity $v = 2\pi r_M / P$ ($P = 1$ sidereal month $= 27.32$ d $= 2.36 \cdot 10^6$ s) and is thus subject to the acceleration (Fig. 2.18):

$$g_M = \frac{v^2}{r_M} = \frac{4\pi^2 r_M}{P^2} = 2.72 \cdot 10^{-3} \text{ m s}^{-2}. \quad (2.40)$$

(g_M should not be confused with the acceleration which an object experiences on the Moon's surface.)

The accelerations g_E and g_M are, in fact, related as the inverse squares of the radii of the Earth, R_E, and of the lunar orbit, r_M, i.e.

$$g_E : g_M = \frac{1}{R_E^2} : \frac{1}{r_M^2} \simeq 3620. \quad (2.41)$$

The numerical value of the universal *gravitation constant* G appears here [cf. (2.38)] only as a product with the initially likewise unknown mass of the Earth, $\mathcal{M}_E$. Similarly, in other astronomical problems, G occurs only in a product with the mass of the attracting celestial object. Thus G can, for fundamental reasons, *not* be determined by astronomical measurements; instead, it must be found from terrestrial experiments.

The first attempt at such a measurement was that of N. Maskelyne in 1774, who investigated the deflection of a plumb line by a large mass (a mountain). In 1798, H. Cavendish used a torsional balance, and in 1881, P. v. Jolly performed measurements with a suitable beam balance. The result of modern determinations is

$$G = (6.673 \pm 0.010) \cdot 10^{-11} \text{ m}^3 \text{ s}^{-2} \text{ kg}^{-1}. \quad (2.42)$$

Gravitational Energy. For applications to celestial mechanics and for the understanding of the structure of planets and stars, we still need the gravitational energy, i.e. the potential energy which results from the action of gravitational forces. We first consider a point mass m at a distance r from a pointlike central mass $\mathcal{M}$; we can readily convince ourselves that the potential or the potential energy associated with Newton's law of gravitation (2.38) is given by:

$$\Phi(r) = -\frac{G\mathcal{M}}{r}, \quad E_{\text{pot}}(r) = -m\frac{G\mathcal{M}}{r}. \quad (2.43)$$

The potential energy can also be considered as the work which must be performed when (in a virtual experiment) we remove the mass $\mathcal{M}$ from the distance r under the influence of the gravitational force $G(\mathcal{M}m/r^2)$ to infinity ($r \to \infty$):

$$E_{\text{pot}}(r) = -G \int_r^\infty \frac{\mathcal{M}m}{r^2} dr = -m\frac{G\mathcal{M}}{r}. \quad (2.44)$$

The negative sign of the gravitational potential indicates a *binding energy* or an attractive interaction: if the distance between the two masses is reduced, then energy is set free from the overall system; e.g. it might be converted into kinetic energy of m. Conversely, energy must be put into the system if we are to increase the distance.

If we now consider a *spherically symmetrical* mass distribution of radius R, then – as we already mentioned – the gravitational action in the *exterior* region $r \geq R$ remains the same. A sphere thus acts as if its mass were concentrated at its center. We shall not consider here the more complicated potential in the interior of the sphere, which in contrast to that outside the sphere is dependent on the mass distribution.

The Acceleration of Gravity. For a sphere of radius R having the mass $\mathcal{M}$, we give the acceleration of gravity: $g(r) = d\Phi(r)/dr$. According to (2.43), outside the sphere we have

$$g(r) = \frac{G\mathcal{M}}{r^2} \quad (r \geq R), \qquad (2.45)$$

i.e. the same value as for a point mass $\mathcal{M}$. However, at a point r within the sphere, the acceleration of gravity is

$$g(r) = \frac{G\mathcal{M}(r)}{r^2} \quad (r \leq R); \qquad (2.46)$$

it thus depends only on the mass within r,

$$\mathcal{M}(r) = 4\pi \int_0^r \varrho(r')r'^2 \mathrm{d}r',$$

$$\mathrm{d}\mathcal{M}(r) = 4\pi r^2 \varrho(r)\mathrm{d}r. \qquad (2.47)$$

Naturally, both expressions give the same value at the surface of the object, R:

$$g \equiv g(R) = \frac{G\mathcal{M}}{R^2}, \qquad (2.48)$$

since $\mathcal{M} \equiv \mathcal{M}(R)$.

We immediately apply (2.48) and calculate from $g_\mathrm{E} = 9.81\,\mathrm{m\,s^{-2}}$, the acceleration of gravity at the Earth's surface, with the known value of the radius of the Earth, $R_\mathrm{E} = 6378$ km, its *mass* $\mathcal{M}_\mathrm{E}$ and – with $\mathcal{M} = (4\pi/3)\overline{\varrho}R^3$ – its *mean density*. We obtain

$$\mathcal{M}_\mathrm{E} = 5.97 \cdot 10^{24}\,\mathrm{kg}, \quad \overline{\varrho}_\mathrm{E} = 5500\,\mathrm{kg\,m^{-3}}. \qquad (2.49)$$

We shall return to a discussion of the geophysical significance of these numbers later.

The Homogeneous Sphere. Finally, we calculate the gravitational energy E_G of a homogeneous sphere of radius R with a constant interior density $\varrho(r) = \varrho_0$. Starting with a sphere of radius r and mass $\mathcal{M}(r)$ – as an "intermediate step" – we add a shell of mass $\mathrm{d}\mathcal{M}(r)$, and find a gain in potential energy $\mathrm{d}E_\mathrm{G} = -G\mathcal{M}(r)\mathrm{d}\mathcal{M}(r)/r$. With

$$\mathcal{M}(r) = \frac{4\pi}{3}r^3 \varrho_0, \quad \mathrm{d}\mathcal{M}(r) = 4\pi r^2 \varrho_0 \mathrm{d}r \qquad (2.50)$$

we immediately obtain

$$E_\mathrm{G} = -G\int_0^R \frac{\mathcal{M}(r)\mathrm{d}\mathcal{M}(r)}{r} = -3G\left(\frac{4\pi}{3}\right)^2 \varrho_0^2 \frac{R^5}{5}$$

$$= -\frac{3}{5}\frac{G\mathcal{M}^2}{R}. \qquad (2.51)$$

For an arbitrary density $\varrho(r)$, only the numerical factor changes, so that in general, the gravitational energy of a sphere of radius R has the order of magnitude

$$E_\mathrm{G} \simeq \frac{G\mathcal{M}^2}{R}. \qquad (2.52)$$

2.4 Celestial Mechanics

Celestial mechanics deals with the application of Newton's law of universal gravitation and the laws of mechanics to the motions of the various bodies which make up our Solar System.

First, we go back to Newton's Principia and derive Kepler's laws from the basic equations of mechanics and the law of gravitation, in order to gain a deeper understanding of the laws themselves and of the numerical values which occur in them. We apply Kepler's laws immediately to the orbits of the planets (Sect. 2.4.1) and to the determination of the masses of the celestial objects (Sect. 2.4.2). We then calculate – with later applications in mind – the escape velotcity (Sect. 2.4.3) and the rotational energy of a rigid body (Sect. 2.4.4). We treat precession (Sect. 2.4.5), the problem of the tides (Sect. 2.4.6), and finally, we consider from the modern point of view the relation between the Ptolemaic and the Copernican worldviews (Sect. 2.4.7). Applications of celestial mechanics to artificial satellites and space probes are treated later, in Sect. 2.5.1.

2.4.1 Kepler's First and Second Laws: Planetary Orbits

The mass of the Sun is clearly very much greater than that of any of the planets, so we start by treating the Sun as fixed and calculate the radius vectors $\boldsymbol{r}$ or $r = |\boldsymbol{r}|$ of the planets from the center of the Sun. The mutual attraction of the planets among themselves, their *perturbations*, will be left out of our calculations.

The motion of *one* planet around the Sun is governed by the central force $G\mathcal{M}_\odot m/r^2$, where $\mathcal{M}_\odot$ is the mass of the Sun, and $m(\ll \mathcal{M}_\odot)$ is that of the planet. Conservation of angular momentum (2.25) thus holds, i.e. the angular momentum vector of the planet

$$\boldsymbol{L} = \boldsymbol{r} \times m\boldsymbol{v} \qquad (2.53)$$

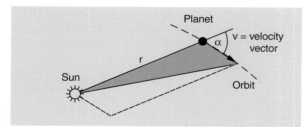

Fig. 2.20. The surface area swept out by a planet per unit time within its orbit, $\frac{1}{2}|\mathbf{r} \times \mathbf{v}| = \frac{1}{2}rv\sin\alpha$

is constant in magnitude and direction. Here, $\mathbf{v}$ is the planet's orbital velocity vector and $\mathbf{r}$ its position vector. Both remain in the same plane, perpendicular to $\mathbf{L}$, the spatially fixed *orbital plane* of the planet. The magnitude of $|\mathbf{r} \times \mathbf{v}| = rv\sin\alpha$ (Fig. 2.20) is twice the area which is swept out by the radius vector $\mathbf{r}$ of the planet per unit time. Conservation of angular momentum is therefore identical with the statement that each planet follows an orbit in a fixed plane with constant areal velocity (Kepler's 2nd and part of his 1st laws).

We shall not reproduce here the somewhat tedious calculations which show that the orbit of a point mass (planet, comet, etc.) under the influence of a central force proportional to $1/r^2$ must be a *conic section*, i.e. a circle (eccentricity $e = 0$), an ellipse ($0 < e < 1$), a parabola ($e = 1$), or a hyperbola ($e > 1$) with the Sun in one focus (Kepler's first law).

2.4.2 Kepler's Third Law: Determination of Masses

We shall carry out the calculation of planetary motion and the derivation of Kepler's 3rd law only for *circular orbits*. The general calculation can be found in any text on classical mechanics. However, in view of future applications, we shall not restrict the mass m of the planet to be much less than the mass $\mathcal{M}_\odot$ of the Sun. We therefore consider the motion of two masses around their common center of gravity, and, on the other hand, the relative motion of the two masses with respect to, for example, the larger one. Let m_1 and m_2 be the masses and a_1 and a_2 their respective distances from the center of gravity S (Fig. 2.21). Then $a = a_1 + a_2$ is their mutual distance. Using the definition of the center of gravity, we have:

$$a_1 : a_2 : a = m_2 : m_1 : (m_1 + m_2) \quad \text{or}$$

$$m_1 a_1 = m_2 a_2 = \frac{m_1 m_2}{m_1 + m_2} a = \mu a, \quad (2.54)$$

where

$$\mu = \frac{m_1 m_2}{m_1 + m_2}, \quad \frac{1}{\mu} = \frac{1}{m_1} + \frac{1}{m_2} \quad (2.55)$$

is termed the *reduced mass*. For each of the two masses m_1 and m_2, the force of gravitational attraction, Gm_1m_2/a^2, must be balanced by the centrifugal force. Calling the orbital period of the system P, we find for the centrifugal force on m_1:

$$\frac{m_1 v_1^2}{a_1} = \left(\frac{2\pi}{P}\right)^2 m_1 a_1. \quad (2.56)$$

Because of action = reaction, as expressed in (2.54), a similar equation and the same value of P naturally hold for m_2. Applying this equation once more, together with (2.54), we obtain after rearranging *Kepler's 3rd law*:

$$\frac{a^3}{P^2} = \frac{G}{4\pi^2}(m_1 + m_2). \quad (2.57)$$

If we relax the requirement of a circular orbit (we shall not carry out the explicit calculation here), the masses m_1 and m_2 are found to move on similar conic sections around the center of gravity S; the relative orbit is also a corresponding conic section. Instead of orbital radii, we then have orbital *semimajor axes*, which are likewise denoted by a_1, a_2 and a. We thus obtain a generalized version of Kepler's 3rd law (2.57). In the Solar System, as we shall see, the mass of e.g. the largest planet, Jupiter, is only about $1/1000$ the mass of the Sun. To this accuracy, we can therefore set the expression $m_1 + m_2$ on the right-hand side of the equation equal to the Sun's mass, $\mathcal{M}_\odot$.

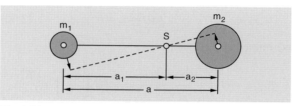

Fig. 2.21. The motion of the masses m_1 and m_2 about their common center of gravity S. From the definition of the center of gravity, $m_1 a_1 = m_2 a_2$

By inserting the numerical values of a and P for the orbit of the Earth or another planet into (2.57), we obtain the *solar mass* $\mathcal{M}_\odot$. Its apparent disk radius of 16′, together with a, gives the solar radius, $R_\odot$. From $\mathcal{M} = (4\pi/3)R^3\bar{\varrho}$, we finally obtain the mean density of the Sun, $\bar{\varrho}_\odot$. The numerical values are:

Mass $\mathcal{M}_\odot = 1.989 \cdot 10^{30}$ kg ,
Radius $R_\odot = 6.960 \cdot 10^8$ m ,
Density $\bar{\varrho}_\odot = 1409$ kg m^{-3} . (2.58)

The uncertainty in these values is in each case about ± one unit in the last place.

In a similar way, we calculate the *masses of the planets* (Table 3.1) by applying Kepler's 3rd law to the orbits of their moons. If the planet has no moons, then mutual perturbations of the planets must be used; this is of course much more difficult. Space travel has opened up further possibilities of determining planetary masses by observing the orbital motions of artificial satellites and space probes.

2.4.3 Conservation of Energy and the Escape Velocity

It is instructive to consider the "Kepler problem" again from the point of view of energy conservation (2.31); this requires that the total energy E, e.g. for a planet,

$$E = E_{\text{kin}} + E_{\text{pot}} = \frac{1}{2}mv^2 - m\frac{G\mathcal{M}_\odot}{r} \quad (2.59)$$

remain constant in time. We can immediately see again from this that the velocity increases on going from the aphelion to the perihelion. For a *circular orbit* ($v = v_0$), the centrifugal force is equal to the attractive force of the two masses, and therefore $mv_0^2/r = m\,G\mathcal{M}_\odot/r^2$, or

$$\frac{1}{2}mv_0^2 = \frac{1}{2}m\frac{G\mathcal{M}_\odot}{r} \ . \quad (2.60)$$

Thus, the kinetic energy E_{kin} is equal to $-E_{\text{pot}}/2$, a special case of the virial theorem (2.36) for the periodic motion of a single point mass attracted by a central force.

If, on the other hand, we consider a *parabolic orbit* (e.g. a nonperiodic comet), which is the limiting case of an elliptical orbit extending to infinity, we find that at infinity, both the kinetic and the potential energy, and thus the total energy, are zero. Since E is a constant of the motion, from $E = 0$ it follows using (2.59) that:

$$\frac{mv^2}{2} = G\frac{\mathcal{M}_\odot m}{r} \quad \text{or}$$
$$E_{\text{kin}} = -E_{\text{pot}} \ . \quad (2.61)$$

At the same distance from the Sun, the parabolic velocity, i.e. v on a parabolic orbit, is larger by a factor of $\sqrt{2}$ than the velocity on a circular orbit. For example, if the mean velocity of the Earth is 30 km s^{-1}, that of a comet or meteor swarm which meets us on a parabolic orbit would be $30\sqrt{2} = 42.4$ km s^{-1}.

Conversely, the conservation of energy in the limiting case of a parabolic orbit, (2.61), also determines the *escape velocity*,

$$v_e = \sqrt{\frac{2G\mathcal{M}_\odot}{r}} = \sqrt{2}\,v_0 \ , \quad (2.62)$$

i.e. the smallest velocity which an object must have in order to leave a circular orbit at a distance r from a central mass $\mathcal{M}$ and to escape its gravitational field. For example, the escape velocity for leaving from the Earth and completely escaping the Solar System is $v_e = 42.4$ km s^{-1}, while that for escaping from the Earth's gravitational field starting from the surface of the Earth is 11.2 km s^{-1} (cf. Sect. 2.5.1).

2.4.4 Rotation and the Moment of Inertia

For later applications, we give the intrinsic or proper angular momentum of a *rigid* rotating body, in which all the mass points m_k have the same angular velocity vector $\boldsymbol{\omega}$, so that their (linear) velocities are:

$$\boldsymbol{v}_k = \boldsymbol{\omega} \times \boldsymbol{r}_k \ . \quad (2.63)$$

The *proper angular momentum* ("spin") of the body follows then from (2.25):

$$S = \sum_k \boldsymbol{r}_k \times m_k(\boldsymbol{\omega} \times \boldsymbol{r}_k) = I\,\boldsymbol{\omega} \ , \quad (2.64)$$

where

$$I = \sum_k m_k r_k^2 \quad (2.65)$$

is the *moment of inertia* of the body relative to the particular axis of rotation.

For a "dumbbell", i.e. two masses m_1 and m_2 which rotate rigidly about their common center of gravity at a fixed distance a (Fig. 2.21), it follows from (2.54) that

$$I = \frac{m_1 m_2}{m_1 + m_2} a^2 = \mu a^2 , \qquad (2.66)$$

where μ again denotes the reduced mass.

For a homogeneous sphere of mass $\mathcal{M}$ and radius R, we have:

$$I = \tfrac{2}{5} \mathcal{M} R^2 . \qquad (2.67)$$

The *rotational energy* of the body, i.e. the total kinetic energy (2.28) which is associated with the (rigid) rotation of all the point masses, is given by

$$E_{\rm rot} = \frac{1}{2} \sum_k m_k v_k^2 = \frac{1}{2} \sum_k m_k r_k^2 \omega^2 \quad \text{or}$$

$$E_{\rm rot} = \frac{1}{2} I \omega^2 = \frac{S^2}{2I} . \qquad (2.68)$$

2.4.5 Precession

Besides the Kepler problem, Newton solved numerous other problems in celestial mechanics. We shall first consider *precession*, at least in outline form.

The "wobble" of the Earth's axis around the pole of the ecliptic is a result of the same physical phenomenon as the wobbling of a top under the influence of gravity: the equatorial bulge of the Earth is pulled into the plane of the ecliptic by the Sun and the Moon, whose mass we can imagine to be distributed around their orbits over the long precession period of the Earth's axis (the Moon has only a small orbital inclination). The angular momentum vector S of the Earth's rotation, which is essentially parallel to the rotation axis, reacts to this torque M (Fig. 2.22) according to (2.24). The rate of change in the Earth's spin S, given by $dS/dt = M$, produces the circling motion of S, i.e. of the Earth's axis, on a cone of constant vertex angle, as can be seen in Fig. 2.22. Numerical calculation yields the correct period for the lunisolar precession (Fig. 2.22).

2.4.6 The Tides

Next, we turn briefly to the old problem of the tides. Galilei became involved in controversies with his contemporaries over a wholly incorrect theory of the 12-hour alternation of ebb and flow (the tides being known to Mediterranean peoples only by hearsay); they were a major contributor leading to his unfortunate trial by the Inquisition. It was again Newton who developed the elements of a *static* theory of the tides (Fig. 2.23).

We assume here for simplicity that the common center of gravity of the Earth and the Moon lies at the midpoint of the Earth. In fact, it is at a distance of 0.73 Earth radii from the midpoint, but still within the Earth (2.54). The gravitational acceleration due to the

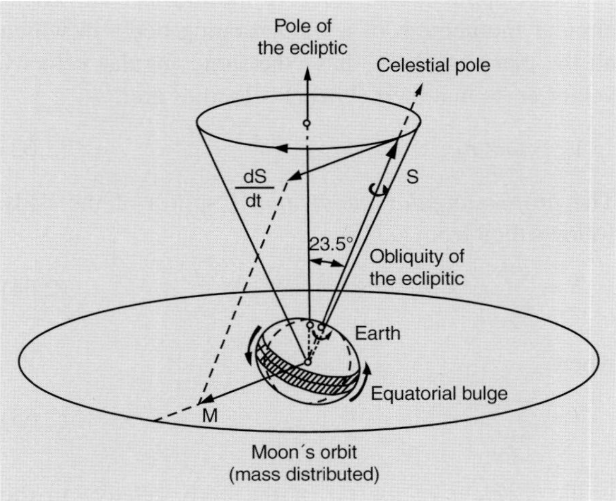

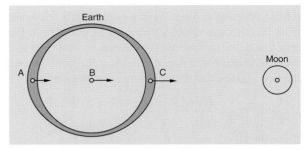

Fig. 2.23. The static theory of the tides. The acceleration of the three points indicated relative to the Moon corresponds to their differing distances from it: A, lower culmination of the Moon: $a - a_{\rm G}$; B, center of the Earth (center of gravity): a; C, upper culmination of the Moon: $a + a_{\rm G}$. The (rigid) Earth takes on, as a whole, the acceleration a. Therefore, at the points A and C, an acceleration $a_{\rm G}$ is left over, and it produces a high water level at both these points

Fig. 2.22. Lunar precession

Moon is exactly compensated by centrifugal acceleration only at the center of gravity (midpoint of the Earth); at every other point, a *tidal acceleration* occurs as their difference. The acceleration vector $\boldsymbol{a}_G$ at the surface of the Earth (at the Earth's radius R_E), acting for example on a water droplet in the ocean, points *upwards* both at the upper and at the lower culmination of the Moon, i.e. at the passages of the Moon through the meridian in the north and the south. We thus expect that, corresponding to the apparent motion of the moon (1 lunar day = 24 h 51 min), two "high-water peaks" and two "low-water valleys" will continually circle the Earth, getting later by 51 minutes every day.

The difference between the accelerations of gravity due to the attractive mass $\mathcal{M}$ of the Moon acting at the Earth's center (at a distance r from $\mathcal{M}$) and that at e.g. the lower culmination (at a distance $r + R_E$, point A in Fig. 2.23) is then

$$a_G = \frac{G\mathcal{M}}{r^2} - \frac{G\mathcal{M}}{(r+R_E)^2} \simeq \frac{2G\mathcal{M}}{r^3} R_E . \quad (2.69)$$

The tidal acceleration is thus only a tiny fraction of the Earth's gravitational acceleration, $a_G \simeq 10^{-7} g_E$. It is a *differential* acceleration and is therefore proportional to $1/r^3$; it is thus a stronger function of the distance than the gravitational acceleration and force. For this reason, the tidal force due to the Sun is only about half as great as that from the Moon, although the Sun's mass is enormously greater. At the new Moon and full Moon, the tidal forces of the Moon and the Sun act together and produce a spring tide; in the first and last quarters of the Moon, they oppose each other and we have a neaptide.

This static theory given here in fact only explains the rough features of tidal phenomena; the exact theory of tides takes into account in detail the Earth's rotation, the fact that the center of gravity is not at the Earth's midpoint, and the spatial and directional variation of the vector $\boldsymbol{a}_G$ on the Earth's surface. It is then found that in reality this vector's *horizontal* components "shove up" the water into tidal peaks and are therefore the real "driving force" behind tidal motion. Furthermore, the deep areas and coastal regions of the oceans play a decisive role in the *dynamic* theory of the tides (G.H. Darwin), in which forced oscillations that are produced in the oceanic basins by tidal forces with the different periods of the apparent solar and lunar motions lead to the propagation of waves.

The average *tidal height* (twice the amplitude of the tidal oscillations) can in the presence of resonances be as large as 16 m. The tidal forces also cause lifting and sinking of the Earth's crust of up to ± 0.3 m.

Tidal friction in the oceans and in the Earth's crust produces a braking efffect on the Earth's rotation (a decrease in the angular velocity, $\Delta\omega_E$) and thus an *increase* in the *length of the day*. The tiny increase of ca. 2 ms per 100 yr is readily measurable over geologic times (Sect. 14.1.4).

According to the law of conservation of angular momentum (2.64), the angular momentum of rotation lost by the Earth, ΔS_E, must be transferred to the orbital motion of the Moon; thus

$$\Delta S_E + \Delta L = I_E \Delta\omega_E + \mathcal{M}\Delta(\omega r^2) = 0 \quad (2.70)$$

holds. The Moon's intrinsic angular momentum can be neglected here. Since furthermore, from Kepler's 3rd law ($\omega^2 r^3 =$ const), the angular momentum L is proportional to the square root of the orbital radius r, the Moon must be gradually moving away from the Earth, i.e. its orbital period $P = 2\pi/\omega$ is increasing. This will continue to occur until the length of the day and the orbital period of the Moon have equalized.

In recent times, the use of Lunar Laser Ranging has permitted the determination of the distance to the Moon with a precision of around 1 cm, so that this effect of tidal friction can be directly measured. The value – averaged over a few years – has been found to be (3.7 ± 0.2) cm yr^{-1}.

2.4.7 The Ptolemaic and the Copernican Worldviews

Before leaving the theory of planetary motions, we want to consider the decisive change which was made in going from the Ptolemaic to the Copernican worldview, looking back from our modern standpoint (Fig. 2.24).

Viewed from the Sun (heliocentric system), let us denote the position vector of a planet by $\boldsymbol{r}_p$ and that of the Earth by $\boldsymbol{r}_E$. Then as seen from the Earth (geocentric system), the position of the planet is given by the difference vector

$$\boldsymbol{R} = \boldsymbol{r}_p - \boldsymbol{r}_E . \quad (2.71)$$

We turn once more to the geocentric view of Ptolemy: (a) for the *outer* planets, e.g. Mars, we begin by first drawing the vector r_P from the Earth and letting it rotate as it did before around the Sun. Following (2.71), we then add the vector $-r_E$ (the position vector of the Sun as seen from the Earth) and thus obtain the position vector R of the planet as seen from the Earth. The imaginary circle which r_P describes with its sidereal period around the Earth is the Ptolemaic deferent. The other circle around the point r_P which is described by the planet at the end of the vector $-r_E$ with the sidereal period of the Earth is the Ptolemaic epicycle. (b) For the *inner* planets, it appeared more reasonable to the Alexandrians to first let the larger vector circle the Earth with a period of one sidereal year as the deferent, and then to allow the smaller vector r_P to circle the point $-r_E$ with the sidereal period of the planet as the epicycle.

So far, the geocentric constructions still correspond exactly to the equation $R = r_P - r_E$. Scheme (b), applied to *all* the planets, would represent the worldview of Tycho Brahe.

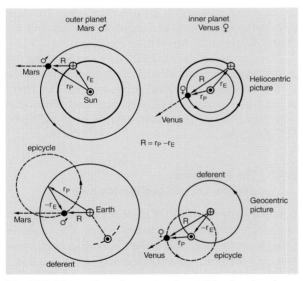

Fig. 2.24. The motion of an outer planet (Mars) and an inner planet (Venus) on the celestal sphere, represented in heliocentric and in geocentric pictures (☉ = Sun, ⊕ = Earth). The dashed arrow indicates in each case the position of the planet in the sky. Correspondingly, for:

	outer planets	inner planets
deferent	r_P	$-r_E$
epicycle	$-r_E$	r_P

In fact, however, we have not yet completed the transition to the Ptolemaic system: as long as only the positions of the planets in the sky, i.e. only their directions but not their distances, could be measured, only the direction, not the magnitude, of the vector R was of importance. One could therefore draw the vector R on a different scale for each planet. This means that the vectors

$$R'_P = A_P \cdot R_P \qquad (2.72)$$

with a fixed but purely arbitrary numerical factor A_P for each planet, give a completely satisfactory representation of the motions of the planets in the sky in the Ptolemaic system.

Now we can see clearly what was lost in our stepwise return from the Copernican to the Ptolemaic system:

1. The change of coordinate system means relinquishing a simple mechanical explanation.
2. The scale factors A_P leave the positions of the planets in the celestial sphere unchanged, but the mutual relations of the planetary positions among themselves are lost.
3. The fact that in the Ptolemaic system the annual period, corresponding to the motion of r_E, is introduced *independently* for *each* planet, is another indication of the "clumsiness" of the ancient worldview.

However, it is important to recognize that a purely *kinematic* consideration of the planetary motions in the sky did not allow a decision to be made between the ancient and the modern views. Only Galileo's observations with his telescope (1609) led to progress: (a) Jupiter, with its freely orbiting moons, could be seen as a "model" of the Copernican solar system. (b) The phases of Venus are determined by the relative positions of the Sun, Earth, and Venus. The smallness of the phase angle e.g. for Jupiter also supports Copernicus. Even the very *idea* of a celestial *mechanics* presupposes, as we should keep in mind, that the basic *similarity* of cosmic and terrestrial matter and its physics be recognized.

2.5 Space Research

Astronomy owes to the advent of space travel a great expansion in available observational methods and with it, an enormous increase of knowledge. On the one hand, the spectral regions which were previously inaccessible due to absorption in the atmosphere have been opened up for observations: the far ultraviolet, X-ray and gamma-ray regions, and at longer wavelengths, the infrared out to the millimeter-wave range and finally the radiofrequencies which are reflected by the ionosphere, with wavelengths longer than ca. 30 m. On the other hand, near flybys of objects in the Solar System and landings onto them have become commonplace, and they offer a wealth of new observational possibilities.

As a conclusion to our discussion of celestial mechanics in the previous Sect. 2.4, we shall in the next section take a brief look at the orbits of artificial satellites and spacecraft (Sect. 2.5.1). Then, in considering the development of space travel in Sect. 2.5.2, we confine ourselves to those aspects relevant to *astronomical* observations. Finally, we give an overview of the space missions which have investigated our nearest neighbor, the Moon (Sect. 2.5.3), and other objects in the Solar System (Sect. 2.5.4).

Here, we will consider only the gravitational field of the Earth, limit ourselves to circular orbits, and neglect the braking effect of the atmosphere.

2.5.1 The Orbits of Artificial Satellites and Space Vehicles

A satellite in a circular orbit near the Earth (Earth's radius $R_E = 6378$ km) must have, according to (2.60), a velocity near $v_0 = 7.9$ km s^{-1} and an orbital period near $P_0 = 84.4$ min. In a larger orbit with a radius r, from Kepler's 3rd law the velocity must be $v = v_0(r/R_E)^{-1/2}$ and the period must be $P = P_0(r/R_E)^{3/2}$. Of particular importance is the fact that the period becomes equal to one sidereal day ($P = 1$) for $r = 6.6\,R_E$. A geostationary satellite at this distance then "stands" at an altitude of nearly 36 000 km above a fixed spot on the Earth's surface.

In order to allow a spacecraft which lacks its own drive motor to escape from the gravitational field of the Earth (alone) and travel to infinite distances, it must start with at least the parabolic or escape velocity $v_0\sqrt{2} = 11.2$ km s^{-1} (cf. (2.61)).

Jules Verne, in his great novel "From the Earth to the Moon" (1865), suggested a solution to this problem by using a gigantic cannon. This would not work, however, since the initial velocity of a cannon shell cannot be much greater than the velocity of sound in the gases from the explosive charge, which is too small.

Higher velocities can be reached by using *rockets*. We shall first deal with the mechanics of rocket propulsion by considering a rocket without the influence of gravity (i.e. on a horizontal test ramp or in space) and without air resistance. Let the mass of the rocket's hull and other parts plus fuel at time t be given by $m(t)$. The change in $m(t)$ per unit time, corresponding to the mass of combustion products expelled from the rocket per unit time, is dm/dt. If we now call the expulsion velocity of the combustion gases u, then a momentum equal to $-(dm/dt)u$ will be transferred to the rocket per unit time. Considering the acceleration of the rocket from the point of view of an astronaut moving along with it, we obtain the Newtonian equation of motion:

$$m(t)\frac{dv}{dt} = -\frac{dm}{dt}u \quad \text{or}$$
$$\frac{dv}{u} = -\frac{dm}{m(t)}. \quad (2.73)$$

By integrating and using the initial condition that for ($t = 0$, $v = 0$) the mass is equal to m_0, we find the *rocket equation*

$$v = u \ln \frac{m_0}{m(t)}. \quad (2.74)$$

If we had taken the (homogeneous) gravitational field in the neighborhood of the Earth into account, for a vertical takeoff we would have had on the right the additional well-known term $-g_E t$.

Taking the relatively favorable values $u = 4$ km s^{-1} and $m_0/m = 10$ at burnout, we find a final velocity (without air resistance!) for our single-stage rocket of $v = 9.2$ km s^{-1}. For real space travel, it is thus necessary to use *multi-stage* rockets, whose basic principle quickly becomes clear on repeatedly applying the rocket equation (2.74).

An extremely effective method which is frequently applied in space travel is the *flyby* (or "slingshot") technique, which produces a change in the orbit of a space probe without consuming fuel, making use of the gravitational field of a massive planet, e.g. Jupiter. We denote the velocity of the probe as v and that of the planet as v_P – both relative to the Sun –, then, according to energy conservation (2.59), the magnitude of the *relative velocity* $|v - v_P|$ at a large distance from the planet before

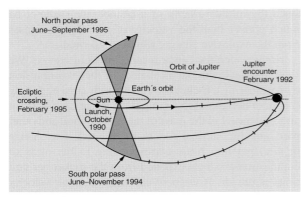

Fig. 2.25. The orbit of the space probe Ulysses, seen from an angle of 15° above the ecliptic. Its second passage through perihelion occurred in May, 2000. The tick marks along the orbit indicate time intervals of 100 days

and after the flyby remains constant, but its *direction* in general will change. More important, however, is the fact that the space probe can be accelerated or decelerated relative to the Sun, since its relative velocity and the velocity of the planet v_P add vectorially. As an example, in Fig. 2.25 we show the redirection of the space probe Ulysses by Jupiter out of the plane of the ecliptic; and in Fig. 2.28, we illustrate the flight paths of the two Voyager probes to the outer planets.

2.5.2 Astronomical Observations from Space

A preliminary phase of space research was the investigation of the celestial objects from simple *rockets*, which can be carried out with a relatively moderate investment; the observational time is, however, limited to a few minutes. The V-2 rockets constructed in Germany during the 2nd World War could attain heights of nearly 200 km. After the war in the USA, the V-2 and improved rockets were employed for the investigation of the highest layers of the atmosphere (the ozone layer and the ionosphere) and of short-wavelength solar radiation.

In 1957, Soviet researchers succeeded in launching the first *artificial satellite* Sputnik 1 into an orbit of altitude between 225 and 950 km. Since then, thousands of satellites for research, television and communications as well as for military purposes have been launched. For long series of astronomical observations, a good stabi-

Table 2.3. Some Astronomical Satellites

Period of operation	Designation		Purpose (wavelength or energy range)
1972–1981	OAO-3 = Copernicus	Orbiting Astronomical Observatory	Stellar UV radiation (95–156 nm) with high spectral resolution, ...
1975–1982	COS-B		Gamma radiation (50–5000 MeV)
1978–1996	IUE	International Ultraviolet Explorer	UV spectroscopy (115–320 nm)
1978–1981	HEAO-2 = Einstein	High Energy Astronomical Observatory	X-rays (0.1–3 keV), first imaging telescope
1983 (10 months)	IRAS	Infrared Astronomical Satellite	Infrared (12–100 μm)
1989–1993	HIPPARCOS	High Precision Parallax Collecting Satellite	Stellar parallaxes
1989–1993	COBE	Cosmic Background Explorer	Microwave radiation (0.1–10 mm)
1990–	HST	Hubble Space Telescope	2.4 m telescope for the visible and UV regions
1990–1998	ROSAT	(German) Röntgen-Satellit	83 cm X-ray telescope (0.1–2 keV)
1991–2000	CGRO	Compton Gamma Ray Observatory	Gamma radiation (0.02–30 000 MeV)
1992–	EUVE	Extreme Ultraviolet Explorer	Extreme ultraviolet (6–76 nm)
1993–2000	ASCA	Advanced Satellite for Cosmology and Astrophysics	X-rays (0.5–12 keV)
1995–1998	ISO	Infrared Space Observatory	Infrared (2.5–240 μm)
1997–	HALCA	Highly Advanced Laboratory for Communications and Astronomy	Radio waves (1.6–22 GHz) für VLBI
1999–	Chandra		X-rays (0.2–10 keV)
1999–	FUSE	Far Ultraviolet Spectroscopic Explorer	Far UV (90.5–118.7 nm) with high spectral resolution
1999–	XMM Newton	X-Ray Multi-Mirror Mission	X-rays (0.1–12 keV)

2.5 Space Research

lization of the satellite to at least $1''$ is important. This was first achieved by NASA (the US National Aeronautics and Space Administration) with the launching of the OSO and OAO series (Orbiting Solar and Astronomical Observatories) in 1962/1975.

Some satellites which have been important in astronomical observations are listed in Table 2.3. We will describe their instrumentation in part in Chap. 5, and give information about the observational results obtained by the various satellites and returned to the Earth by telemetry when we discuss the corresponding celestial objects.

The first *manned* space flight was ventured upon in 1961 by J. Gagarin, who circled the Earth in Vostok 1. The manned space vehicles of the American "Gemini" series (beginning in 1965) and the Soviet "Soyuz" series (beginning in 1967) were able to perform orbital and coupling maneuvers. In 1971, Soyuz 10 coupled to an unmanned vehicle – the first step on the way to a *space station*, on which a crew of several persons could stay in space for a longer period of time and then be replaced by a new crew.

In 1973/74, three different crews in the space laboratory *Skylab* carried out astronomical, biological, and technical experiments. In the joint American-Soviet Apollo-Soyuz mission in 1975, observational data in the extreme ultraviolet spectral region were obtained for the first time. The Soviet/Russian space station *Mir* (1986–2001) was constructed from several modules and has also been used for astronomical observations, among other things.

Since the early 1980's, the launching of satellites and space research missions by NASA has been carried out increasingly by using the manned, reusable *Space Shuttle* instead of multistage launch vehicles. The explosion and loss of the Space Shuttle "Challenger" in 1986 represented a considerable setback for the space program. The Shuttles (Fig. 2.26) are fully maneuverable and can land on a runway.

The vertical takeoff system consists of two solid-fuel rockets whose motors can be recovered and reused, as well as a large, non-reusable fuel tank and the actual Shuttle (orbiter), which resembles an airplane. Depending on the payload, orbits of maximum altitude from 200 to as much as 1000 km can be reached. Satellites such as the geostationary satellites mentioned above, which require orbits of higher altitude, employ an additional rocket motor. Several satellites can be launched

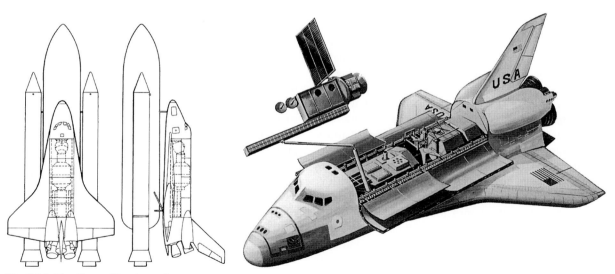

Fig. 2.26. The Space Shuttle. *Left part*: the takeoff configuration, in which the Shuttle is attached to the large fuel tank and two solid-fuel rockets. Its overall height is about 56 m. *Right part*: the Shuttle with its payload transport bay opened. The manipulator arm can be used to launch or retrieve a satellite, for example. (With the kind permission of Octopus Books Ltd., London)

2. Classical Astronomy

on one Shuttle flight, containers with instrumentation may be set out and retrieved, and repairs and maintenance may be performed on satellites. The Shuttle itself can serve as an orbiting laboratory (Spacelab). It will also be used in the construction of the ISS (*I*nternational *S*pace *S*tation), which began in 1998.

Current astronomical research from space is carried out mostly with unmanned satellites and space probes instead of manned observational missions or space stations.

2.5.3 The Exploration of the Moon

The investigation of celestial objects outside the Earth using space vehicles naturally began with the *Moon*, our nearest neighbor in space. In 1959, the Soviet probe Luna 1 approached the Moon to within 5000 km, Luna 2 made a "hard" landing on its surface, and Luna 3 transmitted the first pictures of the back side of the Moon. The unmanned flights of the American Ranger space probes (1961–65) also ended in hard landings. A new era in the exploration of the Moon began in 1966 with the sucessful soft landing of Luna 9 and the placement of Luna 10 in a lunar orbit, making it the Moon's first artificial satellite.

Preparations for *landing men on the Moon*, a difficult and costly undertaking, were carried out by NASA in the period from 1966–69. Possible landing sites were investigated by various Lunar Orbiter and soft-landed Surveyor probes, and manned test flights with Lunar orbiting as well as a number of simulation experiments were carried out. The manned landing itself was accomplished using the following principle: While one of the astronauts circles the Moon a number of times in the command unit, the two others land on the surface in an auxiliary craft (LM: Lunar Module). After completing their mission there they take off with the aid of a rocket motor, couple their LM to the command unit, and start on the return journey after discarding the now-useless LM. The main difficulties in landing result from the heating during reentry into the Earth's atmosphere (heat shield!) and the temporary interruption of radio communication because of ionization of the air.

The first landing on the Moon was made in 1969 with Apollo 11 by the astronauts N. Armstrong, M. Collins,

Fig. 2.27. The Lunar Module "Eagle" of Apollo 11 has landed in the mare tranquillitatis. Astronaut E. Aldrin is setting up the seismic station. Many small craters can be recognized in the foreground; pieces of rock are scattered on the ground, and the shoes of the astronauts' space suits leave sharp imprints

and E. Aldrin in the mare tranquillitatis (Fig. 2.27). They left behind a research station, containing among other things a seismometer, and brought back 22 kg of lunar rocks and loose material. Through 1972, there were five further landings on the Moon as part of the Apollo project.

Soviet researchers in the meantime developed the techniques of unmanned, automated or remote-control exploration of the Moon. In 1970/76, they brought back rock samples from the Moon using *unmanned* spacecraft of the Luna series. The remote-controlled lunar vehicles Lunachod 1 and 2 explored further regions of the lunar landscape.

Only in 1994 was the Moon again the goal of a space mission. The space probe "Clementine" orbited it for 2.5 months and mapped almost the whole lunar surface, taking nearly 2 million digital images; the region near the Moon's south pole was also flown over for the first time. In 1998, the Lunar Prospector, with instruments for the detection of particles and gamma rays, was placed into orbit around the Moon.

2.5.4 Space Probe Missions in the Solar System

The investigation of the *Solar System* using unmanned space probes began in the 1960's with flights to Venus and Mars. The American flights initially made no attempt at landing, but rather gathered data from "flybys" of the probes in the "Mariner" and "Pioneer" series, as close as possible to the planets; in contrast, several of the Soviet space probes in the "Venera" and "Mars" series were launched with the goal of placing instrumentation capsules on the planetary surfaces. We unfortunately cannot describe in detail here the great variety of instrumentation used in the space probes, which includes cameras for direct photography, spectrometers, radiometers, and detectors for particles and magnetic fields.

The first successful flyby of Venus was that of Mariner 2 in 1962, and the first flyby of Mars was by Mariner 4 in 1965. Mariner 10 investigated Venus in 1974 and then continued on to Mercury after its orbit was suitably redirected by passing through the gravitational field of Venus.

Venus. The atmosphere of Venus was investigated in 1967 *in situ* for the first time by a capsule which was ejected from Venera 4 and dropped to the surface on a parachute. In 1970, Venera 7 carried out the first of a series of landings on the surface of Venus. For longer series of observations, artificial satellites (orbiters) were put into orbit around the planet. For example, Pioneer Venus (1978 to 1992) and Magellan (1990 to 1994) carried out mapping of the surface of Venus using radar. Towards the end of its mission, Magellan also measured the gravitational field of Venus, for which purpose its orbit was gradually allowed to drop into the atmosphere until the probe finally burned up.

Mars. Orbiters have also been employed for the study of Mars: Mariner 9, Mars 2 and 3 (1971), Mars 5 (1973) and more recently Viking 1 and 2 (1976). While the instrument capsule landed by Mars 3 transmitted data for only 20 s, the two Vikings have provided us with a large amount of information on the details of the Martian surface. In 1988, two Soviet vehicles of the Phobos mission were launched with the goal of taking closeup photographs of Mars and of its moon Phobos. Radio contact to Phobos 1 broke off after only a few weeks of flight, while Phobos 2 obtained a few images of Mars – along with data about the interplanetary medium – before it also lost contact in 1989. In the 1990's, two further missions failed: the American Mars Observer and the Russian probe Mars 96.

Following these failures, two successful missions, Mars Pathfinder and Mars Global Surveyor, reached the red planet in 1997. Pathfinder landed directly on the martian surface, slowed by a parachute and airbags, without first going into an orbit around the planet. It sent out the ca. 40 cm long, six-wheeled Mars vehicle "Sojourner", which analyzed rock samples with a particle detector and an X-ray spectrometer (Fig. 3.12). Global Surveyor is investigating the surface and the atmosphere of Mars from a circular orbit at an altitude of around 400 km using various instruments, and is also studying the gravitational field and the magnetic fields of the planet.

With the loss of two American probes at nearly the same time (the Mars Climate Orbiter and the Mars Polar Lander), Mars research in 1999 once more suffered a major setback.

The Major Planets. Space missions to the major planets began in 1972 with the launch of Pioneer 10, which reached Jupiter in 1973, and that same year with Pioneer 11, which flew past Jupiter in 1974 and also investigated Saturn during a close approach in 1979. The two space probes Voyager 1 and 2, launched in 1977, have carried out close-up investigations of the Jupiter and Saturn systems; Voyager 2 also reached Uranus in 1986 and finally flew past Neptune in 1989 (Fig. 2.28).

The space probe Galileo was launched in 1989, carrying a number of instruments, with the goal of investigating the Jupiter system. Following near passes by the asteroids 951 Gaspra and 243 Ida, it was able to observe the collision of the comet D/Shoemaker-Levy 9 with Jupiter (Sect. 3.4.1). In 1995, a few months before it reached Jupiter, it ejected a capsule which later entered the Jovian atmosphere. From 1996 until the end

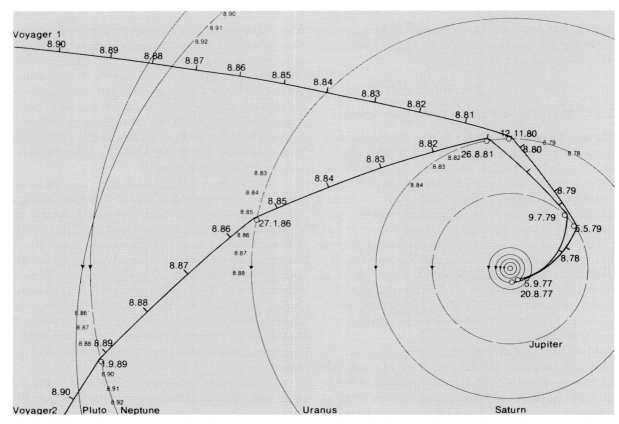

Fig. 2.28. The orbits of the space probes Voyager 1 (launched on Sept. 5, 1977) and Voyager 2 (August 20, 1977) to the major planets. The gravitational field of Jupiter was used to produce the required redirection of the orbit to allow a near passage ("flyby") of Saturn (in 1980 and 1981; "slingshot technique"). The favorable constellation of the planets allowed Voyager 2 to pass near Uranus in 1986, and near Neptune in 1989. (G. Hunt and P. Moore, 1983; with the kind permission of the Herder-Verlag, Freiburg)

of its mission, Galileo passed several times close to the large moons of Jupiter, Ganymede, Europa, Callisto and Io.

Since 1997, the space probe "Cassini" has been underway, en route to study the Saturn system; it is supposed to enter an orbit around the planet in the year 2004.

Comets. Following the passage in 1985 of the International Cometary Explorer through the tail of Comet Giacobi-Zinner 1986, the return of *Halley's Comet* in the same year offered an opportunity for no less than five space probes with a variety of instrumentation to investigate the cometary gas and dust *in situ* and to photograph its head from close up. The European probe Giotto flew past the nucleus of the comet at a distance of only 600 km while the two Japanese probes Sakigake and Suisei and the Soviet probes Vega 1 and 2 – after flybys of Venus (Vega = Venus-Halley, from the Russian *Venera-Ga*llei) passed the comet at considerably larger distances.

Asteroids. In 1991, the space probe Galileo, on its way to Jupiter, passed first by 951 Gaspra and – after another flyby of the Earth – passed near 243 Ida in 1993. Previously, it had twice circled the Sun on "spiral" orbit, passing Venus and the Earth. The probe NEAR (*N*ear *E*arth *A*steroid *R*endezvous) observed the asteroid 253 Mathilde from close up in 1997, before reaching 433 Eros in December, 1998.

The Neighborhood of the Sun. The interplanetary medium in the neighborhood of the Sun was studied in 1974–76 by the two "Sun probes" Helios 1 and 2, whose orbits attained perhelia of around 0.3 AE. The probe Ulysses, launched in 1990, was redirected by the gravitational field of Jupiter into an orbit with a very large inclination (Fig. 2.25), so that for the first time, the properties of the inner *heliosphere* – e.g. magnetic fields and the solar wind – could be measured from *outside* the plane of the ecliptic, even including the poles of the Sun.

The results of the manned and unmanned flights to the Moon as well as those of the interesting missions to other objects within the Solar System will be discussed together with the detailed descriptions of the objects concerned.

3. The Physical Structure of the Objects in the Solar System

The study of the planets and their satellites, exploiting the possibilities of space research, has developed in recent years into one of the most interesting but also most difficult branches of astrophysics. In particular, understanding the observations has required all the available resources of physical chemistry and the Earth sciences (geology, mineralogy, etc.).

We begin this topic by treating the *planets* and their *moons*, considering some of their global properties (Sect. 3.1). Proceeding from these basics, we then discuss the Earth and the Moon, which are the members of the Solar system that have been studied in most detail, and continue with the earthlike planets Mercury, Venus, and Mars (Sect. 3.2); finally, we treat the asteroids (Sect. 3.3). All these celestial bodies have mean densities $\bar{\varrho}$ in the range from 3000 to 5500 kg m^{-3}, corresponding roughly to those of terrestrial rocks or metals. We then turn to the major planets Jupiter, Saturn, Uranus, and Neptune, which have completely different structures, with mean densities $\bar{\varrho}$ in the range 700 to 1600 kg m^{-3}, corresponding roughly to those of liquified gases in the laboratory (Sect. 3.4). Finally, we describe Pluto, which is much smaller than the other outer planets, together with the recently discovered transneptunian objects (Sect. 3.5). We then consider the remaining *small objects* in the Solar System, beginning with the comets (Sect. 3.6), then treating the meteors and meteorites (Sect. 3.7), and concluding with the interplanetary dust (Sect. 3.8), which contains particles of sizes up to those of the smallest meteorites.

The origin and development of the Solar System (cosmogeny) will be dealt with in Sect. 14.1; they can only be treated reasonably after we have learned about the formation and evolution of the stars.

3.1 Global Properties of the Planets and Their Satellites

We begin in Sect. 3.1.1 with some remarks about observational methods; however, we will defer a more detailed desription of some of the instruments used until Chap. 5. We then discuss some general theoretical points of view and the *energy balance* of the planets (Sect. 3.1.2), their inner structures and stability (Sect. 3.1.3), and finally the makeup of the planetary atmospheres (Sect. 3.1.4).

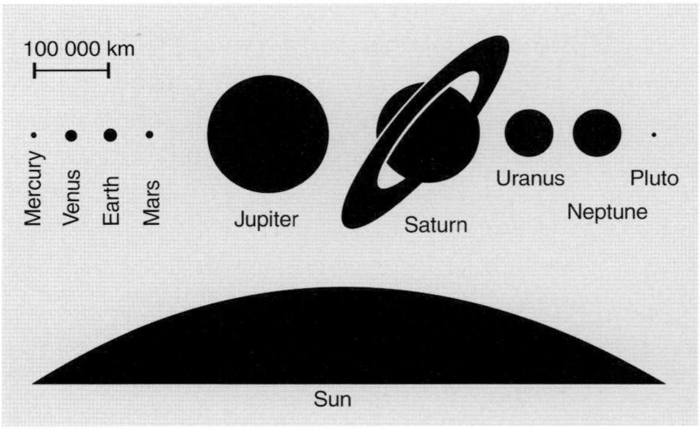

Fig. 3.1. The true relative sizes of the planets and the Sun

3.1.1 Ways of Studying the Planets and Their Satellites

We need make no further remarks about methods of determining the apparent and true diameters of the planets or their masses, and their mean densities $\bar{\varrho}$ which are calculated from those data. Numerical values (relevant to the following sections as well) are collected in Table 3.1; Fig. 3.1 gives some feeling for the true sizes of the planets and the Sun.

The *rotational periods* of celestial objects can be determined by observation of any sufficiently permanent surface features. Another possibility is to use the Doppler effect of the Fraunhofer lines in reflected sunlight or the absorption lines of the atmosphere itself.

Radar methods allow us to discern ring-shaped zones around the midpoint of the planetary disk of nearby planets, by making use of the different propagation times of the radar waves. If the planet rotates, sectors

Table 3.1. Physical Structures of the Planets and of the Moon. Radius in units of $R_E = 6378.1$ km, mass in units of $\mathcal{M}_E = 5.97 \cdot 10^{24}$ kg. The total acceleration g includes the centrifugal acceleration

	Equatorial radius R/R_E	Mass $\mathcal{M}/\mathcal{M}_E$	Mean density $\bar{\varrho}$ [kg m^{-3}]	Sidereal rotational period [d]	Inclination of the equator relative to the orbital plane	Total acceleration at the equator g [m s^{-2}]	Global effective temperature $\bar{T}_{\text{eff}}$ [K]
Mercury	0.38	0.055	5430	58.65	2°	3.7	443
Venus	0.95	0.82	5240	243.0[a]	3°	8.9	230
Earth	1.00	1.00	5520	0.997	23.5°	9.8	255
Moon	0.27	0.012	3340	27.32	6.7°	1.6	274
Mars	0.53	0.11	3930	1.03	23.9°	3.7	216
Jupiter	11.2	317.8	1330	0.41	3.1°	23.2	124
Saturn	9.41	95.2	690	0.45	26.7°	9.3	94
Uranus	4.01	14.6	1260	0.72	97.9°	8.6	59
Neptune	3.81	17.1	1640	0.67	28.8°	11.2	59
Pluto	0.18	0.002	2200	6.39	122°	0.7	42

[a] retrograde rotation

parallel to the projection of the axis of rotation can also be differentiated according to the Doppler shift of the reflected waves. Using the 300 m paraboloid antenna near Arecibo, Puerto Rico in 1964/65, the first clearcut measurements of the rotation of *Venus* and *Mercury* were finally achieved.

Conclusions about the *distribution of mass* in the interiors of the planets and satellites can be reached by measuring the polar flattening due to their rotation, from their precession in the gravitational fields of other masses (see Sect. 2.4.5), as well as from precise measurements of the gravitational potential or acceleration of gravity at their surfaces and (using artificial satellites or space probes) some distance out from the surface. In the case of the Earth and the Moon, investigations of the propagation of seismic waves also give information on the depth dependence of the elastic constants and the densities.

The *reflectivity* is described by quoting the *albedo*, which is defined as the ratio of the intensity of sunlight reflected or scattered in all directions to that of the incident light. Its magnitude and wavelength dependence gives information on the nature of the surface, particularly in the case of the asteroids and moons. More detailed information can be obtained from the brightness as a function of the phase angle, or the surface brightness as a function of the angles of incidence and of reflection, as well as through measurements of the *polarization* of the reflected light.

Some gases, at least, can be identified in the *spectrum* of a planet in terrestrial observations, by utilizing their absorption bands, which occur in addition to the Fraunhofer lines of the reflected solar spectrum. By making the observations outside the Earth's atmosphere, we can avoid interference from the terrestrial bands due to H_2O, CO_2 and O_3, and can also identify other components of the planetary atmosphere by means of spectral lines in the ultraviolet and the infrared.

We obtain information about the *temperatures* in the atmospheres (or on the surface of the planet or moon when the atmosphere is sufficiently transparent) by determining the intensity of the emitted thermal radiation in the infrared or mm- to dm-wave regions of the radio spectrum. However, a theoretical model of the atmosphere is necessary to decide which layers correspond to the measured temperatures.

The many possibilities of *space travel* for investigating the Solar System, such as near approaches of space probes to the planets and their satellites, manned and unmanned landings with *in situ* measurements, or artificial satellites around other planets for longer observations, have already been discussed in Sect. 2.5; there we also gave a summary of the most important missions to the Moon and the planets thus far.

Before we turn to the individual planets and their satellites, we shall first present some theoretical considerations about their energy balances, their interior and atmospheric structures, and their stabilities.

3.1.2 The Global Energy Balance of the Planets

The *input of radiant energy* from the Sun at a distance of 1 AU is given by the solar constant, $S = 1.37\,\text{kW m}^{-2}$ (6.10); for a planet with an orbital radius r, it is

$$S(r) = S \cdot \left(\frac{r}{r_\text{E}}\right)^{-2}. \quad (3.1)$$

The planet (of radius R) absorbs $\pi R^2(1-A)S(r)$ of this power, where A is its mean albedo (mean reflectivity). For the Earth, $A \simeq 0.3$, whereby a major part is due to the clouds (whose reflectivity is $\simeq 0.5$) which cover on the average about 50% of the surface.

Radiant losses occur for the most part in the infrared; for a "black" planet at temperature T, they would be given by the Stefan-Boltzmann radiation law as the surface area $4\pi R^2 \times \sigma T^4$, where the radiation constant $\sigma = 5.67 \cdot 10^{-8}\,\text{W m}^{-2}\,\text{K}^{-4}$ (Sect. 4.3.3).

For objects like the planets, which are not black bodies, a *global effective temperature* $\overline{T}_\text{eff}$ is *defined* by the Stefan-Boltzmann law.

Since the radiant energy input together with the heat input Q from internal energy sources just balances the radiant losses, we have

$$\pi R^2(1-A)S(r) + 4\pi R^2 Q = 4\pi R^2 \sigma \overline{T}_\text{eff}^4. \quad (3.2)$$

For the earthlike planets, the internal energy sources may be neglected in comparison to the radiant energy input from the Sun; in the case of the Earth, the mean internal heat input, which is predominantly due to the energy released by the decay of radioactive elements in the crust, is only $Q = 0.06\,\text{W m}^{-2} \simeq 10^{-4} S$. On the other hand, infrared measurements for the major planets Jupiter, Saturn, and Neptune indicate radiant losses which are 2 to 3 times greater than the absorbed solar radiation:

	Jupiter	Saturn	Uranus	Neptune
$\frac{4\sigma \overline{T}_\text{eff}^4}{(1-A)S(r)}$	1.7 (±0.1)	1.8 (±0.1)	1.1 (±0.1)	2.6 (±0.3)

This energy is due to the release of gravitational energy or to heat remaining from the time of the formation of the planets. Whether or not Uranus also has an internal energy source is still not completely clear.

The global effective temperatures are listed in Table 3.1. As we shall see in the discussions of the individual planets, the actual temperatures which occur in the atmospheres and on the planetary surfaces differ considerably from $\overline{T}_\text{eff}$. On the one hand, rotation and atmospheric flow play a decisive role in equalizing the temperatures in the day and night portions of the planet; on the other, a strong dependence of atmospheric transparency on the wavelength of the radiation can lead to the well-known "greenhouse effect" or to selective heating of particular layers, for example of the ozone layer in the Earth's atmosphere.

3.1.3 Interior Structure and Stability

The pressure distribution in the interior of a planet or moon (or also of a star; see Sect. 8.1.1) is determined by its *hydrostatic equilibrium*. If we consider a volume element with the base area dA and height dr at a distance r from the center, then its mass $\varrho(r)dA\,dr$ (ϱ: density) will be attracted by all the masses which are closer to the center. That part of the mass of a planet $\mathcal{M}(r)$ which is located within a sphere of radius r is given by (2.47):

$$\mathcal{M}(r) = \int_0^r \varrho(r')4\pi r'^2 dr' \quad \text{or}$$

$$\frac{d\mathcal{M}(r)}{dr} = 4\pi r^2 \varrho(r). \quad (3.3)$$

$\mathcal{M}(r)$ produces a gravitational acceleration at its surface according to Newton's law of gravitation:

$$g(r) = \frac{G\mathcal{M}(r)}{r^2}. \quad (3.4)$$

In the region of our volume element, the pressure P thus changes by an amount:

$$-dP\,dA = \varrho(r)\,dA\,dr \cdot g(r)$$
$$\text{force} = \text{mass} \cdot \text{acceleration}. \quad (3.5)$$

The *hydrostatic equation* for the planet is thus given by:

$$\frac{dP}{dr} = -\varrho(r)g(r) = -\varrho(r)\frac{G\mathcal{M}(r)}{r^2}. \quad (3.6)$$

For the special case of a *homogeneous sphere* with $\varrho(r) = \overline{\varrho} = \text{const}$, we can readily integrate this expression and thus obtain an estimate of the pressure P_c at the center of the planet. Let $\mathcal{M}$ be the total mass of the planet and R its radius; then from $\mathcal{M}(r)/\mathcal{M} = (r/R)^3$ (3.6) it follows that:

$$P_c = \int_R^0 \frac{dP}{dr} dr = \overline{\varrho} \int_0^R \frac{G\mathcal{M}}{r^2} \left(\frac{r}{R}\right)^3 dr$$
$$= \frac{1}{2} \overline{\varrho} \frac{G\mathcal{M}}{R} . \quad (3.7)$$

It can be shown that a homogeneous mass distribution gives a minimum value for P_c, as long as we consider only the case that $\varrho(r)$ increases monotonically towards the center.

For the pressure at the center of the Earth, we obtain an estimate from (3.7) and (2.49): $P_c \simeq 170 \, \text{GPa} \simeq 1.7 \, \text{Mbar}$, which is roughly a factor of two smaller than the exact value of 360 GPa (1 GPa = 1 Gigapascal = 10^9 Pa).

In the general case, we shall need the *equation of state* of the material $P = P(\varrho, T; \{\varepsilon\})$, which depends not only on ϱ but also on the temperature T and the chemical composition $\{\varepsilon\}$, and we can therefore obtain the pressure or density distribution only if the temperature profile $T(r)$ is known. This quantity is determined by energy transport. In the interior of an earthlike planet, the equation of state is only weakly dependent on T, so that here $\overline{\varrho}$ and $\mathcal{M}$ together with information about the mass distribution (Sect. 3.3.1) and the chemical composition practically determine the planet's internal structure. The pressures deep in the interior of a planet somewhat exceed those which are at present obtainable in the laboratory ($\lesssim 200$ GPa); the equation of state must therefore be derived from theoretical considerations.

Stability. We now consider the stability of a satellite with respect to tidal forces. A satellite which orbits around its central body at a distance r will tend to be pulled apart by the tidal forces of the latter (Sect. 2.4.6) and, if it approaches too closely, will disintegrate or will never be formed. If the satellite is not too small, we can neglect its internal cohesive forces and can readily estimate the critical tidal force. We assume for the central body (planet) the mass $\mathcal{M}$, radius R, and mean density $\overline{\varrho}$, and correspondingly for the satellite $\mathcal{M}_s$, R_s and $\overline{\varrho}_s$. Now we can estimate the mutual attraction of the parts of the satellite by replacing it by two masses $\mathcal{M}_s/2$ at a distance R_s, i.e.:

$$G \frac{\mathcal{M}_s \mathcal{M}_s}{4 R_s^2} . \quad (3.8)$$

On the other hand, the tidal force which pulls the two fictitious masses $\mathcal{M}_s/2$ apart is, according to (2.69), equal to $G \mathcal{M} \mathcal{M}_s \cdot R_s / r^3$. Centrifugal (i.e. inertial) terms are of the same order of magnitude as these forces. The condition for stability of the satellite is therefore:

$$G \frac{\mathcal{M}_s \mathcal{M}_s}{4 R_s^2} \geq c G \frac{\mathcal{M} \mathcal{M}_s}{r^3} \cdot R_s , \quad (3.9)$$

where c is a constant of the order of one. If we now use the fact that for the satellite $\mathcal{M}_s = (4\pi/3) \overline{\varrho}_s R_s^3$, and correspondingly for the planet, $\mathcal{M} = (4\pi/3) \overline{\varrho} R^3$, we obtain

$$\frac{r}{R} \geq (4c)^{1/3} \left(\frac{\overline{\varrho}}{\overline{\varrho}_s}\right)^{1/3} . \quad (3.10)$$

A more precise calculation due to E. Roche (1850) yields (for a satellite which rotates synchronously, as does our Moon) the *stability limit*:

$$\frac{r}{R} \geq 2.44 \left(\frac{\overline{\varrho}}{\overline{\varrho}_s}\right)^{1/3} . \quad (3.11)$$

A (large) satellite with the same density as its central body is thus not "allowed" to approach the latter closer than 2.44 planetary radii.

For *smaller* satellites, in contrast, it was remarked by H. Jeffreys (1947) that the internal cohesive forces must be considered. These are larger than the intrinsic gravitational forces for objects with radii $R_s \leq R_0$, where R_0 is determined by the *tensile strength* ζ, i.e. by the maximum force per initial cross-section which can pull on the body without tearing it apart. In the framework of our estimate of the attractive force (3.8), we find:

$$\zeta R_0^2 \simeq \frac{G \mathcal{M}_s^2}{4 R_0^2} = \frac{1}{4} \left(\frac{4\pi}{3}\right)^2 G \overline{\varrho}_s^2 R_0^4 \quad (3.12)$$

or, up to a factor of the order of one,

$$R_0 \simeq \left(\frac{\zeta}{G \overline{\varrho}_s^2}\right)^{1/2} . \quad (3.13)$$

For stone, the tensile strength ζ is of the order of 100 MPa; for ices, it is about 10 MPa; and for loose material, for example corresponding to carbonaceous chondrites (Sect. 3.7.3), it is ≤ 1 MPa. Therefore, for a density $\varrho_s \simeq 3000$ kg m^{-3}, R_0 lies in the range 30 to 300 km. From these considerations, we see that the hydrostatic equation (3.6) represents a good approximation for the structure of planets and satellites which are not too small.

Finally, for smaller satellites, we can estimate the stability limit by comparing ζR_0^2 with the tidal force $G\mathcal{M}\mathcal{M}_s R_s/r^3$. We readily see that this limit lies closer to the central body than Roche's limit (3.11) which applies to larger satellites dominated by their intrinsic gravitation. Precise calculations, which also take into account the dynamics of the process of disintegration, yield $r/R \simeq 1.4$.

3.1.4 The Structures of Planetary Atmospheres

The pressure distribution in a planet's atmosphere is, like the pressure distribution in its interior, governed by the hydrostatic equation (3.6). If the thickness of the atmosphere is small compared to the planetary radius R, we can take the gravitational acceleration (3.4) to be constant, $g = G\mathcal{M}/R^2$. Introducing the altitude $h = r - R$, we find

$$\frac{dP}{dh} = -g\varrho(h) . \tag{3.14}$$

The pressure P is related to the density ϱ (or the particle density n), the temperature T, and the mean molecular mass $\bar{\mu}$ through the equation of state of the ideal gas:

$$P = \varrho \frac{kT}{\bar{\mu} m_u} = \varrho \frac{\mathcal{R}T}{M} = nkT \tag{3.15}$$

where $k = 1.38 \cdot 10^{-23}$ J K^{-1} is the Boltzmann constant, $\mathcal{R} = 8.31$ J K^{-1}mol^{-1} the universal gas constant, $m_u = 1.66 \cdot 10^{-27}$ kg the atomic mass unit, and M in [kg mol^{-1}] the molecular weight. Inserting into (3.14) yields:

$$\frac{dP}{P} = -\frac{g\bar{\mu} m_u}{kT} dh = -\frac{dh}{H} \tag{3.16}$$

with the equivalent height or scale height H:

$$H = \frac{kT}{g\bar{\mu} m_u} . \tag{3.17}$$

If H is constant, we can easily integrate and obtain the barometric altitude formula, which is usable for moderate altitudes:

$$\ln P - \ln P_0 = -h/H \quad \text{or}$$
$$P = P_0 e^{-h/H} , \tag{3.18}$$

Here, P_0 is the pressure at the surface or at the initial height $h = 0$. Quite generally, it follows from (3.16) for the region between two levels at h_1 and h_2 and the pressures P_1 and P_2:

$$\ln P_2 - \ln P_1 = -\int_{h_1}^{h_2} \frac{dh}{H(h)} . \tag{3.19}$$

As a result of (3.16), the structure of a planetary atmosphere thus depends on (a) the gravitational acceleration g; (b) the mean molecular mass $\bar{\mu}$, i.e. the chemical composition, and possibly also the dissociation and ionization of the atmospheric gases; and (c) the temperature distribution $T(h)$. This last quantity is determined by the mechanisms of energy transport, that is the input and outflow of thermal energy into each layer from h to $h + dh$, which result from convection and radiation (and in individual layers also from heat conduction).

We can obtain the temperature gradient in a *convective* atmosphere, in which hot matter rises adiabatically (without heat exchange with its environment) and cooler matter falls, by logarithmic differentiation with respect to h of the well-known adiabatic equation:

$$T \propto P^{1-(1/\gamma)} , \tag{3.20}$$

where $\gamma = c_p/c_v$ is the ratio of the specfic heats at constant pressure and constant volume. We obtain:

$$\frac{1}{T} \frac{dT}{dh} = \left(1 - \frac{1}{\gamma}\right) \frac{1}{P} \frac{dP}{dh} . \tag{3.21}$$

If we now use the hydrostatic equation (3.14) together with the equation of state (3.15) and the relation $c_p - c_v = k/\bar{\mu} m_u$, we find the *adiabatic* temperature gradient

$$\frac{dT}{dh} = -\frac{g}{c_p} . \tag{3.22}$$

In the lower, convectively unstable part of the Earth's atmosphere, the troposphere, we obtain with $g = 9.81$ m s^{-2} for dry air ($c_p = 1005$ J kg^{-1} K^{-1}) a gradient equal to 9.8 K km^{-1}; for moist air, the latent heat which is liberated on condensation of moisture leads to a gradient which is only half as large. The measured mean temperature decrease with increasing altitude is 6.5 K km^{-1}.

We will investigate the other limit, that of an atmosphere with energy transport by *radiation*, in Sect. 7.2.1, in connection with radiative energy transport in *stellar* atmospheres. The relations derived there are in principle also applicable to planetary atmospheres. In a planetary atmosphere with surface heating due to unabsorbed solar radiation in the visible range, we observe a temperature increase, the so-called *greenhouse effect*, when the reradiated energy in the infrared spectral region is strongly absorbed by the atmosphere. In the Earth's atmosphere, this effect adds 33 K.

The general case of an interaction between convective and radiative energy transport cannot be treated here, nor can the atmospheric currents and winds, which are due to the variable input of solar energy depending on the time of day and the season. We shall limit ourselves to reporting some particular results in the descriptions of the individual planets.

The Existence of Atmospheres. First, we ask the question as to whether a planet or satellite can retain its *own* atmosphere. The molecules of a gas with mass $\bar{\mu}$ and temperature T have, according to the kinetic theory of gases, a most-probable velocity given by (4.87):

$$\bar{v} = \sqrt{\frac{2kT}{\bar{\mu} m_u}}. \qquad (3.23)$$

On the other hand, a molecule having the velocity v can escape from a celestial body of mass $\mathcal{M}$ and radius R if its velocity is greater than the escape velocity v_e (2.62). If we now introduce the equivalent altitude H (cf. (3.17)) and assume as an approximation that the acceleration of gravity g is constant near the planet's surface, we find $\bar{v} = \sqrt{2gH}$ and $v_e = \sqrt{2gR}$. We thus obtain as a rough criterion for the existence of a planetary atmosphere:

$$\bar{v} < v_e \quad \text{oder} \quad H < R.$$

For the Earth's atmosphere, $T = 288$ K and $\bar{\mu} = 29$ (Table 3.2), so that $H \simeq 8$ km $\ll R_E$. Taking into account the Maxwell-Boltzmann probability distribution of molecular velocities $v > \bar{v}$, we can understand that e.g. Mercury, our Moon, and most of the satellites in the Solar System can have practically no atmospheres, and that on the other hand Saturn's largest moon, Titan, could in fact retain its atmosphere for a very long time.

In the outermost layers of a planetary atmosphere, the *exosphere*, there are only infrequent collisions between the gas particles owing to the low density. The electrically *neutral* particles thus move practically on Keplerian orbits in the gravitational field of the planet; the exosphere cannot be described by the hydrostatic equation. For the motion of *charged* particles, which are formed in the upper atmosphere by ionization processes mainly due to solar UV radiation, the planet's magnetic field is the determining factor. The extent of the *magnetosphere* is determined by the interaction between the planetary magnetic field and the plasma flowing out from the solar wind (Sect. 7.3.7).

3.2 The Earth, the Moon, and the Earthlike Planets

Inside the asteroid belt – which practically divides the Solar System into two physically differing zones – all the planets have masses less than 1 Earth mass and mean densities between 3900 and 5500 kg m^{-3}. They

Table 3.2. The atmospheres of the earthlike planets

	Venus	Earth	Mars
"Solar constant" $S(r)$ [1 kW m^{-2}]	2.6	1.4	0.6
Mean albedo A	0.7	0.3	0.2
Effective temperature T_{eff} [K], (3.2)	230	255	216
Surface temperature T_0 [K]	735	288 (220–310)	210 (145–245)
Pressure at the surface P_0 [10^5 Pa = 1 bar]	93	1	$6 \cdot 10^{-3}$
Relative pressure variations $\Delta P/P$	$\leq 10^{-3}$ (?)	$\simeq 0.01$	0.1
Troposphere:			
Scale height $H = \frac{kT}{g\bar{\mu}m_u}$ [km], (3.17)	14	8	11
Altitude of the tropopause [km]	60	10	15
Mean temperature gradient [K km^{-1}]	8	6.5	3[a]
Chemical composition (Relative volumes or particle densities)			
CO_2	**0.96**	$3 \cdot 10^{-4}$	**0.95**
N_2	0.03	**0.78**	0.03
O_2	$7 \cdot 10^{-5}$	**0.21**	$1 \cdot 10^{-3}$
CO	$2 \cdot 10^{-5}$	$1 \cdot 10^{-7}$	$7 \cdot 10^{-4}$
H_2O	$\simeq 10^{-3}$	$(1-28) \cdot 10^{-3}$	$3 \cdot 10^{-4}$[a]
Ar	$7 \cdot 10^{-5}$	$9 \cdot 10^{-3}$	0.02
Mean molecular mass $\bar{\mu}$	43.4	29.0	43.5

[a] strongly variable

evidently consist essentially of solid matter. Their atmospheres are chemically oxidizing; they contain O_2, CO_2, H_2O, N_2, ... It thus seems justified to collect them under the name *earthlike planets*. We shall examine our Earth somewhat more carefully, as a paradigm for the other planets; of course, we cannot attempt to give a summary of all of geophysics here. Knowledge of our Moon has made a quantum leap within a few years, thanks to the successes of space travel. Mercury, Venus, and Mars have also been examined from close up in recent years.

3.2.1 The Internal Structures of the Earthlike Planets

As a result of its rotation, the *Earth* is to a good approximation a *flattened ellipsoid of rotation*, the so-called terrestrial spheroid, with an equatorial radius of $R_E = a = 6378.1$ km, a polar radius of $b = 6356.8$ km, and a flattening ratio given by

$$\frac{a-b}{a} = \frac{1}{298}. \tag{3.24}$$

Flattening and centrifugal force have the effect that the gravitational acceleration at the equator is $1/189$ weaker than at the poles.

Mars, with its somewhat slower rotation, has a flattening of $1/171$. Precise measurements of the gravitational potential show that both planets have a "pear shape"; for the Earth, the largest deviation from the spheroidal shape is only 17 m, while for Mars it is nearly 2 km. *Venus*, as a result of its extremely slow rotation, exhibits no deviation from a spherical shape.

The *Moon*, due to its rotation which is synchronous with its orbital period around the Earth, is elongated along the axis connecting its midpoint with that of the Earth; its geometrical midpoint lies about 2 km further from Earth than its center of gravity. *Mercury*, owing to the (3 : 2) resonance of its rotation with the (eccentric) orbital motion, is subject to constant deformation by tidal forces ($\propto 1/r^3$).

The *mean densities* of the planets and the Moon have already been discussed (cf. Table 3.1). The density of the Earth's crust (granite, basalt) is 2600 to 3000 kg m^{-3}, and similar densities have been found for the crusts of the earthlike planets.

The *moments of inertia* about their rotational axes, $I = \alpha \mathcal{M} R^2$ ($\mathcal{M}$ mass, R = equatorial radius of the planet), which have been obtained from observations of the precession or determinations of the gravitational field using artificial satellites, indicate that the density increases with depth. In contrast to a homogeneous sphere for which $\alpha = 0.40$ (2.67), the value found for the Earth is $\alpha = 0.33$, and for Mars it is $\alpha = 0.38$; the value for the Moon, $\alpha = 0.39$, differs only slightly from the homogeneous case.

The Earth's Interior. More precise information about the density variations in the interior of the Earth is available from the propagation of *seismic waves*. During an earthquake, longitudinal and transverse elastic waves are generated at the epicenter, which is relatively near to the surface. These propagate through the Earth's interior and will be refracted, reflected, and will interact with each other depending on the depth dependence of the elastic constants and the density. By careful studies of the propagation of seismic waves around 1906, E. Wiechert found that the Earth's interior contains several *surfaces of discontinuity*, at which the elastic constants and the density ϱ change suddenly. The *Earth's crust* has a thickness of about 30–40 km under flat regions, increasing to up to 70 km under recently-formed mountain ranges. Beneath the oceans, it decreases down to $\simeq 10$ km. Its lower boundary is the *Mohorovičić discontinuity* or "Moho". Below it, down to 2900 km depth, is the *mantle*, with a density $\varrho = 3300$ to 5700 kg m^{-3}; it probably consists mainly of silicates. From 2900 km bis 6370 km depth, we have the *core*, with 10 000 to 12 000 kg m^{-3}. In the *outer core*, there is no propagation of transverse waves. In this sense, we can consider it to be liquid, but with an enormous viscosity. Thorough investigations, however, indicate that the *inner core* below 5150 km depth is again solid.

The chemical-mineralogical *composition* of the Earth's core can be only indirectly determined. Laboratory experiments at high pressures and temperatures lead to a phase diagram which is compatible with the geophysical data if the assumption is made that the core, like iron meteorites, consists in the main of Fe and Ni, along with small amounts of S. The *pressure increase* towards the center can be calculated with sufficient accuracy from the hydrostatic equation (3.6). The increase in the *temperature* with increasing depth can be measured in deep wells, which however reach depths of only 10 to 12 km; one obtains a geothermic depth variation of up to 30 K km^{-1}. The temperature distribution at great depths is determined by the heat production due to decay of the radioisotopes ^{238}U, ^{232}Th and – to a lesser extent – ^{40}K; and on the other hand by the slow transport of heat to the exterior by the thermal conductivity and convection of the magma. The temperature of the Earth's core is probably about $T \simeq 5000$ K.

The Moon. Its mean density of 3340 kg m^{-3} is noticeably similar to that of the Earth's mantle; the Moon cannot have the same average composition as the Earth. The seismometers which were set up during the Apollo flights have given us some information about the Moon's inner structure. Surprisingly, seismic waves, which are produced on the Moon by the impacts of meteorites or the crash of the discarded Lunar Lander, can be detected for up to an hour. The damping of lunar quake waves is extremely weak compared to that of earthquake waves, and this may be a result of their scattering or of particular dispersion relations. From recent measurements of the Lunar Prospector and model calculations, the Moon has a metallic *core* with a radius of around 300 to 500 km.

Mercury, Venus, and Mars. Our knowledge of the interiors of these planets is considerably less certain than that of the Earth. Model calculations yield on the whole a similar variation of the density (Fig. 3.2). While the Fe–Ni core makes up about 30% of the total mass of the Earth and Venus, it forms a noticeably smaller core of 15% in the case of Mars, and is considerably larger at 60% for Mercury. (However, the internal structure of Mercury can also be represented using a nearly homogeneous model).

3.2.2 Radioactive Dating. The Earth's History

The methods of radioactive dating give us the most accurate information about the age of the Earth, of the Moon and of meteorites, as well as about the length of the various geologic eras. The radioactive decay of the following sufficiently long-lived isotopes is employed (we list only the stable end products of the whole decay chain in each case):

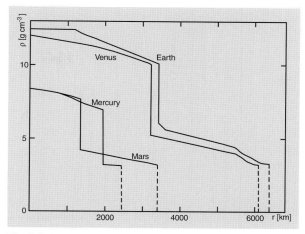

Fig. 3.2. The dependence of the density $\varrho(r)$ on distance from the centers of the earthlike planets Mercury, Venus, Earth, and Mars. ϱ [1 g cm^{-3} = 10^3 kg m^{-3}]

		Half-life T
^{147}Sm	$\rightarrow$ ^{143}Nd $+\,^4$He	$1.06 \cdot 10^{11}$ yr
^{87}Rb	$\rightarrow$ ^{87}Sr $+\,e^-$	$4.88 \cdot 10^{10}$ yr
^{187}Re	$\rightarrow$ ^{187}Os $+\,e^-$	$4.4 \cdot 10^{10}$ yr
^{232}Th	$\rightarrow$ ^{208}Pb $+\,6\,^4$He	$1.40 \cdot 10^{10}$ yr
^{238}U	$\rightarrow$ ^{206}Pb $+\,8\,^4$He	$4.47 \cdot 10^9$ yr
^{235}U	$\rightarrow$ ^{207}Pb $+\,7\,^4$He	$7.04 \cdot 10^8$ yr
^{40}K	$\rightarrow$ ^{40}Ca $+\,e^-$	$1.28 \cdot 10^9$ yr
	$\searrow$ ^{40}Ar (K capture)	$1.1 \cdot 10^{10}$ yr .

In all the methods, the ratio of *end product* to *parent isotope* is determined. This ratio was initially ($t = 0$) equal to zero, and after a time t it is given by $2^{t/T} - 1$ (assuming that the abundance of the parent isotope decreases with a decay constant λ, i.e., a half-life $T = \ln 2 / \lambda$, according to the exponential law $n(t)/n(0) = e^{-\lambda t} = 2^{-t/T}$).

The oldest rocks in the Earth's crust have an age between 3.7 and $3.9 \cdot 10^9$ yr. The age of the Earth since the last mixing – or separation – of its material can be determined by investigating the abundance distribution of the lead isotopes in uranium-free lead minerals or by finding its original value in the "archaic lead" from iron meteorites, which contain practically no uranium or thorium. The relatively exact value obtained for the

Table 3.3. Earth's history

Period Age in 10^6 yr before present	Times of principal mountain formations	Development of flora	Development of fauna	Epochs	First appearance and disappearance	Period Age in 10^6 yr before present
Quaternary		Later angiosperms Neophytic	Cenozoic	Snails Mussels Mammals	First humans	Quaternary
—— 2 ——	Alpine					—— 2 ——
Tertiary						Tertiary
—— 65 ——		Earliest angiosperm		Dinosaurs	Extinction of the ammonites and dinosaurs	—— 65 ——
Cretaceous			Mesozoic		First angiosperms First birds	Cretaceous
—— 144 ——		Later gymnosperms				—— 144 ——
Jurassic					First mammals	Jurassic
—— 213 ——		Mesophytic				—— 213 ——
Triassic		Earliest gymnosperms		Brachiopods	Extinction of many paleozoic animals	Triassic
—— 248 ——				Goniatites Ammonites		—— 248 ——
Permian	Variscan	Later pteridophytes	Palaeozoic		First reptiles	Permian
—— 286 ——						—— 286 ——
Carboniferous		Palaeophytic			First amphibians First spermatophytes	Carboniferous
—— 360 ——				Armored fish		—— 360 ——
Devonian		Earliest pteridophytes			First vascular plants	Devonian
—— 408 ——	Caledonian			Graptolites		—— 408 ——
Silurian					First vertebrates	Silurian
—— 438 ——				Trilobites		—— 438 ——
Ordovician		Eophytic (Age of algae)				Ordovician
—— 505 ——					Development of many phyla of invertebrates	—— 505 ——
Cambrian						Cambrian
—— 590 ——			Proterozoic	Precambrian		—— 590 ——
Precambrian	Assyntic Algomian Laurentian oldest rocks: 3800×10^6 yr		Archean		Oldest traces of life (stromatolites): 3500×10^6 yr	Precambrian

age of the Earth and of the oldest meteorites, and thus for the *age of the Solar System*, is:

$$(4.53 \pm 0.02) \cdot 10^9 \text{ a}. \tag{3.25}$$

The oldest rocks on the Moon likewise are dated at about $4.5 \cdot 10^9$ yr, i.e. the Moon was formed at the same time as the Earth, within the accuracy of the measurements.

The most important *geological periods* of the Earth's history are briefly characterized and their absolute datings are summarized in Table 3.3.

3.2.3 Magnetic Fields. Plate Tectonics

The *Earth* possesses a magnetic field, which corresponds roughly to a dipole field with a magnetic flux density at the equator of $3.1 \cdot 10^{-5}$ Tesla (T) or 0.31 Gauss (G). We can characterize a dipole by its *magnetic moment a*, which points in the direction of the dipole axis and from which we can obtain the vector of the *magnetic induction* or *magnetic flux density B* by taking the gradient:

$$\boldsymbol{B} = -\text{grad}\,\frac{\boldsymbol{a}\cdot\boldsymbol{r}}{r^3} = -\text{grad}\,\frac{a\sin\lambda}{r^2}, \tag{3.26}$$

where $a = |\boldsymbol{a}|$ and λ is the magnetic latitude. The radial and horizontal components (in the directions parallel to r and λ) are given by:

$$B_r = B_p \sin\lambda, \quad B_\lambda = -\tfrac{1}{2} B_p \cos\lambda \tag{3.27}$$

with the flux density at the poles $B_p = 2a/r^3$. The azimuthal component is $B_\phi = 0$. At large distances from the origin, B decreases proportionally to r^{-3}. For a planet of radius R, the field strength at the poles is ($\lambda = 90°$) $2a/R^3$, while at the equator ($\lambda = 0°$), it is just half as large.

In the case of the *geomagnetic field*, $\boldsymbol{a}$ forms an angle of $\alpha = 11.5°$ with the axis of rotation of the Earth, and the dipole axis is shifted from the center of the earth by $0.07\,R_E \simeq 450$ km. The magnetic moment of the Earth's magnetic field is $8 \cdot 10^{15}$ T m^3 or $8 \cdot 10^{25}$ G cm^3; it is currently decreasing by 0.05% per year, and the dipole axis precesses with about $0.04°\,\text{yr}^{-1}$ (with a period of about 9000 yr).

In the case of *Mercury*, the space probe Mariner 10 surprisingly determined a magnetic field with a moment of $5 \cdot 10^{12}$ T m^3 (with $\alpha \simeq 11°$). In contrast, *Venus* and *Mars* probably have no planetary magnetic fields. (The extremely small fields which were measured are likely due to magnetized plasma from the solar wind striking their surfaces.) Our own *Moon* also has no measureable dipole field ($a < 10^9$ T m^3); the weak remanent magnetization of some lunar rocks, corresponding to a few percent of the Earth's field, is an indication of an *earlier* lunar magnetic field.

Variations of the Earth's Magnetic Field. The magnetic field of the Earth and its *secular variation* (the rapid variations caused by the Sun will be treated later) can be interpreted as follows, according to W. M. Elsasser and E. Bullard: the liquid matter in the outer regions of the Earth's core forms large vortices driven by convection and heat exchange. If traces of a magnetic field are present in such a vortex of conducting material, they can be amplified, as in the self-exciting dynamo invented by W. v. Siemens. The details of such dynamos in the Earth's interior are not yet completely clear; however, the dynamo theory offers the only possibility of explaining the terrestrial magnetic field, its secular variations, and its periodic reversals.

The geomagnetic field in past epochs can be determined thanks to the circumstance that when certain minerals are formed, the magnetic field which happens to be present at the time is, so to speak, frozen into them (P. M. S. Blackett, S. K. Runcorn, and others). Such paleomagnetic determinations have shown that the magnetic field vectors in previous times can best be brought into an orderly scheme by applying the hypothesis of *continental drift*, originally deduced from the arrangement of the continents on the globe by A. Wegener (1912). If the reasonable assumption is made that the Earth had essentially a magnetic dipole field in earlier times, we can reconstruct the relative positions of the continents in previous geologic epochs and can show, in good agreement with geological and paleontological observations, how the present continents were formed bit by bit in the course of the Earth's history. For example, the Atlantic ocean had its beginnings about $1.2 \cdot 10^8$ yr ago, i.e. in the Jurassic to Calciferous periods, as a narrow trench, similar to the Red Sea today. On the average, America and Europe have moved apart by a few centimeters per year in the intervening time (Fig. 3.3).

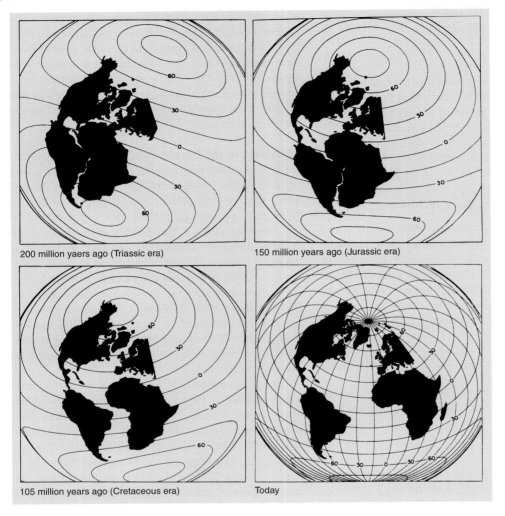

Fig. 3.3. The continental drift. The positions of the continents around the Atlantic ocean (in relation to North America) at different times. The continents move apart with velocities of 2.5 to 4 cm yr^{-1}. (With the kind permission of the Wissenschaftliche Verlagsgesellschaft, Stuttgart)

In the 1950's, paleomagnetic investigations thus provided strong evidence in favor of the continental drift theory, which had previously been the subject of controversy for many years; then, in the 1960's, they provided an insight into the basic mechanisms of the continental motion: it had been noticed from the paleomagnetic measurements that directly adjacent layers of minerals often indicate diametrically opposed directions for the magnetic field. Detailed investigations of precisely dated series of layers showed that this was due not to a spontaneous remagnetization of the rocks, but rather to a reversal of the entire magnetic field of the Earth.

This no longer seems so surprising if one considers that in the self-exciting dynamo, the direction of the current is determined by the random weak magnetic fields present when the machine is started up.

Plate Tectonics. After it had become clear that the geomagnetic field changes its sign in irregular periods of several hundred thousand years, it was discovered by H. H. Hess and others that the floor of the Atlantic ocean shows a whole series of strips with approximately north-south orientation and alternating directions of magnetization. Their magnetic dating indicates that the ocean floor has been spreading for about $1.2 \cdot 10^8$ yr, starting from a mid-atlantic ridge about halfway between America and Europe and moving out on both sides, pushing the two continents apart. This fruitful theory of *sea floor spreading* was soon complemented through the recognition by H. H. Hess, I. T. Wilson and others (around 1965), that in the course of all these processes, large slabs or plates in the upper portion of the Earth's crust are shifted in one piece or are broken along fault lines (this is the origin of tectonic earthquakes); we thus refer to *plate tectonics*.

The driving force behind all of these geologic occurrences is, as had been assumed by A. Holmes as early as 1928, clearly *convection currents*, especially in the upper portion of the mantle; they well up along the mid-oceanic ridges. Here, fresh crust material and fresh plates are formed and forced apart on both sides. Where the oceanic plates are pushed back down into the mantle, deep oceanic trenches are formed, and behind them, young mountain ranges are produced by folding. These and many other basic geological processes have such a simple explanation! The necessary energy is provided by the radioactive decay of minerals within the Earth; from the geothermal temperature profile and the thermal conductivity of the rocks, it can be estimated that the entire Earth obtains about 10^{21} J yr^{-1} of *thermal energy* in this way. About 1/1000 of this energy goes to produce earthquakes. If our "thermodynamic machine" produces mechanical energy with an efficiency of even about 1/1000, it would be quite sufficient to drive the convection in the mantle. Geologic observations furthermore support the idea that the tectonic activity of the Earth has been subject to strong changes in the course of time, with maxima at 0.35, 1.1, 1.8, and $2.7 \cdot 10^9$ yr ago.

Owing to the continual formation of new, hot lithospheric material and its subsequent cooling, plate tectonics or sea floor spreading represents the most important process for cooling the Earth's interior.

We cannot go into the details of oceanic physics here, especially since within the Solar System, oceans are unique to the Earth.

The Earthlike Planets and the Moon. Still very little is known about matter flow, convection, etc. in the interiors of the earthlike planets. On the small planets *Mercury* and *Mars*, as well as on the *Moon*, all tectonic and volcanic activity has now ceased; heat transport from the interiors is accomplished by the thermal conductivity of the lithospheres. It is astonishing that Mercury can generate a magnetic field. Its metallic core is, to be sure, relatively large (Fig. 3.2); but its rotational velocity, which plays an important role in driving the dynamo, is much slower than e.g. that of Mars, which has no magnetic field of its own.

In the case of *Venus*, there are no topographical indications for global plate tectonics, although its inner structure and energy sources are quite similar to those of the Earth. On the contrary, tectonic stresses occur in broad (100 km) border zones as deformations between relatively undisturbed larger blocks. Venus also has no magnetic field. It is possible that differences in the thickness of the crust and its temperature and thermal conductivity, as well as the slow rotation, are responsible for the lack of plate tectonics and of a magnetic field on Venus. Loss of heat from the interior of Venus is probably mostly due to the thermal conductivity of the crust and to cooling of hot magma, which reaches the surface in several volcanic regions.

We shall now turn to the description of the surfaces and the surface-forming processes on the earthlike planets and their moons, beginning with our own Moon.

3.2.4 The Lunar Surface

Selenographic investigations with increasingly larger telescopes, with space probes, and through the landings of the Apollo astronauts have given us detailed information on the formations of the lunar surface (Figs. 3.4, 5).

There can now be no more doubt that all of the circular formations, from the enormous maria (terming them "seas" has only historical significance) through the craters and down to microscopic holes of a few microns diameter in smooth surfaces (Fig. 3.6), were caused by the impact of meteor-like bodies with planetary velocities. Mare Imbrium has a diameter of 1150 km, and the south polar *Aitken basin*, whose status as the largest, deepest, and oldest Mare was recognized only in 1994 as a result of the images sent back by the lunar probe Clementine from the back side of the Moon, has a diameter of 2600 km! The depth of these *impact craters*

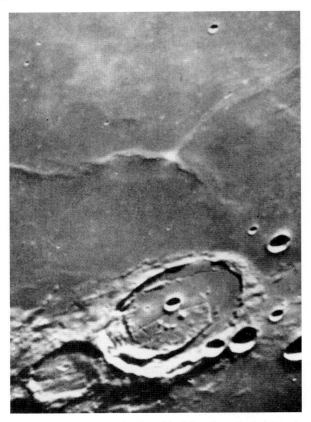

Fig. 3.4. The Moon. At the rim of the Mare Serenitatis (*above*), the crater Posidonius (*lower right*) with a diameter of 100 km, and the smaller crater Chacornac (*lower left*) may be seen. In the Mare Serenitatis, there is a richly structured mountain range of $\simeq 180$ m altitude. Numerous small craters can be seen strewn over the landscape. This photograph was made with the 120″ reflector at the Lick Observatory in 1962

Infrared measurements indicated long ago that the shadow's edge during an eclipse of the Moon changes its temperature only very slowly, and thus the surface material has a very low thermal conductivity. In fact, the surface of the Moon is covered to a large extent with fine dust and loose fragments of stone (regoliths), which were produced by the impacts of the meteorites.

Many of the larger craters are surrounded by a system of bright *rays*, readily seen even with a small telescope. They evidently consist of material which was thrown out when the crater was formed.

Exact measurements by the Clementine probe, carried out using laser beams, indicated that on the Moon, altitude differences up to about 16 km occur. The highest mountains are to be found in the bright, crater-pocked highlands, the *Terra regions*. Their height is limited by the breaking strength of the material and is of the same order as those on the Earth, as was already determined by Galileo, who observed the lengths of their shadows along the terminator or day-night boundary on the Moon.

The *Moon rocks* brought back by the astronauts from the maria are filled with bubble-like hollows, similar to terrestrial lavas which have hardened under low pressure. They are thus *igneous* rocks, i.e. they were formed by hardening from the melt. The *Moon dust* mentioned above was formed by pulverizing of such rocks. The *breccias* consist of dust and other small particles which have become clumped together. Mineralogically, the lunar rocks are roughly similar to terrestrial basalts. In Fig. 3.7, we compare the abundance of some important elements in lunar material with those in terrestrial basalts and in meteorites (eucrites and carbonaceous chondrites; cf. Sect. 3.7.3). This comparison and other studies of rarer (socalled trace) elements show that some elements are enriched in lunar rocks compared to solar matter by up to $\simeq 100$-fold, while the elements which bind to Fe (siderophilic elements), Ni, Co, Cu, ..., are reduced by factors of $\simeq 100$.

When energetic cosmic–ray particles (Sect. 10.4.2) collide with the lunar surface, they produce among other things "fast" neutrons. With the neutron spectrometer of the Lunar Prospector in 1998 in the neighborhood of the two lunar poles, "slow" neutrons were detected, which presumably were *slowed down* by collisions with protons. This is a strong indication of the existence

varies with their diameter in the same way as in terrestrial explosion craters. The largest Maria have depths of the order of 10 km; below them, the crust of the Moon is unusually thin, being only about $\gtrsim 10$ km, while it is otherwise around 100 km thick. The floors of the maria and of many larger craters were evidently later flooded by liquid lava. After it had hardened, the smooth surface was again covered with smaller craters.

The meandering *rills* are probably lava channels which were originally "roofed over" to a considerable extent. Other, longer rills may be explained as cracks in cooling lava.

Fig. 3.5. The Mare Nectaris with the craters Theophilus, Mädler, and Daguerre. The photo was taken from the command module of Apollo 11 at an altitude of 100 km. In the Mare Nectaris, two "ghost craters" can be seen in the foreground, mostly covered over with lava; otherwise, there are only numerous small craters

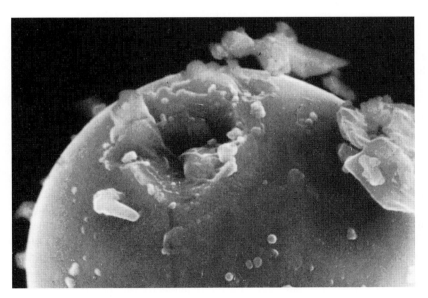

Fig. 3.6. In lunar dust (collected by Apollo 11), glasslike spherules are found; they were formed from stone liquified by meteorite impacts. This electron microscope image by E. Brüche and E. Dick (1970) shows one of these spherules which has a diameter of 0.017 mm; the impact of a micrometeorite has produced a "microcrater" on its surface

of H_2O *ice* in the regions on the Moon which are in perpetual shadow.

While the age of the mountain ranges goes back to $4.5 \cdot 10^9$ yr, the maria are found to be younger: the south polar Aitken Basin is 3.8 to $4.3 \cdot 10^9$ yr old, the Mare Imbrium $3.9 \cdot 10^9$ yr, and the Mare Tranquillitatis only $3.7 \cdot 10^9$ yr. From around 3.3 to $4 \cdot 10^9$ yr ago, the maria were partially filled with basalt lava, which welled up from deeper regions. Their formation was thus completed only about 10^9 yr *after* that of the Moon itself. These data, together with the statistics of the lunar craters and their "covering up", indicate that the cosmic bombardment was initially extremely strong and then decreased in the course of the first 10^9 yr of the Earth's and the Moon's history, at first rapidly and later much more gradually.

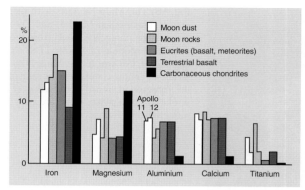

Fig. 3.7. The abundance distribution of the elements iron, magnesium, aluminum, calcium, and titanium (in weight percent) from samples of lunar material (*left half*: Apollo 11, Mare Tranquillitatis; *right half*: Apollo 12, Oceanus Procellarum), from eucrites, i.e. basalt-like achondritic meteorites, and from carbonaceous chondrites of type 1. In the last-named meteorites, the abundance distribution of all non-volatile elements corresponds to that of original solar matter

The great plains of the maria, the "drowned" craters and some remarkable hills on the Moon however demonstrate, as already mentioned, that in addition to meteoric impacts, also volcanic processes, i.e. the melting of rocks and outflows of basaltic lava, must have played an important role in the formation of the lunar surface.

3.2.5 The Surfaces of the Earthlike Planets

Mercury. This planet is very difficult to observe, since it is never more than ±28° away from the Sun. Radar measurements together with older visual observations have shown that the rotational period of Mercury is not, as was previously believed, equal to its period of revolution (88 d); instead, it is 58.65 d, or exactly 2/3 of the period of revolution.

The space probe Mariner 10, which flew past Mercury three times in 1974/75, transmitted numerous high-quality pictures of nearly 50% of the planetary surface with a resolution comparable to that with which the Moon can be seen from the Earth. Mercury's surface, like that of the Moon, is thickly covered with craters. The largest, the *Caloris Basin*, has a diameter of 1300 km, similar to the Mare Imbrium. Other structures (plains, terrae, ...) also exhibit great similarity to the Moon's surface. Because of Mercury's stronger gravitational field, the matter ejected from impact craters is not thrown as far as on the Moon.

The surface formations on Mercury and the Moon have remained essentially intact because tectonic and volcanic activity ceased early on both bodies and since – due to the lack of an atmosphere – no weathering and erosion have occurred. Currently, "weathering" can take place only through the energetic protons of the solar wind, through the impacts of micrometeorites, which produce craters of 1 to 20 mm diameter, and through the strong temperature variations, particularly on Mercury (600 K on the day side compared to 100 K on the night side).

Venus. The surface of Venus cannot be observed visually, because it is completely covered by a thick layer of clouds. Its exploration only became possible with the development of high-sensitivity radar technology and through the soft landings of space probes.

The radar investigations led to the surprising result that the rotation of Venus is *retrograde*, with a sidereal period of 243.0 d. While it was possible to localize some individual mountains on the surface of Venus with earthbound radar, the topographic structure was later surveyed by the Pioneer Venus orbiter, which has been orbiting the planet in a strongly elliptical orbit since 1978, with a horizontal resolution of ≤ 100 km and a vertical resolution which is in part as low as several 100 m. The soviet Venus satellites Venera 15 and 16 provided radar images in 1984/85 with a greatly improved (horizontal) resolution of 1 to 2 km. Finally, starting in 1990 the artificial satellite *Magellan* circled Venus at an altitude of 290 to 8000 km for four years until it dropped down through the atmosphere, and it surveyed practically the entire planetary surface using radar with a resolution down to 120 m.

Venus exhibits no great altitude variations on its surface. About 80% of the surface area shows less than ±1 km deviations from the mean level, i.e. from the radius of the planet (Fig. 3.8). Around 65% is occupied by *rolling plains* with altitudes between 0 and 2 km; about 27% are up to 3 km deep *lowlands*, and only less than 10% are *highlands* (terrae) with altitudes of more than 2 km, which can be compared with Earth's continents in terms of their size. *Ishtar Terra* contains an extended

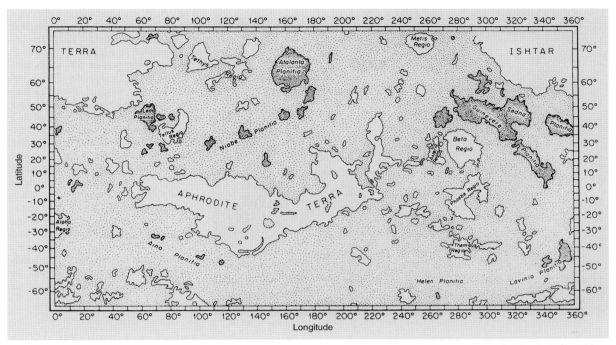

Fig. 3.8. The topography of the surface of Venus in a Mercator projection, from radar observations by the Pioneer Venus Orbiter. The dotted areas are raised up to 1 km from the average altitude level (plains); the light areas are highlands above 1 km altitude; and the dark areas are lowlands below 1 km. G. E. McGill: Nature **296**, 14 (1982). (With the kind permission of Macmillan Magazines Ltd., London, and of the author)

plateau with an altitude of 3 to 4 km and high, steep edges, including the Maxwell Montes, which, at 11 km, have the greatest height of any mountains on Venus.

The surface of Venus is covered by a great variety of tectonic structures: folds, upheavals, rifts, and chains of hills and mountains, which often run in various directions at angles to one another and can be some 100 up to 1000 km long. The surface was shaped almost exclusively by *volcanic activity*; around 85% is taken up by "volcanic plains", covered by lava fields and volcanoes. The latter are mostly flat (basaltic) shield volcanoes with diameters between 100 and 200 km; the smaller (≤ 35 km) round domes with steeper sides are much rarer. The noticeably long (up to several 1000 km) lava channels indicate a rather thin, nonviscous type of lava. Furthermore, we find extended, flat highlands cut through by long trenches more than 1000 km in length, which are similar to terrestrial structures such as the East African rift valley. The circular *coronae*, with diameters of 100 up to 2000 km (Fig. 3.9), in contrast, have no counterparts on the Earth. They are surrounded by a wide trench with a ring-shaped wall; their interiors lie about 100 m above the surrounding regions. The raised *highlands* are also of volcanic origin. Beta regio, with a diameter of 2000 km, is one of the few remaining "hot spots" which are still active today and through which the hot magma from the interior of the planet gives up its heat to the exterior.

Characteristic of the highlands are the *tessera* formations, of which the largest is Alpha Regio. These structures, which are raised above the surrounding volcanic plains by about 1 km, are "striped" and are organized in part into zones in a mosaic pattern, separated by trenches. As a result of their altitude, the tesserae were not flooded with lava.

All together, these structures indicate how the crust of Venus is deformed by upward pressure, compression and stretching as a result of tectonic strains. While the coronae and tesserae are to be found only on Venus, the high mountain ranges and long trenches of the terrae are

Fig. 3.9. A high-resolution radar image of the surface of Venus, recorded by Magellan in January, 1991. The image represents a roughly 300 km wide view of the region at 59° south latitude and 164° longitude within the extended plain south of Aphrodite Terra (Fig. 3.8), with round formations and a complex pattern of ring-shaped and linear fractures. The large circular structure which fills the left half of the image, with a diameter of about 200 km, is *Aine Corona*. Directly north of it, a "pancake dome" of volcanic origin with a diameter of about 35 km may be seen; it was probably formed by an outflow of extremely viscous lava. A smaller dome lies within the western ring-shaped fracture zones of the Corona, and additional small domes (≤ 10 km) are in the southern part of the ring structure. (With the kind permission of NASA/JPL/Caltech)

similar to terrestrial structures which have been formed by plate tectonics. However, on Venus, *no* global system of the apparently tectonic formations is found on the rims of the plates, as is characteristic of the Earth (Sect. 3.2.3). The crust of Venus has probably not broken up into plates.

Considerably more details of the surface structure than yielded by radar observations have been obtained thanks to the pictures and the analyses of the surface material sent back by several soft-landed space probes, first in 1972 by Venera 8, then in 1975 by Venera 9 and 10 with the first pictures of low resolution, and finally in 1982 by Venera 13 and 14, which transmitted color images permitting the recognition of structures of a few mm diameter. In the brief time intervals (1 to 2 h) during which these last-named vehicles survived the inhospitable surface conditions on Venus, with temperatures around 740 K and pressures around 90 bar, among other things rock samples were obtained with a drill and chemically analyzed in the interior of the vehicle using X-ray fluorescence following irradiation by radioactive ^{55}Fe und ^{238}Pu. At both landing sites, on the edge of Beta Regio at an altitude of 2 km (Venera 13) and 960 km distant in the lowlands (Venera 14), we find rocks of volcanic origin, quite similar to the basalts which are widespread on the terrestrial ocean floors and the lunar maria.

The surface of Venus is covered by numerous *impact craters* with diameters of several km up to 300 km. As a result of the strong braking of the impact objects by the heavy atmosphere of Venus, there are hardly craters with diameters ≤ 3 km; the abundance of the craters ≤ 30 km is low, relative to that of the larger ones. Impact craters are distributed statistically and homogeneously over the entire planetary surface, which also – in comparison to the Earth – argues against the presence of plate tectonics on Venus. It is noticeable that nearly 2/3 of the craters show no modification by flooding with lava or by tectonic processes, which are widespread on Venus.

The density of the impact craters offers the only possibility of estimating their age and thus that of the surface, since direct radioactive dating of the stone is not possible. As the mean *age* of the crust of Venus, one finds $0.5 \cdot 10^9$ yr, i.e. only about 1/10 of the age of the planet. (In comparison: the terrestrial continents are around $4 \cdot 10^9$ yr old, the sea floors about $0.1 \cdot 10^9$ yr.) There is still no satisfactory explanation for this result. Was the crust considerably harder at a time $0.5 \cdot 10^9$ yr ago than it is today, so that the structures of the impact craters became rapidly unrecognizable? Or was there some global, catastrophic event at an earlier time – perhaps a mixing up of the entire crust – which destroyed all the older craters or covered them by lava?

Mars. In contrast to Venus, on Mars the surface is readily observable owing to the thin atmosphere. The lively reddish color of the planet is due to a decrease of its reflectivity in the short wavelength region of the visible spectrum. This, along with polarimetric measurements, indicates iron oxides. Visual observations already permit a multiplicity of structures to be seen. However, the long-popular *canals of Mars* have been found to be due to a physiological-optical contrast phenomenon: our eyes have a tendency to "connect" outstanding points and vertices by lines (think of the constellations!). In the telescope, one recognizes two white *polar caps*, which are reduced in size in the Martian summer and increase during the winter.

The first television images from Mariner 4 (1965) showed numerous craters on the surface of Mars, whose diameters ranged from the limit of resolution at a few km up to $\simeq 120$ km. These craters correspond closely to those on our Moon. Mariner 9, which circled Mars from 1971 on, delivered images with a considerably improved resolution. They showed that the surface was shaped not only by meteorite impacts, but also to a large extent by volcanic activity (volcanic craters, shield volcanoes, calderas), by tectonics, by erosion (Fig. 3.10), and by the deposit of minerals.

Since 1976, the two Viking Orbiters 1 and 2 have been circling the planet and have yielded a large number of pictures, confirming the enormous variety of surface structures. The image quality is better than of the images from Mariner 9, whose sharpness was reduced by a screen of dust that was stirred up by a great storm and settled only after some months. Each of the Orbiters released a probe (Viking Lander) to make a soft landing on the surface and investigate more closely two regions of the northern hemisphere at 23° and 48° latitude, separated by 180° longitude (Chryse Planitia and Utopia Planitia). Only 20 years later, in 1997 – with the Global Surveyor – could a satellite again be put into orbit around Mars, and with Pathfinder, a direct landing on the planet was carried out. Ares Vallis in the northern hemisphere at 19° latitude (and 34° west longitude) was chosen as the landing site. The Mars vehicle Sojourner was sent out to analyze the composition of the surface material and of some pieces of rock.

On Mars, we find a large-scale *asymmetry* in the surface formations corresponding to its "pear shape": the northern hemisphere is dominated by lower-lying plains with few impact craters, while the southern hemisphere shows a very high density of craters, comparable to that of the terrae on the Moon. The Tharsis region near the equator is impressive, with several great shield volcanoes of more than 20 km height. Olympus Mons has a diameter of 600 km at its base and rises to 21 km altitude; it has a caldera 80 km in diameter.

The surface contours on Mars are somewhat rounded in comparison to those on the Moon, as a result of erosion. The complex system of canyons and twisted

Fig. 3.10. This Mariner 9 photograph (from 1972) of an area of about $500 \cdot 380$ km^2 on the surface of Mars shows, along with several meteorite craters, a part of the over 2500 km long canyon system of the Coprate region. This canyon was probably formed through a complex interaction of tectonics and erosion. The row of small craters on the right may perhaps be interpreted as having a volcanic origin

crevices, reminiscent of dried-up river beds, indicates that in earlier times there were probably great floods. This idea is supported by the numerous images taken by the Global Surveyor beginning in 1999, at a resolution of a few meters: structures on the walls of craters indicate seepage and water which has flowed underground, and extended regions of layered sediments up to 4 km thick suggest that large areas on Mars were covered with water in earlier times (about $4 \cdot 10^9$ yr ago). Today, water is found only in the form of ice and as traces of water vapor, not as a liquid; a large amount of water may be bound up in the minerals of the crust.

In spite of the numerous sandstorms, wind erosion probably does not now play an important role in forming the surface of Mars, since even very old formations can be found which still retain sharp features. Only around the polar caps do we find deposits of basaltic sand of $\lesssim 100$ m depth left by the wind (transported from equatorial regions?) as well as a belt of dunes (around the northern polar cap only).

The relatively flat, yellow-brown colored *landing sites* of the two Viking Landers remind us of a stony desert on Earth, with their many small stones and fine-grained, windblown sand (Fig. 3.11). During the Martian winter, a thin white layer of frost can sometimes be seen. X-ray fluorescence spectra of samples collected at the two landing places indicate iron-containing clays and hydroxides, along with sulfates and carbonates; furthermore, about 1 wt.-% of water is bound in the surface rocks. The Mars Pathfinder landed in Ares Vallis, a previously-flooded plain with a great variety of rocks lying on the surface. The Mars vehicle Sojourner, which was sent out from the main vehicle, was able to analyze the chemical composition of the surface using an alpha-proton-X-ray spectrometer; moreover, it could also study a number of rocks, which was not possible

Fig. 3.11. This close-up photo of the surface of Mars from the Viking 1 Lander (1976) shows numerous shards of stone ranging from a few cm to several m in size. The ground between the stones is covered by fine-grained material, which is to some extent piled up in the "wind shadows" of the stones. The lower edge of the picture is approximately 4 m away from the camera, and the horizon is at about 3000 m

Fig. 3.12. A detailed image of the martian surface taken in July 1997 by the Mars Pathfinder which landed in Ares Vallis: the Mars vehicle Sojourner (length 40 cm) is empoloying its alpha-proton-X-ray spectrometer to analyze the surface of the rock "Yogi", a primitive, quartz-depleted basalt. (Sojourner (TM), Mars Rover (TM) and space vehicle: design and images © 1996/97 Caltech)

with the Viking Landers (Fig. 3.12). For this purpose, the stones were irradiated with α-particles from a radioactive ^{244}Cm source, and then the emitted protons and X-rays were analyzed. The composition of the surface material in Ares Vallis is no different from that at the Viking landing sites – clearly a result of the global dust storms on the planet. In contrast, the composition of the rocks varies considerably, ranging from primitive basalts to stone very rich in quartz.

Directly over the *northern polar cap* in the summer, a surprisingly high temperature of 205 K was measured, which, together with the low atmospheric pressure of only about 7 mbar near the surface, excludes the existence of CO_2 ice in equilibrium with the atmosphere. Therefore, the main constituent of the polar caps must be *water ice*, which is covered by precipitated CO_2 ice depending on the season. The southern polar cap, in contrast, probably consists of CO_2 ice.

The *temperature on the surface* is found from the radiation temperatures in the radiofrequency region to be 210 K on the average, with variations between about 145 and 245 K.

Both the Viking Landers carried out several experiments to detect the possible presence of life on Mars. Surface samples were heated to 800 K and the liquids driven off were analyzed with gas chromatography and mass spectrometry. The result was negative, since aside from CO_2 and some H_2O, no fragments of organic molecules were detected. On the other hand, the microbiological experiments, which were supposed to react to gas exchange, metabolism and carbon assimilation, have not permitted a clear-cut distinction to be made as yet between biological and chemical activity.

The weak *magnetic fields* detected by the Mars Global Surveyor at several points, ranging up to $4 \cdot 10^{-7}$ Tesla in strength can be interpreted as the remains in the crust of Mars of an earlier global magnetic field. The magnetized strips in the Martian southern hemisphere, which are up to 2000 km long and exhibit polarity, are indicative of the effects of plate tectonics in the past.

Fig. 3.13. Images of the Martian moons Phobos (*above*) and Deimos (*below*), taken by the Viking Orbiter (1977). The diameter of the irregularly-shaped satellites is about 27 and 14 km. They are covered with impact craters having diameters up to 5.3 km. (With the kind permission of NASA/JPL/Caltech)

The Moons of Mars. In 1877, A. Hall discovered the tiny moons of Mars, *Phobos* and *Deimos* (Fig. 3.13), which have diameters of about 27 and 14 km, respectively. The period of revolution of Phobos, 7 h 39 min, is considerably shorter than the rotational period of the planet. Mariner 9 and then later the Viking missions returned photographs showing the traces of strong bombardment by meteorites on the two moons. In 1989, the probe Phobos 2 was able to obtain additional images of Phobos from a distance of 200 to 1000 km with improved resolution.

Surface-forming Processes. Although we have met widely differing surface structures on the earthlike planets and their moons, they may all be understood as the results of the same *basic processes*. For all the objects in the Solar System which have formed solid crusts, the strong bombardment by planetesimals, especially in the first 10^9 yr after their formation, is the most important process. The result is surfaces covered with *impact craters*, which we find all the way out to the satellites of the Neptune system (Sect. 3.4).

In the cases of Mercury, Earth's Moon, and the two small moons of Mars, this original surface structure is for the most part still present, since essentially no processes have occurred which would have destroyed it. In contrast, on Venus, the Earth, and Mars, the crust was reformed on the one hand by *tectonic* and *volcanic* effects, and on the other by *weathering* and *erosion*. While on the Earth, which remains geologically very active, the crust is modified into a variety of forms on a planetray scale mainly by plate tectonics and the associated volcanic activity, on Venus and Mars mostly local tectonic processes such as the rise and fall of mountain ranges, as well as volcanic activity, play a dominant role. On planets with atmospheres and constituents which have condensed from them, we have a variety of *leveling processes*, mechanical and chemical weathering as well as transport and deposit by wind, water, or glaciers. On the Earth, erosion by water (dissolution of minerals, cracking by frost) is dominant; on Venus, owing to the high temperatures and pressures, chemical weathering is probably the most important process. On Mars, we have found indications of water erosion in the past and of wind erosion.

3.2.6 The Atmospheres of the Earthlike Planets

Mercury and Earth's *Moon* have no atmospheres in the strict sense. They merely have *exospheres* (Sect. 3.1.4) with very low concentrations of particles, on the order of 10^9 m^{-3} on the day side and 10^{11} m^{-3} on the night side; they result mainly from an equilibrium between supply of particles by the solar wind and loss as a result of thermal motion.

In contrast, *Venus*, *Earth*, and *Mars* have atmospheres of differing densities (Table 3.2). We have already discussed the basics of their global energy balances in Sect. 3.1.4. The *surface temperatures* of Venus, and to a lesser extent of the Earth, are as a result of the "greenhouse effect" *higher* than the effective temperature, while on Mars with its very thin atmosphere they more closely correspond to the equilibrium value (3.2).

The *chemical composition* (Table 3.2) of the atmospheres of Venus and Mars, which consist mainly of CO_2, is very similar, while the Earth's atmosphere is distinguished by a high proportion of N_2 and O_2. The Earth probably also previously had an atmosphere similar to that of Venus and Mars, formed by volcanic activity and by outgassing. However, through condensation of water into oceans, in which CO_2 dissolves and can then react with silicate minerals to form carbonates, as well as through the presence of living organisms (photosynthesis of oxygen by plants), its composition was fundamentally changed (Sect. 14.2.1).

The Earth's Atmosphere. In particular with a view to comparisons with the other planets, we first consider briefly the structure of the *Earth's* atmosphere. Its pressure profile is, according to (3.19), determined in the main by the temperature profile $T(h)$. The latter is fixed by the mechanisms of energy transport, i.e. the input and output of thermal energy into each layer between h and $h+dh$. In the Earth's atmosphere (Fig. 3.14), the removal of absorbed solar energy in the lowest layer, the *troposphere*, takes place mainly through convection, which leads to a uniform decrease of the temperature on going upwards. Above the socalled *tropopause*, at an altitude of roughly 10 km, radiation becomes the dominant mechanism of energy transport and we arrive at the *stratosphere*, which is almost isothermal. What is now important is the decisive mechanism of absorption of solar radiation on the one hand, and its reradiation into space at longer wavelengths on the other. At an altitude of 25 km we find a warm layer in connection with the formation of ozone, O_3, up to the *stratopause* at an altitude of about 60 km. In the *mesosphere* which lies above it, CO_2 radiates energy in the infrared, while warming through absorption by O_3 no longer occurs, so that a brief decrease in temperature results. Above the *mesopause* at about 85 km altitude, in the *thermosphere*, the temperature rises to about 1000 K (night side) or 2000 K (day side) as a result of the dissociation and ionization of the atmospheric gases N_2 and O_2 by solar UV radiation.

The roughly 20 km thick *ozone layer* has a broad maximum in O_3 partial pressure at about 25 km altitude; the relative particle concentration has a value there of about $5 \cdot 10^{-6}$, the temperature is near 220 K. The ozone in the stratosphere is formed by photolytic dissociation of molecular oxygen, O_2, by the short-wavelength solar radiation (energy $h\nu$ at wavelength $\lambda = c/\nu \leq 242$ nm)

$$\begin{aligned} O_2 + h\nu &\rightarrow O + O, \\ O + O_2 + M &\rightarrow O_3 + M, \end{aligned} \quad (3.28)$$

Here, an additional particle M (e.g. N_2) is required in order to fulfill energy and momentum conservation during the reaction. The O_3 is not destroyed by photolysis reactions $O_3 + h\nu \rightarrow O_2 + O$, which are relatively slow, so that free O preferentially again forms O_3 by combining with O_2. Instead, the destruction of ozone occurs primarily through catalytic reactions:

$$\begin{aligned} O_3 + X &\rightarrow XO + O_2, \\ O + XO &\rightarrow X + O_2 \end{aligned} \quad (3.29)$$

with the net balance $O_3 + O \rightarrow 2O_2$. The most important catalysts X are OH, NO, and Cl, whereby Cl and in part OH and NO are industrial waste products, i.e. they are of anthropogenous origin.

Photoionization produces the electrically conducting layers of the *ionosphere* (with the maximum electron density of the D layer at about 90 km, that of the E layer at about 115 km, and of the F layer at 300 to

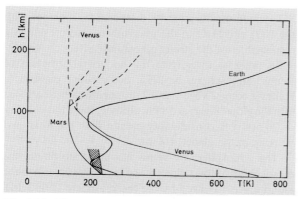

Fig. 3.14. The mean temperature as a function of altitude in the lower atmospheres of the Earth, Venus, and Mars. In the case of Venus, there are large temperature differences between the day and night sides of the planet. The atmosphere of Mars is subject to wide temperature variations; the (smoothed) measurements of the Viking Lander are shown. During major dust storms, the temperature in the lower part of its atmosphere increases (*shaded region*)

400 km altitude). These layers allow the transmission of electromagnetic waves of relatively long wavelengths completely around the globe. The recombination of electrons and ions in the E layer produces the emission lines and bands called the *airglow*.

Up to the socalled *turbopause* around 120 km in altitude, apart from temporal and spatial variations in some socalled trace substances (e.g. H_2O or O_3), the atmosphere is well mixed; above it, the different gaseous constituents separate by diffusion. For example, above about 300 km, atomic oxygen O is predominant, and above about 2000 to 3000 km, atomic hydrogen H is mainly present; it can be observed from satellites through the strong Lyman α emission at $\lambda = 121.6$ nm.

From the *exosphere* (above about 500 km altitude), particles from the atmosphere can escape into space (Sect. 3.1.4). The terrestrial magnetic field is very important for the dynamics of the ionized constituents. The extent of the *magnetosphere* is determined by interactions with the solar wind on the side of the Earth directed towards the Sun. At the subsolar point, its boundary, the *magnetopause*, is at about 10 earth-radii; on the night side, a plasma trail extends to about 1000 earth-radii. At 1.6 and 3.5 earth-radii, we find higher concentrations of charged particles in the "*radiation belts*" discovered in 1958 by J. A. van Allen.

For the global *energy balance* of the Earth's atmosphere, the *clouds* (consisting of H_2O) play a decisive role. On the average, they cover about 50% of the surface; a typical cloud layer has an albedo of about 0.5. The details of both the input of solar radiation and the albedo of Earth's atmosphere and its surface depend on the latitude. The excess of absorbed radiation in the equatorial zones as opposed to the polar regions provides the driving force for atmospheric and oceanic currents. We cannot delve further here into the dynamics of planetary atmospheres and weather, a complex subject in itself.

The Atmosphere of Venus. The discovery in 1932 by W. S. Adams and Th. Dunham based on infrared spectra that Venus possesses a thick atmosphere rich in CO_2 was followed by determinations of the radiation temperatures in the cm- and dm-wavelength spectral regions, and later the flybys of the Mariner satellites, the parachute drops of instruments by the Venera probes, and observations by the Pioneer Venus Orbiter; the last two investigations in particular yielded considerable information about the planet's atmosphere. This dense atmosphere, with a pressure of about 93 bar and a temperature of 735 K at the surface, consists almost entirely of CO_2 and N_2 (Table 3.2). The traces of HCl and HF which have been detected in the lower atmosphere (as well as the drops of sulfuric acid in the clouds) probably result from chemical reactions of the atmosphere with surface minerals. Due to its high density, the lower portion of Venus' atmosphere exhibits only very weak pressure and temperature variations and low wind velocities ($\lesssim 1$ m s^{-1} at the surface).

In the upper part of the troposphere, between about 50 and 70 km altitude, at temperatures of 370 to 220 K, there are dense *cloud layers* which completely block optical observations of the planet's surface and prevent all but a few percent of the incident sunlight from reaching it, owing to scattering. Both above and below the cloud cover there are in addition layers of haze about 20 km thick. Observations with ultraviolet light reveal high-contrast structures in the cloud layer (Fig. 3.15), which are between 10 and 1000 km in size and persist for several days; dark V- or Y-shaped formations are particularly noticeable. The clouds consist mainly of H_2SO_4 droplets a few μm in diameter with a concentration of several 10^8 droplets per m^3; in addition, there are particles of 10 to 15 μm diameter, which apparently consist of solid or liquid sulfur. The sulfuric acid and the SO_2 which results from its dissociation, together with the CO_2 and H_2O in the atmosphere, are presumably responsible for the strong greenhouse effect on Venus.

In order to investigate the horizontal structure of Venus' atmosphere, in 1985, before their mission to Halley's comet, Vega 1 und 2 dropped weather balloons into the cloud layer; at 54 km altitude, the wind velocity of 70 m s^{-1} is considerably greater than the rotational velocity of the planet, 1.8 m s^{-1}.

The atmosphere of Venus has no stratosphere; the mesosphere follows directly above the tropopause or the cloud layer, followed by the thermosphere (Fig. 3.14). The exosphere begins at an altitude of about 160 km. With temperatures in the range 120 to 250 K, it is relatively cool in comparison to the upper atmosphere of the Earth. The turbopause is at about 140 km altitude; above 150 km, atomic oxygen predominates, and above 250 km, helium and hydrogen. The *ionosphere* has its maximum around 140 km with ion densities of several

3.3 Asteroids or Small Planets (Planetoids)

Fig. 3.15. A photograph of the cloud structures of Venus from an altitude of 65 000 km taken with the cloud photopolarimeter of the Pioneer Venus Orbiter (1979). The bright cloud bands near the poles, which are higher than the adjacent background, are set off clearly from the other formations. The north pole is at the top of the picture

10^{11} m^{-3}, comparable to the Earth's E-layer. Its main constituent up to 200 km altitude is the molecule-ion O_2^+, and higher up it is O^+.

The *upper* atmosphere of Venus shows diurnal and annual variations. Furthermore, clearcut thermal structures have been observed, which are to some extent not yet understood. For example, in the mesosphere at around 90 km altitude, the poles are warmest and the subsolar point coolest; at the mesopause, there is a relative maximum in temperature in the middle of the night side. In addition, infrared observations near both poles show long structures ($\simeq$ 4000 km) which are up to 35 K warmer than their surroundings and rotate in the retrograde sense with a period of about 3 d.

The Atmosphere of Mars. Mars has an atmosphere which is very thin in comparison to those of Venus and Earth, and exhibits extensive pressure and temperature variations. Also characteristic are violent *sandstorms* on a planetary scale. The atmosphere reacts sensitively to fluctuations in the absorption of solar radiation, which in turn is strongly dependent on the albedo of the polar caps and of the dust which is stirred up by storms. In particular, the temperature gradient in the lower atmosphere decreases strongly with increasing dust content (Fig. 3.14).

Spectroscopic observations and the mass-spectroscopic measurements of the Viking Landers indicate that Mars' atmosphere, which is well mixed up to an altitude of 120 km, consists mainly of CO_2 and N_2 with some other constituents present in minimal concentrations (Table 3.2). Water vapor plays a special role; it is present only in trace quantities and is subject to very strong local and seasonal variations. (Under the present conditions on Mars, free H_2O can exist as a stable phase only in the form of ice or vapor, however not as a liquid.) In spite of its low concentration, the water vapor in the atmosphere is probably nearly saturated and plays an essential part in the formation of the clouds which have been observed. The *weather on Mars* is characterized by various types of thin ice clouds, surface fog, and by diurnally and annually varying winds and dust storms.

The short-wavelength "airglow" from the upper atmosphere of Mars contains the Lyman α emission line of hydrogen; also, atomic oxygen and carbon, in addition to CO_2^+ and CO, have been detected.

Mars also has an *ionosphere*, onto which – as shown by measurements from the probe Phobos 2 – the solar wind (Sect. 7.3.7) impinges directly. On the side of the planet facing away from the Sun, the ionospheric plasma streams as a kind of "tail" away from Mars.

3.3 Asteroids or Small Planets (Planetoids)

Today, we have catalogued several thousand asteroids, most of them between Mars and Jupiter. The total number of asteroids is estimated to be around 10^6. They are denoted by the year of their discovery and two letters, which give the time when they were first observed (in coded form), and may be followed by a number when more than one were discovered in that time interval. Objects with a known orbit, of which there are more than 20 000, also receive a series number and a name.

3.3.1 The Orbits of the Asteroids

The eccentricities of their orbits have a maximum frequency of occurrence at $e \simeq 0.14$, and the inclinations of the orbits at $i \simeq 10°$; both of these values differ considerably from those of the comets (Sect. 2.2). Their semimajor axes a are mainly in the range between 1.8 and 5.2 AU. Following K. Hirayama and others, we distinguish about 80 *families* of asteroids with similar orbital elements, which can probably be attributed to a larger parent object which has now broken up. The distribution of orbits according to their orbital periods shows several clusters and gaps (named for D. Kirkwood), which correspond to integral ratios with the orbital period of Jupiter; for example, we find a well-developed gap at a ratio of 3 : 1 (i.e. with $a \simeq 2.5$ AU), and a cluster in the group of the Trojans at the socalled libration (or Lagrange) point (1 : 1) of Jupiter's orbit.

433 Eros, for example, has unusual orbital elements, with $e = 0.23$; at opposition, it approaches the Earth to within 0.15 AU and thus permits excellent determinations of the solar parallax. 1566 Icarus ($e = 0.87$, $a = 1.08$ AU) approaches the Sun at perihelion to within the orbit of Mercury. 2102 Tantalus has an unusually large orbital inclination of $i = 64°$. 2060 Chiron, by contrast, remains for the most part between the orbits of Saturn and Uranus ($a = 13.7$ AU); only years after its discovery, when it showed a large increase in brightness and formed a coma (Sect. 3.6.1), was it recognized to be a comet.

We shall treat the recently-discovered small objects outside the orbits of Neptune and Pluto, whose sizes are comparable with those of the asteroids, in Sect. 3.5.2.

Asteroids near the Earth. Particularly interesting for us are those asteroids which cross the Earth's orbit and thus could collide with our planet. Three such families are known: the Aten group, the Apollo group, and the Amor group. Hermes (1937 UB), whose orbit is not known precisely, approached the Earth to within at least 0.005 AU, roughly twice the distance from the Earth to the Moon ($r_M = 384\,400$ km). Also, 1989 FC passed similarly close to the Earth.

Using the 0.9 m Spacewatch telescope in Arizona, a sky survey was begun in 1990 to detect traces of rapidly moving, faint asteroids which might approach the Earth closely. Numerous objects with diameters from around 10 up to 300 m have been found; they thus occupy a position between the smallest asteroids and the largest meteorites (fireballs, Sect. 3.7). Among these was 1994 XM_1, which approached the Earth to within about $0.3 r_M$. It is estimated that an unrecognized object with a diameter of more than 10 m approaches the Earth to within a comparable distance almost daily.

The total number of objects with diameters ≥ 1 km which might approach the Earth closely is estimated to be around 1000, of which not quite half are presently known.

A "Companion" of the Earth. The tiny – still unnamed – asteroid 3753 (1986 TO) with a diameter of only about 5 km has an orbit whose semimajor axis of $a \simeq 1$ AU (varying between 0.977 and 1.003 AU), eccentricity $e = 0.51$ and inclination $i = 19.8°$ at first glance seem not unusual. The orbit reaches from within that of Venus out to somewhat beyond that of Mars and crosses the Earth's orbit with a period of $P \simeq 1$ yr, approximately in (1 : 1) resonance.

However, if one considers the orbit of this object in the coordinate system which rotates with the Earth around the Sun, it can be seen – according to the calculations of P. A. Wiegert and others (1997) – to exhibit an unusual behavior: in the course of a year, 3753 follows a kidney-shaped, not quite closed orbit which shifts along Earth's orbit around the Sun with a cycle of 770 yr. The asteroid never approaches the Earth nearer than about 0.1 AU. Its orbit is sensitive to perturbations by Venus and Mars and thus will probably be stable only over a time of at most 10^8 yr.

Along with the ancient Moon, the Earth thus has another "companion" in the form of the asteroid 3753, which is bound in its motion in a (1 : 1) resonance to the Earth's orbit, at least for some time to come.

3.3.2 Properties of the Asteroids

Mass. The masses of the three largest asteroids can be determined from the perturbations which they cause to the orbits of other asteroids (J. Schubart,

1973). Recent values are $1.2 \cdot 10^{21}$ kg $= 2.0 \cdot 10^{-4} \mathcal{M}_E = 6.0 \cdot 10^{-10} \mathcal{M}_\odot$ for 1 Ceres, $3.5 \cdot 10^{-5} \mathcal{M}_E$ for 2 Pallas and $4.5 \cdot 10^{-5} \mathcal{M}_E$ for 4 Vesta. The overall mass of all the asteroids is estimated to be $5 \cdot 10^{-4} \mathcal{M}_E$ or about 2.5 times the mass of Ceres.

Diameter. The angular diameter can be measured directly only for the largest and brightest asteroids; for the smaller ones, diameters to 0.4 km can be estimated from their brightness or polarization of reflected light by assuming a reflectivity curve. For example, 1 Ceres is found to have a diameter of 940 km, 2 Pallas, of 538 km, and 4 Vesta of 535 km, with an uncertainty of 10 km; their mean densities are thus of the order of 3000 kg m^{-3}, somewhat smaller than those of the earthlike planets.

Rotation. Many asteroids show a periodic variation of brightness, which indicates their rotational periods. Remarkably, they all lie between 3 and 17 h. Brightness variations of 0.1 to 0.3 mag (for the definition of magnitude, cf. Sect. 6.3.1) are typical; however, 433 Eros, for example, has a stronger fluctuation with an amplitude of 1.5 mag and a period of 5.3 h, and 1620 Geographus has an even larger amplitude of 2.0 mag with a 5.2 h period. The light curves of many asteroids indicate that they have elongated, irregular shapes.

Spectrum of Reflected Light. Spectrophotometric measurements of the reflected light from asteroids in the visible and infrared regions allow us to identify various surface characteristics. The reflection spectra of most asteroids are similar to those of the meteorites (Sect. 3.7.3), which are indeed fragments of asteroids. From this relationship, one can reach conclusions about the chemical composition of the asteroids themselves. Most frequently, one finds *type S* ("stony meteorites") with broad absorption bands in the ultraviolet and blue, or *type C* ("carbonaceous chondrites"), with a low albedo (0.03–0.08) and a flat spectrum showing only weak structure. Type S predominates in the inner parts of the asteroid belt, type C in contrast in the outer parts.

Shapes. Using *radar*, in addition to the distance one can determine the size, shape and surface structure, along with the rotation of an asteroid. Frequently, dumbbell or peanut-shaped *"double asteroids"* are observed, which presumably consist of two loosely bound objects. In favorable cases, *stellar occultations* by an asteroid also yield information about its shape.

Close-up Observations. The space probe Galileo while underway to Jupiter in 1991/93 observed the two type S asteroids Gaspra (from a distance of 1600 km), and Ida (from 3100 to 3800 km). Deep Space 1 flew past Braille in 1999 at a distance of only 10 to 15 km. The NEAR-Shoemaker mission (*N*ear *E*arth *A*steroid *R*endezvous), yielded images of Mathilde in 1997 from a distance of 1200 km; then, in 2000, after an unsuccessful maneuver a year before, it was swung into an orbit around Eros and landed on the asteroid in 2001.

951 Gaspra, an elongated, irregular (19 km · 12 km · 11 km) object covered with impact craters (of up to 1.6 km diameter), rotates with a period of 7.04 h; its albedo is 0.23. Its age, i.e. the time since it split off from a larger parent object, lies between 3 and $5 \cdot 10^8$ yr.

243 Ida is at 56 km · 24 km · 21 km more than twice the size of Gaspra, and exhibits larger impact craters with a higher density, which indicates an age of $1 \cdot 10^9$ yr. Its rotation is retrograde, with a period of 4.65 h. Galileo discovered the first *satellite* of an asteroid at a distance of about 100 km from Ida's surface (Fig. 3.16): the tiny (1.6 km · 1.4 km · 1.2 km) moon *Dactyl* revolves around Ida in the same direction as its rotation. It is likewise covered with impact craters (of diameters up to 0.3 km). From its orbit around the asteroid, using Kepler's third law (2.57), one finds for Ida a density of around 2500 kg m^{-3}.

9969 Braille, discovered only in 1992, is indicated by its infrared reflection spectrum to be a 2 km long "splinter" from the large asteroid 4 Vesta. The eucrites, a subclass of the achondritic meteorites (Fig. 3.31), are probably also debris from Vesta.

253 *Mathilde*, of size 66 km · 48 km · 44 km, has a very low albedo of about 0.04 and a mean density of only around 1300 kg m^{-3}. This porous asteroid is likewise thickly covered with craters (of up to 30 km diameter!).

433 Eros, an elongated asteroid (34 km · 11 km · 11 km) of type S, likewise almost completely covered with craters (for the most part of ≤ 1 km diameter), has a density of about 2700 kg m^{-3}. Its chemical composition, obtained by X-ray fluorescence spectroscopy,

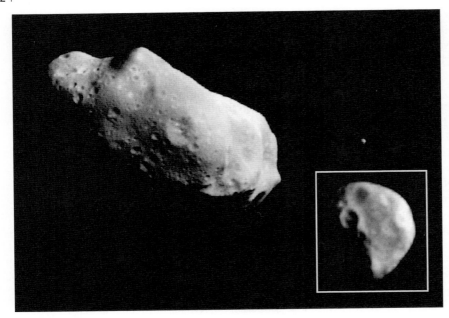

Fig. 3.16. An image taken by the space probe Galileo in August 1993: asteroid 243 Ida with its moon Dactyl (to the right of Ida) from a distance of 10 500 km; *insert:* close-up image of Dactyl at a distance from the probe of only 3900 km. (With the kind permission of NASA/JPL/Caltech)

is similar to that of the common chondrites (Fig. 3.31). This is evidence for the fact that the most common meteorites and the reddish-colored type S asteroids "belong together". The difference in coloring between the grey chondrites and the red-tinted S asteroids could be due to "weathering" of the surfaces of the asteroids, which have spent more time in space.

The unexpectedly low average density of some of the asteroids which have been observed from close up indicates a loose, porous structure (like a "rubble pile"); this idea is supported by their relatively large impact craters as well as the partial covering of their surfaces by regoliths, similar to the Moon (see Sect. 3.2.4). A solid, stony object would break apart as a result of such impacts.

3.4 The Major Planets

After this rather brief survey of the earthlike planets, we turn to the major planets, which, as their masses and mean densities already demonstrate, are of quite different character (Table 3.1).

3.4.1 Jupiter

Jupiter, the largest and most massive of the planets (1/1047 solar masses), exhibits a dense atmosphere on observation from the Earth, with clear *stripes* parallel to the equator (Fig. 3.17), similar to the wind belts on the Earth. The socalled *Great Red Spot*, discovered in 1665 by G. Cassini, is a large oval structure which has persisted for a long time.

Jupiter's Atmosphere. The first close-up pictures and measurements were obtained in 1973 by the space probe Pioneer 10, which flew by Jupiter with a closest approach of 130 000 km ($\simeq 2$ Jupiter radii). In 1979, the two probes Voyager 1 and Voyager 2 transmitted impressive high-resolution images of the flow patterns on the planet (Fig. 3.18): we see among other things cloud bands, convection cells, jetstreams, vortices, white and brown ovals of varying size, and circulation systems. Neighboring structures can have relative velocities of over $100\,\mathrm{m\,s^{-1}}$. In addition to long-lived formations, for example the Red Spot, we can observe changes in large structures on a time scale of a few days.

As early as 1932, R. Wildt identified the strong absorption bands in the spectrum of Jupiter with higher

3.4 The Major Planets

Fig. 3.17. Jupiter. Photograph by B. Lyot and H. Camichel, taken with the 60 cm refractor on the Pic du Midi

overtones of the molecular vibrations of *methane*, CH_4, and of *ammonia*, NH_3; this represented a turning point in the investigation of the major planets. In 1951, G. Herzberg succeeded in detecting some infrared band lines of the hydrogen molecule, H_2, which are weak due to their small quadrupole transition probabilities. During a stellar occultation by the disk of Jupiter in 1971, some insight was gained into the layer structure of the planetary atmosphere, and from the equivalent altitude (3.17), its main components could be concluded to be *hydrogen* and *helium*. Furthermore, especially by means of infrared spectroscopy, numerous *trace compounds* were discovered, such as ethane, C_2H_6, ethyne (acetylene), C_2H_2, water, H_2O, hydrocyanic acid, HCN, phosphine, PH_3, germane, GeH_4, and the deuterated molecules HD and CH_3D.

In the infrared, one can observe into layers at up to 225 K. In this region, the transparency of the atmosphere is more and more limited by *clouds*; only occasionally, thorough gaps in the cloud layer, can deeper layers at up to 280 K (and a pressure of 5 bar) be seen. The high-

Fig. 3.18. This Voyager 1 photograph of Jupiter (1979) from a distance of $4.3 \cdot 10^6$ km shows the region immediately to the southeast of the Great Red Spot, with a vortex-like flow field around one of the "white ovals". The smallest structures which can be distinguished are about 80 km in size

est cloud layer (at about 150 K) consists presumably of NH_3 crystals; the deeper layers are probably NH_4SH and H_2O; the explanation of the red and brown colors requires that trace compounds also be present. The weather on Jupiter appears to be much more complex than in the Earth's atmosphere, particularly because of the occurrence of chemical reactions.

The instrument capsule which was released in 1995 by the Jupiter probe Galileo penetrated into the atmosphere down to about 200 km below the cloud layer before its radio signals ceased after around 10 hours. Along with mass spectrometric determinations of the abundances of various elements, it also studied weather phenomena, in a region relatively free of clouds. The zonal winds around the planet dip with their full strength of around $180\,\mathrm{m\,s^{-1}}$ deep into its atmosphere.

The Inner Structure of Jupiter. A detailed analysis of all the observations together with the basic premises of a theory of inner planetary structures yields the following model: Jupiter consists for the most part of unchanged *solar material*, with hydrogen and helium in the (atomic) ratio $He/H \simeq 0.1$ being the most abundant elements. In spite of the high pressures and densities, the hydrogen in the planetary interior remains for the most part *liquid*, owing to the relatively high temperatures ($\leq 30\,000$ K). At 0.77 jupiter radii, i.e. at a pressure of $3 \cdot 10^{11}$ Pa and a density of $1000\,\mathrm{kg\,m^{-3}}$, a phase transition takes place from liquid molecular hydrogen, H_2, to liquid *metallic* hydrogen. Jupiter in all probability has a *solid core*, which contains about 4% of its mass or about 14 Earth masses, and consists of a mixture of *stones* (SiO_2, MgO, FeO, FeS) and *ices* (CH_4, NH_3, H_2S, H_2O). This core could have served in the early phases of the formation of the Solar System as a "condensation nucleus" for the hydrogen- and helium-rich solar material. The models give a density for the core of $\geq 2 \cdot 10^4\,\mathrm{kg\,m^{-3}}$ and a pressure of about 10^{13} Pa. The excess of thermal radiation from Jupiter compared to the solar irradation has already been discussed in Sect. 3.1.2.

The Magnetic Field of Jupiter. The *radiofrequency emission* of Jupiter is thermal at wavelengths $\lambda \leq 1$ cm, with a radiation temperature of about 120 K, corresponding approximately to that of the tropopause. In the decimeter-wave range, the intensity of the radio emission increases and, due to its partial polarization, must have its origin as nonthermal synchrotron radiation in the planetary magnetic field (see Sect. 12.3.1). In the meter wavelength region, there are are in addition sharply localized radiation *bursts* from well-defined sources on the planetary disk.

The measurements of the Pioneer and Voyager probes indicate that Jupiter has a dipole-like magnetic field with about $4 \cdot 10^{-4}$ Tesla at the equator and a dipole moment (3.27) of $1.5 \cdot 10^{20}$ T m^3, whose axis is tilted away from the planet's axis of rotation by about 10°. Jupiter's *magnetosphere*, similar to the Van Allen belts around the Earth, traps enormous numbers of energetic electrons and protons, as well as thermal plasma. At high altitudes in the Lyman region, one can observe the emission lines of hydrogen and helium, a sort of airglow.

The Crash of a Comet. A spectacular occurrence in 1994 was the crash of the comet D/Shoemaker-Levy 9 (Sect. 2.2.2) onto Jupiter: 21 large fragments impacted one after the other over a period of 6 days at a velocity of $60\,\mathrm{km\,s^{-1}}$ near 45° south latitude on the far side of Jupiter, roughly 5 to 10° behind the rim of the planetary disk. Material which was thrown up to great heights could be observed against the dark background of the sky. Due to the rapid rotation of the planet (rotational period 9.8 h), the impact points themselves became visible about 15 min after each impact. The event was observed by numerous earthbound telescopes, by the Hubble Space Telescope, and from a distance of 1.5 AU by the space probe Galileo.

Two important questions pose themselves: what was the size of the comet fragments, and how did the atmosphere of Jupiter react to the impacts? Since the size or mass of the fragments and thus their kinetic energy were not sufficiently well known, predictions about the phenomena which might be observed remained vague.

Most of the impacts set off a relatively unified chain of events: as soon as the coma of a fragment, with its numerous dust particles, entered the atmosphere, a glow slowly rose up. When the massive nucleus of the fragment entered the atmosphere, within about 10 s a strong flash was seen, much like that which occurs when a meteoroid (bolide) enters the Earth's atmosphere. A few tenths of a second later, the channel which had been "bored" into the atmosphere by the comet's head exploded into a fireball as a result of the input of energy. Hot gas was thrown up out of the channel to a height of about 3000 km. At first, it was so hot (around 7000 to 8000 K) that it emitted light in the optical range; later, as soon as it emerged from Jupiter's shadow, it could be observed in reflected sunlight. Furthermore, ring-shaped sound waves propagated outwards horizontally in the stratosphere, above the cloud layers. After about 10 min, the material which had been ejected fell back onto the upper atmospheric layers and heated them up; this produced a powerful, long-lasting outburst in the infrared range (Fig. 3.19). After a further 10 min, another burst could be seen due to reflected material which had again fallen into the atmosphere. Finally, the falling material formed dark brown structures up to 25 000 km in diameter, which remained visible for weeks and were only gradually pulled apart by Jupiter's rotation. Their dark coloration is caused by dust particles and gas which absorb strongly in the ultraviolet range.

The spectroscopic observations showed that first ammonia – from the uppermost layer of clouds – becomes visible at an increased concentration, then sulfur, S_2, and compounds such as CS_2 and H_2S, which normally are not found in Jupiter's atmosphere. Sulfur and oxygen are abundant elements in comets; however, the low abundance of oxygen compounds observed after the impacts proves that NH_4SH from the middle cloud layers was the main source of the sulfur compounds. Finally, the low concentration of water argues against the impacts having released material from the lowest H_2O-containing cloud layer into the higher visible layers. The H_2O, CO, Mg, Si, and Fe observed in the spectra must have come from the comets themselves.

From the overall picture of the observed phenomena, we can conclude that the largest comet fragments had diameters of only about 700 m, that they exploded at a relatively high altitude in Jupiter's atmosphere, still in the stratosphere, and failed to reach the lowest layer of clouds. The effects of the comet fragments on the magnetosphere of Jupiter cannot be treated here.

Jupiter's Moons. In 1610, Galileo discovered the four brightest satellites J1 to J4. With increasingly larger telescopes, a total of 12 were then detected from the Earth, before the number of known satellites increased to 16 as a result of the Voyager missions (Table 3.4). In 1999/2000 a further moon was discovered with the Spacewatch telescope. Then, in early 2001, S. Shep-

Fig. 3.19. The impact of the comet Shoemaker-Levy 9 onto Jupiter. This image was taken with the radiation from the methane band at $\lambda = 1.7$ μm, in which the planetary disk appears relatively dark; it clearly shows the bright area around the explosion following an impact immediately behind the hemisphere facing the camera (*below left*), as well as the flashes following three other previous impacts of comet fragments – these have been moved into the field of view by the rotation of the planet. Image taken by T. Herbst et al. with the infrared camera MAGIC at the 3.5 m telescope of the Calar Alto Observatory in Spain on the 20th of July, 1994. (With the kind permission of the authors and of Sterne und Weltraum)

Table 3.4. The moons of Jupiter. The distance a from the center of the planet is given in units of Jupiter's radius, $R_J = 71\,400$ km. The satellite radius R (or the semiaxis of the ring) is in [km]

a/R_J	Name	R	a/R_J	Name	R
1.80	J15 Adrastea	10	156	J13 Leda	8
1.80	J16 Metis	20	161	J6 Himalia	95
2.55	J5 Amalthea	135·83·75	164	J10 Lysithea	18
3.11	J14 Thebe	50	165	J7 Elara	38
5.9	J1 Io	1815	291	J12 Ananke	15
9.5	J2 Europa	1570	314	J11 Carme	20
15.1	J3 Ganymede	2630	327	J8 Pasiphae	25
26.6	J4 Callisto	2400	333	J9 Sinope	18

pard and coworkers, using the 2.2 m telescope of the University of Hawaii, found 11 additional tiny, dark moons. The orbital radii of these moons, which are not included in Table 3.4, lie between about 104 and 340 R_J; the radii of the moons themselves are at most 3 to 8 km. The number of presently-known satellites of Jupiter is thus all together 28.

The circular orbits ($e \leq 0.01$) of the eight inner satellites are all nearly in the equatorial plane of the planet (regular orbits). The outer satellites, in contrast, have larger eccentricities (0.1 to 0.5) and orbital inclinations, with the 14 outermost showing a retrograde motion. They are presumably asteroids which have been captured by Jupiter.

Amalthea, the innermost known moon until the Voyager missions (E. E. Barnard, 1892), is a reddish, elongated and irregularly-shaped body, whose orbit lies just outside the Roche stability limit (3.11). A *ring system* was unexpectedly discovered by the Voyager probes; it consists of a flat ($\lesssim$30 km), bright ring with a radius of about 1.8 Jupiter radii, which is embedded in a disk and a weak halo, and of a much weaker ring which extends to about 3 Jupiter radii. At the outer edge of the bright ring, two small moons (J15 and J16) move co-orbitally, i.e. practically on the same orbit, with periods of revolution of only 7 h8 min.

The radii and masses or mean densities of the four *Galilean satellites* are rather precisely known:

Satellite	J1 Io	J2 Europa	J3 Ganymede	J4 Callisto
Mass	4.7	2.6	7.8	5.6 [$10^{-5} \mathcal{M}_J$]
Mean Density	3550	3040	1940	1830 [kg m^{-3}]

Ganymede is the largest moon in the Solar System. While the mean densities at first suggest that Io and Europa mainly consist of silicates, and Ganymede and Callisto in contrast of a mixture of ice and silicates (about 1 : 1), the observations of moments of inertia and the analysis of surface structures during the fly-bys of the Galileo mission have revealed a more detailed structure of the Jupiter moons. Io has a core of Fe and FeS with a radius of $\leq$ 940 km, i.e. $\leq$ 0.52 of the radius of the satellite; outside the core are the partially molten mantle and the crust, made of light silicates. Europa and Ganymede likewise consist mainly of metallic core and a silicate mantle, which on its outside (up to several 100 km) is surrounded by a thick layer of (H_2O) ice.

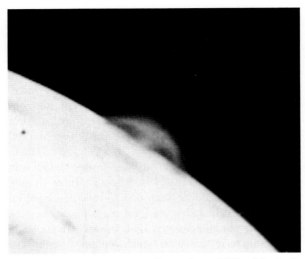

Fig. 3.20. In this Voyager 1 picture from 1979 of Jupiter's moon Io, an eruption of one of the active "volcanoes" (Prometheus) can be readily discerned at the rim; it rises to more than 100 km above the surface

The surface of *Io*, the innermost of the four large moons, is spotted and sprinkled with varying, mainly grayish-yellow colors; its rocks are rich in sodium-potassium compounds and especially in sulfur and sulfur compounds. In a few mountain ranges, the crust rises up to about 10 km altitude. Particularly noticeable is the large number of calderas with diameters of $\geq$ 200 km, indicating dead volcanoes, as is the complete absence of impact craters. The most surprising discovery of the Voyager probes was the strong surface activity on Io: all together, nine active "*volcanoes*" (Fig. 3.20) were observed, which eject sulfur- and oxygen-containing gases with velocities up to 1 km s^{-1} nearly 300 km high, in eruptions lasting several hours. As was shown by the images from the space probe Galileo, nearly 20 years later, some of the volcanoes, e.g. Prometheus (Fig. 3.20), were still active. This volcanic activity continually changes Io's surface and makes the lack of impact craters understandable. A few dark spots, which are about 150 K hotter than their surroundings, can be explained as being recently-solidified lava. The energy source for this activity is probably heating of the interior by strong *tidal forces*, which are generated because Io's orbit is forced to be somewhat eccentric due to resonances with the orbital periods of Europa and Ganymede.

Io has its own thin *atmosphere* (of SO_2), about 120 km thick, and an ionosphere (S, O and Na ions) up to about 700 km altitude, which are constantly renewed from the active surface, since interactions with Jupiter's magnetosphere create a toroidal tube of plasma with particle densities of $\geq 2 \cdot 10^9$ m^{-3} that encloses Io's entire orbit. The interactions with the magnetosphere have a strong influence on the radio bursts from Jupiter, also.

The next Galilean moon, *Europa*, also shows only few impact craters. Its bright, icy surface is covered by a web of dark, interlacing lines. It is astonishingly smooth (altitude differences ≤ 100 m); seen at the resolution of the Voyager mission, the lines seem to be painted on. The markedly improved resolution of the Galileo probe allows us to recognize structures in the thick ice crust (Fig. 3.21), which indicate that it covers a global *ocean*. The stresses caused by rotation and tidal forces break up the "pack ice" and shift the ice blocks, which then once again freeze together. The ocean underneath the ice prevents the formation of higher hills or deeper valleys.

In contrast, the two outermost major moons of Jupiter, *Ganymede* and *Callisto*, are covered with impact craters like our Moon or Mercury. The images of Ganymede from the space probe Galileo reveal, at a resolution down to about 70 m, an icescape of long, closely packed, narrow highlands and rills. For both moons, as for Europa, they give indications of oceans deep below the ice which covers their surfaces. Galileo made the surprising discovery that Ganymede has its own magnetic dipole field with a field strength at the equator of about $8 \cdot 10^{-7}$ T, as well as its own small magnetosphere within that of Jupiter.

3.4.2 Saturn

Saturn (Fig. 3.22) is to a large extent similar to Jupiter. The rings of Saturn were discovered in 1659 by Ch. Huygens with the telescope he constructed (Galileo had already seen indications of them); Huygens also discovered the brightest moon of Saturn, Titan. J. E. Keelers' measurement of the rotational velocity of Saturn's rings in 1895 using the Doppler effect of the reflected sunlight shows that the various zones of the rings revolve at different rates corresponding to Kepler's 3rd law, and thus they must consist of small particles.

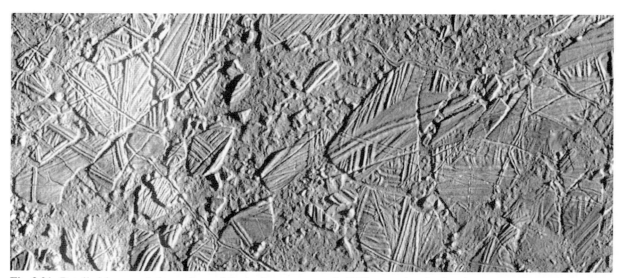

Fig. 3.21. Detailed image of the surface of Jupiter's moon J2 Europa, taken by the space probe Galileo, and put together from several smaller images (1996/97). This photo, representing an area of 70 km · 30 km at 9° north latitude and 274° west longitude, shows a variety of structures ("pack ice") in the thin, fractured ice crust (illuminated by sunlight from the right). (With the kind permission of NASA/JPL/Caltech)

Fig. 3.22. Saturn. A photograph taken by H. Camichel using the 60 cm refractor at Pic du Midi

A quantum leap in our knowledge of the planet, its ring system, and its moons was obtained through the fly-bys of the space probes Pioneer 11 and especially of Voyagers 1 und 2, which reached Saturn in 1979, 1980, and 1981 following their exploration of the Jupiter system.

Saturn's Atmosphere and Interior. The spectrum of Saturn, the chemical composition of its atmosphere, and its cloud structure and flow patterns are all similar to those of Jupiter. In the case of Saturn, the stripes which run parallel to the equator are broader and reach up to higher latitudes. The structures appear more washed out due to a denser layer of haze above the clouds. The temperature of the upper cloud layer, consisting of NH_3 crystals, is about 110 K, and that of the tropopause is about 80 K.

More precise analyses show that the abundance of *helium* in the atmosphere of Saturn is noticeably less than in the case of Jupiter. Apparently a separation of hydrogen and helium occurs on the planet, which is somewhat less massive and cooler than Jupiter; the helium sinks down to lower atmospheric layers. The energy released in this process probably makes a major contribution to the thermal radiation, which in Saturn's case also exceeds the incoming radiation from the Sun (Sect. 3.1.2).

Similar to the case of Jupiter, model calculations for the inner structure indicate a solid ice-silicate core of about 16 Earth masses.

Saturn's Magnetosphere. Like Jupiter, Saturn has a magnetosphere, extending out from 20 to 40 Saturn radii. Corresponding to the dipole moment of $4.6 \cdot 10^{18}$ T m^3, the magnetic field strength at the equator is $2 \cdot 10^{-5}$ T, similar to that of the Earth. It is of interest for the dynamo theory of the generation of planetary magnetic fields that the direction of Saturn's dipole moment is almost precisely ($< 1°$) parallel to its axis of rotation. Finally, Saturn is also surrounded by *radiation belts* containing energetic electrons and protons, with intensity maxima at roughly 7 and 4 Saturn radii from the center of the planet; they evidently contain structure due to the inner moons.

Saturn's Moons. The number of Saturn's known satellites has increased to (at least) 18 following the two Voyager missions (Table 3.5). The data evaluation of this extensive observational material showed after some years ("retroactively") the existence of the moon S18 Pan and some candidates for moons of Saturn. The favorable opportunity in 1995 for observing the *edge* of the ring system from the Earth, as well as earth-bound observations in the following years, showed still further candidates. Currently, the existence of 12 additional moons is considered to be certain, so that Saturn has a total of 30 satellites (as of the year 2001).

In the Saturn system, too, the orbital eccentricities and inclinations of the inner satellites out to and including Titan are small, while those of the outer moons are considerably larger. S9 Phoebe, the darkest outer moon, has a retrograde orbit.

The largest moon of Saturn, S6 *Titan*, is the *only* moon in the Solar System to have its own *atmosphere*. It came as a surprise when in 1944 G. P. Kuiper discovered absorption bands of methane, CH_4, in its spectrum, similar to those from Saturn itself; later, L. Trafton found bands attributable to H_2. According to the estimate in Sect. 3.1.4, the gravitational field of Titan is, indeed, sufficiently strong to retain a (cool) atmosphere.

Voyager 1 approached Titan to within 4000 km and was in particular able to investigate its atmosphere spectroscopically; previously, there had been diverse opinions about its composition. Surprisingly, it was found to contain mostly molecular nitrogen, N_2, with a relative abundance of 90%. Also, in addition to methane, CH_4 ($\simeq 3\%$) and H_2 ($\simeq 0.2\%$), which had long been known, a number of trace constituents were observed in the infrared spectrum, especially hydrocarbons such as ethane, C_2H_6, propane, C_3H_8, ethyne

Table 3.5. The ring system and satellites of Saturn. The distance r from the center of the planet and the semimajor axis a of the orbit are given in units of Saturn's radius, $R_S = 6 \cdot 10^4$ km. The satellite radius R (or the semiaxis of the rings) is in [km]. Roche's stability limit (3.11) is 3.0 R_S (for a mean satellite density of 1300 kg m^{-3})

r/R_s	Ring	Division (gap)	a/R_s	Satellite		R [km]
1.1	D					
1.24	C					
1.50						
		Maxwell				
1.53	B					
1.95						
		Cassini				
2.03	A					
2.28		Encke	2.23	S 18	Pan	10
		Pioneer	2.28	S 15	Atlas	15
2.32	F		2.31	S 16	Prometheus	73 · 43 · 31
			2.35	S 17	Pandora	57 · 42 · 31
			2.51	S 11	Epimetheus	72 · 54 · 49
			2.51	S 10	Janus	98 · 96 · 75
≃ 2.8	G		3.1	S 1	Mimas	196
3.5			4.0	S 2	Enceladus	249
			4.9	S 3	Tethys	530
	E		4.9	S 13	Telesto	17 · 14 · 13
≃ 5.0			4.9	S 14	Calypso	17 · 11 · 11
			6.3	S 4	Dione	560
			6.3	S 12	Helene	18 · 16 · 15
			8.7	S 5	Rhea	765
			20.3	S 6	Titan	2575
			24.6	S 7	Hyperion	205 · 130 · 110
			59	S 8	Iapetus	730
			215	S 9	Phoebe	110

(acetylene), C_2H_2, and ethene, C_2H_4, as well as cyano compounds. In order to explain the mean molecular mass of 28.6, known from observations of occultations, it must be presumed that the second most abundant constituent is argon, with about 10% relative abundance. Dense layers of haze block the view down to the lower levels of the atmosphere and the surface of Titan. Model calculations yield a pressure at the surface of 1.6 bar ($1.6 \cdot 10^5$ Pa) and a surface temperature of 94 K. The density of the atmosphere is thus of the same order as that of Earth's atmosphere. Under these conditions, methane should be present in liquid form. Perhaps methane (together with other hydrocarbons such as ethane) plays a similar role to that of water on the Earth, forming seas, clouds, and rain. The mean density of Titan, at 1940 kg m^{-3}, is similar to that of Jupiter's moon Ganymede, and indicates that Titan also is composed of ice and silicates in roughly equal parts.

The remaining seven *major moons*, S1 Mimas, S2 Enceladus, S3 Tethys, S4 Dione, S5 Rhea, S7 Hyperion, and S8 Iapetus, have icy surfaces covered by numerous impact craters. The density of craters is of the same magnitude as on the Moon, i.e. the rate of impacts of chunks of stone in the first 10^9 yr of the Solar System (Sect. 14.1.1) was comparable out at Saturn's orbit to that near the Earth. From the density of 1200 to 1400 kg m^{-3}, the interior of these moons probably also consists of an ice-silicate mixture.

Within the orbit of Mimas, there are two *minor moons*, S11 Epimetheus and S10 Janus, sharing nearly the same orbit; apparently they exchange their relative positions during their periodic close approaches. An additional small moon (S12 Helene) occupies the same orbit as Dione, in one of the Lagrange points (Fig. 6.15), maintaining a constant angular distance of 60° to the larger satellite. On the orbit of Tethys there are actually two minor moons, S13 Telesto and S14 Calypso, at the socalled Lagrange points L_3 and L_4. The *companion moons* or *"shepherd moons"* in some of Saturn's rings are particularly interesting (see below).

We also note that a larger satellite, according to E. Roche (3.10) could not exist within about three Saturn radii from the midpoint of the planet, as a result of the tidal forces of Saturn.

Saturn's Ring System. Direct images and observations of occultations by the Voyager probes have given us the following picture of the ring system (Table 3.5): the rough structure in the radial direction consists of seven ring zones. In addition to the three brighter rings, which have long been known (denoted from the outermost to the innermost as A, B, and C, with the readily-visible Cassini division between A and B), and the narrow F-ring discovered by Pioneer 11, there are three weaker rings. The D-ring is inside the C-ring and reaches from it to near the planetary surface; the other two (G and E) are further out, near the orbits of the moons Mimas and Enceladus, respectively.

The ring system is extremely thin (< 3 km) in a direction perpendicular to the equatorial plane, and its total mass is probably not more than 10^{-5} of Saturn's mass.

The high spatial resolution of the Voyager images provided some very surprising results concerning the fine structure of the ring system (Fig. 3.23): in the ring zones, several hundred to thousand partial rings can be observed, consisting of thin bright and dark regions with sharp boundaries and apparently irregular spacings. These partial rings are in some cases only a few hundred meters in width. Even within the Cassini division, we find a series of thin rings. The deviations from a circular shape in some of the partial rings are of interest, as are the nearly radial structures ("spokes") in the B-ring, which appear dark in backward-scattered light; they persist for about one rotational period of the planet. The F-ring, which is only about 100 km across, exhibits an unusual, nearly unchanging structure; it consists of a few twisted strands, in which thickenings and kinks can be seen. This ring is accompanied on each side by a small, irregularly-shaped moon (S16 Prometheus and S17 Pandora) of about 100 m diameter. A small companion moon (S15 Atlas) has also been found near the sharp outer edge of the A-ring, and an additional one (S18 Pan) within the narrow Encke division; it "holds open" the division. The companion moons, along with resonances with the orbital periods of the major moons, no doubt play an important role in forming the structure of the rings. The "spoke" phenomenon is possibly caused by the interaction of charged dust particles with the magnetosphere of Saturn.

3.4.3 Uranus

Following Jupiter and Saturn, we turn to the two outermost major planets, Uranus and Neptune. Both are noticeably smaller, but have comparable densities (Table 3.1), and thus must differ in structure from Jupiter and Saturn.

As seen from the Earth, the disk of Uranus hardly shows any recognizable details. The five large, relatively dark moons which were known before the Voyager missions were discovered in 1787 by W. Herschel (U3 Titania, U4 Oberon), in 1851 by W. Lassell (U1 Ariel, U2 Umbriel) and in 1948 by G. P. Kuiper (U5 Miranda). The discovery during observations of a stellar occulation by Uranus in 1977 of a system of nine narrow, *dark rings* came as a complete surprise; they were seen as a series of short, sharp eclipses of the star before and after the expected actual occultation by the planet. In 1986, the fly-by of the space probe Voyager 2 brought a plethora of new information, as it had in the cases of the Jupiter and Saturn systems, also; among them was the discovery of 10 new, small moons and an additional ring.

Planetary Structure. The interior of Uranus – and also of Neptune – is, in contrast to Jupiter and Saturn, not dominated by hydrogen and helium; here, they make up only 15 to 20% of the mass. The major portion consists of stone- and ice-forming heavier elements. The rotational period of Uranus, 17.2 h, is consistent with

Fig. 3.23. A Voyager 1 photograph of the ring system of Saturn from a distance of $8 \cdot 10^6$ km, taken in 1980. The long-known broad ring zones A, B, and C with the Cassini and Encke divisions, as well as the thin F ring with its inner companion satellite (*arrow*) are indicated. The picture shows nearly 100 individual rings, which in turn consist of further narrower rings, as shown by images with better resolution from Voyager 2; thus for Saturn all together at least 100 000 rings are estimated to exist. (From G. Briggs and F. Taylor, 1982)

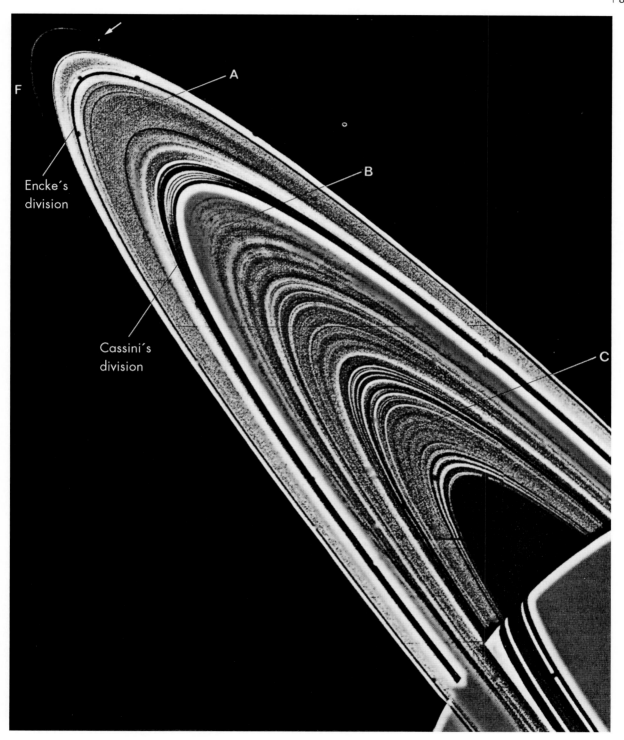

an inner structure consisting of a silicate/iron core (of roughly 7 Earth masses?), surrounded by a mantle of ice (H_2O, CH_4, and NH_3) and a gas shell of H_2 and He.

Uranus' Atmosphere. The cool atmosphere of the planet ($T \simeq 60$ K) is dominated by molecular hydrogen, H_2; its helium abundance of about 12% corresponds to the solar mixture. Methane, CH_4, is relatively abundant in the atmosphere, but ammonia, NH_3, is unexpectedly rare. At a greater depth (corresponding to a pressure of 1.6 bar), there is a layer of CH_4 clouds. The tropopause, above which the temperature again increases, has a pressure of around 0.1 bar. Uranus is surrounded by an extended corona of neutral hydrogen, H; on the sunward side, an intense emission from hydrogen molecules ("dayglow"), whose excitation mechanism is still not understood, is observed in the upper atmosphere.

Uranus' Magnetosphere. Uranus, like Jupiter and Saturn, has a magnetic field and a magnetosphere and emits radiofrequency radiation. The magnetic field at the equator is $2.3 \cdot 10^{-5}$ T; the magnetosphere has an extent of about 20 Uranus radii ($R_U = 25\,600$ km) and forms a long plasma tail on the side away from the sun. The magnetosphere of Uranus has a structure and dynamics which are unique among the planets, as a result of the unusually large angle of about 59° between the magnetic dipole axis and the planetary rotation axis, the latter being nearly in the orbital plane of the planet. (We shall find a similarly large angle also in the case of Neptune.)

Uranus' Moons. The five large, relatively dark moons orbit Uranus practically in its equatorial plane on almost circular, prograde orbits (Table 3.6). Except for the smallest, U5 Miranda, they are comparable in size to e.g. Saturn's moon Rhea. Their densities are of the order of 1600 kg m^{-3}, so that their interiors probably consist of an ice-silicate mixture. Their surfaces, which are covered with ice and dark matter, show in the Voyager pictures not only the expected large number of craters, but also (except for U2 Umbriel) complex geological structures such as rills, valleys, steps, dislocations, etc. (Fig. 3.24).

Voyager 2 discovered 10 small, very dark moons between 2.1 and 3.4 R_U, i.e. within the orbit of U5 Miranda. An additional moon (S/1986 U10) was discovered by E. Karkoschka "retroactively" in 1999 on

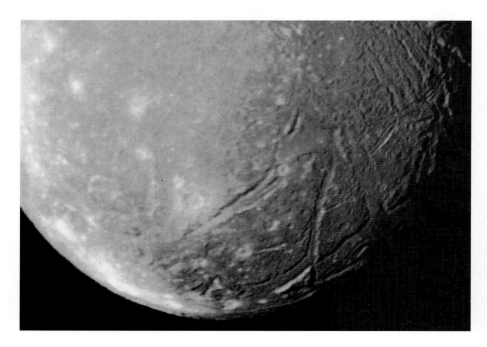

Fig. 3.24. A picture of Uranus' moon U1 Ariel taken by Voyager 2 from a distance of 170 000 km. Along with many craters, the surface shows well-defined rills and valleys

Table 3.6. The satellites of Uranus. The distance a from the center of the planet is given in units of Uranus' radius, $R_U = 25\,600$ km. The satellite radius R is in [km]

a/R_U	Name	R	a/R_U	Name	R
1.95	U6 Cordelia	13	5.12	U5 Miranda	236
2.10	U7 Ophelia	16	7.56	U1 Ariel	579
2.32	U8 Bianca	22	10.5	U2 Umbriel	585
2.42	U9 Cressida	33	17.2	U3 Titania	790
2.45	U10 Desdemona	29	23.1	U4 Oberon	762
2.52	U11 Juliet	42	281	U16 Caliban	30
2.59	U12 Portia	55	311	U20 Stephano	10
2.74	U13 Rosalind	27	477	U17 Sycorax	60
2.94	U14 Belinda	34	631	U18 Prospero	15
2.94	S/1986 U10	20	713	U19 Setebos	15
3.37	U15 Puck	77			

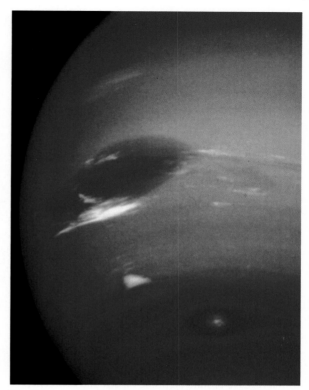

Fig. 3.25. A photograph of Neptune from Voyager 2 (1989). The blue atmosphere shows only a weak striped structure. To the left at 22° south latitude is the "Great Dark Spot" with cirrus clouds in its neighborhood; further south is a smaller dark spot with a bright center. These structures had disappeared in the images taken in 1994 with the Hubble Space Telescope

the Voyager images. In an international collaboration, B.J. Gladman, J.J. Kavelaars, J. Holman et al. have discovered from earthbound observations since 1997 five additional small irregular moons, which orbit Uranus at a large distance.

Uranus' Ring System. The ring system consists of 10 very narrow, extremely dark rings (albedo $\simeq 0.04$) in the equatorial plane at distances between 1.6 and 1.95 R_U from the center of the planet. The outermost, widest ring (ε-ring) is between 20 and 100 km across and is accompanied on either edge by a small shepherd moon; it contains unusually large chunks of stone (≥ 1 m). As in the case of Saturn, some of the Voyager images of Uranus also show finer ring structures and dust-like matter outside the 10 "main rings".

3.4.4 Neptune

The size, mass, and spectrum of Neptune, and therefore its structure, are to a great extent similar to those of Uranus.

Our knowledge of Neptune, whose disk, as seen from the Earth, has a diameter of only $2.3''$, was considerably increased following the fly-by of the space probe Voyager 2, which approached its surface to within 5000 km in 1989.

Neptune's Atmosphere. It consists mainly of hydrogen (H_2), helium, and traces of methane, which causes its bluish color. The atmosphere shows a variety of structures (Fig. 3.25): in addition to stripes and small spots, we can see in the southern hemisphere the "Great Dark Spot", a structure which persists for some months and, with a rotational period of 18.3 h, moves against the direction of rotation of the planet (rotational period 16.1 h) at a velocity of $0.3\,\mathrm{km\,s^{-1}}$. Around 50 km above the blue cloud layer of CH_4 ice, we see the bright fingers of cirrus clouds. Later observations with the Hubble Space Telescope showed that these structures had disappeared a few years after the Voyager mission; by this time, however, other structures, e.g. a larger dark spot, were present in the northern hemisphere. Neptune's brightness appears to have increased steadily in the last 20 years by about 10%.

The (in comparison to Uranus) surprisingly well-developed flow patterns probably have their origin in the high internal energy of the planet, which exceeds the radiation input from the Sun by a factor of 2.6 (Sect. 3.1.2). This by the way balances out the 40% lower solar irradiation as compared to Uranus, so that the surface temperatures of both planets have the same value, 59 K.

Neptune's Magnetic Field. The dipolar magnetic field of Neptune is weaker than those of the other major planets. Its axis is inclined relative to the axis of rotation by 47°, a similarly large angle as in the case of Uranus. The magnetosphere of Neptune shows relatively low particle densities.

Neptune's Moons. Of the Moons of Neptune (Table 3.7), only two were known before the Voyager mission: N1 Triton, discovered in 1846 (by W. Lassell), whose size is nearly as great as that of Earth's Moon and which circles the planet on a circular, *retrograde* orbit; and the considerably smaller N2 Nereid (discovered in 1949 by G. P. Kuiper), which has a strongly elliptical orbit ($e = 0.75$). Voyager 2 discovered six additional moons which revolve around Neptune on circular orbits nearly in the equatorial plane. Like Nereid, they are irregularly shaped, dark objects with radii between 25 and 200 km; their albedo is about 0.06.

The large, reddish moon N1 *Triton*, with a mean density of 2080 kg m^{-3}, consists for the most part of stone and must have a strong source of internal energy owing to its radioactivity. Voyager 2 flew past Triton at a distance of 38 000 km and transmitted information about the variety of surface structures: dislocations and long cracks, "frozen lakes" with terrace-shaped banks, still unexplained flat, dark areas with bright edges, and dark spots more than 50 km long (Fig. 3.26). The latter may consist of dust which was swept out by the eruptions of "geysers" (of nitrogen from deeper liquid layers?). The small number of impact craters on Triton implies a geologically young surface. The region around the south pole is covered at present with frozen methane and nitrogen, whose thickness varies annually. Due to the high albedo of these ices (around 0.85), the temperature here is extremely low, at 38 K.

The atmosphere of Triton, with a surface pressure of only about 15 µbar, is extremely thin. It consists mainly of nitrogen, whose vapor pressure is in equilibrium with N_2 ice, and of traces of methane; occasionally scattered clouds and dust are observed.

Neptune's Ring System. Observations of stellar occultations by Neptune gave indications in 1984/85 of an irregularly shaped ring or fragments of a ring (arcs) at a distance of about 3 R_N. Voyager 2 then discovered a complex ring system. It consists of two narrow, sharply bounded rings (the Leverrier ring at 2.15 and the Adams ring at 2.54 R_N); and of the diffuse, very broad Galle ring within the orbit of the innermost moon (at 1.69 R_N), in addition to an extended disk of finely divided dust particles. The Adams ring contains irregular, large regions of higher density, which must have been responsible for the stellar occultations observed from the Earth.

3.5 Pluto and the Transneptunian Small Planets

Outside the orbit of Neptune, as the only "large" planet we find *Pluto*, which is considerably smaller in terms of its diameter and mass than all of the other planets and is even smaller than our Moon (Table 3.1). Owing to its large orbital eccentricity, it is at times – for example from 1979 to 1999 – closer to the Sun than Neptune is.

Since 1992, in the course of systematic surveys using high-sensitivity detectors, a whole series of *small planets* outside Neptune's and in part outside Pluto's orbit has been discovered. They have diameters of at the most 1/10 that of Pluto and are most comparable to the larger of the asteroids in the belt between Mars and Jupiter. These objects in the *Kuiper belt* (Sect. 2.2.2) are probably "left-over" planetesimals, which were not incorporated into a larger planet.

Table 3.7. The satellites of Neptune. The distance a from the center of the planet is given in units of Neptune's radius, $R_N = 24\,760$ km. The satellite radius R is in [km]

a/R_N	Name	R	a/R_N	Name	R
1.94	N3 Naiade	29	2.97	N7 Larissa	97
2.02	N4 Thalassa	40	4.75	N8 Proteus	209
2.12	N5 Despina	74	14.3	N1 Triton	1350
2.50	N6 Galatea	79	222.7	N2 Nereid	170

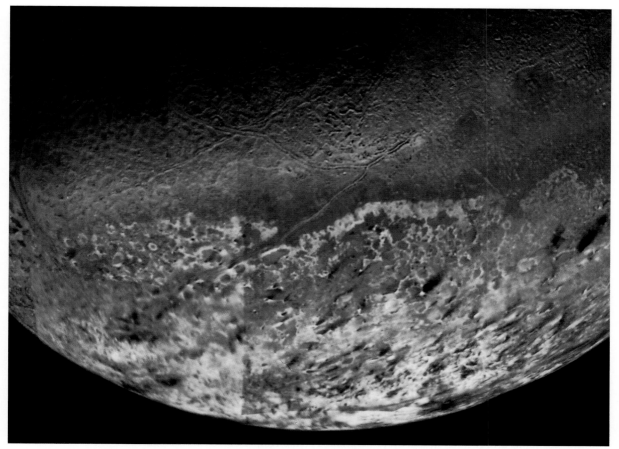

Fig. 3.26. A photograph of Neptune's moon N1 Triton, taken by Voyager 2 (1989). This image, composed of several smaller partial images, shows surface structures in the neighborhod of the south pole

3.5.1 Pluto and Charon

This faint outermost planet of our Solar System is difficult to observe. In the course of a series of astronomical observations at the US Naval Observatory in 1978, J. W. Christy found a systematically-occurring elongation of the planetary disk, whose analysis indicated the presence of a moon; it was later observed directly using speckle interferometry. P1 *Charon*, about 2 mag fainter than Pluto, has an orbital period of 6.39 d, synchronous with the rotation of Pluto. Its mean distance from the planet is only $0.9''$ or 19 600 km. As a result of the discovery of this satellite, the mass and thus the mean density of Pluto can be much more reliably determined than was previously possible from its perturbations of Neptune. In the 1980's, the observation of a series of mutual occultations and then in the 1990's the direct measurement of the diameters of Pluto and Charon with the Hubble Space Telescope permitted an exact determination of the parameters of the system:

	Pluto	P1 Charon	*N1 Triton*
Mass [$\mathcal{M}_E = 5.97 \cdot 10^{24}$ kg]	$2.1 \cdot 10^{-3}$	$3.2 \cdot 10^{-4}$	*$3.7 \cdot 10^{-3}$*
Radius [km]	1137	586	*1350*
Mean Density [kg m^{-3}]	2200	2200	*2100*

For comparison, we have included the largest moon of Neptune, N1 Triton. The mass of the total system is about 1/400 Earth masses, with a mass ratio for Charon/Pluto of 1/7; this mass ratio of a satellite to its planet is by far the largest in the Solar System, so that the Pluto-Charon system can be thought of as a *double planet*.

From its mean density, Pluto consists of about 75% stone and 25% H_2O ice. Little is known about the structure of Charon; it probably resembles that of e.g. Saturn's moon Rhea.

With the Hubble Space Telescope at a resolution of 160 km, one can discern on Pluto's surface large, strongly contrasting light and dark structures, including an ice cap at the north pole. Spectrophotometric investigations show that the surface of Pluto, whose temperature varies between 38 and 63 K, is mainly covered with N_2 ice (absorption at $\lambda = 2.15\,\mu m$); in addition, there is CH_4 ice and traces of CO ice. In contrast, Charon's surface is covered by H_2O ice.

Pluto has an extremely thin atmosphere (pressure at the surface $\leq 10^{-5}$ bar), which has a temperature of about 100 K and consists almost exclusively of N_2, in vapor equilibrium with the frozen gas on the surface. The concentration of methane, CH_4, is only about 1%. The properties of the atmosphere must vary widely, since the ice layer sublimes more or less strongly depending on the distance to the Sun; it probably exists only near perihelion. Charon most likely has no atmosphere at all.

Pluto shows an astonishing similarity to Neptune's moon N1 Triton, both in mass, size, and mean density (see above), as well as in the composition of its surface ices and its atmosphere. This indicates a possible common origin of these two celestial bodies (Sect. 14.1.2).

3.5.2 The Transneptunian Small Planets

At the end of the 1980's, three research groups began the tedious work of a systematic survey of objects in the Kuiper ring (Sect. 2.2.2) in selected regions near the ecliptic.

Then, in 1992, D. Jewitt and J. Luu succeeded in discovering the first faint, reddish "*small planet*" outside the orbit of Neptune: 1992 QB 1 has a brightness in the red of only $m_R = 23$ mag and a diameter of about 200 km. Its orbit has a semimajor axis $a = 43.8$ AU, an eccentricity $e = 0.09$ and an inclination $i = 2.2°$; its orbital period is 290 yr.

For this survey, large–area CCD-detectors (Sect. 5.1.4) with $2048 \cdot 2048$ pixels were used with a 2.2 m telescope. For the identification of a planet at this distance, one essentially makes use of its slow proper motion on the celestial sphere, of not more than $10''\,h^{-1}$. (For comparison: an asteroid in the belt between Mars and Jupiter typically moves at $30''\,h^{-1}$.) These systematic surveys included only 1.2 square degrees down to a limiting brightness of 25 mag in the red spectral region, and initially led to the discovery of 7 transneptunian small planets. Today (as of 2001), we know of more than 300 such objects at 30 to 50 AU from the Sun, with apparent magnitudes of around 21.5 to 24.5 mag in orbits with inclinations around $0°$. Their diameters lie for the most part between 100 and 300 km, and their albedo is estimated to be about 0.04.

An object discovered by J. Luu et al., 1996 TL 66, is – after Pluto and Charon – at $m_R \simeq 21$ mag the brightest transneptunian planet so far. Its orbit, with a semimajor axis of $a = 84$ AU, has (unlike the objects in the Kuiper ring) an unusually large eccentricity of $e = 0.58$ and an orbital inclination of $i = 24°$; thus at aphelion it reaches a distance of 133 AU from the Sun. 1996 TL 66 could be the first of a whole population of small objects between the Kuiper ring and the Oortian cloud (Sect. 2.2.2), which presumably were "scattered" out of the Kuiper belt by larger planets.

With the Hubble Space Telescope, a number of objects down to a brightness (in the red spectral range) of 28 mag, having diameters of 10 to 20 km (at an albedo of 0.04) and proper motions of about $1''\,h^{-1}$ have been identified. They are for the most part candidates for small planets in the Kuiper belt.

Estimates based on the small planets thus far discovered allow us to expect a total of around 35 000 objects with diameters ≥ 100 km in the Kuiper ring, and even $2 \cdot 10^8$ with diameters ≥ 10 km.

The transneptunian small planets which populate the Kuiper ring make up a new group of objects in our Solar System. They are much too small to be considered "genuine" planets and have the wrong orbits for "classical" asteroids (Sect. 3.3). Their connection with the comets, of which the Kuiper ring has been postulated as the source, is still unclear, since comet nuclei have characteristic diameters of only 5 to 10 km.

The relationship of the transneptunian small planets to the asteroids is likewise unclear. A few asteroids such as 944 Hidalgo, 5145 Pholus, 1993 HA 2, 1994 TA, and 2060 Chiron (Sect. 3.3.1), whose orbits reach out well beyond that of Jupiter, could have originally been objects from the Kuiper ring that made their way into the interior of the Solar System as a result of orbital perturbations. Finally, the question arises as to whether even Pluto, along with Charon and Neptune's moon Triton, also could have had their origins in the Kuiper ring – as its most massive objects.

3.6 Comets

We now turn to the physical properties of the comets, a further group of small objects from the outer reaches of the Solar System.

3.6.1 Structure, Spectra, and Chemical Composition

Photographs taken with a suitable exposure time (Figs. 3.27, 28) show that a comet consists of a *nucleus* (which is seldom clearly recognizable), often having a diameter of only a few kilometers. It is surrounded by the *coma* which is like a diffuse, misty shroud that usually takes the form of a series of parabolic shells or rays stretching out from the nucleus. The nucleus and the coma together are called the *head* of the comet; its diameter is in the range of $2 \cdot 10^4$ to $2 \cdot 10^5$ km. Roughly within the orbit of Mars, comets develop the well-known *tail* which, in its visible portion, can attain a length of 10^7 and sometimes even $1.5 \cdot 10^8$ km $= 1$ AU. The brighter comets can be observed in the ultraviolet region of the spectrum from satellites and it is found that the head is surrounded by a *halo* out to a distance of several 10^7 km, consisting of atomic hydrogen which radiates strongly in the Lα line at $\lambda = 121.6$ nm.

The Heads of Comets. The *spectrum* in the visible region shows in part reflected sunlight (Fig. 3.29), whose intensity distribution indicates scattering from dust particles with diameters of the order of the wavelength of visible light (about 0.6 μm). In addition, there are emission bands from numerous molecules and radicals such as CN, CH, C_2, C_3, NH, NH_2, OH, and OH, and from radical ions such as CO^+, CH^+, OH^+, N_2^+, CO_2^+, and H_2O^+. Near the Sun, the [OI] spectral lines of atoms such as Na, Ca, Fe etc. also occur. (Square brackets [ ] indicate "forbidden" transition lines and I denotes neutral atoms; cf. Sect. 7.1.1.)

In the infrared, we can observe the silicate structures at $\lambda = 10$ and 18 μm (Sect. 10.1.5) and, especially, the thermal radiation of the dust particles in the comet. In the microwave region, in addition to OH and CH which were already known from optical spectroscopy, hydrogen cyanide, HCN, hydrogen sulfide, H_2S, methyl cyanide, CH_3CN, and water, H_2O, were discovered. In the ultraviolet, in addition to the atoms and molecules known from other spectral regions, for example C, C^+, O, S, S_2, CO, CS and CN^+ as well as H have been found.

Fig. 3.27. Comet C/1957 P1 Mrkos in a photograph made with the Mount Palomar Schmidt camera (1957). Above, we see the extended, richly structured type I or plasma tail; below, the thicker, nearly featureless type II or dust tail

Fig. 3.28. and Frontispiece, p. 5. Comet C/1995 O1 Hale-Bopp, photographed on April 1st, 1997 at the Puckett Observatory, Mountain Town, Georgia (USA) by D. Lynch with a 30 cm telescopic lens. To the left is the bluish plasma tail, to the right the reddish–yellow dust tail. (With the kind permission of D. Lynch and T. Puckett)

The Nuclei of Comets. The nucleus of roughly 1 to 10 km diameter and mass in the range 10^{12} to 10^{15} kg can be observed only rarely and with difficulty.

According to F. Whipple (1950), it can be considered to be a "dirty snowball", consisting of a mixture of ice and small grains of silicates. This model was verified when, in 1986, the comet 1P/Halley (Sect. 2.2.2) was observed from close up by several space probes. The European probe Giotto passed through the inner coma and approached the nucleus to a distance of 600 km, using among other methods mass spectrometry and impact detectors to measure gas and dust particles *in situ*, and taking closeup images of the nucleus.

The nucleus of the comet 1P/Halley has an elongated, asymmetric shape, similar to a potato or a peanut, with dimensions of about 15 km · 7 km · 7 km. Its mass is around 10^{14} kg and its mean density is thus 500 kg m^{-3}. The surface has an irregular structure and is very dark (albedo about 0.02 to 0.04). The gaseous components, which evaporate as the nucleus approaches the Sun, consist of 80% water, H_2O; CO and CO_2 make up 15 and 4%, respectively, and the lighter volatile compounds such as CH_4, NH_3, and N_2 together contribute less than 1%. The *dust* exceeds the gaseous components in mass and occurs in the form of grains of about 1 cm down to 0.1 μm in size, which are composed of silicates and to a surprisingly large extent of compounds containing only the elements C, H, N, and O.

The nucleus of the bright comet C/Hale-Bopp (Fig. 3.28), which attained its perihelion in 1997, was larger than that of Halley's comet, with a diameter of 30 to 40 km; not only was it larger in size, but also the rate of gas and dust production, around 100 t s^{-1}, was also greater.

The Tails of Comets. The *spectra* of the tails are characteristic in the main of molecular or radical *ions*: N_2^+, CO^+, OH^+, CH^+, CN^+, CO_2^+, and H_2O^+.

The characteristic shapes and motions of the tails of comets require for their explanation that we assume that the Sun produces a repulsive force which must often be stronger than the gravitational force by a large factor.

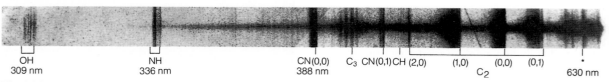

Fig. 3.29. The spectrum of the head of comet C/1940 R2 Cunningham (at a distance of 0.87 AU from the Sun). In the center is the continuous spectrum of reflected sunlight; further out are the emission bands of the molecules OH, NH, CN, C_2, C_3 .... (∗) indicates the nightglow line [OI] at 630 nm

The *narrow, elongated* tails (type I) consist for the most part of molecular ions, as indicated by their spectra. The radiation pressure is in this case not sufficient to explain the observed large ratio of radiation acceleration to gravitational acceleration. According to L. Biermann (1951), these *plasma* tails are instead blown away from the Sun by an ever-present corpuscular radiation, the *solar wind* (Sect. 7.3.7). At a distance equal to that of the Earth's orbit, the solar wind consists of a stream of ionized hydrogen particles, i.e. protons and electrons, with a density of 10^6 to 10^7 particles per m^3 and a velocity of around 500 km s^{-1}. Thus, the often (but not always) observed influence of solar activity on comets can be understood.

The broad *diffuse* and *curved* tails (type II) consist mainly of small *dust particles* ($\lesssim 1$ μm). For such particles, the radiation pressure (each absorbed or scattered light quantum $h\nu$ transfers a momentum $h\nu/c$) can indeed reach a multiple of the gravitational force, as required by the observations.

Chemical Composition. The hydrogen halo as well as the fact that all of the molecules in comets are composed of the light elements H, C, N, and O with their cosmic abundances indicate that comets consist essentially of *solar matter*. The isotopic ratios – insofar as they can be determined – also are the same as in the Solar System. The material of comets has not been "falsified" through chemical or geological processes and thus represents – along with the carbonaceous chondrites (Sect. 3.7.3) – the *original* "primitive" matter of our Solar System.

3.6.2 The Evolution of Comets

On the basis of the spectroscopic and direct observations, we can imagine the evolution of a comet as follows:

At a large distance (> 5 AU) from the Sun, only the *nucleus* is present; its mass is in the range 10^{12} to 10^{15} kg.

As the nucleus approaches the Sun, the "dirty snowball" melts and H_2O, HCN, CH_4, NH_3 etc. evaporate and begin to form the *coma*. These *mother molecules* are dissociated and ionized by solar radiation and by interactions with the solar wind, and stream away at velocities of the order of 1 km s^{-1}. Through multiple chemical reactions, other particles are formed in the outer coma and are excited to fluorescence by the solar radiation. The dust particles embedded in the nucleus are likewise released into the coma during this evaporation process; in the case of Halley's comet, an extremely irregular release of dust from the nucleus, in the form of *jets*, was observed; it originated from only about 20% of the surface of the comet's nucleus. The rest of the surface seems to be covered by a protective layer which prevents the ice within from evaporating. The smaller dust particles ($\lesssim 1$ μm) are then driven away from the Sun by radiation pressure.

The gas in the outer coma is carried along by the solar wind and makes up the plasma tail. On the side facing the Sun, a shock wave forms as a result of the braking of the solar wind by the coma of the comet; in the case of Halley's comet, for example, it was about 10^6 km from the nucleus. The molecules and radicals in the tail are further ionized by energetic solar radiation, while recombination of the positive ions with electrons is rare owing to their low particle density. Therefore, ionic spectral lines dominate in the spectra of comets' tails.

3.7 Meteors and Meteorites

The meteors or "falling stars" represent only a portion of all of the small bodies of our Solar System. A distinction is sometimes made between a *meteor*, which is a brief, luminous trail in the heavens, ranging from "telescopic meteors" to fireballs (bolides) which shine as bright as day, and the body which causes it, the *meteoroid*. If a meteoroid succeeds in passing through the Earth's atmosphere without burning up, and reaches the ground, it is termed a *meteorite*.

As a rule, meteorites, of which nearly 10^4 are known, are named for the place where they were found. Most of them have been found since 1969 on the ice surfaces of the Antarctic, and – as a result of the cool and dry environment – are in a well preserved condition.

3.7.1 Meteorites and Impact Craters

Since celestial bodies on roughly parabolic orbits in the neighborhood of the Earth have a velocity of 42 km s^{-1}, and on the other hand, the orbital velocity of the Earth itself is 30 km s^{-1} (Sect. 2.4.3), depending on the direction of approach (morning or evening), relative velocities between 72 and 12 km s^{-1} can be reached.

On entering the Earth's atmosphere, the objects are heated. In the case of larger objects, the heat cannot penetrate sufficiently rapidly to the interior, and the surface forms melt-pits and burns off; such objects reach the ground as *meteorites*. The largest known meteorite is the Hoba, in Namibia, with a mass of about 60 tons. It must have required very large masses to make some of the *meteoritic craters* on the Earth (as on the Moon and other bodies in the Solar System).

According to E. M. Shoemaker et al. (1979), we can use the diameter D of a (large) impact crater to estimate the associated impact energy E, i.e. the kinetic energy $E = mv^2/2$ of a celestial object of mass m and velocity v, from the empirical formula

$$D \simeq D_0 \cdot \left(\frac{E}{E_0}\right)^{0.294} \quad (3.30)$$

with $D_0 = 15$ km and $E_0 = 10^{20}$ J. For example, for a meteorite with a radius of 1 km and a density of 3000 kg m^{-3} at an incident velocity of 20 km s^{-1}, the energy is $E = 25.1 \cdot 10^{20}$ J and thus the diameter of the resulting crater will be 39 km.

The well-known crater of Canyon Diablo in Arizona has a diameter of about 1.2 km and (today) a depth of 170 m. According to the geologic evidence, it must have been formed about 20 000 years ago by the impact of an iron meteorite of about one million tons mass. The Nördlinger Ries in southern Germany, with a diameter of about 25 km, is probably also a meteoric impact crater, which was produced $15 \cdot 10^6$ yr ago in the Tertiary Era. The crater which has been discovered under thick layers of sediment near Chicxulub in Yucatán, with a diameter of around 180 km, must have been made by the impact of a large object with a diameter of about 10 km, which among other effects led to the extinction of the dinosaurs at the end of the Cretaceous Era $65 \cdot 10^6$ yr ago (Sect. 14.2.4).

3.7.2 Meteors in the Earth's Atmosphere

Small meteorites burn up in the atmosphere as "falling stars" at an altitude of about 100 km. On their paths through the upper atmosphere, they ionize a tube-shaped region of air. Large meteor showers thus contribute to the ionosphere, along with the socalled anomalous E-layer at about 100 km altitude. Furthermore, such a luminous, ionized cylinder emits electromagnetic waves

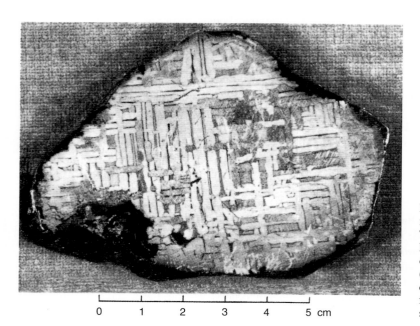

Fig. 3.30. The iron meteorite Toluca, named for the place where it was found. This polished and etched cut surface exhibits Widmannstetter figures, which are due to Fe–Ni crystal layers of kamazite (7% Ni) and taenite (with a large nickel content) which fit together parallel to the four pairs of surfaces of the octahedra; such meteorites are termed *octahedrites*

like a wire, mostly at right angles to its own axis; this is the basis for the enormous impetus which has been given to the study of meteors by *radar technology*. On the radar screen, only the larger objects are themselves visible, but the direction of the "ion-tube" can be determined even for meteors which are well below the level of visual detectability. Velocities can also be measured with radar techniques. Their decisive advantage compared to visual observation, however, is the fact that they are independent of clouds and the time of day.

The most precise information about somewhat brighter meteors is obtained from observations using wide-angle cameras of strong light-gathering power, if possible from two stations a suitable distance apart. In this way, one can determine the precise spatial position of the orbits. Rotating sectors ("choppers") interrupt the image of the orbit and allow the calculation of the orbital velocity. The intensity of the streak records the optical magnitude and its often rapid temporal variations.

Since the air resistance varies as the cross-sectional area (proportional to a^2), but the gravitational force as the mass (proportional to a^3), it is readily seen that for smaller and smaller particles, air resistance becomes so dominant that they never begin to burn up and simply drift slowly to the ground, undamaged. These *micrometeorites* are smaller than a few 100 μm. Using suitable collection apparatus, they can be found in large quantities on the ground and also in deep-sea mud; a difficulty is naturally to distinguish them from terrestrial dirt. In recent times, research with satellites and space probes has yielded important results concerning micrometeorites and the interplanetary dust (Sect. 3.8). Further information has been obtained from the study of microcraters on the Moon (Sect. 3.2.4).

3.7.3 Properties and Origins of Meteorites

The meteorites, being the only cosmic matter which is directly accessible on the Earth, have been carefully investigated, at first using mineralogical and petrographic methods, and more recently for traces of radioactive elements and anomalous isotopes. It was originally believed that the chemical analyses of a large number of meteorites by V. M. Goldschmidt and the Noddacks represented the "cosmic abundance distribution" of the elements; today, these data are used in conjunction with the quantitative analysis of the Sun to give information on the early history of the meteoritic material and of the Solar System.

Classification of Meteorites. A first classification of the meteorites divides them into *iron meteorites* (densities about $7800\,\text{kg}\,\text{m}^{-3}$), whose Fe–Ni crystallites with their characteristic Widmanstetter etch figures (Fig. 3.30) rule out their confusion with terrestrial iron, and *stone meteorites* (densities about $3400\,\text{kg}\,\text{m}^{-3}$). The latter are subdivided into two further classes: the much more common *chondrites*, characterized by silicate spherules of mm size, called chondra or chondrula, and the rarer *achondrites*. The finer classification is given in Fig. 3.31. The *carbonaceous chondrites* of type C1 – similar to the nuclei of comets (Sect. 3.6.2) – have chemical compositions corresponding essentially to unmodified solar matter (Table 7.5); only the noble gases and other readily volatile elements are less abundant or lacking. The matrix in which the chondrula of carbonaceous chondrites are embedded contains many organic compounds, such as amino acids and complex ring compounds, which are *not*, as occasionally presumed, of biogenic origin (Sect. 14.2.3). They even contain traces of Fullerenes ("soccer ball molecules") such as C_{60} and C_{70}.

The matrix of the C1 carbonaceous chondrites must have been formed at temperatures below about 360 K. The formation of the (older) chondrites and in particular the separation of metals (Fe, Ni, ...) and silicates implies complicated separation processes which are only partially understood. H. C. Urey (1952), then later J. W. Larimer, E. Anders, and others have calculated the successive production of various chemical compounds and minerals depending on the pressure and temperature. According to these results, the chondrites required formation temperatures of $\simeq 500$ to $700\,\text{K}$. We shall return to the subject of the origin of meteorites in connection with the development of the Solar System in Sect. 14.1.3.

The *tectites* or glass meteorites are greenish-black objects a few centimeters across which are composed of silicate-rich glass (70%–80% SiO_2 with a density near $2400\,\text{kg}\,\text{m}^{-3}$), whose often rounded or conical shape shows that they must have passed through the air at high speeds in the molten state. Tectites are found only in cer-

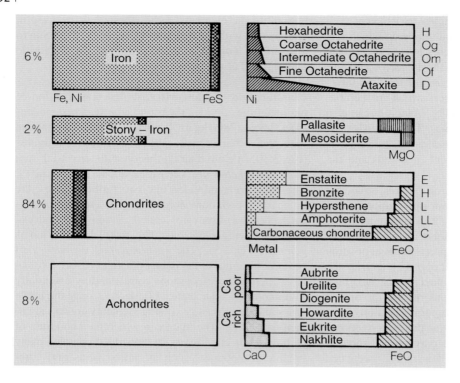

Fig. 3.31. The division of meteorites into four main classes (*left*) according to the ratio of metals (*shaded*) to silicates (*grey*). The finer classification is carried out according to various chemical and structural criteria (*right*). At the far left, the percentage of occurrence is indicated. (E. Anders, 1969)

tain areas, e.g. the *moldavites* in Bohemia. W. Gentner showed that several of these groups can be associated with meteoritic craters of the same age and in neighboring regions; for example, the moldavites are associated with the Nördlinger Ries mentioned above.

Dating of Meteorites. Radioactive dating (Sect. 3.2.2) yields a *maximum age* which is the same as that of the Earth within the error limits, namely $4.5 \cdot 10^9$ yr. On the other hand, it can be determined how long a meteorite was bombarded by energetic protons from the (constant) cosmic radiation background in space. Its main component penetrates about 1 m into matter and produces various stable and radioactive isotopes by spallation of heavy nuclei. The *irradiation age* can be found from the abundancy ratios of these product isotopes. The irradiation ages of the tough iron meteorites are several 10^8 to 10^9 yr, while those of the much more fragile stone meteorites are only 10^6 to $4 \cdot 10^7$ yr. These ages are a measure of the elapsed time since the object in question was broken off from a larger body by a collision in space.

More precise dating (also of the formation of the Solar System, for example) is based not only on the long-lived radioactive elements (Sect. 3.2.2), but also on shorter-lived nuclides which have therefore decayed completely today; their stable decay products can still be found in primitive meteorites such as the carbonaceous chondrites:

	Half-life	Decay products
^{129}I	$1.7 \cdot 10^7$ yr	^{129}Xe
^{244}Pu	$8.2 \cdot 10^7$ yr	$^{131...136}$Xe
^{26}Al	$7.4 \cdot 10^5$ yr	^{26}Mg

Two carbonaceous chondrites: the C3 meteorite *Allende*, which fell in Mexico in 1969, and the *Murchison* chondrite (C2), which impacted in 1969 in Australia, with their numerous, varied components, have been especially thoroughly studied; the abundances of the elements and their isotopes as well as the age of various inclusions were carefully determined.

Isotopic Anomalies. The isotopic mixture in meteorites is not that of solar matter; instead, it has been modified through physical-chemical processes during the formation and evolution of the Solar System, such as radioactive decay, spallation by cosmic radiation, or fractionation as a result of diffusion and other processes. Thus, parts of some meteorites have inclusions containing the *noble gases* neon and xenon. Their isotopes and abundance distribution indicate that, like the noble gases in Moon dust and in the surfaces of Moon rocks, they originated mainly from the solar wind. Investigations of the Xe furthermore show that in some cases, those isotopes are overabundant which result only from the decays of the relatively short-lived iodine and plutonium isotopes ^{129}I and ^{244}Pu.

An isotopic mixture is termed *anomalous* if it cannot be explained as the result of processes in the formation and evolution of the Solar System in terms of isotopically homogeneous matter. In the inclusions of many meteorites, we find not only overabundances of isotopes of Ne and Xe, but also additional anomalies, such as e.g. an enrichment of the isotope ^{16}O (in the case of the Allende meteorite, up to 5% relative to the solar abundances of the oxygen isotopes), as well as an overabundance of ^{26}Mg from the decay of short-lived ^{26}Al in calcium- and aluminum-rich inclusions.

Presumably these anomalies reflect the formation of elements and the condensation of dust in the later phases of stellar evolution, with further modifications within the Solar System. During the formation of the latter, the parent objects of the meteorites, or parts of them (originally at the outer edge of the Solar System?), were *not* chemically and isotopically in equilibrium with the rest of the "normal" solar material.

These investigations, which are very complex in their details, indicate that the origins of all meteorites, including the carbonaceous chondrites, occurred within at most the last 10^7 yr, perhaps even within 10^6 yr.

In the primitive meteorites Allende and Murchison, some inclusions with a high carbon content have been found, which exhibit extremely unusual isotopic ratios that are clearly different from those of solar material and – in terms of their extent – also from the isotopic anomalies just discussed. In these cases, not just trace elements in the inclusions are involved, but also their main constituents. For example, the ratios of the isotopes ^{12}C/^{13}C have been found to exhibit a broad scatter from about 1/10 up to 100 times the normal value of 89 in the Solar System. Furhermore, overabundances of isotopes relative to the values for solar material have been found for the trace elements such as N, O, Mg and others.

Dating of some of the inclusions in Allende yielded $4.9 \cdot 10^9$ yr, an age which is considerably greater than that of the Solar System. This result suggests the existence of *presolar matter*, whose nuclear composition dates from the time *before* the formation of our Solar System. Thus, in these meteorites, small *interstellar* dust grains have survived the formation process of the Solar System unchanged.

Recent investigations of the C-containing inclusions in the primitive meteorites have also shown surprising results, e.g. in the Murchison meteorite. Typically, the mass fraction of carbon is 2 to 3%, with most of this being due to organic compounds; amorphous carbon and graphite are relatively rare. In the *interstellar* dust grains, three fairly heat-resistant carbon forms were found which had survived the formation process of the Solar System: tiny *diamond* grains with diameters of less than 2 nm, small particles of silicon carbide (SiC) with sizes between about 0.2 and 10 μm, and some graphite particles a few μm in size. With a mass fraction of 2% of the carbon, or $4 \cdot 10^{-4}$ of the overall mass, the microdiamonds predominate over SiC and graphite, which occur two orders of magnitude less frequently. These three interstellar components are the carriers of those trace elements which exhibit the isotopic anomalies, including the noble gases Ne and Xe.

Origin of Meteorites. We still know very little about the origin of the meteors and meteorites, and their relation to the comets and the asteroids. From orbital information, especially from radar measurements, we can conclude that a portion of the meteors is of *cometal* origin; another portion, the sporadic meteors, travels on statistically-distributed elliptical orbits with eccentricities very close to one. Hyperbolic orbits or the corresponding velocities do not occur.

On the other hand, a quantitative comparison of their reflectivity over the whole range from the ultraviolet into the infrared for various types of meteorites, with that of the *asteroids* (Sect. 3.3.2), shows that there must be some relationship, at least of the larger meteorites, to the asteroids: we can find asteroids ranging in type from that of the iron meteorites to that of the carbonaceous chondrites. A considerable portion of the meteorites must therefore have had its origin in the asteroid belt, and have been formed there as debris from collisions between larger objects. This hypothesis is also supported by the few available orbital determinations: only in the case of the following four meteorites (named for the places where they were found): Lost City, Pribram, Innisfree and Peekskill – could the luminous trail in the atmosphere be photographed sufficiently accurately to derive the original orbit in the Solar System. The aphelia of these four objects were within the asteroid belt between Mars and Jupiter.

Finally, we know of only a few meteorites which – as shown by their chemical compositions – were broken out of the surfaces of the *Moon* and *Mars* through the impact of a larger object, and then found their way to us on the Earth.

3.8 Interplanetary Matter

Passing from the meteors and meteorites to still smaller particle sizes, i.e. particle radii below 100 μm or masses below 10^{-8} kg, we find the *interplanetary dust*.

The reflection and scattering of sunlight from interplanetary particles with radii of 10 to 80 μm gives rise to the *zodiacal light*, which can be observed as a cone-shaped glow in the sky in the region of the zodiac, in Spring shortly after sunset in the west, or in Autumn just before sunrise in the east. Opposite to the Sun, one can observe the faint *gegenschein* (counterglow). During total eclipses of the Sun, the strong forward scattering by interplanetary dust (Tyndall scattering) gives rise to a continuation of the zodiacal light in the neighborhood of the Sun as an outer part of the solar corona, the so-called *F*- or *Fraunhofer corona*. It is called that because its spectrum, like that of the zodiacal light, contains the dark Fraunhofer lines of the solar spectrum. In both cases, the scattered light is partially polarized.

The *intrinsic radiation* from the dust particles contributing to the zodiacal light, in the far infrared region, was observed by the IRAS satellite. Three narrow emission regions are clearly visible above the background radiation: the brightest coincides with the ecliptic, while the other two are at a distance of about $10°$ above and below the ecliptic. They correspond to bunches of dust at about 2.5 AU from the Sun, i.e. in the asteroid belt (Sect. 3.3.1), and they could have been formed as debris from collisions between asteroids (with orbital inclinations $i \simeq 10°$).

In recent times, it has been possible to *directly* collect and analyze interplanetary dust particles and micrometeorites with satellites and space probes, and from the tiny (diameter ≤ 1 mm) impact craters on the Moon. Flaky, porous micrometeorites can even be collected undamaged by stratospheric aircraft at altitudes around 20 km. Some of the particles contain among other things polycyclic aromatic hydrocarbons with two or more benzene rings (Sect. 10.1.5).

Figure 3.32 gives a summary of the total *flux* of particles of varying mass in the neighborhood of the Earth. In total, a mass of about $4 \cdot 10^7$ kg yr^{-1} falls onto the Earth, i.e. a daily mass of around 10^5 kg, mostly in the form of micrometeorites of diameters in the range of 0.2 μm.

The space probe Ulysses discovered a stream of micrometeorites beyond the orbit of Mars, which

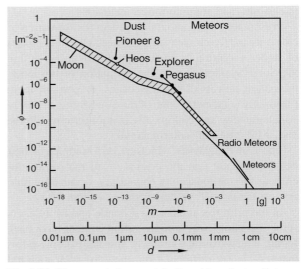

Fig. 3.32. The cumulative particle flux $\phi(>m)$ at a distance from the Sun of 1 AU, as a function of the mass m or diameter d of the particles. $\phi(>m)$ is the total number of particles with masses larger than m which fall on each m^2 per s. Values for the smaller particles are taken from satellite and space–probe observations as well as from the analysis of microcraters on the Moon's surface. (H. Fechtig et al., 1981)

move in a *retrograde* direction (with velocities around 26 km s^{-1}), i.e. against the direction of the planets' motions. Comets are not a possible source of these particles, since they do not occur in the inner parts of the Solar System. Their velocities indicate that they are *interstellar* dust grains.

The comets are probably the most important source of *interplanetary* dust, which is released from them when they are near the Sun. Smaller dust particles are continually produced due to collisions of larger particles and by evaporation of larger objects in the neighborhood of the Sun. The solar radiation pressure drives sufficiently small particles (of diameters $\lesssim 1$ μm) out of the Solar System; these β-meteoroids, on hyperbolic orbits, have been observed from space probes.

In addition to dust, the interplanetary medium includes on the one hand the magnetic plasma of the *solar wind* (Sect. 7.3.7), which at a distance of 1 AU from the sun has a mean particle density of about 10^7 m^{-3} and a magnetic field of about $6 \cdot 10^{-9}$ T, and streams outwards with an average velocity of 470 km s^{-1}. Furthermore, in interplanetary space we also find the highly energetic particles of *cosmic radiation*, both of solar origin and from the galaxy (Sect. 10.4.2).

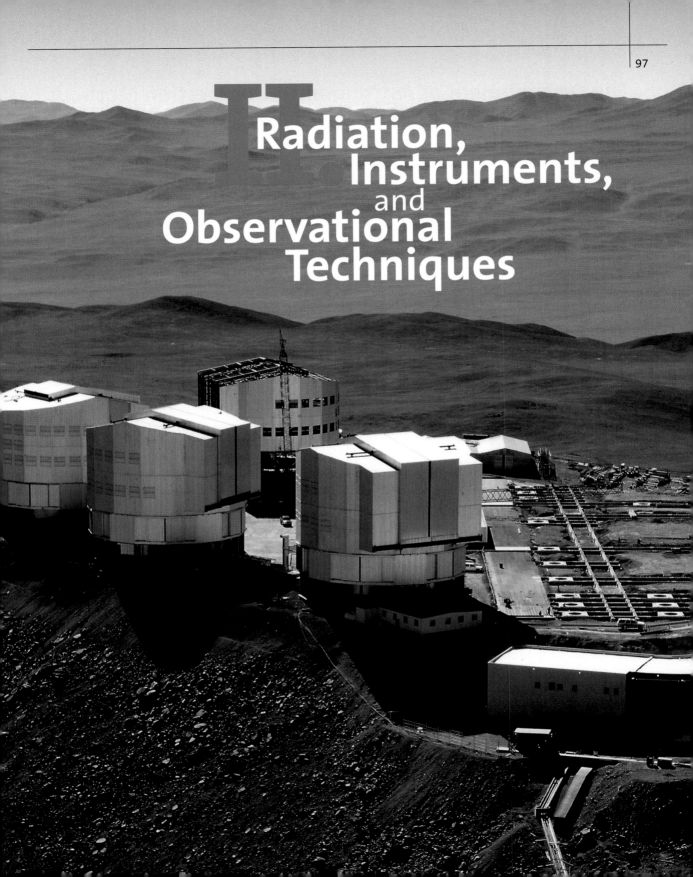

II. Radiation, Instruments, and Observational Techniques

The Development of Astronomical Observation Methods
Historical Introduction to Our Knowledge of the Electromagnetic Spectrum

The great advances in research are often associated with the invention or introduction of new types of *instruments*. The telescope, the clock, the photographic plate, photometer, spectrograph, and finally the whole arsenal of modern electronics and space travel each are associated with an epoch of astronomical research. However, equally important – and we should not forget this – is the creation of new *concepts* and approaches for the analysis of the observations. Scientific attainments of genius are indeed always based upon a combination of the formulation of new concepts and of instrumental developments, which only together can achieve an advance into previously unknown realms of Nature. We are tempted to agree with Simon Stevin (1548–1620), "Wonder en is gheen wonder".

The invention of the *telescope* (G. Galilei, 1609; J. Kepler, 1611) opened a new era for astronomy with previously unsuspected observational possibilities, due to the enormous increase in magnification and light-gathering power.

Beginning with Galileo's refracting telescope with a diameter of only 2 cm, larger and larger optical telescopes were constructed, culminating in those of the 20th century with mirrors (in one piece) of 3.5 up to 6 m in diameter. The 1990's brought the construction and to some extent the use of a new generation of *large telescopes* in which the primary focusing elements are made of several parts and are adjusted by active and adaptive optics, so that their mirror surfaces are equivalent to diameters of 10 to 20 m.

The accessible spectral region of *visual* observations is limited by the sensitivity of the human eye; visible light occupies only a small portion of the electromagnetic spectrum, from about 400 to 750 nm in wavelength, from violet through blue, green, and yellow to red. Only in the 19th century, after the invention of the *photographic plate*, did a light detector for astronomical observations become available which, on the one hand, stores and "integrates over" the incident light, and on the other, possesses sensitivity beyond the visible spectral region. Finally, in the 1970's, highly sensitive *solid-state detectors* were developed, such as the CCD (*C*harge *C*oupled *D*evice), with very high quantum yields, thereby offering a further enhancement of the possibility of observing extremely faint light sources, especially those of extragalactic origin.

The wavelength region which is accessible to astronomical observations from the Earth's surface is limited by the transmission of the *Earth's atmosphere* (Fig. II.1).

The "optical window" includes the near ultraviolet and the near infrared, in addition to the visible region. On the short–wavelength end, it is limited by the absorption due to atmospheric ozone, O_3, near $\lambda = 300$ nm; on the long-wavelength end, by the absorption of water vapor, H_2O, at about $\lambda = 1$ μm. Out to about 20 μm, some observations are still possible in several narrow windows.

Only in the radiofrequency range does the atmosphere again become transparent. The "*radio window*" is bounded on the shortwave end at $\lambda \simeq 1$ to 5 mm or $\nu \simeq 300$ to 60 GHz by the absorption due to atmospheric water vapor and oxygen; on the longwave end, at $\lambda \simeq 50$ m or $\nu \simeq 6$ MHz by reflection from the ionosphere.

Although the propagation of long-wavelength electromagnetic waves in free space was discovered in 1888 by H. Hertz, available radio receivers were for a long time not sufficiently sensitive to detect radio emissions from cosmic sources. Following the fortuitous discovery by K. G. Jansky in 1932 of the radiofrequency emissions of the Milky Way in the meter wavelength region ($\lambda = 12$ to 14 m), G. Reber, beginning in 1939, carried out a survey of the heavens at $\lambda = 1.8$ m using a 9.5 m parabolic mirror antenna (in the garden of his house!). During the 2nd World War, in 1942, J. S. Hey and J. Southworth detected the radio emissions of the active and of the quiet Sun, using receivers which had in the meantime been improved for use in radar apparatus.

In 1951, various researchers in Holland, the USA, and Australia almost simultaneously discovered the 21 cm line of interstellar hydrogen, which had been predicted by H. C. van der Hulst; its Doppler effect opened up enormous possibilities for investigating the motions

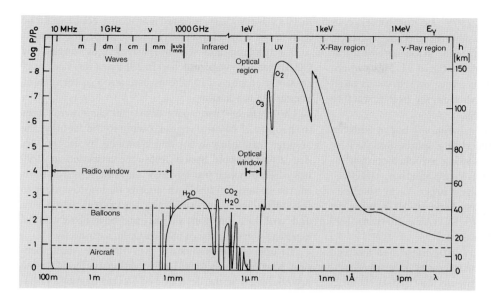

Fig. II.1. The transmission of the Earth's atmosphere for electromagnetic radiation as a function of the wavelength λ (*lower scale*) or the frequency ν and the corresponding photon energy $E_\gamma = h\nu$ (*upper scale*). The altitude h [km] (and the corresponding pressure P in units of the pressure at ground level, $P_0 \simeq 1\,bar$) at which the intensity of the incident radiation is reduced to one-half its initial value is shown, and the maximum altitudes for observations from aircraft and balloons are also indicated

of interstellar matter in our Milky Way galaxy and in other cosmic formations. Since the 1950's, an almost explosive development of radio astronomy has occurred, owing to the construction of individual telescopes and of a wide variety of antenna systems based on the principles of interferometry and of aperture synthesis (M. Ryle) which have yielded better and better *angular resolution*, as well as the improving detection sensitivity, obtained especially through low-noise *amplifiers*. Using long-baseline interferometry with transcontinental base lengths, an angular resolution of better than 10^{-4} seconds of arc has now been attained, by far exceeding the precision of optical observations.

Since about 1970, radio-astronomical observations have been extended to the millimeter and recently to the submillimeter ranges, in particular as a result of progress in amplifier technology.

Observations *outside the atmospheric envelope* of the Earth with rockets and especially with satellites have allowed astronomical observations to extend to all the spectral regions which would otherwise be completely absorbed by the Earth's atmosphere: the medium and far *infrared* between 20 and 350 µm, which is absorbed by atmospheric water vapor and oxygen bands; the *far ultraviolet* past the transmission limit of atmospheric ozone at $\lambda = 285$ nm; the contiguous *Lyman region*, where absorption is mainly due to atmospheric oxygen, O_2; then the *X-ray* and finally the *gamma ray* regions. Although we can investigate the solar spectrum continuously from wavelengths in the radiofrequency region out to the X-ray region, for galactic and extragalactic research we must keep in mind the Lyman continuum of the *interstellar* hydrogen atoms (Sect. 10.2). Their absorption sets in strongly at $\lambda = 91.2$ nm and allows no "viewing" until we reach the X-ray region beyond about $\lambda = 1$ nm.

At the beginning of "space astronomy", immediately after the end of the 2nd World War, observations were made using rockets which were developed in Germany during the war, and later with stabilized research rockets: in 1946, H. Friedman and his group obtained the first ultraviolet spectrum, and in 1948, T. R. Burnight recorded the first X-ray image of the *Sun*. Following the launching of the first artificial satellite Sputnik 1 (1957), a rapid development in the investigation of the Solar System and in astronomical observation using satellites and space probes has occurred, which we have to some extent already described in Sect. 2.5.

From the 1960's on, beginning with the Orbiting Solar Observatory (OSO) and Orbiting Astronomical Observatory (OAO), satellites with sufficient positional stability ($\leq 1''$) to permit longer series of observations of the Sun and other cosmic sources in the ultraviolet, X-ray, and gamma-ray regions became available.

In the *ultraviolet*, from 1979 and for some years thereafter, the IUE satellite with its 0.45 m telescope carried out successful observations. In the *extreme* ultraviolet region below the Lyman edge at 91.2 nm, where observations are blocked in many directions through absorption by interstellar hydrogen, the EUVE satellite has been operating since 1992.

In 1962, (nonsolar) *X-ray astronomy* began with the accidental discovery of the strong source Sco X-1 by R. Giacconi and coworkers using a rocket–carried instrument; in 1970, the first X-ray satellite, UHURU, began its survey of the skies. Larger X-ray satellite–borne telescopes became available in 1978 with the Einstein Observatory, in 1990 (ROSAT), and from 1999 on (Chandra and XMM Newton).

In the 1970's, *gamma-ray astronomy* also attained a sufficient angular resolution to be able to resolve individual cosmic sources. Following the COS-B satellite, launched in 1975, since 1991 the Compton Observatory CGRO has carried out detailed observations in the gamma region. Astronomy with observation of high-energy photons has thus developed within a few years into one of the most interesting areas of research, making use of an arsenal of new and unusual instruments.

In the *infrared* region, the generally weak sources must be detected against a background of thermal radiation from the instrument itself and from the Earth's atmosphere. The development of cooled semiconductor detectors since the 1960's has produced a rapid improvement in our ability to observe infrared sources from stratospheric balloons and aircraft, as well as from the first infrared satellite observatory, IRAS, launched in 1983. In 1995/98, infrared astronomy could again be carried out with a larger satellite telescope, the ISO.

For the mm and sub-mm wavelength ranges, the COBE satellite was launched in 1989 and has been successfully employed for the investigation of the cosmic background radiation.

Observations from space also offer clear advantages in the *optical* spectral region, owing to the fact that the angular resolution is no longer limited by air motion. The 2.4 m Hubble Space Telescope launched in 1990 has offered (since the repair of an imaging error in its mirror system in 1993) an enormous improvement in resolution and detection limits for optical astronomy, and with it a plethora of exciting new observations, especially in the area of extragalactic astronomy.

To end our introductory overview, we shall mention several astronomical observation methods which are not based on the detection of electromagnetic radiation. In 1912, V. F. Hess had already discovered *cosmic radiation* in the course of a balloon ascent; today, it represents an important research area of high-energy astronomy. While the secondary particles resulting from interactions of the primary energetic protons and heavier nuclei in the upper atmosphere can be investigated on the ground, the primary radiations themselves must be observed from outside the atmosphere.

The search for *gravitational waves* from cosmic sources has thus far remained unsuccessful.

The interesting area of *neutrino astronomy*, whose early development was due to R. Davis Jr. (1964), is at present still limited to the Sun and to unusual events such as the explosion of the supernova SN 1987 A in the nearby Large Magellanic Cloud.

4. Radiation and Matter

Having come to know classical astronomy and our Solar System, we have reached what is perhaps the most suitable point to consider the more important astronomical and astrophysical instruments and experimental methods all together, before turning to the astrophysics of the Sun and the stars.

We shall begin with an overview of the electromagnetic spectrum and related basic concepts (Sect. 4.1), and then summarize in Sect. 4.2 the most important results of Special Relativity Theory for later use in the description of particles with high velocities – especially in high-energy astronomy. We then consider the basic elements of radiation theory in Sect. 4.3; it plays a central role in the interpretation of the radiation from cosmic sources. We turn in Sect. 4.4 to a description of the occupation of atomic or molecular energy levels in matter which is in thermodynamic equilibrium. Finally, in Sect. 4.5, we become acquainted with the basics of the interactions of radiation with matter, but leave the treatment of atomic spectroscopy for Sect. 7.1, where we discuss the interpretation of stellar spectra.

4.1 Electromagnetic Radiation

We can describe electromagnetic radiation either in terms of electromagnetic waves, or as a stream of particles (photons).

Electromagnetic Waves. These propagate *in vacuum* at the speed of light, $c = 2.998 \cdot 10^8$ m s^{-1} (and in matter with an index of refraction n, at the velocity c/n). They are a result of the time variation of the electric field vector E or, via the Maxwell equations, by the related magnetic field vector H. For a given direction of propagation, they are characterized by their period of oscillation or their frequency ν, by the phase of the oscillation, by the plane of oscillation (polarization), and by the amplitude of E or of H.

Instead of the frequency ν, we can just as well use the wavelength λ to characterize the electromagnetic radiation. The following relation between λ and ν holds (strictly speaking, only in vacuum):

$$c = \nu\lambda . \tag{4.1}$$

It us usually fulfilled to sufficient accuracy in the plasma of astronomical objects.

The unit of frequency is the Hertz, 1 Hz $= 1$ s^{-1}; wavelengths are quoted in suitable multiples of the meter, depending on the spectral region, e.g. 1 nm $= 10^{-9}$ m in the optical region. In addition, special units such as 1 Å (Ångström) $= 0.1$ nm are also used. Besides ν and λ, the angular frequency $\omega = 2\pi\nu = 2\pi c/\lambda$, the wavenumber $\tilde{\nu} = 1/\lambda = \nu/c$, and the angular wavenumber $k = 2\pi\tilde{\nu} = 2\pi/\lambda$ are also commonly quoted.

The amplitudes of the electric and magnetic fields are related in magnitude by:

$$E = \sqrt{\mu_0/\varepsilon_0}\, H = cB . \tag{4.2}$$

The magnetic flux density in vacuum is given by $\boldsymbol{B} = \mu_0 \boldsymbol{H}$. The electric field constant (permittivity constant of vacuum), $\varepsilon_0 = 8.854 \cdot 10^{-12} \simeq 10^{-9}/(36\pi)$ A s V^{-1} m^{-1}, and the magnetic field constant (permeability constant of vacuum), $\mu_0 = 4\pi \cdot 10^{-7}$ V s A^{-1} m^{-1}, depend on the speed of light via the relation

$$c^2 = \frac{1}{\varepsilon_0 \mu_0} . \tag{4.3}$$

The energy density of an electromagnetic wave in vacuum is:

$$w = \tfrac{1}{2}(\varepsilon_0 E^2 + \mu_0 H^2) = \tfrac{1}{2}(\varepsilon_0 E^2 + B^2/\mu_0) \tag{4.4}$$

and the vector of the energy-current density (Poynting Vector) is defined by:

$$S = E \times H . \tag{4.5}$$

In the *international system of units* (SI), E is quoted in [V m^{-1}], H in [A m^{-1}], B in Tesla [T], w in [J m^{-3}] and S in [W m^{-2}]. For visible electromagnetic radiation, i.e. *light*, there are special SI units derived from the base unit of light intensity, 1 cd (Candela), which corresponds to the radiated power in a unit solid angle [W sr^{-1}]; we shall, however, not make use of them in this book.

In the *Gaussian unit system*, the relations corresponding to equations (4.2, 4 and 5) are:

$$E = H = B, \quad (4.2a)$$

$$w = \frac{1}{8\pi}(E^2 + H^2) = \frac{1}{8\pi}(E^2 + B^2), \quad (4.4a)$$

and

$$S = \frac{c}{4\pi} E \times H, \quad (4.5a)$$

where E is measured in electrostatic units, H in [Oe = Oersted], B in Gauss [G], all three having the same dimensions [cm$^{-1/2}$g$^{1/2}$s^{-1}], while w is expressed in [erg cm^{-3}] and S in [erg s^{-1}cm^{-2}].

In general, electromagnetic radiation from cosmic sources is an *incoherent* superposition of waves of differing frequencies and polarization directions, and the rapid oscillations themselves are not of interest, but only *average values* over many periods of oscillation, for example the average of the energy-current density S, proportional to E^2 or to H^2.

Photons. Especially at high frequencies, it is often expedient to consider electromagnetic radiation not as a wave, but as a stream of photons, which of course move at the velocity c. A photon of frequency ν or wavelength λ has the energy

$$E_\nu = h\nu = \hbar\omega = hc/\lambda. \quad (4.6)$$

Here,

$$h = 2\pi\hbar = 6.626 \cdot 10^{-34} \, \text{J s} \quad (4.7)$$

is Planck's constant. If the photon is moving in the direction $\boldsymbol{n}$ (unit vector, $|\boldsymbol{n}| = 1$), then its momentum is given by

$$\boldsymbol{p}_\nu = \frac{h\nu}{c}\boldsymbol{n} = \hbar k \boldsymbol{n}. \quad (4.8)$$

An often-used unit of energy is the *electron volt*, $1\,\text{eV} = 1.602 \cdot 10^{-19}$ J, which corresponds to a frequency of $2.418 \cdot 10^{14}$ Hz and whose equivalent wavelength is given by

$$E_\nu[\text{eV}] = \frac{1.240 \cdot 10^{-6}}{\lambda[\text{m}]} = \frac{1240}{\lambda[\text{nm}]}. \quad (4.9)$$

We shall return to the basic concepts of radiation theory, such as intensity, radiation current density, etc. in more detail in Sect. 4.3.

4.2 The Theory of Special Relativity

Newtonian mechanics had proved itself very powerful for the classical problems of celestial mechanics, where only velocities which are small compared to the speed of light occur. Towards the end of the 19th century, however, basic difficulties arose, in particular as a result of the experiment of A. A. Michelson and E. W. Morley in 1887, which showed no detectable sign of an "ether wind" relative to the Earth in its orbit. Following initial considerations by H. A. Lorentz, J. H. Poincaré and others, A. Einstein developed his *special relativity theory* in 1905. It starts from the experimental result of Michelson and Morley and requires that the propagation of a light wave, when described in different coordinate systems which may be in (translational, nonaccelerated) motion relative to each other, socalled "inertial systems", be physically identical; i.e. it must obey the same equation. If the light wave moves through a differential distance element of $(dx^2 + dy^2 + dz^2)^{1/2}$ in the time dt at the speed of light in vacuum c, then we have

$$c^2 dt^2 - (dx^2 + dy^2 + dz^2) = 0. \quad (4.10)$$

This quantity can be considered as a line segment ds^2 in a four-dimensional space with the three spacelike coordinates x, y, z and the time coordinate ct (or, more intuitively, ict with $i = \sqrt{-1}$). The light propagation (along a straight line) is then described by

$$ds^2 = 0. \quad (4.11)$$

One now requires generally of a transformation from one Cartesian coordinate system S into another, S', which may be in a (nonaccelerated) state of translational motion with the velocity v relative to S, that the four-dimensional line element be conserved, i.e. it remains *invariant*:

$$ds^2 = c^2 dt^2 - (dx^2 + dy^2 + dz^2) = \text{inv} \quad \text{or}$$
$$ds^2 = d(ict)^2 - (dx^2 + dy^2 + dz^2) = \text{inv}. \quad (4.12)$$

Such a transformation can clearly not be limited to the spacelike coordinates $x, y, z \rightarrow x', y', z'$ alone; rather, the time $t \to t'$ must be transformed, also. This *Lorentz transformation* is – as can be seen from (4.12) – nothing other than a rotation in the four-dimensional space $\{ict, x, y, z\}$, for which by definition the line element ds is invariant. The essential progress of the special relativity theory with respect to the Newtonian theory results from the fact that the particular significance of the speed of light in vacuum c is taken into account from the beginning. It correspondingly leads to the recognition that no material motions and no signal (of any kind) can exceed the velocity $c \simeq 3 \cdot 10^8$ m s^{-1}. Special relativity theory further requires for all of physics that all natural laws be invariant under Lorentz transformations (i.e. be physically independent of translational motions).

4.2.1 The Lorentz Transformation. The Doppler Effect

We give the Lorentz transformations only for the simple case in which a Cartesian system S moves relative to S' with a velocity v *parallel to the x-axis*, whereby at the time $t = t' = 0$, the two spatial coordinate systems are identical:

$$x' = \gamma (x - vt), \quad t' = \gamma \left(t - \frac{vx}{c^2}\right). \quad (4.13)$$

Here, γ is the *Lorentz factor*, which typifies the theory of special relativity:

$$\gamma = \gamma(v) = \frac{1}{\sqrt{1 - v^2/c^2}}. \quad (4.14)$$

For velocities which are small in comparison to the speed of light, we have

$$\gamma \simeq 1 + \frac{1}{2}\frac{v^2}{c^2} + \cdots \quad (v \ll c), \quad (4.15)$$

and the transformation (4.13) becomes equivalent to the Galilean transformation:

$$x' = x - vt, \quad t' = t, \quad (4.16)$$

which forms the basis of Newtonian mechanics, with its concept of absolute time.

From (4.13), it immediately follows that there will be a *contraction* of lengths along the direction of motion (Fitzgerald contraction): a length moving at velocity v appears to be shortened to $l = l_0/\gamma$ in comparison to its value l_0 in the system "at rest" ($v = 0$). A second important result is the *time dilation*: if a clock in the system S gives time marks at intervals of Δt_0, they appear in a system in relative motion to S at velocity v to be longer:

$$\Delta t = \gamma \Delta t_0. \quad (4.17)$$

Time thus passes most rapidly in the system at rest.

Finally, we give the expression for the relativistic *Doppler effect*. A source – considered e.g. to be at rest in the system S' of a radiating atom – is moving in the "laboratory system" S at velocity $\boldsymbol{v}$, with the radial velocity v_{rad} relative to the observer. If the wavelength emitted at rest by the source is λ', i.e. its proper frequency is ν', then the observer measures a wavelength λ or a frequency ν corresponding to

$$\frac{\lambda}{\lambda'} = \frac{\nu'}{\nu} = \gamma \left(1 + \frac{v_{\text{rad}}}{c}\right). \quad (4.18)$$

Compared to the non-relativistic Doppler effect (2.3),

$$\frac{\lambda}{\lambda'} \simeq 1 + \frac{v_{\text{rad}}}{c}, \quad \frac{\nu}{\nu'} \simeq 1 - \frac{v_{\text{rad}}}{c}, \quad (4.19)$$

the relativistic expression (4.18) differs only by the Lorentz factor, which results from the additional time dilation. In contrast to the "simple" Doppler formula, we find in the theory of relativity an effect when the source is moving *transversely* to the observer, i.e. even when $v_{\text{rad}} = 0$.

4.2.2 Relativistic Mechanics

Let us consider the motion of a point mass (in a sufficiently weak gravitational field); the central results of relativistic particle mechanics may be summarized in the equations for conservation of the momentum $\boldsymbol{p}$ and energy E:

$$\boldsymbol{p} = \gamma m \boldsymbol{v}, \quad (4.20)$$

$$E = \gamma m c^2. \quad (4.21)$$

Here, γ is again the Lorentz factor (4.14), and m is the rest mass of the particle, i.e. its mass in the system at rest relative to the particle, where $v = 0$. The relativistic (inertial) mass γm occurs in (4.20); it depends upon the velocity v.

In the non-relativistic case $v \ll c$ ($\gamma \to 1$), (4.20) becomes the Newtonian momentum $\boldsymbol{p} = m\boldsymbol{v}$, while the energy becomes:

$$E = mc^2 + \tfrac{1}{2}mv^2 + \ldots . \tag{4.22}$$

This last expression contains, in addition to the kinetic energy as in the Newtonian case, an additional term, the *rest energy*

$$E_0 = E(v=0) = mc^2 . \tag{4.23}$$

This famous equation of Einstein's, which expresses the equivalence of energy and mass, is valid above and beyond the area of mechanics for *all* types of energy.

The following relations between momentum and energy hold in relativistic mechanics for a particle of mass $m \neq 0$, as we can see from (4.20) and (4.21):

$$\frac{E^2}{c^2} = p^2 + m^2 c^2 , \tag{4.24}$$

$$p = \frac{E}{c^2} v . \tag{4.25}$$

In the *extreme relativistic* limit, i.e. for particles moving at nearly the speed of light, $\gamma \gg 1$ and the energy is $E \gg E_0 = mc^2$, where E_0 e.g. for an electron has the value 511 keV and for a proton, the value 938 MeV. Equation (4.25) is valid also for $v = c$, so that

$$p = \frac{E}{c} . \tag{4.26}$$

In this form, the relation between momentum and energy holds also for particles of rest mass $m = 0$ such as photons (Sect. 4.1). For a photon of energy $E = h\nu$, we then have $p = h\nu/c$, and the inertial mass corresponding to its energy is $\gamma m = h\nu/c^2$.

4.3 The Theory of Radiation

In view of the various radiation fields of cosmic objects, such as the atmospheres and interiors of stars, planetary atmospheres, gas nebulae and interstellar space, we now wish to present the *fundamentals* of radiation theory.

First, in Sect. 4.3.1, we introduce phenomenologically the concepts of intensity, radiation current density and other quantities relating to radiation fields. In Sect. 4.3.2, we present a – likewise phenomenological – description of absorption, extinction, and emission and then introduce the important radiation transport equation. In Sect. 4.3.3, we then describe the special case of thermodynamic equilibrium and introduce the relevant radiation quantities of a cavity radiator or black body.

4.3.1 Phenomenological Description of Radiation

In the radiation field, about which we initially make no special assumptions, we place a surface element dA having a surface normal $\boldsymbol{n}$ (Fig. 4.1); we then consider the radiation energy which passes through dA in a time interval dt at an angle θ to $\boldsymbol{n}$ within a small range of solid angle $d\Omega$ (characterized by the directional angles θ and φ). By spectral decomposition, we separate out the frequency range $(\nu, \nu + d\nu)$ and write:

$$dE = I_\nu(\theta, \varphi) \, d\nu \, \cos\theta \, dA \, d\Omega \, dt , \tag{4.27}$$

where $d\Omega = \sin\theta d\theta d\varphi = -d(\cos\theta)d\varphi$ and $dA\cos\theta$ is the cross-sectional area of the radiation beam.

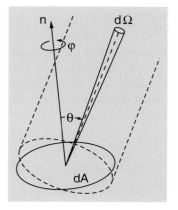

Fig. 4.1. The definition of radiation intensity

Radiation Intensity. The (specific) radiation intensity $I_\nu(\theta, \varphi)$ is then defined as the amount of energy which passes through a unit surface (1 m^2) normal to the direction (θ, φ), per unit solid angle (1 sr = 1 steradiant) and unit frequency range (1 Hz) in one second, corresponding to (4.27).

We can relate the spectral decomposition to the wavelength range instead of the frequency range; since naturally the total amount of energy remains the same, we have

$$I_\nu \, d\nu = -I_\lambda \, d\lambda . \tag{4.28}$$

From $\nu = c/\lambda$ it follows that $d\nu = -(c/\lambda^2) \, d\lambda$ and thus

$$I_\lambda = \frac{c}{\lambda^2} I_\nu \quad \text{or} \quad \nu I_\nu = \lambda I_\lambda . \tag{4.29}$$

The intensity of the *total radiation* is obtained by integrating over all frequencies or wavelengths:

$$I = \int_0^\infty I_\nu \, d\nu = \int_0^\infty I_\lambda \, d\lambda \, . \tag{4.30}$$

We also define the *mean intensity* (averaged over all solid angles):

$$J_\nu = \frac{1}{4\pi} \oint I_\nu \, d\Omega \equiv \frac{1}{4\pi} \int_{\varphi=0}^{2\pi} \int_{\theta=0}^{\pi} I_\nu \sin\theta \, d\theta \, d\varphi \tag{4.31}$$

The average intensity of the total radiation is then $J = \int_0^\infty J_\nu \, d\nu$.

As a simple application, we calculate the energy dE radiated per unit time ($dt = 1$) from one surface element dA into a second surface element dA' at a distance r from the first. The surface normals of dA and dA' are assumed to make the angles θ and θ' with the line r connecting them (Fig. 4.2). dA' as seen from dA subtends a solid angle $d\Omega = \cos\theta' \, dA'/r^2$. We thus have:

$$\begin{aligned} dE &= I_\nu \, d\nu \cos\theta \, dA \, d\Omega \\ &= I_\nu \, d\nu \frac{\cos\theta \, dA \cos\theta' dA'}{r^2} \, . \end{aligned} \tag{4.32}$$

On the other hand, dA as seen from dA' subtends a solid angle $d\Omega' = \cos\theta \, dA/r^2$. We can thus also write, in a manner symmetrical to the first equation in (4.32),

$$dE = I_\nu \, d\nu \cos\theta' \, dA' \, d\Omega' \, . \tag{4.33}$$

Thus, the *intensity* I_ν of, for example, solar radiation, as defined by (4.27), is *the same* in the immediate vicinity of the Sun and somewhere far out in space.

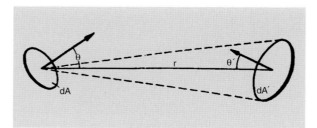

Fig. 4.2. Radiation emitted from one surface element and observed at another one

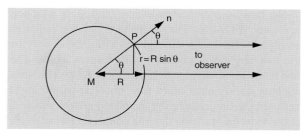

Fig. 4.3. Radiation flux density F_ν from a star. The average intensity $\bar{I}_\nu$ from the disk is equal to F_ν/π

Radiation Flux Density. What is meant in the somewhat fuzzy terms of everyday language by "strength", e.g. of sunshine, is more closely related to the exact concept of radiation flux density or, for short, the *radiation flux*. We define the radiation flux F_ν[1] in the direction $\boldsymbol{n}$ by writing the total energy of radiation with frequency ν which passes through our surface element per unit of time (Fig. 4.1) as:

$$F_\nu = \oint I_\nu \cos\theta \, d\Omega \, . \tag{4.34}$$

For a narrow ray, F_ν corresponds directly to the magnitude of the Poynting vector of an electromagnetic wave, (4.5). In an *isotropic* radiation field (I_ν independent of $\boldsymbol{n}$), $F_\nu = 0$. It is often expedient to decompose F_ν into the emitted radiation ($0 \leq \theta \leq \pi/2$)

$$F_\nu^+ = \int_0^{2\pi} \int_0^{\pi/2} I_\nu \cos\theta \sin\theta \, d\theta \, d\varphi \tag{4.35}$$

and the incident radiation ($\pi/2 \leq \theta \leq \pi$)

$$F_\nu^- = \int_0^{2\pi} \int_\pi^{\pi/2} I_\nu \cos\theta \sin\theta \, d\theta \, d\varphi \, , \tag{4.36}$$

so that $F_\nu = F_\nu^+ - F_\nu^-$ holds.

In analogy to (4.30), the total radiation flux is defined as:

$$F = \int_0^\infty F_\nu \, d\nu = \int_0^\infty F_\lambda \, d\lambda \, . \tag{4.37}$$

---

[1] Often, the radiation flux is also denoted by $\pi \mathcal{F}_\nu$. This socalled astrophysical flux $\mathcal{F}_\nu$ is thus smaller than F_ν by a factor of π.

We now consider the radiation emitted by a (spherical) star (Fig. 4.3). The intensity I_ν of the radiation emitted by the star's atmosphere will depend only on the exit angle θ (calculated with respect to the normal at the particular position), i.e. $I_\nu = I_\nu(\theta)$. The same angle θ occurs once more (see Fig. 4.3) as the angle between the line connecting the observer with the midpoint M of the star, and the line connecting M with the observed point P. The distance of the point P from M, projected onto a plane perpendicular to the direction of observation, in units of the radius R of the star, is thus equal to $\sin\theta$.

Then the mean intensity $\overline{I}_\nu$, averaged over the apparent disk of the star, of the radiation emitted towards the observer is given according to (4.27) by:

$$\pi R^2 \cdot \overline{I}_\nu = \int_0^{2\pi} \int_0^{\pi/2} I_\nu(\theta) \cos\theta \, R^2 \sin\theta \, d\theta \, d\varphi . \quad (4.38)$$

Dividing by R^2 and comparing with (4.35), we see that

$$\overline{I}_\nu = F_\nu^+/\pi , \quad (4.39)$$

i.e. the mean radiation intensity of the stellar disk is equal to $1/\pi$ times the radiation flux F_ν at the star's surface. If the star (of radius R) is at a large distance r from the observer ($\cos\theta \simeq 1$), its disk is seen with a solid angle $d\Omega = \pi R^2/r^2$ and therefore produces, as a result of (4.35), a radiation flux at the observer of

$$f_\nu = \overline{I}_\nu \, d\Omega = F_\nu^+ R^2/r^2 . \quad (4.40)$$

The theory of *stellar* spectra must therefore aim at the calculation of the *radiation flux*. The great advantage of *solar* observations, on the other hand, is due to the fact that here, as first recognized by K. Schwarzschild, we can measure I_ν directly as a function of θ.

In general, the radiation flux is the characteristic measurable quantity for the observation of unresolved sources (point sources), and the intensity or surface brightness is that for the observation of extended sources. Table 4.1 summarizes the units which are currently in common use.

Radiation Density. As a further concept of radiation theory we introduce the (specific) monochromatic *radiation density* u_ν and write the energy density [J m^{-3}] of the radiation field in the frequency range $(\nu, \nu+d\nu)$ as $u_\nu \, d\nu$. The relation between u_ν and the radiation intensity I_ν is found from the following considerations: From (4.27), the basal area dA of a cylindrical volume element of height dl transmits a radiation energy $dA \int I_\nu \, d\Omega$ per unit time; it traverses the volume element in the time dl/c (distance traveled divided by the speed of light). The specific radiation density is then

$$u_\nu = \frac{1}{c} \int I_\nu \, d\Omega = \frac{4\pi}{c} J_\nu , \quad (4.41)$$

where J_ν is the mean intensity (4.31).

If we integrate (4.41) over the whole spectrum, we obtain the *total energy density* of the radiation

$$u = \int_0^\infty u_\nu \, d\nu . \quad (4.42)$$

Photon Density. It is frequently useful to relate the radiation quantities not to the energy, as in (4.27), but rather to the number of the corresponding photons.

Table 4.1. Fundamental concepts and units of radiation

	Radiation quantity		SI-Units	cgs-Units
Extended sources	Intensity or surface-brightness	$\{I_\nu,$ [a] $\{I$	W m^{-2}Hz^{-1}sr^{-1} W m^{-2}sr^{-1}	erg s^{-1}cm^{-2}Hz^{-1}sr^{-1} erg s^{-1}cm^{-2}
Non-resolved (point-) sources	Radiation flux density or radiation current (-flux)	$\{F_\nu, f_\nu,$ [a] $\{F, f$	W m^{-2}Hz^{-1} [b] W m^{-2}	erg s^{-1}cm^{-2}Hz^{-1} erg s^{-1}cm^{-2}

1 SI-Unit = 10^3 cgs-Units

[a] In the λ scale the corresponding quantities of the radiation field, I_λ, F_λ, f_λ, are usually quoted per μm, nm, or Å, according to the particular application. Thus, for example I_λ is given in [W m^{-2}μm^{-1}sr^{-1}], [W m^{-2}nm^{-1}sr^{-1}], [erg s^{-1}cm^{-2}Å^{-1}sr^{-1}] etc.
[b] The unit often used for the flux, particularly in radioastronomy, is 1 Jy (Jansky)= 10^{-26} W m^{-2}Hz^{-1}.

Since the energy of a photon is given by $h\nu$ (4.6), we obtain e.g. the *photon flux* from the radiation flux F_ν (4.34), by simply replacing I_ν with $I_\nu/(h\nu)$.

In the same way, the specific radiation density u_ν corresponds to a *photon density*

$$n_\nu = \frac{4\pi}{c} \frac{J_\nu}{h\nu}, \qquad (4.43)$$

and the total energy density u to a total photon density

$$n = \frac{4\pi}{c} \int_0^\infty \frac{J_\nu}{h\nu} d\nu. \qquad (4.44)$$

Radiation Pressure. Every beam of radiation transports not only energy in the form of the radiation flux F_ν or F, but, from (4.26), also *momentum*. A beam which falls onto a unit surface of surface normal $\boldsymbol{n}$ at an angle θ thus applies a pressure to the surface, which is given by the momentum per unit time and surface area (2.5). This radiation pressure P_r (index r: radiation), which we give as an integral quantity over the whole spectrum of the radiation, is found from (4.34) to be

$$P_\mathrm{r} = \frac{1}{c} \oint I \cos^2\theta \, d\Omega, \qquad (4.45)$$

where I is the total intensity (4.30). The additional factor $\cos\theta$ takes into account the fact that only the perpendicular component contributes to the momentum.

For an approximately isotropic radiation field, we can extract the average value of $\cos^2\theta$ over a sphere, i.e. $1/3$, from the integral, and then using (4.31) obtain the radiation pressure

$$P_\mathrm{r} = \frac{1}{3c} \oint I \, d\Omega = \frac{4\pi}{3c} J. \qquad (4.46)$$

Introducing the total energy density from (4.42), we find

$$P_\mathrm{r} = \frac{1}{3} u. \qquad (4.47)$$

This relation, which shows that the pressure is equal to $1/3$ times the energy density, is valid not only for radiation, but also for a relativistic gas, in which the particle velocities are near the speed of light.

4.3.2 Emission and Absorption. The Radiation Transport Equation

We shall now describe the emission and absorption of radiation, at first phenomenologically: let a volume element dV emit into the solid angle $d\Omega$ and in the frequency interval $(\nu, \nu+d\nu)$ the energy per unit time:

$$j_\nu \, d\nu \, dV \, d\Omega, \qquad (4.48)$$

where j_ν is the *emission coefficient* of dimensions $[\mathrm{W\,m^{-3}\,Hz^{-1}\,sr^{-1}}]$. The total energy output of an isotropically radiating volume element dV per unit time is then given by

$$4\pi \int_0^\infty j_\nu \, d\nu \, dV. \qquad (4.49)$$

We compare this emission to the energy loss by absorption which a beam of radiation of intensity I_ν experiences in passing through a layer of matter of thickness ds with the *absorption coefficient* κ_ν. This is given by:

$$dI_\nu/ds = -\kappa_\nu I_\nu. \qquad (4.50)$$

In the general case, the intensity of the beam is reduced not only by absorption, but also by *scattering* out of the direction of the beam. If we denote the scattering coefficient by σ_ν, then the *extinction coefficient*

$$k_\nu = \kappa_\nu + \sigma_\nu \qquad (4.51)$$

is a measure of the *overall* attenuation of the radiation. The emission coefficient (4.48) may then also contain a corresponding contribution to the intensity from scattering into the direction ds of the beam.

If we combine the changes in the intensity of our radiation beam along the path ds from extinction (4.51) and emission (4.48), we obtain the *transport equation* for radiation of frequency ν:

$$\frac{dI_\nu}{ds} = -k_\nu I_\nu + j_\nu. \qquad (4.52)$$

We can see from this equation that the extinction, the absorption and the scattering coefficients have the dimensions of a reciprocal length $[\mathrm{m}^{-1}]$. The coefficients κ_ν, σ_ν, and j_ν which we have introduced phenomenologically depend in general on the frequency ν as well as on the type and state of the material (chemical composition, temperature, pressure), and possibly on the direction of propagation of the beam relative to a direction within the material (e.g. of a magnetic field applied to a plasma etc.)

If our beam passes through a nonemitting layer ($j_\nu = 0$), we have:

$$\frac{dI_\nu}{I_\nu} = -k_\nu ds, \quad (4.53)$$

and the intensity I_ν of the radiation after passing through this layer of thickness s is related to the intensity I_ν of the incident radiation by:

$$I_\nu = I_{\nu,0} \cdot e^{-\tau_\nu}, \quad (4.54)$$

where the dimensionless quantity

$$\tau_\nu = \int_0^s k_\nu \, ds \quad (4.55)$$

is termed the *optical thickness* of the transmitting layer. For example, a layer of optical thickness $\tau_\nu = 1$ attenuates a beam to $e^{-1} = 0.368$ of its original intensity. In the case that the layer neither emits nor absorbs, i.e. for propagation of radiation in vacuum, we again find that the intensity along the beam stays constant, cf. (4.32, 33).

Using the optical thickness or its differential $d\tau_\nu = k_\nu ds$, we can rewrite the radiation transport equation (4.52) in the form:

$$\frac{dI_\nu}{d\tau_\nu} = -I_\nu + \frac{j_\nu}{k_\nu} = -I_\nu + S_\nu \quad (4.56)$$

where the *source function* has been introduced as the ratio of emission to absorption coefficients:

$$S_\nu = \frac{j_\nu}{k_\nu}. \quad (4.57)$$

As an example, we give here the extinction of the radiation of frequency ν by the *Earth's atmosphere* for a star with the zenith distance z (Fig. 4.4). It corresponds to an *attenuation factor* of:

$$I_\nu/I_{\nu,0} = e^{-\tau_\nu \cdot \sec z}, \quad (4.58)$$

where $\sec z = 1/\cos z$ and τ_ν is the optical thickness of the Earth's atmosphere (as measured in a direction

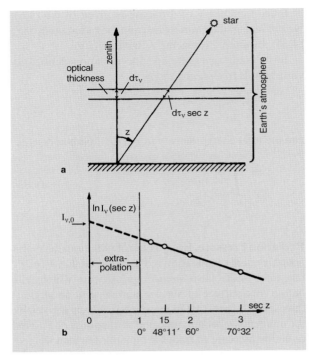

Fig. 4.4. (a) Atmospheric extinction of the radiation from a star at the zenith distance z. (b) Extrapolation to $\sec z \to 0$ yields the extraterrestrial radiation intensity $I_{\nu,0}$

perpendicular to the ground) at the frequency ν. The *extraterrestrial* intensity $I_{\nu,0}$ can, by the way, be obtained by linear extrapolation to $\sec z \to 0$ of the values of $\ln I_\nu$ measured at various zenith distances z, as long as τ_ν is not too large.

As a further important application, we calculate the radiation intensity I_ν emitted by a *layer of matter* having the thickness s in the direction of observation and whose source function S_ν is *constant*, i.e. not dependent on x (Fig. 4.5):

A volume element which extends from x to $x + dx$ in the direction of observation ($0 \leq x \leq s$) and has a unit surface area as its cross section, emits per s and sr the contribution $k_\nu S_\nu \, dx = S_\nu \, d t_\nu$ to the intensity, where $d t_\nu = k_\nu \, dx$ is the optical thickness. Before leaving the layer, this contribution is attenuated by the factor $\exp(-t_\nu)$ with $t_\nu(x) = \int_0^x k_\nu \, dx$. We thus obtain for the overall layer of thickness s with a total optical

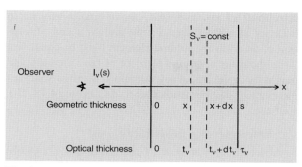

Fig. 4.5. The emission of a layer of thickness s or optical thickness τ_ν with a constant source function S_ν. Geometric scale x, optical scale t_ν

thickness of $\tau_\nu = t_\nu(s)$ the emitted intensity

$$I_\nu(s) = \int_0^{\tau_\nu} S_\nu \, e^{-t_\nu} \, dt_\nu \qquad (4.59)$$

or, since S is supposed to be constant within the layer,

$$I_\nu(s) = S_\nu \left(1 - e^{-\tau_\nu}\right)$$
$$\simeq \begin{cases} \tau_\nu S_\nu & \text{for } \tau_\nu \ll 1 \\ S_\nu & \text{for } \tau_\nu \gg 1 \end{cases}. \qquad (4.60)$$

The radiation from an optically thin layer ($\tau_\nu \ll 1$) is thus equal to its optical thickness times the source function. The radiation intensity from an optically thick layer ($\tau_\nu \gg 1$), in contrast, approaches the value of the source function and cannot exceed it.

4.3.3 Thermodynamic Equilibrium and Black-Body Radiation

We meet a particularly clear situation when we consider a radiation field which is in *thermal equilibrium* or *temperature equilibrium* with its surroundings. Such a radiation field can be generated by immersing an (otherwise arbitrary) cavity in a heat bath at the temperature T. In such a bath, all objects have by definition the same temperature; it thus seems justified to speak of cavity or black-body radiation at a temperature T. In an isothermal cavity, every body or every surface element will emit and absorb the same amount of radiation energy per unit time. Starting from this fact, we can now easily show that the intensity I_ν of the cavity radiation

is *independent* of the material content and the form of the walls of the cavity and of the direction (isotropy). We need only consider a cavity with two different chambers H_1 and H_2 (Fig. 4.6). If $I_{\nu,1} \neq I_{\nu,2}$, we could obtain energy by putting a radiometer (the well-known "light mill" sometimes seen in shop window displays) into the opening between the two cavities; we would then have a perpetual motion machine of the 2nd kind! It is readily shown that $I_{\nu,1} = I_{\nu,2}$ must hold for *all* frequencies, radiation directions, and polarization directions by inserting into the opening between H_1 and H_2 sequentially a colored filter, a tube oriented in a particular direction, or a polarizing filter; in each case, as a result of the 2nd Law of Thermodynamics, one cannot construct a perpetual motion machine of the 2nd kind. The intensity I_ν of the cavity radiation is thus a *universal* function $B_\nu(T)$ *of* ν *and* T.

A small hole in the cavity wall takes up all the incident radiation by multiple reflections and absorption in its interior, and thus represents a socalled *black body*. Cavity radiation is therefore also called black-body radiation or, for short, simply thermal radiation. It can be produced and measured by "sampling" a cavity at the temperature T through a sufficiently small opening.

The Kirchhoff–Planck Function. The important function $B_\nu(T)$, whose existence was recognized in 1860 by G. Kirchhoff and which was first explicitly calculated in 1900 by M. Planck, is called the Kirchhoff–Planck function. Planck recognized that he could find a function which would agree with measurements of the intensity of cavity radiation only by extending the laws of classical mechanics and thermodynamics to include the well-known requirements of *quantum mechanics*,

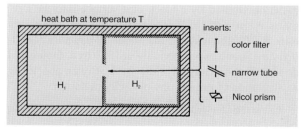

Fig. 4.6. Black-body radiation is unpolarized and isotropic. Its intensity is a universal function of ν and T, the Kirchhoff-Planck function $B_\nu(T)$

whose founder he thus became. He derived the Planck radiation formula, which we give in terms of both frequency ν and of wavelength λ:

$$B_\nu(T) = \frac{2h\nu^3}{c^2} \frac{1}{e^{h\nu/kT} - 1} \quad \text{or}$$

$$B_\lambda(T) = \frac{2hc^2}{\lambda^5} \frac{1}{e^{hc/k\lambda T} - 1}, \quad (4.61)$$

where $k = 1.38 \cdot 10^{-23}$ J K^{-1} is the Boltzmann constant, with the two important limiting cases

$$\frac{h\nu}{kT} \gg 1 : B_\nu(T) \simeq \frac{2h\nu^3}{c^2} e^{-h\nu/kT} \quad (4.62)$$

Wien's Law,

$$\frac{h\nu}{kT} \ll 1 : B_\nu(T) \simeq \frac{2\nu^2 kT}{c^2} \quad (4.63)$$

Rayleigh-Jeans Law.

In the Rayleigh-Jeans radiation law, the quantity h characteristic of quantum mechanics has vanished. For light quanta $h\nu$ whose energy is much smaller than the thermal energy kT, quantum mechanics (quite generally) is transformed into classical mechanics (Bohr's Correspondence Principle). The quantity which occurs in the exponent of the λ-representation (4.61) is summarized in the radiation constant c_2:

$$c_2 = \frac{hc}{k} = 1.439 \cdot 10^{-2} \text{ m K} \quad (4.64)$$

The maximum in the Kirchhoff–Planck function $B_\nu(T)$ or $B_\lambda(T)$ is given by *Wien's displacement rule*:

$$\frac{c}{\nu_{\max}} \cdot T = 5.10 \cdot 10^{-3} \text{ m K} \quad \text{or}$$

$$\lambda_{\max} \cdot T = 2.90 \cdot 10^{-3} \text{ m K}. \quad (4.65)$$

The intensity of the *total radiation* of the black body is obtained by integrating (4.61) over frequencies, $B(T) = \int_0^\infty B_\nu(T) d\nu$. The *total radiation flux*, i.e. the radiation emitted by a black unit surface into the cavity, is $F^+ = \pi B(T)$. If one carries out the integration over $B_\nu(T)$, the result is the *Stefan–Boltzmann Radiation Law* found experimentally in 1879 by J. Stefan and derived theoretically in 1884 by L. Boltzmann using a calculation of the entropy of cavity radiation:

$$F^+ = \pi B(T) = \sigma T^4 \quad (4.66)$$

with the radiation constant

$$\sigma = \frac{2\pi^5 k^4}{15 c^2 h^3} = 5.67 \cdot 10^{-8} \text{ W m}^{-2} \text{ K}^{-4}. \quad (4.67)$$

Kirchhoff's Law. In *thermodynamic equilibrium*, i.e. in a cavity at temperature T, the emission and absorption of an arbitrary volume element must be equal. We write them for a volume element of base area dA and height ds in the frequency domain ν within a solid angle element $d\Omega$ perpendicular to dA. From (4.48, 50), we have:

$$\text{Emission/s} = j_\nu \, dA \, ds \, d\Omega \, d\nu \quad \text{and}$$
$$\text{Absorption/s} = \kappa_\nu \, ds \, B_\nu(T) \, dA \, d\Omega \, d\nu. \quad (4.68)$$

Setting the two quantities equal, we obtain Kirchhoff's Law:

$$j_\nu = \kappa_\nu B_\nu(T). \quad (4.69)$$

This important result means that in a state of thermodynamic equilibrium, the ratio of the emission coefficient j_ν to the absorption coefficient κ_ν is the same universal function $B_\nu(T)$ of ν and T as also describes the intensity of the cavity radiation. For the source function in the radiation transport equation (4.56), we then find $S_\nu = j_\nu/\kappa_\nu = B_\nu(T)$.

The thermal emission of a layer of thickness s with the overall optical thickness τ_ν, in which T (and thus also B_ν) is supposed to be constant, is then found from (4.60):

$$I_\nu(s) = B_\nu(T)(1 - e^{-\tau_\nu}) \quad (4.70)$$

$$\simeq \begin{cases} \tau_\nu B_\nu(T) & \text{for } \tau_\nu \ll 1 \\ B_\nu(T) & \text{for } \tau_\nu \gg 1 \end{cases}.$$

In the optically thick case, the emission thus approaches that predicted by the Kirchhoff–Planck function.

If the matter in the layer is not in thermodynamic equilibrium, then Kirchhoff's law cannot be applied to the radiation transport equation for the resulting *nonthermal* radiation, but the solution (4.60) remains valid if we replace the Kirchhoff–Planck function by the source function $S_\nu = j_\nu/\kappa_\nu \neq B_\nu(T)$. To be sure, its calculation based on the individual processes of interaction of the radiation with the material in the plasma requires a considerably greater effort (Sect. 4.5).

The Black-body Radiation Field. For the isotropic radiation field of a black body, the intensity and the average intensity are given by the Kirchhoff–Planck function, $I_\nu = J_\nu = B_\nu(T)$; therefore, we have

$$u_\nu = \frac{4\pi}{c} B_\nu(T) \,. \tag{4.71}$$

The *total radiation density* is then, from the Stefan-Boltzmann law (4.66), given by

$$u = \frac{4\pi}{c} B(T) = aT^4 \tag{4.72}$$

with the radiation constant

$$a = \frac{4\sigma}{c} = 7.56 \cdot 10^{-16} \,\mathrm{J\,m^{-3}\,K^{-4}} \,. \tag{4.73}$$

The total *photon density* (4.44) for a thermal radiation field is

$$n = \frac{4\pi}{c} \int_0^\infty \frac{B_\nu(T)}{h\nu} \,\mathrm{d}\nu = bT^3 \tag{4.74}$$

where

$$b = 0.370 \frac{a}{k} = 2.02 \cdot 10^7 \,\mathrm{m^{-3}\,K^{-3}} \,. \tag{4.75}$$

We thus finally obtain the *mean energy* of the photons:

$$\langle h\nu \rangle = \frac{u}{n} = 2.70\,kT \,. \tag{4.76}$$

4.4 Matter in Thermodynamic Equilibrium

After having described the radiation field from matter in thermodynamic equilibrium by introducing the Kirchhoff–Planck function in the previous section, we now consider the most important fundamentals needed to describe a *gas* in thermodynamic equilibrium (TE) at a temperature T. This state is attained when the gas density is sufficiently high so that collisions between the particles are frequent. The assumption of global thermodynamic equilibrium – e.g. for a whole star – is, to be sure, not reasonable; but it is often applicable to small volume elements within the star. These in general are at different temperatures. We then speak of a *local thermodynamic equilibrium* (LTE).

4.4.1 Boltzmann Statistics

We consider an (ideal) gas at the temperature T having n particles of one type, e.g. atoms, per unit volume. These particles are in different states i with the energies E_i, which are composed of the excitation energy χ_s of their (internal) states s and their kinetic energies $E = mv^2/2 = p^2/(2m)$:

$$E_i = \chi_s + \frac{p^2}{2m} \,. \tag{4.77}$$

Here, m is the mass, v the velocity, and $p = mv$ the momentum of the particles. (We consider only nonrelativistic particles with $v \ll c$.) The excitation energy χ_s is measured from the ground state 0 of the particle, with $\chi_0 = 0$.

According to the basic principles of statistical thermodynamics developed by L. Boltzmann, for each state i the number of particles in the volume $\mathrm{d}V$ is given by

$$n_i \propto G_i \,\mathrm{e}^{-E_i/kT} \,. \tag{4.78}$$

The multiplicity or degeneracy of the state, i.e. the number of states with the same energy, is taken into account through the *statistical weight* G_i.

By the way, one can consider (4.78) to be a generalization of the barometric pressure formula, (3.18), which gives the density distribution of an isothermal atmosphere at the temperature T as a function of the altitude z: $n(z) \propto \exp(-mgz/(kT))$. Here, g is the acceleration of gravity; the quantity mgz is thus simply the potential energy of a particle at the altitude z above the surface. It corresponds to the energy E_i, which however in quantum mechanics, in contrast to classical statistics, may be quantized.

Since the excitation state s of a particle is independent of its motion p, the overall statistical weight G_i can be written as the product of the statistical weight g_s of the (inner) excitation state with the weight g_p for the translational motion. Statistical mechanics shows that g_p is equal to the number of quantum cells of size h^3 in phase space (configuration times momentum space), where h denotes Planck's constant. If we take the volume element in configuration space to be $\mathrm{d}V$ and in momentum space to be the spherical shell $4\pi p^2 \mathrm{d}p$, we find

$$G_i = g_s \cdot g_p = g_s \frac{4\pi p^2 \mathrm{d}p\,\mathrm{d}V}{h^3} \,. \tag{4.79}$$

For the Boltzmann formula (4.78), using (4.77, 79) we then obtain

$$n_i \propto g_s e^{-\chi_s/kT} \cdot \frac{4\pi p^2}{h^3} \exp\left(-\frac{p^2}{2mkT}\right) dp. \quad (4.80)$$

In this expression and in the following, we consider the particle number per unit volume, i.e. we set $dV = 1$. Integrating over momenta, we obtain the density n_s of all the particles in the excitation state s, independent of their state of motion:

$$n_s \propto g_s e^{-\chi_s/kT} \cdot \frac{(2\pi mkT)^{3/2}}{h^3}. \quad (4.81)$$

For the overall particle number $n = \sum_s n_s$, we then have

$$n \propto Q \cdot \frac{(2\pi mkT)^{3/2}}{h^3}. \quad (4.82)$$

The factor Q in this equation is the important (inner) *partition function*

$$Q = \sum_s g_s e^{-\chi_s/kT}. \quad (4.83)$$

Finally, for the *relative* occupation density of the state i, we find:

$$\frac{n_i}{n} = \frac{g_s}{Q} \cdot e^{-\chi_s/kT} \cdot \frac{4}{\sqrt{\pi}} p^2 (2mkT)^{-3/2}$$
$$\times \exp\left(-\frac{p^2}{2mkT}\right) dp. \quad (4.84)$$

4.4.2 The Velocity Distribution

Leaving the inner energy states out of consideration, the particles are distinguished from one another only through their kinetic energies $mv^2/2 = p^2/(2m)$.

If we now denote the particle number in a unit volume with momentum in the interval $(p, p+dp)$ by $n\Phi(p)dp$, where n is again the total number of all particles in the unit volume, then from (4.84) we find the distribution function $\Phi(p)$ of the momenta:

$$\Phi(p)dp = \frac{4}{\sqrt{\pi}} (2mkT)^{-3/2} p^2$$
$$\times \exp\left(-\frac{p^2}{2mkT}\right) dp \quad (4.85)$$

with the normalization $\int_0^\infty \Phi(p)dp = 1$.

The number $n\Phi(v)dv$ of particles with velocities $(v, v+dv)$ in the unit volume is then given by the *Maxwell-Boltzmann velocity distribution*:

$$\Phi(v)dv = \frac{4}{\sqrt{\pi}} \left(\frac{v}{v_0}\right)^2$$
$$\times \exp\left[-\left(\frac{v}{v_0}\right)^2\right] d\left(\frac{v}{v_0}\right), \quad (4.86)$$

with the most probable velocity

$$v_0 = \sqrt{\frac{2kT}{m}}. \quad (4.87)$$

If the gas consists of a mixture of different types of particles, then due to their differing masses m, their values of v_0 will also differ. In contrast, in thermodynamic equilibrium, all of the types of particles have the *same* temperature T.

4.4.3 Thermal Excitation

We now want to consider the inner energy states s of a particle, e.g. of an atom or ion. In a unit volume, there are all together n atoms, of which n_0 are in their ground state 0 with the energy $\chi_0 = 0$, as well as n_s atoms in excited states s with the excitation energies χ_s. From (4.81), for the occupation densities relative to the ground state, the *Boltzmann formula*

$$\frac{n_s}{n_0} = \frac{g_s}{g_0} e^{-\chi_s/kT} \quad (4.88)$$

holds. The factor for the translational energy cancels here. The statistical weights and excitation energies are calculated in the theory of atomic spectra, which we will discuss in Sect. 7.1.1.

If we refer n_s to the *total number* of all the atoms n, instead of the number in the ground state, we find from (4.82) that

$$\frac{n_s}{n} = \frac{g_s}{Q} \cdot e^{-\chi_s/kT}. \quad (4.89)$$

4.4.4 Thermal Ionization

The thermal excitation of the atoms into quantum states with higher and higher excitation energies χ_s as described by the Boltzmann formula (4.88) always leads finally to *ionization*. For this to occur, the atom must be excited by at least its ionization energy χ_{ion}, i.e. the energy which is just sufficient to remove an electron from the atom. Any additional energy becomes kinetic energy $mv^2/2$ which is carried off by the ejected electron.

How can we calculate n^0/n^+, i.e. the ratio of the numbers of singly ionized to neutral atoms in their respective ground states, or the *degree of ionization*? Clearly, this problem can be reduced to the calculation of the statistical weight of the ionized atom in its ground state *plus* its free electron. The former has the statistical weight Q^+; we do not need to calculate the part due to translational energy, since it cancels with that of the neutral atoms. For a free electron of mass m_e, the statistical weight or partition function is $Q_e = 2$, corresponding to the two possible spin orientations. In addition, for the electron's motion, the volume $(2\pi m_e kT)^{3/2}$ in momentum space, and – taking n_e free electrons per unit volume – the volume $1/n_e$ in configuration space must be taken into account. We thus find for the statistical weight of an ion plus a free electron

$$g = Q^+ \cdot 2 \cdot \frac{(2\pi m_e kT)^{3/2}}{h^3 n_e} . \tag{4.90}$$

Inserting this expression into the Boltzmann formula (4.89), we immediately obtain the *Saha equation* for ionization:

$$\frac{n^+}{n^0} n_e = \frac{Q^+}{Q^0} \cdot 2 \cdot \frac{(2\pi m_e kT)^{3/2}}{h^3} \cdot e^{-\chi_{\text{ion}}/kT} . \tag{4.91}$$

Analogous expressions hold for the transition from the r-th to the $(r+1)$-th ionization state, where the $(r+1)$-th electron is ejected with the ionization energy χ_r, independently of other ionization processes:

$$\frac{n^{(r+1)}}{n^{(r)}} n_e = \frac{Q^{(r+1)}}{Q^{(r)}} \cdot 2 \cdot \frac{(2\pi m_e kT)^{3/2}}{h^3} \cdot e^{-\chi_r/kT} . \tag{4.92}$$

Instead of the number of electrons per unit volume, n_e, the *electron pressure*, i.e. the partial pressure of the free electrons

$$P_e = n_e kT \tag{4.93}$$

could just as well be employed.

M. N. Saha originally derived his formula on the basis of thermodynamic considerations. One considers the process of ionization of an atom A^0 or of the recombination of the ions A^+ with an electron e^- as a *chemical reaction*, in which the reaction

$$A^0 \rightleftharpoons A^+ + e^- \tag{4.94}$$

occurs with equal frequency in both directions. If we limit ourselves to sufficiently low pressures, then the number of recombination processes ($\leftarrow$) will be proportional to the number of collisions between ions and electrons, i.e. $\propto n^+ n_e$. The number of ionization processes ($\rightarrow$) is proportional to the density of the neutral atoms n^0. The proportionality factors depend only on the temperature. Thus, we can understand the form

$$\frac{n^+ \cdot n_e}{n^0} = f(T) \tag{4.95}$$

of the ionization equation (4.91), which can be seen as a special case of the law of mass action (Sect. 4.4.5).

Applications of the Boltzmann and Saha formulas will be given in Sect. 7.1.2 in connection with the quantitative analysis of stellar spectra.

4.4.5 The Law of Mass Action

We shall give here the law of mass action of C. M. Guldberg and P. Waage only for the simple reaction

$$C \rightleftharpoons A + B \tag{4.96}$$

which describes e.g. the *dissociation equilibrium* of a diatomic molecule AB ($=C$):

$$\frac{n_A n_B}{n_C} = K(T) . \tag{4.97}$$

Using (4.82), we can calculate the "equilibrium constant" $K(T)$, obtaining

$$\frac{n_A n_B}{n_C} = \frac{Q_A Q_B}{Q_C} \cdot \frac{(2\pi \mu kT)^{3/2}}{h^3} \cdot e^{-\chi_{\text{diss}}/kT} , \tag{4.98}$$

where χ_{diss} is the dissociation energy and

$$\mu = \frac{m_A m_B}{m_A + m_B} \tag{4.99}$$

is the reduced mass (2.55), using the fact that $m_C = m_A + m_B$.

For the special case of ionization (4.94), we obtain the Saha equation (4.91). Here, $\mu \simeq m_e$, since the electron's mass m_e is small compared to the mass of the atom or the ion.

4.5 The Interaction of Radiation with Matter

In connection with the radiation transport equation in Sect. 4.3.2, we have already met up with phenomenological or macroscopic cross sections in the form of the absorption and emission coefficients κ_ν and j_ν. Now, we will systematically consider the interaction of radiation with matter on the basis of *atomic* processes; its strength will be described by an *atomic cross section* (Sects. 4.5.1, 2). After that, we collect and summarize the formal relations which allow us to calculate the important case of kinetic equilibrium (Sect. 4.5.3). We then give an overview of the fundamental atomic processes in Sect. 4.5.4, followed by a discussion of the extinction and emission coefficients for the radiation transport equation (Sect. 4.5.5). Finally, we shall discuss the interaction of radiation with matter in the high–energy region, which is important for X-ray and gamma astronomy (Sect. 4.5.6).

The basic concepts presented in this Sect. 4.5 will find numerous applications in the discussion of astronomical instruments and the various objects in the cosmos.

4.5.1 Mean Free Paths

We begin by considering a simple *reaction*, the "binary collision":

$$a + b \rightarrow X, \qquad (4.100)$$

in which particles of type a with a relative velocity v collide with particles of type b. In this interaction, particle a is supposed to be "annihilated"; we are initially not concerned with the reaction products X. We consider photons also to be particles in this sense, whose relative velocity is equal to the speed of light c.

Now let τ_a be the mean *time between collisions* of a or the mean lifetime with respect to the reaction (4.100); then the probability of a reaction in the time dt is given by dt/τ_a (insofar as $dt \ll \tau_a$). We could equally well describe the interaction in terms of the mean *free path*

$$l_a = v\tau_a \qquad (4.101)$$

instead of using τ_a, so that the probability of a collision in the time dt or along the path $ds = v\,dt$ is equal to ds/l_a.

We now examine a current of particles $F_a = n_a v$, i.e. the number of particles of type a which passes through a unit surface area per unit time (here n_a is the number of particles in a unit volume). This quantity decreases along a path ds in the target containing a particle density n_b of particles of type b, depending on the number of collisions along the path ds (or in the volume given by ds times the unit surface area), by an amount

$$dF_a = F_a \cdot \frac{ds}{l_a}. \qquad (4.102)$$

On integrating this expression, we find that the particle current decreases exponentially with s from its initial value $F_{a,0}$:

$$F_a = F_{a,0}\,e^{-s/l_a},$$
$$F_{a,0} - F_a = F_{a,0}\left(1 - e^{-s/l_a}\right). \qquad (4.103)$$

4.5.2 Interaction Cross Section and Reaction Rate

How can we now obtain l_a or τ_a for a particular case from atomic physics? To this end, we first introduce the *atomic interaction cross section* σ, with the dimensions of a surface area. We can imagine σ to represent a "target" or a "barrier disk", which is attached to each of the target particles of type b. σ is defined in such a way that a reaction is presumed to occur when a particle of type a hits this disk. For each incoming particle, the fraction $n_b \sigma\,ds$ of the area 1 is "blocked" in its flight along the path ds. Then the probability of a collision, ds/l_a, is equal to $n_b \sigma\,ds$ and thus

$$l_a = v\tau_a = \frac{1}{n_b \sigma}. \qquad (4.104)$$

The *rate* $\mathcal{P}_a$ [s^{-1}] of the reaction with respect to one particle of type a is then given by

$$\mathcal{P}_a = \frac{1}{\tau_a} = \frac{v}{l_a} = n_b \sigma v \qquad (4.105)$$

and the transition probability in the time dt is $\mathcal{P}_a\,dt$.

If we exchange the projectile and target particles in the reaction (4.100), then v and σ naturally remain unchanged, but the mean free path l_b and the reaction rate $\mathcal{P}_b$ for particles of type b are modified as a result of the different densities in the target. We then have

$$\frac{l_a}{n_a} = \frac{l_b}{n_b} \quad \text{and} \quad \frac{n_a}{\tau_a} = \frac{n_b}{\tau_b} = n_a n_b \sigma v. \qquad (4.106)$$

On the other hand, the number of reactions which occur in a unit volume per unit time,

$$\mathcal{P} = n_a \mathcal{P}_a = n_a n_b \sigma v \quad (4.107)$$

is symmetrical with respect to a and b.

In a reaction of the kind described by (4.100), there are often several different possibilities for the product X. For example, the interaction of a photon γ with a neutral atom A^0 can lead to the scattering of the photons (into a different direction), to an excitation (A^*) of the atom through absorption of the photon, or to photoionization:

$$\gamma + A^0 = \begin{cases} A^0 + \gamma' \\ A^* \\ A^+ + e^- \end{cases} \quad . \quad (4.108)$$

We shall describe the probability of each output channel i in terms of a *partial* interaction cross section σ_i, where the sum over all the possible reactions again yields the *total* (integral) interaction cross section, $\sigma = \sum_i \sigma_i$.

Often, one requires of a reaction an exact separation in terms of the properties of the reaction products, e.g. when the number of particles scattered into a particular direction or into a particular energy range is of interest. For this purpose, we introduce *differential* interaction cross sections, such as

$$\frac{d\sigma_i}{d\Omega} \, [m^2 \, sr^{-1}] \, , \quad \frac{d\sigma_i}{dv} \, [m^2 \, Hz^{-1}]$$

or also double-differential cross sections, such as

$$\frac{d^2\sigma_i}{d\Omega \, dv} \, [m^2 \, sr^{-1} \, Hz^{-1}] \, ,$$

which give the number of reactions per unit range of the parameters (solid angle Ω or frequency v). The integral over the whole range of the parameters then again yields the partial interaction cross section, e.g.

$$\int_0^\infty \oint \frac{d^2\sigma_i}{d\Omega \, dv} \, d\Omega \, dv = \sigma_i \, .$$

Instead of the *atomic* interaction cross section – i.e. calculated per particle – $\sigma = \sigma_{at}$, which we now denote for clarity with the index "at", we can also introduce cross sections referred to a unit *volume* or a unit *mass*:

$$\sigma_V \, [m^2 \, m^{-3} = m^{-1}] = n \cdot \sigma_{at} \, ,$$

$$\sigma_M \, [m^2 \, kg^{-1}] = \frac{\sigma_V}{\varrho} = \frac{n}{\varrho} \cdot \sigma_{at} \, , \quad (4.109)$$

where $\varrho \, [kg \, m^{-3}]$ is the density of the material. We have for example already met up with the frequency dependent absorption coefficient $\kappa_v \equiv \kappa_{v,V} [m^{-1}] = \kappa_{v,at} n$ in the radiation transport equation (4.52).

The mean free path l can also be expressed in terms of the *mass column* ϱl [kg m^{-2}] which must be traversed on the average before a reaction takes place.

4.5.3 Collision and Radiation Processes: Kinetic Equations

In general, the interaction cross section of a reaction depends on the relative velocity v or on the corresponding kinetic energy, i.e. $\sigma = \sigma(v)$. We therefore have to take the distribution $\tilde{\Phi}(v)$ of relative particle velocities into account.

In thermodynamic equilibrium, the velocities follow the Maxwell-Boltzmann distribution $\Phi(v)$ (4.86). It can be shown that this distribution also holds for the *relative* velocities if the particle mass m in the expression for the most probable velocity (4.87) is replaced by the reduced mass, $m_a m_b/(m_a + m_b)$.

In the important case that *electrons* interact with atoms, owing to the small mass m_e of the electrons relative to that of the atoms, the relative velocity is practically equal to the velocity of the electrons and the reduced mass is equal to m_e.

After taking the velocity distribution into account, the rate (4.105) becomes

$$\mathcal{P}_a = n_b \int_0^\infty \sigma(v) v \Phi(v) \, dv = n_b \langle \sigma v \rangle \quad (4.110)$$

and the number of reactions per unit volume and unit time, (4.107), is equal to

$$\mathcal{P} = n_a n_b \int_0^\infty \sigma(v) v \Phi(v) \, dv = n_a n_b \langle \sigma v \rangle \, , \quad (4.111)$$

where the angular brackets denote the *thermal* average of the product of $\sigma(v)$ times v.

In most astrophysical applications, an *electron temperature* is established through collisions of the electrons among themselves, so that for the electron component, there is a thermodynamic equilibrium corresponding to T_e. The atoms also usually reach equilibrium with the electrons, so that their kinetic temperature is likewise given by T_e.

The description of the excitation and ionization states of the atoms by the Boltzmann or the Saha formulas (4.88, 91) is based on the validity of the assumption of thermodynamic equilibrium. While in the interiors of stars and for the most part also in their atmospheres (photospheres), the densities are sufficiently high so that equilibrium can be established through mutual collisions of the particles, this is not the case for e.g. the outermost, thin atmospheric layers and in particular also for the gaseous nebulae and the interstellar medium. Here we must return to the detailed *atomic elementary processes* and carry out a calculation in the "non-LTE", i.e. without the assumption of local thermodynamic equilibrium. In such cases, one preferably speaks of a *kinetic* or *statistical* equilibrium.

Usually, we can assume that all the atomic processes take place so frequently that the plasma can be considered to be *stationary*. Then the population density n_i for *every* energy state i of an arbitrary particle averaged over time is constant,

$$\frac{\partial n_i}{\partial t} = 0 \,. \tag{4.112}$$

The kinetic equilibrium is then determined by a system of *rate equations*, which describes the *balance* of the elementary processes leading to all other states $\alpha' = \{a', i'\}$, and from other states $\alpha'' = \{a'', i''\}$ for each energy state i of a particle of type a, which we denote for brevity as $\alpha = \{a, i\}$:

$$\frac{\partial n_\alpha}{\partial t} = -n_\alpha \sum_{\alpha'} \mathcal{P}_{\alpha \to \alpha'} + \sum_{\alpha''} n_{\alpha''} \mathcal{P}_{\alpha'' \to \alpha} = 0 \tag{4.113}$$

or, using (4.107),

$$\sum_{\alpha'} \mathcal{P}_{\alpha \to \alpha'} = \sum_{\alpha''} \mathcal{P}_{\alpha'' \to \alpha} \,. \tag{4.114}$$

The atomic processes include on the one hand *collision processes*, whose rate we denote by $\mathcal{C}$; on the other hand *radiation processes* with the rate $\mathcal{R}$, so that we can assume that

$$\mathcal{P} = \mathcal{C} + \mathcal{R}, \quad \mathcal{P} = \mathcal{C} + \mathcal{R}\,. \tag{4.115}$$

In what follows, we limit ourselves to *excitation* and *ionization* and their reverse processes in atoms and ions

as well as the important case that the *electrons* predominate in the collision processes. The collision rates then have the form

$$\mathcal{C} = n_e \int_0^\infty \sigma(v) v \Phi_e(v) \, dv \,, \tag{4.116}$$

and the radiation rates take the form

$$\mathcal{R} = n_\gamma \int_0^\infty \sigma(v) c \Phi_\gamma(v) \, dv \,, \tag{4.117}$$

where n_e refers to the electron density, n_γ to the photon density, and $\Phi_\gamma(v)$ to the distribution function of the photons over frequency, normalized to one. From (4.43, 44), the photon density

$$n_\gamma = \frac{4\pi}{c} \int_0^\infty \frac{J_\nu}{h\nu} \, dv \tag{4.118}$$

and the number of photons per unit volume and frequency interval $(\nu, \nu + d\nu)$

$$n_\gamma \Phi_\gamma(\nu) \, d\nu = n_\nu \, d\nu = \frac{4\pi}{c} \frac{J_\nu}{h\nu} \, d\nu \tag{4.119}$$

can be expressed in terms of the average radiation intensity J_ν.

4.5.4 Elementary Atomic Processes

The possible energy levels of an atom are represented graphically in an energy-level or *Grotrian diagram* (Fig. 4.7). Every level of energy E is characterized by several quantum numbers (Sect. 7.1.1), which we will represent symbolically here only by a *single* index $i, j, \ldots$. We set the energy of the ground state of the atom equal to 0. The ionization energy, i.e. the minimum energy which is required to remove an electron from an atom (in its ground state) we denote by χ_{ion}. We discriminate between:

discrete energy values $E_i < \chi_{\text{ion}}$, corresponding to the bound or elliptical orbits of the electrons in Bohr's model of the atom; and

continuous energy values $E_\kappa > \chi_{\text{ion}}$, corresponding to the free or hyperbolic orbits of the electrons in Bohr's model. At a large distance from the atom, such an electron has *only* kinetic energy $m_e v^2/2$, where m_e is its mass and v its velocity.

In a *transition* between two energy levels E_i and E_j, a photon of energy

$$h\nu = \hbar\omega = |E_i - E_j| \qquad (4.120)$$

is absorbed (↑) or emitted (↓). Here, h or $\hbar = h/(2\pi)$ refers to Planck's constant and $\omega = 2\pi\nu$ is the angular frequency. In the following, we denote in particular the bound states of an atom by χ (as we have already done in Sect. 4.4.1), and the free states by E.

We can divide the transitions in an atom in a natural way into the following three groups:

1. $\chi_i, \chi_j < \chi_{\text{ion}}$: *bound-bound* transitions, in which a *spectral line* is absorbed or emitted.
2. $\chi_i < \chi_{\text{ion}}, E > \chi_{\text{ion}}$: *free-bound* transitions. At the *absorption edge* where $h\nu_i = \chi_{\text{ion}} - \chi_i$, the free-bound absorption $\nu > \nu_i$ accompanied by ejection of a photoelectron with the kinetic energy $m_e v^2/2 = h\nu - (\chi_{\text{ion}} - \chi_i)$ begins. The atom is ionized in this process.
   The inverse process is *binary recombination*, i.e. the capture of a free electron of energy $m_e v^2/2$, accompanied by the emission of a photon

   $$h\nu = \tfrac{1}{2}m_e v^2 + (\chi_{\text{ion}} - \chi_i) \,. \qquad (4.121)$$

3. $E', E'' > \chi_{\text{ion}}$: *free-free* transitions, in which a photon $h\nu = |E' - E''|$ is absorbed or emitted. The free electron gains or loses the corresponding amount of kinetic energy in the interaction with the atom or ion. In the latter case, we also refer to *bremsstrahlung*.

Line Transitions. We first consider the *bound-bound* transitions. According to A. Einstein, we can describe the interaction of the radiation with matter in terms of the following processes:

1. *Spontaneous emission* of photons with the probability A_{ji}, associated with a transition from an excited state j to a lower-lying state i. The energy emitted per unit time in all directions (with equal probabilities) is $h\nu A_{ji}$.
2. *Stimulated* or *induced emission*. An atom in an excited state j is caused by a photon of frequency ν to emit a photon of the same frequency and direction with a probability proportional to the intensity I_ν of the radiation field, $B_{ji}I_\nu$, per unit time and unit solid angle.
3. *Absorption* of a photon with the probability $B_{ij}I_\nu$ (the inverse process to stimulated emission), associated with the excitation of an atom from the state i into a higher state j. The total energy absorbed from the spectral line per unit time from a unit solid angle is $h\nu B_{ij} I_\nu/(4\pi)$.

The following relations hold between the *Einstein coefficients* A_{ji}, B_{ji} and B_{ij}:

$$A_{ji} = \frac{2h\nu^3}{c^2} B_{ji}\,, \qquad g_i B_{ij} = g_j B_{ji}\,, \qquad (4.122)$$

where g_i and g_j are again the statistical weights of the levels i and j.

We write the frequency-dependent atomic *absorption cross section* as $\sigma_{ij}(\nu) = \hat{\sigma}\phi(\nu)$ with the *profile function* (normalized to one) $\phi(\nu)$; then we find the following relation for the total absorption in the spectral line:

$$\hat{\sigma} = \int_{\text{line}} \sigma_{ij}(\nu)\,\mathrm{d}\nu = \frac{h\nu}{4\pi} B_{ij}\,. \qquad (4.123)$$

We shall discuss the precise form of the profile function, which is resonant at $\nu = |\chi_j - \chi_i|/h$, in Sect. 7.1.4.

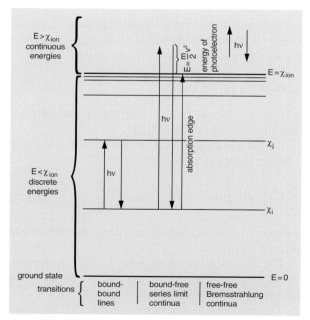

Fig. 4.7. Energy level or Grotrian diagram for an atom (schematic) and transitions (absorption ↑, emission ↓)

An equivalent description of spectral-line emission makes use of the (dimensionless) *oscillator strength* $f \equiv f_{ij}$, which is the "effective number of oscillators" and represents the quantum-mechanical extension of the expression already obtained in the framework of classical electron theory. It attempts to relate the spectral lines to harmonic electron-oscillators of the corresponding frequencies. The f value is related to Einstein's B coefficient by the equation:

$$\int_{\text{line}} \sigma_{ij}(\nu)\, d\nu = \frac{h\nu}{4\pi} B_{ij} = \frac{1}{4\pi\varepsilon_0} \frac{\pi e^2}{m_e c} f, \quad (4.124)$$

where e is the electronic charge and ε_0 the electric field constant. The oscillator strength f can thus also be expressed in terms of the A coefficient using (4.122):

$$g_j A_{ji} = \frac{1}{4\pi\varepsilon_0} \frac{8\pi^2 e^2 \nu^2}{m_e c^3} g_i f. \quad (4.125)$$

The *rates* corresponding to the bound-bound processes are, according to (4.117),

$$R_{ij} = B_{ij} J_\nu,$$
$$R_{ji} = A_{ji} + B_{ji} J_\nu = A_{ji}\left(1 + \frac{c^2}{2h\nu^3} J_\nu\right), \quad (4.126)$$

where for simplicity we have given only the result averaged over all angles.

Along with the bound-bound radiation processes, particle collisions also cause ("radiationless") transitions between the energy states of an atom. We limit ourselves here to collisions with *electrons*, which predominate e.g. in stellar atmospheres at temperatures above some 10^3 K. In other cases, collisions with protons, hydrogen atoms and molecules and other particles would also have to be considered.

For the *excitation* of a level of energy χ_j by electron impact from a lower level χ_i, the electron requires a minimum energy $\chi_j - \chi_i = m_e v_0^2/2$ or a velocity v_0 relative to the atom. The excitation cross section σ_{ij} attains characteristic values of the order of $\pi a_0^2 = 8.80 \cdot 10^{-21}$ m^2 (a_0 is the 1st Bohr radius); it depends in detail on the relative velocity v. The *collision rate* is found from (4.116) to be

$$C_{ij} = n_e \int_{v_0}^{\infty} \sigma_{ij}(v) v \Phi_e(v)\, dv. \quad (4.127)$$

According to the "principle of detailed balance", which cannot be derived here, the rate of each inverse process is determined by that of the corresponding forward process. For example, in the case of bound–bound radiation processes, this yields the relations (4.122).

From the principle of detailed balance, we now obtain the electron collision rate C_{ji} for *deexcitation* $j \to i$:

$$g_j C_{ji} \equiv n_e \int_0^{\infty} g_j \sigma_{ji}(v) v \Phi_e(v)\, dv$$
$$= g_i C_{ij} \exp[-(\chi_j - \chi_i)/kT]. \quad (4.128)$$

The rate for j to i, if we take collisions *and* radiation into account, is $A_{ji} + B_{ji} J_\nu + C_{ji}$.

Ionization. The ionization of an atom or ion from a bound state i, in which the electron is ejected into a free state κ, may be produced by radiation (bound-free transition). Let $\sigma_{i\kappa}(\nu)$ be the interaction cross section for this *photoionization* process; then from (4.117, 119), the corresponding rate is

$$R_{i\kappa} = \int_{\nu_i}^{\infty} \sigma_{i\kappa}(\nu) c\, n_\nu\, d\nu = 4\pi \int_{\nu_i}^{\infty} \sigma_{i\kappa}(\nu) \frac{J_\nu}{h\nu}\, d\nu. \quad (4.129)$$

For hydrogen-like atoms with a nuclear charge Z, $\sigma_{i\kappa}$ decreases from the ionization limit $h\nu_i = \chi_{\text{ion}} - \chi_i$ towards higher frequencies in a way proportional to ν^{-3}:

$$\sigma_{i\kappa}(\nu) = \sigma_i \left(\frac{\nu}{\nu_i}\right)^{-3} \quad (4.130)$$

with $\sigma_i = 7.91 \cdot 10^{-22} n/Z^2$ [m^2] (n is the principal quantum number; cf. Sect. 7.1.1).

On the other hand, an atom can also be ionized by *electron impact* at a rate given by

$$C_{i\kappa} = n_e \int_{v_i}^{\infty} \sigma_{i\kappa}(v) v \Phi_e(v)\, dv, \quad (4.131)$$

where $\sigma_{i\kappa}$ is the collisional ionization cross section and $m_e v_i^2/2 = h\nu_i$.

The cross sections or the rates for the inverse processes of *recombination* can once again be derived from the principle of detailed balance.

The interplay of the various elementary processes will be discussed in terms of a simplified atomic model in Sect. 7.1.2.

Scattering. Another important elementary process is the scattering of a photon by an atom or electron (which is described quantum mechanically as a two-photon process). For astrophysical applications, Rayleigh scattering by abundant atoms or molecules, as well as Compton or Thomson scattering by electrons are significant.

The interaction cross section for scattering of photons by *free* electrons is, for the case of relatively low photon energies ($h\nu \ll m_e c^2$, m_e the rest mass of the electron), given by the frequency-independent *Thomson scattering cross section*

$$\sigma_T = \frac{8\pi}{3} r_0^2 = \frac{8\pi}{3} \left(\frac{e^2}{4\pi\varepsilon_0 m_e c^2} \right)^2$$
$$= 6.65 \cdot 10^{-29} \text{ m}^2 \; ; \qquad (4.132)$$

Here, r_0 is the classical electron radius. Referred to the unit mass of a fully ionized hydrogen plasma, the Thomson cross section is $\sigma_T/m_H = 0.04$ m² kg⁻¹ (m_H: mass of the H atom).

4.5.5 Extinction and Emission Coefficients

The extinction, absorption, scattering and emission coefficients $k_\nu = \kappa_\nu + \sigma_\nu$ and j_ν which occur in the radiation transport equation (4.52) can be obtained from the atomic interaction cross sections for the associated radiation processes by multiplying them with the density of the energy levels in question and summing over all the processes which contribute at the frequency considered.

To calculate the radiation transport, it is usual to include *stimulated* emission, which is proportional to the intensity I_ν, together with the absorption in a single term $\kappa_\nu I_\nu$ in the radiation transport equation, i.e. to treat it as a "negative absorption". Then, if we make the simplification of assuming the same profile function for emission and absorption, the *line absorption coefficient* becomes:

$$\kappa_\nu = \frac{h\nu}{4\pi} (B_{ij} n_i - B_{ji} n_j) \phi(\nu) \qquad (4.133)$$

or, with (4.122, 125),

$$\kappa_\nu = \frac{h\nu}{4\pi} B_{ij} n_i \phi(\nu) \left(1 - \frac{n_j/g_j}{n_i/g_i} \right)$$
$$= \frac{1}{4\pi\varepsilon_0} \frac{\pi e^2}{m_e c} f n_i \phi(\nu) \left(1 - \frac{n_j/g_j}{n_i/g_i} \right) . \qquad (4.134)$$

For the description of *masers*, which also occur in cosmic sources (e.g. OH masers, Sect. 10.2.3), and lasers, a more precise treatment of the stimulated radiation processes would be essential; we cannot include it here, however.

The factor $[1 - (n_j/g_j)/(n_i/g_i)]$ for stimulated emission in thermodynamic equilibrium, according to the Boltzmann formula (4.88) becomes $[1 - \exp(h\nu/kT)]$. It thus produces a strong reduction of the absorption, particularly in the Rayleigh–Jeans region ($h\nu/kT \ll 1$), by a factor of $h\nu/kT$.

The contribution of a spectral line to the emission coefficient is, finally, given by

$$j_\nu = \frac{h\nu}{4\pi} A_{ji} n_j \phi(\nu) . \qquad (4.135)$$

We will dispense with writing down the contributions from photoionization and its inverse process to the absorption and emission coefficients. However, we should mention the contribution of scattering by free electrons with a density of n_e to the *scattering coefficient*: $\sigma_\nu = \sigma_T n_e$.

4.5.6 Energetic Photons and Particles

In the range of high energies, additional processes occur in the interaction of radiation with matter; they are important for X-ray and gamma-ray astronomy and for the understanding of cosmic radiation.

High-Energy Photons. The energy of photons in the optical and ultraviolet regions suffices only to eject electrons from the outer (valence) shells of atoms or ions; in contrast, *X-ray photoionization*, i.e. at energies above about 100 eV or wavelengths below about 10 nm, proceeds mainly by emission of electrons from the *inner* shells (core electrons). Also in contrast to optical spectroscopy, in the X-ray region, except from highly ionized atoms, absorption lines (i.e. transitions between two bound energy states) are seldom observed, since the less strongly-bound states are in general occupied by electrons. Instead, *ionization edges* are characteristic of X-ray absorption spectra; at an edge, the absorption cross section increases abruptly at a frequency ν_n or a wavelength λ_n, corresponding to the binding energy of an electron in a shell of principal quantum number n:

$$h\nu_n = \frac{hc}{\lambda_n} = E_n, \quad (4.136)$$

and then decreases towards higher energies at a rate proportional to E^{-3} or λ^3. E_n is proportional to Z^2/n^2, where Z is the nuclear charge (atomic number) (4.130).

The proportionality between $\sqrt{\nu}$ and Z for corresponding edges of different elements in the X-ray region was first discovered by H. Moseley in 1913.

For the innermost electron of the socalled *K shell*, the ionization energy is given by:

$$E_K = 13.6 \, Z_{\text{eff}}^2 \, [\text{eV}], \quad (4.137)$$

where $Z_{\text{eff}} = Z - s$ is the effective nuclear charge, taking account of the shielding s due to the other electrons. For example, the binding energy in the K shell of hydrogen ($Z=1$) is 13.6 eV; for oxygen ($Z=8$), it is 530 eV; for silicon ($Z=14$), it is 1.84 keV; and for iron ($Z=26$), it is 7.11 keV. We shall not discuss the fine structure of the edges here; it reflects the different "subshells" of the atom. In Fig. 4.8, we show as an example the absorption cross sections in the X-ray region for matter of a composition corresponding to that of the interstellar medium (Sect. 10.2.1).

As long as the energy $h\nu$ of the photons is small relative to the rest energy of an electron, $m_e c^2 = 511$ keV ($m_e = 9.11 \cdot 10^{-31}$ kg is the rest mass of the electron), they give up their energy mainly through photoionization processes. At higher energies, i.e. in the gamma-ray region, another process becomes important: *Compton scattering* of photons by electrons, $\gamma + e^- \rightarrow \gamma' + e^-$. The energy loss on scattering corresponds to a change in wavelength by the amount:

$$\Delta\lambda = \frac{2h}{m_e c} \sin^2 \frac{\Phi}{2}, \quad (4.138)$$

where Φ is the scattering angle of the photon. $\Lambda = h/(m_e c) = 2.43 \cdot 10^{-12}$ m is the Compton wavelength of the electron. It is, by the way, the wavelength corresponding to the rest energy of the electron ($m_e c^2 = h\nu = hc/\Lambda$).

The cross section for Compton scattering in the case of low photon energies, $h\nu \ll m_e c^2$, is given by the *Thomson scattering coefficient* σ_T (4.132). Above $m_e c^2$, $\sigma(\gamma)$ decreases from σ_T roughly in inverse proportion to

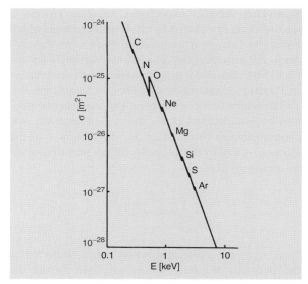

Fig. 4.8. Photoionization in the X-ray region for the elemental mixture of the interstellar medium: dependence of the absorption cross section σ per hydrogen particle on the photon energy $E = h\nu$. The K edges of the most abundant elements are shown; the strongest edge at 0.53 keV is that of oxygen. The absorption coefficient is $\kappa = \sigma \cdot n_H$ with n_H the number density of hydrogen particles (H). (With the kind permission of Cambridge University Press, Cambridge)

the photon energy $h\nu$, as described by the *Klein-Nishina* scattering formula; in the limit $h\nu \gg m_e c^2$, it is given by:

$$\sigma(\gamma) = \frac{3}{8} \frac{m_e c^2}{h\nu} \left(\ln \frac{2h\nu}{m_e c^2} + \frac{1}{2} \right) \cdot \sigma_T . \quad (4.139)$$

If the energy of the photon becomes greater than $2m_e c^2 = 1.02$ MeV, then *pair production* in the Coulomb field of a charge Z, $\gamma \rightarrow e^+ + e^-$, becomes possible. The cross section for this process is of the order of

$$\sigma(e^\pm) \simeq \alpha Z^2 \sigma_T, \quad (4.140)$$

where $\alpha = 1/137.04$ is the fine structure constant. Above some 10 MeV, $\sigma(e^\pm)$ thus becomes larger than the cross section for Compton scattering and pair production is then the dominant process in the interaction of energetic photons with matter. The electron-positron pair takes on the direction of the gamma quantum which produced it, within an angle of the order of $m_e c^2/(h\nu)$.

Energetic Particles. Like the photons, both *protons* and heavy nuclei as well as electrons *ionize* the atoms of the material and thereby lose their energy. They readily destroy the weak bonds within molecules and crystal lattices. The energy loss of an ionizing particle of rest mass m and charge z with the energy $E = \gamma mc^2$ (4.21) or the velocity v (γ: Lorentz factor) on passing through a length x of a material of density ϱ (with atoms having charge Z and atomic number A) is

$$\frac{dE}{d(\varrho x)} = -\frac{z^2}{v^2} f(v) \frac{Z}{A}, \quad (4.141)$$

where $f(v)$ is a function which varies weakly with v, and the relation between velocity and energy for relativistic particles is given by (4.24, 25). The material properties thus enter only weakly (Z/A!). The energy loss at energies of $E \simeq mc^2$ is inversely proportional to E; at higher energies, it increases only slowly with E. The minimum energy loss, at $E \simeq mc^2$, is equal to $0.2z^2$ [MeV] for a path with column thickness of $1\,\text{kg}\,\text{m}^{-2}$. If the ejected electrons gain sufficiently high energies, they can initiate secondary ionization processes.

The Origin of Cosmic Gamma Radiation. A thermal source of these energetic photons is hardly possible, since it would require temperatures above some 10^9 K ($kT \geq 0.5$ MeV), which – apart from the earliest phases of the Universe and perhaps some late phases of stellar evolution – do not occur. Instead, we must consider *nonthermal* sources. In particular, the interactions of relativistic electrons (of energies above $m_e c^2 = 0.511$ MeV) and energetic protons, which become relativistic above 938 MeV, can effectively produce gamma quanta.

1) Gamma radiation can be emitted as *bremsstrahlung* or free-free radiation (Sect. 4.5.4) by energetic electrons in the Coulomb fields of charged particles (of charge Z, atomic number A). In this way, photons with energies of the order of the electron energy E ($\gg m_e c^2$) are emitted; on the one hand, the interaction cross section decreases in a way which varies roughly in inverse proportion to the energy of the bremsstrahlung photon. On the other hand, it is, like that for pair production by a gamma quantum (4.140), proportional to $\alpha Z^2 \sigma_T$. The corresponding energy loss rate (4.105) is then proportional to $\alpha Z^2 \sigma_T c n$, where $n \propto \varrho/A$ is the particle density. Relative to ϱx, the mass column traversed, we find for the energy loss:

$$\frac{dE}{d(\varrho x)} = -\frac{E}{\xi_0}. \quad (4.142)$$

Here, $\xi_0 \propto A(\alpha Z^2 \sigma_T)^{-1}$ is the "radiation length", i.e. that distance over which the initial energy E decreases by a factor of $1/e \simeq 0.37$.

When a relativistic electron passes through matter, there is a high probability that roughly half of its energy E will be converted to one or two gamma quanta. The radiation is emitted predominantly into an angular range given by $m_e c^2/E$ around the direction of travel of the electron. Since however a highly energetic gamma quantum in the field of a charge Z tends to create an electron-positron pair (see above), and both electrons and positrons again yield bremsstrahlung quanta, an *electromagnetic cascade* is produced in the material:

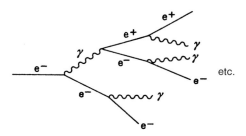

It eventually "runs down", when the energies of the gamma quanta are no longer high enough to initiate pair production.

2) By *inverse Compton scattering*, relativistic electrons can transfer a considerable portion of their energy E to low-energy photons and thus inject them into the X-ray and gamma-ray spectral regions. Photons with the mean energy $\overline{E}_\gamma$ receive on the average in this process an energy given by:

$$\overline{E}_{\gamma'} \simeq \frac{4}{3} \overline{E}_\gamma \left(\frac{E}{m_e c^2} \right)^2. \quad (4.143)$$

In this way, for example, electrons of energy $E \simeq 60$ GeV can transform the very low-energy photons of the cosmic 3 K-background radiation (Sect. 13.2.2), with energies $\overline{E}_\gamma \simeq 6 \cdot 10^{-4}$ eV, into gamma rays of energies in the range of 10 MeV. Inverse Compton scattering can, in the rest system of the electron, be considered as simple Thomson scattering with an interaction cross section σ_T, since the photon energy $E_\gamma E/m_e c^2$ "seen" by the electron is small compared to its own rest energy $m_e c^2$ over a wide range of gamma energies.

3) In the case of inelastic collisions of protons with protons, above a threshold energy of around 300 MeV, neutral *pions* (π mesons) are produced with an interaction cross section of about 10^{-30} m^2. The π^0 meson decays after only around $1.8 \cdot 10^{-16}$ s into two gamma quanta with energies of $m(\pi^0)c^2/2 = 67.5$ MeV each in its rest system. As a result of the velocity distribution of the relativistic pions, the gamma radiation shows a broad maximum at 67.5 MeV.

4) Whenever *positrons* are created, *annihilation* $e^+ + e^- \rightarrow \gamma + \gamma'$ also can occur. The *annihilation radiation* is emitted at an energy of $m_e c^2 = 0.511$ MeV.

5) The *synchrotron radiation* (Sect. 12.3.1), which is emitted by relativistic electrons moving in *magnetic fields*, can extend up to the gamma-ray region. According to (12.25), the electron energy and the strength of the magnetic field must be sufficiently large; e.g. in magnetic fields of $2 \cdot 10^{-10}$ T or $2 \cdot 10^{-6}$ G, typical of the interstellar medium, it would be necessary for the electrons to be extremely energetic, with ca. 10^{16} eV, in order for them to emit gamma radiation of energy 10 MeV as synchrotron radiation. This process thus plays no role in the production of galactic gamma radiation (Sect. 10.4.3). However, it is important in pulsars or neutron stars, with their extremely strong magnetic fields of order 10^6 to 10^8 T (Sect. 7.4.7).

6) Finally, in the low-energy gamma-ray region (≤ 10 MeV), one also observes the emission of nuclear *spectral lines*. These correspond to transitions from excited to lower-lying energy states in nuclei, which are emitted in radioactive decays or following excitation of the nucleus by energetic particles.

5. Astronomical and Astrophysical Instruments

Before we discuss the astrophysics of the Sun and the stars, then the galaxies etc., we give a compact summary in this chapter of the most important astronomical and astrophysical instruments. We start with telescopes and detectors for the optical and the neighboring ultraviolet spectral regions (Sect. 5.1). We then treat instruments for imaging and detecting in the radiofrequency and the infrared regions (Sect. 5.2). Finally, we introduce instruments for high-energy astronomy, the observation of cosmic rays and gamma rays (Sect. 5.3) as well as telescopes and detectors for the X-ray and extreme ultraviolet regions (Sect. 5.4).

More specialized instruments, e.g. those for observing the solar corona, neutrinos, or gravitational radiation, will be most practically described in connection with the corresponding astronomical objects.

At the close of this chapter, we will give a brief account of the role of computers in today's research in astronomy.

5.1 Telescopes and Detectors for the Optical and the Ultraviolet Regions

From the instrumental viewpoint, there is no difference in principle between the ultraviolet and the optical spectral regions, if we exclude the extreme ultraviolet ($\lambda \leq 100$ nm). Instruments intended for ultraviolet astronomy must, however, be designed for observations from space, since absorption by atmospheric ozone prevents observations from the ground for $\lambda \leq 300$ nm (Fig. II.1). In the laboratory, measurements at $\lambda \simeq 200$ nm (vacuum ultraviolet) must be carried out in vacuum owing to absorption by the air. Below $\lambda \simeq 100$ nm, the reflectivity of the usual mirror materials decreases sharply, so that here, as in the X-ray spectral region (Sect. 5.4), systems with grazing incidence mirrors must be used.

To begin, in Sect. 5.1.1 we treat the most important types of telescopes, the refractor and the reflector, with their focusing systems, as well as the Schmidt mirror. In the following Sect. 5.1.2, we discuss some basic concepts such as resolving power, light-gathering power, and the principle of the interferometer. After our discussion of conventional telescopes and interferometers, in Sect. 5.1.3 we introduce the principles of adaptive and active optics and their application to the construction of earthbound telescopes having light-collecting mirror surfaces with effective diameters of more than 8 m. Section 5.1.4 gives an overview of detectors used for observing the radiation collected by the telescope, from the photographic plate to modern high-sensitivity semiconductor detectors. As dispersive element for the spectral decomposition of light, either a prism or, more usually, a diffraction grating is used; the basic principles of the spectrograph will be treated in Sect. 5.1.5. Finally, in Sect. 5.1.6, we introduce some telescopes and their instrumentation designed for carrying out observations from space.

5.1.1 Conventional Telescopes

The principles of Galileo's telescope (1609) and Kepler's telescope (1611) are recalled in Fig. 5.1; in both, the magnifying power is determined by the ratio of the focal lengths of the objective and the ocular lenses. Galileo's arrangement yields an upright image and therefore became the prototype of opera glasses and binoculars. Kepler's telescope, in contrast, casts a real image in the focal plane, which for visual observations is viewed through the ocular, using it like a magnifying glass. This permits a reticle to be inserted in the common focal plane of the objective and the ocular and it can thus be used to fix angles precisely, e.g. on the meridian circle. If the reticle is extended to include a *micrometer scale*, among other things the relative positions of binary stars can be determined.

The disturbing colored borders (chromatic aberration) in the images of telescopes with simple lenses were eliminated by J. Dollond and others around 1758 by the invention of *achromatic lenses*. An achromatic positive

5. Astronomical and Astrophysical Instruments

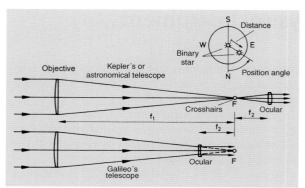

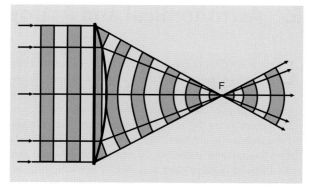

Fig. 5.1. Kepler's and Galileo's telescopes. The former uses a converging lens as ocular, the latter a diverging lens. F denotes the common focal point of the objective and the ocular lenses. The (magnifying) power V is equal to the ratio of the focal lengths of the objective, f_1, and of the ocular, f_2; in this figure, a ratio $V = f_1/f_2 = 5$ has been chosen. In Kepler's telescope, a crosshair or a micrometer scale can be inserted in the focal plane at F to allow the quantitative observation of double stars. (The curvature of the lenses is exaggerated in the drawing)

Fig. 5.2. Image formation by a plane-convex lens. Wavefronts of plane-wave light from a star are incident at the *far left*; the light rays are perpendicular to the wavefronts. The velocity of light in the glass of the lens is a factor n smaller than in vacuum (n = index of refraction); as a result, the wavefronts are bent into spherical surfaces which converge on the focal point F and then diverge behind it

lens consists of a convex lens (positive lens) made of crown glass, whose dispersion compared to its refracting power is relatively small, combined with a concave lens (negative lens) made of flint glass, whose dispersion is large compared to its refracting power. With an objective consisting of two lenses, it is possible to eliminate the change of the focal length f with λ (i.e. to make $df/d\lambda = 0$) at only *one* wavelength λ_0. For a *visual* objective lens, one chooses $\lambda_0 \simeq 529$ nm, corresponding to the wavelength of maximum sensitivity of the eye; for a *photographic* objective lens, in contrast, one takes $\lambda_0 \simeq 425$ nm, the wavelength of maximum sensitivity of a photographic blue plate.

Image Formation. Let us examine in more detail the imaging of a region in the sky by a telescope onto the focal plane of the objective lens. The task of converting the incident plane wave coming from "infinity" into a convergent spherical wave is accomplished by the lens, making use of the fact that in glass (index of refraction $n > 1$), light moves at a speed which is n times slower and therefore the wavelength is n times shorter than in vacuum. As a result, the wavefront is delayed in the middle of the objective lens relative to its outer edges (Fig. 5.2).

What the *lens telescope* or *refractor* accomplishes by inserting layers of material with $n > 1$ of differing thickness into the optical path is achieved in the *mirror telescope* or *reflector* (I. Newton, about 1670) by means of a concave mirror. This arrangement has the *a priori* advantage of suffering no chromatic aberrations. A *spherical mirror* (Fig. 5.3a) focuses a beam of axial rays at a focal length f equal to half its radius of curvature R, as can be seen by simple geometric considerations. Rays which are more distant from the optical axis are focused at somewhat shorter distances from the center of the mirror; this imaging error is called *spherical aberration*. The exact focusing of a beam of axial rays in a *single* focal point is accomplished by a *paraboloid* mirror (Fig. 5.3b); this can be seen most easily by considering the paraboloid as a limiting case of an ellipsoid, whose right focal point has moved away to infinity. Unfortunately, however, a paraboloid mirror yields a good image only in the immediate neighborhood of the optical axis. At a larger *aperture ratio* (the ratio of the mirror diameter D to the focal length f), the usable diameter of the image field is very small, owing to the imaging errors from obliquely incident rays which increase rapidly as one moves away from the axis.

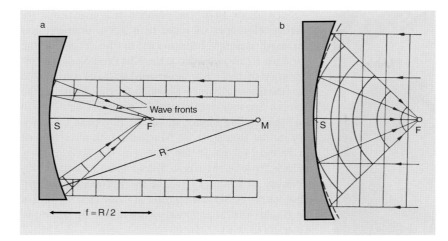

Fig. 5.3. (a) A spherical mirror. A bundle of rays near the optical axis (*upper bundle*) converges at the focal point F, whose distance from the crown of the mirror, S, corresponds to the focal length $f = R/2$, with R the radius of curvature of the mirror. A bundle which is further from the optical axis (*lower bundle*) converges on a point closer to S: this gives rise to spherical aberration. The planar wavefronts incident from the right are converted into spherical wavefronts on reflection by the mirror. (b) A parabolic mirror combines all the incident rays which are parallel to the optical axis precisely at the focal point F, i.e. the plane wave which is incident parallel to the axis is converted into a single convergent spherical wave. The conformal sphere (– – –) has the same curvature at the center point S as the paraboloid

Telescope Mountings. The mounting of a telescope has the task of allowing it to follow the diurnal apparent celestial motion of the Earth with sufficient precision. The commonly-used *equatorial* mount therefore has a "right ascension" axis which is parallel to the axis of the Earth and is driven by a sidereal clock, and perpendicular to it a "declination" axis.

The *azimuthal* mount, with a vertical and a horizontal axis, permits a compact telescope construction with uniform axis loading and is convenient for large, heavy instruments. The problem of the necessary *nonuniform* motion around *both* axes can now be readily solved by a computer-controlled drive system. Azimuthal mounts have long been used for large radiotelescopes but have only recently become widespread in optical astronomy.

Refracting Telescopes. Refractors, with aperture ratios in the range 1:20 to 1:10, are usually equipped with the socalled Fraunhofer or *German Mount*, as for example the largest instrument of this type, the refractor at the Yerkes Observatory of the University of Chicago (Fig. 5.4), with an objective of 1 m and 19.4 m focal length, which was completed in 1897. Large refractors were no longer constructed in the 20th century. For special applications, e.g. in positional astronomy or for the

Fig. 5.4. The refractor of the Yerkes Observatory; it has a Fraunhofer or German mount

Since about 1970, the number of conventional reflecting telescopes of more than 3.5 m diameter has increased sharply. Instruments in this class in the *Northern Hemisphere* are at the Kitt Peak National Observatory of the USA in Arizona (3.8 m), on Mauna Kea (the Canada-France-Hawaii 3.6 m telescope), at the German-Spanish Astronomical Center in Calar Alto in Southern Spain (3.5 m), and at the European Observatory on the Roque de los Muchachos on the island of La Palma (4.2 m). In the *Southern Hemisphere*, they include those at the Cerro Tololo Inter-American Observatory in Chile (4.0 m), at the European Southern Observatory (ESO) on La Silla in Chile (3.6 m), and at the Anglo-Australian Observatory at Sliding Spring in Australia (3.9 m).

In contrast to lens telescopes, the image quality of reflectors is influenced strongly by temperature variations, unless materials of very small thermal expansion coefficients are used. *Glasses*, e.g. Pyrex, from which the 5 m Palomar mirror was constructed, have a linear thermal expansion coefficient $30 \cdot 10^{-7}$ K^{-1}. Quartz has $6 \cdot 10^{-7}$ K^{-1}. The ability since about 1965 to manufacture and work large blocks of high-quality *ceramic glasses*, e.g. Zerodur, has permitted significant progress. This mixture of an amorphous glass and a crystalline ceramic component with opposing thermal properties exhibits no thermal expansion to speak of ($\lesssim 10^{-7}$ K^{-1}). The *reflecting* surface consists of an aluminum layer of only 100 to 200 nm thickness which is vacuum-evaporated onto the mirror.

Fig. 5.5. The Hale-reflector on Mount Palomar

visual observation of binary star systems, however, lens telescopes are even today still preferable to reflecting telescopes.

Reflecting Telescopes. Reflectors are usually constructed with an aperture ratio of 1 : 5 to 1 : 2.5 and equipped either with one of the various types of *fork mounts* (the declination axis passes through the center of gravity of the tube and is reduced to two bosses on either side of the latter), or with an English Mount, in which the north and south bearings of the long right ascension axis rest on separate pillars. The largest reflectors (with the primary mirror in a *single* massive piece) are at present the Hale telescope on Mount Palomar, California, placed in service in 1948, with a diameter of 5 m and a focal length of 16.8 m for the primary mirror (Fig. 5.5), and, since 1976, the (azimuthally mounted) 6 m reflector of the Astrophysical Observatory in Zelenchuk (Caucasus Mountains), whose primary mirror has a 24.0 m focal length.

Focusing Systems. With many reflectors (Fig. 5.6), one can observe either in the *primary* focus of the main mirror or, by means of a plane secondary mirror which reflects the beam out the side of the tube, in the *Newtonian* focus. It is also possible to attach a convex mirror in front of the primary focus and to produce the image at the *Cassegrain* focus (aperture ratio 1 : 20 to 1 : 10) behind an opening in the center of the main mirror. In newer telescopes, the *Ritchey–Chrétien* system (1 : 10 to 1 : 7) is often used; here, *both* mirrors in the Cassegrain system are replaced by mirrors with hyperboloid-like forms in order to obtain a larger field of view (0.5°) and a coma-free image[1]. In both the Cassegrain and the

---
[1]Coma: an image aberration in which rays not parallel to the optical axis are not focused at the same point, so that the image of a point source is drawn out into a "comet's tail" shape.

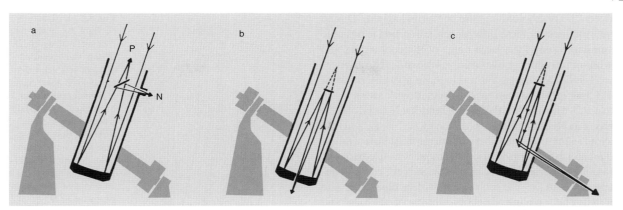

Fig. 5.6a–c. Different focus arrangements, using the example of the 2.5 m-Hooker reflector on Mount Wilson, the first of the large modern telescopes (completed in 1917); it was only recently decomissioned. (**a**) Primary (P)- and Newtonian (N)-focus (aperture ratio 1 : 5), (**b**) Cassegrain focus (1 : 16), (**c**) Coudé focus (1 : 30)

Ritchey–Chrétien systems, boring through the primary mirror can be avoided and the image can be directed out of the tube by a plane mirror to the *Nasmyth* focus. Finally, by means of a complicated arrangement of mirrors, the light can be directed through the hollow pole axis and the image of a star can be projected for example onto the entrance slit of a *fixed* large spectrograph in the *Coudé* focus (1 : 45 to 1 : 30).

Schmidt Cameras and Mirrors. The desire of astronomers for a telescope with a large field of view *and* a large aperture ratio (light-gathering power) was fulfilled by the ingenious invention of the *Schmidt camera* (1930/31). Bernhard Schmidt first noticed that a spherical mirror of radius R will focus small beams of parallel rays, which are incident from any direction but pass roughly through the neighborhood of the spherical center of curvature, onto a concentric sphere of radius $R/2$ – corresponding to the well-known focal length of the spherical mirror, $f = R/2$. With a small aperture ratio, one can thus obtain a good image on a *curved* plate over a *large* angular region, if in addition to the spherical mirror an entrance iris diaphragm is placed at the center of curvature, i.e. at a distance equal to twice the focal length from the mirror (Fig. 5.7). If a high light-gathering power is also desired, so that the entrance iris must be opened wide, spherical aberration becomes apparent as a smearing out of the stellar images. This was eliminated by Schmidt, who placed a thin, aspherically-ground *correction plate* in the entrance iris, compensating the differences in optical path by corresponding thickness variations in the glass and a small shift in the focal plane. These path differences are seen in Fig. 5.3b to correspond to the distance be-

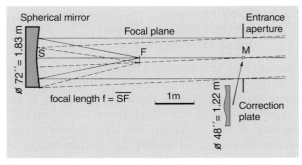

Fig. 5.7. A Schmidt mirror. Bernhard Schmidt based his instrument on a spherical mirror with a radius of curvature $R = MS$. Bundles of parallel rays, even those incident at a considerable angle from the optical axis, are collimated by the entrance aperture at the center point M of the mirror and thus are combined into a spherical wavefront around M with a radius $R/2 = MF$; this is termed the focal surface. The focal length is then $f = FS = R/2$. To remove the spherical aberration, an aspherically-ground, thin correction plate is placed in the aperture. (The dimensions shown correspond to those of the Mount Palomar 48″ Schmidt telescope)

tween the paraboloid and the conformal sphere. Due to the smallness of these differences, correction is possible simultaneously for a large range of incident angles and without introducing disturbing chromatic aberrations.

Schmidt telescopes, as a rule, have aperture ratios of 1:3.5 to 1:2.5, but can be constructed with ratios down to 1:0.3. The 48" Mt. Palomar Schmidt telescope (1:2.5) has a correction plate of diameter $48'' = 1.22$ m. To avoid vignetting, the spherical mirror must have a larger diameter of 1.83 m. This instrument was used for the famous Palomar Observatory Sky Survey; ca. 900 image fields of $7° \cdot 7°$, each taken with a blue plate and a red plate having detection limits of 21 mag and 20 mag, respectively, cover the entire celestial Northern Hemisphere to a declination of $-32°$.

In the case of the *Maksutsov-Bouwers* telescope (or camera), the Schmidt correction plate for avoiding spherical aberrations is replaced by a large *meniscus lens* having spherical surfaces.

Among the special instruments for *positional astronomy*, we must at least mention the *meridian circle* (O. Römer, 1704). The telescope can be moved about an east-west axis in the meridian. The right ascension is determined by the time of transit of a star through the meridian.

5.1.2 Resolving Power and Light Gathering Power. Optical Interferometers

We shall now attempt to develop a clear picture of the usefulness of different telescopes for various applications! The visual observer is primarily interested in the *magnification*. This is, as stated above, simply equal to the ratio of focal lengths of the objective and ocular. A limit to imaging of smaller and smaller sources is however set by the diffraction of light at the entrance aperture of the telescope.

Resolving Power. The smallest angular distance between two stars, e.g. binary stars, which can still just be separated in the image, is termed the resolving power. A square aperture of side D (which is simpler to treat than a circular aperture) produces a diffraction image from the parallel rays of light arriving from a star; it is bright in the center, and has on either side a dark band

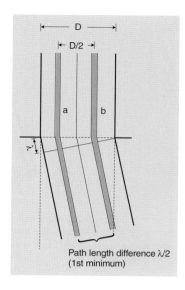

Fig. 5.8. Diffraction of light by a slit or rectangle of width D. When two bundles of rays spaced a distance $D/2$ apart have a path length difference equal to one-half wavelength, i.e. $\lambda/2$, then the first diffraction minimum results from destructive interference; it occurs at an angle of deflection given by λ/D

resulting from interference, where the light excitation from the two halves of the aperture (Fresnel zones) cancels. From Fig. 5.8, this corresponds to an angle (in radians) of λ/D. For a *circular aperture* of radius D, we find for the angular spacing

$$\phi = 1.22 \frac{\lambda}{D}, \qquad (5.1)$$

in which the diffraction disks of two stars half cover each other and can just still be separated. Equation (5.1) thus yields the resolving power of the telescope. If ϕ is calculated in seconds of arc and the diameter of the telescope is in meters, we obtain for $\lambda = 550$ nm a theoretical resolving power of:

$$\phi = \frac{0.14''}{D\,[\text{m}]}, \qquad (5.2)$$

i.e. for example for a 5 m mirror $0.03''$. In the "oldest optical instrument", the *human eye*, the resolving power for a pupil diameter of several mm is found to be of the order of $1'$.

With a focal length f, (5.1) corresponds to a *linear* size of the diffraction disk in the focal plane of $l = f \tan \phi$ or, for small angles,

$$l \simeq 1.22 \lambda \frac{f}{D} \qquad (5.3)$$

(where D/f is the aperture ratio). For $\lambda = 550$ nm we then have $l \simeq 0.67\, f/D\, [\mu\text{m}]$.

Longer exposures made from the ground are subject to a smearing of point images of the order of $1''$ due to "seeing" (scintillation), so that even a small telescope of somewhat more than 10 cm diameter reaches the maximum angular resolution permitted by the atmosphere.

Light Gathering Power. The *aperture* or the *diameter* D of the telescope mirror determines in the first instance the *collecting area* ($\propto D^2$) for the light from a star, a nebula, etc. The collected light or energy flux is imaged onto the receptor in the focal plane; the *light gathering power* I is a measure of how strongly the light is concentrated on the receptor surface. The light from a distant star, i.e. that of a "*point source*", whose diameter is small compared to that of its diffraction disk, is spread in any case over an area given by (5.3), so that its intensity is at the most equal to:

$$I \simeq \left(\frac{D}{\lambda}\right)^2 \cdot \left(\frac{D}{f}\right)^2 \tag{5.4}$$

For a given aperture ratio D/f, I thus increases proportionally to D^2. If the diameter δ of the smallest resolving element of the receptor is now chosen to be larger than the dimension l of the diffraction disk, the intensity of the point source depends only on D^2, not on the aperture ratio.

On the other hand, for an *extended source*, e.g. a comet or a nebula, with an angular diameter whose image ϕf is larger than the diffraction disk or the receptor resolution, the intensity is proportional to $(D/f)^2$. This is also valid for the case that the disk caused by air motion ($\phi \simeq 1''$) is larger than the diffraction disk. The power of an instrument for determining faint area magnitudes therefore depends, like that of a camera, in the first instance only on the aperture ratio, but secondarily on absorption and reflection losses in the optical system. In both respects, the Schmidt mirrors and their variants are unequalled.

A much more complex question is that of how faint a *star* can be and yet still be detected. The *limiting magnitude* of a telescope is clearly determined by the criterion of whether the small stellar disk – its size given by scintillation, diffraction, detector resolution, etc. – can still be distinguished from the background due to airglow and other disturbances. Thus, fainter stars can in the first instance be seen by using an instrument of larger diameter; for a given aperture, a smaller aperture ratio, i.e. a longer focal length, is advantageous. In practice, the exposure times must also be considered!

The Michelson Interferometer. As we have seen, the theoretical resolving power of a telescope is determined by the *interference* of edge rays. A somewhat improved resolution was obtained by A. A. Michelson using his stellar interferometer, by placing two slits at a spacing D in front of the objective (Fig. 5.9a). A "pointlike" star then yields a system of interference fringes with the angular spacing:

$$\phi = n\lambda/D \quad (n = 0, 1, 2, 3, \ldots) \, . \tag{5.5}$$

If a binary star is observed with this instrument, with its two components separated by an angular distance y along the axis of the two slits, then the interference fringes from the two stars are superposed; the brightest fringes are obtained when $y = n\lambda/D$. In between, the interference fringes become undetectable, when the components of the binary star are equally bright; otherwise, they pass through a minimum. Conversely, if one slowly moves the two slits apart, the interference fringes are most readily observable for $y = 0, \lambda/D, 2\lambda/D \ldots$ and least visible for $y = \frac{1}{2}\lambda/D, \frac{3}{2}/D \ldots$.

When a disk of angular diameter y' is observed, it is, as can be shown by an exact calculation, nearly equivalent to two bright points at a spacing $y = 0.41\, y'$, and the first visibility minimum is obtained at a slit distance D_0 corresponding to $0.41\, y' = \frac{1}{2}\lambda/D_0$ or

$$y' = 1.22\, \lambda/D_0 \, . \tag{5.6}$$

Compared to using a normal telescope, it at first may seem that little is to be gained ($D_0 \leq D$ in (5.1)), but in fact the criterion of *visibility* of the interference fringes is less dependent on air motions (seeing) than for example a measurement with a cross-hair micrometer. Michelson and others were thus first able to measure the diameter of Jupiter's moons, closely-spaced binary stars, etc.

Later, however, Michelson inserted a system of mirrors like that of binoculars in front of the mirror of a 2.5 m reflector, and could thus make $D_0 > 2R$ (Fig. 5.9b). It then became possible in the 1920's to measure the diameters of several red giant stars directly (the largest are $\simeq 0.04''$).

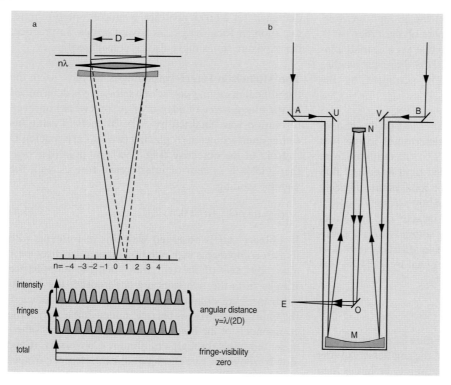

Fig. 5.9. (a) A stellar interferometer as developed by A. A. Michelson. A "point-like" star produces a system of interference fringes whose distances from the optical axis are $\phi = n \cdot \lambda / D$ ($n = 0, \pm 1, \pm 2, \ldots$). The fringes from two (equally bright) stars superpose and give a constant intensity at angular spacings of $y = \frac{1}{2}\lambda/D, \frac{3}{2}\lambda/D, \ldots$, i.e. zero visibility. (**b**) 6.1 m-Michelson interferometer of the Mount Wilson Observatory. A steel beam above the aperture of the 2.5 m reflector carries the inner two 45° mirrors U and V, which are fixed, and the outer, movable mirrors A and B. The distance AB (maximum 6 m) corresponds to the slit width D in (**a**). Observations are made visually at the Cassegrain focus E

Correlation Interferometers. An important problem in the Michelson interferometer is to bring the two rays together with the correct phase. This difficulty, which made the construction of still larger instruments impossible, was overcome by the correlation interferometer developed by R. Hanbury Brown, as follows: two mirrors at a separation D collect the light of a single star, each one onto a photomultiplier. The correlation of the current fluctuations from the two photomultipliers in a particular frequency range is measured. The strength of this correlation is, as can be shown by theory, related to D and y in *exactly* the same way as the visibility of the interference fringes in the Michelson interferometer.

The correlation interferometer in Narrabri, Australia, uses two 6.5 m mirrors, which are each constructed as a mosaic of hexagonal plane mirrors (the image quality is unimportant here); they are mounted on a circle of diameter $D_0 = 188$ m, on which they can be moved, so that their spacing D can be varied in the range $D_0 \leq 188$ m. The maximum resolving power corresponding to D_0, for $\lambda = 420$ nm, is, from (5.1), equal to $6 \cdot 10^{-4}$ seconds of arc. With this instrument, R. Hanbury Brown and R. Q. Twiss in 1962 determined the diameters of about 30 brighter stars, where for each star an integration time of the order of 100 h was required (stable electronics!).

Speckle Interferometry. The limitation of resolving power of a ground-based telescope to roughly 1″ due to air motion can be avoided by applying speckle interferometry, developed by A. Labeyrie in 1970, as well as the Knox-Thompson method (1973) and the speckle masking interferometry technique of G. Weigelt 1977, A. W. Lohmann and others (1983). While speckle interferometry reconstructs only partially the information about the object (the autocorrelation function), the other two methods yield true images with the theoretical diffraction-limited angular resolution. In the usual astronomical exposures, the diffuse seeing-disk is caused by turbulent fluctuations, in particular of the index of refraction in the atmosphere. On the other hand, a "mo-

5.1 Telescopes and Detectors for the Optical and the Ultraviolet Regions

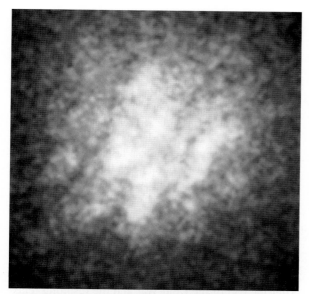

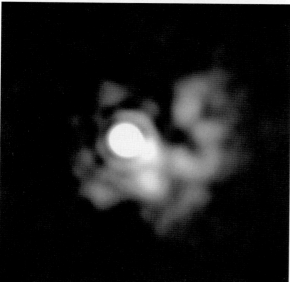

Fig. 5.10. A speckle masking measurement of the star η Carinae with a resolution of $0.076''$. *Upper image:* one of the total of 740 speckle interferograms which were evaluated; they were taken with the 2.2 m ESO-MPG telescope through a 6 nm Hα interference filter at $\lambda = 654$ nm. *Lower image:* the reconstructed picture of η Car at the theoretical diffraction-limited resolution, showing glowing gas in the immediate neighborhood of the star. It was obtained from the 740 speckle interferograms by applying the speckle masking interferometry method. (Images by G. Weigelt and K. Hofmann)

mentary exposure", which is obtained in a time shorter than the mean fluctuation time of the turbulence elements ($\lesssim 0.1$ s), yields a diffraction image composed of numerous small speckles (Fig. 5.10). Although the speckles are spread over an angular range of the order of $1''$, the angular diameter of each individual speckle is on the average $\simeq \lambda/D$, and thus corresponds to the resolving power of the telescope with aperture D (5.1), e.g. $0.03''$ in the visible range for a 4.0 m telescope. Such a *speckle interferogram* contains highly-resolved, coded information about the distribution of brightness of the object, however with a very poor signal to noise ratio. The art of speckle interferometry lies in superimposing 100 to 10^6 images, depending on the brightness of the object, and in then extracting information about the object by Fourier analysis. The three techniques mentioned above can be applied to objects with V magnitudes down to 17 and K magnitudes down to 12 mag.

5.1.3 Adaptive and Active Optics. Large Telescopes

Parallel to the deployment of space telescopes on unmanned satellites and the planning of observations from a space station, the development of large earthbound optical telescopes with effective diameters of more than 8 m has been actively continued. Now that the optical detectors have attained high sensitivity near to the theoretical limit, the principal possibility for further increasing the power of optical telescopes lies in an enlargement of their light-gathering areas. This allows observations to be extended to extremely faint objects while retaining all the advantages of an earthbound telescope: the easy accessibility and the possibility of using complex instruments at a relatively low cost compared to installations in space. The disadvantage of observations from the Earth's surface is the smearing out of the images and the resulting limited resolution due to atmospheric scintillation; this can be overcome to a great extent by correcting the wavefronts with the aid of *adaptive optics*. Another possibility of almost completely eliminating the influence of the atmosphere is through speckle interferometry as we have already mentioned in the previous section.

The construction of conventional telescopes with a stable, rigid primary mirror having a high surface precision ($\leq \lambda/10$) from *one* piece has reached its limit with apertures of 4 to 6 m diameter due to the technical problems which increase drastically with increasing mirror diameter (weight!) and to the cost. The development of large telescopes has therefore taken another direction: on the one hand, the use of lighter and thus *thinner* mirrors, and on the other, the construction of the primary reflecting element in the form of *several* individual mirrors. Since thin mirrors are readily deformed, the shape and direction of the reflecting surfaces must be continuously controlled and adjusted using an automatic feedback system operating on the principle of *active optics*.

Before we introduce some of the modern large telescopes individually, we will briefly describe the principles of adaptive and active optics.

Adaptive Optics. With the aid of this technique, the deviations from planarity of the wavefront from a distant light source (or the corresponding phase deviations) due to turbulence in the atmosphere can be continuously measured and, under computer control, corrected. In this way, the angular resolution is increased by a factor of from 4 to 10 and approaches the diffraction limit; furthermore, all of the light collected by the primary mirror can be focused onto a smaller detector area. In the case of weak (unresolved) objects, this brings a reduction of the sky background and favors in particular spectroscopy with high spectral and spatial resolution as well as interferometric observations.

The *principle* of adaptive optics is simple: the light from the primary mirror is collimated and focused onto a small, thin secondary mirror which is *deformed* via piezoelectric transducers under continuous computer control, so that the effects of atmospheric scintillation are to a large extent eliminated in the light rays reflected from this mirror. The exact surface shape of the mirror, which can be varied by up to a few μm, can be measured interferometrically. In addition, a tiltable plane mirror is also introduced and used to correct errors in the telescope drive motion. The electro-optical correction system has to be able to follow the rapid atmospheric fluctuations, i.e. the corrections must be carried out "online" within a few ms, which is possible only with a relatively small secondary mirror.

The spatial and time-dependent deformations of the wavefront are measured by a wavefront sensor on a *guide star*, which can be the object under observation itself, or a brighter nearby star (at least 10th magnitude). The angular distance between the guide star and the object must be less than about 1 to $2''$, so that its light experiences the same atmospheric disturbances. Such stars are, to be sure, rather rare, so that the development of adaptive optics for astronomy progressed only after it became possible to provide artificial laser "guide stars" by backscattering of a laser beam from the atmosphere above the turbulent layers; these laser guide beams can be positioned in any arbitrary location in the sky. The correction of the wavefront is easier for longer wavelengths, so that the infrared region is more favorable than the visible for this technique.

In practice, the methods of adaptive optics are complex and could be technically realized only in the 1980's; their development for astronomical applications was independent of those in the USA for military purposes. After the end of the Cold War and the resulting relaxation of military secrecy in the early 1990's, a rapid development occurred, so that it is to be expected that adaptive correction systems will be used in practically all optical telescopes in the coming decades.

Active Optics. In this method, the surfaces of the telescope mirrors themselves are corrected for mechanical, thermal and optical defects by computer-controlled transducers (Figs. 5.11, 12). These corrections are much slower than the turbulent variations of the wavefronts, so that the adjustments can be carried out on a time scale of the order of ≥ 1 s, and thus even the primary mirrors of large telescopes can be corrected in this way.

Large Optical Telescopes. Technically, there are several different possibilities for constructing large optical telescopes. One is to use a single, thin mirror of large diameter in connection with active optics; another is a mosaic-like arrangement made of several segments which together form a paraboloid reflecting surface, each of which can be "actively" adjusted. Furthermore, the rays from several independent telescopes in an array can be be collected onto a common focal point; or, finally, several telescopes can be put on a common mount with a common focal point.

The precursor of the new generation of large telescopes was the *multiple-mirror telescope* or MMT on Mount Hopkins in Arizona, which has been used since

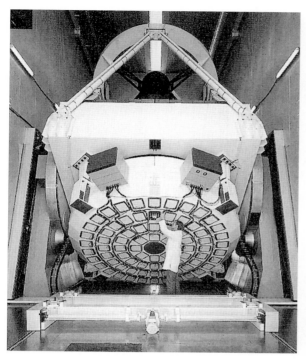

Fig. 5.11. The New Technology Telescope (NTT) of the European Southern Obeservatory (ESO) on La Silla (Chile). In this telescope, completed in 1989, the shape of the 3.5 m mirror is optimized by 78 transducers under computer control. (© European Southern Observatory, ESO)

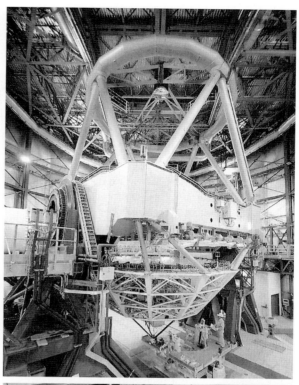

Fig. 5.12. Very Large Telescope (VLT) of the European Southern Observatory (ESO), on the Cerro Paranal, Chile. *Above:* A single telescope with an 8.2 m primary mirror made of Zerodur ceramic glass. *Below:* active optics: the mirror, just 18 cm thick, weighs about 22 t and is supported by a complex steel structure; it rests on 150 force transducers which continuously control and correct its shape. (© European Southern Observatory, ESO)

1979 for optical and infrared observations (Fig. 5.13). Six identical mirrors, each with a diameter of 1.8 m, are mounted together with a common axis; the light from each mirror is redirected by a secondary mirror onto the common quasi-Cassegrain focus. In this way, a light-collecting area is obtained which corresponds to a that of a conventional telescope of $\sqrt{6} \cdot 1.8\,\mathrm{m} = 4.4\,\mathrm{m}$ diameter, while the angular resolution is limited by the diameter of the individual mirrors. The common *azimuthal* mount allows a very compact arrangement of the MMT in a building which weighs "only" about 450 t and is 17 m high; it can be rotated around its vertical axis in the azimuthal direction.

In 1998, the mirrors of the MMT were exchanged for a telescope with a single 6.5 m mirror made of borosilicate (MMT now stands for *Monolithic* Mirror Telescope). A honeycomb structure of glass guarantees its rigidity and also greatly reduces its weight. The

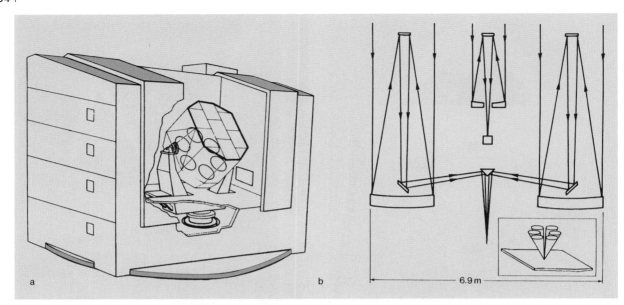

Fig. 5.13a,b. The Multiple Mirror Telescope (MMT) on Mount Hopkins, Arizona. (**a**) Schematic drawing of the telescope with its six mirrors, in mounted azimuthally in its building. (**b**) Light-ray paths for a pair of oppositely-mounted 1.8 m mirrors in a Cassegrain optical system with a common focus. In the central axis, there is a 0.76 m guide telescope for pointing the MMT and steering the main mirrors. These mirrors were replaced in 1998 by a *single* 6.5 m mirror. (With the kind permission of the American Institute of Physics, New York, and of the authors)

Cassegrain secondary mirror serves at the same time as a variable element for adaptive optics and is used to correct the image.

Another forerunner of the new large telescopes is the *New Technology Telescope* (NTT) of the European Southern Observatory on La Silla in Chile, which was completed in 1989 (Fig. 5.11). The 3.5 m primary mirror of its Ritchey–Chrétien optical system is a meniscus mirror of only 24 cm thickness made of Zerodur, whose surface is continuously sensed and shaped by 78 force transducers under computer control. The NTT is mounted azimuthally in a building which rotates with the telescope and has only one Nasmyth focus; the observations can be carried out by remote control.

The first large telescope for the visible and near infrared region (out to about 10 μm wavelength), completed in 1991, is the *Keck I telescope* with a 9.8 m mirror in the Ritchey–Chrétien configuration; it is on Mauna Kea (Hawaii). The mirror consists of 36 hexagonal, thin segments each 1.8 m across and made of Zerodur, which are individually adjusted using active optics. The similar *Keck II telescope*, set up at a distance of 85 m from Keck I, was completed in 1997, so that both instruments can also be used as an interferometer – aided in addition by smaller mirrors with a maximum baseline of 165 m.

The *Very Large Telescope* (VLT) of the European Southern Observatory on the Cerro Paranal in Chile (Figs. 5.12, 14) consists of four identical 8.2 m meniscus mirrors made of Zerodur, which can be operated together and thus provide a light-collecting area corresponding to an effective diameter of 16 m. The four telescopes for observations in the optical and infrared regions ($\lambda \leq 12$ μm) were completed in 1998–2000. They are called Antu (the word for Sun in the language of the Mapuche), Kueyen (Moon), Melipal (Southern Cross), and Yepun (Venus, as the Evening Star). In the coming years, these large telescopes will be complemented by several smaller mirrors in order to obtain a baseline of up to 200 m for long-baseline interferometry (VLTI).

In the Japanese *Subaru telescope*, completed in 1999 on Mauna Kea (Hawaii), the 8.2 m primary mirror is constructed along similar lines to that of the NTT.

Fig. 5.14. and frontispiece, p. 97. The Very Large Telescope (VLT) of the European Southern Observatory (ESO) on the Cerro Paranal in Chile. This aerial photograph, taken in 1995, shows the control building (*in the foreground*) and the concrete apron with the "domes" of the four 8.2 m telescopes, along with the tracks and 30 concrete foundations for positioning several 1.8 m telescopes which together with the large instruments will make up the interferometer system VLTI. The first 8.2 m telescope (Antu, *in front*) saw its "first light" in 1998, the fourth and final one in 2000. (© European Southern Observatory, ESO)

Within the next 5 to 10 years, further large telescopes with effective diameters of the order of 10 m will be put into service; of these, we mention only the following:

The *Large Binocular Telescope* (LBT) on Mount Graham in Arizona consists of two twin telescopes, each with an aperture ratio of 1:1.14 and 8.4 m diameter, and thus an overall effective diameter of $8.4 \,\mathrm{m} \cdot \sqrt{2} = 11.8$ m. The two mirrors are supported at a fixed distance of (23 m) on a common mount and can also be used as an interferometer. As in the "new" MMT, the primary mirrors are made of borosilicate and stabilized by a honeycomb structure.

The 10.4 m mirror of the *Gran Telescopio Canarias* (GTC) on the Canaray Island of La Palma is constructed in a similar manner to the mirrors of the two Keck telescopes; it consists of 36 hexagonal segments, each 1.9 m across.

5.1.4 Optical Detectors

Just as important as the telescopes themselves are the means of detecting and measuring the radiation from stars, nebulae, etc. *Visual* observation today plays practically no role.

The *photographic plate*, one of the most important tools of the astronomer for more than a century, has been replaced in recent times for nearly all applications by semiconductor detectors, in particular the CCD (see below). The (spatial) resolving power of astronomical photographic plates, which is determined by the *grain size* of the emulsion, is in the range between about 5 and 25 μm. The *darkening S* of a plate is defined by $S = -\log I/I_0$, where I/I_0 is its transparency, i.e. the ratio of the intensity I transmitted by the darkened plate to the undarkened intensity I_0. The relationship between the total incident radiation exposure $E = It$ (more precisely: the number of photons incident during the exposure time t) and the darkening is *nonlinear*. The *darkening curve*, S as a function of $\log E$, increases (above a threshold intensity) at first only slowly; it then becomes linear, the slope of this part of the curve giving the *contrast* of the emulsion; and finally, it reaches a saturation region where it flattens out again. A *single* photographic plate covers only a very limited brightness region (*dynamic* range about 1:20, maximum 1:100) before everything "goes black".

Photographic photometry makes it possible to determine the brightnesses of stars with an accuracy of 5 to 10%, provided the darkening curve is determined empirically and individually for each plate.

Photoelectric photometry has, since the 1950's, made use of *photomultiplier* or *electron multiplier* tubes with suitable amplifiers. The incident photons release electrons from the photocathode, as long as their energies are greater than the work function for electron emission (external photoelectric effect). The electrons are then accelerated by a high voltage onto additional electrodes where they cause secondary electron emission, and thus amplify the initial electron current in an avalanche process, giving an amplification ratio of 10^6 to 10^7. The dark current, which represents background noise, can be reduced by cooling the electrodes. Photoelectric instruments cover a similar spectral region to that of photographic plates, but they are more reproducible and precise and respond *linearly* over a larger range of incident light intensity. Their *quantum yield*, i.e. the fraction of the photons which produces a signal, lies for blue light in the range of 20 to 30%, while that of photographic emulsions is at best 2 to 4%. The photomultiplier collects all the electrons at the anode; therefore, image details are lost. In recent times, photometric measurements have been carried out mainly using CCD detectors (see below).

The wish to record more and more distant galaxies and other faint objects as well as their spectra with acceptable exposure times made it attractive to try to combine the advantages of the photographic plate, with its extremely high image resolution, with those of photoelectric detectors: linearity, a large dynamic range with easy amplification, and high quantum yields. This can be achieved by arranging many very small photoelectric detectors in a two-dimensional array, resulting in a *multichannel* detector. As early as the 1930's, A. Lallemand and others developed the first image converters. Progress in microelectronics and solid state physics and their applications to computer manufacture, to television technology, and to military uses (night-vision apparatus, reconnaissance, ...) has led since about 1970 to an enormous development and the rapidly increasing application of various kinds of high-sensitivity photoelectric detectors in optical astronomy. At the end of this development process, the CCD has emerged as *the* detector of choice for the optical and nearby spectral regions. In the following section, we introduce some of the more important of the newer types of detectors, without making a claim to completeness.

In the *image intensifier* or image converter, electrons are released from a photocathode by the incident light and are accelerated, but then – unlike the photomultiplier – they are not all collected by one anode, but rather, using electron optics (applied magnetic or electrostatic fields, combined with glass fibers to make the image plane conform to the shape of the electrodes), they are focused onto a fluorescent screen. This image, having a 50- to 100-times increased intensity compared to the original incident light, is then photographed directly from the screen. By placing several amplifier stages in series (Fig. 5.15), the amplification can be increased up to 10^6 to 10^7, although with reduced image quality.

The *microchannel plate* consists of a thin bundle of parallel glass capillaries ("microchannels", between 1 and 15 μm in diameter and about 1 mm long), which is placed directly behind a photocathode. The photoelectrons are accelerated in the microchannels by a voltage applied to electrodes evaporated on the front and back surfaces of the plate, and impact on the inner channel walls, causing the emission of a cascade of secondary electrons from a thin semiconducting layer applied there. The microchannel plate thus exhibits the high avalanche amplification of a photomultiplier, but at the same time maintains spatial resolution by guiding the electrons within the channels. The image at the anode can be further amplified and enhanced.

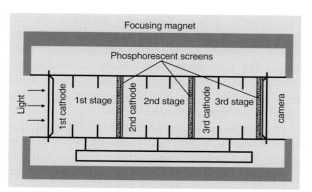

Fig. 5.15. The construction of a three-stage image amplifier with magnetic focusing. (With the kind permission of *Sterne und Weltraum*)

Among *television* imaging systems, in particular the *Vidicons* have been developed for astronomical applications in conjunction with image intensifiers. In the imaging section of the television camera, the photoelectrons are initially accelerated and focused onto a target which consists of a thin foil; the optical image is thus directly converted into an electrical charge distribution, which is then amplified within the target. This image is finally sampled by an electron beam, similar to that in a television tube, in the output section, and, after additional amplification, it is read out, for example into a computer for further treatment.

In *photon-counting* detectors, the amplification is raised to such a high value by a multistage image converter that signals can be discerned on the fluorescent screen which are due to *individual* incident photons. A television camera views this screen and serves as a spatially-resolving "intermediate memory"; it transmits the image to a computer, where it is then digitally reconstructed and further processed.

In a *semiconductor detector* such as the *silicon diode*, the incident photons create electron–hole pairs within the solid semiconductor material, if they have sufficient energy to lift electrons from the valence band into the conduction band (internal photoelectric effect). The free charges can then be collected in potential wells on the boundary layer between p- and n-doped silicon by an externally-applied voltage, and thus build up a charge image during long exposure times. The dark current due to electrons released by thermal lattice vibrations must be suppressed by cooling. Silicon diodes have especially good quantum yields in the red and infrared: at $\lambda = 600$ to 700 nm, they reach values around 80%. Their disadvantage, that avalanche amplification is not readily possible – in contrast to devices using the external photoelectric effect – is offset to a considerable degree by their high quantum yields and the storage ability of semiconductor detectors. The small size of photodiodes permits arraying a large number of them in two-dimensional structures (diode arrays) with their associated electronic control circuits on a single chip.

The silicon diode responds not only to light, but also to (sufficiently energetic) *electrons*. This fact is used in the *Digicon* (E. A. Beaver and C. E. McIlwain, 1971), in which the electrons emitted from a photocathode, after acceleration in a vacuum tube, strike a diode array which serves as detector. Imaging of the photocathode on the diode array is accomplished by electron optics.

To an increasing extent since the mid-1970's, the CCD (Charge-Coupled Device) has found application in optical astronomy. It is a two-dimensional arrangement of extremely small semiconductor detectors, invented at Bell Laboratories in 1970 by W. S. Boyle and G. E. Smith and developed for television applications. A CCD (Fig. 5.16) consists of a thin n-p-doped silicon platelet (chip), on which a two-dimensional arrangement of small electrodes is placed, separated by a thin insulating silicon oxide layer. These electrodes, together with the electronic circuitry, define the smallest independent receptors or image elements (pixels, from *picture el*ements), which have roughly $20\,\mu\mathrm{m} \cdot 20\,\mu\mathrm{m}$ areas. During the exposure, the electrons released, proportional to the number of incident photons, are collected in the potential well of each corresponding pixel. After completion of the exposure, control circuitry carries out a *charge-coupled readout process* by suitable changes in the potentials; in this process, the charge distributions of the pixels are moved to the edge of the picture, line by line, and are read into a computer memory via an amplifier. The readout frequency can be very high ($\gtrsim 1$ MHz), so that the production of new charges during the readout process remains insignificant. In the computer, several charge-images can be digitally combined to give the final picture. CCD chips are in a period of rapid development; at present, chips with up to $8000 \cdot 8000$ pixels on an area of several cm^2 are in use. In order to increase the detector area, several CCD chips can be combined in a mosaic arrangement.

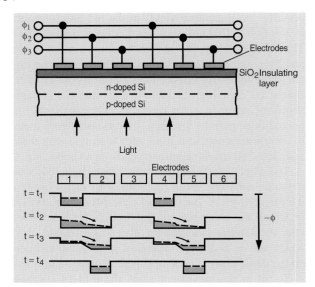

Fig. 5.16. Schematic construction diagram and principle of the readout process of a three-phase CCD (charge-coupled device). The charges collected initially ($t = t_1$) under the electrodes 1 and 4 are shifted in such a way by stepwise changes in the potentials ϕ (at times t_2 and t_3) that they are located under electrodes 2 and 5 at time t_4. (Reproduced by permission from *Sterne und Weltraum*)

The impressive success of CCD cameras in recent years for the observation of extremely faint objects is based in particular on their favorable quantum yields and the large range of their linear response. Their initial disadvantage compared to photographic plates, namely a much lower number of image elements, has been overcome by technological development.

5.1.5 Spectrographs

A precise analysis of the spectral distribution of the light from cosmic sources is the task of *spectroscopy*. We shall here consider only its experimental aspects, by (somewhat artificially) deferring the relevant basic concepts and applications to later sections.

The decomposition of light into its wavelength components requires a prism or a grating as the *dispersive element* of a spectrograph. The telescope has here only the task of focusing the light of a cosmic source, e.g. a star, onto the entrance slit of the spectrograph. Spectrographs of high light-gathering power for the investigation of faint objects are placed at the primary focus of the mirror. Large spectrographs are mounted statically in a separate room behind the Coudé focus. The Cassegrain focus has the advantage that relatively large spectrographs can be mounted there, rigidly attached to the telescope, avoiding light losses by reflection from additional mirrors.

In a *grating spectrograph* (Fig. 5.17), the *collimator* initially selects parallel rays of light and directs them to the grating. The focal length of the collimator mirror is chosen to be as large as allowed by the dimensions of the available grating; its aperture ratio should be equal to that of the telescope, in any case no smaller (full illumination of the grating!). The spectrally-decomposed light is then imaged in the *spectrograph camera*, for which the grating serves as entrance aperture. These cameras are frequently constructed as *Schmidt cameras* (Fig. 5.7), whose advantages we have already seen: large field of view, low absorption and reflection losses, small curvature of the image field, and negligible chromatic aberrations. It is often expedient to combine the grating and the camera mirror into a *concave grating*.

A *diffraction grating* (line grating, e.g. in the form of a reflection grating) yields intensity maxima, i.e. spectral lines, when the optical path difference between neighboring rays is given by:

$$a(\sin\phi - \sin\phi_0) = n\lambda \quad (n = \pm 1, \pm 2, \dots) \quad (5.7)$$

(Fig. 5.17). Here, a is the spacing of the parallel lines of the grating, (the grating constant), ϕ_0 is the fixed angle of incidence onto the grating, ϕ is the angle of the reflected rays, and n is the spectral order. Neighboring orders are separated by $\Delta\sin\phi = \lambda/a$ from one another. Overlapping orders, which would complicate the spectra (for example, the 2nd order at $\lambda = 400$ nm is coincident with the 1st order at $\lambda = 800$ nm) must be separated out using color filters together with the spectral response of the detector.

The angular dispersion of the grating,

$$\frac{d\phi}{d\lambda} = \frac{n}{a\cos\phi} \quad (5.8)$$

is, according to (5.7), *independent* of the wavelength, an advantage compared to a prism with its rather nonlinear dispersion. When the spectrograph camera has a focal length f, (5.8) corresponds to a linear dispersion given

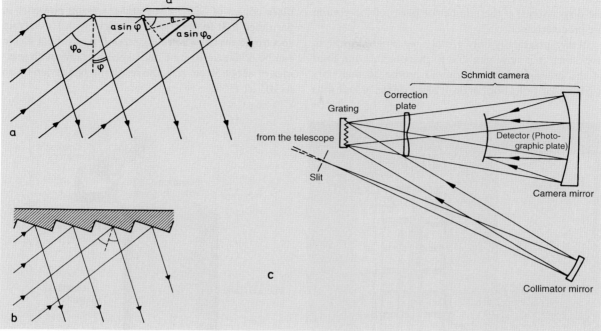

Fig. 5.17a–c. The grating spectrograph. (**a**) Principle of the diffraction grating (reflection grating). Interference maxima, i.e. spectral lines, are obtained when the path difference of neighboring rays is $a(\sin\phi - \sin\phi_0) = n\lambda$ $(n = \pm 1, \pm 2, \ldots)$ according to (5.7), where a is the grating constant. The dispersion (5.8) and resolving power (5.12) are proportional to the spectral order n. (**b**) By vapor deposition of e.g. aluminum onto the grating, called "blazing", specular reflection from the grating steps for certain angles of incidence and emission can be obtained, giving an increased intensity in the spectra. (**c**) A Coudé spectrograph. The star's image is focused onto the entrance slit of the spectrograph by the telescope. The collimator mirror makes the light rays parallel and reflects them onto the grating. The spectrally decomposed light is then imaged in the Schmidt camera; the grating serves simultaneously as the entrance slit of the camera

by:

$$\frac{dx}{d\lambda} = f \cdot \frac{d\phi}{d\lambda} . \tag{5.9}$$

The *spectral resolving power*,

$$A = \frac{\lambda}{\Delta\lambda} \tag{5.10}$$

is defined for a grating by the spacing $\Delta\lambda$ for which two sharp spectral lines are just recognizably separated from each other. For a grating of dimension D, according to the theory of diffraction, (see (5.1)), $\Delta\phi = \lambda/D$ is the smallest resolvable angle, so that (for a small exit angle ϕ), with (5.8) we have:

$$\Delta\lambda = \frac{a}{n}\Delta\phi = \frac{\lambda}{n}\frac{a}{D} . \tag{5.11}$$

If we now introduce the total number of grating lines, $N = D/a$, we find:

$$A = nN , \tag{5.12}$$

i.e. the resolving power depends only on the number of lines and the order.

In astronomical spectroscopy, diffraction effects can be for the most part neglected, since $\Delta\lambda$ is limited by the spatial resolution of the detector, i.e. by the size of the image elements of a CCD (Sect. 5.1.4). The image of the entrance slit produced by the collimator and the camera optics of the spectrograph should, for reasons of economy, be adapted to this resolution. If one wishes, for example, to obtain spectra e.g. of stars of a particular limiting magnitude with a telescope of a given size in a particular exposure time (exposure times > 5 h are inconvenient in practice), then the camera focal length and the dispersion (i.e. the number of lines) of the grating are predetermined.

Today, practically only *grating spectrographs* are in use, since it is now possible to manufacture the lines of the grating with the required profile so that only a particular reflection angle ("blaze angle"), and thus a particular order n, shows a strong light intensity. With blaze angles of $\lesssim 25°$, a spectral resolving power A in the range 500 to 5000 in low orders with Cassegrain spectrographs having grating sizes D up to about 15 cm can be obtained; using Coudé spectrographs with gratings of about 50 cm size, the resolving power can be up to 10^5.

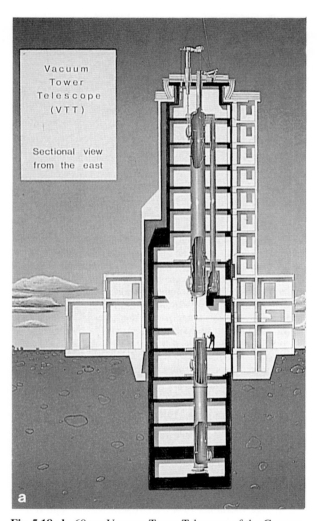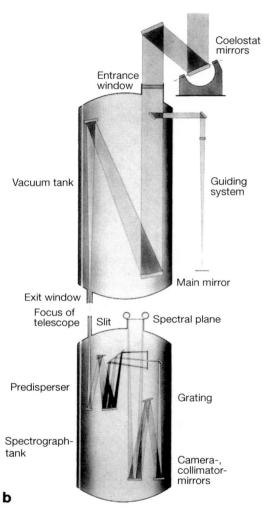

Fig. 5.18a,b. 60 cm Vacuum Tower Telescope of the German Solar Observatory at the Observatorio del Teide of the Instituto de Astrofísica de Canarias on Teneriffa: (**a**) cross section and (**b**) optical path. The two computer-controlled coelostat mirrors direct the sunlight perpendicularly downwards into a tube which is evacuated to 10^{-3} atmospheres to avoid heating; it has a diameter of 2.5 m. At its base is the main mirror with a focal length of 46 m. After exiting the vacuum tube, the light can be directed either horizontally into laboratories or vertically into a high-resolution échelle spectrograph, which is located in a tube 16 m deep under the ground. The 38 m high building consists of two concentric towers; the inner one holds the instruments while the outer prevents vibration, especially from the wind. E. H. Schröter and E. Wiehr (1985)

In recent times, along with "classical" grating spectrographs, the newer *échelle spectrographs* have found increasing application. They permit a high spectral resolution ($A = nN \lesssim 10^5$) with an *échelle grating* (stepped grating) in *high orders* ($n \simeq 10\ldots 1000$) at a large blaze angle (ca. 65°). The overlapping orders are separated at *right angles* to the dispersion direction of the échelle grating by a second grating which need have only a limited resolving power and can be made as a concave grating to serve simultaneously as the camera mirror, so that altogether a "two-dimensionally" arrayed spectrum is obtained. As examples, in Fig. 5.18 we show the échelle spectrograph of the Vacuum-Tower Telescope in the Observatorio del Teide, with which the solar spectrum can be registered at very high resolution; and in Fig. 5.19, we illustrate the high-resolution spectrograph of the Hubble Space Telescope.

For *spectral surveys* of whole starfields, an *objective prism* is used, i.e. a prism is placed in front of the telescope at the angle of minimal deflection, giving the spectrum of each star on the focal plane. For example, the Henry Draper Catalog was prepared in this manner by E. C. Pickering and A. Cannon at the Harvard Observatory; in addition to position and magnitude, it lists the spectral type of about a quarter of a million stars. Instead of an objective prism, an *objective grating* may also be used.

5.1.6 Space Telescopes

In order to observe cosmic sources in the ultraviolet ($\lambda \lesssim 300$ nm), we have seen that the only possibility is to employ a telescope outside the Earth's atmosphere. But even in the visible region, a space telescope offers notable advantages, which must, to be sure, be weighed against the enormous cost of placing an instrument in space. The limitation on resolution due to seeing no longer applies, and the background brightness is reduced by the amount due to the atmosphere, so that for the same mirror diameter, a space telescope can detect fainter objects than an Earth-based instrument. Furthermore, the useful duty cycle of a space telescope is greater, since there in space observations can be carried out day and night, independently of weather conditions.

Telescopes for the Ultraviolet Region. Following initial rocket experiments with very limited observation times, the Copernicus satellite (OAO-3) with an 82 cm telescope offered from 1972 to 1981 the first opportunity to carry out high-resolution *spectroscopy in the ultraviolet* in the range $\lambda \gtrsim 95$ nm as a long series of measurements. The "conventional" spectrograph, in which the spectrum is sampled stepwise by a photomultiplier, limited these spectroscopic observations to relatively bright stars.

The IUE Satellite (International Ultraviolet Explorer), launched in 1978, carried a 45 cm telescope of the Ritchey–Chrétien design, with two spectrographs in its Cassegrain focus, one for the spectral region $\lambda = 115$–195 nm and one for the region 190–320 nm. In each of these regions, either a spectrum of low resolution (5.10), using a concave grating (with $\lambda/\Delta\lambda \simeq$ a few 100) or else a high-resolution spectrum using an

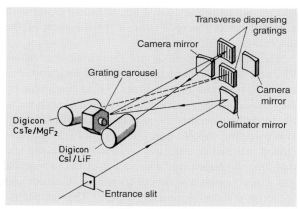

Fig. 5.19. The high-resolution spectrograph of the Hubble Space Telescope (up to 1997). Parallel rays of light from the collimator mirror are first spectrally decomposed by one of the six plane gratings, which are mounted on a carousel so they can be interchanged. For the highest resolution, $\lambda/\Delta\lambda = 10^5$ (the optical path shown), an échelle grating is used, followed by a transverse spectral decomposition using a concave grating which serves simultaneously as the camera mirror. Two digicons with a one-dimensional arrangement of 512 diodes serve as detectors, with either a CsI cathode and a LiF window ($\lambda = 105$–170 nm), or with CsTe/MgF$_2$ (170–320 nm). Spectra of lower resolution ($\lambda/\Delta\lambda = 2 \cdot 10^3$ or $2 \cdot 10^4$) are produced by the remaining five plane gratings on the carousel and, depending on the spectral region, are focused onto the corresponding digicon by one of the two camera mirrors. (From J. Brandt et al., 1982)

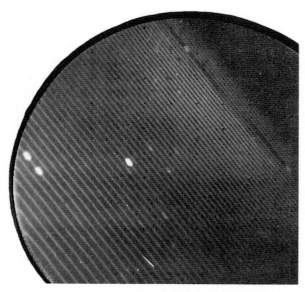

Fig. 5.20. The ultraviolet spectrum of the planetary nebula NGC 3242 in Hydra. The photograph was made by the *International Ultraviolet Explorer* in the range $\lambda = 115$–200 nm by J. Köppen und R. Wehrse in 1979 and shows a section of an échelle spectrum from the 66th to the 125th orders with a resolution of $\lambda/\Delta\lambda = 1.2 \cdot 10^4$, corresponding to $\Delta\lambda \simeq 0.01$ nm. Each order band covers a region of about 2.5 nm. The continuum from the central star is superposed on the emission lines of the nebula, of which the strongest are the C III doublet $\lambda = 190.7/190.9$ nm (*left*) and the He II line at $\lambda = 164.0$ nm (*center*). At the upper right, the resonance line of H I Lα $\lambda = 121.5$ nm, is visible on the one hand as a broad absorption band due to the interstellar medium, and on the other as emission from the geocorona. In the lower part of the picture, the track of a spurious particle ("cosmic") may be seen

échelle grating (with $\lambda/\Delta\lambda = 1.2 \cdot 10^4$; see Fig. 5.20) could be recorded. The detectors were Vidicon television cameras (Sect. 5.1.4), the ultraviolet light first being transformed into visible light by an image converter (CsTe cathode with a MgF_2 entrance window) and guided to the photocathode of the television camera tube using fiber optics. After an exposure, the image, consisting of $768 \cdot 768$ pixels, was read out by an electron beam, and the video signal was digitized and then transmitted to a computer at the ground station, together with information for calibration, image reconstruction, etc. The nearly geostationary, elliptical orbit of the IUE, with an apogee of about 46 000 km, permitted continuous contact with the ground station. By the use of two-dimensional, more sensitive detectors, a considerably weaker limiting magnitude was attained than in the Copernicus satellite, in spite of the smaller telescope. With exposure times of the order of 2 h, high-resolution spectra from 10th magnitude stars could be recorded. The highly successful IUE was decommissioned only in 1996.

Following the Copernicus satellite, since 1999 FUSE (the *F*ar *U*ltraviolet *S*pectroscopic *E*xplorer) has again made possible high-resolution spectroscopy in the *far* ultraviolet (90.5–118.7 nm). The instrument consists of four small telescopes mounted on a common axis, each one with a free aperture of $35 \cdot 39$ cm. Each of these directs the light it collects to a grating spectrograph, attaining a resolution of $\lambda/\Delta\lambda \geq 10^4$. Two of the telescope–spectrograph pairs are coated with Al and LiF, and cover the range from 100 to 119 nm, while the other two, coated with SiC, are for the short-wavelength range. The (two) detectors, each one registering two spectra, are two-dimensional microchannel plates with a spatial resolution of 20 μm in the direction of the dispersion.

The spectral region of the *far* ultraviolet, *below* 91.2 nm, has still been only barely investigated, because the strong absorption in the Lyman continuum by the *interstellar medium* limits possible observations to our immediate neighborhood in the Milky Way Galaxy. However, the Apollo-Sojus mission (1975) demonstrated that earlier estimates were too pessimistic and that the inhomogeneous distribution of the interstellar gas would permit a view out to greater distances (up to about 100 pc) in several "holes". The interstellar medium again becomes transparent only in the X-ray region ($\lambda \leq 10$ nm).

The Hubble Telescope. With NASA's 2.4 m Hubble Space Telescope (HST), a larger instrument for the optical and ultraviolet spectral regions has since its launch in 1990 for the first time been available over an extended period. The telescope, of the Ritchey–Chrétien type, attains an angular resolution of $\leq 0.1''$ at $\lambda = 633$ nm and thus nearly reaches the the diffraction limit (5.1) in its Cassegrain focus (aperture ratio 1 : 24). To be sure, this resolution was not obtained at first, due to a construction error in the mirror system which caused spherical aberration (see Sect. 5.1.1) and thus led to fuzzy images. Through the introduction of several small mirrors (COSTAR: *C*orrective *O*ptics *S*pace *T*elescope *R*eplacement) during a mission of the Space Shuttle "Endeavor", which required several days in 1993, the image errors could be corrected.

In the original instrumentation of the Space Telescope, the following instruments could be selected for use in the focal plane: (a) a *wide-angle camera* with a maximum field of view of $2.7' \cdot 2.7'$, covered by four CCD detectors in a mosaic, each with $800 \cdot 800$ pixels. Its spectral sensitivity ranges from the ultraviolet into the near infrared (115 nm–1.1 µm). (b) A *camera for faint objects* with a more narrow field of view and more limited spectral range (120–600 nm), in which, with a view to image reconstruction, detectors with a very high spatial resolution of $0.008''$ to $0.044''$ are used (image intensifiers with television camera tubes capable of single-photon detection). (c) A *low-resolution spectrograph*, with $\lambda/\Delta\lambda = 250\ldots 1300$ in the range 115–700 nm, using a Digicon detector (a linear diode array with 512 elements, cf. Sect. 5.1.4), for investigation of faint objects; and (d) a *high-resolution spectrograph* for the ultraviolet (110–330 nm), with $\lambda/\Delta\lambda = 2 \cdot 10^3 \ldots 10^5$, also using Digicon detectors; it is illustrated in Fig. 5.19.

The camera for faint objects (b) and the high-resolution spectrograph (d) were replaced with other instruments by the crew of a Shuttle in 1997: (e) NICMOS, a camera and a multiobject spectrometer for the near infrared, as well as (f) an *imaging spectrograph* with two-dimensional diode arrays for both the optical and the ultraviolet spectral regions.

5.2 Telescopes and Detectors for Radiofrequencies and the Infrared

The second major region of transparency of the Earth's atmosphere, besides the "optical window", is the "radiofrequency window"; on the long wavelength end, it is limited by reflection from the ionosphere, mainly in the F-layer (Sect. 3.2.6). The longest wavelength which is transmitted varies strongly, depending on fluctuations in the electron density of the ionosphere, between about $\lambda = 12$ and 100 m (i.e. between 24 and 3 MHz). At the short wavelength end of the region, below about $\lambda = 5$ mm, absorption by atmospheric oxygen and water vapor limits observability more and more, until finally in the submillimeter range below $\lambda = 0.35$ mm (860 GHz), ground-based astronomy is no longer possible. Below about 20 µm, the atmosphere again becomes transparent as we approach the optical window, at first however only in narrow regions between the absorption bands of water vapor (Fig. II.1).

We shall begin with a description of the most important types of radio telescopes (Sect. 5.2.1). In order to obtain high angular resolution in spite of the long wavelengths, both parabolic mirrors with the largest possible diameters, and especially interferometers are employed, the latter consisting of many individual telescopes at a considerable distance from one another. In Sect. 5.2.2, we give a very short summary of the receivers and spectrometers used for the radiofrequency region. Finally, in Sect. 5.2.3, we briefly describe instruments for the infrared spectral region, which lies between the radiofrequency and the optical regions and requires observation techniques combining elements of those used in the two neighboring regions. It is essential for the observation of astronomical sources in the infrared to suppress the intense thermal radiation background from the Earth and from the telescope itself.

5.2.1 Radiotelescopes

The weak radiofrequency signals from space are initially collected by an *antenna*, which should have the largest effective area possible and good directional characteristics (angular resolution); they are then passed to a *receiver* or *radiometer* for amplification and rectification, and on to a *detector*, so that finally their intensity can, for example, be traced by a strip-chart recorder or stored in digital form in a computer for further analysis. The radiation which is collected by the radiometer can also be decomposed into its frequency components with good frequency resolution in a *spectrometer*, or it can be analyzed in terms of its state of polarization in a *polarimeter*.

The most flexible type of antenna, with good directional characteristics over a large frequency range, is the *parabolic mirror*, which, so long as the wavelength to be collected is not too short, can be made of reflecting sheet metal or wire mesh (mesh diameter $\lesssim \lambda/10$). The precision of the reflecting surface determines the shortest wavelength which can be observed. The radiation is collected at the focal point by a *feeder antenna* (a horn or, for $\lambda \gtrsim 20$ cm, a shielded dipole) and input to the receiver via a transmission line. Analogously to an optical telescope, the signal can be taken up by the feeder antenna at the primary focus or, via a secondary reflecting surface, at a secondary focus, e.g. the Cassegrain focus (Fig. 5.6). The latter arrangement offers the possibility of placing several different receivers behind the primary mirror for parallel use.

Resolving Power. The resolving power of a mirror of diameter D in the radiofrequency region is also given by (5.1), as long as $\lambda \ll D$. It is usual in radioastronomy to quote the *beam width* as the angle between the points in the directional characteristic curve at which the energy sensitivity has decreased from its maximum value by one-half (HPBW: half power-beam width). The beam width

$$\varphi = 1.03 \frac{\lambda}{D_{\text{eff}}} \qquad (5.13)$$

(in radians) differs only slightly from (5.1) through its numerical coefficient. Owing to the directional characteristic of the receiver horn, the effective diameter of the primary mirror is $D_{\text{eff}} < D$. Instead of the angle φ, we can just as well describe the resolving power in terms of the *solid angle* $\Omega_A \propto \varphi^2$ of the antenna beam. Since furthermore the (effective) antenna area or aperture A is $A_{\text{eff}} \propto D_{\text{eff}}^2$, the following relation holds (it can be shown to be generally valid):

$$A_{\text{eff}} \cdot \Omega_A = \lambda^2 . \qquad (5.14)$$

Radiation Power. We first recall that the characteristic quantity for radiation from an extended source (Sect. 4.3.1) is the *intensity* or surface brightness I_ν (4.27); in contrast, for an unresolved point source, it is the flux density (4.40):

$$f_\nu = \overline{I}_\nu \Omega_Q . \qquad (5.15)$$

Here, $\Omega_Q = \pi R^2 / r^2 \ll \Omega_A$ is the solid angle which a source Q of radius R subtends at a distance r, and $\overline{I}_\nu$ is the intensity averaged over the source Q. In the radiofrequency region, the Kirchhoff-Planck function (4.61) can furthermore be replaced without loss of accuracy by the Rayleigh–Jeans law (4.63):

$$B_\nu(T) = \frac{2\nu^2 kT}{c^2} = \frac{2kT}{\lambda^2} . \qquad (5.16)$$

To characterize extended sources in radio astronomy, the brightness temperature T_B is also often used. One asks what temperature T_B a black-body radiator would have to be at, in order that it would emit just the intensity I_ν at the frequency ν according to (5.16):

$$I_\nu = \frac{2\nu^2 k T_B}{c^2} = \frac{2k T_B}{\lambda^2} \quad \text{or}$$

$$I_\nu \, [\text{W m}^{-2}\text{Hz}^{-1}\text{sr}^{-1}]$$
$$= 3.08 \cdot 10^{-28} (\nu \, [\text{MHz}])^2 \, T_B[\text{K}] . \qquad (5.17)$$

The brightness temperature is a special case of the radiation temperature (Sect. 6.1.3) for the Rayleigh–Jeans regime; the radiation temperature is generally defined via the exact Planck function.

The radiation flux density of a point source is also often expressed in terms of a temperature, the *antenna temperature* T_A, which however depends not only on the source but also on the properties of the antenna (see below). The common unit which is used in radio astronomy for the "monochromatic" flux F_ν is

$$1 \, \text{Jansky (Jy)} = 10^{-26} \, \text{W m}^{-2} \, \text{Hz}^{-1} , \qquad (5.18)$$

which is frequently denoted by S_ν, a notation which we however do not employ in this book, in order to avoid confusion with the source function (4.56).

Let us now observe a source with the intensity distribution $I_\nu(\theta, \varphi)$; according to (4.27), a planar surface A will receive from it the *radiation power*

$$W_\nu = A \int_\nu^{\nu+\Delta\nu} \oint I_\nu(\theta, \varphi) \cos\theta \, d\Omega \, d\nu \qquad (5.19)$$

in the frequency band $\Delta\nu$ around the frequency ν. We assume that the antenna Ω_A has a well-defined directional characteristic ($\cos\theta \simeq 1$) and that the frequency bandwidth is sufficiently narrow, and take the intensity, which we now denote simply by I, to be the average value over Ω_A and $\Delta\nu$. Then the radio power received by the antenna is

$$W_A = \tfrac{1}{2} A_{\text{eff}} I \Omega_A \Delta\nu . \qquad (5.20)$$

The factor $1/2$ which is usual in radio astronomy in connection with the definition of Ω_A and of the effective antenna area A_{eff} takes into account the fact that a dipole can respond to only *one* of the polarization directions of natural, unpolarized radiation. We now define the *antenna temperature* T_A analogously to the noise power $kT\Delta\nu$ of a resistance at the temperature T according to H. Nyquist (5.26) by

$$W_A = kT_A \Delta\nu \, . \tag{5.21}$$

Then we can also describe the radiation power by

$$kT_A = \tfrac{1}{2} A_{\text{eff}} I \Omega_A \, . \tag{5.22}$$

If we now express I according to (5.16) in terms of the radiation temperature, then for an *extended source*, using (5.14), we find

$$T_A = \frac{A_{\text{eff}} \Omega_A}{\lambda^2} T_B = T_B \, . \tag{5.23}$$

In contrast, for a *point source* of solid angle $\Omega_Q \ll \Omega_A$, with the flux density $f = I\Omega_Q$, we find that the antenna temperature is lower than the radiation temperature:

$$T_A = \frac{\Omega_Q}{\Omega_A} T_B \, . \tag{5.24}$$

The relation between the flux and the antenna temperature is

$$f = \frac{2kT_A}{A_{\text{eff}}} \, . \tag{5.25}$$

Radiotelescope Mirrors. In the cm and lower dm wavelength region, several large radiotelescopes with diameters over 60 m have been available for some time, beginning in the 1950's with the construction of the 76 m telescope at Jodrell Bank near Manchester in England. The currently largest fully-directable radiotelescope, the azimuthally mounted 100 m parabolic mirror of the Bonn Max-Planck-Institute for Radio Astronomy at Effelsberg in the Eifel Mountains (completed in 1972; Fig. 5.21), can be utilized for observations from about 50 cm to 6 mm wavelength (0.6 to 50 GHz).

Fig. 5.21. The 100 m radio telescope of the Max Planck Institute for Radio Astronomy, Bonn, at Effelsberg in the Eifel mountains

In the case of the *fixed* 305 m spherical mirror in a valley at Arecibo, Puerto Rico, an effective variation in the direction of observation within a limited range can be obtained by means of changing the phase of the incident waves by swinging the feeder antenna. The observational capabilities of this instrument (originally commissioned in 1963) were considerably improved by the addition of modern instrumentation in 1997.

The angular resolution of even these largest radiotelescopes is poor compared to that of Galileo's first optical telescope! According to (5.13), for a 100 m telescope at $\lambda = 50$ cm, it is only about $18'$. It is therefore understandable that radio astronomers quite early developed instruments using the *interferometer* principle for the meter wavelength region, in order to obtain better resolution.

Radio Interferometers. The radio interferometer of M. Ryle, which corresponds precisely to Michelson's stellar interferometer (Sect. 5.1.2), attains a high resolving power. In it, the signals from two radiotelescopes are combined, retaining *phase information*, and then further amplified. In contrast to the optical region, the phase of radiofrequency waves can be transmitted over several kilometers by cable and up to several 10 km by a radio link.

The principles of the linear diffraction grating and the two-dimensional grid grating (for a fixed wavelength) have also been applied to antenna technology in order to obtain a high angular resolution with *multi-element* interferometers. In the *Mills Cross* (B. Y. Mills, 1953), two long, cylindrically-parabolic antennas, each of which has a planar directional characteristic (i.e. good angular resolution in only *one* dimension), are mounted in a cross-shaped arrangement and are alternately connected to the receiver in phase and then with a $\lambda/2$ phase shift in one of the transmission lines. Taking the difference of the two signals yields a synthetic, rod-shaped power beam in the center of the cross, with an angular resolution of the order of λ/D, where D is here the length of the crossarms. Thus, for example, with the crossed antenna in Molonglo, Australia, whose arm lengths are 1.6 km, an angular resolution at $\lambda = 73.5$ cm (408 MHz) of $1.4' \cdot 1.4'$ is obtained.

The British interferometer MERLIN (Multi-Element Radio-Linked Interferometer Network; 1990), which is composed of seven telescopes between Jodrell Bank and Cambridge with a maximum baseline of 230 km, achieves a resolution under $0.01''$ at 22 to 24 GHz.

The principle of the *correlation interferometer*, first applied to radio astronomy by R. Hanbury Brown and R. Q. Twiss, has already been described for its "optical" version.

Aperture Synthesis. It has furthermore been shown by M. Ryle how the principles of aperture synthesis can be applied to use the information on amplitude and phase obtained *successively* from *several* small antennas at suitable, preselected positions, rather than that collected by a single large instrument during a particular time interval. For this purpose, the individual antennas, usually in an X-, T-, or Y-shaped arrangement, are moved with respect to each other – and the Earth's rotation is also used – to change their relative distances (projected onto a sphere). The angular resolution according to (5.13) corresponds to that of a single antenna whose aperture would be equal to the area covered in the course of time by all the small antennas taken together.

Among the large aperture-synthesis telescopes for the cm and dm ranges, we mention the Synthesis-Radiotelescope at Westerbork, in the Netherlands, with 12 fixed and 2 movable 25 m mirrors in an east-west arrangement with a maximum baseline of 1.6 km; and the Very Large Array at Socorro, NM, USA, with 27 movable 25 m telescopes in a Y-shaped arrangement, the lengths of the three arms being up to 21 km (Fig. 5.22). With these antenna systems, angular resolutions of a few $0.1''$ to $1''$ are obtained.

Very Long-Baseline Interferometry. In order to attain a resolution much better than $1''$, since about 1970 the technique of very long baseline interferometry (VLBI) has been developed: two radiotelescopes which are very far apart, up to distances of order of the Earth's diameter, are used. They detect the radiation from the same source at exactly the same frequency independently of one another as a function of time, and thus of the effective baseline. The signals (i.e. the intermediate frequencies) are recorded digitally on videotape; they are accompanied by extremely precise time markers from two atomic clocks, so that their interference patterns can later be analyzed with *known phases* in a computer (correlator). Instrumental phase fluctuations and disturbances in the atmosphere, which are different for the two sites

Fig. 5.22. The Very Large Array (VLA) in Socorro, New Mexico (USA). This is a radio synthesis-telescope with 27 movable 25 m diameter mirrors in a Y-shaped arrangement (maximum arm lengths 21, 21, and 19 km). (With the kind permission of the American Institute of Physics, New York, and of the author)

at the large distances used, can be nearly eliminated by suitable combination of the signals from three or more telescopes (arranged along a "closed loop"). Thus, using a worldwide network of the largest radiotelescopes with D on the order of $\gtrsim 10^4$ km, the obtainable angular resolution and positional precision according to (5.13) is reduced to $0.0001''$; i.e. about 10^4 times better than that achievable in optical measurements, which is limited by the seeing!

With the *Very Long Baseline Array* (VLBA), radio astronomers have had access since 1995 to a further development in long-baseline interferometry: it is an arrangement of 10 *identical* 25 m telescopes with identical antennas, receivers and computers which spans the North American continent from Hawaii via California to New Hampshire on the East Coast and St. Croix (Virgin Islands) in the Caribbean, giving a maximum distance of 9600 km. The array can be used to investigate both continuum and line sources in the range from $\lambda = 7$ mm (43 GHz) to 90 cm (0.33 GHz). Its angular resolution reaches $2 \cdot 10^{-4''}$ at the shortest wavelengths. The advantage of this system as compared to previous VLBI lies on the one hand in the optimal usage of angular resolution and dynamic range even for the shortest wavelengths as a result of the homogeneous instrumentation; on the other hand, observing time on the VLBA is available only for interferometric observations.

Using an additional telescope on a *satellite* in a strongly elliptical orbit, the angular resolution of long-baseline interferometry could be further improved to several 10^{-5} seconds of arc. Since 1997, the Japanese radio satellite HALCA (Table 2.3), together with the associated ground stations, has been available for this purpose; it circles the Earth on a strongly elliptical orbit at altitudes between 560 and 21 000 km every 6.1 h.

Millimeter and Submillimeter Telescopes. In the mm and sub-millimeter wavelength region, the resolution becomes more favorable, according to (5.13), but the requirements for surface precision of the mirrors become considerably more demanding, especially considering thermal deformations due to differential heating during the day- and night cycle. Besides steel, carbon-fiber strengthened plastics with a reflecting foil surface are used in this region. In particular for the submillimeter region, telescope and receiver technology are in a state of rapid development.

Since the Earth's atmosphere becomes opaque towards the submillimeter range, the telescopes for this spectral region are set up on the highest possible sites.

Among the largest telescopes for the mm region are the 45 m mirror at Nobeyama, Japan, built in 1985 and usable down to the 3 mm range; and the French-German 30 m mirror of the IRAM (Institut de Radio Astronomie

Millimetrique) on the Pico Valeta near Granada, Spain, for wavelengths down to 1.3 mm. Among the larger instruments for the sub-mm region down to wavelengths of 0.85 mm, we mention the 15 m mirror on La Silla in Chile (SEST: Sweden-ESO Sub-mm Telescope), and, down to 0.35 mm, the 15 m James Clerk Maxwell telescope on Mauna Kea, Hawaii with the SCUBA detector (*Submillimeter Common-User Bolometer Array*), as well as the 10 m Heinrich Hertz telescope (Fig. 5.23) on Mount Graham, Arizona, which has been operated jointly since 1995 by the Max Planck Institute for Radio Astronomy in Bonn and the University of Arizona.

Since 1998, NASA's SWAS satellite (*Submillimeter Wave Astronomy Satellite*) has been investigating the spectral lines of important components such as O_2, H_2O, ^{13}CO and CI in star-formation regions.

As with longer wavelengths, interferometry and long baseline interferometry (VLBI) are carried out in the mm region. As examples, we mention firstly the Array of the IRAM, consisting of four 15 m mirrors on the Plateau de Bure in the French Alps: it is an interferometer with only a few large mirrors; and secondly, the BIMA (Berkeley-Illinois-Maryland Array) on Hat Creek, California: it has more (nine), but smaller (6 m) mirrors.

COBE. Precise measurements of the isotropy and the spectral distribution of the cosmic 3 K background radiation (Sect. 13.2.2) were carried out in the microwave (and infrared) regions from 1989 to 1993 by the COBE (*Cosmic Background Explorer*) *satellite* from its circular orbit which passed over the poles at an altitude of 900 km. Its three instruments were cooled by superfluid helium to temperatures below 2 K.

A *differential* microwave radiometer measured the intensity differences between two different points on the sphere at wavelengths of 3.3, 5.7, and 9.6 mm using two horn antennas, and thus determined the isotropy

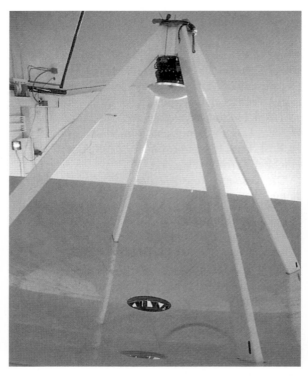

Fig. 5.23. *Left:* The 10 m Heinrich-Hertz Telescope of the Submillimeter Telescope Observatory on Mount Graham, Arizona (USA). *Right:* a portion of the primary mirror, made of carbon-fiber strengthened polymer and covered with a vapor-deposited aluminum layer, along with the movable secondary mirror (69 cm diameter) at the primary focus. Photo: D. Fiebig

of the background radiation. With a spectrometer, *absolute* measurements of the far-infrared radiation in the range of 0.1 to 10 mm were carried out, in order to detect possible deviations from the Planck distribution. In addition, the diffuse infrared radiation from 1 to 300 μm was observed, in an attempt to find indications of galaxies from the early stages of the cosmos whose radiation has been shifted into this region.

5.2.2 Receivers and Spectrometers for the Radiofrequency Region

We cannot treat amplifier technology here, but we shall mention at least a few principles and the most important types of amplifiers.

Noise Power. The in general weak, incoherent radiofrequency radiation from outer space ("noise") has to be detected and measured in the receiver or radiometer against the background of antenna noise and that of the receiver components, as well as the background radiation from the Earth's atmosphere in the case of shorter wavelengths ($\lambda \lesssim 1$ cm).

In the receiver, a noise power (H. Nyquist, 1928) is produced by the *thermal* motion of electrons in every (ohmic) resistance at temperature T in the frequency bandwidth $\Delta \nu$:

$$W_R = kT \Delta \nu \tag{5.26}$$

where $k = 1.38 \cdot 10^{-23}$ J K^{-1} is Boltzmann's constant. In addition, there is "shot noise" and "semiconductor noise" in some components such as transistors, diodes, etc. The overall noise in the receiver, W_N ($> W_R$) can, analogously to (5.26), be represented by an effective *receiver noise temperature* T_N:

$$W_N = kT_N \Delta \nu , \tag{5.27}$$

where T_N is higher than the physical temperature T. The main contribution to W_N or T_N comes in the *first* amplifier stage and the components before it, so that particularly *low-noise* amplifiers are used for this function. Since the contribution due to thermal noise is proportional to T, W_R and thus W_N can frequently be considerably reduced by *cooling*.

The *total noise power* W of the receiver system is the sum of antenna noise, W_A, and the receiver noise W_N itself:

$$W = W_A + W_N = k(T_A + T_N) \Delta \nu , \tag{5.28}$$

where W_A, again analogously to (5.21), can be expressed in terms of the *antenna temperature* T_A. In the ideal case, W_A comprises only the noise power of the cosmic object being observed. In practice, W_A also includes a noise contribution from the antenna itself, and, for $\lambda \lesssim 1$ cm, radiation from the atmosphere and possibly from the ground, or for $\lambda \gtrsim 30$ cm, the background radiation from our Milky Way galaxy (nonthermal synchrotron radiation, Sect. 12.3.2).

Sensitivity Limit. The limit of sensitivity, i.e. the weakest radiation power ΔW which can still be detected above the overall noise power W, is given for a measurement in a frequency bandwidth $\Delta \nu$ with an integration time τ by:

$$\Delta W = \frac{W}{\sqrt{\tau \Delta \nu}} \tag{5.29}$$

since in this case, the number of independent measured values is $N = \tau \Delta \nu$ and the relative accuracy, according to the law of statistical fluctuations, is $\Delta W/W = 1/\sqrt{N}$.

For the detection of weak cosmic signals it is thus important to have a stable amplifier (constant amplification) over the integration time τ. Furthermore, W_N must be kept as low as possible, since all contributions to W are of stochastic nature and in principle cannot be separated from one another. The frequency bandwidth is determined to a large extent by the problem at hand (a continuum measurement or observation of a single spectral line).

Amplifiers. Fluctuations in the amplification can be eliminated to a large extent by the method of R. H. Dicke (1946), employing a *Dicke* or *modulation receiver*, in which the receiver input is periodically switched between the antenna and a resistance at a fixed temperature, and the *difference signal* is then analyzed. If the resistor is replaced by a second antenna, which is pointed at a neighboring area of the sky, the noise contributions from celestial background and atmospheric radiations can be largely eliminated by taking the difference between the signals from the two antennas (beam switching). Another possibility for reducing background radiation is either to shift the radiotelescope mirror at a low "wagging frequency" between the source of interest and a neighboring area in the sky, and again to use only the difference signal in the receiver (on-off technique), or else to swing the secondary mirror back and forth relative to the telescope's optical axis (wobbling).

Following the first amplifier stage, the high frequency radio signal is usually converted to a (low) *intermediate frequency* (difference frequency) in a *superheterodyne* receiver using a *mixer* to combine with a fixed frequency from a stable oscillator; the intermediate frequency signal is then further analyzed.

In the case of *dm-* and *meter-waves* ($\lambda \gtrsim 30$ cm or $\nu \lesssim 1$ GHz), the nonthermal background noise radiation from our own galaxy dominates all other noise contributions; its radiation temperature increases with increasing λ from a few 10 K to well above 1000 K. Thus, in this range, there is no need for particularly low-noise amplifiers, and *conventional* amplifiers from radio technology, e.g. silicon transistor amplifiers and, more recently, (uncooled) field effect transistors ($T_N \simeq 300$ K) are used.

In the *cm-wave region* (1 to 30 cm, 30 to 1 GHz), the celestial background is "cold" (radiation temperature 3 to 10 K: microwave background radiation, Sect. 13.2.2), so that it is worth the effort to use *low-noise* amplifiers. With HEMT amplifiers (*H*igh *E*lectron *M*obility *T*ransistor) cooled to about 15 K, noise temperatures T_N of several 10 K are obtained; with *masers* cooled to still lower temperatures ($T \simeq 4$ K), one obtains T_N even below 1 K and thus an enormous increase in the precision of measurements of faint radio sources, to be sure at the price of reduced bandwidth.

Going into the *mm-wave region*, the thermal radiation from the Earth's atmosphere increases rapidly, and radiation temperatures soon reach about 300 K. For the longer-mm-wavelength region, SIS (*s*uperconductor–*I*nsulator–*S*uperconductor) amplifiers are used. Below about $\lambda = 3$ mm ($\nu > 100$ GHz), there are currently no low-noise amplifiers available. Here, the signal is fed *directly* into a *mixer* and reduced to an intermediate frequency in the 1 GHz range without preamplification; the latter can then be amplified conventionally. With very low-temperature ($T \lesssim 4$ K) Josephson mixers and SIS mixers, $T_N \simeq 100$ K is achieved at 100 GHz.

Receiver technology for the observationally difficult *submillimeter wavelength region* ($\nu \gtrsim 700$ GHz) is currently undergoing rapid development; it uses "quasi-optical" components in heterodyne receivers, based on a "hybridization" of waveguide techniques as used for microwaves, and the methods of geometric optics as used in the infrared and optical regions.

Spectrometers. For the investigation of radiofrequency *spectral lines*, as for example the 21 cm line of neutral hydrogen or the many molecular lines in the cm- and mm-regions, the broadband signal from the receiver is decomposed into its component frequencies in a spectrometer and their intensities are determined. In a *filter spectrometer*, the original frequency band (in some cases after conversion to another frequency range using the heterodyne principle) is divided up by a large number (several hundred) of narrow bandpass filters with fixed center frequencies. The signals from the individual spectrometer channels are then analyzed in parallel. In the *acousto-optical spectrometer*, the electromagnetic waves are first converted to ultrasonic oscillations using the piezoelectric effect. These then produce standing waves in a (transparent) crystal, corresponding to fluctuations in the density and the index of refraction of the material, and thus forming a "diffraction grating" which is irradiated with monochromatic light. The spectral energy distribution of the radio signal is thus finally converted into an optical diffraction image, whose intensity distribution can be measured, e.g. with a CCD.

5.2.3 Observation Methods in the Infrared

Observations from the ground can be made only in the *near* infrared ($0.8 \lesssim \lambda \lesssim 1.2$ μm) and in limited "windows" through the absorption by atmospheric water vapor in the *medium* infrared ($1.2 \lesssim \lambda \lesssim 20$ μm); owing to the strong decrease in water vapor content of the air with increasing altitude, they are best carried out from high mountains ($\gtrsim 4$ km). Astronomical observations in the *far* infrared ($\lambda \gtrsim 20$ μm), in contrast, have to be made from aircraft (altitude range $\lesssim 15$ km) and balloons ($\lesssim 40$ km) or from space vehicles and satellites. Only in the very-far infrared or submillimeter region ($\lambda \gtrsim 350$ μm $= 0.35$ mm) does the atmosphere again become transparent in several windows (Fig. II.1).

The infrared *celestial background* contains contributions from the zodiacal light (Sect. 3.8) for $\lambda \leq 50\,\mu\text{m}$ and from the diffuse galactic emission for $\lambda \geq 100\,\mu\text{m}$.

Telescopes. In the entire infrared range, optical telescopes can in principle be used (Sects. 5.1.1, 3). However, since at $T \simeq 300$ K the background radiation from not only the atmosphere but also from the telescope parts themselves is strong in the infrared, this bothersome thermal radiation has to be reduced as far as possible, by limiting the field of view, by keeping to a minimum the number of reflecting surfaces, and by cooling the telescope and the detectors, among other methods. In order to detect the faint infrared radiation from cosmic sources against the background of intense, variable emissions from the Earth's atmosphere and from the sky, a difference signal between the source and a neighboring area of the heavens is obtained by periodic "wobbling" as in the radiofrequency region (for example by wobbling the secondary mirror in a Cassegrain arrangement at about 10 Hz; this is termed chopping or beam switching).

Along with temporarily "modified" optical telescopes, especially constructed *infrared telescopes* are employed, for example the 3.2 m mirror of NASA and the British 3.8 m telescope (UKIRT), both on Mauna Kea (Hawaii).

Longer observations from altitudes of 12 to 15 km were possible from 1974 to 1995 with the *Kuiper Airborne Observatory*, an aircraft equipped with a 91 cm mirror for infrared astronomy.

Infrared Satellites. The first infrared satellite, IRAS (*I*nfrared *A*stronomical *S*atellite) carried out a sky survey in 1983 with a 57 cm telescope in four broad wavelength regions between 12 and 100 μm with an angular resolution on the order of 1′. By using a tank filled with liquid helium, the entire telescope was maintained for 10 months at a temperature $\lesssim 10$ K, and the focal plane with its detectors at a still lower temperature of $\lesssim 2$ K, until the cooling agent was spent. IRAS found about 250 000 infrared "point sources", among them many in distant galaxies whose light is primarily in the infrared, and made some surprising discoveries (among others new comets, interstellar dust clouds, and a dust ring around Vega).

The ISO satellite (*I*nfrared *S*pace *O*bservatory, launched in 1995) carried out observations for around 2 years from its strongly elliptical orbit – with an altitude of 70 600 km at its apogee – and detected infrared objects in the extended wavelength range of $\lambda = 2.5\,\mu\text{m}$ to 240 μm, before the more than 2000 l of superfluid helium coolant were used up. On board, it carried a Ritchey-Chrétien telescope of 60 cm diameter with four instruments cooled to between 2 and 8 K: – for broadband and narrow-band observations – a camera together with a polarimeter and an imaging photopolarimeter, furthermore two spectrometers with a spectral resolution $\lambda/\Delta\lambda$ of their gratings in the range of 1000 to 2500 in the short wavelength region ($\lambda \leq 45\,\mu\text{m}$) and 200 in the long wavelength region. By using Fabry-Pérot interferometers, resolutions up to $3 \cdot 10^4$ or 10^4 could be obtained. In particular, with ISO the widely-spread distribution of the cold ($T \simeq 20$ K) interstellar dust in numerous galaxies could be observed.

Detectors. For the infrared, which lies between the visible and the radiofrequency regions, techniques for detection of both the radiation from the visible (Sects. 5.1.4, 5) and from the submillimeter ranges (Sect. 5.2.2) can be applied.

As detectors for the *near* infrared region ($\lambda \leq 1.2\,\mu\text{m}$), photomultipliers and image converters, as in the optical region, but with special cooled photocathodes (AgOCs or GaAsP) are used; silicon photodiodes are also employed. Out to about $\lambda = 5.5\,\mu\text{m}$, indium antimonide (InSb) photodiodes cooled with liquid nitrogen at 77 K or liquid helium at 4.2 K are used; lead sulfide cells are also employed to about 4 μm. In the *medium* and *far* infrared, cooled semiconductor detectors predominate; they can also be used in CCD cameras. In a *photoconductive detector*, the absorption of a light quantum in a semiconductor creates an electron-hole pair by exciting an electron from the valence band through the energy band gap which separates it from the conduction band. The briefly increased electrical conductivity (until recombination of the pairs occurs) is then used for detection. Compared to the corresponding optical detectors, the photon energies in the infrared are considerably lower, so that the detector material must have a small band gap. In pure semiconductors such as germanium or silicon, the band gap is of the order of 1 eV, corresponding to $\lambda \simeq 1.2\,\mu\text{m}$. Doping with suitable materials, however, creates "impurity states" within the band gap, with energy spacings down to about 0.01 eV, so that the response limit of the photoconductor is extended to much longer wavelengths. The semiconductor must be cooled to a temperature at which these energy states are not populated thermally, i.e. cooling to lower temperatures is necessary as the limiting wavelength increases. By doping silicon with In, Ga, As, or Sb or germanium with Cu, and cooling to below about 10 K, detectors for infrared radiation out to the 20 to 30 µm range can be fabricated; germanium detectors doped with Ga and cooled to liquid helium temperatures ($\lesssim 4$ K) are sensitive even as far out as 120 µm.

In the whole infrared region, *thermal* detectors are also used, in which the absorbed radiation is transformed into the heat energy of a crystal lattice: in the *bolometer*, one makes use of the strong temperature dependence of the electrical resistance of a suitably doped cooled semiconductor crystal for the detection of infrared radiation. At liquid helium temperature, using ^{4}He or ^{3}He (4.2, 0.3 K), a range from 1 to about 1000 µm is accessible with a gallium-doped *germanium bolometer*.

Spectroscopy. For spectroscopy in the infrared, besides cooled prisms and gratings of medium resolving power, high-resolution *Fourier spectrometers* are also employed. The *heterodyne spectrometers* developed for the microwave region (Sect. 5.2.2) permit spectroscopy of line sources in the far infrared for wavelengths greater than about 100 µm.

5.3 Instruments for High-Energy Astronomy

From the long-wavelength region of the electromagnetic spectrum, we now jump to the region of the shortest wavelengths, or the highest frequencies and energies: to the *gamma-ray* and *X-ray* regions. Reflection, e.g. from the surface of a mirror, is as a result of the strong penetrating power of these high-energy photons possible only for soft X-rays (energies below a few keV) at grazing incidence, i.e. at angles $\lesssim 1°$ to the surface. At higher energies, the detection of the photons is carried out with instruments developed in high-energy physics for counting energetic particles, making use of the *energetic electrons* released from the detector material by the photons.

Along with high-energy electromagnetic radiation, the radiation consisting of *energetic particles* has become increasingly important: the *solar wind* blows solar matter, i.e. mainly high-energy hydrogen and helium ions and electrons, with velocities of about 200 to 1000 km s^{-1} into space. Protons of velocity around 500 km s^{-1} have an energy of 1 keV. The range of even higher energies is taken up by *cosmic radiation*, which reaches us with energies of up to 10^{10} eV from outbursts on the Sun, and in addition from the Milky Way (and possibly from other galaxies) with energies of a few 10^8 up to 10^{20} eV; it continually bombards the Earth's atmosphere (Sect. 10.4.2).

Many of the physical processes behind the interactions of energetic photons and particles with matter, which are important for the understanding of the particle detectors and the telescopes used in high-energy astronomy, as well as for the description of the propagation of gamma rays or cosmic rays from their sources to the point of detection, have already been introduced in Sect. 4.5.6.

We thus begin immediately in Sect. 5.3.1 with a discussion of the most important types of particle detectors, and continue in Sect. 5.3.2 by treating telescopes used to observe cosmic rays. Finally, in Sect. 5.3.3, we describe the instrumentation for the detection of high-energy photons in the gamma-ray region.

5.3.1 Particle Detectors

When energetic (charged) particles pass through matter, they give up their energy mainly through *ionization* of the atoms (4.141) in the material or by interactions with the *nucleons* in their nuclei. For electrons, the *bremsstrahlung* (4.142) which is produced as a result of their acceleration by the charges of the ions and electrons in the material can lead in addition to a considerable energy loss. For ions, in contrast, energy losses through bremsstrahlung play no role, since the interaction cross section for its production is inversely proportional to the square of the particle mass.

While energetic protons and heavy ions maintain their original directions practically unchanged in matter, the lighter electrons are strongly deflected.

When energetic protons or heavy ions approach the nuclei of atoms upon penetrating matter to within a distance of $\leq 10^{-15} A^{1/3}$ m (A: atomic number), i.e. to within the range of the nuclear force, they undergo nuclear interactions. In this process, 40 to 50% of their energy is converted into a multiplicity of particles, in particular into π *mesons*, but also into nucleons, hyperons, etc. The charged particles lose a large part of their energy by ionization of the material; the secondary nucleons, if they have sufficient energy, can interact with other nuclei and lead to a *nucleon cascade*. The charged poins, $\pi^{\pm}$, decay within a few 10^{-8} s into muons that have a high penetrating power, and into neutrinos or antineutrinos. In contrast, the neutral pions decay after their average lifetime of only about $1.8 \cdot 10^{-16}$ s into two gamma quanta each; these then initiate electromagnetic cascades (Sect. 4.5.6).

The orbit of an energetic particle can be directly obseved in a *bubble chamber*. The ions which are produced along its track act as nucleation points for vaporization in a superheated liquid with boiling hysteresis, so that the path becomes visible as a string of small bubbles of vapor. The particle's energy can be derived from the intensity of the track; a magnetic field can be applied to affect the particle orbits and allow the determination of their charge and mass. Bubble chambers are, however, not suitable for use in satellites or space vehicles due to their size and complexity.

In *plastic detectors*, the tracks of strongly ionizing particles can be detected using the permanent damage to the material which they produce. Since the damaged areas have an increased chemical reactivity, the particle tracks can be made visible by etching.

In *nuclear emulsions*, which consist of a gelatine layer containing suspended AgBr crystals much like a conventional photographic emulsion, the crystals are first activated by secondary electrons. After development, the tracks of ionizing particles become visible as a result of precipitation of silver grains. The density of grains at each point is proportional to the energy loss dE/dx.

A *scintillation detector* converts the energy of charged particles or energetic photons to *visible light flashes* via secondary electrons released by the energetic particle in a suitable crystal or liquid; the yield is a few percent, and the light flashes are detected using photomultipliers. The intensity of the current pulse from the photomuliplier is proportional to the energy loss of the primary particle. Materials for scintillation detectors are NaI and CsI crystals, thallium doped NaI, and also organic liquids and plastics.

In order to register energetic particles above a particular threshold energy or velocity, *Cherenkov detectors* are used; they are made of a transparent, solid dielectric material such as plexiglas, or employ a gas. If a particle moves through a medium with a velocity v which is *greater* than the velocity of light c/n in the medium (of index of refraction n), it will emit Cherenkov radiation at an angle to the direction of flight of the particle which is determined by v and n. This weak light in the optical region can once again be detected by photomultipliers.

Individual charged particles can be detected by a *gas counter tube*, in which they ionize the gas charge, e.g. Ne or Xe, along their paths. A high voltage is applied between the outer metal wall of the tube and a thin wire along its axis, giving rise to an electron avalanche due to ionization of the gas, which can be detected as a current pulse on the wire electrode. In the *gas proportional counter*, the voltage is set so that the total number of electrons detected is directly proportional to the number of electrons produced by the primary incident particles. Gas proportional counters are currently not so much used as particle detectors, but instead as detectors for X-radiation (Sect. 5.4.1).

By increasing the applied voltage, one can operate a gas counter tube, as in the *Geiger–Müller detector*, in the saturation region, in which each particle ionizes the whole gas and produces a spark. Counters of this type can be made with very small dimensions and assembled into a *spark chamber*. Because of their good positional resolution, spark chambers are suitable for determining the orbits of particles.

In *solid state detectors*, *electron–hole pairs* are created in a semiconducting material, which may be fabricated as an extremely pure single crystal. The pairs, similarly to the electron–ion pairs in a gas counter, then make the material conducting and can be registered as a voltage pulse after amplification. The disadvantages of the relatively small detector areas and the need for cooling are compensated by the high energy resolution of semiconductor detectors, which is due to the well-defined energy of the electron–hole pairs and to the short stopping ranges of the primary particles in the material.

5.3.2 Telescopes for Cosmic Radiation

Now that we have briefly introduced the numerous types of detectors with which energetic particles can be observed, we may ask how a *particle telescope* for observation of cosmic radiation can be constructed. Since these energetic particles can neither be reflected nor focused by matter, owing to their high penetrating power, it is necessary to construct a particle telescope in a completely different manner from, for example, an optical telescope. To select a particular type of energetic particles, characterized by their charge z and mass m, at the same time discriminating against other types of particles, and to determine their direction of incidence and energy E, one assembles several detectors having different sensitivities, together with shielding and absorbing elements, into a telescope. Its geometric arrangement, together with its electronic logic, which includes coincidence and anticoincidence circuits that operate on the detector signals, are designed to yield the desired information (Figs. 5.24, 25). The observable *angular range* of the telescope can, for example, be geometrically defined by the transmission of a particle through two particular detectors; details of particle orbits can be obtained from numerous small detectors arranged, for example, into a spark chamber. The determination of the *charge z* requires the analysis of the energy loss $dE/d(\varrho x)$, e.g. using the degree of ionization in a detector (4.141), with as many variously-responding detectors as possible. Finally, in order to find the particle's *mass m*, the total kinetic energy $(\gamma - 1)mc^2$ ($\gamma =$ Lorentz factor) must be precisely measured. For this purpose, the particle must be stopped in the telescope. This is verified by an "exit detector", in which the particular particle is *not* allowed to produce a signal and which is connected via an anticoincidence circuit to the other detectors with their positive detection signals.

The particles of galactic cosmic radiation interact strongly even with the upper layers of the Earth's atmosphere, since their average distance of travel corresponds to a mass column for strong interactions of about 100 kg m^{-2}, i.e. less than one-hundredth of the whole atmosphere. Thus, complex apparatus can be placed on the ground to investigate the cascades or air showers of *secondary particles*, from which conclusions can be drawn about the primary radiation. The secondary particles excite the air molecules (N$_2$) to fluorescence; the short flashes of light can then be detected.

We list some of the sensitive detectors which were put into operation during the 1990's:

1. The *"Fly's Eye"* near Dugway in the Arizona desert, which consists of a large number of 1.5 m mirrors, each of which focuses the light onto about a dozen photomultipliers, each one "seeing" a hexagon on the celestial sphere. The mirrors are arranged in such a way that all of the detectors together cover the *whole* sky – like the eye of a fly. Since the year 2000, its successor, the High Resolution Fly's Eye (HiRes), also near Dugway, has been available for observations of cosmic radiation. It attains an angular resolution of $1° \cdot 1°$.

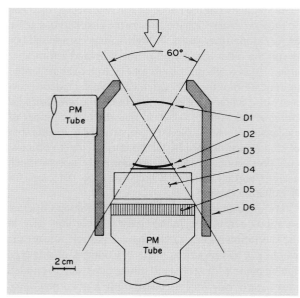

Fig. 5.24. A telescope for detecting cosmic-radiation particles. A cross-section through the telescope constructed by the University of Chicago for the IMP-7 satellite (launched in 1977), and used to measure beryllium isotopes from cosmic radiation in the energy range around 100 MeV per nucleon. The various detectors D1 to D6, together with logic circuits, detect the type, energy, and direction of incidence of the particles and eliminate signals from secondary and background radiations; the silicon drift counters D1 through D3 determine the angular resolution of the telescope. "Desired" particles must be stopped in the Tl-doped CsI counter D4, and therefore may not produce a signal in the Cherenkov detector D5 or in the plastic scintillator shielding D6 (PM tube = photomultiplier tube). M. Garcia-Munoz et al., Astrophys. J. **217**, 859 (1977). (Reproduced by permission of the Unversity of Chicago Press, the American Astronomical Society, and the authors)

Fig. 5.25. A large-area particle telescope for observing cosmic-radiation isotopes at several 100 MeV per nucleon from a stratospheric balloon. The dimensions are roughly 1.4 m · 1.4 m · 1.4 m (a collaboration of the University of Siegen, Germany, with the Goddard Space Flight Center of NASA; photograph by M. Simon)

2. The *Akeno* Giant Air Shower Array (AGASA) near Akeno in Japan, with over 100 scintillation detectors, each with an area of 2.2 m², which are set up at distances of about 1 km apart.

On the other hand, however, the *primary* cosmic radiation, which consists mainly of protons of ≥ 1 GeV energy and, to a lesser extent, of heavier nuclei and electrons, must be observed outside the atmosphere. This places heavy demands on the technology, since in particular for the most energetic particles, which have relatively low intensities, large, heavy telescopes must be employed over long periods of time. Satellites and space vehicles are used, as well as stratospheric balloons; it must be remembered, however, that even at an altitude of 50 km, a layer of atmosphere of about 20 kg m^{-2} thickness (Fig. II.1) still remains above a balloon-carried telescope.

5.3.3 Gamma-ray Telescopes

Gamma rays from space are absorbed in the upper layers of the Earth's atmosphere; in addition, large numbers of gamma quanta are produced by the impact of highly energetic cosmic ray particles on the upper atmosphere. Therefore, astronomy in the gamma-ray region (above a few 100 keV photon energy) can be carried out only from balloons, satellites, or space vehicles. Only extremely high-energy gamma rays can be investigated by ground-based instruments using the air showers which they produce in the atmosphere.

In the 1960's, instruments in balloons and satellites such as Explorer XI (1961) and the Orbiting Solar Observatory OSO-3 (1967) were able to detect the gamma radiation from the Sun and our Milky Way galaxy, although only limited angular and energy resolution was achieved. For example, the angular resolution of OSO-3 was only about 20°. At the beginning of the 1970's, longer series of observations with sensitive gamma-ray telescopes could be carried out; they first attained an angular resolution worthy of mention. Thus, the surveys performed by the satellites SAS-2 and COS-B (launched in 1972 and 1975) at energies $\gtrsim 100$ MeV had an angular resolution of a few degrees and an energy resolution $\Delta E/E$ of about 50% (Fig. 10.20). Using the instruments of the large gamma-ray satellite CGRO (*C*ompton *G*amma *R*ay *O*bservatory), observations were carried out at an angular resolution around 10′ and an energy resolution of about 20% in the high-energy range (100 to 2000 MeV) and of 4 to 10% at several MeV, during the period from 1991–2000.

Detectors. In principle, the detection of a gamma quantum is performed using the particles released by it in the detector, so that all of the varieties of *particle detectors* used in high energy physics (Sect. 5.3.1) can be employed for this purpose. At low energies, $E_\gamma \lesssim 30$ MeV, detection is performed using Compton backscattering electrons; at higher energies, electron–positron pair formation is used (Sect. 4.5.6). In the *low-energy* gamma-ray region, the determination of the direction of incidence of the gamma radiation is particularly difficult. It can be accomplished e.g. by using two large-area scintillator detector systems placed a certain distance apart. The relatively good angular resolution achieved at *high* energies is based upon the fact that the $e^\pm$ pair retains the direction of the gamma quantum which produced it to within an angle of $\simeq m_e c^2/E_\gamma$ (the rest energy of the electron is $m_e c^2 = 0.511$ MeV).

Telescopes. Cosmic gamma radiation with its low photon flux has to be detected against a strong background of secondary gamma radiation generated by cosmic rays in the atmosphere and in the detector itself. Gamma-ray telescopes therefore consist of a suitable geometric arrangement of several types of detectors with differing response functions for charged particles and photons, together with electronic logic circuits.

In the case of the gamma-ray telescope on COS-B (Fig. 5.26), a spark chamber serves as the primary detector; it contains layers of wire mesh and tungsten plates, and is filled with a mixture of neon and ethane gases. The incident gamma quantum creates an electron–positron pair in the tungsten, which then ionizes the gas along its path and is detected on passing through a scintillation or Cherenkov counter. The system of wires in the spark chamber localizes the orbits of the electrons and positrons and thus determines the *direction* of the gamma quantum. To this end, a discharge in the ionized gas along the tracks is initiated by a high-voltage pulse triggered from the detection counter, leading to current pulses in the "targeted" wires. In order to verify that the electron–positron pair was in fact produced by a gamma quantum and not by a cosmic-ray particle, the spark chamber is surrounded by a large-area scintillation counter which responds to cosmic radiation, but not to gamma radiation. An *anticoincidence circuit* opposite to the detection counter eliminates triggering of the spark chamber by the unwanted particles. The *energy* of the gamma quantum is finally determined from the energy of the electron–positron pair, which produces flashes of light upon absorption in an additional detector system consisting of a cesium iodide and a plastic scintillator; their intensity depends on the gamma energy.

The *Compton Observatory* CGRO carried four instruments which could detect cosmic gamma radiation in different energy ranges at the same time. The detector system BATSE (*B*urst *a*nd *T*ransient *S*ource *E*xperiment) used NaI scintillation crystals and photomultipliers and "saw" the *entire* celestial sphere (except for the part covered by the Earth); it was used in particular for the investigation of the brief, rapidly varying gamma bursts which were for a long time unexplained (Sect. 7.4.9), and of other transient sources at relatively low energies ($\lesssim 2$ MeV). Its resolving time was less than 1 ms and its positional accuracy (for strong sources) was about $1°$. The *spectrometer* OSSE (*O*riented *S*cintillation *S*pectroscopy *E*xperiment) consisted of a system of NaI and CsI scintillation counters and allowed spectroscopic observations between 0.1 and 10 MeV with an energy resolution $\Delta E/E$ of 4 to 12% and angular resolution of around $10'$. In this range of energies, there are many gamma-ray lines

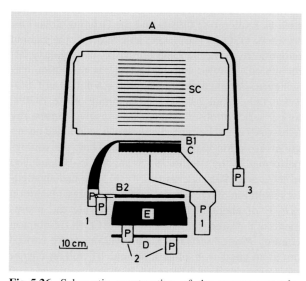

Fig. 5.26. Schematic construction of the gamma-ray telescope on the COS-B satellite. SC: wire spark chamber; B1,2: scintillation counters; C: Cherenkov detector; A: anticoincidence shield; D: plastic scintillators; E: CsI crystal; P: photomultipliers. (With the kind permission of Sterne und Weltraum)

from radioactive nuclides, which give information about nucleosynthesis in stars (Sect. 10.4.3). At somewhat higher energies (1–30 MeV), the imaging *Compton Telescope* COMPTEL achieved an energy resolution of about 7%, a positional accuracy of 5 to 30′, and an angular resolution of a few degrees. In this telescope, gamma quanta which were scattered in a liquid scintillator were detected by scintillation crystal detectors at a suitable distance. Finally, for the high-energy region (20 MeV–30 GeV), the CGRO carried the EGRET telescope(*E*nergetic *G*amma-*R*ay *E*xperiment *T*elescope), constructed with a number of spark chambers and a large NaI scintillation crystal. It permitted the determination of the positions of gamma sources to a precision of 5 to 10′, but with only moderate energy resolution ($\simeq 20\%$).

The Observation of Air Showers. At energies above 10^{12} eV = 1 TeV, gamma rays can be detected indirectly from the *ground* by means of their electromagnetic cascades, which result from their interactions with the particles of the upper atmosphere. At the very highest energies, ≥ 100 TeV, the large air showers reach the ground over an area of a few 100 m diameter and can be registered by the usual particle telescopes. Below several 10 TeV, the *Cherenkov radiation* from the energetic particles can be detected in the optical region with relatively small telescopes. For example, the Crab-nebula pulsar, the binary star system Cyg X-3, and the radio galaxy Cen A were discovered in the high-energy region in this way.

At the Whipple Obervatory on Mount Hopkins (Arizona), since 1968 showers have been observed with a 10 m mirror which is assembled out of a mosaic of 250 hexagonal segments. Recently, a new telescope was set up there at a distance of 140 m from the first. Other observatories for the observation of gamma radiation in the TeV range using Cherenkov radiation are for example HEGRA, an assembly of five telescopes for *H*igh *E*nergy *G*amma *R*ay Astronomy, built by a German-Spanish-Armenian collaboration on the Canary Island of La Palma, and its successor HESS (*H*igh-*E*nergy *S*tereoscopic *S*ystem) in Namibia, as well as CAT, the French Cherenkov Array Telescope near Thémis, in the Pyrenees.

The difficulty of making air shower observations is to differentiate the electromagnetic cascades originating with gamma quanta from the over 1000 times more frequent "nuclear" cascades due to cosmic ray particles, which contain considerably more muons.

5.4 Instruments for X-rays and the Extreme Ultraviolet

We now turn to the last region in the electromagnetic spectrum, the X-ray and its neighboring extreme ultraviolet (EUV). The X-ray region includes photon energies from around 100 eV to several 100 keV, i.e. wavelengths from about 10 to a few 10^{-3} nm; the EUV lies in the energy range above about 10 eV or at wavelengths below 100 nm and borders on the X-ray region.

Corresponding to their position between the visible and the gamma-ray ranges, observations in the X-ray and EUV require instrumentation which is intermediate between those used for the neighboring spectral regions. Up to photon energies of a few keV, one can still use *imaging* telescopes with grazing incidence at the reflecting surfaces. At still higher energies, it is necessary to use detectors and telescope arrangements similar to those for the gamma-ray region.

In the X-ray and EUV regions, astronomical observations must again be carried out from above the Earth's atmosphere. The first cosmic X-ray sources (apart from the Sun) were discovered in 1962 by R. Giacconi and coworkers using a rocket-borne detector. This discovery was followed by a number of observations of X-ray sources from rockets and balloons. The first *X-ray satellite*, UHURU, was able to carry out a complete sky survey and discovered around 340 sources. Within the following decade, the number of known X-ray sources increased through further satellite observations to well over 1000, of which many could be identified with a variety of different celestial objects. The most surprising result of nonsolar X-ray astronomy was the discovery of the great multiplicity of *temporal* variations of the cosmic sources.

The first satellite able to carry out systematic observations in the EUV was launched only in 1992: the Extreme Ultraviolet Explorer (EUVE). There were no earlier attempts, because it was generally assumed that absorption by neutral hydrogen in the interstellar medium would make observations in this spectral region practically impossible.

5.4.1 Detectors and Spectrometers for the X-ray Region

The most important detector for the X-ray region is the *gas proportional counter*, which can be constructed with a large surface area and is usually filled with Ar or Xe; such counters have yields of 30 to 40% up to energies of $\lesssim 20$ keV. In order to distinguish whether the ionization in the counter was due to an X-ray quantum or a cosmic-ray particle, it is necessary to use anticoincidence circuits or electronic discrimination methods based on the pulse shape of the electron avalanche pulses, as with all types of X-ray detectors. In gas-filled (Xe) *scintillation proportional counters*, the electrons which are set free by the X-ray quanta are accelerated so gently that they cause no secondary ionization of the gas atoms, but instead just excite them to fluorescence, which is then detected in the visible by the usual photomultiplier tubes.

Solid-state detectors, which are fabricated of lithium doped germanium or high-purity germanium and cooled with liquid nitrogen or helium, have the best currently available energy resolution, but can be manufactured only with a relatively small detector area.

In order to make use of the imaging properties of an X-ray telescope, *position sensitive* detectors are required. In the *imaging proportional counter* (IPC), the localization of the X-ray photons is carried out by using crossed wire grids together with electronic discriminator methods. Arrangements using multichannel plates and and CCD cameras have also recently been applied.

The *Sun* is the only X-ray source whose radiation flux in the X-ray region is sufficiently high to permit images to be captured on photographic film, as for example during the Skylab mission in 1973 (Fig. 7.19).

In the range of hard X-radiation ($\gtrsim 10$ keV), one can also employ *scintillation detectors* using crystals of NaI(Tl) or CsI(Na), where once again, the optical light flashes are detected by photomultipliers.

Spectral decomposition into more or less well-defined wavelength intervals in the X-ray range is accomplished with absorption *filters*, making use of the sharp absorption cutoff on the long-wavelength side of the absorption edge of the elements in the filter (Sect. 4.5.6). The spectral resolution or energy resolution of the detectors themselves is not very high. The best resolution is offered by *solid-state spectrometers*, which, depending on the energy range, reach 1 to 10%. In the soft X-ray region, using *transmission gratings* or Bragg *crystal spectrometers*, one can obtain high resolutions of the order of $\lambda/\Delta\lambda \simeq 10^3$ to 10^4.

5.4.2 Telescopes and Satellites for the X-ray Region

Images in the X-ray spectral region were first achieved with the Sun as source, using a primitive *pinhole camera*, then with mechanical collimators behind which e.g. a large-area scintillation detector counts the X-ray quanta. The moderate angular resolution of around 1° can be improved by using *modulation collimation*, in which the X-ray intensity is modulated by e.g. moving wire grids.

Only in 1964 did R. Giacconi take up the method developed by H. Wolter in 1951 (initially for X-ray microscopy), the *X-ray mirror telescope*. It is based on the following considerations: a metal surface, no matter how well polished it is, can reflect X-rays like a mirror only at grazing angles, i.e. for angles between the beam and the surface of less than a few degrees. The first attempt to produce an image of a distant object thus used a metal ring cut from a long *paraboloid*, while the unused portion of the entrance aperture was covered with a round metal disk (Fig. 5.27a). A telescope of this design would however yield only very poor images owing to the strong aberrations (coma). These can, according to Wolter, be considerably reduced by following the first reflection from the paraboloid by a second reflection from a coaxial and confocal *hyperboloid* (Fig. 5.27b). In order to increase the radiation-gathering area of the telescope, a number of these confocal double mirrors can be stacked one inside the other.

On the *Einstein Observatory* (HEAO-2) operated by NASA from 1978 to 1981, the first *imaging* Wolter telescope, developed under the leadership of R. Giacconi, with an aperture of 56 cm, was available for high-resolution observations (angular resolution $\lesssim 2''$) of long duration in the energy range from 0.1 to 4 keV. It was able to discover new, very weak X-ray sources. The European EXOSAT satellite (1983–1986) carried two small 27 cm Wolter telescopes in a strongly eccentric orbit (perigee 190 000 km); it permitted long series

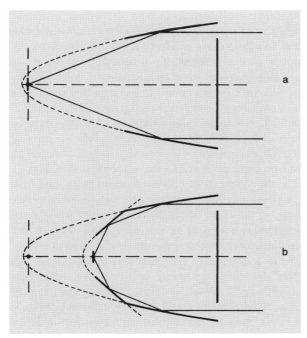

Fig. 5.27a,b. The X-ray mirror telescope designed by H. Wolter (1951) is based on a paraboloid mirror used at grazing incidence with a ring-shaped entrance slit (**a**). The extreme aberrations (coma) in this arrangement are practically eliminated by a further reflection of the rays from a confocal and coaxial hyperboloid (**b**). Still more effective reduction of the aberrations can be obtained by using more complex rotationally-symmetric mirror systems

of observations (without interruption by the Earth's shadow).

The ROSAT satellite, developed by J. Trümper and coworkers and launched in 1990, first carried out a complete sky survey with its 83 cm Wolter telescope from a circular orbit at an altitude of 570 km; thereafter, interesting objects were individually investigated. The image detectors are sensitive in the energy range of 0.1–2.5 keV, and in this range – without further energy resolution – an angular resolution of 3″ can be attained; alternatively, four spectral regions ($E/\Delta E \simeq 2.5$) can be observed at a reduced angular resolution of 25″.

The four identical telescopes on the Japanese satellite ASCA (*A*dvanced *S*atellite for *C*osmology and *A*strophysics) have since 1993 allowed for the first time in the X-ray region the simultaneous acquisition of images and spectra (0.5 to 12 keV) with good resolution.

Each of the telescopes consists of 120 closely fitted concentric pairs of hyperbolic and parabolic mirror segments made of thin aluminum foil with a surface layer of acrylic and gold; they thus represent the optimum realization of the Wolter-Giacconi design (Fig. 5.27), i.e. the largest possible radiation-collecting area with low weight. However, since the foils cannot be manufactured with sufficient surface smoothness, these telescopes have the disadvantage of a limited angular resolution of only about 3′.

Two of the telescopes on ASCA are each equipped with a spectrometer and a gas proportional counter as detector; the other two have imaging solid-state spectrometers (CCD cameras), which permit an energy resolution of $E/\Delta E \simeq 20$ at energies of 1.5 keV or $\simeq 50$ at 6 keV.

For the observation of strongly variable and eruptive X-ray sources, the RXTE satellite (*R*ossi *X*-ray *T*iming *E*xplorer) was launched in 1995 by NASA. Since 1996, the Italian-Dutch satellite BeppoSAX has served to identify and investigate gamma bursters (Sect. 7.4.9).

In the year 1999, two large X-ray observatories of a new generation were put into strongly eccentric orbits around the Earth: Chandra (named for S. Chandrasekhar) was launched by the American space agency NASA, and XMM Newton (*X*-ray *M*ultimirror *M*ission) by the European ESA. Both satellites are equipped with very efficient, high-resolution X-ray spectrometers which are sensitive over a large range of energies.

Chandra carries a Wolter telescope with a very high angular resolution of 0.5″, which is constructed with four pairs of concentric confocal mirror shells. This instrument covers the energy range from about 0.2 to 10 keV with an effective mirror area of 400 cm². Along with a high-resolution camera constructed of microchannel plates and having a field of view of 30′ · 30′, an imaging spectrometer and two transmission gratings can be employed for the spectral decomposition of the X-radiation.

XXM Newton has three X-ray telescopes of the Wolter type, each with a diameter of 70 cm. Each telescope consists of 58 concentric, thin (0.5 to 1 mm thick) mirror shells made from nickel covered by a 200 nm layer of gold. The high precision of the surfaces and the stiffness of the mirrors leads to an angular resolution in the range of 5 to 15″. The X-radiation from one of the telescopes is imaged at its primary focus by a CCD camera; that from the other two is spectrally decomposed by a reflection grating in a spectrometer for each telescope, and then also registered by CCD cameras. The radiation which is not diffracted is also detected by a CCD camera. Observations in the range from about 0.1 to 12 keV are carried out using all three instruments simultaneously, giving an overall effective area (5 keV) of around 1800 cm^2.

5.4.3 Telescopes for the Extreme Ultraviolet

In the extreme ultraviolet, beginning below about 50 nm, telescopes with "normal" mirrors can no longer be used, since in this region, light which falls nearly perpendicularly onto the reflecting surfaces will be completely absorbed. Instead, reflection at a *grazing angle*, as in the X-ray region, must be employed.

A first survey in the limited extreme ultraviolet region from 6–20 nm was carried out by the X-ray satellite ROSAT.

The EUVE satellite (*Extreme Ultraviolet Explorer*) was launched in 1992 as the first satellite intended exclusively for observations in the extreme ultraviolet; its instruments were developed at the University of California in Berkeley by S. Bowyer and coworkers. EUVE carries three 40 cm telescopes, three spectrometers, and a telescope for spectrometric surveys. The energetic light is imaged by *metal* mirrors with grazing angles of reflection; the detectors are multichannel plates. EUVE can carry out photometric observations in four bands: 6–18, 16–24, 35–60, and 50–74 nm; spectroscopy is possible in the range of 7–76 nm at a resolution of $\lambda/\Delta\lambda \simeq 250$.

Computers in Astronomy. To conclude our introduction to observational methods in the various spectral or energy regions, we should mention the rapid development of electronic data processing devices (computers), which – one can say without exaggeration – have opened up a new era in astronomical observation and measurement techniques. With the steady and rapid progress of semiconductor technology, miniaturization and low-cost mass production of components, computers with ever-increasing processing capacities and growing memory size have taken on such an important role in astronomy that today, the acquisition and evaluation of observational data without the computer has become simply unimaginable. We mention here only a few examples, such as the steering of telescopes (azimuthal mounting) and the control of mirror surfaces applying the principles of active optics, the synthesis of enormous amounts of data from long-baseline interferometry into high-resolution radio images, the logic operations required by particle and gamma-ray detectors, the readout control of CCD cameras and the resulting generation of a "digital image" in the computer, or the reduction of incoming "raw data" by either automatic or interactive, often very complicated software systems. Finally, "space astronomy" from satellites and space vehicles only became feasible through the application of powerful digital computers.

Astronomers have become accustomed to the fact that their *observational material* today for the most part exists only in the form of *digital records or arrays* on suitable data media (magnetic tapes or disks, semiconductor memories, ...). Even the information from photographic plates is often digitized for direct input into a computer.

We should also mention here that not only observational astronomy, but also *theoretical astrophysics*, with its enormously complex model calculations (e.g. of radiation transport, stellar structure and stellar evolution, or of magnetohydrodynamic processes) would be unthinkable without powerful computers with large memory capacities.

The fantastic "working capacity" of digital computers opens up unimagined new possibilities for research on the one hand; on the other, it puts a heavy responsibility on today's astronomers.

III. The Sun and Stars: The Astrophysics of Individual Stars

Astronomy + Physics = Astrophysics
Historical Introduction

The first measurements of trigonometric star parallaxes by F. W. Bessel, T. Henderson and F. G. W. Struve in 1838 provided the final confirmation of the Copernican worldview (whose validity was by then hardly doubted). But in addition, they provided a secure foundation for the determination of all cosmic *distances*. Bessel's parallax of 61 Cygni, $p = 0.293''$, indicated that this star is at a distance of $1/p = 3.4$ parsec or 11.1 light years. With this value, we can, for example, compare the luminosity of this star directly with that of the Sun. However, determination of *trigonometric* parallaxes became an important method in astrophysics only after about 1903, when F. Schlesinger developed their photographic determination to a tool of unbelievable precision ($\simeq 0.01''$)

Information about the *masses* of stars is obtained from *binary star systems*. W. Herschel's observations of Castor (1803) left no doubt that here, two stars were revolving about one another on elliptical orbits under the influence of their mutual attraction. As early as 1782, J. Goodricke had observed the first *eclipsing variable* star, Algol. The application of the variety of information obtained from binary star systems is the work of H. N. Russell and H. Shapley (1912). The first *spectroscopic* binary system (determination of the motion by means of the Doppler effect) was Mizar, whose binary nature was discovered by E. C. Pickering in 1889.

Stellar photometry, the measurement of the apparent brightness of stars, obtained a firm basis in the middle of the 19th century, after its beginnings in the 18th century (Bouguer 1729, Lambert 1760 and others). For one thing, in 1850 N. Pogson introduced the definition of a *magnitude*: it was taken to correspond to a decrease in the logarithmic brightness of 0.400, i.e. a brightness ratio of $10^{0.4} = 2.512$. Furthermore, J. F. Zöllner in 1861 constructed the first *visual* spectrophotometer, with which even the colors of stars could be determined. About the same time, the great catalogues of stellar magnitudes and positions made knowledge of the stars available on a widespread basis: e.g. in 1852/59, the *Bonner Durchmusterung*, a sky survey with data for 324 000 stars down to about 9.5 mag, was prepared by F. Argelander and coworkers; later followed the *Córdoba Durchmusterung* for the southern hemisphere.

Photographic stellar photometry was founded by K. Schwarzschild with the Göttinger actinometrie in 1904/08. He also recognized that the color index (= photographic minus visual brightness) can be used as a measure of the color and thus of the temperature of a star. Soon after the invention of the photoelectric cell, from 1911, H. Rosenberg, P. Guthnick and J. Stebbins began the development of *photoelectric* photometry. Since the invention of the photomultiplier shortly after the second World War, it has become possible to measure the brightnesses of stars with a precision of a few thousandths of a magnitude in expediently chosen wavelength ranges, such as the system of UBV magnitudes (*U*ltraviolet, *B*lue, *V*isual) of H. L. Johnson and W. W. Morgan (1951). The corresponding color indices can then be derived.

Parallel to stellar photometry, the *spectroscopy* of the Sun and stars was developed. In 1814, J. Fraunhofer discovered the dark lines in the solar spectrum which bear his name. In 1823, with an extremely modest apparatus, he was able to observe similar lines in the spectra of several stars and noted their differences. Modern *astrophysics*, i.e. the investigation of the stars with physical methods, began in 1859, when G. Kirchhoff and R. Bunsen in Heidelberg discovered *spectral analysis* and explained the Fraunhofer lines in the solar spectrum (Fig. III.1). Beginning in 1860, G. Kirchhoff formulated the basic principles of a theory of radiation, in particular Kirchhoff's Law, which deals with the relation between the absorption and emission of radiation in thermal equilibrium. This theorem, together with Doppler's principle ($\Delta\lambda/\lambda = v/c$), represented for forty years the entire theoretical basis of astrophysics. The spectroscopy of the Sun and of stars was at first directed to the following applications:

1. recording of the spectra and determination of the spectral-line wavelengths of all the elements in the *laboratory*. Identification of the lines from

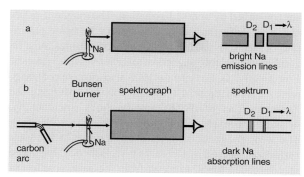

Fig. III.1a,b. The basic experiment of spectral analysis, after G. Kirchhoff and R. Bunsen, 1859. (**a**) A Bunsen-burner flame to which some sodium (salt) has been added shows bright Na emission lines in its spectrum (*right*). (**b**) When the light of a carbon-arc lamp, which is at a much higher temperature, is passed through the sodium flame, a continuous spectrum is obtained with dark Na absorption lines, similar to the solar spectrum

  stars and other cosmic light sources (W. Huggins, F. E. Baxandall, N. Lockyer, H. Kayser, Ch. E. Moore-Sitterly and many others);
2. *photographic* registration and increasingly precise analysis of the spectra of *stars* (H. Draper 1872; H. C. Vogel and J. Scheiner 1890, and others); and of the *Sun* (H. A. Rowland 1888/98).
3. *classification* of the stellar spectra, initially in a one-dimensional series, essentially as a function of (decreasing) temperature. Following early works by W. Huggins, A. Secchi, H. C. Vogel and others, E. C. Pickering and A. Cannon et al. created the Harvard Classification (starting in 1885) and the Henry Draper Catalogue. Later progress was made as a result of the discovery of the *luminosity* as the second classification parameter, and thus the determination of *spectroscopic parallaxes* by A. Kohlschütter and W. S. Adams in 1914; much later, under "modern" aspects, the Atlas of Stellar Spectra was compiled by W. W. Morgan, P. C. Keenan and E. Kellman in 1943 using the MKK classification.
4. However, for a long time the main interest of astronomers was held by the determination of the *radial velocities* of stars using Doppler's principle. Following visual attempts by W. Huggins in 1867, first usable photographic radial velocity measurements were obtained H. C. Vogel in 1888. The rotation of the Sun had already been verified spectroscopically

by that time. The further development of the technique of precise radial velocity determination was in particular the achievement of W. W. Campbell at the Lick Observatory.

What might be termed the conclusion of this period in astrophysics came with the discovery in 1913 of the *Hertzsprung–Russell Diagram*. As early as 1905, E. Hertzsprung had recognized the difference between giant and dwarf stars. H. N. Russell drew the well-known diagram based on improved trigonometric parallaxes, with the spectral types on the abscissa and the absolute magnitudes on the ordinate; it showed that most of the stars in our neighborhood fall within the narrow band of the *main sequence* (Fig. 6.8), while a smaller number occur in the region of the giant stars. Russell at first based a theory of stellar evolution on this diagram (a star begins as a red giant, undergoes compression and heating to join the main sequence, then cools along the main sequence), which however had to be abandoned ten years later. We shall take up further developments in this theme later on.

What astrophysics most needed at the beginning of the 20th century was an extension of its physical and theoretical *fundamentals*. The theory of *cavity radiation*, i.e. of the radiation field in thermal equilibrium, begun by G. Kirchhoff in 1860, had been completed in 1900 by M. Planck with the discovery of the *quantum theory* and the law of spectral energy distributions for black-body radiation. Astronomers then attempted to apply this theory to the *continuous spectra of stars* in order to estimate their temperatures. However, the brilliant Karl Schwarzschild (1873–1916) soon developed one of the future pillars of the theory of stars, his *theory of stationary radiation fields*. He showed in 1906 that in the photosphere of the Sun (i.e. in the layers which emit the main portion of the solar radiation), energy transport takes place from within to the outer layers by means of radiation. He calculated the increase in temperature with increasing depth under the assumption of radiation equilibrium, and was able to obtain the correct *center–limb darkening* for the solar disk. Schwarzschild's paper on the solar eclipse of 1905 is a masterpiece of the mutual interaction of observation and theory. In 1914, he investigated radiation exchange theoretically and spectroscopically in the broad H- and K-lines of the solar spectrum. It was then clear to him that the further development of his ideas required first of all an atomic theory of absorption coefficients, i.e. of the interaction of radiation and matter; he thus accepted with great enthusiasm the quantum theory of atomic structure founded by N. Bohr in 1913. He produced his famous papers on the quantum theory of the Stark effect and on band spectra. In 1916, K. Schwarzschild died, much too soon, at the age of only 43.

The relationship between the theory of radiation equilibrium and the new atomic physics was clarified on the one hand by A. S. Eddington in 1916/26 with his *theory of the inner structure* of stars. On the other hand, J. Eggert (1919) and M. N. Saha (1920) sought the key to the interpretation of the spectra of the Sun and of stars in the theory of *thermal ionization* and *excitation*. The fundamental principles of the modern *theory of stellar atmospheres* and of the solar and stellar spectra thus were rapidly developed. We should mention also the works of R. H. Fowler, E. A. Milne and C. H. Payne on *ionization* in stellar atmospheres (1922/25); the measurement and calculation of *multiplet intensities* by L. S. Ornstein, R. de L. Kronig, A. Sommerfeld, H. Hönl, and H. N. Russell; and then the important contributions of B. Lindblad, A. Pannekoek, M. Minnaert and others. By 1927, a serious attempt could be made to combine the theory of *radiation transport*, which had been further developed, with the *quantum theory of absorption coefficients*, and to construct a rational theory of solar and stellar spectra (M. Minnaert, O. Struve, and A. Unsöld). This made it possible to determine the *chemical composition* of stellar atmospheres from their spectra and thus to study *stellar evolution* empirically in relation to energy production by nuclear processes in the interior of stars.

The theory of *convective flow* in stellar atmospheres and especially in the Sun was founded in 1930 by A. Unsöld with the discovery of the hydrogen convection zone. The connection to hydrodynamics (theory of mixing lengths) was soon thereafter made by H. Siedentopf and L. Biermann, after S. Rosseland had already pointed out the astrophysical importance of turbulence.

We know that the ionized gases in stellar atmospheres and in other cosmic objects have a high electrical conductivity. T. G. Cowling and H. Alfvén pointed out in the 1940's that *cosmic magnetic fields* might persist for very long times until the currents associated with them are consumed by ohmic dissipation. The magnetic fields and flow fields interact continuously; the basic principles of electrodynamics and of hydrodynamics must be combined into the science of *magnetohydrodynamics*. Empirically, G. E. Hale in 1908 had discovered magnetic fields of up to 4 T in sunspots by observing the Zeeman effect of the Fraunhofer lines (incidentally, he started from physically quite incorrect hypotheses!). The weak magnetic field on the remainder of the solar surface, measured in 1952 by H. W. Babcock, has more recently been found to result from the superposition of numerous, thin flux tubes, each having a magnetic field intensity of a few tenths of a Tesla. The sunspots, the faculae which surround them, prominences, flares, and many other solar phenomena are statistically correlated in the $2 \cdot 11$-year cycle of *solar activity*.

Only by studying the flow phenomena and magnetic fields in the Sun can we arrive at a certain degree of understanding of the structure and excitation of the *corona*, the outermost shell of the Sun with a temperature of some 10^6 degrees. Here, very complex processes occur which lead to the production of X-rays and the radiofrequency radiation of the Sun, as well as various types of corpuscular radiation and the solar wind.

The investigation of the chromospheres, coronas, and actvity phenomena of other *stars* has become possible through observations in the ultraviolet and X-ray spectral regions made from space. The ultraviolet resonance lines of common elements from hot giant and supergiant stars exhibit evidence in their short-wavelength absorption components of massive *stellar winds* with velocities of several 1000 km s^{-1}. The *loss of mass* connected with these stellar winds is orders of magnitude higher than in the Sun and has an important influence on the evolution of these stars. A. Deutsch found already in the 1950's that cool giant stars are surrounded by extended, slowly expanding gas shells; these could later be observed using the infrared emissions of the dust particles and molecular emission lines in the radiofrequency region.

Since the 1920's, the *theory of the internal structure* of the Sun and of other stars has developed in parallel with progress in the understanding of the outer, "visible" stellar layers. In relation to the age of the Earth and the Sun of about $4.5 \cdot 10^9$ yr, known from radioactivity measurements, J. Perrin and A. S. Eddington suggested already in 1919/20 that the energy which is continuously radiated by the Sun is generated by conversion of hydrogen to helium. In 1938, H. Bethe and C. F. v. Weizsäcker were then able to show, based on the understanding of *nuclear physics* which had in the meantime been developed, just *which* thermonuclear reactions would, for example, be able to slowly "burn" hydrogen to helium in the interior of main-sequence stars at temperatures of about 10^7 K.

In the 1950's, in particular investigations of the color-magnitude diagrams of star clusters (Sects. 9.1.3, 2.3), together with the theory of the internal structure of stars, increased our understanding of *stellar evolution*: A star remains on the main sequence in the phase of hydrogen burning until it has consumed around 10% of its hydrogen. It then moves to the right in the color-magnitude diagram (M. Schönberg and S. Chandrasekhar, 1942) and becomes a red giant star. The theoretical investigations begun by F. Hoyle and M. Schwarzschild in 1955 thus represent a fundamental modification of A. S. Eddington's theory of stellar structure, since the assumption of a continuous mixing of the matter in the interiors of stars had to be abandoned under the pressure of observations. The new assumption is that a burned-out helium zone forms in the center of a star. The resulting forced displacement of the nuclear burning zone away from the center is the cause of the expansion of the star into a red giant.

The burning phases which follow hydrogen burning, i.e. helium burning, carbon burning, etc. take place at higher and higher temperatures and densities and on increasingly short time scales. Characteristic of these *later* evolutionary stages is a strong loss of mass through stellar winds and expulsion of outer layers, a major energy loss by emission of neutrinos, and a complex interplay of hydrodynamic and nuclear processes, so that models of these phases can be calculated only with the aid of the most powerful supercomputers. Our knowledge of the nuclear reactions which are important for energy release in stars was enormously increased by the measurements of nuclear interaction cross sections by W. A. Fowler. A closely related question to stellar evolution and the nuclear reactions which occur in it is that of the formation of the chemical elements. Stars produce heavy elements and – in part through supernova explosions – enrich the interstellar matter in these elements. Form these, a new generation of "metal rich" stars is formed, etc. Although many detailed questions are still open, we now understand the basic facts of stellar evolution including its final phases: stars which begin with less than about 8 solar masses end – often after a considerable loss of mass – as white dwarfs, while stars with originally larger masses end as neutron stars or black holes.

With regard to the compact *final stages* of stellar evolution after all energy sources have been exhausted, R. H. Fowler recognized as early as 1926 that *white dwarfs*, for example the star accompanying Sirius, have a structure which is completely different from that of normal stars; their high densities are due to the Fermi degeneracy of their electrons. A star of e.g. 0.5 solar masses may shrink down to the size of the Earth. Even its stored thermal energy suffices to supply its weak radiation losses over billions of years. According to S. Chandrasekhar (1931), white dwarfs are stable only below a mass limit of about 1.4 solar masses.

Near the end of the evolution of massive stars, the stellar matter can become even more dense (to about 10^{17} kg m^{-3}), causing the protons and electrons to combine into neutrons. These *neutron stars*, which were already predicted in 1932 by L. Landau, shortly after the discovery of the neutron, were first treated theoretically in the framework of general relativity theory by J. R. Oppenheimer and G. M. Volkoff in 1939. Only in 1967 were they discovered experimentally as *pulsars* by A. Hewish using their radio emissions. W. Baade and F. Zwicky had realized by 1934 that the collapse of a star into a neutron star releases sufficient gravitational energy to explain the violent outburst of a *supernova*. Still stronger compression leads finally, according to general relativity, to the *black holes*, whose enormous gravitational fields can prevent the escape even of light quanta.

The discoveries of *X-ray binary stars*, of X-ray and gamma-ray bursters, etc., with their intense, time-varying emissions, have provided an additional opportunity since the 1970's to observe neutron stars and, possibly, black holes. In a *close* binary star system, the components exchange matter in the course of their evolution. If a compact object, e.g. a neutron star, accretes the matter which is sucked out of its companion star, the gravitational energy which is released can be converted into X- and gamma-radiation, giving rise to unusual *high-energy* phenomena. For many years, the origin of the energetic bursts in the gamma-ray region remained one of the great unsolved riddles of astronomy, until in 1997 the *gamma bursters* could be attributed to distant galaxies as a result of their "aftgerglow" in other spectral regions.

A very interesting possibility for testing Einstein's theory of general relativity was offered by the "binary pulsar" PSR 1913+16, discovered in 1974. The lengthening of its period corresponds to the predicted energy loss rate through emission of *gravitational waves* as a result of the movements of mass within the system.

A "view" into the interior of stars is in general possible only in the case of the *Sun*: the *neutrinos* which are generated in the solar interior by the fusion of hydrogen reach the outside with practically no interactions. Their detection was achieved by R. Davis and coworkers in an experiment which has been running since the mid-1960's. Besides the Sun, the only other cosmic source of neutrinos which could thus far be identified has been the supernova SN 1987 A, relatively near to the Earth in the Large Magellanic Cloud.

Information about the interior of the Sun is also obtained from "*helioseismology*", in which the various types of *solar oscillations* (R. B. Leighton 1961, F. L. Deubner 1975, and others) are analyzed on the basis of the periodic, large-area components in the photospheric flow patterns. The oscillations of stars have recently also been spectroscopically investigated, opening up the new field of "asteroseismology".

Summarizing briefly, we can characterize the development of the *physics of stars* as follows: (1) the theory of radiation: interactions between radiation and matter; (2) the thermodynamics and hydrodynamics of flow processes; (3) magnetohydrodynamics and plasma physics. Parallel to these for stellar evolution: (4) nuclear physics and energy generation; matter at high densities; finally, in more recent times: (5) high-energy astrophysics: cosmic corpuscular radiation, X- and gamma-rays, neutrinos, and gravitational waves.

6. The Distances and Fundamental Properties of the Stars

We shall begin with a discussion of our nearest star, the *Sun*. In Sect. 6.1, we collect the data concerning the Sun, which are often used to define *units* for describing the properties of other stars, and introduce the energy distribution of the solar radiation from the photosphere: a continuum with numerous absorption lines.

Before turning to the fundamental properties of the *stars*, we consider in Sect. 6.2 their parallaxes or distances, and describe their motions relative to the Sun. Only when the distance to a star is known can we convert e.g. its apparent magnitude into an absolute radiation flux. In Sect. 6.3, we then treat the fundamental properties of the stars, among others their magnitudes, colors and radii, and in Sect. 6.4 we discuss the great variety of stellar spectra; here, we introduce the Hertzsprung–Russell diagram, which is an important way of representing the data for many stars. Finally, in Sect. 6.5, we discuss the stellar masses in connection with an overview of binary star systems.

6.1 The Sun

We briefly summarize once more the data relevant to the Sun; when dealing with the stars, we shall often use them as intuitively apparent *units of measure*.

Making use of the solar parallax of $8.794''$, we first obtained the mean distance from the Earth to the Sun, the astronomical unit:

$$1 \text{ AU} = 23\,456 \text{ equatorial Earth radii}$$
$$= 149.6 \cdot 10^6 \text{ km} . \qquad (6.1)$$

The corresponding apparent radius of the solar disk is

$$15'\, 59.63'' = 959.63'' = 0.004652 \text{ rad} . \qquad (6.2)$$

We obtain from this the radius of the Sun,

$$R_\odot = 696\,000 \text{ km} . \qquad (6.3)$$

Therefore, on the Sun, $1'' = 725$ km. Owing to atmospheric scintillation, the limit of resolving power for most observations made from the Earth is about 500 km.

The *rotation* of the Sun can be determined e.g. by observing sunspots (Sect. 7.3.3). At the solar equator, its period has the value 24.8 d, corresponding to a velocity of only 2 km s^{-1}. The equator of the Sun has an inclination angle relative to the ecliptic of $7°15'$.

The flattening of the Sun as a result of its rotation is at the limit of current detectability ($\lesssim 0.01''$).

From Kepler's 3rd Law, we obtain the *mass* of the Sun:

$$\mathcal{M}_\odot = 1.989 \cdot 10^{30} \text{ kg} . \qquad (6.4)$$

With it, we can calculate the mean density of the Sun, $\bar{\varrho}_\odot = 1409 \text{ kg m}^{-3}$; from the density, we readily obtain the gravitational acceleration at the Sun's surface,

$$g_\odot = 274 \text{ m s}^{-2} \qquad (6.5)$$

or 27.9 times greater than the acceleration of gravity on the Earth's surface.

The *spectrum* of the Sun appears on observation to consist of a *continuum*, if we initially disregard the far ultraviolet and the radiofrequency regions; it contains many dark *absorption lines*, the *Fraunhofer lines*. Those layers of the solar atmosphere which give rise to the continuous radiation and the major portion of the Fraunhofer lines are termed the *photosphere*.

In Sect. 6.1.1, we give an overview of the photospheric spectrum and its intensity curve from the center of the Sun's disk to the limb. Then, in Sect. 6.1.2, we discuss observations of the absolute energy distribution in the solar spectrum and derive in Sect. 6.1.3 the luminosity and the effective temperature of the Sun. We delay the analysis of the Fraunhofer spectrum until Sect. 7.3.

During total solar eclipses, the higher layers of the atmosphere can be observed at the limb of the solar disk; they do not emit a continuum, but rather practically only the emission lines corresponding to the Fraunhofer spectrum. This layer is called the *chromosphere*; it yields only a small contribution to the intensity of the absorption lines. Above it comes the *corona*, which stretches outwards into the interplanetary medium. We shall deal with these outermost layers of the Sun, and likewise with all the various phenonena of *solar activity* (sunspots, prominences, eruptions, etc.) or, as it is also called, with the disturbed Sun, in Sect. 7.3.

6.1.1 The Spectrum of the Photosphere. Center–Limb Variation

Using large grating spectrometers, for example in a tower telescope (Fig. 5.18), the solar spectrum can be recorded with extremely high resolution, so that the Fraunhofer lines can be investigated in great detail (Fig. 6.1).

A quantity of observational data concerning the wavelengths λ, the identifications, and the intensities of the photospheric Fraunhofer lines are collected in atlases, tables and, more recently, in digital form. Here, we mention only the tables of H. A. Rowland (1895/97) and their (second) revised edition by C. E. Moore, M. G. J. Minnaert, G. F. W. Mulders, and J. Houtgast (1966) for the range from 293.5 to 877 nm, as well as the Utrecht Photometric Atlas of the Solar Spectrum, by M. G. J. Minnaert, G. F. W. Mulders, and J. Houtgast (1940).

Many of the measurements refer to the *center* of the solar disk. Towards the limb, the solar disk exhibits a considerable *center–limb darkening*; along with the continuous spectrum, the line spectrum also shows a center–limb variation.

We specify the position of observation on the solar disk (Fig. 4.3) by giving its distance from the center in units of the solar radius, $r/R_\odot$, or, for theoretical purposes, by giving the angle of emission θ relative to the normal to the solar surface. We thus have:

$$\frac{r}{R_\odot} = \sin\theta, \quad \mu = \cos\theta = \sqrt{1-\left(\frac{r}{R_\odot}\right)^2}. \quad (6.6)$$

The center of the Sun corresponds to $\sin\theta = 0$, $\cos\theta = 1$ and its limb to $\sin\theta = 1$, $\cos\theta = 0$. The *radiation intensity* at the surface of the Sun (optical thickness $\tau_0 = 0$, Sect. 7.2.1) at a distance $r/R_\odot = \sin\theta$ from the center of the disk is denoted by $I_\lambda(0,\theta)$ or $I_\nu(0,\theta)$ referred to a wavelength or a frequency scale, (4.28). The *center–limb variation* of the radiation intensity, $I_\lambda(0,\theta)/I_\lambda(0,0)$ is measured by setting the entrance slit of the spectrograph at various positions on the solar image. The main difficulty in this and all other measure-

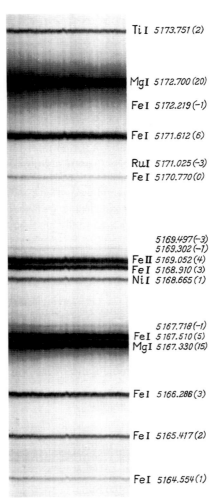

Fig. 6.1. A solar spectrum (from the center of the Sun's disk), $\lambda = 516$–518 nm recorded in the 5th order of the vacuum grating spectrograph of the McMath–Hulbert Observatory. The wavelength in Å (1 Å = 0.1 nm), the identification, and the (estimated) Rowland intensity of the Fraunhofer lines are noted at the right

ments of details on the solar disk lies in the elimination of scattered light from the instrument and the Earth's atmosphere. A summary representation of measurements of the center–limb variation at different wavelengths is given in Fig. 6.2. The ratio of the mean intensity of the solar disk, $\bar{I}_\lambda = F_\lambda/\pi$ (4.39) to the intensity at the center of the disk, $I_\lambda(0,0)$, can be readily calculated using (4.35).

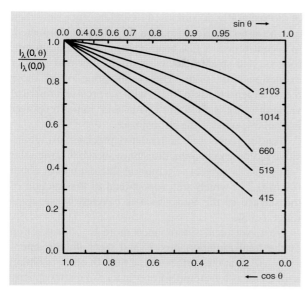

Fig. 6.2. The center–limb darkening of the Sun. The ratio of the radiation intensities $I_\lambda(0, \theta)/I_\lambda(0, 0)$ is plotted against $\cos \theta$ for several wavelengths λ in [nm] (at which there are no spectral lines) from the blue into the infrared. The non-uniform upper $\sin \theta$ scale gives the distance from the center of the disk in units of the Sun's radius

6.1.2 The Energy Distribution

Absolute values for the radiation intensity, for example at the center of the Sun's disk, $I_\lambda(0, 0)$, are obtained by reference to a cavity radiator (black body) at a known temperature. The highest fixed point on the temperature scale which has been precisely determined by gas thermometry is the melting point or *solidification point of gold*, T_{Au}. The temperature scale above T_{Au} is based on this fixed point and on pyrometer measurements using Planck's radiation formula or the radiation constant c_2 (4.61, 64). The International Practical Temperature Scale of 1968 uses the value $T_{Au} = 1337.58$ K or $1064.43\,°$C and the radiation constant $c_2 = 1.4388 \cdot 10^{-2}$ mK.

The extinction of the Earth's atmosphere must of course be precisely determined at a number of wavelengths (Fig. 4.4) and corrected for.

Since the solar spectrum (Fig. 6.1) contains many Fraunhofer lines, the best procedure is to measure the intensity $I_\lambda^m(0, 0)$ of the spectrum *including* the lines, averaged within sharply-defined wavelength regions of e.g. $\Delta \lambda = 2$ nm (superscript m). The integration over the solar disk using the center–limb variation measured in the same wavelength regions then gives directly the corresponding mean values of the radiation flux, $F_\lambda^m(0)$. Using absolute measurements of this kind, high-resolution spectra of both the center of the disk and of the radiation integrated over the disk can be calibrated.

The *continuum* between the lines, $I_\lambda^c(0, 0)$ or $F_\lambda^c(0)$, is also of interest for the physics of the Sun. Its determination in the long-wavelength region $\lambda > 450$ nm can be carried out without difficulties. In the blue to violet spectral region, however, the lines are so closely spaced that the determination of the "true" continuum becomes problematic below $\lambda = 450$ nm. In this region, only a "local quasi-continuum" can be defined using those places in the spectrum ("windows") which are the least influenced by the lines. In this part of the spectrum, the "true" continuum can only be determined by reference to a well-developed theory.

The results of the most precise absolute measurements and center–limb observations up to the present, by D. Labs and H. Neckel, as well as the high-resolution Fourier spectra by J. Brault, are shown in Fig. 6.3.

The fraction η of the radiation absorbed by the lines from the continuum, averaged over well-defined regions $\Delta \lambda$ of about 1 to 10 nm width,

$$\eta_\lambda^m = 1 - I_\lambda^m(0, 0)/I_\lambda^c(0, 0) \tag{6.7}$$

or the corresponding expression for $F_\lambda^m(0)$, is determined from high-resolution spectra. η_λ^m in the ultraviolet (300 to 400 nm), determined in 1 nm wide regions and relative to the local continuum, lies between 25 and 80%; only above $\lambda > 550$ nm does it drop down to a few percent.

6.1.3 Luminosity and Effective Temperature

By integration of the radiation flux $F_\lambda(0)$ over all wavelengths, whereby the end regions in the ultraviolet and infrared which are cut off by absorption in the Earth's atmosphere must be supplemented using measurements made outside the atmosphere, the *total radiation flux* from the Sun's surface can be obtained:

$$F = \int_0^\infty F_\lambda \mathrm{d}\lambda = 6.33 \cdot 10^7 \, \mathrm{W\,m^{-2}} \ . \tag{6.8}$$

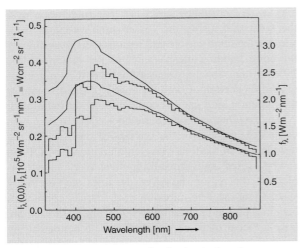

Fig. 6.3. The absolute energy distribution in the solar spectrum after H. Neckel and D. Labs (1984). The intensity $I_\lambda(0,0)$ at the center of the disk (*upper curves*) and the intensity $\bar{I}_\lambda = F_\lambda/\pi$ averaged over the disk (*lower curves*) are shown. The stepped curves represent intensities I_λ^m or $\overline{I_\lambda^m}$ averaged over intervals of $\Delta\lambda = 10$ nm, including the Fraunhofer lines. The smooth curves show the dependence of the quasi-continuum, I_λ^c or $\overline{I_\lambda^c}$. The right-hand scale shows the radiation flux (lower curves) at the Earth's position ($r = 1$ AU), but outside the atmosphere, $f_\lambda = F_\lambda(R_\odot/r)^2$. Integration of f_λ over all wavelengths yields the solar constant S (6.10)

From this value, one readily calculates the total emission of the Sun per unit time, i.e. its *luminosity*:

$$L_\odot = 4\pi R_\odot^2 F = 3.85 \cdot 10^{26} \text{ W} . \tag{6.9}$$

On the other hand, we can obtain the radiation flux S at the distance of the Earth ($r = 1$ AU) according to (4.40) by multiplying the mean radiation intensity of the solar disk by its solid angle (as seen from the Earth), $\pi R_\odot^2/r^2 = 6.800 \cdot 10^{-5}$ sr, or by multiplying the radiation flux F by the attenuation factor $(R_\odot/r)^2$. In this way, we obtain the *solar constant*:

$$S = 1.37 \text{ kW m}^{-2} . \tag{6.10}$$

This important quantity was first determined precisely by K. Ångström (about 1883) and by C. G. Abbot (about 1908), following initial experiments by C. S. Pouillet in 1837. They started by measuring the total solar radiation arriving at the Earth's surface using a "black" receiver, a pyrheliometer. This measurement must then be complemented by relative determinations of the spectrally decomposed radiation, since only then can the atmospheric extinction (Fig. 4.4) be eliminated. In recent times, S has been determined from aircraft, rockets, and space vehicles, with smaller and smaller extinction corrections. The combination of *all* these measurements leads to (6.10) with the precise numerical value of (1366 ± 3) W m^{-2}. This solar radiation power is available outside the atmosphere.

Measurements with high relative precision using sensitive radiometers on satellites have shown that the solar "constant" is subject to temporal fluctuations of about 0.1 to 0.2%, which are proportional to the fraction of the surface of the Sun which is covered by sunspots (Sect. 7.3.3).

Next, we shall return once more to consideration of the total radiation flux F and the radiation intensity $I_\lambda(0,0)$ at the surface of the Sun. In order to obtain a rough idea of the temperatures in the solar atmosphere, at first in a rather formal and preliminary way, we interpret the total flux F in terms of the Stefan–Boltzmann radiation law and thus define the *effective temperature* of the Sun:

$$F = \sigma T_{\text{eff}}^4 , \quad T_{\text{eff}} = 5780 \text{ K} . \tag{6.11}$$

Furthermore, we interpret the radiation intensity $I_\lambda(0,0)$ using Planck's radiation law (4.61) and thus define the *radiation temperature* T_λ for the center of the solar disk as a function of the wavelength λ. Analogously, a wavelength-dependent radiation temperature can also be defined by means of the monochromatic radiation flux $F_\lambda(0)$.

However, the Sun does *not* radiate as a black body (otherwise, for one thing, $I_\lambda(0,\theta)$ would be independent of θ, i.e. the Sun would exhibit no center–limb darkening; for another, the Fraunhofer lines would not occur); thus we cannot interpret T_{eff} and T_λ all too literally. T_{eff} or T_λ will at least be a reasonable approximation to the temperature in *those* layers of the solar atmosphere which give rise to the overall observed radiation or the radiation of wavelength λ, respectively. The effective temperature T_{eff} is furthermore an important characteristic of the solar atmosphere (and, correspondingly, of the atmospheres of stars), since it yields by definition, together with the Stefan–Boltzmann law, the total radiation flux F, i.e. the total energy per unit time which arrives at a unit surface of the Sun from its interior.

The analysis of the solar spectrum, the temperature and density profiles in the photosphere, and the abundances of the chemical elements will be treated in connection with the theory of stellar atmospheres in Sect. 7.2.

6.2 Distances and Velocities of the Stars

Having dealt with the Sun, we now turn to the much more distant *stars* and concern ourselves first, in Sect. 6.2.1, with the trigonometric measurement of parallaxes and thus the determination of distances to the stars, with which their true or absolute magnitudes can be calculated from the apparent magnitudes. Then, in Sect. 6.2.2, we consider the radial velocities and the proper motions of the stars, and in Sect. 6.2.3, we treat the stream or cluster parallax as an additional method of distance determination. In Sect. 6.2.4, we give an overview of the positions of the stars on the celestial sphere and of the most important star catalogs, and finally, in Sect. 6.3–5, we turn to the fundamental properties of the stars.

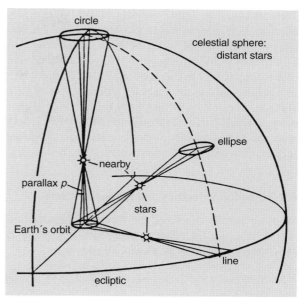

Fig. 6.4. The star parallax p. A nearby star describes a circle of radius p at the pole of the ecliptic relative to the much more distant background stars; in the plane of the ecliptic, it appears to move on a line of length $\pm p$, and in between it moves on an ellipse

6.2.1 Trigonometric Parallaxes

Due to the orbital motion of the Earth around the Sun, a nearby star describes a small ellipse in the course of a year relative to the more distant, fainter stars (Fig. 6.4; not to be confused with the aberration, Fig. 2.17!). Its semimajor axis, i.e. the angle which the radius of the Earth's orbit would subtend as seen from the star, is called the heliocentric or annual *parallax* p of the star (from $\pi\alpha\varrho\acute{\alpha}\lambda\lambda\alpha\xi\iota\varsigma$ = back-and-forth motion).

In the year 1838, F. W. Bessel in Königsberg succeeded (with the aid of a Fraunhofer heliometer) in measuring the direct trigonometric parallax of the star 61 Cyg to be $p = 0.293''$. We should also mention the observations carried out at the same time by F. G. W. Struve in Dorpat and by T. Henderson at the Cape Observatory. The star α Cen, observed by Henderson, is, along with its companion Proxima Centauri, our nearest neighbor in interstellar space. Their parallaxes, according to recent measurements, are $0.75''$ and $0.76''$. A fundamental advance was made by F. Schlesinger in 1903, who succeeded in making photographic determinations of trigonometric parallaxes with a precision of about $\pm 0.01''$.

An improved precision down to $0.001''$ for brighter stars and $0.002''$ for fainter ones has recently been achieved with the extremely precise position measurements from the European astronomical satellite HIPPARCOS (*Hi*gh *P*recision *Par*allax *Co*llecting *S*atellite), launched in 1984.

The first measurements of star parallaxes meant not only a confirmation of the Copernican world view (at that time hardly any longer necessary), but more importantly the first step towards measurements of the distances to the stars. We define suitable units of measure:

A *distance* of $360 \cdot 60 \cdot 60/2\pi = 206\,265$ astronomical units or radii of the Earth's orbit corresponds to a parallax of $p = 1''$. This distance is termed 1 parsec (from *par*allax and *sec*ond), abbreviated 1 pc. We thus have:

$$1 \text{ pc} = 3.086 \cdot 10^{16} \text{ m} = 3.26 \text{ lightyears},$$
$$1 \text{ lightyear} = 0.946 \cdot 10^{16} \text{ m} = 0.31 \text{ pc}. \quad (6.12)$$

This means that light, moving at $300\,000\,\mathrm{km\,s^{-1}}$, requires 3.26 years to travel a distance of 1 pc. A parallax p (in seconds of arc) corresponds to a distance of $1/p$ [pc] or $3.26/p$ [light years]. Given the precision of measurement of trigonometric parallaxes, these take us out into space to distances of "only" 500 or at most 1000 pc. The *stream* or *cluster parallaxes*, for groups of stars which "stream" through the Milky Way galaxy with the same velocity vectors, reach out somewhat further; they will be discussed later in Sect. 6.2.3.

6.2.2 Radial Velocities and Proper Motions

We have already described the spectroscopic determination of *radial velocities* V_r using the *Doppler effect* (H. C. Vogel, 1888; W. W. Campbell and others):

$$V_r = c\frac{\Delta\lambda}{\lambda_0}, \qquad (6.13)$$

which gives us V_r in absolute units, e.g. in [km s^{-1}], (2.3). The sign is defined in such a way that a positive (negative) radial velocity means a red shift (blue shift), i.e. moving away from the Sun (approaching the Sun). The changing influences of the orbital motion and the rotation of the Earth are avoided by using the *Sun* as reference point.

In order to describe completely the motions of the stars in space, we still require their *proper motions* μ, (PM = Proper Motion) on the celestial sphere, which were discovered much earlier by E. Halley (1718). They are usually given in seconds of arc per year. One measures relative proper motion (relative to distant stars with small PM) by comparing two photographs taken when possible with the same instrument at a time interval of 10 to 50 yr. The reduction to *absolute* proper motion presumes that the absolute positions of some stars have been determined for various epochs. Following C. D. Shane, one uses distant galaxies or also quasars as an extragalactic reference system. From 1989 to 1993 using the astrometry satellite HIPPARCOS, the proper motions of around 100 000 stars were determined with a precision of a few thousandths of an arc-second per year.

The proper motion μ is related to the *tangential component* V_t [km s^{-1}] of the star's velocity as follows: If p is the parallax of the star in seconds of arc, then μ/p is equal to V_t in astronomical units per year. The latter is equal to $1/(2\pi)$ times the orbital velocity of the Earth or $4.74\,\mathrm{km\,s^{-1}}$. We thus have

$$V_t\,[\mathrm{km\,s^{-1}}] = 4.74\,\frac{\mu\,[''\mathrm{yr}^{-1}]}{p\,['']}. \qquad (6.14)$$

The *spatial velocity* v of the star is then

$$v = \sqrt{V_r^2 + V_t^2}. \qquad (6.15)$$

The angle θ with which the star moves relative to the line of sight is given by the relations

$$V_r = v\cos\theta \quad \text{and} \quad V_t = v\sin\theta. \qquad (6.16)$$

6.2.3 Stream Parallaxes

L. Boss discovered in 1908 that e.g. for a group of many stars in Taurus, which are collected around the open star cluster called the *Hyades*, the proper motions on the celestial sphere all point to a *convergence point* at $\alpha = 6\,\mathrm{h}\,31\,\mathrm{min}$, $\delta = 5°59'$. The stars of this star stream or *moving cluster* thus apparently carry out parallel motions in space like a school of fish, with directions all tending towards the convergence point. Let the velocity of the cluster (relative to the Sun) be v_C. If we now know the proper motion μ (Fig. 6.5) of one star in the cluster, and its radial velocity V_r relative to the Sun, and if furthermore θ is the angle of the star in the heavens from the convergence point, then we can apply the considerations of (6.14) and (6.16), and find

$$\begin{aligned}V_r &= v_C\cos\theta,\\ V_t &= v_C\sin\theta = 4.74\,\mu/p,\end{aligned} \qquad (6.17)$$

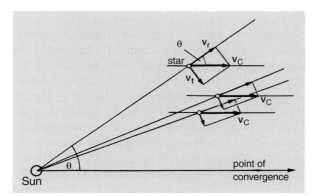

Fig. 6.5. A moving cluster or star stream

from which one immediately obtains the parallax of the star:

$$p['']=\frac{4.74\,\mu\,['' \text{ yr}^{-1}]}{V_r\,[\text{km s}^{-1}]\cdot\tan\theta}\,. \tag{6.18}$$

The Hyades cluster is at a distance of 46 pc from the Earth, and most of its stars are within a region about 10 pc in diameter. This method of stream parallaxes is – in terms of range and often in terms of accuracy – comparable to the measurements of trigonometric parallaxes by the Hipparcos satellite.

6.2.4 Star Positions and Catalogs

Every observation of stellar properties presumes that the star under consideration has been "found and identified" with a telescope, i.e. that its position e.g. in the equatorial coordinate system, (α, δ) (Sect. 2.1.1) is sufficiently precisely known.

We term the measured position of a star its *apparent position*; it is initially corrected for instrumental errors, refraction in the Earth's atmosphere (Sect. 2.1.1) and for the daily aberration. From this position relative to the Earth's center, we then compute the *true* position relative to the center of the Sun, by taking into account the annual aberration (Sect. 2.2.3) and the parallax (Sect. 6.2.1). A star catalog finally contains the *mean* position, which is obtained from the true position after correction for precession and nutation (Sect. 2.1.2) and is referred to a particular time (equinox).

The Almagest of Ptolemy (around 150 A.D.) contains about 1000 stars, whose positions are given to a precision of 10 to 15′, together with their magnitudes. Among newer catalogs, we mention the Bonner Durchmusterung (BD) (1855), with around 460 000 stars, and its extension to the southern sky (1875), the Córdoba-Durchmusterung (CD), with 610 000 stars. The Henry Draper Catalogue (HD) of the Harvard Survey (around 1900) contains about 225 000 stars with their spectral types. Among the most important tools of the astronomer is also the General Catalogue of Trigonometric Stellar Parallaxes by L. F. Jenkins (1952, 1963) and the Catalogue of Bright Stars (D. Hoffleit and C. Jaschek, 1982).

The stars are, properly considered, not "fixed", but rather change their relative positions due to their *proper motions* μ (Sect. 6.2.2). Thus for example our nearest neighbor star Proxima Centauri moves by $3.85''\text{yr}^{-1}$ and Barnard's star by as much as $10.3''\text{yr}^{-1}$ tangentially on the celestial sphere. In order to calculate precise stellar positions for any given time, it is thus necessary to know μ.

The laws of mechanics are formulated for *inertial systems* (Sect. 4.2), so that we require at least a good approximation to such a system in order to investigate the motions of the celestial objects. To fulfill this requirement for an equatorial coordinate system (α, δ), besides exact determinations of the positions and proper motions we need to apply the constants of celestial mechanics and the theory of motions of the Earth, the Moon and the planets; we thus require a "dynamic" system.

A catalog which contains especially precise star positions based on *absolute* measurements is termed a *fundamental catalog*. Since however the preparation of such a catalog is very time-consuming, for practical purposes a *reference catalog* is used, which contains the coordinates and proper motions of a large number of stars from all over the celestial sphere – for a particular equinox. From it, the data for any desired time and for other stars can be obtained by interpolation. In the 1960's, the rapid development of satellites and space probes required a modern, complete reference catalog; this need was filled by the SAO (Smithsonian Astrophysical Observatory) catalog with the positions and proper motions of 260 000 stars.

Since the 1990's, the standard for the equatorial coordinate system has been the FK5/J2000.0 System, which was set in place by the International Astronomical Union. It is represented on the celestial sphere by the Fundamental Catalogue FK5 (1988), with more than 1500 bright stars whose positions have been determined absolutely. The coordinates α and δ are referred to the vernal equinox of the Julian year 2000. This new coordinate system replaces the FK4/B1950 System, in which α and δ were calculated from the vernal equinox at the beginning of the Bessel year 1950. It is distinguished from the earlier system by – among other things – the improved position of the equinox and the precession constant ($50''\text{yr}^{-1}$). This system is very near to an inertial system. Deviations occur mainly because of the fact that the Sun moves around the center of the Milky Way galaxy with the small angular velocity of $0.55''$ per century (Sect. 11.2.1).

The reference star catalog for the new system is the PPM Star Catalogue (*Positions* and *Proper Motions*), completed in 1991/92, with about 380 000 stars distributed over the entire celestial sphere, whose positions are exact to ca. 0.3″ and the proper motions to about 0.4″ per 100 yr.

A further increase in precision, e.g. of the parallaxes to around one-thousandth of an arc second, was obtained through the measurements of the astrometric satellite HIPPARCOS from 1989 to 1993. Parallaxes for more than 20 000 stars with uncertainties of less than 10% are available. The data are collected in two catalogs of the European Space Agency (ESA) (1997): *Hipparcos*, with parallaxes and proper motions of around 120 000 stars, and *Tycho*, with more than 10^6 astrometrically investigated stars with their magnitudes and their colors.

6.3 Magnitudes, Colors and Radii of the Stars

We first take up in Sect. 6.3.1 the apparent magnitudes of the stars, then in Sect. 6.3.2 their colors and spectral energy distributions. With knowledge of their distances, we can compute the absolute magnitudes of stars from their apparent magnitudes (Sect. 6.3.3). In Sect. 6.3.4, we introduce the bolometric magnitude and the luminosity, and finally we give an overview of the radii of stars and the methods of determining them in Sect. 6.3.5.

6.3.1 Apparent Magnitudes

Hipparchus and many of the earlier astronomers had already begun to make catalogues of the brightnesses of stars. *Stellar photometry* in the modern sense dates from the definition of the magnitude scale by V. Pogson in 1850 and the construction of a visual photometer in 1861 by J. F. Zöllner. With this instrument, the brightness of a star could be precisely compared with that of an artificial star image by using two Nicol prisms; when their directions of polarization are rotated through an angle α, they attenuate the transmitted light by a factor $\cos^2 \alpha$.

If the radiation fluxes measured from two different stars are in the ratio s_1/s_2, then according to Pogson's definition, their "*apparent magnitudes*" differ by an amount

$$m_1 - m_2 = -2.5 \log(s_1/s_2) \quad \text{[mag]} \quad (6.19)$$

or, conversely,

$$s_1/s_2 = 10^{-0.4(m_1-m_2)} . \quad (6.20)$$

Thus, a difference of:

$-\Delta m = 1 \quad 2.5 \quad 5 \quad 10 \quad 15 \quad 20$ [mag]

corresponds to a ratio of the radiation fluxes (or, for short, a brightness ratio) of:

$s_1/s_2 = 2.51 \quad 10 \quad 10^2 \quad 10^4 \quad 10^6 \quad 10^8$.

The unit of measure of m is the *magnitude* and is abbrevated as mag or m.

The zero point of the magnitude scale was originally based on the international *pole sequence*, a series of stars in the neighborhood of the pole, whose magnitudes were precisely measured and checked for their constancy. Since brighter stars correspond to smaller and finally to negative magnitudes, it is reasonable to say, for example, "The star α Lyr (Vega) with an apparent (visual) magnitude of 0.14 mag is 1.19 magnitudes *brighter* than α Cyg (Deneb), which has 1.33 mag", or, "α Cyg is 1.19 magnitudes *fainter* than α Lyr".

The early photometric measurements were carried out visually. Photographic photometry came into practice in 1904/08, beginning with K. Schwarzschild's Göttinger Aktinometrie, initially using conventional blue-sensitive plates. It was soon discovered that by using plates sensitized in the yellow region and placing a yellow filter in front of them, one could imitate the spectral sensitivity of the human eye. Thus, in addition to visual magnitudes m_V, photographic magnitudes m_{pg} and then photovisual magnitudes m_{pv} could be determined. Today, it is possible to adjust the sensitivity maximum of the apparatus to any desired wavelength region by using a suitable combination of detectors with corresponding color or interference filters.

We begin by clarifying our concepts and asking the question, "What do the various magnitudes actually mean?"

Assume that a star of radius R emits at its surface a radiation flux F_λ at the wavelength λ. If it is at a distance r from the Earth, we obtain a radiation flux outside the Earth's atmosphere (we shall initially neglect interstellar extinction; see (4.40)) given by:

$$f_\lambda = R^2 F_\lambda / r^2 \,. \tag{6.21}$$

Now assume that the value which our apparatus indicates for a normal spectrum $f_\lambda \equiv 1$ can be described as a function of wavelength by a sensitivity function E_λ, which is determined by the transmission of the instrument and the sensitivity of the detector. The indicated value is then proportional to the integral:

$$s = \frac{1}{r^2} \int_0^\infty R^2 F_\lambda E_\lambda \, d\lambda \,, \tag{6.22}$$

and the apparent magnitude m of our star is given by the following expression (with a constant of integration which must be fixed by convention):

$$m = -2.5 \log \frac{1}{r^2} \int_0^\infty R^2 F_\lambda E_\lambda \, d\lambda + \text{const} \,. \tag{6.23}$$

In (6.23), we have left out the extinction due to the Earth's atmosphere. Since E_λ in practice takes on reasonably large values only within a limited wavelength range, the *extinction determination* can indeed be carried out as in Fig. 4.4 without further problems, as if for monochromatic radiation at the sensitivity maximum of E_λ.

The *visual* magnitudes m_V are, from (6.23), thus defined by the sensitivity function of the human eye *times* the transmission of the instrument; the sensitivity maximum, called the isophotic wavelength, lies at about 540 nm, in the green. In a corresponding manner, the sensitivity maximum for the *photographic* magnitudes m_{pg} lies near $\lambda = 420$ nm.

6.3.2 Color Index and Energy Distribution

The difference of two apparent magnitudes measured in different wavelength ranges X and Y is termed the

$$\text{color index} = m_X - m_Y \,. \tag{6.24}$$

As a standard system for the determination of stellar magnitudes and colors, the UBV system developed by H. L. Johnson and W. W. Morgan in 1951 is currently

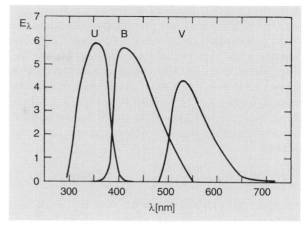

Fig. 6.6. Relative sensitivity functions E_λ (referred to a light source with $f_\lambda = \text{const}$) for UBV photometry. After H. J. Johnson and W. W. Morgan

most often used (U = ultraviolet, B = blue, V = visual); the resulting magnitudes are denoted in brief as:

$$U = m_U \,, \quad B = m_B \,, \quad V = m_V \,. \tag{6.25}$$

The corresponding sensitivity functions E_λ, which can be obtained photographically or photoelectrically, are displayed in Fig. 6.6. Their sensitivity maxima for average star colors are:

$$\lambda_U \simeq 365 \text{ nm} \,, \quad \lambda_B \simeq 440 \text{ nm} \,,$$
$$\lambda_V \simeq 548 \text{ nm} \,. \tag{6.26}$$

For hot (blue) or cool (red) stars, these are shifted to shorter or longer wavelengths, respectively.

The three magnitudes U, B, and V are by definition related to each other in such a way (i.e. the three constants in (6.23) are chosen in such a way) that for A0 V stars (e.g. α Lyr = Vega; Sect. 6.4)

$$U = B = V \,,$$
$$U - B = B - V = 0 \,. \quad \text{(for A0 V)} \tag{6.27}$$

For practical purposes, including transfer from one instrument to another with a (possibly) somewhat different sensitivity function, the UBV system is referred to a number of precisely measured *standard stars*, whose magnitudes and colors cover a large range.

The color indices give a measure of the *energy distribution* in the stellar spectra, i.e. of the wavelength dependence of the radiation flux f_λ or F_λ, as first recognized by K. Schwarzschild; they are thus also a measure of the temperatures of the stellar atmospheres. As a first approximation, we obtain the *color temperature* T_C, when we fit the energy distribution, in a limited spectral region, to Planck's radiation law. In Wien's approximation ($c_2/(\lambda T) \ll 1$), we have $F_\lambda \propto \exp(-c_2/(\lambda T_C))$; if we limit the integrations in (6.23) to the sensitivity maxima of the sensitivity functions, we obtain for example:

$$B - V = \frac{2.5 \, c_2 \log e}{T_C}\left(\frac{1}{\lambda_B} - \frac{1}{\lambda_V}\right) + \text{const.} \qquad (6.28)$$

or, with the radiation constant $c_2 = 0.014$ mK and (6.26),

$$B - V \simeq \frac{0.7 \cdot 10^4}{T_C} + \text{const.} \qquad (6.29)$$

The numerical values of the color temperatures calculated e.g. from the color index B − V have no great significance, owing to the considerable deviations of the stars from black body behavior (absorption constants, Fraunhofer lines, ...; see Fig. 6.7). The importance of the color indices lies elsewhere: they (or the color temperatures) are related to fundamental parameters, in particular to the effective temperature T_{eff} (total radiation flux!), the gravitational acceleration g, and the absolute magnitudes of the stars (Sect. 7.2), as we shall see from the theory of stellar atmospheres. Since the color indices can be measured with a precision of 0.01 mag without difficulty, it can be expected from (6.29) that e.g. around 7000 K, temperature *differences* can be determined with an accuracy of about 1%, which is not obtainable with any other method. The temperatures themselves are of course not nearly so accurately determined.

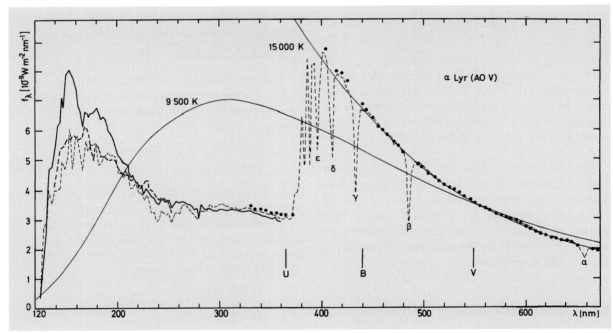

Fig. 6.7. The energy distribution in the spectrum of the A0 V star α Lyr (Vega), showing the observed radiation flux density f_λ in units of $[10^{-11}\,\text{W}\,\text{m}^{-2}\,\text{nm}^{-1}]$. In the *optical region*: absolute measurements by H. Tüg et al. (1977), corrected for extinction in the Earth's atmosphere excluding the regions where there are strong hydrogen lines (points ●●●). In the *ultraviolet*, $\lambda \leq 350$ nm: satellite observations from TD 1 (C. Jamar et al., 1976; (– – –) and from Copernicus (A. D. Code and M. Meade 1979) (———). *For comparison*: the Kirchhoff-Planck function $B_\lambda(T)$, fitted to the data at $\lambda = 555$ nm for 9500 K, corresponding to the overall radiation flux or the effective temperature, and for 15 000 K, the color temperature in the blue. Also shown is a theoretical model (– – –) by R. L. Kurucz (1979) for $T_{\text{eff}} = 9400$ K and $g = 89.1$ m s^{-2}, taking into account absorption lines

In addition to the UBV system, the UGR system of W. Becker, with sensitivity maxima at $\lambda = 366$, 463, and 638 nm and the *six color system* of J. Stebbins, A. E. Whitford, and G. Kron, which reaches from the ultraviolet to the infrared ($\lambda_U = 355$ nm ... $\lambda_I = 1.03\,\mu$m), have attained some importance. The mutual interrelation of magnitudes and color indices among the various systems with not too different isophotic wavelengths is calculated with the aid of empirically derived relations which are for the most part linear.

H. L. Johnson and others have extended the UBV system into the red and infrared by adding the following filters and sensitivity-maximum wavelengths, determined essentially by the transmission of the Earth's atmosphere:

$$\begin{array}{c|ccccc}
 & R & I & J & H & K \\
\lambda: & 0.7 & 0.9 & 1.25 & 1.63 & 2.2 \\
 & L & M & N & Q & \\
 & 3.6 & 5.0 & 10.6 & 21 & [\mu\text{m}]
\end{array} \quad (6.30)$$

If the colors of the faintest stars are not required, broadband photometry (with halfwidths of the sensitivity functions $\Delta\lambda \gtrsim 50$ nm) is not necessary, and measurements can be carried out with narrow band filters, which offer advantages with respect to a theoretical calibration as a function of the fundamental stellar parameters. The *medium bandwidth photometry* developed by B. Strömgren ($\Delta\lambda \simeq 10$–30 nm) is often used; it employs the following filters:

$$\begin{array}{c|cccc}
 & u & v & b & y \\
\lambda: & 350 & 411 & 467 & 547 & [\text{nm}]
\end{array} \quad (6.31)$$

The color indices used in this system are $u - b$ and $b - y$ (analogous to $U - B$ and $B - V$) as well as $c_1 = (u - v) - (v - b)$ and the "metal index" $m_1 = (v - b) - (b - y)$.

Furthermore, *narrow band filters* having a halfwidth of a few nm can be employed to investigate individual strong lines or groups of lines in the spectra. For example, Strömgren's uvby System is extended to include the β-index for brighter stars; this measures the intensity of the hydrogen line $H\beta$ $\lambda = 486$ nm relative to the neighboring continuum through a filter with $\Delta\lambda \simeq 3$ nm.

For measurements in spectral regions of only a few nm width, photoelectric *spectrophotometry* has become increasingly important since the 1960's. In this method, the narrow spectral regions are not defined by color or interference filters, but rather are defined by a slit or by the width of the detector (for example a CCD chip) in the focal plane of a spectrograph. The goal of spectrophotometry, to investigate as much of the entire accessible spectrum as possible, can be accomplished with *one* detector which measures the various parts of the spectrum *sequentially* (scan technique), or with *several* detectors which register the energy distribution in the spectrum by simultaneous measurements at different wavelengths (multichannel spectrophotometry).

The Standard Star α Lyrae. Although it is relatively easy to determine relative radiation fluxes with spectrophotometry, relating them to carefully investigated standard stars, it is difficult to obtain from these the *absolute* fluxes. The A0 V-star α Lyra (Vega) serves as the primary photometric standard. We show in Fig. 6.7 the absolute measurements of α Lyra by H. Tüg, N. M. White and G. W. Lockwood (1977), in which the calibration was accomplished by comparison to a cavity radiator (black body) of precisely known temperature. The melting point of platinum at 2042.1 K was one of the fxed points used to determine the temperature scale. The radiation fluxes obtained in this manner should be accurate to 1 to 2%. In the far ultraviolet, there are observations of α Lyra using, for example, the European astronomical satellite TD 1 and the Copernicus satellite, with bandwidths of a few nm. In this range, absolute calibration is very difficult, and the accuracy of the fluxes obtained is probably not better than 10 to 20%.

The infrared satellite IRAS discovered in the case of α Lyr in the 60 µm and 100 µm bands a radiation flux which was around 10 times greater than expected on the basis of the theoretical atmosphere models, which however reproduce the optical and ultraviolet regions well. From the observed diameter of the infrared source of about $35''$, this excess radiation can be traced to the thermal emission from a *circumstellar disk of dust* about 140 AU in diameter with a temperature of only 85 K, making use of the radiation balance of the particles (10.12) as in the case of the interstellar dust. The dust particles, with radii mainly in the range of 0.1 to 10 µm, are somewhat larger than those in the usual interstellar medium (Sect. 10.1.4). The mass of the disk is of the order of magnitude of the mass of our Solar System, so that this could be an example of a "protoplanetary system".

6.3.3 Absolute Magnitudes

The observed apparent magnitude of a star (in some arbitrary photometric system) depends, according to (6.23), on its true magnitude *and* on its distance. We now define the *absolute* magnitude M of a star by imagining it to be displaced from its true distance r to a *standard distance* of 10 pc. From the $1/r^2$ law of photometry (6.21), its magnitude is then modified by a factor of $(r/10)^2$. In magnitudes, we thus find

$$m - M = 5 \log \frac{r\,[\text{pc}]}{10} = 5 \log r\,[\text{pc}] - 5$$
$$= -5 - 5 \log p[''] \,. \quad (6.32)$$

The quantity $(m - M)$ is termed the *distance modulus*. A modulus $(m - M)$ of

$$-5 \quad 0 \quad +5 \quad +10 \quad +25 \quad [\text{mag}] \quad (6.33)$$

corresponds to the distance r

$$1\,\text{pc} \quad 10\,\text{pc} \quad 100\,\text{pc} \quad 10^3\,\text{pc} \quad 10^6\,\text{pc}$$
$$= 1\,\text{kpc} \quad = 1\,\text{Mpc} \,.$$

In our calculation using the $1/r^2$ law, we initially disregarded *interstellar extinction*. We must recognize that it becomes quite important at distances of more than 10 pc. Then the relation (6.23) between the distance modulus and the distance or parallax must be corrected:

$$m - M = 5 \log r\,[\text{pc}] - 5 + A \,, \quad (6.34)$$

where A gives the interstellar extinction in magnitudes. $A = 1$ mag corresponds to an optical thickness $\tau \simeq 1$ for extinction in the wavelength region under consideration, since from (4.54), $A = -2.5 \log e^{-\tau} = 1.08\,\tau$.

In principle, absolute magnitudes can be quoted in any photometric system; they are then denoted by M with the corresponding subscript, for example M_V = absolute visual magnitude. If no subscript is given, M_V is always to be understood.

We shall now collect the data for the *Sun as a star*, which are important as a point of reference. We need hardly mention that it is metrologically extremely difficult to make a photometric comparison of the Sun with the fainter stars, which have brightnesses at least 10 orders of magnitude fainter. The *distance modulus* of the Sun is obtained from the definition of the parsec and (6.32) and is $(m - M)_\odot = -31.57$ mag. With this, in units of magnitude, we obtain:

Apparent magnitude	Color indices	Absolute magnitude
$U_\odot = -25.85$	$(U-B)_\odot = 0.18$	$M_{U,\odot} = +5.72$
$B_\odot = -26.03$	$(U-V)_\odot = 0.67$	$M_{B,\odot} = +5.54$
$V_\odot = -26.70$		$M_{V,\odot} = +4.87$

The limits of error of these data lie in the range of several hundreths of a magnitude.

6.3.4 Bolometric Magnitudes and Luminosities

Along with the radiation in different wavelength ranges, the *total radiation* of the stars is of interest. In analogy to the solar constant, we therefore define the *apparent bolometric* magnitude in the sense of (6.23):

$$m_{\text{bol}} = -2.5 \log \frac{1}{r^2} \int_0^\infty R^2 F_\lambda \, d\lambda + \text{const}$$

$$= -2.5 \log \frac{R^2 F}{r^2} + \text{const} \,, \quad (6.35)$$

where

$$F = \sigma T_{\text{eff}}^4 \quad (6.36)$$

(T_{eff}: effective temperature) again denotes the total radiation flux at the stellar surface. The constant is defined in such a manner that the bolometric correction, B.C.

$$\text{B.C.} = m_V - m_{\text{bol}} \quad (6.37)$$

corresponding to the color indices, never becomes negative, i.e. its minimum, at $T_{\text{eff}} \simeq 7000$ K, is set to zero. (A definition with reversed sign can also be found, so that B.C. = $m_{\text{bol}} - m_V$ never becomes positive.) Since the Earth's atmosphere completely absorbs wide spectral regions, the bolometric magnitude *cannot* be measured directly from the ground; its name is in fact misleading. It must be determined by making use of measurements from satellites, for example. With the aid of theory, m_{bol} or B.C. can be calculated as a function of other, *measureable* parameters of stellar atmospheres.

For the Sun, we obtain

$$m_{\text{bol},\odot} = -26.83 \text{ mag} \,,$$
$$\text{B.C.}_\odot = +0.13 \text{ mag} \,. \quad (6.38)$$

The *absolute* bolometric magnitude $M_{\rm bol}$ of a star is a measure of its total radiation energy output per unit time, i.e. of its *luminosity L*. The latter quantity is generally referred to the solar luminosity $L_\odot$ as unit of measure. With $(m-M)_\odot = -31.57$ mag, we obtain $M_{\rm bol,\odot} = 4.74$ mag and thus:

$$M_{\rm bol} = 4.74 - 2.5 \log\left(\frac{L}{L_\odot}\right) \quad [\text{mag}],$$
$$L_\odot = 3.85 \cdot 10^{26} \text{ W}. \qquad (6.39)$$

Since the energy radiated from the stars is generated in their interiors by nuclear processes, the luminosity belongs among the most important starting data values for investigations into internal stellar structure.

6.3.5 Stellar Radii

If a star of radius R is located at a distance r, then the very small angle which its radius subtends at this distance is $\alpha = R/r$ in angular units (rad) or

$$\alpha['']= 206\,265 \cdot \frac{R}{r} = 4.65 \cdot 10^{-3}\,\frac{R/R_\odot}{r\,[\text{pc}]} \qquad (6.40)$$

in seconds of arc. Even for the star system which is nearest to us (α Cen), at a distance of $r = 1.3$ pc, the angular radii are thus only a few milliseconds of arc and place heavy demands on the accuracy of measurements.

Only in the case of the Sun and a few other stars can α be measured *directly*; thus, using the Hubble Space Telescope, it was possible to determine the angular radius in the ultraviolet of the red giant α Ori (Betelgeuse), and with the New Technology Telescope, α of the Mira variable R Dor could be measured in the near infrared.

For several bright stars, *interferometric* methods allow a determination (Sect. 5.1.2). Although measurements with the Michelson interferometer are limited to a very few red giant stars, R. Hanbury Brown und R. Q. Twiss were able to use their correlation interferometer in Narrabri (Australia) to determine the angular radii of about 30 stars, including dwarfs, to about 0.001″. With the Sydney University Stellar Interferometer (SUSI), commissioned in 1996 and also near Narrabri, J. Davis and coworkers have been able to determine the radii of stars down to the 8th magnitude with a precision between 0.020″ and just under 0.0001″. This instrument has a base line between 5 and 640 m at an effective aperture of 14 cm.

Speckle interferometers on individual telescopes have angular resolutions from (5.1) given only by the diameter of the primary mirror, but they can be used to observe stars of apparent magnitudes in the range of 14 to 17 mag (with good seeing) and to make very precise measurements.

Large telescopes (Sect. 5.1.3), such as the Keck telescopes or the VLT(I), which can be used as interferometers with base lines of up to nearly 100 m, will in the near future likewise attain angular resolutions of around 0.001″.

For stars in the neighborhood of the ecliptic, *lunar occultations* (Sect. 2.1.5) can be used to measure angular radii, also to about 0.001″. In this case, an arrangement with a sufficiently short time constant (≤ 1 ms) registers the intensity fluctuations from Fresnel diffraction of the starlight by the sharp edge of the Moon; these pass by a particular observer in a few tens of ms. The angular radius of the stellar disk can then be derived from the contrast of the intensity bands compared to that expected from a point source.

In the rare cases of *eclipsing variable systems* (Sect. 6.5.2), the radii can be derived very precisely from the observed light curves.

For the majority of stars, the radii must be estimated *indirectly* using photometric methods from their magnitudes or radiation fluxes. From (4.40), we obtain the monochromatic radiation flux f_λ to be expected from a star at a distance r:

$$f_\lambda = F_\lambda \left(\frac{R}{r}\right)^2 = F_\lambda \alpha^2 \qquad (6.41)$$

and the total radiation flux (6.36)

$$f = F\alpha^2 = \sigma T_{\rm eff}^4 \alpha^2. \qquad (6.42)$$

Thus, from the measured radiation fluxes f_λ or f, or using (6.23) with the apparent magnitude m measured in one of the photometric standard regions (e.g. U,B, or V), the angular radii can be calculated when the radiation flux F_λ or F at the star's surface, or its effective temperature $T_{\rm eff}$, are known. These latter quantities can be derived from the color or the energy distribution by making use of the theory of stellar atmospheres.

When greater precision is required in determining the angular radii, the *center–limb variation* (Sect. 6.1.1) must be taken into account. Especially in the case of cool giant stars, with their extended atmospheres, the radius depends also on the wavelength or color which is used for the observations.

In order to obtain the *stellar radii R* themselves from the angular radii α, we require the distance r or the parallax p, or else the modulus of distance or the absolute magnitude (6.32), or, because of the relation

$$4\pi r^2 f = 4\pi R^2 F = 4\pi R^2 \sigma T_{\text{eff}}^4 = L \qquad (6.43)$$

we can use the luminosity L (derived e.g. from the star's spectrum).

The largest observed angular radii are found for the red giant stars and the Mira variables. Thus, for α Ori in the optical range $\alpha = 0.028''$ was measured, and in the ultraviolet (with the Hubble Space Telescope) $0.062''$, while for R Dor, in the near infrared, $\alpha = 0.028''$. Angular radii larger than $0.02''$ were also found for, among others, W Hya, R Leo, o Cet (Mira) and α Sco (Antares). The radii of the red giants are for the most part somewhat variable.

From these angular radii, we find – using the none too precise distances r in (6.40) – radii R of 400 to 500 $R_\odot$, or around 2 AU; for α Ori (M2 Iab, V = 0.5 mag, $r = 200$ pc) even $R \simeq 1000\,R_\odot \simeq 4.6$ AU. The diameters of these stars are thus greater than that of Mars' orbit!

In the case of α Ori and some Mira variables such as o Cet, W Hya and R Cas, *non-spherical* brightness distributions are found, which indicate flattening as a result of rapid rotation, or else star spots.

6.4 Classification of the Stellar Spectra. The Hertzsprung–Russell Diagram

When, as a result of the discoveries of J. Fraunhofer, G. Kirchhoff and R. Bunsen, the observation of *stellar spectra* was begun, it quickly became clear that they can for the most part be ordered according to their spectral type (Sect. 6.4.1) in a *one-parameter* sequence. As early as 1905, E. Hertzsprung recognized the difference between giant and dwarf stars, and H. N. Russell by 1913 had begun to investigate the dependence of the absolute magnitude M_V on the spectral type Sp of the star. In Sect. 6.4.2, we discuss the connection between M_V and Sp, which is represented in the Hertzsprung–Russell diagram, one of the most important diagrams in astronomy, along with the luminosity classes of the stars. The modern classification of W. W. Morgan, P. C. Keenan and E. Kellman will be treated in Sect. 6.4.3. In the following Sect. 6.4.4, we introduce the two-color diagram, and in Sect. 6.4.5 we give a summary of the rotation of stars.

6.4.1 The Spectral Type

Based on the work of W. Huggins, A. Secchi, H. C. Vogel, and others, E. C. Pickering and A. Cannon developed the *Harvard Classification* of stellar spectra in the 1880's; it formed the basis of the Henry Draper Catalogue. The series of *spectral classes* ("Harvard types") which are ascribed to the colors of the stars:

$$\begin{array}{c} \nearrow S \\ O - B - A - F - G - K - M \\ \searrow R - N \end{array} \qquad (6.44)$$

blue yellow red

resulted from older classification schemes after numerous modifications and simplifications. H. N. Russell's students in Princeton invented the well-known mnemonic for this series: *O Be A Fine Girl, Kiss Me Right Now* (Note, for experts only): S stands for "Smack" (or *Sweetheart*!). The spectral classes R and N today are usually combined into the class C (for *c*arbon stars).

We still refer to O and B as *early*, A, F, and G as *middle*, and K and the classes beyond as *late* spectral classes; this terminology dates from earlier, incorrect ideas about the evolutionary stages of the stars.

The changes in the colors of the stars, or their color indices, which occur parallel to the single-parameter sequence of the Sp (also indicated in (6.44)) showed that the stars had been ordered according to decreasing *temperature*.

Between each of these letters, a finer subdivision is indicated by a following number, 0 to 9. For example, a B5 star is between B0 and A0 and has about an equal amount in common with each of these classes. The definition of the Harvard sequence is accomplished principally by comparison of the spectral profiles of certain standard stars (Fig. 6.10).

The MK Classification (Sect. 6.4.3), which is in general use today, has adopted the nomenclature of the Harvard Classification to a large extent.

6.4 Classification of the Stellar Spectra. The Hertzsprung–Russell Diagram

Table 6.1. The classification of stellar spectra

Spectral class	Temperature [K]	Classification criteria
O	50 000	Lines from highly ionized atoms: He II, Si IV, N III, …; hydrogen H relatively weak; occasional emission lines.
B0	25 000	He II missing; He I strong; Si III, O II; H stronger.
A0	10 000	He I missing; H at maximum; Mg II, Si II strong; Fe II, Ti II weak; Ca II weak.
F0	7600	H weaker; Ca II stark; the ionized metals, e.g. Fe II, Ti II had their maxima at ∼ A 5; the neutral metals, e.g. Fe I, Ca I now have about the same strength.
G0	6000	Ca II very strong; neutral metals Fe I, … very strong.
K0	5100	H relatively weak, neutral atomic lines strong; molecular bands.
M0	3600	Neutral atomic lines, z. B. Ca I, very strong; TiO bands.
M5	3000	Ca I very strong, TiO bands stronger.
C	3000	Strong CN-, CH-, C_2-bands, TiO missing. neutral metals as with K and M.
S	3000	Strong ZrO-, YO-, LaO-bands; neutral atoms as with K and M.

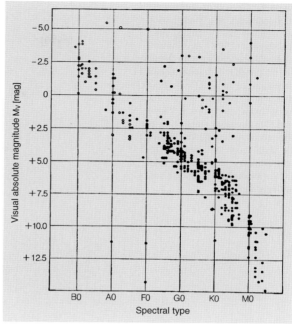

Fig. 6.8. Hertzsprung–Russell diagram (H. N. Russell 1927). The visual absolute magnitude M_V is plotted against spectral type. The Sun corresponds to $M_V = 4.8$ mag and class G2

In Table 6.1, we give the spectral classes together with the behavior of their more important classification criteria. Instead of the spectral lines used for the classification, we give here only their corresponding elements and ionization states (I = neutral atom, II = singly ionized atom, III = doubly ionized atom, …). The quoted temperatures correspond *approximately* to the colors of the stars and are intended only to give an initial orientation.

6.4.2 The Hertzsprung–Russell Diagram. Luminosity Classes

In the year 1913, H. N. Russell had the happy thought of investigating the relationship between the *spectral type* Sp and the *absolute magnitude* M_V of the stars, by constructing a diagram with Sp as abscissa and M_V as the ordinate, and plotting all stars for which the parallaxes were known with sufficient accuracy. Figure 6.8 shows such a diagram, which Russell drew in 1927 using considerably improved observational results, for his textbook; it served a whole generation as "astronomical Bible".

Most stars populate the narrow band of the *main sequence*, which stretches diagonally from the (absolutely) bright blue-white B and A stars (e.g. the belt stars in Orion) through the yellow stars (e.g. the Sun, G2 and $M_V = +4.8$ mag) out to the faint red M stars (e.g. Barnard's star, M5 and $M_V = +13.2$ mag).

On the upper right, we find the group of *giant stars*; in contrast, those stars of the same spectral class which have much smaller luminosities are termed *dwarf stars*. Since, at the same temperatures, the difference in absolute magnitude can only result from a corresponding difference in *radii*, these names seem quite appropriate. The distinction and classification of the giant and dwarf stars dates back to older work (1905) of E. Hertzsprung, so that today we refer to the (Sp, M_V) diagram as a *Hertzsprung–Russell Diagram* (HRD).

Instead of the spectral type, a color index, e.g. B − V, can be plotted; in this way, one obtains a Color–Magnitude Diagram (CMD) which is equivalent to the HRD. Figure 6.9 shows the CMD (B − V, M_V) with its now very sharply defined main sequence, yellow and red giant stars (upper right) and white dwarfs (lower left); it contains individual and cluster stars from our neighborhood with precisely determined parallaxes. Since color indices can be precisely measured even for faint stars,

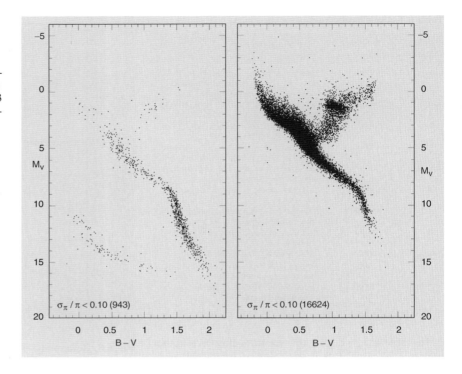

Fig. 6.9. The color–luminosity diagram with M_V [mag] plotted against $B-V$ [mag] after H. Jahreiss (1999); it contains main sequence stars with trigonometric parallaxes having relative uncertainties of $\leq 10\%$. *Left:* 943 stars whose parallaxes were measured from the ground; *Right:* 16 623 individual stars from the Hipparcos Catalog, for which the data were obtained by the astrometry satellite HIPPARCOS. (With the kind permission of the author)

the CMD has become the most important tool of stellar astronomy.

The extremely bright stars along the upper edge of the HRD or CMD are called supergiants. For example, α Cyg (Deneb; A2) has an absolute magnitude $M_V = -7.2$ mag; it thus exceeds the luminosity of the Sun ($M_V = +4.8$ mag) by 12.0 magnitudes, i.e. by a factor of 60 000!

A further clearly recognizable group are the *white dwarfs* at the lower left. Since they have only weak luminosities in spite of their relatively high temperatures, they must be very small; their radii are readily calculated from $L \propto R^2 T_{\text{eff}}^4$ (6.43) to be barely larger than the Earth's radius. For Sirius' companion α CMa B and a few other white dwarfs in binary star systems (Sect. 6.5.1), the masses are also known, so that mean densities of the order of 10^8 to 10^9 kg m^{-3} can be calculated. The internal structure of such stars must therefore be quite different from that of other stars. R. H. Fowler showed in 1926 that in white dwarfs, the matter (more precisely, the electrons) is *degenerate* in the sense of Fermi statistics, in the same way as was demonstrated soon thereafter by W. Pauli and A. Sommerfeld for the conduction electrons in a metal. This means that nearly all the quantum states are completely occupied, taking into account the Pauli Principle, as in the inner shells of a heavy atom.

We shall return to further groups of stars in the HRD, which for the most part are relatively small and specialized, in connection with other topics.

It was noted by E. Hertzsprung in 1905 that stars with sharp hydrogen lines, for example the A2 star α Cyg, are remarkable for their high luminosities. In 1914, W. S. Adams and A. Kohlschütter then showed that the stars of a particular spectral class can be further subdivided, corresponding to their luminosities or *absolute magnitudes* M_V, on the basis of new spectroscopic criteria. Among the absolutely bright stars, for example, the lines from ionized atoms are stronger relative to those from neutral atoms; among the A stars, as mentioned, the hydrogen lines can be thus used as a criterion for luminosity.

If such a *luminosity criterion* (which can hold only in a particular range of spectral classes) is calibrated using stars of known absolute magnitude, one can determine *absolute* magnitudes spectroscopically. If the interstellar absorption (which was completely unknown in 1914!) can be neglected or corrected for, one can determine *spectroscopic* parallaxes by combining with the known apparent magnitudes of the stars (6.34). We shall have more to say in Chap. 11 about their importance for the investigation of the Milky Way galaxy. Here, we follow the significant insight that the majority of stars can be classified using *two* parameters, Sp and L (or M_V).

6.4.3 The MK Classification

From the Harvard classification, W. W. Morgan and P. C. Keenan developed the two dimensional MK Classification, which is today in general use, and is summarized in "An Atlas of Stellar Spectra. With an Outline of Spectral Classification." Its general principles hold for *any* classification:

1. The classification is based only upon *empirical* criteria, i.e. directly observable absorption and emission phenomena.
2. The observational data are *unified*. In order on the one hand to be able to define sufficiently fine spectral criteria, but on the other to penetrate far enough into the galaxy, a unified dispersion of about 12.5 nm/mm for Hγ is employed, *even* for bright stars.
3. The *transferability* of the classification system to other instruments is guaranteed by a list of suitable *standard stars*, i.e. by direct comparison, not by (possibly semitheoretical) descriptions.
4. The classification is carried out according to *spectral type* Sp and *luminosity class* LC.

A series of standard stars for defining the spectral classes is shown in Fig. 6.10, together with the spectral lines used for their classification.

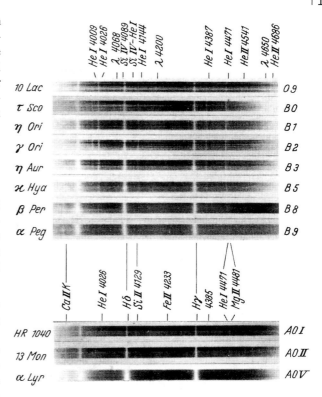

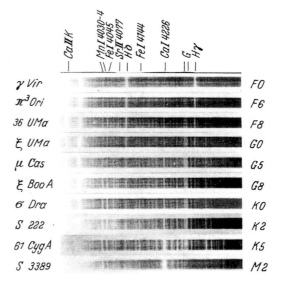

Fig. 6.10. The MK classification of stellar spectra, from "An Atlas of Stellar Spectra" by W. W. Morgan, P. C. Keenan und E. Kellmann (1943). Along the main sequence (luminosity class V), above, O9-B9, then A0 V and below, F0-M2. In the case of A0, the luminosity classes I (supergiants) and II (bright giants) are also shown, in order to make clear the significance of the absolute magnitudes (spectroscopic parallaxes!). The spectral lines used in the classification are indicated with their identification and wavelengths in [Å] (1 Å = 0.1 nm)

The luminosity criteria should depend to first order and as sensitively as possible only on the luminosity of the star, over the whole range of Sp. W. W. Morgan's luminosity classes LC at the same time indicate the position of the star in the HRD; they are (with their proper names):

Ia-0	Hypergiants
I	Supergiants
II	Bright Giants
III	Giants
IV	Subgiants
V	Main Sequence (Dwarfs)
VI	Subdwarfs

As needed, the luminosity classes I to V can be subdivided using the suffixes a, ab, and b. Figure 6.10 shows the important spectral classes of the main sequence stars (LC = V) and, for the spectral type A0, the division into luminosity classes I to V (depending on the width of the hydrogen lines).

About 90% of all stellar spectra can be accounted for in the MK classification; those remaining are in part composite spectra of unresolved binary stars, and in part the peculiar spectra of pathological individuals.

The *relationship* between the parameters of the MK classification Sp and LC on the one hand, and the color indices B – V and absolute magnitudes M_V on the other, is shown in Fig. 6.11.

Some of the peculiarities of stellar spectra which cannot be taken into account in a two-parameter classification scheme are denoted by the following abbreviations: e.g. the suffix n (nebulous) indicates a particularly diffuse appearance of the lines, e denotes emission lines, v means variable spectrum, and p (peculiar) characterizes *any sort of* unusual feature, e.g. an anomalous intensity of the lines of a certain element.

Supplementary Information. The MK classification system has been added to and refined in a number of ways since its introduction. We mention the "Revised Spectral Atlas for Stars earlier than the Sun" by W. W. Morgan, H. A. Abt and J. W. Tapscott (1976), the supplement for cooler stars (G, K, M, S and C) by P. C. Keenan and R. C. McNeil (1976 and later supplements), which takes into account the most important frequency anomalies of the giant stars, and the new classification of the stars in the Henry–Draper Catalog by N. Houk (1976/78).

Spectral atlases of the MK classes are "An Atlas of Objective Prism Spectra" (Michigan 1974) by N. Houk, N. J. Irvine and D. Rosenbush, the Bonn Atlas for Objective Prism Spectra by W. C. Seitter (1970/75) and "An Atlas of Representative Stellar Spectra" by Y. Yamashita, K. Nariai and Y. Norimoto (Tokyo 1978).

Through observations with the IUE Satellite, it was possible to set up a classification system for the earlier spectral types in the *ultraviolet* region ($\lambda = 115$ to 320 nm) as well, with a resolution of around 0.7 nm (A. Heck, D. Egret, M. und C. Jaschek, 1984).

The *white dwarfs* are denoted by a D written before the spectral class. Most of the white dwarfs are of type DA (with only H I in their spectra, temperature range 5000 to 70 000 K). For the helium-rich white dwarfs, which fall in the temperature range from 5000 to 120 000 K, the types DO (He II lines present, $T \geq 50\,000$ K), DB (only He I), DQ (atomic or molecular carbon) and DZ (only lines from metals) are used

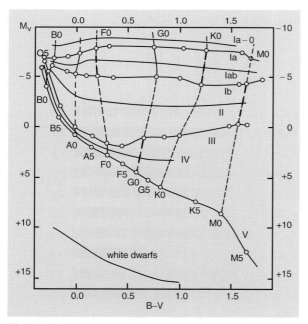

Fig. 6.11. Spectral class Sp and luminosity class LC of the MK classification as functions of the color index B – V and the absolute magnitude M_V [mag]

– ordered here according to decreasing temperature. The PG 1159 stars, with extremely high surface temperatures up to 180 000 K, are related to the hydrogen-poor white dwarfs.

We mention also the *Wolf-Rayet stars*, which are notable for clear emission lines; they are denoted by WC or WN, depending on whether their spectra contain the lines of carbon or of nitrogen. We shall discuss these and other special groups of stars in another connection in Sect. 8.2.5.

In recent times, the *spectral classes L and T*, which follow M, have been introduced to describe cooler stars. A lack of TiO bands is characteristic of type L stars ($T_{\text{eff}} \lesssim 2000$ K); instead, one observes CrH bands and especially strong lines of Li I and the other neutral alkali atoms Na I and K I. Class T is defined by the occurrence (at $T \lesssim 1400$ K) of bands from methane, CH_4, which are also to be seen in the spectra of the major planets Jupiter and Saturn.

6.4.4 Two-Color Diagrams

Along with the color–luminosity diagram, the two color diagram developed by W. Becker (1942) plays an important role. Here (Fig. 6.12), the short-wavelength color index $U - B$ is plotted (downwards) against $B - V$ as the abscissa. For *black body* radiators one obtains in this diagram approximately a straight line inclined at 45°, as can be readily calculated using (6.29) and the corresponding expression for $U - B$. In Fig. 6.12, the relationship between $U - B$ and $B - V$ is plotted for the *main sequence* stars, with their spectral classes and absolute magnitudes indicated.

The great differences between the spectral energy distributions of the stars and of a black body radiator will be clarified in Sect. 7.2.3, where we make use of the theory of stellar spectra. The applications of the two-color diagram for the determination of interstellar reddening, as well as for identifying particular types of stars, will be discussed in Chaps. 9 and 10.

6.4.5 Rotation of the Stars

We consider a star whose rotational velocity at the equator is v and whose rotational axis is inclined to the line of sight by an angle i. The projection of the equato-

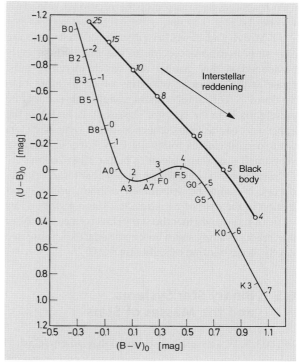

Fig. 6.12. A two-color diagram for main sequence stars using the (interstellar-reddening-independent) indices $(U - B)_0$ and $(B - V)_0$ in [mag] (after H. L. Johnson, W. W. Morgan and others). Along the curves, the MK spectral types and the absolute magnitudes of the stars are indicated. Black-body radiators having the $T_{\text{eff}} \cdot 10^{-3}$ [K] values shown would give the nearly straight line above the main sequence curve. Interstellar reddening displaces the data points for stars parallel to the arrow drawn at the upper right; this line refers directly to O stars

rial velocity on the line of sight is then $v \sin i$, and the Doppler shift at the wavelength λ, according to (2.3), is

$$\Delta\lambda = \pm\lambda \frac{v \sin i}{c} . \qquad (6.45)$$

A spectral line which would be sharp in the case of a star at rest now appears to be spread out into a band of width $2\Delta\lambda$, whose profile reflects the brightness distribution of the "stellar disk". If the latter exhibits no center–limb darkening, then the line profile is elliptical.

If, for example, a B star of radius $5R_\odot$ which rotates with a period of 5 d and has $i = 90°$, its projected equatorial velocity will be $v \sin i = 250$ km s^{-1} and the halfwidth of e.g. the Mg II $\lambda = 448.1$ nm line – often used for the determination of the rotational velocity – will be $\Delta\lambda = \pm 0.37$ nm.

O. Struve and coworkers discovered that there are also stars in whose spectra all the lines are strongly broadened in this manner and which therefore must rotate with equatorial velocities of up to more than 300 km s^{-1}. The rapidly rotating single stars occur quite preferentially in the spectral classes O, B, and A in the upper part of the main sequence. Main sequence stars with spectral classes following F5 V have, in contrast, very small rotational velocities below 10 to 20 km s^{-1}.

Since in (6.45) only the *projected* equatorial velocity occurs, it cannot be determined without further information in the case of an *individual* star e.g. with a low value of $v \sin i$, whether the star in fact rotates slowly, or is a rapidly rotating star whose rotational axis is directed nearly along our line of sight ($\sin i \simeq 0$).

6.5 Binary Star Systems and the Masses of Stars

In 1803, F. W. Herschel, discovered that α Gem (Castor) is a *visual binary star*, whose components move around each other under the influence of their mutual gravitational attraction. The observation of binary stars thus offers the possibility of extracting the stellar masses $\mathcal{M}$. Since the stellar masses are accessible *only* through their gravitational interactions, the investigation of binary star systems remains of fundamental importance for all of astrophysics today.

We begin with an overview of the types of binary star systems: visual binary star systems (Sect. 6.5.1), the spectroscopic binary systems and the eclipsing variables (Sect. 6.5.2). In Sect. 6.5.3, we consider briefly the rotational periods of binary star systems and the rotation of stars. Stellar masses, determined in various ways, are summarized in Sect. 6.5.4, where we also give the mass-luminosity relation, important for the theory of stellar structure. Particularly interesting are also the *close* binary systems (Sect. 6.5.5), in which exchange of stellar matter can occur between the components, and those few systems which contain a radio pulsar (Sect. 6.5.6). Finally, we apply the methods of mass determination in Sect. 6.5.7 to faint objects of low mass, in particular in connection with the search for *planet*-like companions of nearby stars.

6.5.1 Visual Binary Stars

We first separate the optical (apparent) pairs using statistical criteria, possibly also taking account of their individual proper motions and radial velocities, from the *physical* pairs, i.e. the true binary stars. The *apparent* orbit of the fainter component (the "companion") around the brighter component is observed and its separation (in seconds of arc) and position angle (N $0°$ – E $90°$ – S $180°$ – W $270°$) are recorded. Along with direct observation, in recent times speckle interferometry (Sect. 5.1.2) has been applied to the determination of the distance ($\gtrsim 0.001''$) of the two components.

If the apparent orbit is plotted, the result is an *ellipse*. If we were to look down onto the orbital plane from a perpendicular direction, we would necessarily find the brighter component at one focus of the orbit. This is, in general, not the case, since the orbital plane subtends an *inclination angle i* with the celestial plane (perpendicular to the direction of observation). Conversely, the orbital inclination i can clearly be determined in such a way that the true orbit fulfills Kepler's laws. Then, using Kepler's 3rd law (2.57), we can determine the total mass of the two stars:

$$\mathcal{M}_1 + \mathcal{M}_2 = \frac{4\pi^2}{G} \frac{a^3}{P^2} . \tag{6.46}$$

Let a'' be the semimajor axis of the (relative) true orbit in seconds of arc, and p'' be the parallax of the binary star system (also in seconds of arc); then $a = a''/p''$ is the semimajor axis of the true orbit in astronomical units AU (radii of the Earth's orbit). If, furthermore, P is the orbital period in years, we can write down the total mass of the two stars (in units of the solar mass):

$$\frac{\mathcal{M}_1 + \mathcal{M}_2}{\mathcal{M}_\odot} = \frac{(a \,[\mathrm{AU}])^3}{(P \,[\mathrm{yr}])^2} . \tag{6.47}$$

If the motion of the two components has been measured *absolutely* (i.e. relative to the background stars, after subtracting the parallactic and the proper motions), then the semimajor axes a_1 and a_2 of their true orbits about the common center of gravity can be obtained, and from (2.54), we find:

$$a_1/a_2 = \mathcal{M}_2/\mathcal{M}_1 , \qquad a = a_1 + a_2 , \tag{6.48}$$

so that now the individual masses $\mathcal{M}_1$ and $\mathcal{M}_2$ can be calculated.

When the fainter component is *not* directly observable, its presence can still be inferred from the (absolutely measured) motion of the brighter component about the center of gravity (astrometric binaries). If a_1 is its semimajor axis (again in seconds of arc), we obtain the following relation from $a_1/a = \mathcal{M}_2/(\mathcal{M}_1 + \mathcal{M}_2)$:

$$(\mathcal{M}_1 + \mathcal{M}_2) \cdot \left(\frac{\mathcal{M}_2}{\mathcal{M}_1 + \mathcal{M}_2}\right)^3 = \frac{4\pi^2}{G} \frac{a_1^3}{P^2}. \quad (6.49)$$

In this way, F. W. Bessel in 1844 using meridian circle observations discovered that Sirius (α CMa, A1 V, $\mathcal{M}_1 = 2.2 \mathcal{M}_\odot$) must have a "dark" companion. In 1862, A. Clark indeed discovered Sirius B, 9.8 mag fainter. It has an absolute magnitude of only $M_V = 11.2$ mag, although its mass is $\mathcal{M}_2 = 0.94 \mathcal{M}_\odot$. Since the surface temperature of this little star is quite "normal" at about 23 000 K, it must be extremely *small* (as already noted). In 1923, F. Bottlinger came to the conclusion "that it is a question of something new here", namely a white dwarf star.

A further interesting application is the precise measurement of the orbits of nearby stars, in order to find dark companions, which either represent a transitional stage from stars to planets, or are indeed already *planets* (Sect. 6.5.7).

6.5.2 Spectroscopic Binary Stars and Eclipsing Variables

In 1889, E. C. Pickering observed that in the spectrum of Mizar (ζ UMa) the lines become double (twice) within a period of $P = 20.54$ d. Mizar was thus shown to be a *spectroscopic binary star*. In this particular system, two similar A2 stars move around each other; their angular separation is too small to be resolved telescopically. In other systems, only *one* component can be recognized in the spectrum; the other is evidently too faint.

If the radial velocity of one or both components, obtained from the Doppler effect, is plotted against time, a *velocity curve* (cf. also Fig. 6.17) is obtained. After subtracting the mean or center-of-gravity motion, one can read off the component(s) of the orbital velocity in the direction of observation. From these, the semimajor axis of the orbit itself cannot be determined; however, using methods which we shall not treat in detail here, the quantity $a_1 \sin i$ can be calculated (i is the unknown orbital inclination angle) for component 1 if only it can be observed, and $a_2 \sin i$ as well, if component 2 is also observable.

If only *one* spectrum is visible, Kepler's 3rd Law and the Center of Gravity theorem immediately yield from (6.49) the *mass function*:

$$(\mathcal{M}_1 + \mathcal{M}_2) \cdot \left(\frac{\mathcal{M}_2}{\mathcal{M}_1 + \mathcal{M}_2}\right)^3 \sin^3 i$$
$$= \frac{\mathcal{M}_2^3 \sin^3 i}{(\mathcal{M}_1 + \mathcal{M}_2)^2} = \frac{4\pi^2}{G} \frac{(a_1 \sin i)^3}{P^2}. \quad (6.50)$$

For statistical purposes, one can make use of the fact that the mean value of $\sin^3 i$ over a sphere is equal to 0.59, or, taking the probability of discovery into account, about 2/3.

When both spectra are visible, one obtains $\mathcal{M}_1 \sin^3 i$ and $\mathcal{M}_2 \sin^3 i$ and thus the mass ratio $\mathcal{M}_1/\mathcal{M}_2$.

In Sect. 6.5.7, we shall discuss the observations of *planet-like* companions of nearby stars based on periodic line shifts in their spectra.

If the orbital inclination of a spectroscopic binary system is near to 90°, then "eclipses" occur and the system is called an *eclipsing variable*.

The classic example is β Per (Algol), identified by J. Goodricke in 1782, with a period of $P = 2.87$ d.

From the brightness of an eclipsing variable, measured over a long period of time, its *period P* and then its *light curve* is determined. From the latter (Fig. 6.13),

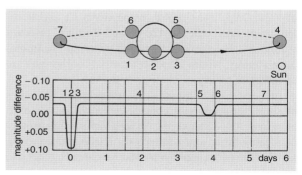

Fig. 6.13. Apparent relative orbit and light curve of the eclipsing binary variable IH Cas. Corresponding points on the light curve and the orbit are indicated by numbers. The main eclipse of the brighter component by the fainter and smaller one is in this case ring-shaped

the radii of the two stars can be obtained, relative to the radius of the relative orbit, as well as the orbital inclination i. If, in addition, the velocity curve can be determined spectroscopically for one or even both components, then the absolute dimensions of the system and its masses, and thus the mean densities of both stars can be calculated. In favorable cases, even the ellipticity (flattening) and the center–limb darkening of the two stars can be extracted. Thanks to the methods for determination of the elements of eclipsing variables developed by H. N. Russell and H. Shapley, they are today among the most precisely described stars. In *close pairs* (Sect. 6.5.5), the two components also interact physically, as first shown by O. Struve from a thorough analysis of their spectra.

6.5.3 Periods and Rotation in Binary Systems

Let us now attempt to gain an overview of the general features of binary star systems.

The visual binaries, spectroscopic binaries, and eclipsing variables, which differ only in the manner of their observation, form a continuum, with some overlap. Their periods range from a few hours up to many millenia. Binary stars with short periods mostly have circular orbits; systems of long period prefer large orbital eccentricities. In addition to binary systems, *multiple* systems also occur frequently; they usually contain one or more close pairs. The "binary star" α Gem (Castor) discovered by F. W. Herschel consists of *three pairs* A, B, and C, each spectroscopic binaries with periods of 9.21, 2.93, and 0.814 d, respectively. A and B orbit around each other in 420 yr, and Castor C revolves around A + B in several thousand years. In our immediate vicinity, within 20 pc, at least half of the stars are members of binary or multiple systems.

In the spectra of binary stars and eclipsing variables of short period, whose components circle about each other at a close distance, the *Fraunhofer lines* are usually noticeably broad and washed out. This is related to the fact that the two components rotate about each other like a rigid body due to tidal friction, in a manner similar to the Earth-Moon system. Their periods of rotation are equal to the orbital (revolution) periods. The distribution of their spectral lines due to the rotation is given by (6.45).

We shall return later to the significance of the rotation of single and binary stars and the role of angular momentum in the problems of stellar evolution.

6.5.4 The Stellar Masses

We will now try to gain an overview of the masses $\mathcal{M}$ of the stars, obtained from all types of binary star systems. Their numerical values range from about 0.07 $\mathcal{M}_\odot$, the smallest mass found for a "visible" star, to 100 $\mathcal{M}_\odot$, with the majority of stellar masses falling in the region from 0.3 to 3 $\mathcal{M}_\odot$. The connection to the other quantities of state of the stars remained unclear until A. S. Eddington in 1924 discovered the *mass-luminosity relation* in connection with his theory of the inner stellar structure. From a modern standpoint, we can understand the essential features as follows: the stars on the main sequence are evidently in analogous stages of development (their energy requirements are supplied by the fusion of hydrogen to helium) and therefore, as a rule, they have structures formed "according to the same recipe". A particular mass $\mathcal{M}$ will correspond to internal energy sources of a particular magnitude, which in turn determine the luminosity L of the star. We should thus expect a relationship between the mass $\mathcal{M}$ and the luminosity L, or the absolute bolometric magnitude M_{bol}, of these stars. In fact, as an analysis of all the observational data (Fig. 6.14) shows, such a relation holds for the *main sequence stars*; in the upper mass range ($\mathcal{M} \gtrsim 0.2\,\mathcal{M}_\odot$), it can be approximated by the empirical formula

$$\log \frac{L}{L_\odot} = 3.8 \log \frac{\mathcal{M}}{\mathcal{M}_\odot} + 0.08 \quad (6.51)$$

The stars which have evolved away from the main sequence (Sects. 8.2.3, 4) do not obey this relation, as is to be expected; for example, the white dwarfs with an average mass of about 0.58 $\mathcal{M}_\odot$, and especially the neutron stars with masses of the order of 1 $\mathcal{M}_\odot$, (Sect. 6.5.6) show deviations. In the case of the red giants, for which there are hardly any reliable empirical mass determinations, the mass-luminosity relation is also inapplicable, since they form a heterogeneous group in terms of their evolution. As a statistical average, the masses of the red giants are about 1.1 $\mathcal{M}_\odot$.

Considering the radii R of the stars (Sect. 6.3.5) as given we can calculate an important quantity for the

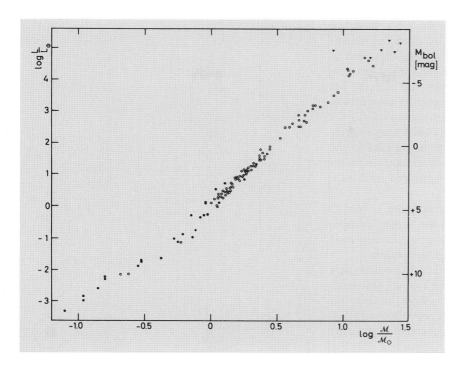

Fig. 6.14. The empirical mass-luminosity relation for main sequence stars. The luminosity L or the absolute bolometric magnitude M_{bol} of the stars is plotted as a function of their masses $\mathcal{M}$ (D. M. Popper, 1980). • visual binary stars; ○ spectroscopic binary stars: optically resolved systems and eclipsing variables (detached systems); ▼ OB eclipsing variables (presumably contact systems)

theory of stellar spectra, the *gravitational acceleration* on the stellar surface:

$$g = \frac{G\mathcal{M}}{R^2}. \quad (6.52)$$

We find that it has a numerical value which is constant within a factor of 2, $g \simeq 2 \cdot 10^2$ m s^{-2} to $\simeq 4 \cdot 10^2$ cm s^{-2}, for the stars of the main sequence out to the spectral class M2 V. For giants and supergiants, it is considerably smaller (ca. 10^{-2} m s^{-2}), for white dwarfs much larger (about 10^6 m s^{-2}).

Finally, we summarize the important results for the relations between the luminosity L, the absolute bolometric magnitude M_{bol}, the effective temperature T_{eff}, the stellar radius R and the gravitational acceleration on the surface of the star, g; here, we have made use of (6.43) and (6.52) and have related all the quantities to their corresponding *solar* values:

$$\frac{L}{L_\odot} = \left(\frac{R}{R_\odot}\right)^2 \cdot \left(\frac{T_{\text{eff}}}{T_{\text{eff},\odot}}\right)^4,$$

$$M_{\text{bol}} - M_{\text{bol},\odot} = -2.5 \log L/L_\odot,$$

$$\frac{g}{g_\odot} = \frac{\mathcal{M}}{\mathcal{M}_\odot} \cdot \left(\frac{R}{R_\odot}\right)^{-2} \quad (6.53)$$

with $L_\odot = 3.85 \cdot 10^{26}$ W, $M_{\text{bol},\odot} = 4.74$ mag, $T_{\text{eff},\odot} = 5780$ K, $R_\odot = 6.96 \cdot 10^8$ m and $g_\odot = 274$ m s^{-2}.

As we shall see in Sect. 7.2, the analysis of stellar spectra permits the extraction of the effective temperature T_{eff} and the acceleration of gravity g. It is then possible using (6.53) to calculate the *ratio* of mass to luminosity:

$$\frac{\mathcal{M}}{L} = \frac{1}{4\pi G\sigma} \frac{g}{T_{\text{eff}}^4}. \quad (6.54)$$

Should one wish to determine $\mathcal{M}$ and L separately, it is necessary either to refer to the theory of internal stellar structures (Chap. 8) or to use corresponding empirical data.

6.5.5 Close Binary Star Systems

In the case of close pairs of stars, there are for one thing very strong tidal forces acting between the two components, which tend to synchronize their rotational periods with the orbital periods. Furthermore, there is often a direct physical interaction between them; as was first shown by O. Struve in the 1940's and 50's using spectroscopic analyses, there are common gas shells and gas flows from one component to the other. In recent times, it has become clear from, for example, investigations of novas and nova-like variable stars, and of galactic X-ray sources (Sects. 7.4.5, 6), that a gas flow often does not enter the companion star directly, but rather, owing to conservation of angular momentum, forms a rotating disk (accretion disk) around it.

The cause of matter exchange in close binary systems is to be found in the changes in stellar radii during the course of the stars' evolution, in particular the enormous increase in radius upon approaching the red giant stage (Sect. 8.2.3).

In a binary star system, we denote the *originally more massive* star consistently as the *primary* component, independently of whether the mass ratio may have reversed in the course of the system's evolution.

We first consider the *equipotential surface* of a binary star system, whose components are initially still separated. We take a point at distance r_1 from the mass $\mathcal{M}_1$ and distance r_2 from the mass $\mathcal{M}_2$; according to (2.44), at that point there acts a gravitational potential given by:

$$\Phi_G = -G\left(\frac{\mathcal{M}_1}{r_1} + \frac{\mathcal{M}_2}{r_2}\right) . \quad (6.55)$$

If the system rotates with the angular velocity ω, we can represent the centrifugal acceleration (z is the distance from the axis of rotation) by an additional potential $\Phi_z = -z^2\omega^2/2$. On a surface given by:

$$\begin{aligned}\Phi &= \Phi_G + \Phi_z \\ &= -G\left(\frac{\mathcal{M}_1}{r_1} + \frac{\mathcal{M}_2}{r_2}\right) - \frac{1}{2}z^2\omega^2\end{aligned} \quad (6.56)$$

a test mass can be moved freely without performing work; therefore, the surface of e.g. an ocean, or of a celestial body, corresponds to an equipotential surface $\Phi = $ const.

In our binary star system, each component is at first surrounded by "its own" closed equipotential surfaces, out to the point where the first common equipotential surface begins; it surrounds both objects in the shape of an hourglass, and is termed the *Roche surface*. Further out, all the equipotential surfaces surround both objects (Fig. 6.15).

If now the object of larger mass, let us say $\mathcal{M}_1$, evolves into a giant star, it can grow out beyond the innermost common equipotential surface. We then have a *semidetached system*; gas flows from component 1 to component 2, so that the mass ratio can even be reversed.

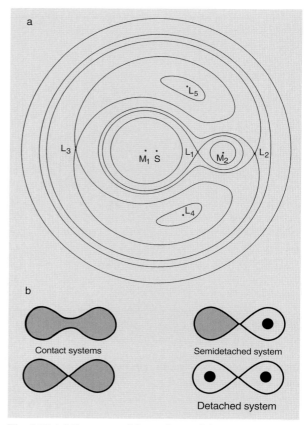

Fig. 6.15. (**a**) Geometry of the equipotential surfaces in a close binary star system. The curves for $\Phi = $ const according to (6.56) in the orbital plane are drawn for a mass ratio $\mathcal{M}_2/\mathcal{M}_1 = 0.17$. The *Roche* limit curve meets itself at the Lagrange point L_1. S is the center of gravity of the system, through which the axis of rotation passes. (**b**) Types of spectroscopic binary systems

What can happen in detail will be described in Sect. 8.2.6, after we have treated the various phases of stellar evolution.

If the two components completely fill a common equipotential surface, we speak of a *contact system* or W Ursae Maioris system (Fig. 6.15).

6.5.6 Pulsars in Binary Star Systems

A further possibility for determining stellar masses, besides the spectroscopic measurements of the changing radial velocities in binary star systems discussed above (Sect. 6.5.2), is offered by the regular signals sent out at short time intervals by the pulsars (Sect. 7.4.7).

We have already met up with the frequency shift $\Delta\nu$ from a moving radiation source due to the Doppler effect (2.3). The Doppler formula

$$\Delta\nu/\nu_0 = v/c \quad (v \ll c), \tag{6.57}$$

is applicable not only to electromagnetic waves (or acoustic waves; c is then the velocity of sound), but also to *every* regular sequence of signals such as, for example, the "ticking" of a pulsar with a pulse frequency ν_0 or a period $1/\nu_0$. If a pulsar is approaching us with a relative velocity v owing to its motion in a binary star system, the time interval between arriving signals is shortened, or the pulse frequency ν_0 is lengthened, according to (6.57).

The exceptionally high precision with which the arrival of radio pulses can be measured (ca. 1 µs) permits a very precise determination of the orbital elements and, above all, of the masses of the pulsars (neutron stars), whereby we can simply apply the formulas from Sect. 6.5.2. To be sure, the socalled *radio pulsars* (Sect. 7.4.7) are for the most part single stars, but in 1974, R. A. Hulse and J. H. Taylor, in the course of a sky survey with the 300 m radio telescope at Arecibo in Puerto Rico, discovered fluctuations in the pulse frequency of the pulsar PSR 1913 + 16 (= PSR J1915 + 1606). This pulsar, which has a period of 0.059 s or a frequency of $\nu_0 = 17$ Hz, exhibits changes in its period of up to 80 µs with a modulation period of 0.323 d; the modulation was interpreted by its discoverers as due to orbital motion in a binary system, with radial velocities of the pulsar between 60 and 330 km s^{-1} (Fig. 6.16).

Longer series of observations have yielded unusual orbital elements for this "*binary pulsar*": the semimajor axis of the orbital ellipse of the pulsar, projected onto the sphere, is is found to be $a \sin i = 702\,000$ km; with an inclination of $i \simeq 50°$, this corresponds to a semimajor axis of only 1.3 $R_\odot$! The orbital period is $P = 0.32$ d, and the eccentricity of the orbit is $e = 0.62$. The mass of the pulsar or neutron star is 1.39 $M_\odot$, and that of its companion, also a neutron star, is 1.44 $M_\odot$. These orbital elements are so extreme that Newton's theory of gravitation is not sufficient to describe them; instead, due to the strong gravitational fields which occur, Ein-

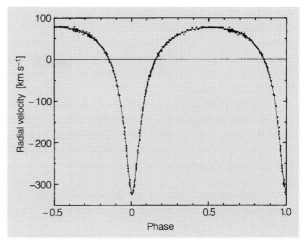

Fig. 6.16. The observed radial velocity curve of the binary pulsar PSR 1913 + 16 (PSR J1915 + 1606) over one orbital period. The large orbital eccentricity $e = 0.62$ expresses itself as a noticeable deviation from a sine curve. From R. A. Hulse and J. H. Taylor: Astrophys. J. **195**, L51 (1975). (Reproduced with the kind permission of The University of Chicago Press, The American Astronomcal Society, and the authors)

stein's *General Relativity Theory* (Sect. 8.4) must be employed. Particularly notable is the large *precession of the periastron* of the orbit, 4.23° yr^{-1}, corresponding to a complete rotation in 85 yr. The effect of General Relativity is much stronger in this binary-pulsar system than in the analogous precession of the perihelion of Mercury around the Sun, which is 43″ per 100 yr (see (8.85)).

Einstein's gravitational theory predicts that the motions of the masses in a binary system such as PSR J1915 + 1606 will give rise to the emission of *gravitational waves* and therefore lead to an energy loss by the system. Indeed, a small systematic *decrease* $\dot{P}$ in the *orbital period*, $\dot{P}/P = -2.4 \cdot 10^{-12}$, has been observed; it agrees well with the theoretically predicted value and can thus be considered to represent an indirect observation of gravitational radiation.

At present, we know of more than 50 binary star systems with radio pulsars; most of these are "millisecond pulsars" with periods of less than 25 ms (Sect. 7.4.7). On the average over all radio pulsars, only about 10% are in a binary star system. The systems within the galactic disk appear to fall into two classes: one of these, including PSR J1915 + 1606, contains somewhat more massive stars ($\geq 0.7\,\mathcal{M}_\odot$) on orbits with short periods and strong eccentricities; the other includes systems having lower masses. Furthermore, a surprising result is that more than 50% of the binary-star pulsars are to be found in globular star clusters (Sect. 9.2.1).

The *X-ray pulsars* (Sect. 7.4.6), such as Her X-1, differ from the radio pulsars in that they are, as a rule, members of a *close binary system*. Frequently, in addition to the modulation of the X-ray pulses, the variations in radial velocity can also be observed in the spectrum of the other component.

Analysis of the orbits of the known X-ray pulsars yields masses for the neutron stars in the range 1.2 to 1.6 $\mathcal{M}_\odot$.

6.5.7 Companions of Substellar Mass: Brown Dwarfs and Exoplanets

An interesting application of the techniques for mass determination is offered by the faint "substellar" objects with masses of $\leq 0.08\,\mathcal{M}_\odot$ or ≤ 80 Jupiter masses, which represent a transitional stage from stars to planets (brown dwarfs, Sect. 8.3.1) or are in fact planets. These objects in the course of their evolution never pass through the phase of central hydrogen burning characteristic of a star (Sect. 8.2.3). Of particular interest to us is naturally the search for *planet-like* companions and *planetary systems* around nearby stars. This search demands an extremely high precision from the measurements, which tests the limits of performance of current observational methods.

Until the 1980's, indirect *astrometric* observation according to (6.49) was predominant. Numerous measurements taken over a period of years by P. van de Kamp and others of the motions of our next-nearest neighbor, Barnard's star (BD+4°3561), which is at a distance of 1.8 pc, in the spectral class M5 V and has a mass of $0.15\,\mathcal{M}_\odot$, indicated two companions with about 0.7 and 0.5 Jupiter masses and periods of 12 and 20 yr. Later independent analyses, however, did *not* confirm the existence of this "planetary system". Other candidates for substellar objects, which were found by applying speckle interferometry or from an excess of infrared radiation of the main component, also failed to stand up to more exact examination.

Only after the development of *spectroscopic* observation methods of very high precision, with which today radial velocities can be determined with errors of a few m s^{-1}, did a breakthrough occur. The modern methods are based on the one hand on recording of the stellar spectral lines together with very sharp comparison lines, e.g. from an iodine vapor source in the optical path using a high-resolution CCD detector; and on the other, on the application of computers, allowing the correlation of numerous lines and a comparison with synthetic spectra.

The Jupiter–Sun System. Before we discuss individual observations of objects with substellar masses, we will first try to estimate the order of magnitude of the effects which are to be expected by considering *our own* Solar System (Table 2.2, 3.1); we examine the motions of the Sun (of mass $\mathcal{M}_\odot$) and of Jupiter ($\mathcal{M}_J = \mathcal{M}_\odot/1047$) around their common center of gravity with the period $P \equiv P_J = 11.9$ yr. According to (2.54), the distance of the Sun from the center of gravity is $a_\odot = a_J \cdot (\mathcal{M}_J/\mathcal{M}_\odot)$ with $a_J = 5.2$ AU, and thus, due to $v = 2\pi a/P$, the orbital velocity of the Sun around the center of gravity is $v_\odot = v_J \cdot (\mathcal{M}_J/\mathcal{M}_\odot)$ with $v_J = 13.1$ km s^{-1}. The "reflex" of the Sun on Jupiter's motion is thus $a_\odot = 0.005$ AU or $v_\odot = 12.5$ m s^{-1}.

If we were to observe the Jupiter–Sun system e.g. from a distance of 5 pc at an angle of inclination i, then the Sun would be "shifted" by only a tiny angle of at most around $0.001'' \cdot \sin i$; the largest amplitude of its radial velocity would be only (12.5 m s^{-1})$\cdot \sin i$, which, from the Doppler formula (2.3), corresponds to an extremely small shift of the wavelength of $\Delta\lambda/\lambda \simeq 3 \cdot 10^{-8}$.

The *radiation* from Jupiter in the visible range is only about 10^{-9} as intense as that from the Sun; in the infrared, where its own radiation is more intense than the reflected sunlight, this fraction is higher, about 10^{-6} to 10^{-4}. The *direct* observation of a planet-like companion to a star at a large distance from us is thus rendered enormously difficult by the great difference in luminosities together with the small apparent angular distance between the two components.

Spectroscopic Observation. The maximum amplitude of the radial velocity $V_{r,\max} = (2\pi a/P)\sin i$ from (6.50) for a star of mass $\mathcal{M}$ with a planet of mass $\mathcal{M}_{pl} \ll \mathcal{M}$

(not visible in the spectrum) is given by

$$V_{r,\max} \, [\text{m s}^{-1}] \quad (6.58)$$
$$= 12.5 \cdot \frac{\mathcal{M}_{\text{pl}}}{\mathcal{M}_{\text{J}}} \left(\frac{\mathcal{M}}{\mathcal{M}_\odot}\right)^{-2/3} \left(\frac{P}{P_{\text{J}}}\right)^{-1/3} \cdot \sin i$$

where we express $\mathcal{M}_{\text{pl}}$ in Jupiter masses, P in units of $P_{\text{J}} = 11.9$ yr and $\mathcal{M}$ in solar masses. According to Kepler's 3rd law, the factor $(P/P_{\text{J}})^{-1/3}$ in (6.58) can be replaced by $(a/a_{\text{J}})^{-1/2}$.

Since – owing to the projection factor $\sin i$ – the true radial velocity is larger than the observed velocity (or at most equal to it), we obtain only a *lower limit* for $\mathcal{M}_{\text{pl}}$ from (6.58).

D. W. Latham, B. Campbell and others found in 1988 for HD 114762 variations of the radial velocity ≤ 700 m s^{-1} with a period of 84 d, which suggested a companion with the order of ≥ 9 Jupiter masses at a distance of $a_{\text{pl}} \simeq 0.3$ AU from the star. The eccentric orbit of HD 114762B, to be sure, indicates a brown dwarf with greater likelihood than a planet.

The first convincing discovery of an exoplanet on the basis of its radial velocity curve (Fig. 6.17) was made in 1995 by M. Mayor and D. A. Queloz, observing the G3 IV star *51 Peg* ($=$ HD 217014, $m_V = 5.5$ mag) at a distance of 15.4 pc from the Earth. From the nearly sinusoidal curve with a maximum amplitude of 59 m s^{-1} and the surprisingly short period of $P = 4.23$ d, one calculates a nearly circular orbit for the companion with $a_{\text{pl}} = 0.05$ AU. The distance from 51 Peg B to the main component is thus considerably less than that from Mercury to the Sun (0.39 AU). Its mass is $\mathcal{M}_{\text{pl}} \simeq 0.5 \mathcal{M}_{\text{J}}/\sin i$. We can also set an upper limit for the mass by *assuming* that the rotational axis of the star is nearly (i.e. within about 10°) perpendicular to the orbital plane of the companion. From the measured rotational velocity at the equator, $v \sin i = (2\pi R/P) \sin i \simeq 2.2$ km s^{-1}, we then find using the known stellar radius R and the period obtained from the chromospheric activity, $P_{\text{rot}} \simeq 37$ d, a mass limit of $\mathcal{M}_{\text{pl}} \leq (1.2 \ldots 2) \mathcal{M}_{\text{J}}$. The small mass and the circular orbit are strong arguments for the assumption that 51 Peg B is a *planet*.

In the following years, G. Marcy, R. P. Butler and others have found numerous further planets and candidates for planets with masses usually similar to that of Jupiter, around sunlike stars. The discovery of a planet of the size of our *Earth*, even around a very near star, stretches current observational techniques beyond their limits; it could however succeed in the near future.

Planet-like companions have, by the way, also been detected around pulsars (Sect. 7.4.7).

We shall return to the detailed consideration of planets around other stars in Sect. 14.1.5, where we also discuss the problem of distinguishing between planets and brown dwarfs.

Direct Observation. The first clearcut direct detection of the *substellar* component of a binary star system was accomplished only in 1995 by T. Nakajima and coworkers in the course of their search for faint companions of nearby ($r \leq 15$ pc) stars with the 1.5 m telescope on Mt. Palomar. They covered the bright component with a stellar coronagraph (cf. Fig. 7.15) and employed adaptive optics to supress motion in the image due to atmospheric turbulence (Sect. 5.1.3):

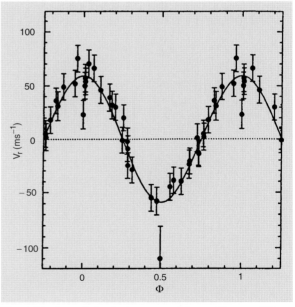

Fig. 6.17. The radial velocity curve of the sunlike star 51 Peg. Observations and orbital calculations (*full curve*) by M. Mayor and D. Queloz. The amplitude of 59 m s^{-1} and the period of 4.23 d indicate a planet-like companion of with an orbital eccentricity of $e = 0$ and ≥ 0.5 Jupiter masses. (With the kind permission of Nature **345**, 779, © Macmillan Magazines Ltd., and the authors)

The M1 V star *Gliese 229* ($=$ HD 42581, $m_V = 8.1$ mag) in our near neighborhood (distance $r = 5.7$ pc) has a faint companion at a distance of 7.8″ or 44 AU, whose effective temperature – as can be seen by the occurrence of methane bands in its spectrum – is about 950 K. The luminosity of Gliese 229 B is $L \simeq 6 \cdot 10^{-6} L_\odot$, and its mass is found from model calculations to be 40 to 50 M_J. We are thus dealing here with an old brown dwarf.

Using direct observational methods, several other substellar companions around nearby stars have also been detected.

7. The Spectra and Atmospheres of Stars

A stellar atmosphere is defined as that layer of a star from which electromagnetic radiation can escape; it is thus the "visible" outer part of the star. Its structure is closely connected with the spectra it emits.

Now that we have dealt with some general theoretical fundamentals in Chap. 4, in the following Sect. 7.1 we will consider further basic concepts of atomic physics and its application to stellar spectra. In Sect. 7.2, we then treat the theory of stellar atmospheres (photospheres), which forms the basis for the quantitative analysis of stellar spectra. Sect. 7.3 is devoted to the outermost layers of the atmosphere of our Sun, the chromosphere and the corona, and to the solar wind; in this section, we also treat the role of magnetic fields and matter currents and the phenomena of solar activity and solar oscillations. In Sect. 7.4, we turn to the corresponding phenomena on other stars, and try to get an overview of the great variety of variable stars.

7.1 Spectra and Atoms

The interpretation of stellar spectra and their classification led to M. N. Saha's theory of thermal excitation and ionization in the year 1920, following important preliminary work by N. Lockyer. This theory is based essentially on the quantum theory of atoms and atomic spectra developed by N. Bohr, A. Sommerfeld and others after 1913. We shall permit ourselves here to recall some fundamentals without giving a complete justification: the basic concepts of atomic spectroscopy (Sect. 7.1.1), the excitation and ionization of atoms (Sect. 7.1.2), the absorption coefficients for spectral lines (Sect. 7.1.3), and spectral line broadening (Sect. 7.1.4). We close this subchapter in Sect. 7.1.5 with some remarks on molecular spectra.

7.1.1 Basic Concepts of Atomic Spectroscopy

We recall Sect. 4.5.4 and remind ourselves that in a transition between the energy levels of an atom, characterized by their quantum numbers i and j and having the energies E_i and E_j, a photon can be emitted or absorbed, whose frequency ν or angular frequency ω is given by

$$h\nu = \hbar\omega = |E_i - E_j| \,. \tag{7.1}$$

Instead of ν or ω, we can also use the wavenumber $\tilde{\nu}$, with the dimensions of a reciprocal length [m^{-1}, or cm^{-1} = Kayser], or the wavelength $\lambda = 1/\tilde{\nu} = c/\nu$, quoted in convenient fractions of a meter (cf. also Sect. 4.1).

The energy values of an atom are calculated with reference to its ground state. The energy unit used is generally not the Joule, but rather 1 cm^{-1} (Kayser), corresponding to the formula $E = hc \cdot \tilde{\nu}$. We then speak of the energy *terms* and the term scheme, represented in a Grotrian diagram (Fig. 7.2). The unit electron volt [eV] is also often used; it corresponds to the energy which an electron gains on passing through a potential difference of 1 Volt. 1 eV is equal to $1.602 \cdot 10^{-19}$ J or 8066 cm^{-1}.

In thermal equilibrium, we are dealing with energies of the order of kT. In this sense, we can quote the temperature T which corresponds to a given energy E: 1 eV equals 11 605 K.

We shall limit ourselves here to the *bound* states of the atoms and ions, whose energy values will be denoted by χ.

One-electron Systems. A particular energy level of an atom or ion with *one* valence electron (i.e. the other electrons do not participate in transitions) is described by the four *quantum numbers* $\{n, l, s, j\}$:

1. n is the *principal quantum number*. In *hydrogen-like* atoms (i.e. in a Coulomb field), in the language of Bohr's theory, $a_n = n^2 a_0/Z$ is the semimajor axis of the electron's orbit, where e is the elementary electronic charge, m_e the electron's mass, and ε_0 the electric field constant (permittivity of vacuum). $a_0 = 4\pi\varepsilon_0 \hbar^2/(e^2 m_e) = 5.29 \cdot 10^{-11}$ m is the 1st Bohr radius of the hydrogen atom (i.e. of the innermost electronic orbit) and Z is the effective nu-

clear charge number ($Z = 1$ for a neutral, $Z = 2$ for a singly-ionized atom, etc.).
The energy of the state with principal quantum number n is

$$\chi_n = Z^2 \cdot \chi_{\text{ion}}(\text{H}) \cdot \left(1 - \frac{1}{n^2}\right), \tag{7.2}$$

and $-Z^2 \chi_{\text{ion}}(\text{H})/n^2$ is its binding energy. The ionization energy of the hydrogen atom is

$$\chi_{\text{ion}}(\text{H}) = \frac{e^4 m_e}{32\pi^2 \varepsilon_0^2 \hbar^2} = \frac{1}{4\pi\varepsilon_0} \frac{e^2}{2a_0} \tag{7.3}$$

and is equal to

$$\chi_{\text{ion}}(\text{H}) = hc \cdot R_\infty = 13.59 \text{ eV}, \tag{7.4}$$

where $R_\infty = 1.097 \cdot 10^7 \text{ m}^{-1}$ is the Rydberg constant. (The small shift due to the motion of the nucleus will be neglected here.)

2. l is the *orbital angular momentum* of the electron, quoted in the quantum unit $\hbar$. It can take on the integral values $0, 1, \ldots, (n-1)$.

$$\begin{array}{cccccc} l = 0 & 1 & 2 & 3 & 4 & 5 \quad \text{gives an} \\ \text{s} & \text{p} & \text{d} & \text{f} & \text{g} & \text{h} \quad \text{electron}. \end{array} \tag{7.5}$$

This notation originally referred to the upper term of the series (s: sharp subseries, p: principal series, d: diffuse subseries, f: fundamental series).

3. $s = \pm 1/2$ is the *spin* of the electron, also in units of $\hbar$.

4. j is the *total angular momentum* in units of $\hbar$. It is formed by vector coupling of l and s and can take on only the two values $l \pm 1/2$.

An electron with the quantum numbers $n = 2$, $l = 1$ and $j = 3/2$ is, for example, referred to as a $2p_{3/2}$-electron.

Several-electron Systems. In atoms and ions with more than one electron, the angular momentum vectors are often coupled according to the Russell–Saunders or *LS-coupling* scheme: the orbital angular momentum vectors l add vectorially to give the total orbital angular momentum $\boldsymbol{L} = \sum \boldsymbol{l}$; likewise the spin angular momenta s add to give the total spin $\boldsymbol{S} = \sum \boldsymbol{s}$. The vectors $\boldsymbol{L}$ and $\boldsymbol{S}$ then combine to give the total angular momentum $\boldsymbol{J}$ (Fig. 7.1), where for the corresponding quantum numbers we have:

$$|L - S| \leq J \leq L + S. \tag{7.6}$$

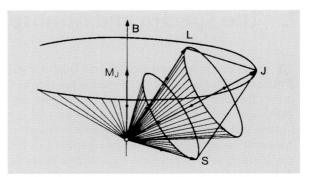

Fig. 7.1. Vector diagram for an atom with Russell–Saunders coupling. The vectors of the total orbital angular momentum $\boldsymbol{L}$ and the spin angular momentum $\boldsymbol{S}$ add to give the total angular momentum $\boldsymbol{J}$ (all in units of $\hbar$). M_J is the component of J in the direction of an applied field $\boldsymbol{B}$. $\boldsymbol{L}$ and $\boldsymbol{S}$ precess about $\boldsymbol{J}$, and $\boldsymbol{J}$ precesses about $\boldsymbol{B}$. The drawing corresponds to an energy level $L = 3$, $S = 2$, $J = 3$, i.e. $^5F_3$

L is always an integer; S and J are half-integral or integral for atoms with odd or even numbers of electrons, respectively.

A given pair of values of S and L yields a *term*. As in the case of a one-electron system, the orbital angular momentum quantum number corresponds to

$$\begin{array}{cccccc} L = 0 & 1 & 2 & 3 & 4 & 5 \quad \text{an} \\ \text{S} & \text{P} & \text{D} & \text{F} & \text{G} & \text{H term} \end{array} \tag{7.7}$$

As long as $L \geq S$, this term can be decomposed into $r = 2S + 1$ energy levels corresponding to different values of J. The number r is referred to (even when $L < S$) as the *multiplicity* of the term; it is written at the upper left of the term symbol. J is written at the lower right as an index which characterizes the individual values of the term. An overview of the possible terms of different multiplicities, their energy levels and the usual notation is given in Table 7.1.

In an *external* field (e.g. a magnetic field), the vector of the total angular momentum J takes on an orientation such that its components M_J parallel to the field likewise have integral or half-integral values. M_J can thus take on the values $J, J-1, \ldots, -J$; i.e. the directional quantization of J yields $2J + 1$ possible orientations (Fig. 7.1). When the external field is vanishingly small, these $2J + 1$ energy levels collapse into a single level; one then speaks of a $(2J + 1)$-fold *degenerate* level J. Furthermore, we divide the terms according to their *parity* into two groups, the even and odd terms, depending

Table 7.1. The terms and the J values of their levels for various quantum numbers L and S in Russell–Saunders coupling

		$S=0$ $r=2S+1=1$ Singlet	$1/2$ 2 Doublet	1 3 Triplet	$3/2$ 4 Quartet
$L=0$	S term	$J=0$	$J=1/2$	$J=1$	$J=3/2$
1	P term	1	1/2 3/2	0 1 2	1/2 3/2 5/2
2	D term	2	3/2 5/2	1 2 3	1/2 3/2 5/2 7/2
3	F term	3	5/2 7/2	2 3 4	3/2 5/2 7/2 9/2

Example: ▭ quartet P term with the energy levels $^4P_{1/2}$, $^4P_{3/2}$, $^4P_{5/2}$. Statistical weight of the term $g(^4P) = 4 \cdot 3 = 2+4+6$.

on whether the arithmetic sum of the l values of the electrons is even or odd, respectively. *Odd* terms are denoted by a symbol $^\circ$.

In a transition between two energy levels, a spectral line is emitted or absorbed. All of the possible transitions between all levels of a term generate a group of neighboring lines, called a *multiplet*.

Selection Rules. The possible transitions (with emission or absorption of electric dipole radiation, in analogy to the well-known Hertzian dipole radiation) are limited by *selection rules*:

1. There are transitions only between even and odd levels.
2. J changes only by $\Delta J = 0$ or ± 1. The transition $0 \leftrightarrow 0$ is forbidden.

For Russell–Saunders coupling, two additional rules hold:

1. $\Delta L = 0, \pm 1$.
2. $\Delta S = 0$, i.e. no intercombinations (e.g. singlet–triplet) are allowed.

Russell–Saunders or LS coupling can be recognized for example by the fact that the multiplet splittings arising from the magnetic interaction between the orbital and spin momenta (i.e. their associated magnetic moments) are small in relation to the splittings between neighboring terms or multiplets.

If the selection rules (1) and (2) are *not* fulfilled, there can still be *forbidden* transitions involving electric quadrupole radiation or magnetic dipole radiation with much smaller transition probabilities.

An example of a term scheme or Grotrian diagram is given in Fig. 7.2, which shows the term scheme for neutral calcium, Ca I.

Identification. The theory of atomic spectra outlined in the previous paragraphs allowed the classification of the wavelengths λ or the wavenumbers $\tilde{\nu}$ of most of the chemical elements and their ionization states. This means that for each spectral line, the lower and upper term (usually referred to the ground state) and the term classifications can be identified. For example, Fraunhofer's K line, $\lambda = 393.37$ nm, the strongest line in the visible solar spectrum, can be identified as Ca II $4s\,^2S_{1/2} - 4p\,^2P^\circ_{1/2}$.

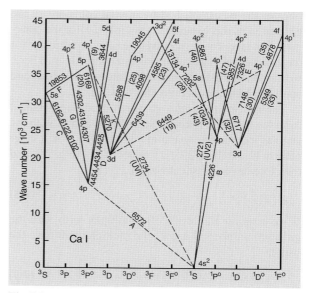

Fig. 7.2. A term scheme or Grotrian diagram for the spectrum of neutral calcium, Ca I. The more important multiplets are denoted by their numbers in "A Multiplet Table of Astrophysical Interest" or "An Ultraviolet Multiplet Table" by C. E. Moore, as well as by the wavelengths of their most intense lines in [Å]. 10^4 cm^{-1} corresponds to 1.24 eV

The *intensity* of a line (where we at this point use the word in a qualitative sense without precise definition) will depend on the one hand on the Einstein coefficients A_{ji} (4.122) or the oscillator strength f (4.124) of the transition as well as on the abundance of the element in question. On the other hand, it also depends on what fraction of the atoms of the element are in the excitation state (energy level) from which the line can be absorbed.

7.1.2 Excitation and Ionization

We now consider the interaction of excitation and ionization, in order to interpret the line intensities of the atoms and ions quantitatively. As long as we can assume that the stellar gas is in thermal equilibrium, i.e. that it – at least locally – to a good approximation represents the conditions in a closed cavity radiator at the temperature T, then the Boltzmann formula (4.88, 89) describes the excitation and the Saha equation, (4.92), describes the ionization. While in the interiors of stars and for the most part also in their atmospheres (photospheres), the densities are sufficiently high that thermodynamic equilibrium can be established by mutual collisions of the gas particles, this condition is not met e.g. for the outermost thin layers of the atmospheres and in particular for the gaseous nebulae and the interstellar medium. There, we have to return to the individual atomic processes and apply the kinetic equations (Sects. 4.5.3, 4).

The *statistical weights* g_s which occur are obtained from the theory of spectra (Sect. 7.1.1): a level with the angular momentum quantum number J, e.g. in a magnetic field, exhibits $(2J+1)$ values of M_J (Fig. 7.2) and therefore has a statistical weight

$$g_J = 2J + 1 \ . \tag{7.8}$$

If we *combine* the levels of a multiplet term with the quantum numbers S and L, the multiplet will have the weight

$$g_{S,L} = (2S+1)(2L+1) \ . \tag{7.9}$$

The addition of the corresponding g_J values of course leads to the same result (Table 7.1).

Thermal Equilibrium. We denote the total number of neutral atoms per unit volume by n_0, the number of singly ionized atoms by n_1 etc. Within each ionization state r, we denote the excitation state by an additional second index s, s' etc.:

Ionization state:	Neutral	Singly ionized	...r-fold ionized
Free electrons per atom:	0	1	...r
Ionization energy:	χ_0	χ_1	...χ_{r-1}
Spectra, e.g. iron:	Fe I	Fe II	...Fe$(r+1)$
All atoms in ionization state:	n_0	n_1	...n_r
Atoms in level $s, s', \ldots$:	$n_{0,s}$	$n_{1,s'}$	...$n_{r,s}$

Corresponding notations are used also for the statistical weights and the excitation energies.

The ratio of the number of particles in the excitation level s' of the ionization state $r+1$ to that in the level s of the ionization state r is given by a combination of the Saha and the Boltzmann equations:

$$\frac{n_{r+1,s'}}{n_{r,s}} n_e = \frac{g_{r+1,s'}}{g_{r,s}} \cdot 2 \cdot \frac{(2\pi m_e kT)^{3/2}}{h^3} \\ \times \exp\left(-\frac{\chi_r + \chi_{r+1,s'} - \chi_{r,s}}{kT}\right) . \tag{7.10}$$

By summing over all the states s, s' etc., one again obtains (4.92).

Sometimes it is expedient to use the measure of temperature suggested by H. N. Russell:

$$\Theta = \frac{5040}{T \text{ [K]}} \ ; \tag{7.11}$$

after inserting the numerical constants we then obtain the Saha equation (4.92) in logarithmic form

$$\log\left(\frac{n_{r+1}}{n_r} \cdot P_e \text{ [Pa]}\right) = -\chi_r \text{ [eV]} \cdot \Theta + \frac{5}{2}\log T \text{ [K]} \\ + \log\frac{2\,Q_{r+1}}{Q_r} - 1.48 \ . \tag{7.12}$$

In Fig. 7.3, we have plotted the fraction of the atoms H, He, Mg, and Ca in several ionization and excitation states whose absorption lines play a role in the MK classification of stellar spectra. The results in the figure correspond to (7.10) for an electron pressure $P_e = n_e \cdot kT = 10$ Pa (which can be considered roughly as an average value for the atmospheres of the main-sequence stars) and for temperatures of 3000 to 50 000 K.

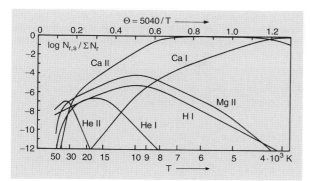

Fig. 7.3. Thermal ionization and excitation (4.10) as functions of the temperature T or $\Theta = 5040/T$ for an electron pressure $P_e = 10$ Pa ($\simeq$ average value for the atmospheres of main-sequence stars). The temperature scale covers the whole region from the O stars (*left*) to the M stars (*right*). The Sun (G2) should be placed at about $T = 5800$ K

Spectrum	Ionization energy χ_r [eV]	State	Excitation energy χ_r [eV]
H I	13.60	$n=2$	10.20
He I	24.59	$2\,^3P^\circ$	20.96
He II	54.42	$n=3$	48.37
Mg II	15.04	$3\,^2D$	8.86
Ca I	6.11	$4\,^1S$	0.00
Ca II	11.87	$4\,^2S$	0.00

The maxima of the curves, which correspond roughly to the maximum strengths of the spectral lines, occur because within a given ionization state, with increasing T at first the excitation also increases. If T continues to increase, this state is "ionized away", so that the fraction of active atoms again decreases. For example, hydrogen is predominantly neutral up to around 10 000 K; the excitation of the 2nd quantum state, from which the Balmer lines in the visible spectrum are absorbed, increases with increasing T. Above 10 000 K, the hydrogen quickly becomes ionized. Thus we can understand that the Balmer lines have their intensity maximum in the A0 stars at $T \simeq 10\,000$ K. R. H. Fowler and E. A. Milne in 1923 were first able to estimate the average electron pressure P_e in the stellar atmospheres by taking the temperatures of the maxima as known.

The Saha theory could in addition explain the increase of the intensity ratios of the lines of singly ionized atoms to those of neutral atoms on going from main-sequence stars to giant stars as a result of increasing ionization, i.e. of n_1/n_0, due to the low pressure.

Mixtures of Elements. The ionization of a mixture of several elements can be most simply calculated by considering the temperature T and the electron pressure $P_e = n_e \cdot kT$ or n_e as independent parameters and then applying the Saha equation to each element and its ionization. The gas pressure

$$P_g = n_{\text{total}} \cdot kT \qquad (7.13)$$

is then readily obtained from the sum of *all* particles n_{total}, including the electrons.

Analogously, the average molecular weight $\bar{\mu}$ can be calculated. For example, completely ionized hydrogen corresponds to $\bar{\mu} = 0.5$, since upon ionization, one proton and one electron are produced.

For application to the theory of stellar atmospheres and of the interiors of stars, we show in Table 7.2 the composition of "*normal*" stellar matter, of which – as we shall see – the Sun and most of the stars are composed. We consider here only those elements which make notable contributions to the electron pressure P_e. The relative abundances of the elements of this mixture are denoted by $\varepsilon_\odot$.

Table 7.3 contains the relation $P_e(P_g, T)$ for the solar or normal mixture, also taking into account the formation of hydrogen molecules at lower temperatures. At an electron pressure characteristic of the atmospheres of main-sequence stars, $P_e \simeq 10$ Pa and at $T \simeq 10\,000$ K, the stellar matter is nearly completely ionized and thus $P_g/P_e \simeq 2$. At the temperature of the Sun, $T \simeq 6000$ K, essentially only the metals are ionized; corresponding to their relative abundance of around 10^{-4}, we find $P_g/P_e \simeq 10^4$. At still lower temperatures, the electrons are provided only by the most readily ionizable group of elements, Na, K, and Ca.

Kinetic Equations. For the general case, when local thermodynamic equilibrium (LTE) cannot be assumed, we will limit ourselves here to explaining the principle of the calculation of the occupation probabilities from the kinetic equations, i.e. the "non-LTE problem", using a greatly simplified atom. The interplay of the individual elementary processes will be represented as in Fig. 7.4 for an atom which consists of only two bound states 0 and 1 and the continuum, κ. The index κ implies a summation over the Maxwell-Boltzmann velocity distribution, (4.86), for the free electrons. The occupation probabilities of the atomic

Table 7.2. These elements (nuclear charge number Z, atomic weight μ, solar abundance relative to hydrogen = 100) make important contributions to the electron pressure $P_e = n_e \cdot kT$ in stellar atmospheres ($P_e \simeq 10$ Pa). We also give the ionization energy and statistical weight of the ground state of the first ionization states. The three groups – grouped according to the ionization energy χ_0 of the neutral atom – become active in different temperature ranges

Z	Element		Atomic weight μ	Abundance $\varepsilon_\odot$ (H = 100)	Neutral atom		Singly ionized		Doubly ionized		At $P_e = 10$ Pa important for
					χ_0 [eV]	g_0	χ_1 [eV]	g_1	χ_2 [eV]	g_2	
1	H	Hydrogen	1.008	100	13.60	2	–	–	–	–	$T > 5700$ K
2	He	Helium	4.003	8.5	24.59	1	54.42	2	–	–	
12	Mg	Magnesium	24.31	$2.6 \cdot 10^{-3}$	7.65	1	15.04	2	80.14	1	
14	Si	Silicon	28.09	$3.3 \cdot 10^{-3}$	8.15	9	16.35	6	33.49	1	$6000 > T > 4500$ K
26	Fe	Iron	55.85	$4.0 \cdot 10^{-3}$	7.87	25	16.16	30	30.65	25	
11	Na	Sodium	23.00	$1.8 \cdot 10^{-4}$	5.14	2	47.29	1	71.64	6	
19	K	Potassium	39.10	$8.9 \cdot 10^{-6}$	4.34	2	31.63	1	45.72	6	$T < 4700$ K
20	Ca	Calcium	40.08	$2.0 \cdot 10^{-4}$	6.11	1	11.87	2	50.91	1	

Table 7.3. The relation between electron pressure P_e, gas pressure P_g and temperature T for stellar matter with the normal chemical composition (cf. Table 7.2). P_e and P_g are given in [Pascal]

P_g [Pa] \ T [K]	10^{-2}	10^{-1}	1	10^1	10^2	10^3	10^4	10^5
3000	$7.01 \cdot 10^{-7}$	$4.12 \cdot 10^{-6}$	$1.96 \cdot 10^{-5}$	$9.33 \cdot 10^{-5}$	$5.33 \cdot 10^{-4}$	$3.47 \cdot 10^{-3}$	$2.18 \cdot 10^{-2}$	$1.02 \cdot 10^{-1}$
4000	$2.36 \cdot 10^{-6}$	$1.26 \cdot 10^{-5}$	$1.00 \cdot 10^{-4}$	$8.69 \cdot 10^{-4}$	$5.98 \cdot 10^{-3}$	$3.09 \cdot 10^{-2}$	$1.38 \cdot 10^{-1}$	$6.78 \cdot 10^{-1}$
5000	$1.04 \cdot 10^{-4}$	$3.39 \cdot 10^{-4}$	$1.12 \cdot 10^{-3}$	$3.96 \cdot 10^{-3}$	$1.68 \cdot 10^{-2}$	$1.02 \cdot 10^{-1}$	$6.85 \cdot 10^{-1}$	3.64
6000	$1.49 \cdot 10^{-3}$	$5.38 \cdot 10^{-3}$	$1.78 \cdot 10^{-2}$	$5.77 \cdot 10^{-2}$	$1.88 \cdot 10^{-1}$	$6.31 \cdot 10^{-1}$	2.36	$1.09 \cdot 10^1$
8000	$4.77 \cdot 10^{-3}$	$4.58 \cdot 10^{-2}$	$3.54 \cdot 10^{-1}$	1.74	6.43	$2.14 \cdot 10^1$	$6.93 \cdot 10^1$	$2.24 \cdot 10^2$
10 000	$4.82 \cdot 10^{-3}$	$4.80 \cdot 10^{-2}$	$4.77 \cdot 10^{-1}$	4.56	$3.46 \cdot 10^1$	$1.67 \cdot 10^2$	$6.13 \cdot 10^2$	$2.04 \cdot 10^3$
15 000	$5.00 \cdot 10^{-3}$	$5.00 \cdot 10^{-2}$	$4.99 \cdot 10^{-1}$	4.96	$4.85 \cdot 10^1$	$4.74 \cdot 10^2$	$4.37 \cdot 10^3$	$2.94 \cdot 10^4$
20 000	$5.00 \cdot 10^{-3}$	$5.00 \cdot 10^{-2}$	$5.00 \cdot 10^{-1}$	5.00	$5.00 \cdot 10^1$	$4.97 \cdot 10^2$	$4.86 \cdot 10^3$	$4.62 \cdot 10^4$
30 000	$5.19 \cdot 10^{-3}$	$5.19 \cdot 10^{-2}$	$5.17 \cdot 10^{-1}$	5.08	$5.01 \cdot 10^1$	$5.00 \cdot 10^2$	$4.99 \cdot 10^3$	$4.96 \cdot 10^4$
40 000	$5.19 \cdot 10^{-3}$	$5.19 \cdot 10^{-2}$	$5.19 \cdot 10^{-1}$	5.19	$5.18 \cdot 10^1$	$5.14 \cdot 10^2$	$5.04 \cdot 10^3$	$4.99 \cdot 10^4$

states are then determined by three rate equations of the form of (4.113) along with conservation of the total particle density, $n = n_0 + n_1 + n_\kappa = $ const. For example, the rate equation for the excited level is given by:

$$n_1(R_{10} + C_{10} + R_{1\kappa} + C_{1\kappa})$$
$$= n_0(R_{01} + C_{01}) + n_\kappa(R_{\kappa 1} + C_{\kappa 1}) . \quad (7.14)$$

Its average lifetime for transitions into both the ground state 0 and into the continuum κ is given by $(R_{10} + C_{10} + R_{1\kappa} + C_{1\kappa})^{-1}$.

To conclude these rather formal considerations, we give the *source function* S_ν, i.e. the ratio of emission to absorption coefficients (4.57) for the line transition $0 \leftrightarrow 1$ in a simple *two-level atom*, whereby we leave off the continuum, in contrast to Fig. 7.4. According to (4.133, 135), we have

$$S_\nu = \frac{A_{10} n_1}{B_{01} n_0 - B_{10} n_1} . \quad (7.15)$$

Here, if we insert n_1/n_0 from the rate equation (7.14) in the correspondingly simplified form

$$n_1(A_{10} + B_{10} J_\nu + C_{10}) = n_0(B_{01} J_\nu + C_{01}) \quad (7.16)$$

and express C_{01} using (4.128) in terms of C_{10} and B_{01}, B_{10} from (4.122) in terms of A_{10}, and finally introduce the Kirchhoff–Planck function $B_\nu(T)$ (4.61), we obtain the source function in the straightforward form

$$S_\nu = (1 - \varepsilon) J_\nu + \varepsilon B_\nu(T) , \quad (7.17)$$

where the coefficient

$$\varepsilon = \frac{C_{01}(1 - e^{-h\nu/kT})}{A_{10} + C_{10}(1 - e^{-h\nu/kT})} \quad (7.18)$$

is a measure of the ratio of the collision rate for deexcitation of level 1 to the rate of spontaneous emission.

An important application of this two-level model, the explanation of the strength of "forbidden" lines in gaseous nebulae, will be introduced in Sect. 10.3.1.

The system of kinetic equations that we have set up here naturally contains – as a limiting case – the well-known relations for thermodynamic equilibrium; e.g. for n_1/n_0 from (7.14), the Boltzmann formula (4.88) must be obtained, and for $n_\kappa/(n_0+n_1)$, the Saha formula (4.91) results.

Using the expression (7.17) for the source function, the conditions for validity of the limiting case of *thermodynamic equilibrium* can readily be shown: we obtain Kirchhoff's rule (4.69), i.e. $S_\nu = B_\nu(T)$, which holds in thermodynamic equilibrium, either when the collisions predominate over radiation processes ($C_{10} \gg A_{10}$ or $\varepsilon \simeq 1$) or when the radiation field is that of a black body ($J_\nu = B_\nu$), or when both conditions are fulfilled.

7.1.3 Line Absorption Coefficients

In the quantitative spectral analysis and determination of elemental abundances, the absorption coefficient of a spectral line plays a central role. We have already met up with the general expressions for the bound-bound radiation processes and for the line absorption coefficients κ_ν [m^{-1}] for the transition $i \rightarrow j$ in Sects. 4.5.4, 5. Applying (4.134), we write:

$$\kappa_\nu = \tilde{\kappa}_\nu \cdot [1 - (n_j/g_j)/(n_i/g_i)] \quad (7.19)$$

and in what follows, we leave off stimulated emission, i.e. we discuss only the contribution

$$\tilde{\kappa}_\nu = \frac{h\nu}{4\pi} B_{ij} \phi(\nu) n_i = \frac{1}{4\pi\varepsilon_0} \frac{\pi e^2}{m_e c} f \phi(\nu) n_i \quad (7.20)$$

with

$$\frac{1}{4\pi\varepsilon_0} \frac{\pi e^2}{m_e c} = 2.65 \cdot 10^{-6} \quad [\text{m}^2 \, \text{s}^{-1}]. \quad (7.21)$$

In the Gaussian unit system, instead of this expression, we would have $\pi e^2/(m_e c) = 2.65 \cdot 10^{-2}$ cm^2 s^{-1}.

Within the line, the *profile function* describes the precise frequency dependence of the resonant absorption coefficient (Sect. 7.1.4) at $\nu = |\chi_j - \chi_i|/h$; it is normalized to

$$\int_{\text{Line}} \phi(\nu) \mathrm{d}\nu = 1. \quad (7.22)$$

As an atomic constant to determine the strength of the spectral line, we choose here the f value, which, according to (4.125), could also be replaced by the Einstein coefficient A_{ji}. Numerically, relation (4.125) yields

$$g_j A_{ji} = \frac{6.670 \cdot 10^{13}}{(\lambda \, [\text{nm}])^2} g_i f \quad [\text{s}^{-1}]. \quad (7.23)$$

Since the occupation probability n_i of the lower (absorbing) level in (7.20) is proportional to its statistical weight g_i according to the Boltzmann formula, (4.88), one often quotes the gf value directly, instead of f.

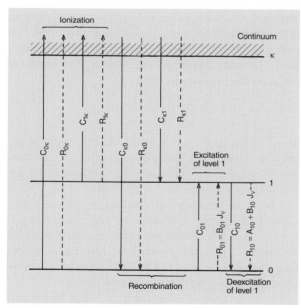

Fig. 7.4. Elementary processes in an atom which is schematically represented by two bound energy states 0 and 1 and the continuum κ (averaged over the Maxwell–Boltzmann distribution for free electrons). The dashed arrows indicate the radiative processes with their rates R_{ij}, and the fully-drawn arrows show collision processes with rates C_{ij}. A_{10}, B_{10} and B_{01} are the Einstein coefficients for the line transition $0 \leftrightarrow 1$, and J_ν is the angle-averaged intensity in the corresponding line

Oscillator Strengths. A first overview of the *absolute values* of the oscillator strengths can be gained from the f sum rule of W. Kuhn and W. Thomas: *from* a given level i of an atom or ions with z electrons (as an approximation, one considers only the "valence electrons" which take part in the transition considered), the absorption transitions $i \rightarrow j$ with oscillator strengths f_{ij} are possible, while from lower-lying levels, the transitions $j' \rightarrow i$ with $f_{j'i}$ lead *to* i. Then we have

$$\sum_j f_{ij} - \sum_{j'} \frac{g_{j'}}{g_i} f_{j'i} = z \, . \qquad (7.24)$$

If essentially a *single* strong transition leads from the ground-state term to the next highest term, then as a good approximation we can set $f \simeq 1$ for the whole resonance multiplet. Thus, for example, for the two Na I D lines ($^2S - {}^2P^\circ$), we estimate for both lines together $f \simeq 1$. The exact value is $f = 0.98$; from (7.23), it corresponds to an Einstein coefficient of $A = 0.64 \cdot 10^8 \, \text{s}^{-1}$.

The *relative* f values within a multiplet are found from quantum-mechanical formulas derived by A. Sommerfeld and H. Hönl, H. N. Russell, and others. For example, the ratios of the gf values for a doublet, like the Na D lines $^2S_{1/2} - {}^2P^\circ_{1/2, 3/2}$, are given by 1:2; for a triplet, like Ca I $4p^3P^\circ_{2,1,0} - 5s\,^3S_1$ ($\lambda = 616, 612, 610$ nm; Fig. 7.2), they are 5:3:1.

Exact f or gf values can either be calculated from quantum mechanics or experimentally determined. We cannot take up the various methods in detail here. The voluminous data on the transition probabilities and the energy levels of atoms – ordered by element – is collected in *compilations* and *bibliographies*, which are issued for the most part by the National Institute of Standards and Technology in Washington, D.C. Term schemes are given in several volumes of the "Atomic Energy-Level and Grotrian Diagrams" by S. Bashkin and J. O. Stoner (1975 and later). We can find f values in the "Tables on Atomic Transition Probabilities" by W. L. Wiese, J. R. Fuhr and others (1966 and later), and in additional works.

7.1.4 Broadening of Spectral Lines

We now turn to the calculation of the profile function $\phi(\nu)$ of the line absorption coefficients as a function of the temperature T, electron pressure P_e or the gas pressure P_g, and of the distance from the center of the line, $\Delta \nu$ in frequency units (or $\Delta \lambda$ in wavelength units). Here, we first consider the broadening of the lines of a single atom or ion by radiation and collision damping. This then has to be superimposed onto the thermal motion of the atoms.

Lorentz Line Profiles. In classical optics, a spectral line corresponds to a wave train with a characteristic length (on the time axis) of τ (according to a well-known theorem of Fourier analysis); its absorption coefficient exhibits a typical *Lorentz* or *damping profile*

$$L(\Delta \nu) = \frac{\gamma}{(2\pi \Delta \nu)^2 + (\gamma/2)^2} \qquad (7.25)$$

with the damping constant $\gamma \, [\text{s}^{-1}] = 1/\tau$, which is at the same time the full width at half maximum of the absorption coefficient in units of the circular frequency, $\omega = 2\pi \nu$. The integral $\int L(\Delta \nu) \text{d}(\Delta \nu)$ over the entire line is normalized to one.

Depending on whether the temporal length of the radiation process is limited by the emission of the atom itself or by collisions with other particles, we speak of *radiation* or *collision damping*. From quantum mechanics, we also obtain a line profile of the form of (7.25), with

$$\gamma = \gamma_{\text{rad}} + \gamma_{\text{coll}} \, . \qquad (7.26)$$

The *radiation damping* constant γ_{rad} is equal to the sum of the reciprocal lifetimes (decay constants) of the two energy levels i und j between which the transition takes place. In many cases, we can neglect the stimulated emission process; then we find

$$\gamma_{\text{rad}} = \sum_{k<i} A_{ik} + \sum_{k<j} A_{jk} \, . \qquad (7.27)$$

Since, for allowed transitions, γ_{rad} is of the order of 10^7 to $10^9 \, \text{s}^{-1}$, we expect from this mechanism (full) widths at half maximum for the absorption coefficients, e.g. for $\lambda = 400$ nm, of about 10^{-6} to 10^{-4} nm.

The *collision damping* constant is $\gamma_{\text{coll}} = 2 \cdot$ (the number of effective collisions per unit time). According to W. Lenz, V. Weisskopf and others, these are close passages of a perturbing particle near the radiating particle, in which the phase of the light wave is shifted by more than one-tenth of an oscillation period.

In cooler stars, such as the Sun, where the hydrogen is for the most part neutral, collision damping by neutral hydrogen atoms predominates; these influence the emitting atom through *van der Waals forces*, whose interaction energy is $\propto r^{-6}$ (r is the distance of the collision partners). In addition to the van der Waals forces, at smaller r values, shorter-ranged repulsive forces can also become important. The damping constant γ_{coll}, as long as it is due to the interaction of the emitting atom with only one H atom at a time, is proportional to the gas pressure P_g.

In the case of spectral lines which exhibit a large *quadratic Stark effect*, and in predominantly ionized atmospheres, collision damping by free electrons is sometimes the most important effect. The interaction energy is then proportional to the square of the field strength produced by the electron at the site of the emitting atom, i.e. $\propto r^{-4}$. The damping constant is then proportional to the electron pressure P_e.

At about 10^9 effective collisions per s, we can expect halfwidths of the line absorption coefficients of a few 10^{-4} nm. Here, it is unimportant whether the collisions – as in the atmosphere of the Sun – are predominantly with hydrogen atoms, or – as in hot stars – predominantly with free electrons.

The Linear Stark Effect. The lines of *hydrogen* and of singly ionized *helium* require special consideration. Since they show particularly large linear Stark splittings in an electric field, their line broadening in partially ionized gases is due for the most part to the quasi-static Stark effect from the statistically distributed electric fields which are produced by the (slow-moving) ions. Beginning with the consideration that in the distance range $r \ldots r + dr$ from an H atom, a perturbing ion can be found with a probability proportional to $4\pi r^2 dr$ and then produces a field $\propto 1/r^2$ (to which the line splitting is in turn proportional), one can show that in the wings of the absorption line, the absorption coefficient becomes approximately proportional to $(\Delta v)^{-5/2}$ or $(\Delta \lambda)^{-5/2}$. This theory, originally developed by J. Holtsmark, requires refinement by including nonadiabatic effects, collision damping by electrons, and an improved calculation of the microfield in the plasma.

The Doppler Profile. In addition to broadening by radiation and collision damping, there is a contribution to broadening due to the Doppler effect as a result of the thermal velocities and possibly of turbulent flows. Corresponding to the Maxwell-Boltzmann velocity distribution for the atoms (4.86), the Doppler profile, again normalized to one, is given by

$$D(\Delta v) = \frac{1}{\sqrt{\pi} \Delta v_D} \exp\left[-(\Delta v/\Delta v_D)^2\right] \quad (7.28)$$

with a full width at half maximum of $2\Delta v_D \sqrt{\ln 2}$ (in frequency units). The Doppler width Δv_D or $\Delta \lambda_D$ is defined by the most probable velocity (4.87), $v_0 = (2kT/m)^{1/2}$, m = mass of the absorbing atom:

$$\frac{\Delta v_D}{v_0} = \frac{\Delta \lambda_D}{\lambda_0} = \frac{v_0}{c} \quad (7.29)$$

(v_0 or λ_0 refer to the center of the line). The thermal velocity, e.g. of the Fe atoms in the atmosphere of the Sun at $T \simeq 5700$ K, is 1.3 km s^{-1}, which leads e.g. for the Fe I-Linie $\lambda = 386$ nm to a Doppler width of $\Delta \lambda = 1.7 \cdot 10^{-3}$ nm. Turbulent flows in the stellar atmospheres ("microturbulence") often produce similar velocities and a corresponding contribution to $\Delta \lambda_D$.

The Voigt Profile. Finally, we consider the *interaction* of the Doppler effect and damping, which will lead us to the profile function via the convolution integral

$$\varphi(\Delta v) = \int_{-\infty}^{+\infty} L(\Delta v - \Delta v') D(\Delta v') \, d(\Delta v') \, . \quad (7.30)$$

It is termed a Voigt profile when the normalization to one refers not to the line area, but rather to the maximum value at the center of the line. Since the ratio of half the damping constant, $\gamma/2$, to the Doppler width $\Delta \omega_D = 2\pi \Delta v_D$ (likewise in circular frequency units)

$$\alpha = \gamma/2\Delta \omega_D \quad (7.31)$$

in stellar atmospheres is almost without exception < 0.1, one might at first assume that the broadening through damping could be neglected in comparison to the Doppler broadening. This is, however, not correct, because the Doppler distribution, (7.28), decreases exponentially outwards from the line center, while the damping distribution, (7.25), decreases only proportionally to $(\Delta v)^{-2}$. Each moving atom produces a damping distribution with a sharp center and broad wings, which as a whole is Doppler shifted, depending on its velocity. Thus, one obtains a profile function with a rather sharply limited Doppler core which is connected almost directly to the damping wings (Fig. 7.5).

To conclude this section, we give the two limiting cases of the Voigt profile for the absorption coefficient $\tilde{\kappa}_v = \tilde{\kappa}_\lambda$ in [m^{-1}] (7.20) at a distance $\Delta \lambda$ from the center of a line at wavelength λ_0. We convert the frequency scale into a wavelength scale using $|\Delta v| = (c/\lambda_0^2)\Delta \lambda$,

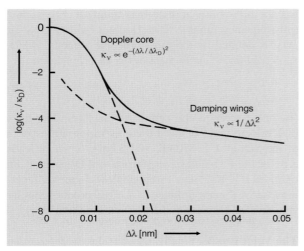

Fig. 7.5. Line absorption coefficient κ_ν (referred to the line center). Doppler core and damping wings of the Na D lines calculated according to (7.25) and (7.28) for $T = 5700$ K and purely radiative damping. Their superposition (———) gives the Voigt profile

and obtain the Doppler core:

$$\tilde{\kappa}_\lambda = \frac{1}{4\pi\varepsilon_0} \frac{e^2}{m_e c^2} f n_i$$
$$\times \sqrt{\pi} \frac{\lambda_0^2}{\Delta\lambda_D} \exp\left[-\left(\frac{\Delta\lambda}{\Delta\lambda_D}\right)^2\right] \quad (7.32)$$

and for the damping wings:

$$\tilde{\kappa}_\lambda = \frac{1}{4\pi\varepsilon_0} \frac{e^2}{m_e c^2} f n_i \cdot \frac{\lambda_0^4}{4\pi c} \frac{\gamma}{(\Delta\lambda)^2} \,. \quad (7.33)$$

The quantity $e^2/(4\pi\varepsilon_0 m_e c^2) = 2.82 \cdot 10^{-15}$ m is the classical electron radius, and n_i [m^{-3}] is the occupation density of the lower level. As we have already mentioned, (7.32, 33) are not valid for the lines of H I and He II, which are broadened by the linear Stark effect.

7.1.5 Remarks on Molecular Spectroscopy

Molecules become important for the cooler ($T \leq$ 4000 K) stellar atmospheres and play a central role in particular in the coolest and densest components of interstellar matter, the molecular clouds (Sect. 10.2.3). We cannot give a complete treatment of the fundamentals of molecular spectroscopy here, but will make some remarks on the subject.

As in the case of atoms and ions (Sect. 7.1.1), the energy states of molecules are essentially determined by their electrons, depending on their quantum numbers; here however, the electric fields of the nuclei are not Coulomb fields, but rather – in the simplest case of a *diatomic* molecule AB – a dipole field. The transitions between the different electronic states occur as a rule in the optical or ultraviolet spectral ranges.

Every electronic state is split as a result of the vibrations of the nuclei into a number of *vibrational* levels (quantum number v), which are in turn split into a series of *rotational* levels due to the rotation of the molecule (quantum number J). The transitions between (pure) vibrational or rotational states are generally in the infrared and radiofrequency ranges. For the coolest molecular clouds, mainly the transitions from the lowest rotational states are important.

In the simplified case of the *rigid rotor* ("dumbbell"), in which the two nuclei with masses m_A and m_B rotate about their common center of gravity (Fig. 2.21) at a fixed distance a from a rotation axis which is perpendicular to the line connecting them, we find the moment of inertia (2.65), using (2.54), to be:

$$I = \frac{m_A m_B}{m_A + m_B} a^2 \quad (7.34)$$

and therefore the rotational energy is

$$E_{\text{rot}} = \frac{h^2}{8\pi^2 I} J(J+1) = hcB\, J(J+1)\,, \quad (7.35)$$

where $J = 0, 1, 2, \ldots$ is the rotation quantum number and B is the rotational constant, usually given in units of [cm^{-1}]. In classical mechanics, we would have $E_{\text{rot}} = I\omega^2/2$ (ω: angular velocity) or, if we introduce the angular momentum $S = I\omega$ (2.64), $E_{\text{rot}} = S^2/(2I)$ (2.68). In the quantum-mechanical expression, S enters in units of $\hbar$, more precisely $S = \hbar\sqrt{J(J+1)}$.

The selection rule for electric dipole radiation is $\Delta J = 0, \pm 1$ (except for $0 \leftrightarrow 0$).

The occurrence of $m_A m_B/(m_A + m_B)$, the reduced mass, in I and E_{rot} shows that molecular spectra are well suited for observing different *isotopes* of an element. For example, between the lines of $^{12}C^{16}O$ and $^{13}C^{16}O$, corresponding to the reduced masses of 6.86 or 7.17, according to (7.35) there will be an energy or frequency difference giving a factor of 1.046 (cf. Sect. 10.2.3). In contrast, isotope effects in atomic lines are considerably smaller, since here the reduced mass is determined mainly by the electron mass m_e, which is much smaller than that of the nucleus; thus the isotope lines can be resolved only for the lightest atoms, e.g. for ^{1}H/^{2}D or ^{6}Li/^{7}Li.

Expression (7.20) for the line absorption coefficients of atoms can be applied also to molecules.

The fraction of the elements which is bound in molecular form can – in thermodynamic equilibrium – be calculated from the law of mass action (Sect. 4.4.5).

7.2 The Physics of Stellar Atmospheres

With an eye to the quantitative explanation and evaluation of spectra from the Sun and the stars, we turn now to the physics of stellar atmospheres. The *atmosphere* by definition includes those layers of a star which send us radiation directly. In Sect. 7.2.1, we discuss the characteristic parameters of a stellar atmosphere (photosphere) and the basic equations for its structure. In Sect. 7.2.2, we consider the absorption coefficients in stellar atmospheres in more detail. Then in Sect. 7.2.3, we treat the construction of a model atmosphere and show calculated spectral energy distributions for some stars. In Sect. 7.2.4 we treat the theory of Fraunhofer lines, and in Sect. 7.2.5 the growth curve of a spectral line. Finally, in Sect. 7.2.6, we turn to our main task, the quantitative analysis of stellar spectra, and introduce in Sect. 7.2.7 the element abundances derived for the Sun and selected stars.

7.2.1 The Structure of Stellar Atmospheres

We limit ourselves here to *compact* atmospheres, whose thickness is small compared to the radius of the star, so that we can approximate the layer structure as a series of parallel planes and can assume the acceleration of gravity (cf. (3.4), (6.52)) to be constant.

A compact stellar atmosphere is characterized by the following *parameters*:

1. The *effective temperature* T_{eff}. According to the Stefan-Boltzmann radiation law (4.66), the radiation energy flux at the star's surface is given by

$$F = \int_0^\infty F_\lambda d\lambda = \int_0^\infty F_\nu d\nu = \sigma T_{\text{eff}}^4 \,. \quad (7.36)$$

For the Sun, we obtained in (6.8) and (6.11) directly from the solar constant $T_{\text{eff},\odot} = 5780$ K or $F_\odot = 6.33 \cdot 10^7$ Wm^{-2}.

2. The *acceleration of gravity* g at the surface of the star. For the Sun, we found in (6.5) $g_\odot = 274$ m s^{-2}; the acceleration of gravity of the main sequence stars is, according to (6.53), not very different from this value.

3. The *chemical composition* of the atmosphere, i.e. the abundance distribution of the elements. In Table 7.2, we have given some values in advance.

Possible additional parameters, such as rotation or oscillation of the star, stellar magnetic fields, etc., will not be considered here.

One can readily reach the conclusion that with a given T_{eff}, g, and chemical composition, the *structure*, i.e. the temperature and pressure distribution in a static stellar atmosphere, can be completely calculated. For this purpose, we need two equations (in the case of local thermodynamic equilibrium, see below): (a) an equation for *energy transport*, by radiation, convection, heat conduction, mechanical or magnetic energy, which determines the temperature distribution in the atmosphere; and (b) the *hydrostatic equation*, or, generally speaking, the basic equations of hydrodynamics or magnetohydrodynamics which determine the pressure distribution. The degree of ionization, the equation of state, and all the material constants of stellar atmospheres can in principle be calculated from the results of atomic physics, so long as we know the chemical composition. The theory of stellar atmospheres thus makes it possible, beginning with parameters 1, 2, and 3, to calculate a *model atmosphere*, and furthermore to find out how certain *measurable* quantities, e.g. the intensity distribution in the continuum (the color indices), or the intensities of the Fraunhofer lines of certain elements in particular ionization and excitation states, depend on these parameters.

Once we have solved this problem, we can (and this is the decisive point!) reverse the process and, in a series of successive approximations, find the answer to the question, "Which T_{eff}, g, and abundance distribution of the chemical elements does the atmosphere of a particular star have, when its color indices (energy distribution in the continuum), intensities of various Fraunhofer lines, etc. have given measured values?" In this way, we arrive at a procedure for the *quantitative analysis* of the spectra of the Sun and stars.

The Temperature Distribution. The temperature distribution in a stellar atmosphere is determined, as mentioned above, by the type of energy transport. As was recognized in 1905 by K. Schwarzschild, the energy transport in most stellar atmospheres – more precisely photospheres, cf. Sect. 6.1 – takes place predominantly through radiation; we thus speak of *radiation equilibrium*.

In order to describe the *radiation field* (Fig. 7.6), we imagine an element of unit surface area at a depth t (measured from an arbitrarily chosen zero level); its surface normal makes an angle θ with that of the stellar surface ($0 \leq \theta \leq \pi$). Then, within a solid angle element $d\Omega$, the radiation energy $I_\nu(t, \theta) d\nu d\Omega$ in the frequency interval ν to $\nu + d\nu$ passes through this surface element in the direction of its normal per unit time (cf. Sect. 4.3.1). Along a path element $ds = -dt/\cos\theta$, the radiation intensity I_ν suffers an overall change, according to the radiation transport equation (4.52) and assuming local thermodynamic equilibrium (LTE), i.e. applying Kirchhoff's law (4.69) to each volume element in the atmosphere, of

$$dI_\nu(t, \theta) = I_\nu(t, \theta) \kappa_\nu dt / \cos\theta - B_\nu(T(t)) \kappa_\nu dt / \cos\theta , \quad (7.37)$$

where $B_\nu(T)$ is the Kirchhoff–Planck function (4.61) for the local temperature T at the depth t and the frequency ν.

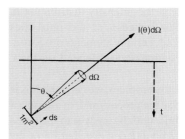

Fig. 7.6. Radiation equilibrium

In order to describe the different layers of the atmosphere, it is preferable to use the optical depth for radiation of frequency ν instead of the geometric depth t:

$$\tau_\nu = \int_{-\infty}^{t} \kappa_\nu dt , \quad d\tau_\nu = \kappa_\nu dt . \quad (7.38)$$

It is calculated from the viewpoint of the observer ($t = -\infty$). We thus obtain for a compact atmosphere, which can be approximated by *planar, parallel* layers,

$$\cos\theta \frac{dI_\nu(\tau_\nu, \theta)}{d\tau_\nu} = I_\nu(\tau_\nu, \theta) - B_\nu(T(\tau_\nu)) . \quad (7.39)$$

If *radiation equilibrium* holds, i.e. if the overall energy transport takes place via radiation, we can apply energy conservation, which tells us that the *total* radiation flux must be independent of the depth t, i.e

$$F = \int_{\theta=0}^{\pi} \int_{\nu=0}^{\infty} I_\nu(t, \theta) \cos\theta \, 2\pi \sin\theta \, d\theta \, d\nu = \sigma T_{\text{eff}}^4 . \quad (7.40)$$

The boundary conditions of the system of equations (7.39) and (7.40) are: (a) the incident radiation vanishes at the stellar surface, i.e. $I_\nu(0, \theta)$ for $0 < \theta < \pi/2$, and (b) at great depths, the "enclosed" radiation field, which approximates that of a cavity radiator, cannot increase more rapidly than exponentially with increasing depth (we cannot treat this point in detail here).

The solution of the system of equations now becomes relatively simple, if we first assume that the absorption coefficient is *in*dependent of the frequency ν; that is, we insert in (7.39) in place of κ_ν its harmonic mean value (with suitable weighting factors) averaged over all frequencies ν, i.e. the *Rosseland* opacity coefficient $\bar{\kappa}$ (8.1, 2) and the corresponding optical depth:

$$\bar{\tau} = \int_{-\infty}^{t} \bar{\kappa} dt . \quad (7.41)$$

For such a "grey atmosphere", according to E. A. Milne and others, the solution gives to a good approximation the temperature distribution:

$$T^4(\bar{\tau}) = \frac{3}{4} T_{\text{eff}}^4 (\bar{\tau} + 2/3) . \quad (7.42)$$

The effective temperature T_{eff} is thus obtained at an optical depth $\bar{\tau} = 2/3$. At the stellar surface $\bar{\tau} = 0$, T approaches a finite limiting temperature

$T_0 = T_{\text{eff}}/\sqrt[4]{2} = 0.84 \, T_{\text{eff}}$. It is frequently expedient to use an optical depth τ_0 instead of $\overline{\tau}$; it refers to the absorption coefficient κ_0 at a suitably chosen wavelength λ_0 (e.g. $\lambda_0 = 500$ nm). The relation between $\overline{\tau}$ and τ_0 (and correspondingly also between the optical depths for two different wavelengths) is readily found from $d\overline{\tau} = \overline{\kappa} dt$ and $d\tau_0 = \kappa_0 dt$ to be

$$\frac{d\overline{\tau}}{d\tau_0} = \frac{\overline{\kappa}}{\kappa_0} \tag{7.43}$$

for a given depth.

The Pressure Distribution. If the temperature distribution $T(\overline{\tau})$ or $T(\tau_0)$ in a stellar atmosphere is known, the calculation of the pressure distribution offers no difficulties. The increase of gas pressure P_g with depth t in a static atmosphere is determined by the *hydrostatic equation* (3.6):

$$dP_g/dt = g\varrho, \tag{7.44}$$

where the density ϱ is related to P_g and T by the equation of state for an ideal gas; g once again denotes the acceleration of gravity. Dividing both sides by κ_0 and using $\kappa_0 dt = d\tau_0$, we obtain the relation

$$\frac{dP_g}{d\tau_0} = \frac{g\varrho}{\kappa_0}. \tag{7.45}$$

Since the right-hand side of this equation is known as a function of P_g and T, and since T is known as a function of τ_0 from the theory of radiation equilibrium, (7.45) can be (numerically) integrated without difficulty.

In hot stars, the radiation pressure must be taken into account in addition to the gas pressure. If currents are present in the stellar atmosphere, and their velocity v is no longer small compared to the local velocity of sound (in atomic hydrogen at 10 000 K, for example, 12 km s^{-1}), then the dynamic pressure $\varrho v^2/2$ also plays a role. In sunspots and in the Ap stars with their strong magnetic fields, the magnetic forces must also be taken into consideration.

Non-grey Atmospheres. For a real analysis of stellar spectra, the "grey approximation" (7.42) is too imprecise; we need to take into account the frequency dependence of the absorption coefficient κ_ν, which is composed of the continuous absorption coefficient and of the very strongly frequency-dependent line absorption (Sect. 7.2.2). In the radiation equilibrium, the total radiation flux F must then be equal at all depths t. The calculation of such "non-grey" atmospheres can be carried out only by applying successive approximation methods.

For the calculation of stellar atmospheric models, besides the absorption coefficient κ_ν or $\overline{\kappa}$ as a function of temperature and pressure (Sect. 7.2.2), we still need the relation between gas and electron pressure for various mixtures of elements, $P_e(P_g, T)$ (Table 7.3), and, in the equation of state, the mean molecular mass $\overline{\mu}$. For neutral (atomic) or fully ionized stellar matter having the solar elemental composition (Table 7.2), $\overline{\mu} = 1.26$ or 0.60.

In the deep layers of the atmospheres of later spectral classes, energy transport by *convection* is predominant over radiative transport (Sect. 7.3.1). In this case, the condition of radiation equilibrium (7.40) must be extended to include the *total* energy flux from radiation *plus* convection, which must be independent of depth.

The assumptions of radiation equilibrium (possibly including convective transport) and of local thermodynamic equilibrium are relatively well fulfilled in the *photospheres* of the Sun and of most stars, i.e. in those layers from which the absorption line spectrum arises. The thinner layers above the photosphere, where the LTE is no longer a reasonable approximation and where (in addition to the radiation field) matter flows and magnetic fields play an important role, will be treated in Sect. 7.3. Here, we first consider the fundamentals of the quantitative analysis of photospheric spectra.

7.2.2 Absorption Coefficients in Stellar Atmospheres

In calculating the temperature distribution, besides the continuous absorption, the line absorption must be taken into account. Initially, the optical spectra (Fig. 6.10) give the impression that line absorption plays a role only for cooler stars, such as the Sun, with their numerous Fraunhofer lines, but not for the hotter stars. However, satellite observations have shown that even in the O and B stars in the range of their maximum energy flux, in the ultraviolet near $\lambda \leq 200$ nm, the absorption lines are similarly dense as in the case of the Sun in the blue region.

We first discuss the *continuous* absorption coefficients, which are due to bound–free and free–free transitions (Sect. 4.5.4) of various atoms and ions, and then line absorption or bound–bound transitions.

Continuous Absorption. In the case of neutral *hydrogen*, a series-limit continuum occurs on the short-wavelength side adjacent to the limits of the Lyman series at $\lambda = 91.2$ nm, the Balmer series at $\lambda = 364.7$ nm, and the Paschen series at $\lambda = 820.6$ nm, etc. Their absorption decreases as $\propto 1/\nu^3$ (see (4.130)). At long wavelengths, the series-limit continua overlap and pass over smoothly into the free–free continuum of H I.

As was noticed in 1938 by R. Wildt, in cooler stellar atmospheres, the bound–free and free–free transitions in the *negative hydrogen ion* H$^-$, which can be formed from a neutral H atom by electron attachment, play an important role. The H$^-$ ion, with only one bound state (the ground state ^{1}S) has a binding energy or ionization energy of 0.75 eV. Corresponding to this small energy, the long-wavelength limit of the bound–free continuum lies in the infrared at 1.655 μm. The free–free absorption also increases (as in the H atom) at longer wavelengths.

The *atomic* coefficients for absorption from a particular energy level have been calculated quantum-mechanically for the H atom and the H$^-$ ion with great precision. In order to obtain from them the absorption coefficient κ_ν of stellar matter of a given composition (in local thermodynamic equilibrium) as a function of the frequency ν, the temperature T, and the electron pressure P_e (or the gas pressure P_g), the ionization of the various elements and the excitation of their energy levels must first be calculated using the formulas of Boltzmann and Saha.

For the "normal" elemental mixture (Table 7.2), one first finds that in hot stars ($T \gtrsim 7000$ K), the continuous absorption of H atoms, in cooler stars that of the H$^-$ ions, is predominant. Although for example in the Sun's atmosphere (at roughly $\tau_0 = 0.1$, Fig. 7.7) the particle density of the H$^-$ ion is only about 10^{-8} of that of H I, it nevertheless dominates the continuous absorption in the visible region, because the relative occupation density of the second excited H I level (12.1 eV), from which the "competing" absorption in the Paschen continuum originates, is much lower (about 10^{-11}), and the atomic cross sections are of comparable magnitude.

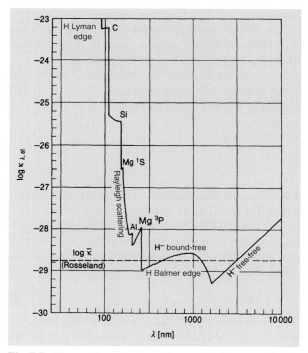

Fig. 7.7. The continuous absorption coefficient $\kappa_{\lambda,\text{at}}$ in the solar atmosphere (G2 V, [m^2] per nucleus) at $\tau_0 = 0.1$ (τ_0 corresponds to $\lambda = 500$ nm), i.e. $T = 5400$ K, $P_e \simeq 0.32$ Pa, $P_g \simeq 5.8 \cdot 10^3$ Pa

In addition to the absorption of H I and H$^-$, the following must be considered, in particular: free–bound and free–free absorption of He I and He II (in hot stars) and of the heavier elements C I, Mg I, Si I (in cooler stars); also, in the coolest stars, the continuum of H$_2^-$. Along with these absorption processes, the *scattering* of light is important: in hot stars, Thomson scattering σ_T by free electrons, (4.132); and in cooler stars, Rayleigh scattering by neutral hydrogen atoms.

In Figs. 7.7 and 7.8, we give as examples the continuous absorption coefficients as functions of the wavelength for mean values of the state variables (Sect. 7.2.3) in the atmosphere of a cooler star: the Sun; and of a hotter star: τ Sco (B0 V).

Line Absorption. We have already treated the theory of line absorption coefficients in Sects. 7.1.3, 4. For the calculation of a model atmosphere, in general a very large number ($\geq 10^6$) of lines have to be taken into con-

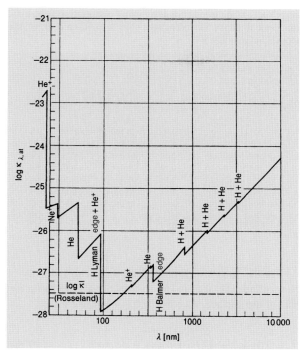

Fig. 7.8. The continuous absorption coefficient $\kappa_{\lambda,\text{at}}$ in the atmosphere of τ Scorpii (B0 V) at $\bar{\tau} \simeq 0.1$, i.e. $T = 28\,300$ K, $P_e = 320$ Pa, $P_g \simeq 640$ Pa

sideration. In this process, only a few strong lines can be treated individually and thus sufficiently precisely; the majority of the medium-strong and weak lines is taken into account by means of *statistical* methods, e.g. sampling techniques or distribution functions. Here, a high precision for the individual lines is not important.

Since the lines reduce the outflow of radiation from the depths, but increase its intensity in the higher layers, they tend to cause a steeper temperature decrease in the outer layers as compared with the temperature distribution calculated on the basis of continuous absorption only ("blanketing effect").

7.2.3 Model Atmospheres. The Spectral Energy Distribution

Having described the equations which govern the structure of a stellar atmosphere, the absorption coefficients, and other material functions, we present in Table 7.4 the *models* for atmospheres at various effective temperatures calculated according to the current state of the art (and with the assumption of local thermodynamic equilibrium). The details of the stellar atmosphere model calculations go beyond the scope of this book. Table 7.4 contains the model of a cool star (the Sun, G2 V), a star of medium temperature (α Lyr, A0 V), and a hot main-sequence star (B0 V), represented by τ *Sco*, which has been intensively investigated following the pioneering work by A. Unsöld (1941/44).

The radiation which is emitted at the surface of a star originates from various layers within its atmosphere; that from the deeper layers naturally undergoes more attenuation by absorption before emerging at the stellar surface than does radiation from further up. Radiation of frequency ν, emitted from a depth τ_ν at an angle θ relative to the surface normal of the atmosphere, is attenuated by the factor $\exp(-\tau_\nu/\cos\theta)$ before leaving the atmosphere. We thus obtain from (7.39) the *radiation intensity* at the surface, $\tau_\nu = 0$:

$$I_\nu(0, \theta) = \int_0^\infty S_\nu(\tau_\nu) e^{-\tau_\nu/\cos\theta} d\tau_\nu / \cos\theta , \quad (7.46)$$

where the source function S_ν, defined as the emission coefficient j_ν divided by the absorption coefficient κ_ν (4.57) in local thermodynamic equilibrium, is equal to the Kirchhoff–Planck function at the local temperature, $B_\nu(T(\tau_\nu))$.

Since we can, on the one hand, calculate the *temperature* T as a function of the optical depth (e.g. τ_ν) using the theory of radiation equilibrium, and since on the other, the relation between the differently-defined optical depths $\bar{\tau}, \tau_\nu, \tau_0, \ldots$ can be established with (7.43) using the absorption coefficients κ_ν, the entire theory of stellar spectra is contained in (7.46), including (for the Sun) the *center–limb variation* ($\theta = 0°$ corresponds to the center of the solar disk, $\theta = 90°$ to its limb).

Indeed, calculations (which we shall not describe in detail here), carried out on the basis of the model of the *solar atmosphere* in Table 7.4, yield the solar spectrum $I_\nu(0, 0)$ (e.g. for the center of the solar disk (Fig. 6.3)), and the center–limb variation in the continuum (Fig. 6.2) with good accuracy.

The Eddington–Barbier Approximation. In order to determine which layers of the atmosphere make

Table 7.4. Models of stellar atmospheres after R. L. Kurucz (1979) for the solar element mixture and different effective temperatures T_{eff} and gravitational accelerations g. Line absorption is taken into account by using distribution functions; the optical depth τ_0 refers to κ_λ at $\lambda = 500$ nm, $\bar{\tau}$ to the Rosseland average $\bar{\kappa}$

	Sun G2 V $T_{\text{eff}} = 5770$ K, $g = 274$ m s^{-2}				α Lyr A0 V $T_{\text{eff}} = 9400$ K, $g = 89$ m s^{-2}				B0 V $T_{\text{eff}} = 30\,000$ K, $g = 100$ m s^{-2} [a]			
$\bar{\tau}$	τ_0	T [K]	P_g [Pa]	P_e [Pa]	τ_0	T [K]	P_g [Pa]	P_e [Pa]	τ_0	T [K]	P_g [Pa]	P_e [Pa]
10^{-3}	$1.1 \cdot 10^{-3}$	4485	$3.46 \cdot 10^2$	$2.84 \cdot 10^{-2}$	$0.6 \cdot 10^{-3}$	7140	6.52	$4.31 \cdot 10^{-1}$	$0.8 \cdot 10^{-3}$	19 680	2.13	1.07
0.01	0.01	4710	$1.29 \cdot 10^3$	$1.03 \cdot 10^{-1}$	$0.5 \cdot 10^{-2}$	7510	$2.70 \cdot 10^1$	1.61	$0.9 \cdot 10^{-2}$	21 450	$1.60 \cdot 10^1$	7.98
0.10	0.09	5070	$4.36 \cdot 10^3$	$3.78 \cdot 10^{-1}$	0.05	8150	$9.13 \cdot 10^1$	7.33	0.14	24 880	$1.01 \cdot 10^2$	$5.03 \cdot 10^1$
0.22	0.19	5300	$6.51 \cdot 10^3$	$6.43 \cdot 10^{-1}$	0.11	8590	$1.22 \cdot 10^2$	$1.40 \cdot 10^1$	0.37	27 030	$1.86 \cdot 10^2$	$9.31 \cdot 10^1$
0.47	0.40	5675	$9.55 \cdot 10^3$	1.34	0.24	9240	$1.53 \cdot 10^2$	$2.94 \cdot 10^1$	0.92	29 840	$3.33 \cdot 10^2$	$1.66 \cdot 10^2$
1.0	0.84	6300	$1.29 \cdot 10^4$	4.77	0.53	10 190	$1.79 \cdot 10^2$	$5.81 \cdot 10^1$	2.2	33 490	$5.87 \cdot 10^2$	$2.95 \cdot 10^2$
2.2	1.8	7085	$1.52 \cdot 10^4$	$2.13 \cdot 10^1$	1.3	11 560	$2.12 \cdot 10^2$	$9.21 \cdot 10^1$	5.5	38 310	$1.04 \cdot 10^3$	$5.29 \cdot 10^2$
4.7	3.5	7675	$1.71 \cdot 10^4$	$5.86 \cdot 10^1$	3.6	13 480	$2.99 \cdot 10^2$	$1.40 \cdot 10^2$	13.3	43 940	$1.81 \cdot 10^3$	$9.43 \cdot 10^2$
10	7.1	8180	$1.89 \cdot 10^4$	$1.27 \cdot 10^2$	11.5	16 000	$5.81 \cdot 10^2$	$2.77 \cdot 10^2$	37	51 310	$3.60 \cdot 10^3$	$1.88 \cdot 10^3$

[a] Corresponds roughly to the parameters of τ Sco (B0 V) $T_{\text{eff}} = 31\,500$ K and $g = 140$ m s^{-2}.

the *principal contribution* to the emitted intensity at a frequency ν, we can carry out a greatly simplified calculation of $I_\nu(0, \theta)$ by expressing the source function $S_\nu(\tau_\nu)$ or $B_\nu(\tau_\nu)$ at a particular optical depth τ^*, which we shall initially leave undetermined, in the form of a series expansion:

$$S_\nu(\tau_\nu) = S_\nu(\tau^*) + (\tau_\nu - \tau^*)\left(\frac{dS_\nu}{d\tau_\nu}\right)_{\tau^*} + \ldots \quad (7.47)$$

If we carry out the integration (7.46) using this approach, we readily obtain the result:

$$I_\nu(0, \theta) = S_\nu(\tau^*) + (\cos\theta - \tau^*)\left(\frac{dS_\nu}{d\tau_\nu}\right)_{\tau^*} + \ldots . \quad (7.48)$$

Now, setting $\tau^* = \cos\theta$, the second term on the right vanishes, and we obtain the *Eddington–Barbier approximation*:

$$I_\nu(0, \theta) \simeq S_\nu(\tau_\nu = \cos\theta) . \quad (7.49)$$

This means that on the surface of e.g. the Sun, the emitted intensity corresponds to the source function or, in the sense of Planck's radiation law, to the temperature at the optical depth $\tau_\nu = \cos\theta$, measured perpendicular to the Sun's surface; or $\tau_\nu / \cos\theta = 1$, measured along the line of sight from the observer. This can be understood intuitively: we can say (with a grain of salt) that the radiation emerges at unit optical depth, as measured along the line of sight.

For the stars we can now find the mean intensity of the stellar disk, that is (up to a factor π) the *radiation flux* at the surface. According to (4.35), we have:

$$F_\nu(0) = 2\pi \int_0^{\pi/2} I_\nu(0, \theta) \cos\theta \sin\theta \, d\theta . \quad (7.50)$$

With the approach of (7.48), we obtain from (7.50) the approximation:

$$F_\nu(0) \simeq \pi S_\nu(\tau_\nu = \tfrac{2}{3}) . \quad (7.51)$$

With local thermodynamic equilibrium, the stellar radiation at a frequency ν thus corresponds to the local temperature at an optical depth $\tau_\nu = 2/3$.

The Energy Distribution of the Stars. To illustrate the energy distribution of stellar radiation, we show in Figs. 6.7 and 7.9 the radiation fluxes calculated for main-sequence stars with $T_{\text{eff}} = 9400$ K (A0 V) and 30 000 K (B0 V), using the models in Table 7.4. All similarity to the spectrum of a black-body radiator has now vanished. Instead, the F_λ curve represents more or less a mirror image of the wavelength dependence of the absorption coefficients κ_λ or κ_ν (compare the wavelength dependence of the continuous absorption coefficient in Fig. 7.8). Since $d\tau_\lambda/d\bar{\tau} = \kappa_\lambda/\bar{\kappa}$, for *large* κ_λ, the radiation comes from layers near the surface with *small* $\bar{\tau}$; these layers are relatively *cool*, and the radiation curve F_λ is decreased.

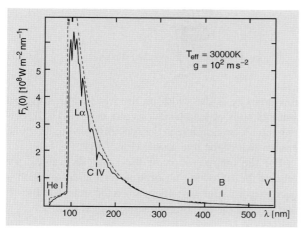

Fig. 7.9. The calculated spectral energy distribution $F_\lambda(0)$ of a main sequence star with an effective temperature of $T_{\text{eff}} = 30\,000$ K (after R. L. Kurucz, 1979). (———) Radiation flux including the Fraunhofer lines, which were calculated using a distribution function with a spectral resolution of 2.5 nm. (– – –) Radiation flux in the continuum, maximum at $\lambda = 92$ nm: $10.1 \cdot 10^8$ W m^{-2}nm^{-1}

We can therefore understand that at each absorption edge of hydrogen ($\lambda = 91.2$ nm, Lyman edge; $\lambda = 364.7$ nm, Balmer edge, ...), F_λ shows a sudden decrease towards shorter wavelengths. (We shall discuss the energy flux in the Fraunhofer lines later, in Sect. 7.2.4). The hotter of the two stars emits its radiation flux mainly between the Balmer and the Lyman edges; in the case of the cooler star, the region between the Paschen and the Balmer edges contributes about the same order of magnitude to the emitted radiation as does the ultraviolet. It is clear that for hot stars, observations in the far ultraviolet from satellites are of decisive importance; measurements in the optical region, which for a long time formed the exclusive basis of our knowledge of these stars, must be extremely precise in order to make a reasonable contribution to the adjustment of theoretical calculations to the observations. This is particularly true of the *bolometric correction*, B.C. (6.37), which is obtained by integration of F_λ over the whole spectrum and comparison with the radiation flux in the visual region; for example, for a B0 V star, B.C. $\simeq 3.2$ mag.

Finally, Fig. 7.9 clarifies the role of *spectral line absorption* in the far ultraviolet for hot stars: the energy distribution taking into account the Fraunhofer lines (they are "smeared out" over 2.5 nm wide wavelength regions) lies well below the "true" continuum. For the A stars (Fig. 6.7), the ultraviolet region is dominated by numerous lines from metallic elements, while in the optical region, the broad Balmer lines from hydrogen must be taken into account. In cooler stars, roughly below F5, the metal lines block out a major portion of the spectrum, especially in the blue and near ultraviolet regions (Fig. 6.3); in the K and M stars, the absorption from molecular bands also comes into play (Table 6.1).

Departures from LTE. When greater precision is required, departures from local thermodynamic equilibrium must be considered within the framework of the kinetic equations described in Sect. 4.5.3 for calculating the model atmospheres. These *"non LTE"* calculations are much more tedious than those which assume local thermodynamic equilibrium.

Departures from LTE become ever more important the hotter and more luminous the star is, since then radiation processes become more important relative to collisional processes. For example, on the main sequence, the O and B stars show notable departures from LTE, in particular for the occupation numbers of the lower energy states of hydrogen and helium.

7.2.4 Radiation Transport in the Fraunhofer Lines

The spectral distribution from the Sun, where one can observe the center–limb variation, is described in terms of the intensity $I_\nu(0,\theta)$ or $I_\lambda(0,\theta)$ which emerges at the surface ($\tau = 0$) in the direction θ; in the case of the stars, whose disks as a rule cannot be resolved, we use the radiation flux $F_\nu(0)$ or $F_\lambda(0)$. With modern spectrographs, in the optical region one can obtain resolutions in the solar spectrum up to $\lambda/\Delta\lambda \simeq 10^6$; for brighter stars, up to $\lambda/\Delta\lambda \geq 10^4 \ldots 10^5$. For example, the intensity distribution recorded with a CCD detector is available in "bite-sized" digital form to be analyzed with a computer. Photographically recorded spectra are analyzed with a microphotometer, which automatically corrects for the darkening curve (Sect. 5.1.4), calibrated by intensity marks (cf. Fig. 7.10).

In spectral regions with relatively few Fraunhofer lines, it is not difficult to determine a *"true" continuum*

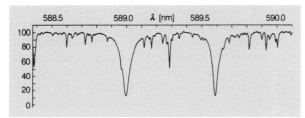

Fig. 7.10. The intensity distribution of the NaD lines in the solar spectrum

which is normalized to 1 or to 100% and from which the intensity reduction in a line

$$r_\ell = \frac{I - I_\ell}{I} \quad \text{or} \quad R_\ell = \frac{F - F_\ell}{F} \qquad (7.52)$$

i.e. the *line profile*, can be measured as a function of the distance to the center of the line, $\Delta\nu$ or $\Delta\lambda$. We denote here all quantities in the range of a spectral *line* by the index ℓ. Quantities which refer to the neighboring *continuum*, which can be taken to be constant over the line profile, will be written *without* an index.

In spectral regions with a high density of lines, the determination of a continuum level and thus the calculation of an isolated spectral line is in general not possible. Here, a longer section of the spectrum must be calculated "in one piece", and then compared with the observed intensity distribution. This technique of *spectrum synthesis* however requires precise knowledge of atomic data for a large number of spectral line transitions. In the following, we shall limit ourselves to the discussion of individual lines, for which the neighboring continuum (background) can be determined.

Since the spectrograph itself reproduces even an infinitely sharp line with a finite width which is called the *instrumental profile*, it must be kept in mind that the width and structure of weak Fraunhofer lines always contains an instrumental contribution. This can be determined, for example, by measuring the profiles of sharp lines from a laboratory source. The wings of strong lines are only slightly influenced by the instrumental profile; the center portions, as well as weak lines, are, however, strongly affected. The absorbed energy is clearly independent of the instrumental distortion of the line profile. It is therefore frequently advantageous, as pointed out by M. Minnaert, to measure the *equivalent width*

$$W_\lambda = \int_{\text{Line}} r_\ell \, d\lambda = \int_{-\infty}^{\infty} r_\ell(\Delta\lambda) \, d(\Delta\lambda) ; \qquad (7.53)$$

this quantity is the width on the wavelength scale (in suitable units of length such as pm, mÅ, ...) of a rectangular absorption line in the spectrum having the same area as that of the line profile (Fig. 7.11).

For the interpretation of the r_ℓ or W_λ values, we require the following absorption coefficients, which – each in a different way – depend on the state variables T and P_e, and on the corresponding optical depths:

Line: κ_ℓ

$$\tau_\ell(t) = \int_{-\infty}^{t} \kappa_\ell(t') dt'$$

Continuum: κ

$$\tau(t) = \int_{-\infty}^{t} \kappa(t') dt'$$

Total: $\kappa + \kappa_\ell$

$$x_\ell(t) = \int_{-\infty}^{t} \big(\kappa(t') + \kappa_\ell(t')\big) dt' . \qquad (7.54)$$

We can now apply the methods and calculations from Sect. 7.2.3 to compute line profiles, if we replace κ_ν there by $\kappa + \kappa_\ell$.

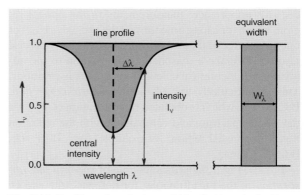

Fig. 7.11. Profile and equivalent width of a Fraunhofer line

The *intensity* of the radiation emitted at the angle θ for the line is then given by:

$$I_\ell(\theta) = \int_0^\infty S_\ell(x_\ell) e^{-x_\ell/\cos\theta} dx_\ell / \cos\theta \qquad (7.55)$$

and for the neighboring continuum by:

$$I(\theta) = \int_0^\infty S(\tau) e^{-\tau/\cos\theta} d\tau / \cos\theta \,. \qquad (7.56)$$

Here, in contrast to (7.46), we have abbreviated $I(0, \theta)$ as $I(\theta)$, etc. S_ℓ or S is again the source function (4.57), the ratio of emission to absorption coefficients. The relation between the two optical depths is given by

$$\frac{dx_\ell}{d\tau} = \frac{\kappa + \kappa_\ell}{\kappa}, \qquad x_\ell = \int \frac{\kappa + \kappa_\ell}{\kappa} d\tau \,. \qquad (7.57)$$

The intensity reduction in the spectral line is then, corresponding to (7.52),

$$r_\ell(\theta) = \frac{I(\theta) - I_\ell(\theta)}{I(\theta)} \,. \qquad (7.58)$$

If we again make the assumption of local thermodynamic equilibrium (LTE) with regard to radiation transport, which is a realistic approximation in particular for atoms or ions with more or less complex term schemes in atmospheres which are not too hot, then the source function becomes identical to the Kirchhoff–Planck function $B_\nu(T)$. The small frequency range within a spectral line is naturally unimportant in this function.

To get a rough feeling for these quantities, we can again employ the Eddington–Barbier approximation (7.49) and obtain:

$$r_\ell(\theta) = \frac{B_\nu(\tau = \cos\theta) - B_\nu(x_\ell = \cos\theta)}{B_\nu(\tau = \cos\theta)} \,. \qquad (7.59)$$

The important point here is that the radiation in the line comes from a higher layer, $x_\ell = \cos\theta$, with correspondingly lower temperature, than the adjacent continuum, which originates from $\tau = \cos\theta$. If the absorption coefficient at the center of the line is very much stronger than in the continuum ($\kappa_\ell \gg \kappa$), then the central intensity of the line simply corresponds to the Kirchhoff–Planck function at the surface temperature T_0 of the atmosphere. All of the sufficiently strong lines in a particular region of the spectrum then have the same, i.e. the largest possible, intensity reduction, $r_c(\theta)$.

For applications to *stars* we require the corresponding expressions for the radiation flux F_ℓ in the line and F in the adjacent continuum. The intensity reduction in the line (7.52) then becomes (again assuming local thermodynamic equilibrium):

$$R_\ell = \frac{B_\nu(\tau = \tfrac{2}{3}) - B_\nu(x_\ell = \tfrac{2}{3})}{B_\nu(\tau = \tfrac{2}{3})} \,. \qquad (7.60)$$

For weak lines and in the wings of the stronger lines, where $\kappa_\ell \ll \kappa$ holds, it follows from (7.57) in the spirit of the Eddington–Barbier approximation that:

$$x_\ell = \frac{2}{3}\left(1 + \frac{\kappa_\ell}{\kappa}\right) ; \qquad (7.61)$$

then by series expansion of $B_\nu(x_\nu = 2/3)$ in (7.60) we obtain for the intensity reduction

$$R_\ell = \frac{2}{3} \frac{\kappa_\ell}{\kappa} \left(\frac{d \ln B_\nu}{d\tau}\right)_{\tau = 2/3}, \qquad \kappa_\ell \ll \kappa \,. \qquad (7.62)$$

We see for one thing that in a stellar spectrum, the continuum as well as the wings of the Fraunhofer lines originates mainly in *those* layers of the atmosphere whose optical depth is given by $\tau \simeq 2/3$ for the neighboring continuum. Secondly, the line intensities depend strongly on the temperature *gradient* in the atmosphere. We obtain *absorption* lines when the temperature *increases towards the stellar interior*.

The central portions of the stronger lines, where $\kappa_\ell \gg \kappa$, originate in correspondingly higher layers, $\tau \simeq (2/3)\kappa/\kappa_\ell$.

Employing the Eddington–Barbier approximation, we were able to get an overall picture of the origin of Fraunhofer lines; however, the correct calculation of the lines must proceed by direct integration of (7.52) with (7.46, 50).

7.2.5 The Curve of Growth

We shall now investigate how the profile and the equivalent width W_λ of an absorption line increase as we increase the concentration of the absorbing atom, n, or its product with the oscillator strength f. The line absorption coefficient κ_ℓ is dominated in the central part of the line by the Doppler effect, and in the wings by damping (Fig. 7.5), the damping constant being determined by radiative and collision processes.

The relationship between the intensity reduction in the line, R_ℓ, and the line absorption coefficient κ_ℓ, using an *absorption tube* (without reemission) of length H and optical thickness $\tau_\ell = \kappa_\ell H$ in the laboratory, would be (from (4.50)):

$$R_\ell = 1 - e^{-\kappa_\ell H} . \tag{7.63}$$

For a *stellar atmosphere*, it is given by (7.46, 50, 52) and can, in general, be calculated only by direct integration. When the requirements for accuracy are not too demanding, these tedious calculations can often be avoided by using the approximate interpolation formula

$$R_\ell = \left(\frac{1}{\kappa_\ell H} + \frac{1}{R_c} \right)^{-1} . \tag{7.64}$$

Here, H is an effective height, or $N = nH$ an *effective number* of absorbing atoms over a unit area on the star's surface (column density). For $(\kappa_\ell H \ll 1)$ (absorption in an optically thin layer), $R_\ell = \kappa_\ell H$; for $(\kappa_\ell H \gg 1)$ (optically thick layer), R_ℓ tends towards the limiting value for very strong lines, R_c.

H or N can be calculated by comparing (7.64) with the formulas (7.46, 50, 52); within a limited wavelength region, the effective layer thickness of an atmosphere may be treated as constant. We shall immediately give the result of this calculation: in the lower part of Fig. 7.12, we have drawn the *line profiles* for various values of the quantity Nf. The reductions for the weak lines, $R_\ell = \kappa_\ell H$, simply reflect the Doppler broadening of the absorption coefficient in the core of the line. With increasing Nf, the value at the line center approaches the maximum depth R_c, since here only radiation from the uppermost layers with the limiting temperature T_0 contributes. On the other hand, the line is initially not strongly broadened, since the absorption coefficient drops off sharply with increasing $\Delta\lambda$. This changes only with further increasing Nf, when the effective optical depth becomes important in the damping wings also. Since now $\kappa_\ell H \propto Nf\gamma/(\Delta\lambda)^2$, the line acquires broad "damping wings" and its width for a given reduction R_ℓ becomes $\propto \sqrt{Nf\gamma}$.

By integrating over the line profile, we readily obtain the *equivalent widths* W_λ of the lines. We have plotted

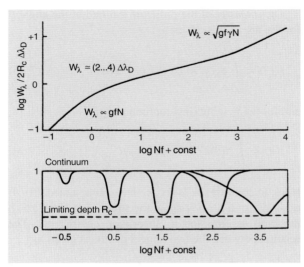

Fig. 7.12. Curve of growth (*above*). The equivalent width W_λ, referred to a rectangular profile of width equal to twice the Doppler broadening $\Delta\lambda_D$ and height given by the limiting depth R_c, is plotted as a function of the effective concentration of absorbing atoms. The line profiles (*below*) show how the growth curve is generated

them in units of $2R_c\Delta\lambda_D$, i.e. relative to a rectangular absorption line whose depth corresponds to the maximum intensity reduction and whose width corresponds to twice the Doppler width. We thus obtain *universal curves of growth* (Fig. 7.12, upper part), which are important for the evaluation of stellar spectra, giving $\log(W_\lambda/(2R_c\Delta\lambda_D))$ as a function of $\log Nf + \text{const}$.

As can be easily understood in the light of our discussion of line profiles, for weak lines (at left in the figure), W_λ increases proportionally to Nf; we are in the *linear* part of the growth curve. This is followed by the *flat* or Doppler region, in which the equivalent width is equal to 2–4 times the Doppler width $\Delta\lambda_D$. For strong lines (right), corresponding to the increase in width of the profile, $W_\lambda \propto \sqrt{Nf\gamma}$; we have come to the *damping* or *square-root region* of the growth curve. Here, in general, the damping constant γ is also of importance. In Fig. 7.12, we have used 0.03 for the ratio of damping to Doppler width, corresponding to an average value for the metal lines in the solar spectrum. For example, the strong D lines from Na I and the H and K lines from Ca II in the solar spectrum lie in the damping region of the growth curve.

7.2.6 Quantitative Analysis of Stellar Spectra

From the measurements of the lines and the continuum, we now have to determine the fundamental parameters which characterize the stellar atmosphere, i.e. the effective temperature T_{eff}, the acceleration of gravity g, and the chemical composition $\{\varepsilon\}$.

The complexity of this task suggests a method based on stepwise or successive approximations. We first ask: which equivalent widths or line profiles for various lines would result from a *model* atmosphere with *predetermined* T_{eff}, g and $\{\varepsilon\}$? We then discuss the question of how the observed quantities, especially the W_λ of particular elements and ionization or excitation states, are related to the chosen parameters. Finally, we refine the values of these parameters until the best possible agreement has been obtained between the calculated and the measured spectrum of the star being investigated.

Coarse Analysis. In the *quantitative* analysis of stellar spectra, one often begins with an initial orientation based on a simple approximation method, called the coarse analysis, in which calculations are carried out for a whole atmosphere using *constant* average values of the temperature T, the electron pressure P_e, the effective layer thickness H (over a large spectral range), etc. The universal growth curve calculated with the interpolation formula (7.64) can then be immediately applied, in order to determine the column density $N = nH$ of absorbing particles in the lower energy level of any given atom or ion, by comparison to the measured equivalent width W_λ of the corresponding Fraunhofer line. For this purpose, the f value and often the damping constant γ for the line must be known.

The *Doppler width* $\Delta\lambda_D$ or $\Delta\nu_D$ also enters the growth curve (Fig. 7.12); it depends via the thermal motion v_0 on the temperature. Furthermore, turbulent flow fields often contribute the same order of magnitude as v_0 to the Doppler width. The corresponding *microturbulent velocity* ξ adds quadratically in (7.29) to the thermal velocity, so that

$$\frac{\Delta\lambda_D}{\lambda_0} = \frac{1}{c}\sqrt{v_0^2 + \xi^2} \quad (7.65)$$

holds. Since it is currently not possible to derive this "microturbulence" from the stellar parameters from first principles using a hydrodynamic theory, ξ must be determined in the analysis as an additional parameter along with T, P_e, N, ... This is done using lines from the flat portion of the growth curve, which, as we saw, depend sensitively on $\Delta\lambda_D$.

By comparison of N for energy levels with various excitation energies and for various ionization states of the same element (e.g. Ca I and Ca II), it is finally possible using the formulas of Boltzmann and Saha (Sect. 7.1.2) to calculate the temperature T and the electron pressure P_e. Knowing them, however, we are led immediately from the numbers of atoms in particular energy levels to the total number of all ε particles of the element considered (independently of the ionization and excitation state) and thus to the *abundance distribution of the elements*. When the latter, as well as the degree of ionization of the different elements, is known with reasonable completeness, the step from the electron pressure P_e to the gas pressure P_g can be taken and the acceleration of gravity g can be calculated from the hydrostatic equation: the gas pressure P_g is just the weight, i.e. the mass times the gravitational acceleration g, of all particles above a unit area on the stellar surface.

Fine Analysis. Based on the coarse analysis, one can carry out the more precise but much more tedious fine analysis of the stellar spectrum: a *model* of the stellar atmosphere under investigation as described in Sect. 7.2.3 is constructed using the most plausible values of T_{eff}, g and the chemical composition $\{\varepsilon\}$. In this model, using the theory of radiation transport and of the continuous and line absorption coefficients (including their depth dependence), the profiles or equivalent widths W_λ of the lines are calculated, taking the microturbulence velocity initially as an undetermined parameter (7.65). The results of the model calculations are then compared with spectral measurements for those elements which are represented in the spectrum in various states of ionization and excitation. The hydrogen lines also play an important role, since hydrogen is *the* most abundant element in nearly all stars; thus the elemental abundance is not relevant for this element. Furthermore, the energy distribution or color indices can be taken into account. It can be seen from our previous considerations which criteria are more strongly dependent on T_{eff}, g, ξ, or the abundance ε of a certain element, so the initial values of these quantities can now be improved

7.2.7 Element Abundances in the Sun and in Other Stars

Having illustrated briefly the methods of a quantitative analysis of stellar spectra, we now turn to the *results* of such an analysis:

The Sun. For the Sun (spectral class G2 V), we have incomparably better observational data than for any other star; furthermore, T_{eff} and g are known independently. Since the first quantitative analyses by C. H. Payne (1925), A. Unsöld (1928) und H. N. Russell (1929), the solar elemental abundance distribution has been derived repeatedly using more precise observations and improved theoretical models. In the case of the Sun, we can determine the abundances even of rare elements, whose weak lines can be seen only in low-noise spectra having high dispersion. In addition to the Fraunhofer spectrum of the photosphere, the emission spectra of the solar chromosphere and the corona (and of the prominences) from the optical out to the X-ray region (Sect. 7.3.4) permit the determination of abundances of many elements, especially of the noble gases He, Ne and Ar. They are not seen in the Fraunhofer absorption spectrum of the photosphere, due to its relatively low temperatures ($T \lesssim 8000\,\text{K}$) and to their term structures (states of high excitation energy). To determine abundances of the more common elements, measurements of the solar wind (Sect. 7.3.7) and of solar cosmic rays (Sect. 10.4.2) have also been evaluated. The abundances obtained by different methods agree with each other well (for the elements where several methods can be applied).

Table 7.5. Element abundances $\log n$ in the Solar System: the *Sun* ($\odot$) after H. Holweger (1985) and *carbonaceous chondrites of Type* C 1 after E. Anders and M. Ebihara (1982). Normalized to hydrogen $\log n(\text{H}) = 12.0$, adjustment of the solar and meteoritic abundance distributions for $\log n(\text{Si}) = 7.6$. Determination of the solar abundances from the photosphere with the exceptions of He, Ne, and Ar (from the corona or prominences) and Tl (from sunspots). Meteorites: C 1-chondrites except for Be, B, Br, Rh, I, for which other chondrites were used. For Kr, Xe, and Hg, the values were estimated from interpolations. Radioactive elements: for Th and U, the current abundances are quoted; at the time of the formation of the Solar System $4.5 \cdot 10^9$ yr ago (3.26), the abundances were $\delta \log n = 0.2$ (Th) or 0.3 (U) higher, respectively

		⊙	C1			⊙	C1			⊙	C1			⊙	C1
1	H	12.0	–	22	Ti	5.1	5.0	44	Ru	1.8	1.9	66	Dy	1.1	1.2
2	He	*11.0*	–	23	V	4.1	4.1	45	Rh	1.1	*1.1*	67	Ho	0.3	0.6
3	Li	1.1	3.4	24	Cr	5.8	5.7	46	Pd	1.7	1.7	68	Er	0.9	1.0
4	Be	1.2	*1.5*	25	Mn	5.4	5.6	47	Ag	0.9	1.3	69	Tm	0.3	0.1
5	B	2.5	*3.0*	26	Fe	7.5	7.5	48	Cd	1.9	1.8	70	Yb	1.1	1.0
6	C	8.6	–	27	Co	4.9	5.0	49	In	1.7	0.9	71	Lu	0.8	0.2
7	N	8.0	–	28	Ni	6.2	6.3	50	Sn	1.9	2.2	72	Hf	0.9	0.8
8	O	8.9	–	29	Cu	4.2	4.3	51	Sb	1.0	1.1	73	Ta	–	0.0
9	F	4.6	4.5	30	Zn	4.6	4.7	52	Te	–	2.3	74	W	1.1	0.7
10	Ne	7.6	–	31	Ga	2.9	3.2	53	I	–	*1.6*	75	Re	–	0.3
11	Na	6.3	6.4	32	Ge	3.5	3.7	54	Xe	–	(2.2)	76	Os	1.4	1.5
12	Mg	7.5	7.6	33	As	–	2.4	55	Cs	–	1.2	77	Ir	1.4	1.4
13	Al	6.4	6.5	34	Se	–	3.4	56	Ba	2.1	2.2	78	Pt	1.8	1.7
14	Si	7.6	7.6	35	Br	–	2.7	57	La	1.1	1.3	79	Au	1.1	0.9
15	P	5.4	5.6	36	Kr	–	(3.3)	58	Ce	1.6	1.7	80	Hg	–	(1.3)
16	S	7.2	7.3	37	Rb	2.6	2.5	59	Pr	0.7	0.8	81	Tl	*0.9*	0.9
17	Cl	–	5.3	38	Sr	3.0	3.0	60	Nd	1.4	1.5	82	Pb	1.9	2.1
18	Ar	*6.7*	–	39	Y	2.2	2.3	62	Sm	0.8	1.0	83	Bi	–	0.8
19	K	5.1	5.2	40	Zr	2.6	2.6	63	Eu	0.5	0.6	90	Th	0.2	0.1
20	Ca	6.4	6.4	41	Nb	1.4	*1.5*	64	Gd	1.1	1.1	92	U	–	0.4
21	Sc	3.1	3.1	42	Mo	1.9	2.0	65	Tb	0.2	0.4				

In Table 7.5, the relative abundances n (ordered according to atomic number) in the Sun are set out; they are taken mainly from the analysis of the Fraunhofer spectrum. As usual, all abundances are given relative to that of hydrogen, H, and are normalized to

$$n(\mathrm{H}) = 10^{12} \quad \text{or} \quad \log n(\mathrm{H}) = 12.0 \ . \tag{7.66}$$

For example, from this relation, a ratio $n(\mathrm{C})/n(\mathrm{H}) = 4.2 \cdot 10^{-4}$ corresponds to $\log n(\mathrm{C}) = 8.6$. The imprecision of the solar abundance determination is about $|\Delta \log n| \lesssim \pm 0.1$ for most elements.

For comparison, in Table 7.5 we also give the abundances for the *carbonaceous chondrites* of type C 1 (Sect. 3.7), whose precision is $\Delta \log n \simeq \pm 0.05$ with a few exceptions. The agreement with the values for the Sun is quite good. These meteorites thus represent (except for the very volatile elements) nearly unchanged solar matter. The deviations for Li and in part for Be and B can be explained by the fact that these light elements are destroyed in the lower hydrogen convection zone (Sect. 7.3.1) by nuclear reactions with protons, and their abundances at the solar surface have decreased over long periods of time due to the convective mixing of the outer layers of the Sun.

In Table 7.6, we list some selected *isotope* ratios for the material in the Solar System; in most cases, they are again obtained with the greatest precision from the analysis of meteorite data.

Stars. Since the pioneering analysis of the B0 V star τ Sco by A. Unsöld in 1941/44, precise elemental abundance analyses were initially carried out for a number of selected bright stars, representing the different spectral classes in the Hertzsprung–Russell diagram, on the basis of high-resolution spectra with the help of model atmospheres. We mention as examples the main-sequence stars 10 Lac (O9 V), ι Her (B3 V), α Lyr (A0 V), β Vir (F8 V), the giants and supergiants ζ Per (B1 Ib), α Cyg (A2 Ia), ε Vir (G8 III), β Gem (K0 III), α Boo (K2 III), the white dwarf van Maanen 2 (DZ) and the "metal deficient" stars HD 140283 (G0 V), HD 122563 (K0 III) and HD 161817 ($\simeq$A2).

The *precision* with which the abundance of an element in stars of different temperatures can be determined is naturally variable. For one thing, it depends on the quality of the spectrum (signal–noise ratio); for another, on the number of lines from the element seen in the spectrum and in which region of the growth curve they lie. Furthermore, it must be remembered that the abundances can only be determined together with T_{eff} and g; for medium spectral classes, for example, a value for T_{eff} chosen too low would lead to abundances of metallic elements which are too small! In general it can be said that, relative to hydrogen, helium and the lighter elements are more precisely determined in the hot stars, owing to their high ionization energies, while the more readily ionized metallic elements can be more precisely determined in cooler stars. The uncertainty of a careful abundance determination is at present of the order of $\Delta \log n = \pm 0.2$. In recent times, a number of metal abundances have been obtained even in hot stars by using lines in the far ultraviolet observed from satellites.

The quantitative analysis of stars is expediently performed *relative* to comparison stars whose temperature and gravitational acceleration are not too different, thus avoiding many difficulties; for example the uncertainty in the f values and damping constants is avoided. Relative accuracies of $\Delta \log n \simeq \pm 0.1$ for the element abundances can be obtained in this way.

For stars ($*$), the abundances are often referred to the *solar* or *normal* ($\odot$) abundances. We denote the abundance of an element El by

$$\varepsilon(\mathrm{El}) = \frac{\bigl(n(\mathrm{El})/n(\mathrm{H})\bigr)_*}{\bigl(n(\mathrm{El})/n(\mathrm{H})\bigr)_\odot} \tag{7.67}$$

or in logarithmic form by

$$[\mathrm{El/H}] \equiv \log \varepsilon(\mathrm{El}) \ . \tag{7.68}$$

In stellar spectroscopy, it is usual to denote *all* elements heavier than carbon, i.e. with atomic numbers $Z \geq 6$, without reference to their chemical properties,

Table 7.6. Some isotope ratios in the Solar System

^{2}D/^{1}H	$2.0 \cdot 10^{-5}$	^{15}N/^{14}N	$3.7 \cdot 10^{-3}$
^{3}He/^{4}He	$1.4 \cdot 10^{-4}$	^{17}O/^{16}O	$3.8 \cdot 10^{-4}$
^{6}Li/^{7}Li	$8.1 \cdot 10^{-2}$	^{18}O/^{16}O	$2.0 \cdot 10^{-3}$
^{10}B/^{11}B	0.25	^{25}Mg/^{24}Mg	0.13
^{13}C/^{12}C	$1.1 \cdot 10^{-2}$	^{26}Mg/^{24}Mg	0.14

as *"metals"*. The mean abundance of the metals M is given as [M/H] = $\log \varepsilon$(M).

Abundance Distributions. The abundance distribution $\{\varepsilon\}$ of the elements is clearly related to the history of their formation and to their transmutation by *nuclear processes* in the course of *stellar evolution*. Without anticipating later discussions (Chap. 8, Chap. 12), we summarize the most important results of the quantitative analysis of the spectra of the Sun and stars, limiting ourselves to the Milky Way galaxy:

The chemical composition of the *normal* stars of the spiral arm Population I and the disk population of the Milky Way; (Sect. 11.2.6) is nearly constant; in particular, it agrees with that of *solar* matter. For historical reasons, this mixture of elements is also termed *"cosmic"* matter. We denote those stars by definition as "normal" which can be uniquely positioned in the MK classification. The feasibility of a two-dimensional classification in fact presumes that no further parameters besides T_{eff} and g are necessary, i.e. that these stars all have the same chemical compositions.

The *interstellar matter* in which young, hot stars such as τ Sco (B0 V) are continually being formed has nearly the same composition as these stars (Sects. 10.2.1, 3.1). Many (but, as we shall see, by no means all) stars which are in advanced stages of development, such as the red giant ε Vir (G8 III), have atmospheres or shells of unmodified normal matter.

The *oldest* stars, such as the high-velocity stars (e.g. the horizontal-branch-A star HD 161817 or the G-subdwarf HD 140283) and members of globular clusters, which belong to the socalled stellar Population II of the galactic halo (Sect. 11.2.6), are characterized by an *underabundance* of all heavier elements ("metals") relative to hydrogen. We find reduction factors for the metals in the whole range from 1 to at least around 1000, or mean metal abundances [M/H] = $\log \varepsilon$(M) of 0 to below -3; however, stars with extreme metal deficiencies are rare. Conversely, there are also "metal-rich" stars, e.g. β Vir, to be sure with only moderate metal enrichment, [M/H] $\leq +0.4$.

The *relative* abundance distribution of the metals among themselves in the metal-deficient stars (except for some of the most extreme cases) is very similar to that in the Sun and in the Population I stars. Significant departures from this rule, which clearly exceed a factor of 2 or 3, are found for only a few elements, e.g. for C, N, O, and Al, as well as for the elements beyond the iron group which are most difficult to assess spectroscopically, such as Ba und Y.

The abundance distribution in a stellar atmosphere is determined in part by the composition of the *interstellar* matter at the time when the star was formed in it. Not yet mature (main-sequence) stars, both from the young and from the old stellar populations, still exhibit these original chemical compositions in their spectra, except for a few special cases. From the observed deficiency of the metals in the oldest stars, we can conclude that an enrichment in the heavy elements (> He) has occurred in the interstellar medium in the Milky Way since its formation (Sects. 8.2.4 and 12.5.4). Furthermore, the chemical composition of a stellar atmosphere can also change in the course of the star's evolution, e.g. when products of the *nuclear processes* which take place in its interior (Sect. 8.1.3) reach the surface through convection or flow processes, possibly aided by mass loss (stellar winds). The first clear evidence for nuclear processes was obtained with the discovery in 1952 of technetium lines in the spectra of some cool giant stars by P. W. Merrill. The most long-lived isotope of this element (^{98}Tc) has a half-life of only $4 \cdot 10^6$ yr, which is much shorter than the time required for evolution of these stars. As we shall see in Sect. 8.2.5, abundance anomalies of He, C, N, O and heavy elements such as Y, Ba, and La are also indicators for nuclear processes during stellar evolution, accompanied by transport processes to the surface.

Not all abundance anomalies in stellar atmospheres originate, however, with nuclear processes. In particularly stable, fixed or unmixed stellar layers, a *sedimentation* of the elements can occur, the heavier elements diffusing out of the observable atmosphere into the layers below. Here, radiation pressure can oppose the sinking of certain atoms or ions as a sort of selective buoyancy when they have "favorable" absorption properties. Thus, the anomalies in the spectra of the *Ap* or *Bp stars* and of the metallic-line stars (Sect. 7.4.2) can be explained by this kind of *diffusion processes*. The extreme metal deficiencies in the atmospheres of the *white dwarf stars*, which nearly all show either only hydrogen lines or only helium lines, have also been explained by diffusion of the heavier elements into the layers lying below the atmosphere.

7.3 The Sun: Its Chromosphere and Corona. Flow Fields, Magnetic Fields and Activity

We begin in Sect. 7.3.1 with the flow fields in the photosphere, the granulation, which we treat together with its "driving force", the hydrogen convection zone lying beneath the photosphere. The existence of the extremely hot outer layers of the solar atmosphere, the chromosphere and the corona, as well as solar activity phenomena, also have their origins in the convection zone and are furthermore strongly influenced by magnetic fields. Therefore, in Sect. 7.3.2 we first collect the most important relations of magnetohydrodynamics and cosmic magnetic fields, before turning to sunspots and the activity cycle together with photospheric magnetic fields in Sect. 7.3.3. Section 7.3.4 is devoted to the chromosphere and the corona. Further phenomena of solar activity are the prominences (Sect. 7.3.5) and powerful eruptions, the flares (Sect. 7.3.6). In Sect. 7.3.7 we describe the plasma which streams out from the Sun into interplanetary space: the solar wind. Finally, in Sect. 7.3.8, we introduce the oscillations of the Sun. Their effects, which can be observed on the solar surface, are the basis of the young discipline of "helioseismology".

7.3.1 Granulation and Convection

If one studies the solar surface carefully, it appears (in white light) to have a grainy structure. This *granulation* consists of brighter "granula", whose temperature is 100 to 200 K higher than that of the darker areas in between (Fig. 7.13). The granulation pattern changes constantly; photographs taken in series show that the cells have a mean lifetime of about 8 min before they divide or dissolve. The diameters of the granulation cells range from the largest of about 3500 km down to the telescopic resolution limit of about 100 km. In photospheric *spectra* with good spatial and time resolution, one can distinguish vertical motions of the individual cells with velocities of about $0.5 \, \mathrm{km \, s^{-1}}$ from the Doppler shift (2.3) of the Fraunhofer lines. Since the entrance slit of the spectrograph admits light from many granulation elements, the absorption lines have a sawtooth-like structure (Fig. 6.1).

In addition to the granulation, there is a second network of cells on a larger scale, whose meshes have diameters from 15 000 to 40 000 km. The flow in this *supergranulation* rises in the center of the cells, with radial velocities of about $0.4 \, \mathrm{km \, s^{-1}}$, and sinks at the edges, with velocities of $\lesssim 0.2 \, \mathrm{km \, s^{-1}}$. The mean lifetime of the cells (roughly equal to their turnover times) is about 36 h. The supergranulation also reaches into the chromosphere, the layer of the solar atmosphere which lies above the photosphere, and is particularly visible in Ca II spectroheliograms (Sect. 7.3.4).

The flow fields of the granulation and supergranulation are closely related to the *magnetic fields* on the Sun. We shall come back to this point in connection with the phenomenon of solar activity in Sect. 7.3.3.

First, however, we ask the question: What "thermodynamic machine" produces the mechanical energy which is necessary to maintain these flows, accompanied (as required by the 2nd Law of thermodynamics)

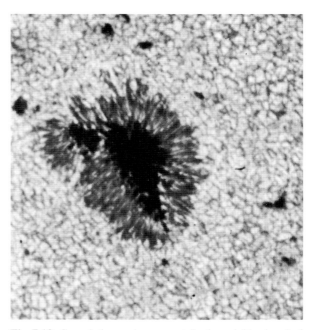

Fig. 7.13. Granulation and a sunspot. In the neighborhood of the spot, there are several dark "pores" with diameters of a few seconds of arc. The photograph was made with the 30 cm stratosphere telescope of M. Schwarzschild (1959) at an altitude of 24 km. The exposure time was 0.0015 s, the spectral range of sensitivity (547 ± 37) nm

by a flow of heat energy from a higher to a lower temperature region? This task is performed by the *hydrogen convection zone* (A. Unsöld, 1931): from the deeper layers of the photosphere, i.e. from an optical depth $\tau_0 \simeq 2$ (referred to $\lambda = 500$ nm) at a gas pressure of $P_g \simeq 1.5 \cdot 10^4$ Pa and a temperature of $T \simeq 7000$ K downwards to about $P_g \simeq 10^{13}$ Pa and $T \simeq 2 \cdot 10^6$ K, the Sun's atmosphere is *convectively unstable*. The thickness of this layer of about 200 000 km corresponds to 0.29 solar radii. Above the layer, hydrogen (the most abundant element) is practically neutral, while within the layer, it is partially ionized, and below the layer, its ionization is complete. Now the following happens: when a volume element containing partially ionized gas rises, the hydrogen ions begin to recombine. In each recombination process, 13.6 eV (this corresponds to $16\,kT$ at a temperature of 10 000 K) is added to the thermal energy ($3kT/2$ per particle). In this way, the adiabatic cooling of the gas is reduced to such an extent that the effective ratio of the specific heats, c_p/c_v, approaches one and the rising gas volume becomes *warmer* than its new surroundings, which are in radiation equilibrium. The volume element thus continues to rise. A sinking volume element experiences precisely the reverse process. This effect is amplified still further by the opposite influence of ionization on the radiation temperature gradient, and we obtain a zone with *convective flow*. In this zone, convection is responsible for practically the whole energy transport; radiation energy transport becomes unimportant in the deeper layers of the convection zone.

While the thermodynamics of the hydrogen convection zone are relatively simple and straightforward, its *hydrodynamics* are among the most difficult problems in the theory of flow phenomena. Usually, models can be calculated only with the rather rough *mixing-length* approach following W. Schmidt and L. Prandtl: a gas volume of dimension l moves along a sort of mean free path of the same order of magnitude, l, and then its excess temperature, its momentum etc. are suddenly given up to its surroundings by mixing. This is clearly a very rough schematic description of the complicated convective flow process. Detailed calculations for the convection zone are still in the early stages.

The *granulation* can now be attributed to the particularly strong instability of a surface layer of the hydrogen convection zone which is only a few hundred km thick. The granula are of the order of the thickness of this zone, but also not much larger than the equivalent height of the atmosphere at that point. The coarser network of *supergranulation* consists of cells whose diameters are up to about 40 000 km. It thus appears tempting to associate it with a flow which includes a considerable part of the depth of the whole convection zone.

In the solar spectrum, the flows which we have discussed so far, with velocities of 0.2 to 3 km s^{-1}, contribute to the fact that the Fraunhofer lines take on a sawtooth shape due to the Doppler effect and are on the average broadened. As long as the flowing volume elements are optically thin, their Doppler effects simply superpose on those due to the thermal motion: one thus obtains the socalled *microturbulence*, which increases the purely thermal $\Delta\lambda_D$ by factors of the order of magnitude of 1.2 to 2 (7.65).

7.3.2 Magnetic Fields and Magnetohydrodynamics

Since the structure of the outer layers and the activity phenomena of the Sun, as well as many other astrophysical phenomena, are affected in an essential way by magnetic fields, it seems useful at this point to summarize some quantities related to magnetic fields and some important results of magnetohydrodynamics.

Magnetic Fields. We describe a magnetic field by the paths of the field lines of the magnetic *flux density* ("field strength") **B**, whose magnitude B is quoted in [Tesla] (1 T = 10^4 Gauss). The magnetic *flux* through an area A is given by

$$\Phi = \int_A B_n \, dA , \qquad (7.69)$$

where B_n is the field component parallel to the surface normal. The unit of magnetic flux is 1 T m^2 ($= 10^8$ G cm^2).

The *energy density* of the magnetic field, (4.4), is

$$w_B = \frac{B^2}{2\mu_0} \qquad (7.70)$$

with the magnetic field constant $\mu_0 = 4\pi \cdot 10^{-7}$ V s A^{-1}m^{-1}; in the Gaussian system, it is $B^2/(8\pi)$. The magnetic *pressure* depends upon the details of the field configuration; it is of the same order of magnitude as the energy density, i.e. $P_B \simeq B^2/(2\mu_0)$.

The Zeeman Effect. By observing the splitting or broadening and the polarization of spectral lines by magnetic fields, one can determine the field strength spectroscopically. In a homogeneous magnetic field which is not too strong, with a flux density of $B \leq 10$ Tesla, each energy level, described by the quantum numbers L, S and J (Russell–Saunders coupling, Sect. 7.1.1), splits proportionally to B into $(2J+1)$ equidistant energy substates, which are characterized by their magnetic quantum numbers M_J with $-J \leq M_J \leq +J$ (Fig. 7.1). The energy shifts are given by

$$\Delta E = g_{LSJ} M_J \cdot \frac{\hbar e B}{2 m_e} = g_{LSJ} M_J \cdot \hbar \omega_L , \quad (7.71)$$

where g_{LSJ} is the Landé factor or g factor, which depends on L, S and J, and $\omega_L = 2\pi \nu_L = eB/(2m_e)$ is the Larmor frequency. With the selection rules $\Delta M_J = 0, \pm 1$, we find from the splitting of the lower and upper level of a line transition a set of frequency components which are shifted from their "zero-field" position ν_0 or λ_0 by

$$|\Delta \nu_B| = \frac{c}{\lambda_0^2} \Delta \lambda_B = \hat{g} \frac{eB}{4\pi m_e} = \hat{g} \frac{\omega_L}{2\pi} . \quad (7.72)$$

The factor $\hat{g}$ is composed of the Landé factors of the lower and upper levels. While the frequeny shift is independent of λ_0, $\Delta \lambda_B$ increases $\propto \lambda_0^2$.

In the simplest case that both the upper and the lower term are singlet states ($S = 0$), we find $\hat{g} = 1$ and obtain the classical Lorentz triplet with three components at wavelengths λ_0 and $\lambda_0 \pm \Delta \lambda_B$. Here, the shifted components are circularly polarized in opposite senses, while the unshifted component is linearly polarized.

Numerically, we find for the frequency or wavelength shifts

$$\Delta \nu_B \, [\text{Hz}] = 1.40 \cdot 10^{10} \, \hat{g} \cdot B \, [\text{T}] ; \quad (7.73)$$

$$\Delta \lambda_B \, [\text{nm}] = 4.67 \cdot 10^{-8} \, \hat{g} (\lambda_0 \, [\text{nm}])^2 \cdot B \, [\text{T}] .$$

For a magnetic field of 0.1 T (1 kGauss), which is typical of a larger sunspot (Sect. 7.3.3), at a wavelength of $\lambda_0 = 500$ nm the shift of a line component with $\hat{g} = 1$ is around 1 pm ($= 10^{-3}$ nm). It is thus only of the same order of magnitude as the linewidth e.g. of an Fe line in the solar spectrum, whose Doppler width $\Delta \lambda_D$ at a temperature of 5700 K is around 2 pm, according to (7.29). If the magnetic field is weaker than 0.1 T, a line broadening can still be observed. Furthermore, much smaller shifts can be detected if one makes use of the different polarization of the components, e.g. with polarization optics which measures alternately the right-hand and the left-hand circularly polarized outer line components.

Magnetohydrodynamics. We now turn to the theory of the flow of conducting matter together with magnetic fields, i.e. to magnetohydrodynamics (MHD); it will provide us with the basic physics needed to understand the phenomena of solar activity and many other astrophysical processes. We must of course leave the complex mathematical apparatus by the wayside. Initially, however, an intuitive understanding of the physical fundamentals should be more important, in any case.

From the basic work of H. Alfvén, T. G. Cowling, and others, we take the following ideas:

Practically all cosmic plasmas have a very high electrical *conductivity* σ (the reciprocal of the electrical resistivity). For fully ionized hydrogen gas, we find:

$$\sigma \, [\Omega^{-1} \text{m}^{-1}] \simeq 10^{-3} (T \, [\text{K}])^{3/2} \quad (7.74)$$

(T = temperature in K); e.g. at corona temperatures of 10^6 K, σ is about $10^6 \, \Omega^{-1} \text{m}^{-1} = 10^6 \, \text{V}^{-1} \text{A m}^{-1}$, thus only an order of magnitude less than for pure metallic copper ($6 \cdot 10^7 \, \Omega^{-1} \, \text{m}^{-1}$). For a conductor *at rest*, the current density $\boldsymbol{j}$ (current per cross-sectional area) depends on the electric field strength $\boldsymbol{E}$ through *Ohm's Law*:

$$\boldsymbol{j} = \sigma \boldsymbol{E} . \quad (7.75)$$

If in addition a magnetic field of flux density $\boldsymbol{B}$ is present in the conducting medium, a change in its strength will induce currents (according to the law of induction); on the other hand, the induced current density $\boldsymbol{j}$ again produces a magnetic field (Ampère's law). The relations between the electric and magnetic vortex fields and $\boldsymbol{B}$ or $\boldsymbol{j}$ are contained in *Maxwell's equations*. Here, we shall content ourselves with an estimate of the order of magnitude, by replacing the curl operator simply by the change over the characteristic dimension x of the volume of conducting plasma considered:

$$\text{curl } \boldsymbol{E} = -\frac{\partial \boldsymbol{B}}{\partial t}, \qquad \frac{E}{x} \simeq \frac{B}{\tau}$$

$$\text{curl } \boldsymbol{B} = \mu_0 \boldsymbol{j}, \qquad \frac{B}{x} \simeq \mu_0 j \simeq \mu_0 \sigma E. \quad (7.76)$$

Here, τ refers to a characteristic time scale and μ_0 is the magnetic field constant (permeability of vacuum). In the equation for curl $\boldsymbol{B}$, we have neglected the displacement current, an approximation which is justified for the relatively slow flow processes in the Sun.

According to (7.76), on the one hand the time-dependent change of the field, $\partial \boldsymbol{B}/\partial t \simeq B/\tau$, produces an electric field and a current density, $E \simeq j/\sigma \simeq xB/\tau$. (The induced voltage $U = Ex$ is equal to the change of the magnetic flux $\Phi \simeq Bx^2/\tau$). On the other hand, the magnetic field produced by $\boldsymbol{j}$ has a field strength $H = B/\mu_0 \simeq jx$ (Ampère-turns per m!). If we eliminate B/j from both equations, we obtain the characteristic decay time

$$\tau \simeq \mu_0 \sigma x^2. \quad (7.77)$$

The dependence $\tau \propto x^2$ demonstrates that the propagation and decay of a magnetic field in a conductor at rest has the character of a diffusion process.

If we now permit *motions* of the conducting medium with velocity $\boldsymbol{v}$, there will be an additional term in Ohm's law (7.75) due to the Lorentz force, so that in the moving system we have

$$\boldsymbol{j} = \sigma(\boldsymbol{E} + \boldsymbol{v} \times \boldsymbol{B}), \quad (7.78)$$

and the equations of hydrodynamics have to be solved together with the Maxwell equations. The result will depend upon whether the magnetic field $\boldsymbol{B}$ can move more quickly by diffusion, independently of the motions of the matter, or is "*frozen into* the matter".

Magnetohydrodynamic waves or disturbances in a medium of density ϱ propagate with the Alfvén velocity:

$$v_A = \frac{B}{\sqrt{\mu_0 \varrho}}. \quad (7.79)$$

When the associated propagation time x/v_A is shorter than the diffusion time (7.77), i.e. when *Alfvén's condition*:

$$\sqrt{\frac{\mu_0}{\varrho}} \sigma Bx > 1 \quad (7.80)$$

is fulfilled (and this is frequently the case in cosmic plasmas) the magnetic field remains frozen into the plasma matter. The matter can then move essentially only *parallel* to the magnetic lines of force, like glass beads on a string. Since the magnetic pressure is of the order of $B^2/(2\mu_0)$, the dynamic pressure $\varrho v^2/2$ will frequently be of the same order (in magnetohydrodynamic flows which are not too peculiar).

7.3.3 Sunspots and the Activity Cycle. Magnetic Flux Tubes

The *sunspots*, discovered already by Galileo and his contemporaries, appear for the most part in two zones having the same heliographic north and south latitudes. A typical sunspot has roughly the following structure and dimensions:

	Diameter	Area in millionths of the ☉ hemisphere
Umbra (dark center)	18 000 km	80
Penumbra (somewhat brighter rim)	37 000 km	350

The reduced brightness in the spots is due to a reduced temperature. In the largest spots, the effective temperature decreases from 5780 K for the normal solar surface to 3700 K. As a result, the spectrum of a large sunspot is on the whole similar to that of a K star; we have already anticipated the explanation based on Saha's theory of ionization in Sect. 7.1.2.

The outer regions of a sunspot, the *penumbra*, show bright and dark, radially directed filaments (Fig. 7.13), which at a very high spatial resolution appear as a series of bright elements (granules) strung together on a darker background. Observations of the Doppler effect in the spectral lines near the edge of the solar disk show that the matter in the dark zones is moving radially outwards at about 6 km s^{-1}, i.e. parallel to the solar surface, while the brighter granules are moving more slowly inwards.

Sunspots usually occur on the solar surface in groups. A *sunspot group* (Fig. 7.13) is surrounded by brighter *faculae* ("torches"). Furthermore, there are socalled polar faculae, independent of sunspots. The brightness of the faculae is a few percent above that of the normal surface only at the perimeter of the Sun. Applying (7.49), we conclude that in the faculae, only the layers nearest the surface (about $\tau \leq 0.2$) are overheated by a few hundred degrees.

Using the sunspots and, at higher heliographic latitudes, the faculae, the *rotation* of the Sun can be observed. Heliographic latitude is measured from the equator. It is found that the Sun does not rotate as a rigid body, but rather that higher latitudes rotate more slowly than the equator:

Heliographic latitude:	0°	20°	40°	70°
Mean sidereal rotation:	14.5°	14.2°	13.5°	$\simeq 12°\,\mathrm{d}^{-1}$
Sidereal period:	24.8	25.4	26.7	$\simeq 31$ d

Spectroscopic measurements of the Doppler effect at the Sun's perimeter (equatorial velocity about $2\,\mathrm{km\,s^{-1}}$) confirm this picture. The synodic period (as seen from the Earth) is correspondingly longer; for the *sunspot zones*, a (rounded) value of 27 d is obtained; it determines the quasiperiodic behavior of many geophysical phenomena. Another method of determining the rotation of the Sun, in particular in the invisible deeper layers down into the hydrogen convection zone, is provided by "helioseismology" (Sect. 7.3.8), through observation of the solar oscillations.

Activity Cycle. As was first shown by the pharmacist H. Schwabe about 1834, the abundance of sunspots varies with an average period of 11 years. All other phenomena of *solar activity*, to which we shall return later, follow this sunspot cycle; one therefore refers to the 11-year solar activity cycle. As a measure for the activity cycle, we use the

Relative number (of sunspots) R
$= k \cdot (10 \times$ Number of visible sunspot groups
$+$ Total number of spots) (7.81)

which was introduced by R. Wolf in Zurich. Here, k is a constant which depends on the size of the telescope used. Another measure, in use at the Greenwich Observatory after R. Carrington, is the photographically determined area of the umbrae, the overall spots, and the faculae, either directly in projection (in units of 1 millionth of a solar disk) or corrected for foreshortening (in units of 1 millionth of a solar hemisphere).

The activity cycles are numbered consecutively; the maximum of the arbitrarily chosen 1st cycle was in 1761.5, and that of the 22nd cycle was in 1990.3; the 23rd cycle began in 1995. There are occasionally longer "inactive" periods, most recently between 1645 and 1715, when only a few sunspots occurred (Maunder minimum).

Evolution of Sunspots. In each cycle, the area where new sunspots appear moves from high latitudes (±30 to 40°) at the maximum to low latitudes (±5°) at the minimum (Fig. 7.14). New spots or sunspot groups appear preferentially again and again in the same regions, called the *activity centers*.

The evolution in time of a (larger) sunspot group exhibits the following characteristic pattern: at first, two main sunspots form, each surrounded by several smaller spots. Their axis lies practically parallel to the equator. Gradually, the smaller spots disappear, and a "double spot" (bipolar group, see below) remains. The spot which trails behind with respect to the Sun's rotation becomes smaller and finally vanishes, while the leading sunspot remains visible for a long time.

Magnetic Fields. Structures like the radial filaments of the penumbra, and also the vortex-shaped flow fields which are observed on Hα spectroheliograms (Sect. 7.3.4) in the neighborhood of sunspots, the filamentary composition of the prominences (Fig. 7.20), and the polar bundles and rays of the corona at minimum (Fig. 7.18), all suggested some time ago that in solar physics, hydrodynamics alone is insufficient; rather, in addition, *magnetic fields* must play an important role. Thus G. E. Hale in 1908 searched for the Zeeman effect (7.73) and discovered magnetic fields which attain roughly 0.4 T in the largest sunspots. It could also be shown that the two spots (or halves) of a *bipolar* sunspot group always represent a north and a south pole, like the poles of a horseshoe magnet. Long series of observations have since revealed that the leading sunspot (relative to the Sun's rotation) in such a group always has the opposite sign in the northern or in the southern hemisphere, and that this sign alternates from cycle to cycle, so that the true period of a sunspot cycle is $2 \cdot 11$ years (Hale cycle).

The magnetic fields are a more sensitive indicator of disturbances in the activity centers than are the visible sunspots themselves, since they are frequently observable before the appearance of a spot or after its disappearance. Within the umbra of a sunspot, the magnetic field lines are perpendicular to the Sun's surface,

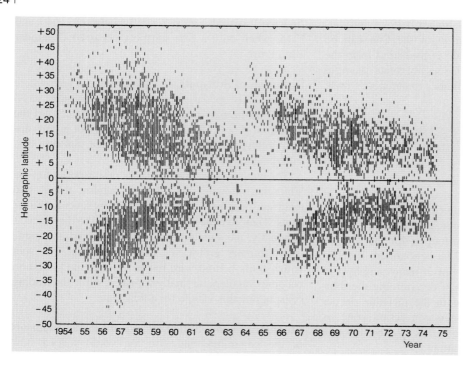

Fig. 7.14. A butterfly diagram (Spörer's law) showing the movement of sunspot zones in an activity cycle, from observations made at the Mt. Wilson Observatory. Each vertical dash represents a sunspot group, which is observed at the corresponding heliographic latitude within a synodic rotational period of about 27 d. R. Howard (1977). (Reproduced with the kind permission of D. Reidel Publishing Company, Dordrecht, Holland)

i.e. they form a *flux tube*. The magnetic flux (7.69), the integral of the magnetic flux density over the area through which field lines pass, lies in a range between about $5 \cdot 10^{12}$ to $3 \cdot 10^{14}$ T m². The energy density of the magnetic field, which gives also the lateral magnetic pressure of the flux tubes, is $B^2/2\mu_0$ (7.70). For $B = 0.1$ T (1 kG) as a typical value for a larger spot, we obtain a pressure of about 10^4 Pa, which is comparable with the pressure of the surrounding photosphere (cf. Table 7.4). In order to maintain pressure equilibrium between the sunspot and the photosphere, the gas pressure within the spot must be lower than in the gas surrounding the spot. In the upper regions of the solar atmosphere, the magnetic field lines fan out as a result of the rapidly decreasing gas pressure.

An explanation of the lower temperature in the spots using purely thermodynamic approaches has remained unsuccessful; we must assume that in the deeper regions of the sunspots, the convective energy flow (Sect. 7.3.1) is strongly impeded. Clearly, this energy flow is not simply deflected to reach the surface at another point as an area of higher temperature. Instead, it is stored for long periods of time (ca. 10^5 yr) below the photosphere in the convection zone. This idea has been suggested by recent precise measurements of the solar luminosity (Sect. 6.1.3), which is found to vary by 0.1 to 0.3%. The variation corresponds precisely to the dark areas of the sunspots present at the time of the measurement.

In 1952, H. W. and H. D. Babcock, using a considerably more sensitive apparatus, succeeded in recording Zeeman effects on the Sun corresponding to 1 to $2 \cdot 10^{-4}$ T (with a moderate spatial resolution of several seconds of arc); the spectral line splitting is only a small fraction of the linewidth. Among other things, it was found that a weak, spatially extended magnetic field of the order of a few 10^{-4} T exists at high heliographic latitudes (above 55°), having a total flux of $3 \cdot 10^{14}$ T m², and its polarity reverses with the 11-year period. In this polarity-reversal process, both hemispheres may exhibit the same polarity for up to several months. The field is composed of a number of magnetic regions, which are the remains of the fields in the activity centers and which drift towards the poles.

Within the experimental precision of about 10^{-4} T, the Sun has *no* general magnetic field. Instead, it has become clear since the 1970's as a result of spatially highly-resolved observations that the magnetic fields are concentrated in *thin flux tubes* at the limit of resolution (diameter $\lesssim 300$ km), each having a magnetic flux of about $3 \cdot 10^9$ T m^2 and a flux density of 0.1 to 0.2 T. They thus attain nearly the field strength found in large sunspots. These flux tubes, which are embedded in the photosphere and are directed perpendicular to the solar surface, are spread over the entire Sun, but are more numerous at the rims of flow cells of the supergranulation (Sect. 7.3.1), and naturally also in the activity centers. All together, they take up about 1% of the Sun's surface area. The regions between the flux tubes are probably free of magnetic fields. Above the photosphere, the flux tubes broaden out and merge to some extent.

Magnetohydrodynamics gives us the physical fundamentals for understanding the sunspots and the solar activity cycle:

As early as 1946, T. G. Cowling pointed out that the magnetic field of a sunspot would decay in a solar atmosphere *at rest* only over a time of the order of 1000 years by "diffusion", owing to the very high electrical conductivity σ of the plasma (7.77). In fact, however, sunspots have lifetimes of only several days up to a few months. Only in 1969 was it noted by M. Steenbeck and F. Krause that the turbulence in the hydrogen convection zone contributes strongly to vortex formation in the magnetic fields and in the electrical current and matter velocity fields which are associated with them; in the solar atmosphere, this produces a reduction of the effective electrical conductivity by a factor of 10^4, so that now (7.77) yields the right order of magnitude for the lifetimes of the sunspots.

The formation of a *bipolar sunspot group* (others can be traced back to this case) was proposed by V. Bjerknes (1926) to proceed as follows: in the Sun, there are always toroidal "tubes" of magnetic field lines (i.e. parallel to the lines of latitude). Since in these tubes the pressure is partly of magnetic origin, the gas pressure and the density are lower than in the surrounding medium. The tubes are therefore pushed up towards the surface ("magnetic buoyancy") and are "cut open" there. The two open ends of such a field tube form a bipolar sunspot group.

The Solar Dynamo. As is shown by Spörer's law of sunspot zone motion (Fig. 7.14), and Hale's law for the magnetic polarity of the sunspot groups, the entire activity cycle is based upon a reversal of the overall flow and magnetic fields in the interior of the Sun (mainly in the lower portion of the hydrogen convection zone) with a period of $2 \cdot 11$ years, and their stability (on the average) in the intervening time. The theory of this solar dynamo, whose driving force is the hydrogen convection zone together with the Sun's rotation, has become clearer piece by piece, beginning with the work of H. W. Babcock, R. B. Leighton, M. Steenbeck and F. Krause in the 1960's. The toroidal magnetic field must, in the appropriate phase of the cycle, give rise to a meridional magnetic field, and *vice versa*. This is made possible by the turbulence; it generates a current parallel to the field lines (in a manner which we cannot explain in detail here), giving rise to a meridional field component, i.e. perpendicular to the original field lines. For our dynamo, not only is induction due to the *differential rotation* of the Sun essential, but also that due to the statistically distributed *turbulent flow*. The dynamo theory must finally take into account the observation that the magnetic flux of the photosphere is almost completely concentrated into thin *flux tubes* at the rims of the supergranulation convection cells. The extremely complex calculation resulting from these assumptions appears to be able, in principle, to reproduce the phenomena of solar activity in terms of their order of magnitude.

7.3.4 The Chromosphere and the Corona

During total solar eclipses, the radiation from the Sun's disk (the photosphere) is blocked by the Moon, so that the *outer* layers of the solar atmosphere, which emit much less light in the optical region of the spectrum, become visible. They are called the chromosphere and, further out and extending far out into space, the corona.

A view of the higher layers of the solar atmosphere is also permitted by observations of the light of its *spectral lines*. As we saw in (7.49), this light arises at the optical depth for continuous *plus* line absorption:

$$x_\ell = \int_{-\infty}^{t} (\kappa + \kappa_\ell) dt' \simeq \cos\theta \qquad (7.82)$$

(7.54). Thus, layers whose optical depth in the continuum is only $\tau \simeq 10^{-3}$ and less, and which therefore do not appear at all in ordinary photographs, can be observed separately.

Instruments. We first introduce some types of instruments which are used in particular for the observation of the *outer* layers of the Sun:

The *spectroheliograph* (G. E. Hale and H. Deslandres, 1891), a large grating monochromator with which the image of the Sun in a precisely defined wavelength range of ≤ 0.01 nm within a Fraunhofer line can be photographed bit by bit. The most interesting images are obtained using the intense K line of Ca II ($\lambda = 393.3$ nm) or the hydrogen line Hα ($\lambda = 656.3$ nm). These calcium or hydrogen *spectroheliograms*, which originate in rather high layers of the solar atmosphere (chromosphere), are one of the most important tools for the investigation of solar activity.

The *Lyot polarization filter* (B. Lyot, 1933/38), which permits photography of the entire solar image in *one* short exposure, e.g. using the red Hα line of hydrogen, with a wavelength bandwidth (which is to some extent adjustable) of only about 0.05 to 0.2 nm. The "Hα filtergrams" thus attain a somewhat better image definition than the corresponding spectroheliograms. Lyot also constructed similar filters for the lines of the corona.

The observation of the highest atmospheric layers at the rim of the Sun is made difficult by *scattered light* which comes to some extent from impurities in the Earth's atmosphere and in the optics, but in large part is due to diffraction of the light at the entrance slit of the instrument. *Instrumental* light scattering is avoided for the most part by the *Lyot coronagraph*, whose principle is illustrated in Fig. 7.15.

Furthermore, total solar eclipses still provide an indispensable tool for research related to the outermost layers of the Sun.

In recent years, observations made from *satellites* and from *space vehicles* have become increasingly important for the investigation of the outer solar atmosphere. Here, scattered light from the Earth's atmosphere is avoided, and also the ultraviolet and X-ray spectral regions are accessible, which would otherwise be absorbed by the air. In these wavelength regions, and in the radiofrequency region, the radiation from the chromosphere and the corona predominates completely over that from the photosphere.

We mention the manned Skylab mission (1973/74), the satellite observations by OSO-8 (launched in 1975) and the Solar Maximum mission (1980). Since 1996, SOHO (*So*lar and *H*eliospheric *O*bservatory) has observed the Sun and the solar wind over several years with a variety of instruments from the inner Lagrange point L_1 (Fig. 6.15) at a distance of about $1.5 \cdot 10^6$ km from the Earth, where the gravitational attraction and centrifugal force in the Earth-Sun system compensate each other. SOHO was "lost" for several months in 1998 as a result of control errors at the ground station, but was later found and put back into operation. The 30 cm telescope of the TRACE satellite (*T*ransition *R*egion *A*nd *C*oronal *E*xplorer), launched in 1998, yields images of the transition region and the corona which are highly resolved in space and time, and is particularly dedicated to the exploration of the fine structure of the solar

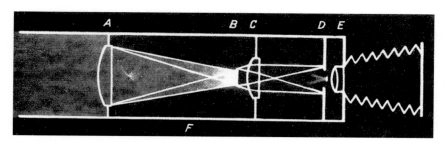

Fig. 7.15. A coronagraph as designed by B. Lyot (about 1930). The objective A, a simple plane-convex lens made of particularly homogeneous glass, projects an image of the Sun onto a circular plate at B; the plate extends $10''$ to $20''$ beyond the rim of the Sun's image and blocks the light coming from the solar disk. The field lens C forms an image of the objective aperture at D; here, a ring-shaped collimator removes the light diffracted from the edge of A. The objective E finally forms an image of the corona, prominences, etc. on the detector or the entrance slit of a spectrograph

magnetic fields. The observations are carried out in part simultaneously with those of SOHO.

The Chromosphere. If we observe the perimeter of the Sun, at first in the optical continuum, its brightness decreases rapidly on going outwards, as soon as the optical thickness along the direction of observation becomes < 1. The corresponding optical thickness for a strong Fraunhofer line however remains ≥ 1 for hundreds or thousands of kilometers further outwards. This means that the Fraunhofer spectrum with its absorption lines becomes an *emission* spectrum in the highest layers of the solar atmosphere, the *chromosphere*. This was first noticed by J. Janssen and N. Lockyer during the total solar eclipse of 1868: when the Moon covers the Sun out to its outer edge, the emission spectrum of the chromosphere "flashes" for a few seconds, producing the *flash spectrum* (Fig. 7.16). Spectrographic observations with moving-picture cameras (time resolution about 1/20 s) yield excellent information about the structure of the solar chromosphere, when combined with the relative motion of the Moon with respect to the Sun, calculated from the ephemerides. The scale height is larger than expected for an isothermal atmosphere at the limiting surface temperature of $T_0 \simeq 4000$ K, and it even increases on going outwards.

The lower chromosphere up to about 2000 km above the rim of the Sun was for a long time considered to be nearly homogeneously layered, like the photosphere, while the upper chromosphere shows strong spatial and temporal density fluctuations. Only with the high-resolution UV images taken by the TRACE satellite has it become clear that the lower chromosphere is also characterized by a variable and extremely inhomogeneous temperature and pressure layering. In Ca II spectroheliograms and in the light of other chromospheric spectral lines, bright and dark graininess can be

Fig. 7.17. Spicules at the rim of the Sun's disk (which itself is blocked out) photographed in Hα light. The altitude of these structures, which were first described by A. Secchi, attains up to 10 000 km above the rim of the Sun. Their thickness is about 900 km; they move with velocities of around 25 km s^{-1} up- (and more rarely down-)wards; their lifetimes are about 5 min. The direction of the spicules follows local magnetic fields

seen. The bright grains are arranged into the *chromospheric network* and are contiguous with the rims of the supergranulation cells (Sect. 7.3.1). Observed from the perimeter of the Sun, that is seen from the side, the higher layers of the chromosphere, e.g. in Hα light, look like a "burning prairie" with small "flamelets", the socalled spicules, (from Latin spiculum, "spike" or "point") which move with velocities of about 25 km s^{-1} upwards or downwards (Fig. 7.17). The spicules are not distributed uniformly, but rather are also concentrated at the rims of the flow cells of the supergranulation.

In the *active regions* near sunspots, we find in spectroheliograms extended bright emission regions called the *chromospheric plages*, which lie above the photospheric faculae (Figs. 7.22, 23). Here we can also observe an intensification of the magnetic fields, which are in the range of 10^{-2} T averaged over large areas.

In the chromosphere, the *temperature* increases up to about 20 000 K; in the corona, which lies further up, the temperatures are considerably higher, about 10^6 K, as we shall see in the next section. Since the temperature must somehow increase within a thin transition layer (≲ 15 000 km) up to the higher value in the corona, it does not seem surprising that lines of greater ionization and excitation energies are already observed with strong intensities in the upper chromosphere: among others the Balmer lines of hydrogen (excitation energy 10.2 eV) and the lines of neutral and ionized helium such as the He I D$_3$ $\lambda = 587.6$ nm (20.9 eV) and He II $\lambda = 468.6$ nm (48.2 eV) lines. Incidentally, the D$_3$ line of the "Sun element" helium was first observed in the solar spectrum by J. Janssen in 1868; only in 1895 was W. Ramsay able to isolate helium from terrestrial minerals.

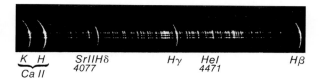

Fig. 7.16. A flash spectrum = emission spectrum from the solar chromosphere. Taken by J. Houtgast during the total solar eclipse in Khartoum in 1952 with an objective-prism camera

In the *ultraviolet*, the solar spectrum below $\lambda \simeq 160$ nm is dominated by the numerous emission lines from the chromosphere. The strongest line by far is the H I Lα line at $\lambda = 121.6$ nm. Additional strong chromospheric lines are the resonance lines of He I at $\lambda = 58.4$ nm and of He II at $\lambda = 30.4$ nm, as well as that of Mg II at $\lambda = 279.5/280.2$ nm.

The Corona. During the totality of a solar eclipse or using the Lyot coronagraph on a high mountain with the clearest air possible, the solar corona can be observed out to several solar radii (Fig. 7.18). Its form (flattening, radial structure, etc.) and brightness are functions of the 11 year cycle. Spectroscopic analysis distinguishes the following phenomena, which we shall in part attempt to explain immediately:

The *inner* corona ($r \simeq 1$ to 3 solar radii) exhibits a completely *continuous* spectrum in the visible region; its energy distribution corresponds to that of normal sunlight (K corona). The light is partially linearly polarized. We attribute this to Thompson scattering of photospheric light by the free electrons of the completely ionized gas (plasma) in the corona. The Fraunhofer lines are therefore completely smeared out by the Doppler effect, corresponding to the high electron velocities.

From the brightness distribution of the K corona in white light, the *average* (i.e. neglecting inhomogeneities) electron density n_e as a function of the distance r from the Sun's center can be calculated, since the distribution of the luminosity and the Thompson scattering coefficient per free electron are known (4.132). For e.g. the (round) corona at maximum, we find:

$r =$	1.03	1.5	2.0	3.0	Solar radii
$n_e \simeq$	$3 \cdot 10^{14}$	$2 \cdot 10^{13}$	$3 \cdot 10^{12}$	$3 \cdot 10^{11}$	Electrons [m^{-3}]

These values should, however, be taken only as a rough indication of the radial density distribution, due to the extremely inhomogeneous structure of the corona.

In the *outer* corona, which builds up rapidly on going outwards from the K corona, scattered photospheric light with *unmodified* Fraunhofer lines can be observed. Following W. Grotrian, who termed this component the F corona (Fraunhofer corona), it was pointed out in 1946/47 by C. W. Allen and H. C. van de Hulst that the light was due to Tyndall scattering, i.e. mainly forward scattering by small particles (somewhat larger than the wavelength of the light); they are so far from the Sun that they are not heated sufficiently to be vaporized. Measurements of the distributions of brightness and polarization in the outer corona have shown that the F corona is simply the innermost portion of the zodiacal light (Sect. 3.8). The F or dust corona or the zodiacal light, thus do not belong to the Sun at all and are influenced by it only to a relatively small degree.

In the *inner* corona, observations of the emission of the socalled *corona lines* were continued in the optical region; their identification however remained one of the great riddles of astrophysics, until in 1941, B. Edlén succeeded in explaining them as arising from forbidden transitions (Sect. 7.1.1) of metastable levels of the ground states of *highly ionized* atoms. The strongest and most important are ($\chi =$ ionization energy):

Fig. 7.18. The solar corona, near the sunspot minimum. Taken during the solar eclipse in Khartoum in 1952 by G. van Biesbroeck. The corona at minimum exhibits extended "rays" in the region of the sunspot zones; above the polar regions, there are finer "polar brushes". The corona at maximum has a more rounded shape

	λ [nm]	Identification		χ [eV]
Red corona line	637.4	[Fe X]	$3s^2 3p^5\, ^2P_{3/2} - {}^2P_{1/2}$	235
Green corona line	530.2	[Fe XIV]	$3s^2 3p\, ^2P_{1/2} - {}^2P_{3/2}$	355
Yellow corona line	569.4	[Ca XV]	$2s^2 2p^2\, ^3P_0 - {}^3P_1$	820

Additional, similar lines of the elements Ar, K, Ca, V, Cr, Mn, Fe, and Co have been identified with certainty. The ionization energies of several hundred eV indicate clearly that the electron temperature is of the order of a million degrees.

The following facts also support corona temperatures of several 10^6 K:

1. the density distribution $n_e(r)$, together with the hydrostatic equation (3.16) or the scale height (3.17);
2. the linewidths of the corona lines, which are due to the thermal Doppler effect (7.29);
3. the spectra in the far ultraviolet and X-ray regions, which in the main reflect the emission spectrum of the inner corona, with numerous allowed and forbidden lines from *higher* ionization states of the more abundant elements; and
4. the continua of the coronal plasma in the X-ray region.

The *ionization* and *excitation* in the corona can*not* be calculated under the assumption of local thermodynamic equilibrium (Saha and Boltzmann formulas), since in the extremely tenuous plasma, there is no radiation field corresponding to 10^6 K. Instead, the ionization, recombination, and excitation processes must be considered individually (Sects. 4.5.3, 4). From detailed calculations, temperatures of $(1$ to $5) \cdot 10^6$ K are found; within about this range, the temperature in the corona varies with time and place. The *abundances* of the elements relative to hydrogen, determined from the line intensities compared with the electron-scattering continuum of the K corona, agree well with the values determined using the absorption lines of the photosphere (Table 7.5).

Images of the Sun in the *soft X-ray region* using a Wolter telescope (Fig. 7.19) clearly show the structure of the corona and its changes with time: relative to the thermal emission of the corona at several 10^6 K, the continuum of the photosphere at about 6000 K, which is predominant in the optical region, is not at all visible in the X-ray region. The X-ray emission of the corona is extremely variable and is non-uniformly distributed. While spatially extended structures can

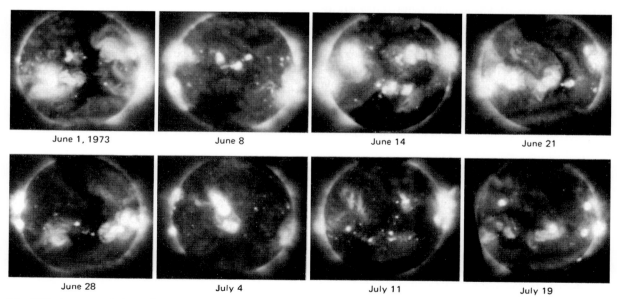

Fig. 7.19. Images of the solar corona in the soft X-ray region (using filters to limit the spectral range) taken at intervals of 7 days over nearly two rotational periods. Photographs made with the X-ray telescope of American Science and Engineering, Inc. on Skylab, 1973 (From J. A. Eddy, 1979)

be followed through several rotational periods, other smaller emission regions change form in days or even hours.

Images taken at higher resolution show that nearly all the X-ray emissions consist of a number of loop- or arch-shaped structures, which occur both in quiet regions and in active regions. These *coronal loops* represent *closed* magnetic flux tubes. Outside the active regions, we find long, extended arches, which reach up to several tenths of a solar radius above the perimeter of the Sun, with plasma densities of about $2 \cdot 10^{14}$ to 10^{15} m^{-3} and temperatures of 1.5 to $2 \cdot 10^6$ K. Active regions are often connected together by long arches ($\lesssim 700\,000$ km), in which somewhat higher temperatures (2 to $6 \cdot 10^6$ K) are found.

Above sunspot groups or activity centers, both in the optical and in the ultraviolet, as well as with the light from the corona lines and in the X-ray region, we can observe the hot and more dense *coronal condensations*, whose area is for the most part the same as that of the chromospheric faculae regions or the plages. In a dense ($\lesssim 10^{16}$ m^{-3}) core region with a lifetime of several days, temperatures of $\gtrsim 3 \cdot 10^6$ K are attained. At a high spatial resolution it becomes clear that the corona above the active regions consists of bundles of small coronal arches (of 10^4 to 10^5 km length), tightly crowded together and having temperatures of $\lesssim 2.5 \cdot 10^6$ K. Above them, we find roughly radially-directed *coronal rays* in a brush or fan-shaped arrangement, whose densities are about 3 to 10 times higher than those of the surrounding medium and which reach out as far as 10 solar radii into space.

The coronal rays above the active regions, as well as the polar brushes of the corona at minimum, represent long, for the most part *open* magnetic field lines. The extended *coronal holes*, which appear dark in X-ray pictures and occur preferentially at higher heliographic latitudes, also exhibit open magnetic field structures. From them, relatively cool matter ($T \simeq 10^6$ K) – the solar wind – streams outwards with velocities up to 800 km s^{-1} (cf. Sect. 7.3.7).

Finally, the X-ray images also show small *bright spots* (diameter $\lesssim 20\,000$ km), which are distributed over the whole Sun and have lifetimes of only several hours up to a few days. They are formed (at a rate of about 1500 per day) by small flux tube loops rising up to the surface of the Sun from the depths, and become recognizable on photospheric magnetograms as a small bipolar region. The magnetic flux in all the bright spots is greater than that in the active centers at sunspot minimum; at maximum, it is still almost 50% of the total flux which emerges from the solar surface.

Because the magnetic energy density in the corona is greater than the thermal energy density, magnetic field structures dominate its density distribution and flow properties (Sect. 7.3.2). In X-ray photographs, we can see the magnetic field lines more or less directly, so to speak. Not only material flows, but also energy transport by thermal conduction and magnetohydrodynamic waves follow the field lines or flux tubes and are strongly inhibited in their motion perpendicular to them. A coronal arch is thus well "filled up", so that its density is higher than that of its surroundings. Heating by waves can take place along the curved flux tubes. Within an "undisturbed" arch, there is on the one hand hydrostatic equilibrium, on the other thermal equilibrium between heating, radiation and thermal conduction. The magnetic field ensures that the "overly dense" matter cannot escape to the sides. The precise physics of the various density concentrations in the solar atmosphere, its heating mechanisms, dynamics and stability certainly requires further investigations.

Radiofrequency Radiation. The *thermal* radiofrequency radiation of the Sun can be explained in terms of free–free radiation from the chromosphere (millimeter and centimeter waves) and the corona (meter waves). In the radiofrequency range, the free–free absorption coefficient of the plasma at frequency ν is proportional to $n_e^2 T^{-3/2} \nu^{-2}$. With increasing frequency, one can thus "look" deeper and deeper into the solar atmosphere. Thus, the radio spectrum of the quiet Sun in the decimeter-wave range gives practically an exact image of the temperature and pressure distribution in the transition layer between the chromosphere and the corona.

The fluctuations in the intensity of the thermal radiation in the course of days and months allow us to decompose them into the ever-present radiation from the quiet Sun and the slowly varying radiation which comes from the active regions (S component in the range 1.5 cm $\lesssim \lambda \lesssim 70$ cm). At the rim of the Sun and in the denser parts of the corona, an optical thickness of > 1 for radiofrequency radiation is even reached in some places, so that we can measure the black-body radiation directly. Its temperature, which corresponds to the electron temperature, is several 10^6 K.

Heating. Finally, we turn to a more exciting question: Why does the temperature *increase* from ca. 4000 K at the upper edge of the photosphere to around 20 000 K in the chromosphere directly above it and then from the chromosphere to the corona, i.e. within the region of the transition layer which is only $\leq 15\,000$ km thick, does it rise even to values $\geq 10^6$ K? From the Second Law of thermodynamics, this heating of the highest layers of the solar atmosphere, opposing the "natural" (i.e. entropy-increasing) temperature gradient from within the Sun to the outside, can arise only through *mechanical* energy or other "ordered" energy forms which have a large negative entropy.

Indeed, following M. Schwarzschild and L. Biermann, we can show that in the upper layers of a partially convective atmosphere, mechanical energy transport becomes increasingly important relative to radiative energy transport. In the turbulent flow fields of the uppermost layer of the hydrogen convection zone (Sect. 7.3.1), only a few 100 km thick and directly below the photosphere, *acoustic waves* (with periods in the range from 20 to 100 s) are produced; this was first investigated in detail by I. Proudman and M. J. Lighthill. Furthermore, due to interactions with the magnetic fields present in the plasma of the solar atmosphere, *magnetohydrodynamic* waves are also generated; they couple the oscillations of the plasma to correspondingly oscillating magnetic fields.

All of these waves penetrate into the higher and thinner layers of the solar atmosphere. As long as the velocity amplitude Δv e.g. of an acoustic wave remains small compared to the velocity of sound c_s, the wave is practically not damped. However, since the energy flux of an acoustic wave in a plasma of undisturbed density ϱ is given by

$$F_s = \tfrac{1}{2} \varrho (\Delta v)^2 \cdot c_s \qquad (|\Delta v| \ll c_s) \,, \qquad (7.83)$$

we see that $(\Delta v)^2$ increases $\propto 1/\varrho$ when the wave moves outwards into a less dense medium. When Δv approaches the velocity of sound, *shock waves* are formed. Their energy is relatively rapidly dissipated, i.e. it is converted back to heat. The temperature then increases until an equilibrium between energy input and energy loss through radiation has been established. While in the chromosphere, energy losses via radiation dominate, in the corona several processes contribute to losses, and their relative importance depends strongly on the magnetic fields and densities. Thus, in the coronal arches, thermal conductivity by free electrons in the plasma and radiation in the X-ray region are significant; in the coronal holes, in contrast, losses through convective energy to the solar wind are predominant.

While the physical processes which lead to the steep temperature increase in going outwards to the chromosphere and the corona are by no means understood in detail, we can estimate the required (minimal) *mechanical* energy current from observations of the total radiation flux from the chromosphere and the corona in connection with model calculations. If we neglect the energy losses of the acoustic and magnetohydrodynamic waves in the convection zone and the photosphere, we find that for heating of the chromosphere, about 5 kW m^{-2}, and for the corona, about 0.4 kW m^{-2} are required; i.e. it is only 10^{-4} or 10^{-5} of the overall energy flux of the Sun (6.8) which in the end need be tapped from the convection zone.

7.3.5 Prominences

At the rim of the Sun during eclipses or using the coronagraph, the prominences can be seen as bright "long, extended clouds" in the corona. In front of the Sun's disk, e.g. in Hα light, they appear as thin, dark filaments (Figs. 7.20, 23).

The spectra of the prominences are for the most part similar to those of the upper chromosphere (spicules). They exhibit strong Ca II H and K lines along with the He I D$_3$ line at $\lambda = 587.6$ nm. The ultraviolet region is dominated by numerous emission lines. We see here a dense, cool plasma with about 10^{17} particles m^{-3} and excitation temperatures in the range from 5000 to 12 000 K, which is embedded in the less dense, hotter corona.

The *quiescent* prominences (filaments) retain their form with minor changes (flow fields of the order of 10 km s^{-1}) often for weeks at a time. They are characterized by a peculiar threadlike structure (Fig. 7.20). The transition from the cool threads to the surrounding hot corona at about $\geq 10^6$ K takes place in an extremely thin

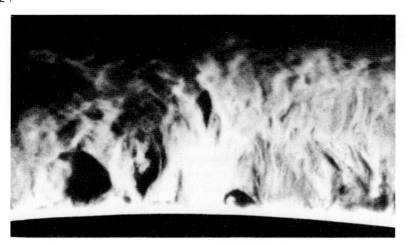

Fig. 7.20. The quiescent prominences (filaments) have, on the whole, the form of a thin sheet standing upright on the Sun's surface on several "feet": their thickness is about 7000 km (4000 to 15 000 km), height around 45 000 km (15 000 to 120 000 km), length about 200 000 km (up to 10^6 km). This detail photograph from the Sacramento Peak Observatory (in Hα light) shows threadlike structures in which matter flows upwards or downwards with velocities on the order of 10 to 20 km s^{-1}. The shape of the prominences is clearly determined in part by solar magnetic fields

"skin" only 10 to 100 km thick, as can be seen in images made with the ultraviolet light from highly excited or highly ionized states. Magnetic fields of $\lesssim 10^{-3}$ T have been measured in the filaments.

By applying magnetohydrodynamics, we can understand how the *quiescent* prominences can "float" in the much hotter corona that surrounds them. Only the cool matter can effectively radiate away energy, i.e. cool matter remains cool and hot matter remains hot, even when a certain energy transport occurs. Pressure equalization between the prominences and the corona requires that $P \propto \varrho T$ be about the same in both. The density of the prominences must therefore be about 300 times greater than that of the corona. The fact that the quiescent prominences do not fall down is a result, according to R. Kippenhahn and A. Schlüter (1957), of their being supported on a cushion of magnetic force lines (the matter cannot penetrate the lines of force), similar to rain water in the depressions of the protective cover over a haystack. However, we must still explain that fact that individual prominences do not fall along the force lines of their magnetic guide fields according to Galilei's laws, but instead float slowly downwards with velocities which are constant over long times. This behavior, which is reminiscent of that of clouds in the Earth's atmosphere, apparently results from the high *viscosity* of the corona, which is in turn due to the high velocities and long mean free paths of its electrons. That the prominences do not ever fall suddenly down into the photosphere, but instead float like real clouds is (in both cases) due to the predominance of viscous forces over pressure and inertial forces; we have here an example of "creeping flow" with small Reynolds numbers.

Sometimes, the prominences or parts of them are accelerated more or less suddenly to velocities of 100 km s^{-1} or occasionally up to 600 km s^{-1}, without any previous external indications. These *eruptive* or *ascending* prominences (Figs. 7.21, 22) can then escape into interplanetary space. In other cases, a disturbed prominence "rains" back onto the solar surface, its material streaming down along arch-shaped paths (following the magnetic lines of force).

Above sunspot groups or activity centers, we observe manifold forms of *active* prominences as jets, sprays, or arches with lifetimes of only a few minutes to hours; some of them are accelerated to above the escape velocity from the Sun. The temperatures and magnetic fields in the active prominences are considerably higher than in the static prominences.

The activation of quiescent prominences and the occurrence of active prominences are often connected with solar eruptions or flares (Sect. 7.3.6). Along with eruptive prominences, expanding clouds of matter or arches with radially-directed velocities up to 1200 km s^{-1} can be observed in the outer corona (coronal transients). The corona above an active region can thus be "blown away" within about an hour.

We cannot give a more detailed treatment of the dynamics of prominences here.

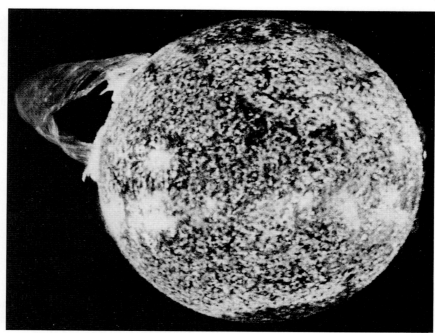

Fig. 7.21. An eruptive or ascending prominence. The time interval between the two photographs was 71 min. Maximum altitude: 900 000 km above the rim of the Sun; maximum velocity: 230 km s^{-1}

Fig. 7.22. An enormous eruptive prominence photographed on December 19th, 1973 with the light of the He II resonance line at $\lambda = 30.4$ nm, using the extreme-ultraviolet spectrograph of the U. S. Naval Research Laboratory on Skylab (from J. A. Eddy, 1979)

7.3.6 Solar Eruptions or Flares

The impressive phenomenon of *eruptions* (not to be confused with the eruptive prominences!) or *flares* can be most readily observed as a "brightening" in the Hα line; they are, however, accompanied by an intensification of the radiation in the whole electromagnetic spectrum as well as by the production of energetic particles.

When an active region is observed with the Hα or the Ca II-K line (Fig. 7.23), the chromospheric torch areas or plages between the sunspots and in their neighborhood show structures of the order of several 1000 km across with smaller irregular motions and variations in brightness. Suddenly, many of the brighter structures merge together; then, in a major eruption, a region of 2 to $3 \cdot 10^{-3}$ of the solar hemisphere, corresponding to a whole sunspot group, flares up in the Hα and Ca II K lines. The Hα emission in the middle attains an intensity up to about 3 times that of the normal continuum and a width of several tenths of a nanometer. The lifetime of the flares varies from the order of one second in the case of the tiny "microflares" of about 1″ diameter, up to several hours in the case of major flares. The "importance" of the eruptions is classified according to the area of the Hα emission at the maximum as S (subflare, $< 10^{-4}$ hemisphere), 1, 2, 3 and 4 ($> 10^{-3}$ hemisphere).

The increase of intensity in Hα is accompanied by increased emission in the whole spectrum, which, except for the hard X-ray and the radiofrequency regions, is of *thermal* origin. The flaring up of the upper chromosphere in Hα and other lines is however clearly only a secondary phenomenon. In the far ultraviolet and especially in the soft X-ray regions ($\gtrsim 10$ eV), an increase in the radiation intensity can be observed about 10 min before the beginning of the Hα flare; it indicates heating of the lower corona. In the case of large flares, at the beginning of the intensity rise in Hα a burst occurs in the microwave and hard X-ray regions, lasting $\lesssim 5$ min. This non-thermal radiation is produced by the interaction of electrons (which are accelerated to high energies

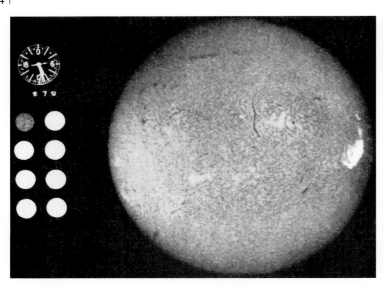

Fig. 7.23. A large solar eruption (type 4 flare), taken on July 18th, 1961 (*right*); it was accompanied by strong cosmic-ray particle emissions. This is an Hα solar observation photo taken at the Cape Observatory. Somewhat above the midpoint of the solar disk, an extended filament can be seen, i.e. a prominence in absorption (compare Fig. 7.20). *Left*: Photometric intensity calibrations and a time marker

during the flare) with the solar plasma. In this phase, in the soft X-ray region a bright, dense, and hot core (diameter $\lesssim 4000$ km) with temperatures up to $3 \cdot 10^7$ K can be observed; it is often at the highest point of a corona arch (Fig. 7.24).

Corpuscular Radiation. The emission of corpuscular radiation over a large energy range accompanies major flares. Plasma clouds or magnetohydrodynamic shock waves, which reach the Earth about one day later and thus have a velocity of about 2000 km s^{-1}, cause magnetic storms and auroras. Major flares also make a solar contribution to cosmic rays, as found by S. E. Forbush and A. Ehmert, with particle energies of 10^8 to 10^{10} eV.

Gamma Rays. The interaction of the energetic particles with atomic *nuclei* in the solar matter also causes the emission of low-energy gamma radiation. Aside from a continuum, several strong emission *lines* are observed: nuclear transitions in the abundant nuclei ^{12}C and ^{16}O at 4.43 and 6.14 MeV, and, with a time delay, lines at 0.511 and 2.23 MeV. The 0.511 MeV line is produced through annihilation of positrons, which are generated by the flare, and electrons; the 2.23 MeV line comes from the formation of deuterium, ^{2}D, through reaction of neutrons which are also produced in the flare with protons (hydrogen nuclei) in the solar atmosphere.

Since 1991, the gamma radiation from flares has been observed with the Compton Gamma Ray Observatory over a wide range of energies both in the continuum and in the low energy lines. From the strongest outbursts, gamma radiation up to the GeV range is emitted. The time correlation between the decay of the 2.23 MeV line and the high-energy radiation leaves no doubt concerning the common origin of the neutrons and the gamma quanta.

Radiofrequency Emissions. Accompanying the flares and also the less spectacular phenomena of solar activity is a *non-thermal* contribution to the radiofrequency radiation of the Sun. Its analysis with the aid of a radiofrequency spectrometer, registering the intensity in a large frequency range as a function of time, has been carried out by J. P. Wild and coworkers in order to distinguish between several different types of "bursts" or intense radiofrequency emissions, initially in the range below 400 MHz (Fig. 7.25).

In order to understand this analysis, we must first remark that electromagnetic waves of frequencies below the critical or *plasma frequency* can*not* be emitted from a plasma of electron density n_e:

$$\nu_p = \sqrt{\frac{1}{4\pi\varepsilon_0} \frac{e^2}{\pi m_e} n_e} ,$$

$$\nu_p \, [\text{MHz}] = 9.0 \cdot 10^{-6} \sqrt{n_e \, [\text{m}^{-3}]} , \qquad (7.84)$$

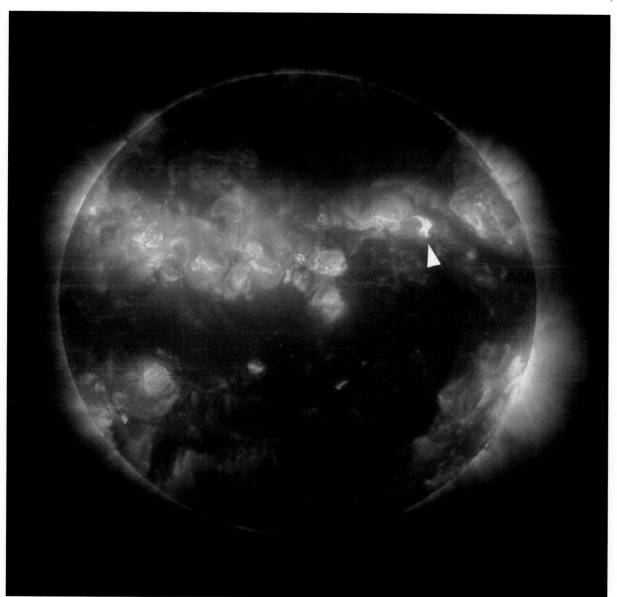

Fig. 7.24. A high-resolution X-ray image of the solar corona, showing a flare (indicated by the arrow) and active regions. The angular resolution is 0.75″, and the temperature of the hottest regions shown is about $3 \cdot 10^6$ K. The image was made by L. Golub et al. (1989) using a rocket carrying the 0.25 m X-ray telescope NIXT (*N*ormal *I*ncidence *X*-ray *T*elescope) in a limited wavelength region around $\lambda = 6.35$ nm which includes the emission lines of FeXVI at $\lambda = 6.37/6.29$ nm and MgX at $\lambda = 6.33/6.32$ nm. In contrast to an image-forming X-ray telescope of the Wolter type, with grazing incidence (Fig. 5.27), here the image is produced at normal incidence as with an ordinary optical mirror. Reflection of the X-rays is made possible by vapor-deposition of alternating thin layers of cobalt and carbon, so that constructive interference results for the wavelength 6.35 nm. NIXT was developed by L. Golub and coworkers at the Smithsonian Astrophysical Observatory, Cambridge, together with the IBM Thomas J. Watson Research Center, Yorktown Heights, New York. (With the kind permission of the SAO and IBM)

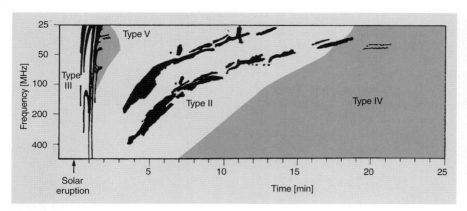

Fig. 7.25. The dynamic (i.e. time-dependent) radio spectrum in the meter wave region (schematic), after J. P. Wild: the time evolution of "bursts" of different types following a major solar eruption. Lower/higher frequencies are emitted in general by higher/lower layers of the corona. Type II and type III bursts are often accompanied by 2 : 1 harmonics

since then the index of refraction is < 0. Thus, radiofrequency radiation of a particular frequency in the corona can have had its origin only from *above* a certain layer.

The *type II* and *type III* bursts (Fig. 7.25) exhibit a slow or a rapid shift of their frequency bands to lower frequencies, respectively. From this it can be concluded that the phenomenon which produces them sweeps through the corona with velocities of the order of $1000\,\mathrm{km\,s^{-1}}$ for type II and up to 0.4 of the velocity of light for type III. Simultaneously with the type III bursts, in the *microwave region* above $\nu \gtrsim 1.5\,\mathrm{GHz}$ ($\lambda \le 20\,\mathrm{cm}$), pulses of 1 to 5 min length (with a smoother time envelope) are detected. The *type IV* events emit a continuum over a long period of time, covering a broad frequency band; this is due to synchrotron radiation from the fast electrons.

In 1967, J. P. Wild and coworkers at the Culgoora Observatory in Australia succeeded in observing the motion of various types of burst sources at 80 MHz ($\lambda = 3.75\,\mathrm{m}$) on and outside the solar disk, directly on the display monitor of the *radioheliograph*. Using 96 parabolic antennas each having a diameter of 13 m and mounted in a circle of 3 km diameter, the Sun and the solar corona could be probed over a field of view of $2°$ diameter with 2 to $3'$ resolution. The observations verify the results from the radio spectra. We unfortunately cannot discuss here the complicated plasma-physics treatment of the different types of bursts.

The total *energy* released in a flare ranges from $10^{22}\,\mathrm{J}$ in the case of a subflare up to $3 \cdot 10^{25}\,\mathrm{J}$ in a type 4 flare; it is divided about equally among electromagnetic radiation, kinetic energy of the ejected plasma, and high-energy particles. Although the complex processes which are responsible for solar flares are not understood in detail, we can assume that in the end, they involve *magnetic energy* (energy density $B^2/2\mu_0$), which is transformed into thermal energy which is then radiated, and which also causes the acceleration of particles. In order to produce the energy of $3 \cdot 10^{25}\,\mathrm{J}$ in a major flare, for example in a volume with a diameter of about 30 000 km, a magnetic field of $5 \cdot 10^{-2}\,\mathrm{T}$ would have to be "annihilated", perhaps by an instability which leads to a regrouping of field lines (magnetic reconnection). Measurements of the solar magnetic fields support this idea: flares occur preferentially in activity regions with complex field configurations and steep field gradients, and there mostly in regions where the polarity of the magnetic field changes its sign.

7.3.7 The Solar Wind

As we already mentioned in Sect. 3.6.1, it was suggested by L. Biermann in 1951 that the plasma tails of comets are blown away from the Sun not by radiation pressure, but rather by an everpresent corpuscular solar radiation. Measurements from satellites and space vehicles later showed that this plasma, which has roughly the composition of solar matter, streams away from the Sun with velocities – in the neighborhood of the Earth's orbit – of about $500\,\mathrm{km\,s^{-1}}$ (with variations between about 300 and $750\,\mathrm{km\,s^{-1}}$). Its density corresponds to about $5 \cdot 10^6$ protons and electrons per m^3 or $\varrho = 8 \cdot 10^{-21}\,\mathrm{kg\,m^{-3}}$, also with large variations. The (more or less) statistical part of the particle velocities corresponds to a temperature of about $10^5\,\mathrm{K}$.

Associated with the plasma are magnetic fields of the order of $6 \cdot 10^{-9}$ Tesla. In 1959, E. N. Parker named this phenomenon the *solar wind* and suggested that it could be explained on the basis of hydrodynamics or of magnetohydrodynamics. If we calculate the pressure distribution $P(r)$ of the corona (schematically assumed to be isothermal as a first approximation) at a large distance r from the Sun, we find that the finite limiting value of the pressure there, for corona temperatures $T < 5 \cdot 10^5$ K, lies *below* that of interstellar matter (Chap. 10). In this case, the interstellar matter would stream into the Sun. However, for $T > 5 \cdot 10^5$ K, as holds for the real corona, matter streams continually outwards: this is just the solar wind.

To calculate the properties of the solar wind, we need the basic equations of hydrodynamics (here, we will neglect the effect of the magnetic fields), together with the equation of state of the matter. We limit ourselves to *spherically symmetrical* flows of matter outwards from a central mass $\mathcal{M}$.

Equation of State. For the matter in the solar wind, we assume the ideal gas equation (3.15), which, for this application, we write in the form

$$P = \frac{kT}{\bar{\mu} m_u} \varrho = \frac{\gamma - 1}{\gamma} c_p T \cdot \varrho \qquad (7.85)$$

by introducing the specific heat at constant pressure, c_p, via the relations

$$\gamma = \frac{c_p}{c_v}, \quad c_p - c_v = \frac{\gamma - 1}{\gamma} c_p = \frac{k}{\bar{\mu} m_u} \qquad (7.86)$$

(Sect. 3.1.4). Here, k again refers to Boltzmann's constant, $\bar{\mu}$ to the mean molecular weight, m_u to the atomic mass unit, and c_v to the specific heat at constant volume. For a completely ionized plasma consisting of 90% H and 10% He, we have $\bar{\mu} = 0.65$.

Hydrodynamics. The motion of a volume element in a gas is described by the *Euler equation*, analogously to the equation of motion "mass × acceleration = force" from mechanics:

$$\varrho \frac{dv}{dt} = \varrho \frac{\partial v}{\partial t} + \varrho v \frac{\partial v}{\partial r} = -\frac{\partial P}{\partial r} - \varrho \frac{G\mathcal{M}}{r^2} . \qquad (7.87)$$

Here, we take into account only the force caused by the pressure gradient and the gravitational force $\varrho G \mathcal{M}/r^2$ of the central mass (2.38). The *total* acceleration dv/dt at the position r and time t includes both the local acceleration $\partial v / \partial t$ and the rate of change of v as a result of the motion of the volume element ("advection of the acceleration"). In the following, we assume *stationary* flow, i.e. we require that $\partial / \partial t = 0$, but *not* that the overall acceleration vanishes. For example, v increases in a tube even for stationary flow, if the diameter of the tube becomes smaller (7.88). In the case that the overall acceleration is also 0, we again obtain from (7.87) as a special case the equation for hydrostatic equilibrium (3.6).

The *equation of continuity* – in the stationary case – is given by the condition that the same amount of matter flows out through each spherical shell $4\pi r^2$ in a unit time, i.e.

$$4\pi r^2 \varrho(r) v(r) = \dot{\mathcal{M}} = \text{const}. \qquad (7.88)$$

The constant is thus equal to the rate of *mass loss* $\dot{\mathcal{M}}$ by the solar wind.

These statements must be drastically modified if the matter is guided by a magnetic field. Then, a particular matter current must move along a particular magnetic force tube, as long as the pressure of the matter on its "walls" does not exceed the magnetic pressure. We cannot pursue the solution of this extremely difficult problem in magnetohydrodynamics (Sect. 7.3.2) any further here.

Rather, we first ask what conclusions we can draw from Bernoulli's equation, i.e. the *conservation of energy*.

Energy Balance. By integration of the stationary Euler equation (7.87), divided by ϱ, we find that along a streamline, i.e. a line in the velocity field v, the sum of the kinetic energy $v^2/2$, pressure energy $\int dP/\varrho$, and the potential energy or potential Φ (all energies calculated per unit mass) is constant. The gravitational potential of the mass $\mathcal{M}$ (2.43) can be expressed using (2.62) in terms of the escape velocity $v_e(r)$ at the position r in the region of attraction of the mass $\mathcal{M}$:

$$\Phi(r) = -\frac{G\mathcal{M}}{r} = -\frac{1}{2} v_e^2(r) . \qquad (7.89)$$

We thus obtain the *Bernoulli equation*

$$\frac{1}{2} v^2(r) + \int^r \frac{dP}{\varrho} - \frac{1}{2} v_e^2(r) = \text{const}, \qquad (7.90)$$

where the pressure integral extends up to the value of r being considered.

We now apply (7.90) to the solar wind. In order to calculate its velocity $v(r_E)$ in the neighborhood of Earth's orbit $r = r_E$, we assume that the solar wind begins at the base of the corona, i.e. at $r \simeq R_\odot$, with $v(R_\odot) \simeq 0$, and consider the acceleration out to r_E. We then obtain

$$v^2(r_E) = 2 \int_{r_E}^{R_\odot} \frac{\mathrm{d}P}{\varrho} - v_e^2(R_\odot), \qquad (7.91)$$

where we can neglect $v_e(r_E)$ in comparison to $v_e(R_\odot) = 620 \text{ km s}^{-1}$.

The pressure integral depends decisively on the temperature distribution $T(r)$ between the Sun and the Earth. If we rely on the observation that T decreases relatively little from $T_\odot \simeq 10^6$ K in the corona to $T_E \simeq 10^5$ K near the Earth's orbit, we can as an approximation use the constant mean temperature $\overline{T} \simeq T_\odot$ and then readily carry out the integration, using $\mathrm{d}P = \text{const} \cdot \mathrm{d}\varrho$ (7.85):

$$\int_{r_E}^{R_\odot} \frac{\mathrm{d}P}{\varrho} = \frac{2k\overline{T}}{\bar{\mu} m_u} \ln\left(\frac{\varrho_\odot}{\varrho_E}\right) = \frac{\gamma - 1}{\gamma} c_p \overline{T} \ln\left(\frac{\varrho_\odot}{\varrho_E}\right). \qquad (7.92)$$

Here, $\varrho_\odot$ is the density at the base of the corona and ϱ_E is the density at the position of the Earth. From (2.23), $\sqrt{2k\overline{T}/(\bar{\mu} m_u)} = 165 \text{ km s}^{-1}$ corresponds to the thermal velocity of the particles in the corona, or – with an accuracy of about 10% – the velocity of sound

$$c_s = \sqrt{\gamma \frac{kT}{\bar{\mu} m_u}}, \qquad (7.93)$$

so that we obtain as a good approximation

$$v^2(r_E) \simeq c_s^2 \ln\left(\frac{\varrho_\odot}{\varrho_E}\right) - v_e^2(R_\odot). \qquad (7.94)$$

Taking the observed value of $\varrho_\odot/\varrho_E$, we can readily convince ourselves that the right order of magnitude results for the velocity, $v(r_E) \simeq 500 \text{ km s}^{-1}$. As can be seen by comparing with (7.93), the solar wind is a *supersonic flow*.

It is important to realize that the fact that our calculation leads to a value which is in agreement with the observations is a result of our *implicit* assumption, in the numerical value chosen for $\overline{T}$, that the outer corona or the solar wind is *heated* by some process or other

(dissipation of wave energy?) and has only very small radiation losses.

If we had for example calculated the case of an *adiabatic* outflow from the corona, we would have obtained from the adiabatic equation for an ideal gas, (3.20),

$$\frac{P}{P_0} = \left(\frac{\varrho}{\varrho_0}\right)^\gamma, \quad \frac{T}{T_0} = \left(\frac{P}{P_0}\right)^{1-\frac{1}{\gamma}} = \left(\frac{\varrho}{\varrho_0}\right)^{1-\gamma}, \qquad (7.95)$$

with $T_0 = T_\odot$ and $\varrho_0 = \varrho_\odot$ for the pressure integral

$$\int_{r_E}^{R_\odot} \frac{\mathrm{d}P}{\varrho}\bigg|_{\text{ad}} = c_p T_\odot \left[1 - \left(\frac{\varrho_E}{\varrho_\odot}\right)^{\gamma-1}\right] \simeq c_p T_\odot ; \qquad (7.96)$$

i.e. instead of (7.92), for the maximum value only $c_p T_\odot$, the enthalpy per unit mass of the coronal matter. The matter streaming out adiabatically would cool rapidly and remain "stuck" near the Sun.

Our hydrodynamic theory of the solar wind indeed gives the correct order of magnitude of the observed result, but it is to some extent not satisfying since, as one can easily show, the mean free path is of the same order as the characteristic lengths in the model. Therefore, a calculation based on the kinetic theory of gases, in which the magnetic field would necessarily be taken into account ("collision-free plasma"), is actually required.

Structure. When the solar wind streams out into interplanetary space, it takes magnetic field lines along with it. We first consider the motion of particles in the neighborhood of the Sun's equatorial plane, where we introduce polar coordinates r and φ. A particle with velocity v arrives in the time t at a distance $r = vt$ from the Sun. The Sun has, in the meantime, rotated through an angle $\varphi = \omega t$, where the angular velocity $\omega = 2\pi/(\text{sidereal rotational period})$. The particles which are emitted from a particular point on the Sun thus lie along an Archimedian spiral at a particular time:

$$\varphi = \frac{\omega r}{v}, \qquad (7.97)$$

which traverses a radius vector everywhere at the same angle, determined by

$$\tan \alpha = \frac{r \mathrm{d}\varphi}{\mathrm{d}r} = \frac{\omega r}{v}. \qquad (7.98)$$

With $v = 500 \text{ km s}^{-1}$, one obtains $\alpha \simeq 45°$ at the position of the Earth, in good agreement with observations.

The magnetic field lines follow these spirals. Since they must always be closed curves, they form loops whose beginnings or ends lie in regions of opposite magnetic polarity on the Sun. Such regions fill up a considerable fraction of the solar surface area, so that enormous field loops are continually pulled out into interplanetary space by the solar wind. Using observations from the satellite IMP-1 (*I*nterplanetary *M*onitoring *P*latform), J. M. Wilcox and N. F. Ness in 1965 first detected in the neighborhood of the Earth's orbit the *sector structure* of the interplanetary magnetic field (Fig. 7.26), which corresponds to the field distribution on the Sun. The number (usually two or four) and the distribution of the sectors is variable and corresponds to the arrangement of the field configuration on the Sun.

It is essential for an understanding of the variations in the sector structure to have a knowledge of the *three dimensional* distribution of the solar or interplanetary magnetic field in the *heliosphere*, the region in space filled by the solar wind. The first direct measurements *outside* the plane of the ecliptic, beginning in the mid-1970's with the space probes Pioneer 11 and Voyager 1 on their way to the outer Solar System, showed no signs of a sector structure at around 16° above the ecliptic, but rather magnetic field lines directed away from the Sun. With Ulysses, it then became possible in 1994/95 for the first time to observe the structure of the solar wind above the poles of the Sun and at other locations well outside the ecliptic.

Near the Sun's equator, there are many active regions with closed field lines; however, at higher heliographic latitudes, we find mainly *open* field lines, along which the solar wind streams outwards (Sect. 7.3.4). These field lines, which emerge from or enter the Sun's surface roughly perpendicular to it, are deflected towards the equatorial plane at a few solar radii away from the surface and thereafter run essentially parallel to it. Since the magnetic field (except for localized regions) exhibits opposite magnetic polarity in the northern and southern solar hemispheres, field lines of opposite direction run close to each other in the equatorial plane. They are separated by a thin neutral layer (in which, according to the law of induction, (7.76), electric currents flow). This *heliospheric current layer* has, particularly at maximum activity, an extremely complex folded and twisted structure, comparable to the "whirling skirt of a ballerina" (Fig. 7.27). Its shape is determined on the one hand by the inclination of the solar magnetic equator relative to the rotational equator, by asymmetries in the field distribution, and by field reversals in the course of the activity cycles; on the other hand, the layer is twisted by the rotation of the Sun and is transported outwards by the solar wind.

The boundaries of the sectors, as for example observed in the neighborhood of the Earth (Fig. 7.26), result finally from the intersection of the plane of the ecliptic, which is inclined by 7.25° relative to the equatorial plane of the Sun, with this rotating, deformed neutral current layer.

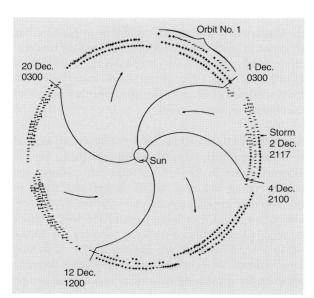

Fig. 7.26. The sector structure of the interplanetary magnetic field. Observations with IMP-1, from J. M. Wilcox and N. F. Ness (1965). The + and − signs indicate magnetic fields which are directed outwards or inwards, respectively. The portion of the field near the Sun is extrapolated schematically

The Heliopause. The heliosphere reaches out to roughly the point where the dynamic pressure of the wind becomes equal to the pressure of the interstellar medium. It is approximately spherical in shape, with a gentle "wasp waist" in the equatorial plane. The transition zone, the heliopause, lies at a distance of about 100 AU from the Sun and is reached by the solar wind, with its velocity of around 500 km s^{-1}, in around a year. Since the solar wind strikes the interstellar medium at

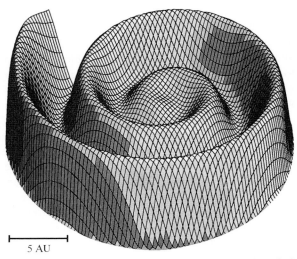

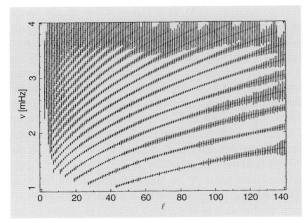

Fig. 7.27. The heliospheric current layer and the solar wind. The diameter of the region shown is approximately 25 AU and the distance between the turns is about 5 AU. The extended magnetic field of the Sun is directed e.g. outwards above the layer, inwards below it. These are calculations of the current layer for a constant solar wind, flowing outwards with a velocity of 400 km s^{-1}, according to the formula of J. R. Jokipii and B. Thomas (1981)

Fig. 7.28. Oscillations on the Sun: the power spectrum of the p-modes (acoustic waves) from observations by K. G. Libbrecht and M.F. Woodard. The frequency ν in [mHz = 10^{-3} Hz] is plotted as a function of the rank l of the spherical harmonic functions (number of node lines) or the horizontal wavenumber $k_h = \sqrt{l(l+1)}/R_\odot \simeq l/R_\odot$ (7.101). The discrete ridge lines each correspond to a particular number n of radial nodes with $n = 1$ for the lowest frequencies; the vertical marks represent 1000-fold enlarged error bars (1 σ). (With the kind permission of Nature **378**, 355, © Macmillan Magazines Ltd., and of the authors)

supersonic velocities, a shock wave forms: in a relatively thin, turbulent transition zone the density, temperature, and magnetic field change abruptly.

7.3.8 Oscillations: Helioseismology

Along with the irregular flow patterns of the granulation (Sect. 7.3.1), we can see in the Doppler shifts of the Fraunhofer lines nearly periodic, large-area *wave motions* in the photosphere (and the chromosphere). In 1960, R. B. Leighton discovered oscillations in the *radial* velocities with periods of 5 min or frequencies of $3.3 \cdot 10^{-3}$ Hz = 3.3 mHz (millihertz) and amplitudes from 0.1 to 500 m s^{-1}, where frequently regions of the surface of sizes $\leq 0.04\, R_\odot$ oscillate in phase. These Doppler motions are associated with small intensity variations.

The oscillations can be interpreted as acoustic waves, which are treated as standing waves in a limited layer of the Sun below the photosphere, like waves in a resonator. In 1975, F. L. Deubner succeeded in observing the characteristic patterns of oscillation as predicted by R. Ulrich in 1970, resolved with respect to their frequencies ν and (horizontal) wavelengths λ_h or wavenumbers $k_h = 2\pi/\lambda_h$ (Fig. 7.28). Since then, using more refined methods of observation, many further modes of oscillation have been found.

Methods of Observation. The most important method of observing solar oscillations is still based on the Doppler effect in the spectral lines; in addition, the corresponding intensity variations are also used. In order to measure frequencies and amplitudes very precisely, averaging over large parts of the solar surface is required, and also very long observation times, up to several months, are important. The first long observational series made use of the long antarctic summer. By combining the observations made at several stations distributed over the Earth (GONG: *G*lobal *O*scillation *N*etwork *G*roup), the precision could also be increased. Finally, since 1996, the solar observatory SOHO (Sect. 7.3.4) has permitted observations over long periods of time from space.

Acoustic Waves. As we saw in Sect. 7.3.1, in the upper layers of the turbulent convection zone, acoustic waves are produced. These cannot propagate in the interior of the Sun, since they are on the one side reflected by the steep density gradient on going outwards towards the photosphere; and on the other, in the lower convection zone, where the sound velocity c_s increases strongly on going inwards, they are again directed outwards due to refraction. This leads to the result that the layer between these two boundaries acts as a resonant cavity, in which for *particular* frequencies – below about 5.5 mHz – *standing* waves can be formed. For waves of longer wavelength, the "bottom" of the resonator lies deeper in the solar interior than for those of shorter wavelengths.

The power spectrum in Fig. 7.28 shows that the energy in the oscillations is not evenly distributed over ν and k_h, but instead is concentrated into discrete "ridge lines" which are determined by the standing wave condition.

Since in the case of acoustic waves, the compression of the gaseous medium is opposed by the pressure P, they are also termed *p-modes*. The velocity amplitude of the individual p-modes is extremely small, typically $0.01\,\mathrm{m\,s^{-1}}$; the corresponding relative intensity variation is just 10^{-7}. Only by superposition of a large number (around 10^7) of modes can stronger amplitudes of up to $500\,\mathrm{m\,s^{-1}}$ occur.

Gravity Waves. In addition to acoustic waves, gravity waves or g-modes can be excited in the Sun, whereby buoyancy or gravitational force provides the return force. The waves from the solar interior below the convection zone are, however, strongly damped by the time they reach the surface. Their amplitudes are only about 1/10 of those of the acoustic waves and they have not yet been detected with certainty.

Normal Modes. The analysis of the oscillations as standing and also the propagating waves is carried out in the form of a superposition of the *normal* oscillation modes or proper oscillations of a sphere of gas.

We first determine the period Π_0 of the fundamental oscillation or radial pulsation of a sphere of gas, e.g. of the Sun or a star, having the radius R, a mean density $\bar{\varrho}$ and a mass $\mathcal{M}$. Here, we cannot give the derivation from the basic equations of hydrodynamics, but will limit ourselves to an estimate of the order of magnitude. We interpret the pulsation as a *standing acoustic wave*. Its velocity is given according to (7.93) by $c_s = \sqrt{\gamma P/\varrho}$, where γ is the specific heat ratio, P the pressure, and ϱ the density. The mean pressure $\overline{P}$ in the gaseous sphere can be estimated by the central pressure P_c (3.7) from the hydrostatic equation (3.6), so that we obtain the order of magnitude $\overline{P} \simeq \bar{\varrho}\,G\mathcal{M}/R$. The period of oscillation is, again to an order of magnitude, given by

$$\Pi_0 \simeq \frac{R}{c_s} \simeq R\left(\gamma \frac{G\mathcal{M}}{R}\right)^{-1/2}. \quad (7.99)$$

With $\mathcal{M} = (4\pi/3)\bar{\varrho}R^3$, we find from this the important relation between the period and the mean density:

$$\Pi_0 \simeq \frac{1}{\sqrt{G\bar{\varrho}}}. \quad (7.100)$$

For the Sun, $\bar{\varrho}_\odot = 1410\,\mathrm{kg\,m^{-3}}$ (2.58), and thus $\Pi_{0,\odot} \simeq 1\,\mathrm{h}$. In contrast to the pulsating variable stars (Sect. 7.4.1), in the Sun the excitation of numerous high harmonics clearly predominates, with wavelengths $\ll R_\odot$ and with very small amplitudes.

Radial harmonics are also treated as standing acoustic waves and are characterized by the number n of oscillation nodes between the surface and the center of the Sun. Furthermore, *non-radial* oscillations are also possible. These are described in spherical polar coordinates θ and φ – corresponding to an expansion in spherical harmonic functions $Y_{l,m}(\theta,\varphi) = P_{l,m}(\cos\theta)\cdot\cos(m\varphi)$ – by the overall number l of node circles on the spherical surface and by the number m of node circles which pass through the poles (Fig. 7.29). Independently of this, the number n of radial nodes enters the description. There are correspondingly $l=0$ *radial* pulsations ($n=0$: fundamental oscillation), $l=1$ *dipole* and $l=2$ *quadrupole* oscillations.

For values of l which are not too small, we obtain for the horizontal wavenumber or the wavelength

$$k_h = \frac{\sqrt{l(l+1)}}{R_\odot} \simeq \frac{l}{R_\odot}, \quad \lambda_h \simeq \frac{2\pi R_\odot}{l}. \quad (7.101)$$

The energy in an oscillation $\{n,l,m\}$ in the spherically symmetric case does *not* depend on the azimuthal number of nodes m; there is thus an energy degeneracy analogous to that in the energy states of an atom with

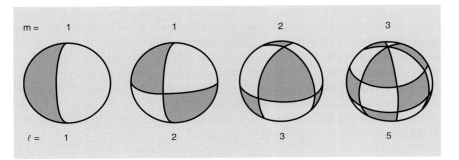

Fig. 7.29. Nonradial oscillations, represented by the spherical harmonic functions $Y_{l,m}$ for some selected l (number of node lines) and m (number of node lines through the poles) values. When e.g. the grey zones are oscillating outwards, the white regions are moving inwards and vice versa. (With the kind permission of R. Kippenhahn)

respect to the magnetic quantum number for the angular momentum (Sect. 7.1.1). This degeneracy is lifted by the rotation of the Sun (see below).

The Seismology of the Sun. In contrast to electromagnetic waves, the interior of the Sun is largely transparent to the mechanical waves produced in the convection zone. The propagation conditions for acoustic waves (velocity c_s, reflection, refraction) and thus also the "walls" of the resonant cavity are determined by the temperature and density layering within the Sun (Sect. 8.2.2).

Therefore, by conversely applying *helioseismology* (so termed by D. Gough in 1983), in analogy to earthquake waves, one can obtain information about the *interior* of the Sun below the photosphere, from which no electromagnetic radiation can reach us, by an analysis of the complicated temporally and spatially variable pattern of oscillation nodes and maxima on the Sun's surface, produced by the superposition of around 10^7 individual modes of differing wavelengths and frequencies. Thus, solar seismology led to the correction of our ideas about the lower limit of the hydrogen convection zone, which lies considerably deeper than was previously assumed. Furthermore, the analysis of the solar oscillations allows the determination of the *rotational* velocities in the interior of the Sun and of the extended, slow *matter flows* (of several $10 \, \text{m s}^{-1}$) down to 40 000 km below the solar surface.

The observation of oscillations with periods of about 5 min reveals p-modes with l in the range of around 10 to 2000 and n up to about 40, i.e. according to (7.101), of waves with horizontal wavelengths of about 0.003 to 0.6 $R_\odot$. A particularly deep view into the solar interior is offered by radial oscillations with *small l*, and thus with long wavelengths or periods. These can be observed by integrating over large areas of the solar surface, so that oscillations with shorter wavelengths are for the most part averaged out.

Rotation. At the Sun's *surface*, we can determine its rotation from the motion of sunspots (Sect. 7.3.3). The period Π_{rot} increases from 24.8 d at the equator to about 31 d at higher latitides; this corresponds to a decrease of the angular velocity Ω or the frequency $\nu_{\text{rot}} = \Omega/(2\pi) = 1/\Pi_{\text{rot}}$ from about 460 to 370 nHz (nanohertz). The rotational frequency is thus considerably lower than the frequencies of the oscillations. A wave propagating in the direction of the rotation has a higher frequency relative to the non-rotating case by $m \cdot \nu_{\text{rot}}$; a wave propagating in the opposite direction has a correspondingly lower frequency (m is the number of azimuthal nodes). Since these very small splittings in the modes can still be measured, it is possible by means of seismology to derive the variation of the rotation even in the *interior* of the Sun. However, measurements at small m, which would correspond to deep layers and high latitudes φ, are extremely difficult.

The rotation of the Sun in the whole hydrogen convection zone (Sect. 7.3.1) is very similar to that on the surface and shows a similar variation with latitude (Fig. 7.30). Just below the convection zone, the Sun rotates with an approximately constant angular velocity of 440 nHz, corresponding to an average of the values in the outer layers.

Asteroseismology. In principle, the methods of solar seismology can also be applied to the *stars*; the first of the extremely difficult observations (such as e.g. those of the G0 IV star η Boo) have, to be sure, not yet been

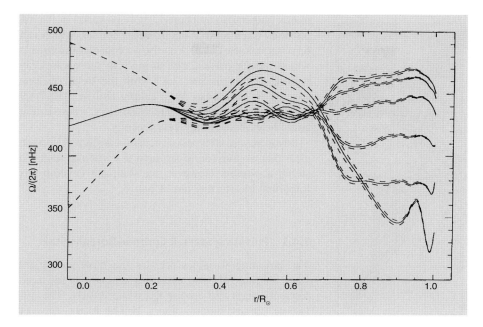

Fig. 7.30. Solar rotation: the dependence of the rotational frequency $\nu_{\rm rot} = \Omega/(2\pi)$ on the radius $r/R_\odot$ and on the heliographic latitude, after J. Schou and the SOI Internal Rotation Team (1998). $\nu_{\rm rot}$ is in [nHz = 10^{-9} Hz]; curves are shown (*from above to below*) for the latitudes 0°, 15°, 30°, 45°, 60° and 75°. The *dashed* lines give the 1σ error limits. (With the kind permission of the authors and of the International Astronomical Union)

verified. With asteroseismology, in particular the observation of the p-modes of low rank, we are standing on the threshold of a promising new area of stellar astrophysics.

7.4 Variable Stars. Flow Fields, Magnetic Fields and Activity of the Stars

The first observations of variable stars at the turn of the 16th to the 17th centuries represented at the time a weighty argument against the Aristotelian dogma of the immutability of the heavens. Tycho Brahe's and Kepler's discoveries of the supernovae of 1572 and 1604 have continued to contribute to knowledge of these mysterious objects even in our day and have permitted the radioastronomical identification of their remnants. Fabricius' discovery of the Mira Ceti should also be mentioned in this connection.

Variable stars are denoted by capital letters, R, S, T, ... Z, and the genitive case of the name of the constellation; these are followed by RR, RS, ... ZZ, AA ... AZ, BB ... QZ, without using J. After these 334 possible combinations have been used, one employs V335 etc. (V = variable star), together with the genitive of the constellation.

It should be clear from the outset that the investigation of variable stars promises us much deeper insights into stellar structure and evolution than that of the static stars, which are "eternally the same". On the other hand, the observation and theory of variable stars presents much greater difficulties. A warning against easy, *ad hoc* hypotheses is not inappropriate here.

It is impossible within the scope of this book for us to describe the numerous classes of variable stars (usually named for a prototype) with any degree of completeness. We shall leave aside the eclipsing variables, which have already been discussed, and consider a few interesting and important types of *physically* variable stars which we group according to common physical aspects of their descriptions. To gain a better overview, we summarize the most important groups in Fig. 7.31 schematically with their positions in the Hertzsprung–Russell diagram.

We first treat the pulsating variables (Sect. 7.4.1) and the magnetic variables with their variable spectra (Sect. 7.4.2). After having introduced the outer layers of the solar atmosphere and solar activity in Sect. 7.3,

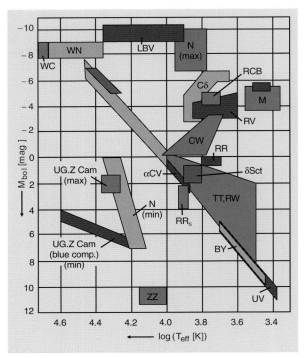

Fig. 7.31. Schematic positions of some types of variable stars in the Hertzsprung–Russell diagram (M_{bol}, T_{eff}), after H. W. Duerbeck and W. C. Seitter

αCV	α CVn variable	RRs	Dwarf Cepheids
BY	BY Dra stars	RV	RV Tau stars
Cδ	Classiscal Cepheids	RW	RW Aur stars
CW	W Vir stars	TT	T Tau stars
δSct	δ Sct stars	UG	U Gem stars
LBV	Luminous Blue Variables	UV	UV Cet stars
M	Mira variables	WC, WN	Wolf-Rayet stars
N	Novae	Z Cam	Z Cam stars
RCB	R CrB stars	ZZ	ZZ Cet stars

we now discuss in Sects. 7.4.3, 4 the indications for activity, chromospheres, coronas, etc. in *stars* of different classes. The cataclysmic variables, to which the novae belong (Sect. 7.4.5) and the very diverse stellar X-ray sources (Sect. 7.4.6) represent quite different groups of variable stars. The phenomena observed in these two groups are due to matter flows within close binary systems, which are "pulled in" by a compact star component (a white dwarf or neutron star). Then, in Sect. 7.4.7, we turn to the spectacular phenomenon of the burst of brightness observed in a supernova, which is accompanied by the casting off of a shell of stellar matter (supernova remnants). These powerful stellar explosions often leave a neutron star behind, which can be observed as a pulsar. Finally, we consider the objects which radiate particularly strongly in the gamma-ray region, dealing in Sect. 7.4.8 with the stellar gamma sources in our Milky Way galaxy and in Sect. 7.4.9 with the extremely bright gamma bursters, whose nature could be explained only very recently.

In discussing all types of variable stars, it will become clear that we must consider them as particular stages in stellar *evolution*. However, we shall not develop this important idea further until Chap. 8.

7.4.1 Pulsating Stars. R Coronae Borealis Stars

The pulsating variables are, for the most part, giant stars, but there are also pulsating stars among the main sequence stars and among the white dwarfs. The following groups, among others, belong to the class of pulsating variables; we also note here the star populations in the Milky Way Galaxy to which they belong (Sect. 11.2.6):

RR Lyrae stars or *cluster variables*. These are stars with regular changes in brightness having periods of about 0.2 to 1.2 d, brightness amplitudes of the order of 1 mag (ranging from about 0.4 to 2 mag), spectral types A and F, and masses from 0.5 to 0.6 $\mathcal{M}_\odot$. They are found in the halo and core of the Milky Way galaxy and are important in the globular clusters.

δ Cephei Stars (classical Cepheids). Stars of high luminosities (class Ia to II) which also exhibit very regular brightness changes having periods from 1 to 50 d and about the same brightness amplitudes as the cluster variables (0.1 to 2 mag), but belonging to later spectral types (F5–K5) and having masses between about 5 to 15 $\mathcal{M}_\odot$. They occur in the spiral arms of the galaxy.

W Virginis stars with very similar properties to the δ Cep stars, but weaker in absolute magnitude (by 1 to 2 mag) and with low masses (0.4 to 0.6 $\mathcal{M}_\odot$). They are found in the halo and core regions of the Milky Way.

Dwarf Cepheids and *δ Scuti stars*. Short-period variables near the main sequence, of spectral types A and F (masses 1 to 2 $\mathcal{M}_\odot$), having periods between 0.03 and 0.2 d, and brightness amplitudes of 0.3 to 0.8 mag in the case of the dwarf Cepheids and, as a rule, ≲ 0.1 mag in the δ Sct variables.

ZZ Ceti stars. White dwarfs of spectral type DA with very short periods in the range from 3 to 20 min and small amplitudes between 0.01 and 0.3 mag.

Mira variables or *long-period variables* are all giant stars of the later spectral types (M, C, and S), usually with emission lines; Mira Ceti = o Cet (M7IIIe) is a member of this group. The light curves are not as stable as in the case of the Cepheids; they have periods of from 80 d up to more than 500 d, and large brightness amplitudes of more than 2.5 to about 8.8 mag in the visible region. Their masses are of the order of 1 $\mathcal{M}_\odot$, and their radii range from 100 to 1000 $R_\odot$. Mira variables occur both in the young and in the old stellar populations of the Milky Way.

RV Tauri stars. Bright giants and supergiants with spectral types F to K having alternately deep and shallow minima in their light curves, periods between about 30 and 150 d, and amplitudes up to 3 mag.

Semiregular variables. Giants to supergiants of medium or late spectral types ($\gtrsim$ F) with quasi-periods in the range of 30 to over 1000 d.

Radial Velocity Curves. The first indication of the physical nature of the groups of variable stars described here was given by their radial velocity curves. These are closely connected to their light curves. Initially, it was attempted to attribute the very regular velocity fluctuations of e.g. the classical Cepheids to a binary star motion. Integration over the radial velocity yields (without further hypotheses) the dimensions of the "orbit", since (x = coordinate in the direction of the line of sight):

$$\int_{t_1}^{t_2} V_r \, dt = \int_{t_1}^{t_2} \frac{dx}{dt} \, dt = x_2 - x_1 . \qquad (7.102)$$

It soon became apparent, however, that the star would have no room in this orbit alongside the postulated companion star. Thus, in 1914, H. Shapley returned to the possibility of a radial *pulsation* of stars, which had been discussed in the 1880's by A. Ritter as a purely theoretical problem. The pulsation theory of the Cepheids (and related variables) was then developed further in 1917 by A. S. Eddington. This, in turn, gave the impulse for his pioneering work on the internal structure of stars (Chap. 8).

The period Π_0 of the pulsation of a sphere of gas was already estimated in Sect. 7.3.8 and the important relation $\Pi_0 = 1/\sqrt{G\overline{\varrho}}$ (7.100) with the mean density $\overline{\varrho}$ was obtained there. This relation has been well verified by comparison with extensive observational results.

Light Curves. A second test of the pulsation theory was suggested by W. Baade: the brightness of a star is proportional to the area of its "disk", πR^2, times the radiation flux F_λ at its surface. The temporal variation of the stellar radius R can be obtained directly by integrating the radial velocity curve as in (7.102). On the other hand, F_λ can be determined independently from the theory of stellar atmospheres using the color indices or other spectroscopic criteria. (We shall apply this method later in Sect. 12.1.1 to the determination of the distances of galaxies).

Figure 7.32 summarizes some of the fundamental properties of δ Cep stars, with their time variations. The expected proportionality of the observed magnitudes to $R^2 F_\lambda$ is indeed well fulfilled.

Henrietta S. Leavitt at the Harvard Observatory in 1912 discovered a relation between the periods and, initially, the apparent magnitudes m_V of the many hundred cepheid variables in the Magellanic Clouds. Since all these stars have the same distance modulus, she had actually discovered a *period–luminosity relation*. H. Shapley determined the zero point of the scale of *absolute* magnitudes using the modest amount of observational data of proper motions then available (Sect. 6.2.2). It thus became possible to determine the *distance* to every cosmic object in which Cepheids could be found (the problem of interstellar absorption was as yet unknown). For example, H. Shapley in 1918 was able to determine the distances to many RR Lyr stars in the globular clusters for the first time and thus to fix the boundaries of our galaxy in the modern sense. Then in 1924, E. Hubble, using the same methods, but employing classical Cepheids (with longer periods), determined the distances to some of the neighboring spiral nebulae and showed definitively that they are *galaxies* of a similar scale to our Milky Way. We shall discuss this "penetration of deep space" in Sects. 12.1.1 and 13.1.1. Here, however, we should consider an important correction to the fundamentals of the cepheid method, discovered about 1952 by W. Baade. He was able to show that the zero point of the period–magnitude relation is different for different types of Cepheids. In particular, the classical Cepheids of Population I (cf. Sect. 11.2.6) are 1 to 2 magnitudes brighter than the W Virginis stars of Population II with the same period.

The *amplitude* of the light variation, measured e.g. in visual magnitudes m_V, increases systematically on going to cooler stars. This is essentially based on Planck's radiation law. If we write m_V in Wien's approximation, analogously to (6.28), we find:

$$m_V = \frac{1.56 \cdot 10^7}{\lambda_V \, [\text{nm}] \, T \, [\text{K}]} + \text{const}_V \, . \tag{7.103}$$

A given temperature variation ΔT thus corresponds to a brightness amplitude of

$$\Delta m_V = -\frac{1.56 \cdot 10^7}{\lambda_V T^2} \Delta T \, , \tag{7.104}$$

which for cooler stars increases $\propto 1/T^2$.

Pulsation Theory. An interesting theoretical problem is the *conservation* of pulsation: what "valve" guarantees that the stellar oscillations, like the pistons of a thermal engine, are always pushed with the right phase? The production of thermal energy by nuclear processes near the center of the star is practically not influenced by the pulsation. Rather, the critical factor is the temperature and pressure dependence of the opacity (7.41), which regulates the flow of radiation energy and thus determines the temperature of a particular layer. This "κ mechanism" is found to be particularly effective, together with the change in the adiabatic temperature gradient, in the region of the second ionization of helium. Similar (very difficult) calculations can make it clear theoretically which combination of fundamental properties, i.e. which regions of the Hertzsprung–Russell diagram, permit the existence of pulsation. Thus, one can understand that the δ Sct stars and the dwarf Cepheids as well as the R R Lyr, W Vir- and δ Cep stars are all found close to an *instability band* which stretches from an effective temperature of about 8000 K near the main sequence "upwards and to the right" to 5000 K at the Cepheids (Fig. 7.31).

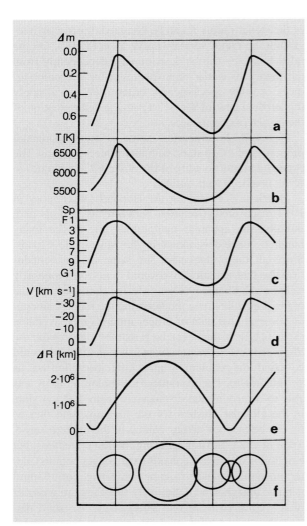

Fig. 7.32a–f. Periodic variations of δ Cephei. From above to below, the curves represent (**a**) the brightness (light curve in [mag]), (**b**) the color temperature, (**c**) the spectral type, (**d**) the radial velocity, (**e**) the changes in radius $\Delta R = R - R_{\min}$, and (**f**) the stellar disk

Long-Period Variables. The cooler pulsation variable stars with longer periods of light variation, such as the R V Tauri variables and long-period variables, have an increasingly irregular light variation. The theory of inner stellar structure (Chap. 8) shows that in cooler stars, the hydrogen convection zone becomes more and more extended. It is therefore tempting to attribute the observed semi-regular light variation to a coupling of the pulsation with the turbulent flow processes of convection.

The shells of the *Mira variables* expand with velocities of the order of 10 km s^{-1} as can be determined from measurements of the radial velocities of absorption and emission lines. The stars lose a corresponding amount of mass at a rate of about 10^{-8} to $10^{-6}\,\mathcal{M}_\odot$ yr^{-1}. From some Mira variables, lines in the radiofrequency region can also be observed, in particular the *maser* emission of the *OH radical* at $\lambda = 18$ cm (Sect. 10.2.3), where the 1612 MHz component is strongest (type II maser). The characteristic double structure of this component also indicates an expanding motion of the gas shell.

Radio surveys at $\lambda = 18$ cm, searching for OH sources, together with infrared observations, have led to the discovery of optically invisible *OH/IR stars*. These cool stars, which are hidden behind a thick shell of dust, exhibit brightness variations in the infrared corresponding to the type of the Mira variables; the periods extend up to 2000 d. The period–luminosity relation of the OH/IR stars indicates that we can take them to represent a continuation of the Mira variables. Their mass loss rates of about 10^{-5} to $10^{-4}\,\mathcal{M}_\odot$ yr^{-1} are greater than those of the Mira stars.

R Coronae Borealis Stars. Within the group of the red giant stars, there is quite a different kind of slow variables, the RCrB stars. Their brightness decreases suddenly by several magnitudes from a constant normal value, and then slowly returns to the normal value. Spectral analysis of these relatively cool stars shows that their atmospheres have low hydrogen contents, but high contents of carbon (and probably helium). One might suspect that the R CrB stars at times emit clouds of colloidal carbon, i.e. a sort of soot cloud, which darken the star for a while.

7.4.2 Magnetic or Spectrum Variables. Ap Stars and Metallic-Line Stars

In the region of the main sequence, various types of stars are found which do *not* fit into the two-dimensional MK classification. They are all characterized by peculiarities in their spectra. The hotter Ap(Bp) stars are for the most part variable, while the cooler metallic-line stars are not. Whether these two groups have anything in common remains an open question.

Ap Stars. The (cooler) Ap stars (peculiar A stars) or *spectrum variables* exhibit anomalous intensities and a periodic variation in the intensities of certain spectral lines, with different spectral lines showing different behaviors. The prototype is α^2 *Canum Venaticorum* with a period of 5.5 d, in which the lines of Eu II and Cr II change with reversed phase, while e.g. the Si II and Mg II lines remain nearly constant. Along with the spectral changes, there are usually brightness variations of around 0.1 mag. H. W. Babcock was able to show by Zeeman effect measurements (Sect. 7.3.2) that these stars have magnetic fields of 0.1 to 1 T (10^3 to 10^4 G), which exhibit periodic changes in intensity and often even in sign. According to A. Deutsch, at least the major part of the observations can be explained by the hypothesis that these stars possess enormous magnetic *spots*, in which, depending on their polarity, one or another group of spectral lines is intensified. The variations observed are attributed to the *rotation* of the stars.

In the color–magnitude diagram, the Ap stars lie on or near the main sequence roughly in the region $-0.20 \lesssim (B-V) \lesssim +0.20$ mag corresponding to effective temperatures between about 18 000 to 8000 K. A large number of the Ap stars are thus in fact Bp stars, which were originally classified as A stars due to their characteristic weak He I lines. In general, however, the classification as Ap stars has been retained for all the members of this group.

Among the late B and A stars, there are at least 10 to 15% Ap(Bp) stars. These are subdivided into the following groups, according to their most noticeable spectral anomalies: (a) at the cooler end of the region are the *Eu-Cr-Sr stars*, which extend towards the higher effective temperatures ($T_{\text{eff}} \gtrsim 12\,000$ K) of the *Si stars*; among the latter are e.g. α^2 CVn. (b) At temperatures similar to those of the Si stars are the *Hg-Mn stars*, which however, in contrast to the magnetic Eu-Cr-Sr- and Si stars, show no observable magnetic fields and are not spectral variables. (c) The *weak-helium-line stars*, with less noticeable anomalies, represent the continuation of both groups towards the highest effective temperatures up to about $\lesssim 20\,000$ K.

In some Ap stars, unusual anomalies are observed: for example, a subgroup of the weak-helium-line stars exhibits the lines of ^{3}He with comparable intensities to those of ^{4}He, but somewhat shifted in wavelength.

In some of the Eu-Cr-Sr stars, lines of otherwise rare elements such as Os I and II and Pt II and perhaps also U II occur with unexpectedly strong intensities. In the spectrum of HR 465, the radioactive promethium isotope ^{145}Pm, with a half-life of 17.7 yr, has been identified. This can be regarded as an indication that the anomalous element abundances could in part be a result of neutron irradiation. On the other hand, most of the abundance anomalies of the Ap stars, and of the metallic-line stars (see below), can hardly be understood as resulting from nuclear processes. F. Praderie, E. Schatzman and G. Michaud (1967/70) have therefore suggested an explanation on the basis of *diffuson* processes, by which the elements can be separated by the interplay of selectively-acting radiation pressure as a "buoyancy force", and sedimentation resulting from the gravitational acceleration, in layers of these stars near to their surfaces. The details, which are by no means understood, probably depend sensitively on the strength of turbulence or convective mixing and of magnetic fields.

Metallic-Line Stars. Along the main sequence going to cooler temperatures, the non-variable metallic-line stars (Am stars) border on the Ap stars, with a considerable region of overlap. They are mostly members of binary star systems, and their rotational velocities are lower ($\lesssim 100$ km s^{-1}) than those of the normal A stars. Strong magnetic fields are not observed. About 10% of the (brighter) A stars are metallic-line stars. They are classified according to their hydrogen lines, for example as A0 and F1. On the basis of their hydrogen types, these stars lie on the main sequence. However, the lines of calcium (especially H and K) and/or scandium are too weak, while the metal lines of the iron group and the heavier elements are too strong. Precise analysis yields effective temperatures between about 7000 and 10 000 K and show, by comparison of lines from various ionization and excitation stages, that the *abundances* of the elements are indeed anomalous, rather than the observed line intensities being due to some sort of anomalous excitation processes (deviations from LTE).

Both the Ap and the Am stars occur on the main sequence of relatively young (only 10^6 to 10^7 yr old) star clusters and groups. This also indicates that the anomalies in these stars must have been produced quickly and in the immediate neighborhood of the main sequence.

7.4.3 Activity, Chromospheres, and Coronas of Cool Stars

In Sect. 7.3, we found that the Sun can be considered to be a variable star, with its $2 \cdot 11$ year magnetic activity cycle. The manifold phenomena of solar activity are, in the final analysis, based on the interaction of the *rotation* of the Sun with the velocity and magnetic fields of the *hydrogen convection zone*; this interaction gives rise to a dynamo process which maintains the activity cycle. The temperature increase towards the outer layers of the solar atmosphere, the chromosphere and the corona, is also due to the convection zone, in which acoustic and magnetohydrodynamic waves are generated which then give rise to heating of the upper layers. On the other hand, the theory of the inner structure of stars shows that all cooler stars, later than about F0 ($T_{\text{eff}} \lesssim 6500$ K), must have more or less extensive hydrogen convection zones near their surfaces. It is thus tempting to search for indications of chromospheres and coronas and for solar-like activity phenomena in other stars from this region of the Hertzsprung–Russell diagram.

Since about 1970, it has been possible through observations in the ultraviolet and X-ray regions from satellites to investigate the chromospheres, coronas and stellar winds of various types of stars on a large scale.

Chromospheric Activity. On the Sun, the active regions or their torch spots are characterized by Ca II H and K emission lines, sometimes superposed on a finer absorption structure. Exactly the same spectral features are observed in many main sequence and giant stars of types G through M. This chromospheric emission or stellar activity, discovered by K. Schwarzschild and G. Eberhard in 1913 and later thoroughly investigated by O. C. Wilson, often exhibits variations with time, which are an indication for the rotation or an activity cycle in the star.

Then in 1957, a completely unexpected phenonenon was observed by O. C. Wilson and M. K. V. Bappu: the *width* of the Ca II emission lines was found (independently of their intensity) to be a function of the *absolute* magnitude of the stars. This provides an excellent method for the determination of the spectroscopic parallax of cooler stars, which is valid over 15 magnitudes or 6 powers of ten in the luminosity. Just why the velocity of the turbulence in the chromospheres of the cooler main sequence and giant stars is related to their other atmospheric parameters in the observed manner remains for the most part an unanswered question.

From the work of O. C. Wilson, R. P. Kraft, A. Skumanich and others it has become clear since the 1960's that there is a connection between the chromospheric activity or Ca II emission with the *rotation* of the stars, and, through comparisons of the stars in stellar clusters of differing ages, with their *ages* also. With increasing stellar ages (on the average), both the rotational velocities and the activities of stars decrease. This is true not only of chromospheric emission, but also of all the other indicators of stellar activity, such as flares and X-ray emission (see below). Young, rapidly rotating stars show a very irregular activity; in contrast, older, slowly rotating stars exhibit activity in cycles similar to those observed on the Sun. In the course of the evolution of a cool star, angular momentum is transferred to the interstellar medium by the outward streaming of plasma with its magnetic fields, similar to the solar wind, so that the star's rotation is slowed down. This in turn causes the dynamo process which maintains the stellar activity cycle and produces the star's magnetic field to become less effective, and the activity phenomena, which depend essentially on the magnetic field, thus decrease.

Starspots. In several cool dwarf stars, G. E. Kron (1952) observed small periodic brightness variations, which he attributed to starspots, analogous to sunspots. Following B. V. Kukarkin and others, we refer to this type of rotating stars as *BY Draconis variables*. They include main sequence stars of the later spectral types (with emission lines) and quasiperiodic brightness variations ($\lesssim 0.6$ mag). Their periods lie in the range from several tenths of a day up to a few days; the amplitudes may change in the course of some years.

Flare Stars. Many cool dwarf stars, usually of type M with hydrogen emisson lines, also show flares at irregular intervals; these differ from those observed on the Sun only in their sometimes greater brightness. The optical brightness of the star increases within 3 to 100 s by amounts of 6 to 7 mag; the decrease back to normal magnitude is considerably slower. Smaller fluctuations in brightness merge into nearly continuous variations. The spectra of these flare stars, or, as they are named for their prototype, the *UV Ceti stars*, show a superposed continuum in the ultraviolet during a flare outburst, so that the brightness in this region may increase up to 10 mag. They also exhibit strong emission lines from Ca II, H I, He I, and even He II as in solar eruptions. However, while on the Sun the continuous spectrum of the flare reaches at most a few percent relative to the brightness of the photosphere, this ratio reverses in cool stars ($\lesssim 4000$ K), since the radiation flux from their photospheres is weaker by several magnitudes. Accompanying an optical flare, B. Lovell et al. discovered in 1963 the corresponding radio flares in the meter wavelength range. Likewise, in the X-ray region, outbursts of flare stars amounting to 10^{23} to 10^{24} W were observed from the Einstein satellite; they showed a great similarity to solar flares.

Ultraviolet Spectra. In many stars of the later spectral types, we find, as in the Sun, ultraviolet emission lines such as H I Lα at $\lambda = 121.6$, O I at $\lambda = 130.4$, C I at $\lambda = 165.7/156.1$, Si II at $\lambda = 180.8/181.7$ and Mg II (h and k) at $\lambda = 279.6/280.3$ nm, which are of chromospheric origin. These occur together with lines of higher ionization states such as Si IV at $\lambda = 139.4/140.3$, C III at $\lambda = 97.7/117.5$, C IV at $\lambda = 154.8/155.1$ or N V at $\lambda = 123.9/124.3$ nm, which indicate temperatures above $3 \cdot 10^4$ K and are formed in the transition layer to a corona or in a cool corona (Fig. 7.33). In contrast, in the cooler, more luminous stars such as α Ori (M2 I ab), the lines from highly excited states are lacking, so that their chromospheres at temperatures $\lesssim 2 \cdot 10^4$ K probably merge directly, with no hot transition zone, into cool, massive *stellar winds* having relatively low velocities (see below).

Mass Losses. Based on the *blue* shifted circumstellar absorption components of strong lines such as the Ca II H- and K lines in the spectra of cool, luminous giants, A. J. Deutsch recognized in 1956 that the expansion velocities of the extended shells are greater than the escape velocities; thus, these stars give off matter into the interstellar medium. The rate of mass loss $\dot{M}$ can now be derived not only from the optical circumstellar lines, but also among other possibilities from the infrared radiation of the dust in the shell as well as from lines of OH, H$_2$O or CO in the radiofrequency region.

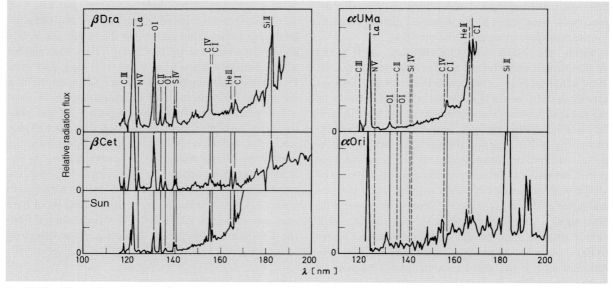

Fig. 7.33. Ultraviolet spectra of some cooler stars (observations with the International Ultraviolet Explorer by J. L. Linsky and B. M. Haisch, 1979): β Dra (G2 II), β Cet (K1 III), α UMa (K0 II–III), α Ori (M2 Iab) and, for comparison, the quiet Sun. The positions of solar spectral lines which are not present in the stellar spectra are indicated by dashed vertical lines

According to D. Reimers (1975), the order of magnitude of the mass loss from cool giants and supergiants is given by:

$$\dot{\mathcal{M}} [\mathcal{M}_\odot \text{ yr}^{-1}] = 4 \cdot 10^{-13} \frac{L/L_\odot}{(g/g_\odot)(R/R_\odot)} \quad (7.105)$$

which relates it to the fundamental stellar parameters luminosity L, radius R, and surface gravity, $g = G\mathcal{M}/R^2$. The rates of mass loss for the coolest supergiants attain 10^{-7} to $10^{-5}\,\mathcal{M}_\odot$ yr^{-1}; for the OH/IR stars, they go as high as $10^{-4}\,\mathcal{M}_\odot$ yr^{-1}. On the other hand, a weak stellar wind comparable to the solar wind ($\dot{\mathcal{M}} \simeq 2 \cdot 10^{-14}\,\mathcal{M}_\odot$ yr^{-1}), as expected for cool main sequence stars, lies below the current detection limit.

The (generally variable) stellar winds are expected to be due, as in the Sun, to the transport of energy by acoustic and magnetohydrodynamic waves from the hydrogen convection zone. In the case of the highest mass loss rates, acceleration by radiation pressure from the star probably also plays a role; it acts on the circumstellar dust particles with their large absorption cross sections.

Coronas and Stellar Winds. In the X-ray region, corresponding to its corona temperature of several 10^6 K, the Sun emits at a rate of 10^{20} W (quiet corona) up to $2 \cdot 10^{22}$ W (many active regions). Only with the sensitivity of the Einstein satellite and following X-ray satellites did it become possible to detect *stellar* coronas of a similar kind by means of their X-ray emissions. The sky survey carried out by ROSAT indicated coronas for several 10 000 stars.

The X-ray luminosity L_X of the cool *main sequence stars* of spectral types G to M is in the range 10^{19} to 10^{21} W, and is thus quite comparable to that of the Sun at an "average" activity level. While the X-ray emission of a G dwarf makes up only about 10^{-7} of the overall radiation L, the relative amount in the case of M dwarfs is noticeably higher, 10^{-3} to 10^{-2}, due to their much smaller total luminosities. The strong scatter of the observed values for a given spectral type indicates that L_X depends not only on the effective temperature, i.e. on the existence of a hydrogen convection zone. Indeed, we find, as for the chromospheric activity, that the X-ray luminosity depends strongly on the *rotation* of the star, as well as on its *age*. Independently of L, the following approximate relation holds:

$$L_X \, [\text{W}] \simeq 10^{21} \cdot (v \sin i \, [\text{km s}^{-1}])^2 \quad (7.106)$$

up to a "saturation value" $L_X/L \simeq 10^{-3}$ ($v \sin i$ is the projected rotational velocity, cf. (6.45)).

The stars in the group of the cooler *giants* and supergiants can be divided into *two classes* depending on their X-ray emissions; these are separated by a relatively sharp dividing line in the Hertzsprung–Russell diagram at early K stars or $T_{\text{eff}} \simeq 4600$ K. To the left of this dividing line, we find stars with ultraviolet emission lines corresponding to a temperature around 10^5 K (Fig. 7.33) and with very high luminosities L_X in the soft X-ray region, of up to $3 \cdot 10^{23}$ W. In contrast, stars to the right of the line show no corresponding emission lines in the ultraviolet and *no* X-ray emission ($L_X < 3 \cdot 10^{18}$ W). The limit of detection corresponds to a value of L_X/L which is well below that of the quiet Sun ("coronal holes"). On the other hand, these stars have massive, cool ($T \leq 2 \cdot 10^4$ K) *stellar winds* with high rates of mass loss $\dot{M}$ in the range of 10^{-10} to $10^{-7} \, \mathcal{M}_\odot \, \text{yr}^{-1}$ and velocities of only 20 to 100 km s^{-1}. The structure of these stellar winds is thus clearly different from that of the solar wind.

Many variables of the type *RS Canum Venaticorum* are detached, close binary systems, mainly having periods of less than 30 d and with late supergiants and dwarfs as components. Although from their age, only a weak coronal emission would be expected, the RS CVn stars exhibit variable, strong emissions in the soft X-ray region of $L_X \simeq 10^{22}$ up to $3 \cdot 10^{24}$ W as well as indications of chromospheric activity and starspots. Their activity is apparently also based on their rapid rotation, which is syncronized with their orbital motions by tidal interactions.

Pre-Main-Sequence Stars. Strong activity or variablilty, which shows similarities with those of the Sun and other cooler dwarf stars, are also found in *very young* stars of spectral types F to M, which have not yet evolved onto the main sequence and lie above it in the Hertzsprung–Russell diagram (Sect. 10.5.1). The *T Tauri* or *RW Aurigae stars*, named for their prototypes, are characterized spectroscopically by strong emission lines, especially from Ca II H and K, Hα and other Balmer lines of hydrogen. Their luminosities "flicker" in an irregular manner, of the order of 1 mag within a few days; many are strong X-ray sources ($\leq 5 \cdot 10^{24}$ W), whereby the somewhat older, "classical" T Tau stars, which are still surrounded by an accretion disk dating from their formation, have roughly a factor of 3 weaker X-ray luminosities than the younger, "weak-line" T Tau stars. W. A. Ambarzumian, G. Haro, and others have shown that these variables are to be found in the sky particularly in the neighborhood of the dark clouds and young star clusters. These are stars which were formed from interstellar matter relatively recently ($\leq 4 \cdot 10^8$ yr). Along with them, a large number of flare or UV Cet stars are found.

In rare cases, strong luminosity outbursts can occur in T Tau stars, which are accompanied by ejection of matter (FU Orionis phonomenon). For example, in 1969, the luminosity of V 1057 Cyg increased by 6 mag in about 300 d, and then dropped back very slowly. Shortly after the outburst, the spectrum of V 1057 Cyg was similar to that of an A supergiant.

7.4.4 Coronas, Stellar Winds and Variability of Hot Stars

Many, probably in fact all *supergiants* show irregular brightness variations of the order of 0.1 to 0.2 mag as well as variations in the radial velocities of their Fraunhofer lines up to several km s^{-1} with typical time scales of days to months; for example, α Cyg (A2 Ia) has been known since 1896 to be a spectral variable. In the brighter supergiants, emission lines are also observed and are also variable, such as Hα and other Balmer lines, He II at $\lambda = 468.6$ nm, or (in the Of stars) N III lines around $\lambda = 463$ nm.

The rare supergiants of highest absolute magnitudes, with M_v in the range from -7.5 to -9.5 mag and spectral types between B and F (hypergiants), the *Luminous Blue Variables* (LBV), which include the Hubble-Sandage variables as well as variables of the S Dor type (in the Large Magellanic Cloud), η Car (Fig. 10.8) and P Cyg, exhibit a wide range of brightness variations, ranging from rapid, small variations (of a few 0.01 mag) on a time scale of hours, up to slow changes of the order of 1 mag in years to decades and occasional stronger bursts of brightness. Their spectra are rich in emission lines, which frequently show a *P Cygni profile*: a broad emission with an absorption component

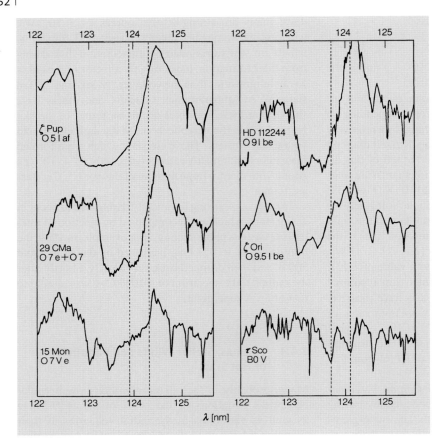

Fig. 7.34. Stellar winds and mass losses in OB stars: as an example, the P Cyg profile of the resonance doublet of N V at $\lambda = 123.88/124.28$ nm is shown (the wavelengths at rest are indicated by dashed lines). Observations made with the Copernicus Satellite by T. P. Snow and D. C. Morton (1976). (With the kind permission of the International Astronomical Union)

shifted to the short-wavelength side (Fig. 7.34), which can be attributed to *ejected* matter. The velocities reach several $100\,\mathrm{km\,s^{-1}}$, and the mass loss caused by this stellar wind is considerable ($\gtrsim 10^{-5}\,\mathcal{M}_\odot\,\mathrm{yr}^{-1}$).

Stellar Winds. Although in the optical region, only the particularly dense, massive stellar winds from the most luminous stars can be recognized on the basis of their emission lines or P Cygni profiles, the far *ultraviolet* offers the advantage that here the *resonance lines* of many ions of abundant elements are to be found; therefore, relatively low densities of matter can be detected. The first ultraviolet observations with rockets by D. C. Morton and coworkers (1967) already indicated strong P Cygni profiles from some brighter stars. Using more sensitive spectrographs, in particular those of the Copernicus and the IUE satellites, it became clear that stellar winds or *mass losses* occur in all OB stars with luminosities $L \gtrsim 10^4\,L_\odot$.

The mass-loss rate $\dot{\mathcal{M}}$ due to a stationary wind is given by the equation of continuity, (7.88):

$$\dot{\mathcal{M}} = 4\pi r^2 \varrho(r) v(r) = \mathrm{const}\,, \qquad (7.107)$$

where $\varrho(r)$ is the density and $v(r)$ the wind velocity at a distance r from the center of the star. The Doppler shift of the observed absorption components of a P Cyg profile gives us information about the maximum velocity v_∞, while the equivalent width W_λ, which in the optically thin case is proportional to the column density $\int_R^\infty \varrho(r)\,\mathrm{d}r$ (R = star's radius), essentially determines the density. The rate of mass loss is thus proportional to the product $R v_\infty W_\lambda$.

In the region of the OB stars, $\dot{\mathcal{M}}$ increases for the most part with increasing luminosity; empirically, the relation

$\dot{M} \propto L^{1.7}$ is found. On the one hand, the supergiants with their well-developed P Cyg profiles show mass losses up to several $10^{-5} \mathcal{M}_\odot$ yr^{-1} at wind velocities up to 3500 km s^{-1}. The mass losses of the Wolf–Rayet stars, (Sects. 6.4.3 and 8.2.5), $4 \cdot 10^{-5} \mathcal{M}_\odot$ yr^{-1}, surpass even those of the OB stars of *comparable* luminosity by about a factor of 10. Mass losses of this magnitude during the "dwell time" in the supergiant phase make up a considerable portion of the original mass of a star and must be taken into account in considering stellar evolution (Sects. 8.2.3, 4). On the other hand, in the early B main sequence stars, such as τ Sco (B0 V), $\dot{M}$ is a few $10^{-8} \mathcal{M}_\odot$ yr^{-1} or less and can be recognized only through the *asymmetric* absorption lines with an extended wing on the short wavelength side (Fig. 7.34).

The extended, expanding shells of the supergiants can also be detected by their free–free radiation in the *radiofrequency* and infrared regions. The first of these stars to be observed in the radio range was P Cyg, by H. J. Wendker et al. (1973) at 5 and 11 GHz.

Coronas. The occurrence of higher ionized states such as C IV, N V and O VI in the ultraviolet spectra of the stellar winds indicates temperatures from 10^5 to 10^6 K. The discovery by the Einstein satellite in 1979 that OB stars belong among the strongest *X-ray sources*, excepting close binary systems (Sect. 7.4.6), likewise indicates temperatures of 10^6 K. The X-ray emission of the coronas of the O and B stars is to a large extent independent of the spectral type and luminosity and is in the range $L_X \simeq 10^{23}$ up to several 10^{27} W. In the A stars, we observe no or only very weak X-ray emission.

The existence of coronas in hot stars is surprising, since here, in contrast to the cooler stars, *no* extended hydrogen convection zones which could serve as a source of heating for a corona by mechanical transport are present. The characteristics of X-ray emission of the hotter stars differ clearly from those of the cooler ones: nearly independently of the spectral type, there is a close correlation between L_X and the total luminosity L:

$$L_X \text{[W]} \simeq 1.5 \cdot 10^{-9} (L \text{[W]})^{1.08}, \quad (7.108)$$

as well as with the rate of momentum $\dot{M} v_\infty$ and kinetic energy $\dot{M} v_\infty^2 / 2$ transport by the stellar wind; on the other hand, L_X is correlated neither with the wind velocity v_∞ itself nor with the rotation $v \sin i$. The acceleration of the stellar wind is probably due to the high *radiation pressure* of the hot stars. The physical processes which lead to the coronal temperatures, high degrees of ionization, and X-ray emission are still for the most part unexplained; the X-rays may originate in compressed areas of the stellar wind, which are heated by shock waves. At any rate, the "mechanical power" which must be generated by the star to drive the stellar wind is considerable; e.g. in the case of an O supergiant such as ζ Pup, with $\dot{M} = 6 \cdot 10^{-6} \mathcal{M}_\odot$ yr^{-1} and $v_\infty = 2700$ km s^{-1}, it is nearly $10^4 L_\odot$, i.e. about 1% of its total luminous power.

7.4.5 Cataclysmic Variables: Novae and Dwarf Novae

The group of the *eruptive variables* offers quite different perspectives; they are characterized by single or multiple, sudden increases in brightness. This group includes the novae, with an increase in brightness of about 7 to 20 mag within a few days; the dwarf novae, with weaker outbursts (2 to 6 mag) and a more or less regular rhythm; and some nova-like variables. These all belong to close, semidetached binary star systems, in which matter from a cool main sequence star (secondary component) that fills up its Roche surface (Sect. 6.5.5) continually overflows onto a somewhat more massive white dwarf (primary component, with about $1 \mathcal{M}_\odot$). They are therefore called *cataclysmic* variables or binaries (from the Greek κατακλυσμός, flood). The X-ray binary stars, which we shall discuss in the next section, are related to them. In contrast, the supernovae, considering the strength of their outbursts ($\gtrsim 20$ mag) and their causes, represent a completely different phenomenon.

Novae. A nova outburst occurs roughly as follows: the initial state, the *prenova*, is a hot object with an absolute magnitude of $M_V \simeq +4$ mag, consisting of a white dwarf surrounded by an accretion disk (see below). Within a very short time, its brightness increases to a maximum at about $M_V \simeq -8$ mag ($L \simeq 10^5 L_\odot$) i.e. by about 5 powers of ten. The luminosities of individual novae vary considerably; the least and most bright range from 9 to 19 mag, corresponding to around 3 up to more than 7 powers of 10.

Since novae are usually discovered only when – or after – they have reached their intensity maximum, there are just a few observations of the initial phase of increasing brightness. At its maximum, the spectrum of a nova is similar to that of an A or F supergiant such as α Cyg (A2 Ia), while the radial velocities of around 1000 km s^{-1} indicate an enormous expansion of the star.

Depending on whether the decrease of the visual brightness by 3 magnitudes occurs in less or more than 100 d, one terms the nova fast or slow. The former also reach their maxima faster, within a few days, and show larger amplitudes.

The novae with the greatest apparent magnitudes in recent times were V 1500 Cyg (Nova Cyg 1975), with an apparent magnitude at its maximum of $m_{V,\text{max}} = 1.6$ mag and a decay time (by 3 mag) of 6 d; and V 1974 Cyg (Nova Cyg 1992) with 4.4 mag and 42 d.

During the continuing decrease in brightness, sometimes fluctuations similar to those of the Cepheids occur. In some (slow) novae, around 40 to 150 d after the maximum, a strong decrease in the optical light curve is observed, which is correlated with an increase in the infrared brightness: in the expanding shell, small *dust particles* condense out and absorb the radiation from the nova, re-emitting it in the infrared.

After its passage through the brightness maximum, broad emission lines are observed, which often exhibit a P Cyg profile (Fig. 7.34). As the brightness decrease continues, the emission line spectrum becomes similar to the socalled Orion spectrum, like that of the luminous gaseous nebula (Sect. 10.3.1). The nova is now ejecting a shell of gas with velocities of the order of 2000 km s^{-1} and masses in the range of 10^{-5} to 10^{-3} $\mathcal{M}_\odot$. In several cases, these shells, consisting of individual condensation regions, and their expansion could be observed for more than a decade in direct images.

In the course of many years up to decades (in the case of the slow novae), the nova returns to its initial brightness and finally to its original state.

The *recurrent novae* exhibit several nova-like outbursts with amplitudes of 7 to 9 mag at intervals of about 10 to 100 yr; in between, their luminosities are nearly constant. In the case of the novae U Sco and T Pyx, the outbursts occur at particularly short intervals of 9 or 12 yr, respectively. Based on theoretical models, it is presumed that the outbursts of "common" novae also repeat themselves, to be sure at longer intervals of perhaps 10^3 to 10^6 yr, so that in historical time periods

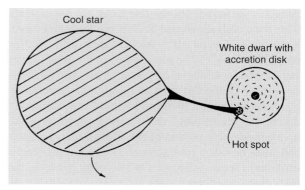

Fig. 7.35. A model of a cataclysmic binary star system. The geometric relationships correspond to those of the dwarf nova Z Cam. E. L. Robinson, Ann. Rev. Astron. Astrophys. **14**, 119 (1976). (Reproduced with the kind permission of Annual Reviews Inc., Palo Alto)

only one outburst can be observed. All together, it is estimated that in our galaxy there are about 100 nova outbursts per year, taking the interstellar extinction into account.

The energy which is released by a nova is of the order of 10^{38} J; an overall *energy release* of this magnitude is observed as the kinetic energy of the ejected shell. This corresponds to the thermal energy content of a thin layer of matter, at for example several 10^6 K and having only 1/1000 to 1/100 the mass of the Sun. All indications are that novae represent a sort of "skin disease" of stars. More precise observations, especially of several favorable cases of eclipsing binaries (Sect. 6.5.2), yield the following picture (Fig. 7.35): in the cataclysmic binary star systems whose periods lie in the range of about 1.3 to 15 h, the matter ejected from the secondary component, due to its angular momentum, does not fall directly into the white dwarf, but instead forms a rapidly rotating *accretion disk* around the dwarf. In this disk, the matter loses angular momentum through friction and thus gradually moves inwards to the surface of the white dwarf. At the point where the gas flow meets the disk, a "hot spot" is formed. The main contribution to the luminosity comes not from the white dwarf, but, depending on the particular case, from the disk or the hot spot.

The hydrogen-rich material collects until nuclear hydrogen burning (Sect. 8.1.3) is initiated *explosively*, leading to a burst of radiation intensity and to the ejection of a shell of matter. This process will be discussed in more detail in Sect. 8.2.6, after we have studied the evolution and nuclear energy production of the stars.

In the accretion disk and especially in the explosion on the white dwarf, high temperatures (of more than several 10^5 K) occur, so that a major part of the radiation is emitted in the *ultraviolet* and soft X-ray regions. Especially through observations with the IUE satellite, important knowledge about the physics of nova outbursts has been obtained.

Referring to the ultraviolet emission lines, one can distinguish two classes of novae: the carbon–oxygen novae, with overabundances of the elements C, N and O; and the neon-novae, in which O, Ne and Mg are overabundant. These classes represent white dwarfs of differing chemical compositions.

The bright nova Cyg 1992 was observed in a very early phase of its outburst by the IUE satellite. Its luminosity in the ultraviolet decreased at first within 2 d, due to rapid cooling of the expanding ejected matter, and then increased during several weeks. This increase can be explained by the fact that as a result of the changing temperatures and densities during the expansion, radiation from deeper and deeper layers is able to reach the surface and be detected. Indeed, the *total* luminosity, i.e. in the ultraviolet plus the optical region during the first 50 d following the outburst is practically constant and only after this time begins to decrease.

Dwarf Novae. The *U Geminorum* and *Z Camelopardalis* variables, again named for their prototypes, are like the "classical" novae, hot, blue stars below the main sequence. These dwarf novae exhibit less violent outbursts (up to 6 mag) at irregular intervals from about 10 d up to several months. In the Z Cam stars, the sequence of the outbursts is occasionally interrupted by a period of relative stability at an intermediate level of brightness.

In the outbursts of the dwarf novae, the luminosity of the accretion disk increases strongly, but no matter is ejected. In contrast to the novae proper, the outbursts are caused here by *instabilities* of the disk. The energy source in the final analysis is the release of gravitational energy corresponding to an average rate of around $10^{-10}\,\mathcal{M}_\odot\,\mathrm{yr}^{-1}$ of accreted material.

Magnetic Binary Star Systems. An interesting group of cataclysmic variables are the *AM Herculis* stars, which are notable on the one hand for the high (and variable) degree of circular and linear polarization of their radiation in the optical and ultraviolet regions, and on the other for their strong emissions in the soft X-ray region. The cause is a strong *magnetic field* of the white dwarf amounting to several 10^3 Tesla, which prevents the formation of an accretion disk and, instead, concentrates the gas flow onto a spot around one (or sometimes both) of the magnetic poles. There, in a stationary shockwave front, its kinetic energy is converted to thermal energy and emitted as radiation. The temperature in the accretion column above the poles is 10^7 to 10^8 K, so that the emission is primarily in the X-ray and ultraviolet regions. In the magnetic field of several thousand Tesla, the plasma also emits *polarized* cyclotron radiation of higher harmonics in the infrared and optical regions, corresponding to the cyclotron frequency ω_C (10.31).

Due to the strong magnetic field, the rotational period of the white dwarf is synchronized with the orbital period of the binary system. The orbital periods of the AM Her stars are in the range of 1.3 to 9 h; the distance between the two stars is of the order of 1 $R_\odot$.

In the *DQ Herculis* systems, whose prototype is DQ Her = Nova Her 1934, the primary component is a magnetic, rapidly rotating white dwarf with a rotational period in the range of 0.5 min up to 2 h. Its magnetic field is somewhat weaker than in the case of the AM Her systems, so that the rotation is not synchronized with the orbital motion and the formation of the accretion disk is prevented by the magnetic field only in the neighborhood of the white dwarf.

7.4.6 X-ray Binary Systems: Accretion onto Neutron Stars

As a result of the progress in X-ray astronomy, a large number of "point sources" in our Milky Way galaxy have been discovered by satellite observations in recent years; they radiate *mainly* in the X-ray region and are characterized by a strong *variability*. Intensity fluctuations and bursts of radiation are observed, in which the X-ray emission increases by factors of 10 to 100; they show an impressive variety of behaviors, including irregular flickering on a time scale of a few milliseconds, series of short bursts lasting several seconds, nova-like outbursts followed by a slow decrease of the intensity back to the original value, X-ray flares which occur at irregular intervals and last for several days, and regular or irregular pulses on a time scale of seconds. Some sources exhibit more or less constant emission, in which often several different periods can be discerned, with active and quiet phases succeeding each other on a time scale of weeks or months.

In spite of this mutiplicity of phenomena, it is possible to describe practically all of these variable galactic X-ray sources within the framework of a *single* theory, that of *accretion* of matter in close *binary star* systems. They thus differ from the cataclysmic variables (Sect. 7.4.5), where X-rays can, to be sure, also be observed, but the main contributions to the luminosity are in the visible and ultraviolet spectral regions; by contrast, in the *X-ray binary systems*, the luminosity L_X in the X-ray region is in the range from 10^{27} to 10^{32} W, and matter is not accreted by a white dwarf star, but rather by a magnetic *neutron star* (Sect. 8.3.4) or possibly by a black hole (Sect. 8.4.4). This theoretical model is in many cases supported by optical identification of one of the binary system components and by the occurrence of X-ray pulses of short periods, which can be explained only by the rotation of a neutron star ("searchlight effect", Sect. 7.4.7).

Pulsed X-ray emission from *single* (non-binary) stars will be treated in connection with the rotating *radio* pulsars in Sect. 7.4.7.

Energy Source. The energy source for the X-ray emissions is the gravitational potential energy released by the accreting gas. If matter of mass m falls from a large distance to a distance R from the center of a (spherical) mass $\mathcal{M}$, according to (2.44), an energy equal to $m\, G\mathcal{M}/R$ is released. If $\dot{m}$ is the *rate* at which the matter falls, then acceleration by the strong gravitational field of a compact object, e.g. a white dwarf, neutron star, or black hole, and then braking and heating near its surface, generates X-rays very effectively, producing a luminosity of the order of:

$$L_X \simeq \dot{m}\, \frac{G\mathcal{M}}{R}. \qquad (7.109)$$

For example, in order to obtain 10^{31} W, a relatively modest gas flow of $\dot{m} \simeq 10^{-8}\, \mathcal{M}_\odot\, \mathrm{yr}^{-1}$ in the gravitational field of a neutron star with $\mathcal{M} \simeq 1\, \mathcal{M}_\odot$ and $R \simeq 10$ km is sufficient. This rate can readily be produced either by mass flow through the Roche surface or by stellar winds.

Hercules X-1. The well-studied pulsating X-ray source Her X-1 has been identified with the eclipsing binary system HZ Her discovered by C. Hoffmeister in 1936. Its X-radiation first of all exhibits *pulses* with a period of 1.24 s, which are attributed to the rotation of a neutron star. This period is then modulated by the Doppler effect due to orbital motion (Sect. 6.5.6), with a period of 1.7 d. The latter period is seen both as an eclipsing of the X-ray source and as an intensification of the visible light, which occurs whenever the side of the optical component facing us is irradiated by the X-ray source. An additional period of 35 d is related to the precession of the axis of the neutron star. Her X-1 can also "switch off" its X-ray emissions over periods of several months, as a result of occlusion by the accretion disk surrounding the neutron star. The mass of the latter is estimated to be about $1.3\, \mathcal{M}_\odot$, while that of the other component is about $2.2\, \mathcal{M}_\odot$.

The line at 58 keV which was discovered in 1976 by J. Trümper and coworkers in the spectrum of Her X-1 (Fig. 7.36) and represents an emitted power of around $2 \cdot 10^{28}$ W, allows a direct estimate of the magnetic flux density B in the neighborhood of the surface of the neutron star. This line is too strong to be explained as an atomic or nuclear transition of a heavy element. Instead, it must be regarded as *cyclotron emission*, which is produced by the spiral motion of nonrelativistic electrons in magnetic fields at multiples of the cyclotron frequency $\nu_C = (1/2\pi)\, eB/m_e$ (10.31). For relativistic electrons, by the way, the cyclotron emissions become continuous synchrotron radiation (Sect. 12.3.1), which is important for the explanation of nonthermal radio emissions. In the extremely strong magnetic fields of pulsars or neutron stars, the quantization of the gyrating motion into discrete energy states (Landau levels), which are separated by the energy $h\nu_C$, must be taken into account. For Her X-1, the observed energy of 58 keV corresponds to an estimated magnetic field of about $5 \cdot 10^8$ Tesla.

The details of the dynamics of the inflowing gases in a close binary system with a neutron star which has a strong magnetic field are very complex. The dipole-like magnetic field hinders the formation of an accretion disk and influences its structure. It directs and concen-

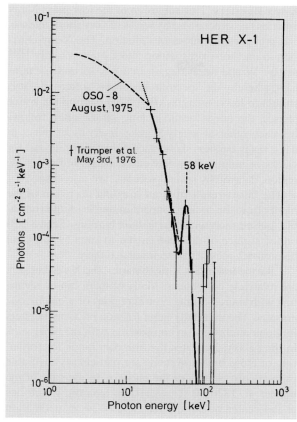

Fig. 7.36. The cyclotron emission line at 58 keV in the pulsed X-ray spectrum of Her X-1; from observations by J. Trümper et al. using balloons (as well as observations of Her X-1 from the OSO-8 satellite)

trates the flowing material towards the poles and has important effects on the shape of the X-ray pulses.

The mass determinations of the components of the X-ray binary systems (Sect. 6.5.6) yield a relatively limited range of 1.2 to 1.6 $\mathcal{M}_\odot$ for the neutron stars.

The X-ray binary systems can be classified according to the mass of their secondary component essentially into *two groups*:

High-Mass X-ray Binary Systems. In these systems, the secondary component is usually a young OB star which is bright in the optical region and has a mass of $\mathcal{M} \geq 10\,\mathcal{M}_\odot$; it dominates the overall luminosity of the system. The ratio of X-ray emission to emissions in the optical region is in the range $L_X/L_{\mathrm{opt}} \simeq 10^{-3} \ldots 10$.

The neutron star has a strong magnetic field of 10^8 to 10^9 T, which guides the matter flow to the magnetic poles of the star. The temperature of the hot spot at the pole attains values of $T \simeq 10^8$ K, so that the "X-ray pulsar" emits radiation in the hard X-ray range.

The pulse periods which reflect the rotation of the neutron star range from 0.7 s up to several 100 s. In some cases, a secular *decrease* in the period, (interrupted by brief fluctuations), at a rate of $\dot{\Pi}/\Pi \simeq -(10^{-2} \ldots 10^{-6})\,\mathrm{yr}^{-1}$ is observed. This is apparently due to an acceleration of the rotation ("spin-up") owing to transfer of angular momentum from the matter flowing onto the star.

The orbital periods can be derived both from the modulation of the X-ray emission and from the optical spectrum of the OB star; they are in the range of a few days.

Members of this group are e.g. Her X-1, whose secondary component has, to be sure, a mass of only 2.2 $\mathcal{M}_\odot$, and Cen X-3 (O6.5 III). In these systems, the accretion takes place via an overflow through the Roche surface, while in the case of e.g. Vel X-1 = HD 77 581 (B0.5 Ib), the neutron star accretes matter from the strong stellar wind of the supergiant.

Low-Mass X-ray Binary Systems. This group includes sources which radiate intensely ($L_X > 10^{27}$ W), mainly in the soft X-ray range, and whose emission is as a rule not pulsed. One subgroup, the X-ray bursters (see below), exhibits irregular outbursts. The companion of the neutron star is a dwarf star of spectral type G to M and mass $\mathcal{M} \leq 1\,\mathcal{M}_\odot$, so that the radiation in the X-ray region is much stronger than that in the optical region ($L_X/L_{\mathrm{opt}} \simeq 10^2 \ldots 10^4$). These systems are related to the cataclysmic variables (Sect. 7.4.5), with the difference that here a neutron star replaces the white dwarf.

The low-mass X-ray binaries are concentrated towards the center of our galaxy, and therefore represent *old* objects ($\geq 10^9$ yr). The magnetic field of the neutron star will for the most part have decayed away, and at values around 10^4 to 10^6 T, it is considerably weaker than that in the younger, more massive systems. The gas stream from the cool companion star can distribute itself into an accretion disk or over the whole surface of the neutron star and is not concentrated onto the polar regions as in the X-ray pulsars. A strong continuum in the blue as well as optical emission lines indicate the presence of an accretion disk.

Some of the *transient* X-ray sources, in which the intensity rises above the detection limit for only short periods of time, can be classified as *X-ray novae* on the basis of their light curves: a rapid increase in brightness (to $L_X \leq 10^5 \, L_\odot$) is followed by an essentially exponential decay which can be observed for several months and is interrupted by irregular flares (lasting a few days and sometimes observable in the optical region as well).

In certain cases it has been possible to observe also *gamma* radiation from X-ray binary systems: Her X-1 emits pulsed gamma radiation at energies above 10^{12} eV, which occurs with the same period of 1.24 d known from the optical and X-ray regions. Cyg X-3 = V1521 Cyg is a strongly variable pulsed gamma source, for which a modulation with the orbital period of 4.8 h can be distinguished. By means of air-shower observations, extremely energetic gamma rays from Cyg X-3 with energies from 10^{15} to 10^{16} eV have been detected, with a luminosity in this region alone of at least $10^3 \, L_\odot$. Presumably, in this system the neutron star is a rapidly rotating, active pulsar, which accelerates protons to extreme relativistic energies of 10^{17} to 10^{18} eV; these then produce gamma quanta through interactions with the matter in the system. The system Cyg X-3 is thus probably also a strong source of energetic cosmic radiation (Sect. 10.4.2).

Some binary systems are particularly notable through their intense radiation in the gamma-ray region (Sect. 7.4.8), but otherwise are not essentially different from the X-ray binaries, so that the transition between the X-ray and the gamma-ray sources is continuous.

Black Holes? On the basis of the properties of the X-ray emission alone, one can practically not distinguish between accretion of matter by a neutron star or by a black hole; it is possible that an irregular variation in the millisecond range could be an indication of a matter flow into a black hole. Since a stable neutron star however has an upper *limiting mass* (which is not very precisely known) of around $2 \, \mathcal{M}_\odot$ (Sect. 8.3.4), we can in principle consider all objects with greater masses to be *candidates* for black holes. Among others, these are Cyg X-1 = HDE 226 868 with a mass of $(3 \ldots 10) \, \mathcal{M}_\odot$, LMC X-1, LMC X-3, and the X-ray nova Muscae 1991.

Additional candidates for black holes can be found by observing strong galactic sources in the *gamma-ray* range (Sect. 7.4.8).

X-ray Bursters. This subgroup of the low-mass X-ray binaries is characterized by series of very short, intense, and non-periodic bursts of radiation. The X-ray emission in one burst rises within about 1 s to roughly 10^{31} to 10^{32} W and decays again in a few seconds or minutes, whereby the spectrum becomes "softer", i.e. it is shifted to lower energies (Fig. 7.37). During the active phases, the bursts occur in long series, often following each other at nearly regular intervals of hours to days; many bursters show inactive phases which last for weeks or months. In a few cases, bursts in the visible region, delayed by several seconds, are also observed.

The bursts result from an irregular accretion of the matter from the cooler companion star, leading to unstable, pulsed thermonuclear *helium* burning (Sect. 8.2.3) on the surface of the neutron star as soon as sufficient helium-rich matter has accumulated there.

Like the low-mass X-ray binaries, the X-ray bursters are relatively old objects. Many of them can be identified with sources in the central regions of *globular star clusters*. The unusually large fraction of X-ray bursters in globular clusters, in comparison with the other stars in the galaxy, indicates that the conditions for formation of close binary systems are particularly favorable in the dense centers of the clusters. In the case of the burster 4U 1820-30 (U = UHURU survey), which lies at the center of the cluster NGC 6624, a modulation

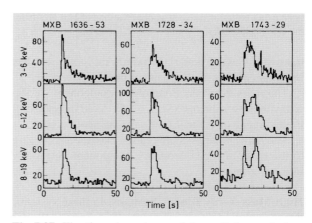

Fig. 7.37. The time dependence of the radiation emissions from three X-ray bursters in different energy ranges; observations from the SAS-3 satellite. The intensity is expressed as the number of counts in 0.4 s (*upper row*) or in 0.8 s (*lower two rows*)

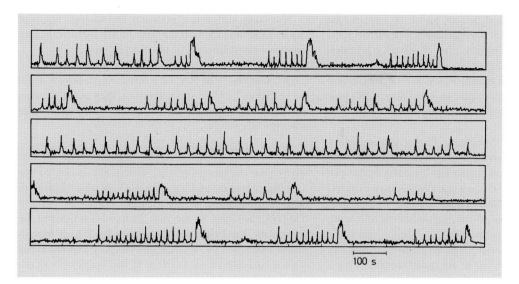

Fig. 7.38. The time dependence of radiation emissions from the "rapid X-ray burster" MXB 1730-335. Portions of observations made from the SAS-3 satellite in March, 1976

of its X-ray emissions with a period of only 11.4 min was discovered in 1984/85 in EXOSAT observations; it has been interpreted as the *orbital* period of a system consisting of a neutron star and a very low-mass (0.055 $\mathcal{M}_\odot$) white dwarf.

The remarkable "rapid burster" MXB 1730-335 in the globular cluster Liller 1 at a few degrees from the galactic center was unknown before the discovery of the X-ray source in 1976. During its active periods, which last for a few weeks and follow each other with a period of about 6 months, it produces several 1000 bursts per day (Fig. 7.38). Only in 1996, using the Compton Observatory, was another object of this type discovered: GRO J1744-28, likewise roughly in the direction of the galactic center. Its outbursts, which last for several 10 s in the energy range from 25 to 60 keV, at first occurred at intervals of about 200 s, but these have gradually lengthened.

SS 433 = V 1343 Aql (SS refers to the list of stars with Hα-emission by C. B. Stephenson and N. Sanduleak, 1977). The brightness of this unusual system with an orbital period of 13.1 d shows irregular fluctuations in the optical, X-ray, and radiofrequency spectral regions. The components of the strong, broad emission lines from H I and He I are notable: they exhibit Doppler shifts which are unusually large for stars, up to 50 000 km s^{-1} with a period of 164 d. SS 433 consists of an O star and a neutron star (or a black hole) with an accretion disk, which makes the main contribution to the emitted radiation. Roughly perpendicular to the orbital plane, two collimated beams of matter emerge in opposite directions with nearly relativistic velocities of 80 000 km s^{-1} (i.e. about 0.26 of the velocity of light); they show a precessional motion with a period of 164 d. SS 433, with its two highly energetic jets, represents a rare stellar "miniature version" of a phenomenon which we shall meet later in many galaxies (Sects. 12.3.3, 4), with much larger dimensions.

A few additional objects related to SS 433 have been discovered in our galaxy on the basis of their intense gamma radiation (Sect. 7.4.8).

7.4.7 Supernovae and Pulsars

Among the eruptive variables, the *supernovae* represent cosmic explosions of a much greater magnitude than the novae, as was recognized by W. Baade and F. Zwicky in 1934. A supernova at its maximum can attain the luminosity of an entire galaxy (to which it belongs).

The last supernova in our Milky Way galaxy was observed in 1604 by J. Kepler. Only in 1987 was a supernova again visible to the naked eye, in our neighboring galaxy, the Large Magellanic Cloud. Except for a few "historical" events in our galaxy, all of the supernovae have been discovered in *other* galaxies. The rate of their discovery increased abruptly after F. Zwicky initiated the systematic Mount Palomar survey in 1934, and has continued to rise as a result of the use of larger and larger telescopes; for example, in 1999 about 200 new supernovae were discovered.

Classification and Spectra. Early classification schemes of the supernovae based on their spectra were developed by R. Minkowski (in 1941) and by F. Zwicky (in 1965). Since about 1985, a classification system has been in use which is based on the optical spectra near the *brightness maximum* and distinguishes the following three main types of supernovae:

We first term all supernovae in whose spectra the hydrogen lines are lacking or are quite weak as *Type I* supernovae. Supernovae with strong Balmer lines are classed as *Type II*.

The SN I are then subdivided into *Type Ia*, in which an absorption structure occurs at $\lambda \simeq 610$ nm, due to blueshifted Si II lines; and *Types Ib/c*, where this structure is lacking. We will not discuss here the fine details which distinguish types Ib und Ic.

The spectra of the SN Ia show many lines with broad emission and absorption features. The brightness maximum is dominated by lines from singly ionized Si, S, Mg and Ca; some weeks later, lines from singly ionized elements of the iron group appear. The spectra of the Ia supernovae exhibit individual variations at the beginning, but after several months they become increasingly similar.

The spectrum of a Type II supernova and its time development are astonishingly similar to those of ordinary novae. Along with the Balmer lines, lines of He and of the metals such as Ca II and Fe II occur; in later phases there are also forbidden lines such as those of [O I] and [O III].

The spectra of the SN Ib are initially, near the intensity maximum, similar to those of type SN Ia; but in later phases, they resemble those of the supernovae II.

The emission lines of both types of supernovae are often accompanied by short-wavelength absorption components (P Cyg profile), whose Doppler shifts indicate *ejection velocities* of up to $2 \cdot 10^4$ km s^{-1}; on the average, the SN I show higher velocities than those of type II.

The basic difference between the various types of supernovae thus seems to be determined by the different chemical compositions of the ejected matter.

Light Curves. The *SN Ia* exhibit very similar light curves (Fig. 7.39): at their maxima, they attain a mean absolute magnitude in the blue (corrected for interstellar absorption) of

$$M_{B,\max} = -(19.2 \pm 0.1) \text{ mag},\quad (7.110)$$

where the distance to the supernova is based on a Hubble constant of $H_0 = 65$ km s^{-1} Mpc^{-1} (13.4); for other values of H_0, we would have $M_{B,\max} = -19.8 + 5 \log(H_0/50)$ mag. In the first 20 to 30 d following the maximum, the brightness decreases by 2 to 3 magnitudes; from then on, it decreases roughly exponentially with a half-life of 40 to 70 d. If the brightness is expressed in magnitudes, this corresponds to a linear decrease with time. Due to their uniform light curves, the supernovae Ia are suitable as "standard candles" for distance determinations in the cosmos (Sect. 13.2.5).

Supernovae Ib/c have similar (optical) light curves to those of the SN Ia, but at maximum they attain a brightness which is less by about 1.5 mag.

The maximum brightnesses of the *SN II* exhibit strong variations; on the average, they reach "only" $M_{B,\max} \simeq -17$ to -18 mag. This still corresponds to around the 10^4-fold luminosity of an ordinary nova. The light curves show strong individual variations. We distinguish two subtypes: the light curves of the SN II-P are relatively flat from about 30 to 80 d following the maximum (*p*lateau), before the linear brightness decrease – in [mag] – sets in, while in type SN II-L, the magnitude decreases *l*inearly with time from the beginning (Fig. 7.39).

The colors and the intensity distribution in the continuum, which is particularly difficult to determine for supernovae I owing to the large number of spectral lines, indicate that the temperature of the emitting layers decreases from $\geq 10^4$ K at the maximum to about 6000 K. These temperatures, which are not very high, correspond to photospheric radii of the order of $10^4\ R_\odot$ using the relation $L = 4\pi R^2 \sigma T_{\text{eff}}^4$ (6.43), due to the extremely high luminosities of supernovae.

Historical Supernovae. In our *Milky Way galaxy*, the following supernovae have been identified with cer-

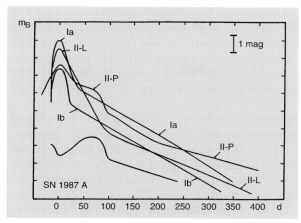

Fig. 7.39. Schematic mean light curves in the blue for supernovae of types Ia, Ib (including Ic), II-L and II-P, and the light curve of SN 1987A after J. C. Wheeler, 1990. (With the kind permission of the author)

tainty in historic times: the very bright supernova of 1006 A.D. in Lupus reached the brightness of the half-full Moon and is mentioned in East Asiatic, European, and Arabian sources. The supernova of 1054 A.D. in Taurus is described in Chinese and Japanese annals as a "guest star" but, remarkably, it is not mentioned in European sources, although its brightness at maximum was greater than that of Venus. It was the source of the Crab Nebula (Fig. 7.41), which was identified by J. G. Bolton as the radio source Tau A. Tycho *Brahe* described the supernova of 1572 in Cassiopeia, whose maximum brightness was comparable to that of Venus, and J. *Kepler* observed the supernova 1604 in Ophiuchus, with a maximum brightness corresponding to about -2.5 mag; these two supernovae are also mentioned in Chinese and Korean sources. Kepler's SN 1604 is the last supernova to be directly observed in our Milky Way galaxy.

The strong radio source Cas A and the nebula at the same position, which is expanding with a velocity of 7400 km s^{-1}, have been clearly identified as the remains of a supernova which must have exploded around 1680, but was not observed then.

On the average, therefore, *one* outburst is observed every few hundred years. Based on extensive statistics of supernovae in *other* galaxies, G. A. Tammann in 1981 estimated that in a galaxy like ours, a supernova occurs on the average every 25 yr, with an uncertainty in this rate of a factor of 2. Due to the interstellar extinction in the disk of our galaxy, we can, to be sure, observe only a tenth of these explosions.

Origins. A first indication of the origin of supernovae is provided by their occurrence in galaxies of different types (Sect. 12.1.2). Supernovae Ia are the only type which occur in *all* kinds of galaxies, in particular also in elliptical galaxies and in the halos of spiral galaxies. In contrast, the SN II and SN Ib/c are observed only in spiral and irregular galaxies, where they are mainly concentrated in the spiral arms. It therefore can be presumed that supernovae of types SN II and SN Ib/c are to be attributed to young, massive Population I stars. The origin of the SN Ia must be from old objects.

Causes of the Outbursts. The total *energy* release E from the explosion of a supernova can be estimated on the one hand by integration of the light curve with an approximate bolometric correction, in favorable cases also using observations in the ultraviolet and infrared, to obtain the *total emitted radiation*; on the other hand, the *kinetic energy* of the ejected gas shell (0.1 to 10 $\mathcal{M}_\odot$ at around 10^4 km s^{-1}) can be taken into account. The order of magnitude is $E \simeq 10^{44}$ J. In comparison, one can readily calculate that the energy released e.g. by the fusion of 1 $\mathcal{M}_\odot$ of hydrogen to helium is around 10^{45} J (Sect. 8.1.3). Furthermore, the collapse of a star, e.g. of 1 $\mathcal{M}_\odot$, to a *neutron star* of $R = 10$ km radius, with a density comparable to that of nuclear matter (10^{17} kg m^{-3}), can, as was recognized already in 1934 (!) by W. Baade and F. Zwicky, release about $3 \cdot 10^{46}$ J of gravitational energy (Sect. 8.1.4), and thus could readily provide the required energy for a supernova. This estimate does not, to be sure, take into account the fact that, as observed from the nearby supernova SN 1987A of type II, about 10^{45} to 10^{46} J in the form of *neutrinos* are emitted, and therefore the electromagnetic radiation and the kinetic energy of the gas shell make up only around one percent of the total required energy. However, the collapse of a massive star of about 10 $\mathcal{M}_\odot$ can readily release even this amount of energy.

A supernova (of type II) emits light when the shock wave moving outwards from the central collapse reaches the surface of the star and heats it up. The maximum of the light curve depends on the size of the pre-supernova: red giants attain greater brightnesses when they explode than do the more compact blue supergiants (SN 1987A, see below). The decrease of the light curve, both for the SN I as well as for the SN II – in their later phases – reflects the energy release by the radioactive isotope ^{56}Ni. This nuclide, which is formed during the collapse, decays with a half-life of 6.1 d to ^{56}Co, which in turn decays with 77.3 d to ^{56}Fe. In the cases of a few bright supernovae, this model is supported by the observation of *gamma-ray lines* in the MeV range from this decay chain, which occur several weeks after the brightness maximum.

The further question of what actually causes the explosion of a supernova will be treated in Sect. 8.2.4, together with the topic of instabilities in stellar evolution.

Supernova 1987A. This bright type II supernova has particular significance; it was observed on February 23rd, 1987, in our neighboring galaxy, the Large Magellanic Cloud (Fig. 7.40), which is only about 50 kpc away, and reached its maximum (apparent) visual magnitude $m_V = 2.9$ mag after about three months. After 120 d, it entered the phase of exponential decrease in light intensity. Its absolute magnitude at maximum, $M_V = -15.5$ mag however remained well below the average value for supernovae of type II. This first bright supernova in modern times could be observed from early on with a high spectral resolution in a wide wavelength range. For SN 1987A, for the first time, the *pre-supernova* could be clearly identified; it was a *blue* B3-supergiant with an absolute magnitude of $M_V = -6.6$ mag.

In the case of SN 1987A, it was possible to detect the *neutrino emission*, for the first time for any object excepting the Sun (see also Sect. 8.2.2); the first neutrino signals were in fact detected several hours *before* the initial optical observation of the supernova was registered. The observation of neutrinos with a total energy of 10^{45} to 10^{46} J and an average particle energy of the order of 10 MeV supports our basic theoretical understanding of supernova explosions of type II as the collapse of a massive star, in which gravitational energy is released almost exclusively in the form of neutrinos, with only about 1% as electromagnetic radiation and kinetic energy of the ejected gas shell (Sect. 8.2.4).

Measurement of the time during which the neutrinos were emitted offers the possibility in principle (by means of a time-of-flight analysis) of determining the rest mass of the neutrino, a fundamental quantity both for elementary particle theory and for cosmology (Sect. 13.3.2). The current models for a supernova explosion indicate an upper limit for the mass of the (electron) neutrino in the range of 10 to 30 eV c^{-2}.

Pulsars. In the last part of this section, we shall concern ourselves with the situation following a supernova explosion, i.e. with the observation of the *remnant stars*. The fate of the ejected shells, the *supernova remnants*, SNR, will be treated in Sect. 10.3.3 together with interstellar matter.

Although the structure of *neutron stars* was already investigated theoretically in the 1930's (Sect. 8.3.4), the first one was discovered only in 1967 in a completely unexpected manner by A. Hewish and J. Bell. At the Cambridge radiotelescope, they noticed signals in the meter wavelength range which repeated themselves very regularly with a period of 1.337 s. Observations of the apparent motion in the sky showed that these were not due to terrestrial disturbances, but rather to a cosmic object.

These *pulsars* or *radio pulsars* are denoted by PSR and their (approximate) right ascension and declination, whereby the coordinates in the new system for the epoch 2000.0 (Sect. 6.2.4) are preceded by a J, and those in the old system (1950.0) by a B, for clarity. For example, the first pulsar discovered at $\alpha = 19$ h 19 min and $\delta = 21°\,47'$ is called PSR 1919+21 = PSR B1919+21 or PSR J1921+2153. Their pulse periods lie roughly in the range from 1 ms to 8 s, mostly between 0.3 and 2 s.

The signals, whose length is about 3% of the period, all arrive somewhat later at lower frequencies than at higher ones, because the propagation velocity of electromagnetic waves in the interstellar plasma decreases at lower frequencies. The delay is proportional to the num-

Fig. 7.40a,b. and frontispiece, p. 161: The Large Magellanic Cloud and the supernova 1987A: a color photograph made by C. Madsen, ESO; (**a**) a few hours before the outburst on Feburary 23rd, 1987; and (**b**) 2 d after the outburst. The H II regions in the galaxy appear reddish due to their strong Hα emissions. SN 1987A can be clearly recognized to the southwest of the giant H II complex 30 Doradus (*at the left*). The distance to the Large Magellanic Cloud is 50 kpc. (With the kind permission of the European Southern Observatory)

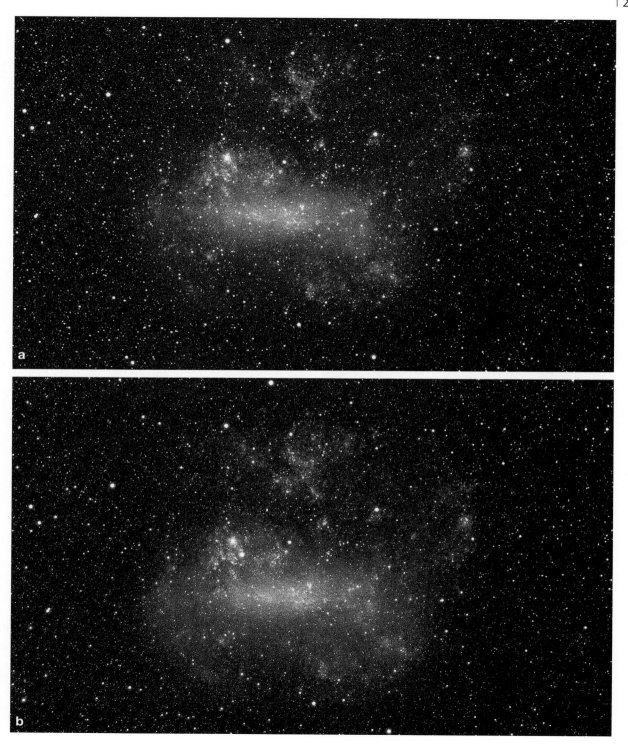

ber of electrons in a column (of unit cross-sectional area) reaching from the observer to the pulsar, the socalled *dispersion measure*:

$$DM = \int n_e \, dl \, . \tag{7.111}$$

If the value $n_e \simeq 3 \cdot 10^4$ m^{-3} is used as a rough measure of the electron density, one can, conversely, estimate the *distances* of the pulsars.

It represented a great step forward when in 1968/69 the pulsar PSR B0531+21, with the shortest known period at the time, $P = 33.2$ ms, could be identified *optically* with the *central star* of the *Crab Nebula* (Fig. 7.41). It was then soon possible to demonstrate that this star emits similar "pulses" in the optical and even in the X-ray and gamma-ray regions (Fig. 7.42), using observations through a chopper (rotating collimator) adjusted to the period. The pulse shape is about the same from the gamma region out to wavelengths of around 1 m; at still longer wavelengths, the pulses are broadened due to rapid local fluctuations in the interstellar electron density n_e, the "interstellar scintillation". The high luminosity in the gamma region is surprising; it is about 10^5 to 10^6 times greater than in the rediofrequency region. We know of only a few pulsars which emit out to the gamma-ray region, including the Vela-Pulsar PSR B0833-45, with $P = 89$ ms, and Geminga, with $P = 237$ ms (see below).

In the few cases where a pulsar could be observed within the remains of a supernova, a more precise distance determination becomes possible; e.g. the Crab pulsar is at a distance of 2000 pc, and the Vela pulsar at 450 pc.

The extraordinarily short periods and the regularity of the pulses limits possible explanations of pulsars to *rotating neutron stars* with the period P, emitting a beam with an opening angle of about 20° like a lighthouse, over the enormous frequency range from meter waves to gamma rays. The requirement that the surface of the star cannot exceed the velocity of light already implies a considerable restriction on its possible radius.

The extraordinary precision of the observations of pulsars quickly made it possible to determine that the period P of all pulsars, neglecting occasional small "jumps", is *increasing*, i.e. the rotation is slowing down ($\dot{P} = dP/dt > 0$). As a readily-understandable measure of the slowing down of their "ticking", we define the time $P/\dot{P}$, after which the period would double for a constant rate $\dot{P}$. On the basis of models for the braking process, we define the characteristic time $\tau = P/(2\dot{P})$, which indicates a lifetime or an *age* of the same order of magnitude. The pulsars which are changing the most rapidly are PSR J1846-0258, at a distance of about 20 kpc, which was discovered only in the year 2000 and has $P = 324$ ms and $\tau = 720$ yr, and the Crab pulsar, with $\tau = 1250$ yr; for most pulsars, times τ or ages of 10^6 to 10^7 yr are found.

The pulsars obtain the energy for their radiation emissions in the end from the rotational energy of the neutron star. The *mechanism of emission* in pulsars is not yet clear in all its details, but it can be considered that a rapidly rotating neutron star with a strong *magnetic field* ejects plasma outwards in the region of the magnetic poles. Recently, with the X-ray satellite Chandra,

Fig. 7.41. The Crab Nebula, M 1 = NGC 1952: a photograph in red light made with the 5 m Hale telescope by W. Baade. In 1949, J. G. Bolton first succeeded in optically identifying a radio source, and showed that Tau A is identical to the Crab Nebula, M 1. The inner, homogeneous part emits continuous synchrotron radiation, while the outer parts, the "legs of the crab", emit a nebula spectrum, particularly the red hydrogen line Hα. The Crab Nebula was formed in 1054 A.D. in a supernova explosion. The remnant star is the pulsar PSR B0531 + 21, a neutron star which "ticks" with a period $P = 33.2$ ms

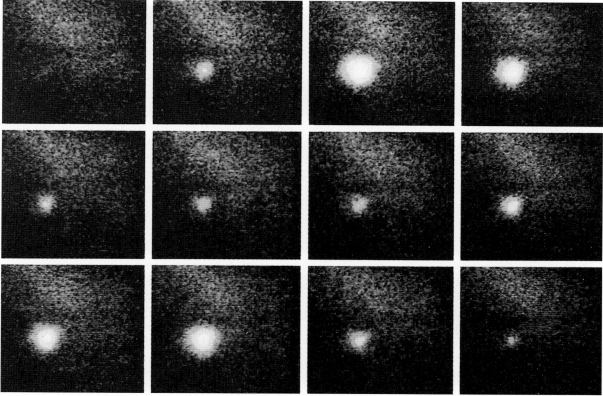

Fig. 7.42. X-ray emissions of the Crab Pulsar and the Crab Nebula in the range from 0.1–4.5 keV; observations from the Einstein Satellite with an angular resolution of about 4″. The pulsar period of 33.2 ms is time-resolved in 16 images, each comprising about 2′·2′; 12 of these are shown here (3rd picture: main pulse; 10th picture: secondary pulse). The large-area X-ray emissions of the Crab Nebula are shifted to the NW relative to the pulsar. From F. R. Harnden and F. D. Seward: Astrophys. J. **283**, 279 (1984). (Reproduced by permission of The University of Chicago Press, The American Astronomical Society, and the authors)

it was possible to resolve a sharply defined, bright ring of about 0.2 pc (0.4′) diameter around a pulsar, a broad "vortex", and two jets emitted by the pulsar. These structures no doubt represent the transition from the pulsar to an extended nebula.

The ejected plasma initially rotates with the star, reaching the velocity of light c on a cylindrical surface of radius $cP/(2\pi)$ – in the case of the Crab pulsar, for example, at 1580 km. The conditions for emission of a *beam* of radiation are then favorable. The finer details of the pulse *structure* are not yet understood satisfactorily. The opening angle of the radiation cone appears to become larger with increasing rotation. The Crab pulsar emits a weaker secondary pulse between every two main pulses, while other pulsars do not. Presumably, two pulses occur when the radio emission within the cone is not uniform.

If the pulse period increases as a result of the braking to more than about 5 s, the signals emitted become so weak that they apparently can no longer be detected.

The radius of a neutron star is of the order of 10 km (Sect. 8.3.4). If a star like the Sun initially rotates with a period of 25 d and collapses to form a neutron star, then this period will become 1 millisecond(!), assuming that angular momentum is completely conserved. (The surface of the star will still be far from attaining the velocity of light). If the star initially has a reasonably ordered magnetic field of about $5 \cdot 10^{-4}$ Tesla, its lines of force will be compressed corresponding approximately to the cross-section of the star, and the neutron star will obtain an enormous magnetic field of about 10^6 T. Further-reaching considerations, based on models for the emission of radiation, require even higher fields in the range of 10^7 to 10^9 T at the surface of the rotating neutron stars.

In a few cases, the *proper motions* of the pulsars have been determined, and they show *high* spatial velocities of (≥ 100 km s^{-1}). According to A. Blaauw, these result when – in a rapidly rotating binary system – the main component is ejected out into space by a supernova explosion, apart from a low-mass remnant star, and the gravitational binding of the system is broken.

With a very few exceptions, observations show that the radio pulsars, in which the rotation of the neutron star provides the energy for the radio emissions, are *individual* stars. Exceptions are a few systems such as the binary star pulsar PSR B1913+16 (Sect. 6.5.6) as well as the class of the millisecond-pulsars (see below).

In contrast to the radio pulsars, the related X-ray pulsars (see Sect. 7.4.6) are all neutron stars in close binary systems, whose coherence was evidently not destroyed by the supernova explosion. Their energy comes from the accretion of matter from the companion star.

Millisecond Pulsars. The first member of this group, PSR B1937+21, with a period of $P = 1.558$ ms, was discovered by D. C. Backer et al. in 1982; it at first seemed to be an extremely young object because of its very rapid rotation. However, the extremely small rate of change of its period ($\dot{P}/P$ corresponding to an age of about 10^{10} yr), its magnetic field which is weaker by a factor of about 10^4, and the lack of a supernova remnant argue that it is an *old* pulsar, which probably obtained its high angular momentum during an earlier phase by matter accretion as a close (low-mass) *X-ray binary system* ("recycled pulsar"; cf. Sect. 7.4.6). Today, it is an individual object, presumably because its outflux of energetic particles together with the strong tidal forces have destroyed its companion in the course of time.

Around 10% of the radio pulsars belong in the class of the (old) "millisecond" pulsars with periods of 1.6 ms up to about 25 ms. Among them, in contrast to the "normal" radio pulsars, is a large number, about half, which are members of close binary star systems. An astonishingly large proportion of more than 50% of all millisecond pulsars furthermore occurs in *globular clusters* such as M 15 and 47 Tuc (Sect. 9.2.1). Here, the conditions for the formation, but also for the destruction of binary star systems are particularly favorable.

Precise observations of the pulsar PSR B1257+12, which has a period of 6.22 ms, led A. Wolszczan and D. A. Frail in 1992 to the discovery of a *"planetary system"*: Two bodies with a few Earth masses orbit with periods of 0.18 and 0.26 yr at a distance of about 0.5 AU on nearly circular paths around the pulsar. This system was probably formed when the pulsar captured some "fragments" after the supernova explosion of the massive principal component of the original binary system.

Geminga. This interesting object, discovered in 1972 with the SAS-2 satellite, is, according to observations with COS-B, one of the strongest gamma-ray sources above about 50 MeV in the sky (Fig. 10.20). Geminga (2 CG 195+04) was identified in 1983 with an X-ray source, which finally in 1992 by means of observations with ROSAT was found to have a pulse *period* of 237 ms. With this knowledge, it was then possible to detect the same period in the gamma-ray spectral range using the Compton observatory.

Geminga, at a distance of about 30 pc, is indeed the closest neutron star to the Earth. From the increase in its pulse period, its age can be estimated to be about $3.7 \cdot 10^5$ yr.

Geminga has the typical characteristics of the young radio pulsars in the Crab Nebula and in Vela, with one exception: the source is *not* observed in the radio region! The reason for this could be that the radiation cone of this pulsar is not pointing in our direction.

7.4.8 Stellar Gamma-ray Sources

We have already met up with stellar gamma-ray sources in the case of the Sun with its flare outbursts, and of the supernovae. As we saw, the emissions of some X-ray binary stars (Sect. 7.4.6) and pulsars (Sect. 7.4.7) also extend beyond the X-ray range into the gamma-ray region.

Here, we discuss a few selected gamma-ray sources, i.e. objects which emit strongly in the gamma-ray range; distinguishing from the X-ray range is often somewhat arbitrary, since the transition from the hard X-ray to the soft gamma-ray region is not well defined. The bursters, which were unexplained for a long time, are characterized by short, intense radiation flashes in the gamma-ray region; we treat them separately in the following section (7.4.9).

In particular in the case of the transient gamma sources, the temporal structure of the emission is often very similar to that of the X-ray binary stars, so that we can assume a common origin, namely the accretion of matter onto a neutron star with a strong magnetic field, or also onto a black hole, in a close binary system. For example, the brightness decrease of the gamma source GRO J0422+32 a few days after its outburst showed great similarity to that of the X-ray nova Muscae (Sect. 7.4.6), so that we can classify this object also as a candidate for a black hole. GRO J0422+32, by the way, attained three times the brightness of the Crab Nebula and was for a short time the brightest gamma source in the sky.

The notation for gamma sources gives the right ascension and the declination, analogously to that for the pulsars (Sect. 7.4.7); their rough *galactic* coordinates can then be read out (e.g. using Fig. 11.1).

In particular from the direction of the *galactic center* we find a number of bright, noticeably variable gamma sources which, as a result of the strong interstellar extinction, cannot be observed in the optical region. When in the late 1970's, using satellites and balloons (with an angular resolution of only $\simeq 15°$), gamma radiation from the direction of the galactic center at 0.511 MeV with a strong time variation on a scale of several months was observed, the center of the Milky Way itself was at first assumed to be the source of this *annihilation radiation* (Sect. 4.5.6) from positrons and electrons, $e^+ + e^- \rightarrow \gamma + \gamma$. Only with the improved resolution ($\simeq 15'$) of the imaging gamma telescope SIGMA on the GRANAT satellite did it become clear in the 1990's that the annihilation radiation was coming from a variable "point source" at about $50'$ (100 pc) distance from the center of the galaxy, which coincided with the X-ray source 1E 1740.7-2942 observed by the Einstein satellite.

Outside the region of the galactic center, two other strong gamma sources have been discovered: GRS 1915+105 by GRANAT, and GRO J1655-40 by the Compton observatory. Both sources are close to the plane of the galaxy, about 40° and 15° from the center, respectively, so that they are likewise unobservable in the optical spectral region. Their outbursts can, however, be seen to some extent in the X-ray, infrared, and radiofrequency ranges.

High-resolution *radio* observations with the VLA at $\lambda = 6$ and 20 cm wavelengths show for 1E 1740.7-2942, along with a variable, compact central source, also two *jets* which extend in opposite directions and are about 1 pc long (Fig. 7.43). The radio emission of these jets is probably synchrotron radiation from relativistic electrons and also from positrons. The structure of this stellar galactic source in the radio range is similar to that of many active galaxies and quasars (Sects. 12.3.3, 4), although their dimensions and magnitudes are different. Observations with the VLA showed also that the two strong gamma sources GRS 1915+105 and GRO J1655-40 are *"microquasars"*, an unusual structure for stellar sources. Another object with energetic jets, which we have already described, is the X-ray source SS 433 (Sect. 7.4.6).

VLA observations over longer periods of time have shown that the radio jets from GRS 1915+105 and GRO J1655-40 move away from the centers of the sources with velocities apparently greater than the velocity of light, c. This phenomenon of *apparent superluminal velocity* was first found in the jets of active galaxies; we thus explain it in connection with these objects in Sect. 12.3.4.

7.4.9 Gamma Bursters

In the 1970's, the American Vela satellites (which were intended to detect explosions of nuclear weapons) observed short, intense bursts of radiation in the gamma-ray range. In the following years, several hundred gamma bursts were observed from satellites and space probes, until from 1991 on, after the launch of the Compton observatory (CGRO), with its enormous increase of sensitivity and resolution, the rate of discovery rose to about *one* burster *per day*. In the meantime, the instruments of the CGRO, especially BATSE (Burst and Transient Source Experiment, Sect. 5.3.3), have yielded the main contribution to the roughly 3000 known gamma bursters (as of 2001).

The gamma bursters are, during the short time of their outbursts (of the order of 10 s), by far the brightest

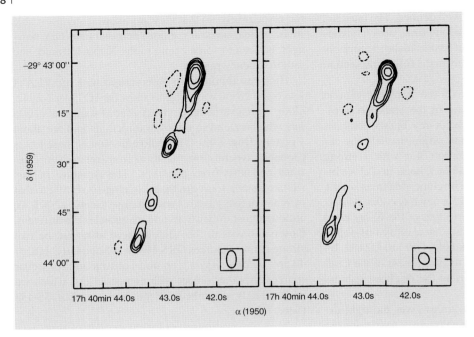

Fig. 7.43. Radio image of the X-ray and gamma-ray source 1E 1740.7−2942 after I. F. Mirabel et al.. The observations with the VLA at $\lambda = 6$ and 20 cm show, along with a compact central source, two jets in opposite directions. (With the kind permission of Nature **358**, 215 © 1992 Macmillan Magazines Ltd., and of the authors)

sources in the gamma-ray range. Their emission occurs mainly at energies above about 50 keV and exhibits a great variability of its intensity structure and spectrum. Since the position of a short burst in the gamma range cannot be determined with great precision and its occurrence is not predictable, it was for nearly three decades not possible to observe the emissions from the gamma bursters in other spectral regions. Therefore, no direct identification with any other known objects and no distance determination could be made, so that the nature of the gamma bursters remained unknown.

The only exception was the tiny group of SGR's (*Soft Gamma-ray Repeaters*), which at present (2001) has only four members. These objects exhibit characteristically shorter bursts with a softer gamma-ray spectrum than the "normal" gamma bursters; they occur in multiple bunches. They also show stronger outbursts in other spectral regions. Two of these bursters are in young supernova remnants (Sect. 10.3.3) of our galaxy and thus are probably associated with neutron stars. The unusually intense gamma bursts from the third source (SGR 0526-66) on March 5, 1979 were registered by eleven space probes. The short rise time (≤ 0.25 ms) in connection with the large distances of the probes from each other made it possible to carry out an exact position determination by "time-of-flight triangulation": the position of this burster coincides with that of the supernova remnant N 49 in our neighboring galaxy, the Large Magellanic Cloud. The gamma flashes from SGR 0526-66 were exceeded in intensity by the outburst at 1900+14 on August 27, 1998. Here too, observations from several space probes made a precise position determination possible and allowed the identification with an X-ray source.

Only in 1997 was it possible for E. Costa and coworkers, using the Italian-Dutch satellite *BeppoSAX*, launched in 1996, to detect the outbursts from several of the *"classical"* gamma bursters in the X-ray range as well. This then opened up the possibility of observing their "afterglow" in the optical and radiofrequency regions and thus finally of clarifying the origin of the gamma bursts.

Properties of the Bursts. The *duration* of the outbursts lies between about 10 ms and more than 1000 s; the most probable value lies around 20 s. It is to be sure quite difficult to determine, since it depends sensitively on the intensity of the burst and on the spectral background. Single pulses without additional time structure can be observed, often with a very short risetime, followed by an exponential decay. Other bursters show smooth emissions with a number of peaks. There are also outbursts which are separated by quiet intervals of up to 100 s; and finally, complex, chaotic series of sharp intensity maxima with differing strengths can be observed.

There is *no* clear indication of *repeated* outbursts.

The *spectra* exhibit a smooth continuum, which stretches over a very wide energy range from a few keV up to 10 GeV, whereby the main emissions occur between about 100 keV and 1 MeV. The occurrence of spectral features is very rare. It is furthermore interesting that the arrival times of the high-energy photons relative to that of the low-energy photons are delayed by up to about 1 h.

Distribution. A first indication of the nature of the gamma bursters is given by their distribution on the celestial sphere: they occur randomly, and their distribution is highly *isotropic*.

Further information is given by the (integrated) number $N(>S)$ of those bursts whose *maximum* flux, integrated over the whole energy range, exceeds the value S. If the sources were distributed homogeneously in (Euclidian) space and their luminosity functions were everywhere the same, then we would expect

$$N(>S) \propto S^{-3/2}, \qquad (7.112)$$

since on the one hand, $N(>S) \propto r^3$, and on the other, $S \propto 1/r^2$ where r is the distance to the source (cf. also the brightness distribution of the stars in Fig. 11.2). However, the observations clearly show fewer weak gamma bursts in comparison to the prediction of (7.112) (see Fig. 7.44). The deviation of the intensities from the $S^{-3/2}$-function, together with the isotropic distribution on the celestial sphere, restricts the possible interpretations severely: the gamma bursters must be nearly homogeneously distributed within a *limited*, roughly spherical volume.

For this distribution, there are in principle two models, which could be distinguished decisively only after the identification of several gamma bursters on the basis of their optical afterglow:

(a) a very extended (≥ 100 kpc) spherical *halo* around our *Milky Way Galaxy*. To be sure, no suitable objects with such a distribution are known. For neutron stars, which are under discussion as sources of the gamma bursts owing to their properties (see below), we

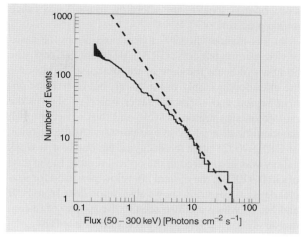

Fig. 7.44. The cumulative distribution $N(>S)$ of gamma bursters with a maximum flux $>S$ in the energy range from 50 to 300 keV, from observations using the detector system BATSE of the Compton observatory. The dashed curve gives the expected function for a homogeneous distribution, $N(>S) \propto S^{-3/2}$ (7.112). (With the kind permission of *Sterne und Weltraum*)

expect (for those belonging to the Milky Way) a concentration within the plane of the galaxy and not an isotropic distribution.

(b) a *"cosmological"* origin of the gamma bursters, i.e. a distribution which extends far past our nearest neighboring galaxies, since no concentration of the bursters e.g. in the direction of the Andromeda galaxy can be detected. In order to explain the deviations from the $S^{-3/2}$-function (Fig. 7.44) through effects of the red shift and of the temporal development of the sources, the weakest sources would then have to have red shifts of at least $z \simeq 1$ (i.e. distances of the order of 1 Gpc $= 10^9$ pc; cf. Sect. 13.1.1).

Identification. The precision with which their positions can be determined, of "only" about 2° for the stronger gamma bursts, did not initially allow their identification with known objects. Only with the aid of the BeppoSAX satellite, which can observe the outbursts both in the gamma-ray and in the X-ray regions, was it possible to carry out an improved position determination on the basis of the X-ray source. With this information, an imaging X-ray telescope on BeppoSAX can – within a few hours – localize the afterglow in the X-ray range with a precision of $\leq 1'$.

In this manner, in 1997 it was possible in the cases of two gamma bursters, GRB 970228 and 970508 (the numbers indicate the date of the outburst), to observe the afterglow not only in the X-ray, but also in the *optical* region and – for the latter object – also in the radiofrequency region as well, and thus to accomplish the identification of these two bursters. At the position of GRB 970228 after the decay of the outburst, a weak galaxy was discovered in the optical region. In the case of GRB 970508, the optical afterglow persisted for several days; its spectrum shows absorption lines from Fe II and Mg II, which are also seen in the spectra of distant quasars (Sect. 12.3.4), with a red shift of $z = 0.84$.

In the meantime (2001), observations of the afterglow have become available for more than 30 gamma bursters; for the most part, they can be identified with faint, distant mother galaxies (of brightness range 23–26 mag in the red). Often, these galaxies show indications of a high rate of star formation.

We list here several of the particularly interesting gamma bursters: GRB 990123, with a red shift of $z = 1.6$, was discovered in the optical region while it was still emitting gamma rays. Assuming isotropic emission, for this burster the gamma flash was found to contain an unusually high energy of $4 \cdot 10^{47}$ J (corresponding to $2\mathcal{M}_\odot c^2$); probably the gamma-ray emission in this case is not isotropic, but instead is focused (in jets?). In the optical afterglow from GRB 000301C ($z = 2.04$), the brightness increased for a short time by more than a factor of 2. This increase can be attributed to a star in the foreground acting as a mivro-gravitational lens (Sect. 8.4.3). GRB 980425 was the first burster which could be connected with a supernova explosion; its position agrees sufficiently well with that of SN 1998 BW, which flared up in a relatively nearby galaxy (at a distance of 40 Mpc). GRB 000131, with $z = 4.50$, has a particularly high red shift.

All these observations clearly support a *cosmological* origin for the gamma bursters.

Energy Source. From the shortest time variations of the signals, amounting to a few tenths of a millisecond, we can estimate the spatial extent of the emission region to be at most 100 km, since the signals can propagate with at most the velocity of light. This supports the idea that, as in the case of the X-ray bursters, processes in neutron stars or black holes are responsible for the radiation flashes.

At "cosmological" distances $z \simeq 1$, the energy in a gamma burst, assuming isotropic emission, would be about 10^{44} to 10^{46} J; at a typical duration of 10 s, this corresponds to a luminosity of $10^{19} L_\odot$. The required energy is thus not essentially greater than that which is released in a supernova explosion in the form of neutrino emissions; however, the luminosity of a gamma burster exceeds that of a supernova by an enormous factor (Sect. 7.4.7).

The *physical process* which leads to a radiation flash mainly in the gamma-ray region, followed by an afterglow in the longer wavelength spectral range, can be described in terms of a "fireball": when an enormous amount of energy is released in a short time within a small volume, which is initially optically dense with respect to gamma rays, then nearly all the energy of the explosion is converted into kinetic energy of highly relativistic particles. This leads to a rapid expansion of the source, whereby a considerable portion of energy is emitted in the form of gamma radiation. When the relativistic particles collide with the surrounding matter, e.g. the interstellar medium of the mother galaxy, the afterglow results.

As the *source* of the energy of around 10^{45} J, on the one hand the *fusion* of two *neutron stars* in a close binary system is a possibility; this is a process which in an average galaxy might occur every 10^6 yr. Through radiation of gravitational energy, the orbital angular momentum of the system is reduced, analogously to the model for supernovae of type Ia, where two white dwarf stars fuse (Sect. 8.2.4); as a result, the two neutron stars approach each other more and more closely. This model, to be sure, is applicable only to the short-duration gamma bursters (with bursts of ≤ 1 s).

The longer-lasting gamma bursts (≥ 1 s) are probably related to the *hypernovae*, an unusual type of supernova explosion in which a very massive, rapidly rotating individual star collapses to form a black hole.

8. The Structure and Evolution of Stars

The significance of his diagram for investigating stellar evolution was clearly realized by H. N. Russell as early as 1913. But its deeper understanding, and thus a theory of stellar evolution founded on observational facts, became possible only in connection with the study of the *inner structure* of stars. The older works of J. H. Lane (1870), A. Ritter (1878–89), R. Emden (the "Spheres of Gas" appeared in 1907) and others could be based only on classical thermodynamics. A. S. Eddington succeeded in combining these approaches with the theory of radiation equilibrium and with Bohr's theory of *atomic structure*, which had in the meantime been formulated; his book "The Internal Constitution of Stars" (1926) gave the start signal for the whole development of modern astrophysics.

Starting from the knowledge that the *Sun* has essentially not changed its luminosity since the formation of the Earth $4.5 \cdot 10^9$ ago, J. Perrin and A. S. Eddington recognized in 1919/20 that the mechanical or radioactive energy sources which had been considered up to that point could not have been sufficient to supply the Sun's radiative energy. They made the assumption that *nuclear energy* is released in the interior of the Sun and the stars, by transmutation, or, as it is also termed, "burning" of hydrogen into helium. The rapid developments in nuclear physics in the 1930's then made it possible in 1938 for H. Bethe and others to find out which *nuclear reactions* would be possible at temperatures of 10^6 to 10^8 K in solar matter and other element mixtures, and to calculate their consequences. Experimental investigations of the reaction cross-sections at low proton energies, especially those of W. A. Fowler, made essential contributions to our understanding of nuclear energy release in stars.

Although we can understand the basic ideas of Eddington's theory with a very modest mathematical effort, the solution of the fundamental equations, together with the complicated material equations (for energy release, etc.) and the equation of state of stellar matter, requires a considerable numerical effort. The application of increasingly powerful computers has given the field of stellar structure and evolution a strong impulse since the mid-1950's.

We shall first consider in Sect. 8.1 the basic equations for the structure of a star, along with the associated matter equations; we then apply these in Sect. 8.2 to understand the evolution of the Sun and the stars, i.e. the changes in their structures with time. The explanation of the color–magnitude diagrams of star clusters, which indeed provided the initial impulse for the development of the basic ideas of stellar evolution, will be left for Chap. 9. As a further application here, we consider in Sect. 8.3 the final stages of stellar evolution, with their extremely high matter densities: the white dwarfs and neutron stars. For the calculation of the precise structure of a neutron star, Newton's Theory of Gravitation is no longer sufficient, and instead Einstein's General Theory of Relativity must be employed. It is thus appropriate that in Sect. 8.4, we give an overview of the fundamental assumptions and the most important results of this theory, particularly in view of later applications to the activity of galactic nuclei and especially to cosmology.

8.1 The Fundamental Equations of Stellar Structure

In Sect. 8.1.1 we treat the hydrostatic equation together with the equation of state of stellar matter. In Sect. 8.1.2, we follow this with a discussion of energy transport by radiation and convection, important for the determination of the temperature distribution within a star. Energy production can take place on the one hand through thermonuclear reactions, of which we give an overview in Sect. 8.1.3, and on the other through release of gravitational binding energy; the latter will be treated in Sect. 8.1.4, in connection with the thermal energy content of a star. In Sect. 8.1.5, we turn to the most important question of the stability of a star. Finally, in Sect. 8.1.6, we summarize the system of fundamental equations and derive their general conclusions for stellar structure.

8.1.1 Hydrostatic Equilibrium and the Equation of State of Matter

We first consider a star as a sphere with the mass $\mathcal{M}$ and radius R in hydrostatic equilibrium: the gravitational force of the mass $\mathcal{M}(r)$ which lies within r acts on each volume element with density $\varrho(r)$ at a distance r from the star's center, and produces a radial variation in the pressure P within the volume element. The *hydrostatic equation*, which we have already derived in connection with planetary structures in Sect. 3.1.3, is then:

$$\frac{dP}{dr} = -\varrho(r)\frac{G\mathcal{M}(r)}{r^2}, \quad (8.1)$$

where $\mathcal{M}(r)$ is given by

$$\frac{d\mathcal{M}(r)}{dr} = 4\pi r^2 \varrho(r). \quad (8.2)$$

In most stars, P is practically equal to the gas pressure P_g; only in very hot and massive stars is it necessary to include the radiation pressure P_r explicitly, i.e. $P = P_g + P_r$. Using (8.1), we can readily estimate the pressure, for example at the center of the Sun ($P_{c\odot}$), by setting $\varrho = $ const (3.7). We thus obtain $P_{c\odot} \simeq 1.3 \cdot 10^{14}$ Pa; exact model calculations yield a central pressure which is about a factor of 100 larger ($2.5 \cdot 10^{16}$ Pa).

The relation (at each point) between the pressure P, the density ϱ, and the temperature T as the third state variable is contained in the *equation of state*. Following Eddington, we begin with the equation of state of the *ideal gas* (3.15):

$$P_g = \varrho \frac{kT}{\bar{\mu} m_u}, \quad (8.3)$$

where $k = 1.38 \cdot 10^{-23}$ J K^{-1} is the Boltzmann constant, $m_u = 1.66 \cdot 10^{-27}$ kg the atomic mass unit ($\simeq$ the proton's mass), and $\bar{\mu}$ the average molecular weight. Equation (8.3) can be used so long as the interactions of neighboring particles are sufficiently small relative to their thermal (kinetic) energies. On Earth, we are accustomed to the fact that this no longer applies, i.e. that condensation begins, roughly at densities of $\varrho \gtrsim 500$ to 1000 kg m^{-3}. In stars, this limit is shifted up to much higher values by *ionization*. In particular, the most abundant elements H and He are completely ionized at relatively shallow depths in all stars.

In the range of higher densities (and lower temperatures), at $T = 10^7$ K and roughly above $\varrho \simeq 10^6$ kg m^{-3}, Fermi–Dirac degeneracy of the *electrons* arises, so that (8.3) is no longer applicable. The equation of state of these degenerate electron components, which are important in the later phases of stellar evolution, will be discussed in Sect. 8.3.2.

The *average molecular weight* $\bar{\mu}$ is, for complete ionization, equal to the atomic mass divided by the total number of particles, i.e. nuclei plus electrons. We thus obtain for

Hydrogen	Helium	Heavy elements
$\bar{\mu} = 1/2$	4/3	$\simeq 2$

From these values, the average molecular weight can easily be calculated for any mixture of elements.

The average temperature in the interior of the Sun can be estimated from (8.3) with $\varrho = \bar{\varrho}_\odot$, $\bar{\mu} = \mu_H = 0.5$ and $\overline{P}_g \simeq P_{c\odot}/2$ (see above) to be $\overline{T}_\odot \simeq 6 \cdot 10^6$ K.

8.1.2 Temperature Distribution and Energy Transport

To calculate the temperature distribution $T(r)$ in the interior of a star, we must first investigate the mechanisms of energy transport. Poor energy transport leads to a steep temperature gradient, good energy transport yields a flat temperature gradient. (Whoever doubts this can test it by holding first a wooden stick and then a nail in a flame with bare fingers!).

Radiation Equilibrium. We first consider the energy transport via *radiation*, i.e. radiation equilibrium (cf. Sect. 7.2.1, also for nomenclature); we immediately integrate all the radiation quantities such as I_ν, F_ν and B_ν over the *total* frequency spectrum. These integrals are denoted as I, F, and B. Furthermore, we again introduce Rosseland's opacity $\bar{\kappa}$ (7.41) as an average over all frequencies, here using *mass* absorption coefficients $\bar{\kappa}_M$ [m^2 kg^{-1}] (4.109), as is usual in the theory of stellar structure; for simplicity, we denote it simply by κ. The radiation transport equation (7.37) then takes on the form:

$$\cos\theta \frac{dI}{\kappa\varrho\, dr} = -I + B. \quad (8.4)$$

From the radiation intensity I, we again calculate the radiation flux F (4.34) by multiplying with $\cos\theta$ and integrating over all directions. For the total radiation we thus obtain from (8.4):

$$F = -\int_0^\pi \frac{dI}{\kappa\varrho\, dr} \cos^2\theta \cdot 2\pi \sin\theta\, d\theta. \quad (8.5)$$

(Since the Kirchhoff–Planck function B is isotropic, the integral of $B\cos\theta$ over all directions vanishes.)

In a *stellar interior*, the radiation field I is nearly isotropic, so that we can extract it from the integral and carry out the integration over θ on the right side of (8.5); this gives

$$\int_0^\pi \cos^2\theta \cdot 2\pi \sin\theta\, d\theta = 4\pi/3 \,. \tag{8.6}$$

Furthermore, we can write down I in this case as a function of temperature T using the Stefan–Boltzmann law. It is

$$I = \frac{\sigma}{\pi} T^4 = \frac{ac}{4\pi} T^4 \,, \tag{8.7}$$

and we thus obtain

$$F = -\frac{c}{3\kappa\varrho}\frac{d}{dr}(aT^4) \,, \tag{8.8}$$

i.e. the radiation flux is proportional to the gradient of the energy density $u = aT^4$ (4.72) and to the "conductivity" $(\kappa\varrho)^{-1}$ of the stellar matter with respect to photons.

The total radiation energy which passes outwards per unit time through a spherical surface of radius r is then given by:

$$L(r) = 4\pi r^2 F = -\frac{16\pi acr^2}{3}\frac{T^3}{\kappa\varrho}\frac{dT}{dr} \,. \tag{8.9}$$

At the surface of the star, $r = R$, the shell luminosity $L(r)$ becomes the directly measurable stellar luminosity $L = L(R)$. At temperatures near $\simeq 6 \cdot 10^6$ K, the maximum in Planck's radiation function according to (4.65) lies at $\lambda_{max} = 0.5$ nm, i.e. in the X-ray range. The *absorption coefficient* is determined here by bound–free and free–free transitions of the atomic states which are not yet completely "ionized away". In an element mixture such as we find in the Sun and in Population I stars, these are higher ionization states of the more abundant, heavier elements such as O, Ne, … In the "metal poor" stars of Population II, the contributions of H and He must also be taken into account.

Convective Energy Transport. Besides energy transport by radiation, a hydrogen and helium convection zone can occur in stellar interiors, as we saw in the example of the Sun; the conditions for its occurrence were discussed in Sect. 7.3.1. In addition, further convection zones can be formed in connection with nuclear energy release. In all such convection zones, energy transport by convection (rising of hotter and sinking of cooler matter) is strongly predominant over transport by radiation, except at the the boundary regions of the zone. The relation between the temperature T and the pressure P is then given by the adiabatic equation $T \propto P^{1-1/\gamma}$, which we have already used in treating planetary atmospheres (3.20) and in connection with the solar wind (7.95). $\gamma = c_p/c_v$ is the ratio of specific heats at constant pressure and at constant volume. By logarithmic differentiation with respect to r, we obtain (3.21) the *temperature gradient* in a convection zone:

$$\frac{dT}{dr} = \left(1 - \frac{1}{\gamma}\right)\frac{T}{P}\frac{dP}{dr} \,. \tag{8.10}$$

The methods for calculating convective energy transport are still very unsatisfactory in quantitative terms.

8.1.3 Energy Production Through Nuclear Reactions

In *hydrogen fusion*, the overall reaction consists of the combination of four hydrogen nuclei to form one helium nucleus, $4\,^1\text{H} \rightarrow {}^4\text{He}$. The energy released, ΔE, can be readily calculated using the mass–energy relation discovered by A. Einstein in 1905,

$$\Delta E = \Delta mc^2 \tag{8.11}$$

from the mass difference Δm between the products and the reactants in the above nuclear reaction. The mass of 4 hydrogen atoms is $4 \cdot 1.007825$ atomic mass units m_u, that of the helium atom is $4.0026\,m_u$, so that the mass difference is given by $0.0287\,m_u$ ($1\,m_u = 1.6605 \cdot 10^{-27}$ kg $= 931.49$ MeV/c^2). This corresponds to an energy difference of:

$$\Delta E(4\,^1\text{H} \rightarrow {}^4\text{He}) = 4.28 \cdot 10^{-12}\text{ J}$$
$$= 26.73 \text{ MeV} \,. \tag{8.12}$$

In fact, we require here the masses of the atomic *nuclei*. However, since the number of electrons remains unchanged in the reaction, we can use the atomic masses, whose experimental values are more precisely known, risking only very small errors due to differences in the electronic binding energies.

The "usable" energy released by a nuclear reaction, Q, is still smaller than calculated from the mass defect by the amount which escapes through neutrino emission, since neutrinos, owing to their extremely small interaction cross-sections, escape even from the interior of a star practically without energy loss (unless we consider the extremely dense final stages of stellar evolution). In the case of hydrogen fusion, we thus find instead of 26.73, only at most 26.23 MeV which will be available within the star (cf. (8.18)).

If we imagine that e.g. one solar mass of pure hydrogen were to be converted into helium, then an energy of $1.28 \cdot 10^{45}$ J would be released; this would supply the luminosity of the Sun at the current level $L_\odot$ for a period of $1.05 \cdot 10^{11}$ yr.

The continued formation of elements heavier than ^{4}He from lighter elements allows further energy releases in accord with (8.11), up to ^{56}Fe, at which point the nuclear binding energy is a maximum, equal to 8.4 MeV per nucleon. The main portion of this energy gain is already released in the fusion of hydrogen to helium, which yields 6.7 MeV per nucleon (8.12).

Energy Production Rate. In order to calculate the rate of nuclear energy release in stars, we must consider the individual nuclear reactions in detail. In the following, we summarize the most important reactions, employing the usual notation:

$x(a, b)y$

Reactant nucleus x (reacts with a, ejects b) Product nucleus y.

Here, we use the symbols

p:	Proton	α:	^{4}He nucleus
n:	Neutron	γ:	Quantum of radiation
e^-:	Electron	ν:	Electron neutrino
e^+:	Positron		

A nucleus (nuclide) is denoted by adding its mass number A to the upper left of the chemical symbol X and, where needed, its atomic number (number of protons) Z to the lower left: ^{A}X or ^A_ZX. For example, we write the two stable helium isotopes as ^{3}He and ^{4}He or $^3_2$He and $^4_2$He. A nucleus in an excited state is denoted by an asterisk (*).

A positron e^+ which is released in a nuclear reaction immediately forms two gamma quanta by annihilating with a thermal electron e^-. The neutrinos escape from the star practically without energy loss (see above).

The number of reactions $x(a, b)y$ in a unit volume per unit time (cf. (4.111)) is now:

$$\mathcal{P} = n_x n_a \langle \sigma v \rangle , \qquad (8.13)$$

where n_x and n_a are the particle densities of the reaction partners x and a, v is their relative velocity, and σ the reaction cross-section for the fusion reaction considered. The thermal average $\langle \sigma v \rangle$ is obtained by integration over the Maxwell–Boltzmann distribution (4.86), which also holds for the relative velocities (Sect. 4.5.3).

The energy release rate ε [W kg^{-1}], i.e. the energy released per unit time and mass, is then

$$\varepsilon = Q \frac{\mathcal{P}}{\varrho} . \qquad (8.14)$$

Here, ϱ is the density and Q the energy released by *one* reaction.

Due to the short range of the nuclear forces, the positively-charged reactant nuclei, which repel each other corresponding to the Coulomb potential

$$V(r) = \frac{1}{4\pi\varepsilon_0} \frac{Z_x Z_a e^2}{r} , \qquad (8.15)$$

must approach to within a distance of the order of a nuclear diameter, $r_0 \simeq 10^{-15} A^{1/3}$ [m], in order for the reaction to occur. They must first overcome a "Coulomb barrier" of height $B \simeq V(r_0)$ (Fig. 8.1). While B is in the range of several MeV, the average mean kinetic or thermal energy $E = mv^2/2 = (3/2)kT$, e.g. at a temperature of 10^8 K, is only about 0.01 MeV. The Coulomb barrier can thus only be "tunneled" through, as predicted by quantum mechanics. The probability for this process, as calculated by G. Gamow, is

$$P(v) = e^{-2\pi\eta} , \qquad \eta = \frac{1}{4\pi\varepsilon_0} \frac{Z_x Z_a e^2}{\hbar v} . \qquad (8.16)$$

This leads to a very strong temperature dependence of the thermonuclear reactions; one can speak of an "ignition temperature" for the initiation of energy production, particularly in the case of nuclei with higher proton numbers Z.

In a more precise calculation, the *screening* of the nuclear charges by the electrons in the plasma must be considered; it causes a lowering of the Coulomb barrier,

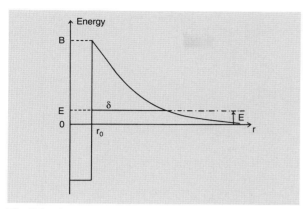

Fig. 8.1. Thermonuclear reactions: the radial potential function $V(r)$ (schematic) for the approach of two positively charged particles. For $r \geq r_0$, i.e. outside the range of the attractive nuclear forces, the Coulomb repulsion is predominant. A particle with the (relative) kinetic energy E at infinity, corresponding to a mean thermal energy of $(3/2)kT$, must tunnel through the width δ of the Coulomb barrier of height $B \gg E$, before a nuclear reaction can take place

A second possibility for the conversion of hydrogen into helium is offered by a reaction *cycle* investigated in 1938 by H. Bethe and C. F. von Weizsäcker, in which the elements C, N and O are involved and essentially determine the reaction rate, but are quantitatively "returned" at the end. Energy release by the *CNO triple cycle* (Fig. 8.2) is mainly due to the principal or CN cycle (24.97 MeV), whose slowest reaction is ^{14}N$(p,\gamma)^{15}$O. The two subcycles take place when a γ quantum instead of an α particle is emitted following the proton capture by ^{15}N; they occur e.g. in the Sun about 1000 times less frequently than the principal cycle.

In the stationary state, the *abundance ratios* of the isotopes involved are determined by the rates of the individual reactions as in a "radioactive equilibrium", although in each *single* cycle, the C, N, and O nuclei, as "catalysts", are not consumed. At temperatures from about 10^7 to 10^8 K, most of the matter originally present as C, N and O is converted by the CNO cycle into ^{14}N.

Starting from the normal element mixture (Tables 7.5, 6), the nitrogen abundance increases by about a factor of 10. The abundance ratio of the two carbon isotopes ^{12}C/^{13}C, which is about 90 in terrestrial and solar matter and between 40 and 90 in interstellar matter, settles down to a much smaller value $\simeq 4$. This is an important indicator for (possibly previous) hydrogen combustion by the CNO cycle.

In Fig. 8.3, the mean *energy production* ε from hydrogen fusion for an element mixture as found in the Sun and in Population I stars is plotted as a function of temperature over the range 5 to $50 \cdot 10^6$ K. Due to the higher electric charges of the reactant nuclei, the CNO cycle is initiated only at higher temperatures than the pp chain. Cool main sequence stars with central temperatures of up to $1.8 \cdot 10^7$ K, in particular also the Sun, with $T_{c\odot} \simeq 1.5 \cdot 10^7$ K, therefore extract their energy mainly from the pp chain; hotter stars in the upper portion of the main sequence derive theirs mainly from the CNO cycle. For the old stars of Population II, the

since according to the Debye theory, the electrostatic potential (8.15) is modified according to

$$\tilde{V}(r) = V(r)\, e^{-r/r_D} . \qquad (8.17)$$

Here, $r_D = \sqrt{\varepsilon_0 kT/(e^2 n)}$ is the Debye length for a completely ionized hydrogen plasma with a particle density of n.

Hydrogen Burning. We now consider the reactions of hydrogen burning, i.e. the combination of four protons to a helium nucleus, in detail. A first possibility is the *proton–proton reaction* or *pp-chain* (8.18).

The slowest reaction, which represents the rate-limiting step for the whole chain, is ^{1}H$(p, e^+\nu)^2$D. The neutrinos which are released in this step carry away on the average 0.50 MeV of energy, so that 26.23 MeV of energy is available to the star from the main pp I chain. The branches pp II and pp III become more significant with increasing temperature.

$$^1\text{H}(p, e^+\nu)\,^2\text{D}(p,\gamma)\,^3\text{He} \begin{cases} ^3\text{He}(^3\text{He}, 2p)\,^4\text{He} & \text{pp I} \quad 26.23\ \text{MeV} \\ ^3\text{He}(\alpha, \gamma)\,^7\text{Be} \begin{cases} ^7\text{Be}(e^-, \nu)\,^7\text{Li}(p, \alpha)\,^4\text{He} & \text{pp II} \quad 25.67\ \text{MeV} \\ ^7\text{Be}(p, \gamma)\,^8\text{B}(e^+\nu)\,^8\text{Be}(2\alpha) & \text{pp III} \quad 19.28\ \text{MeV} \end{cases} \end{cases} \qquad (8.18)$$

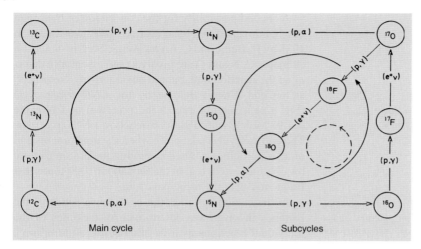

Fig. 8.2. Reactions of the CNO triple cycle

contribution of the CNO cycle relative to the pp chain is proportional to their underabundances of C, N, and O relative to H.

Helium Burning. If a major portion of the hydrogen is used up in the interior of a star, and its temperature increases to above 10^8 K (as a result of contraction), then the combustion of helium is initiated, as noted by E. J. Öpik and E. E. Salpeter in 1951/52, leading initially to ^{12}C by the *3α process*: $3\alpha \rightarrow {}^{12}$C.

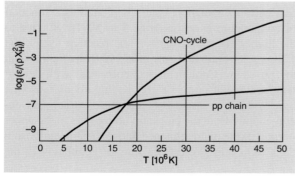

Fig. 8.3. Temperature dependence of the energy release ε by hydrogen fusion in equilibrium for the element mixture of the Sun and Population I stars. ε in [W kg^{-1}], ϱ in [kg m^{-3}], X_H = relative mass abundance of hydrogen. $\varepsilon/\varrho X_\text{H}^2$ depends only on T in the pp chain, but for the CNO cycle, it is also proportional to the abundance $X_{\text{C,N}}/X_\text{H}$. At the center of the *Sun*, $X_\text{H} = 0.36$ (originally 0.73), $\varrho = 1.6 \cdot 10^5$ kg m^{-3}, $T = 1.5 \cdot 10^7$ K and thus $\varepsilon \simeq 1.8 \cdot 10^{-3}$ W kg^{-1}

This process begins with the slightly endothermic fusion of two He nuclei:

$$^4\text{He} + {}^4\text{He} + 95 \text{ keV} = {}^8\text{Be} + \gamma \, . \tag{8.19}$$

The ^{8}Be is present in thermal equilibrium (^{4}He + ^{4}He $\rightleftharpoons$ ^{8}Be) at a very small concentration which is then further decreased by the reaction

$$^8\text{Be} + {}^4\text{He} \rightleftharpoons {}^{12}\text{C}^* \rightarrow {}^{12}\text{C} + \gamma \tag{8.20}$$

with an energy release of 7.28 MeV per ^{12}C nucleus. The decay of the excited ^{12}C*, releasing 7.65 MeV of energy, back to ^{4}He and ^{8}Be is more than 1000 times more rapid than the transition back to its ground state, ^{12}C. Only 2.4 MeV are released per helium nucleus by its fusion, i.e. about 10% of the energy which was obtained from its formation from hydrogen. Starting with ^{12}C, the following nuclei (with A a multiple of four) can be formed by further (α, γ) reactions:

$$^{12}\text{C}(\alpha,\gamma){}^{16}\text{O}(\alpha,\gamma){}^{20}\text{Ne}(\alpha,\gamma){}^{24}\text{Mg}(\alpha,\gamma){}^{28}\text{Si} \, , \tag{8.21}$$

whereby the last two reactions hardly contribute to the energy release, due to the slow reaction rate of ^{16}O$(\alpha, \gamma)^{20}$Ne.

At somewhat higher temperatures than for the the 3α process, ^{14}N, the main product of the CNO cycle, is consumed by the reaction series

$$^{14}\text{N}(\alpha,\gamma){}^{18}\text{F}(e^+\nu){}^{18}\text{O}(\alpha,\gamma){}^{22}\text{Ne} \, . \tag{8.22}$$

The reaction

$$^{22}\text{Ne}(\alpha, n)^{25}\text{Mg} \quad (8.23)$$

can continue this series; it provides free *neutrons* for the synthesis of the *heavy elements* of $A \gtrsim 60$ (through the socalled s-process in red giant stars; see Sect. 8.2.5). This synthesis could in no case occur by means of charged-particle reactions, due to the strong Coulomb fields of the heavy nuclei.

Carbon Burning and Further Burning Phases. After the cessation of helium burning and the reactions which follow it, at temperatures of 6 to $7 \cdot 10^8$ K, in the course of the evolution of a star *carbon burning* can take place; its most important reactions:

$$^{12}\text{C} + {}^{12}\text{C} \rightarrow {}^{23}\text{Na} + p, \quad {}^{23}\text{Na}(p, \alpha)^{20}\text{Ne},$$
$$^{12}\text{C} + {}^{12}\text{C} \rightarrow {}^{20}\text{Ne} + \alpha \quad (8.24)$$

lead to ^{20}Ne and release 2.3 MeV of energy per ^{12}C nucleus. Finally, above 1.5 to $2 \cdot 10^9$ K, the energies of thermal photons are sufficiently high to destroy ^{20}Ne by *photodisintegration* and, by further reactions with the helium nuclei formed in the disintegration, to convert it to ^{24}Mg and ^{28}Si:

$$^{20}\text{Ne}(\gamma, \alpha)^{16}\text{O},$$
$$^{16}\text{O}(\alpha, \gamma)^{20}\text{Ne}(\alpha, \gamma)^{24}\text{Mg}(\alpha, \gamma)^{28}\text{Si}. \quad (8.25)$$

This *neon burning* is followed at somewhat higher temperatures by oxygen burning (^{16}O + ^{16}O) and silicon burning, a quasi-equilibrium between photodisintegration and α capture, which leads to the formation of heavier elements up to the nuclei of the iron group.

In the carbon burning and the later burning phases, the temperatures and densities in stellar interiors are so high that *neutrinos* are produced in large numbers at the expense of the thermal energy of the stellar matter, via the weak interaction between electrons, positrons, and photons; they leave the star practically without hindrance. The energy loss from these neutrinos is of the same order of magnitude as the energy release from thermonuclear processes (in contrast, the neutrinos formed by β-decay of the nuclear reaction products represent only a moderate energy loss). Formally, the neutrinos can be included in the reactions as a negative energy release.

Thermal Equilibrium. The application of nuclear energy release in the theory of the internal structure of stars is (fundamentally) simple: if $\varepsilon(\varrho, T)$ is the energy released per unit of time and mass by *all* the nuclear reactions which occur at the corresponding temperatures T and densities ϱ, then in a spherical shell $r \ldots (r + dr)$ in unit time, the energy $\varrho\varepsilon 4\pi r^2 dr$ is released and the energy flux $L(r)$ increases according to the equation:

$$\frac{dL(r)}{dr} = 4\pi r^2 \varrho(r)\, \varepsilon(r)\,. \quad (8.26)$$

8.1.4 Gravitational Energy and Thermal Energy

When the internal temperature of a star (which has a particular chemical composition) becomes too low to permit thermonuclear processes to continue, the only energy source in the intermediate period until nuclear reactions can again be ignited is the *gravitational energy* E_G, i.e. energy of contraction.

It seems appropriate to mention here that as early as 1846, soon after his discovery of the law of energy conservation, J. R. Mayer raised the question of the origin of the radiation energy emitted by the Sun. He considered the possibility that a mass m of meteorites which fall into the Sun would release their energy as heat according to (2.44)

$$m G \mathcal{M} / R \quad (8.27)$$

(where G is again the gravitational constant, $\mathcal{M}$ and R the mass and radius of the Sun). Since in fact the mass of meteorites falling into the Sun is rather small, H. von Helmholtz pointed out in 1854, and Lord Kelvin in 1861, that the *contraction of the Sun itself* would be a more effective source of gravitational energy.

As one can readily see from (8.27), the energy which is released by the formation of a sphere of gas of radius R from matter which was originally far apart, or also by the strong contraction of a star to the radius R, is given by:

$$E_G \simeq \frac{G \mathcal{M}^2}{R}\,. \quad (8.28)$$

For the Sun, this is $\simeq 3.8 \cdot 10^{41}$ J; the Sun's thermal energy content is of the same order of magnitude. The energy E_G could supply the Sun's radiant power at the current level of $L_\odot = 3.85 \cdot 10^{26}$ W for only $E_G/L_\odot \simeq 3 \cdot 10^7$ yr.

8. The Structure and Evolution of Stars

The Virial Theorem. The close connection between the gravitational energy of a star and its thermal energy becomes clear from the virial theorem (2.36), which states that in a system of massive objects with mutual gravitational attraction, on the average over time the kinetic energy E_{kin} and the potential energy E_{pot} are related by:

$$\overline{E}_{\text{kin}} = -\tfrac{1}{2}\overline{E}_{\text{pot}} \; . \tag{8.29}$$

This relation is presented more clearly in Fig. 8.4. Since the total energy is $E = E_{\text{kin}} + E_{\text{pot}}$, we also have $\overline{E}_{\text{kin}} = -\overline{E}$.

We apply the virial theorem initially to a star (in hydrostatic equilibrium) which we consider for simplicity to be a *homogeneous* sphere (of mass $\mathcal{M}$, radius R, density ϱ). If we further assume it to consist of a monatomic gas of atomic mass μ, then the only kinetic energy present is the *thermal energy* E_{T}. It is on the average $(3/2)kT/(\mu m_{\text{u}})$ per unit of mass ($m_{\text{u}} =$ atomic mass unit), and for the whole star

$$E_{\text{T}} = \frac{3}{2}\frac{kT}{\mu m_{\text{u}}}\mathcal{M} \; . \tag{8.30}$$

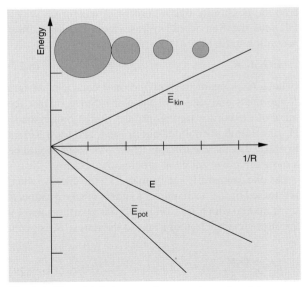

Fig. 8.4. The virial theorem for a system of point masses under the influence of mutual gravitational attraction, $\overline{E}_{\text{kin}} = -\tfrac{1}{2}\overline{E}_{\text{pot}}$. A contracting star loses potential energy; half of the amount lost is, however, added to its reserves of kinetc energy. $2\overline{E}_{\text{kin}} < -\overline{E}_{\text{pot}}$ leads to gravitational instability, according to J. Jeans (Sect. 10.5.3)

The potential energy or the *gravitational energy* of a homogeneous sphere has already been calculated in Sect. 2.3.6; from (2.51):

$$E_{\text{G}} = -\frac{3}{5}\frac{G\mathcal{M}^2}{R} \; . \tag{8.31}$$

Using the *virial theorem*

$$2E_{\text{T}} + E_{\text{G}} = 0 \tag{8.32}$$

we can now derive directly the fact that the temperature in the stellar interior must be 10^6 to 10^7 K, using the known data for main sequence stars, so long as radiation pressure plays no essential role. (We had already estimated this temperature in Sect. 8.1.1 by a different route.)

The Contraction of a Star. However, a star's energy does *not* remain constant, due to the energy which is *radiated away* (luminosity L); instead, it decreases in a time interval δt according to $\delta E = L \delta t$. If no nuclear energy sources are available to the star, it can replace this energy loss only through contraction; from (8.28), in a time δt, contraction provides the quantity $\delta E_{\text{G}} = G\mathcal{M}^2 \delta(1/R)$. If this contraction takes place sufficiently slowly, so that hydrostatic equilibrium can be maintained in the star, then from the virial theorem (8.32), we find

$$2\delta E_{\text{T}} + \delta E_{\text{G}} = 0 \quad \text{and}$$
$$\delta E_{\text{T}} + \delta E_{\text{G}} = L\delta t \; . \tag{8.33}$$

This result shows that one-half of the gravitational energy δE_{G} is required to increase the thermal energy, and the other half to supply the radiation energy (Fig. 8.4). When energy is *extracted* (by radiation), the star thus reacts by *increasing* its temperature, i.e. it represents a system with "negative specific heat". The star is stabilized by the initiation of nuclear energy release as its temperature rises, the prerequisite for this being that the pressure increase with increasing temperature, i.e. that the equation of state (8.3) be valid. The contraction is halted and the luminosity is then maintained from nuclear energy sources.

The time interval during which a star can supply its radiation energy at the expense of gravitational energy E_{G}, the *Helmholtz–Kelvin time*, is of the order of

$$t_{\text{HK}} \simeq \left|\frac{E_{\text{G}}}{L}\right| \simeq \frac{G\mathcal{M}^2}{RL} \simeq \frac{E_{\text{T}}}{L} \; . \tag{8.34}$$

From the virial theorem, it is comparable to the thermal relaxation time. In general, t_{HK} is much shorter than the nuclear evolution time t_E (8.45).

The Rate of Gravitational Energy Release. For those evolutionary stages of a star in which gravitational energy plays a role due to mass configuration changes, we must thus extend (8.26) to include the rate of release of gravitational energy. There is then no longer an equilibrium in a mass shell simply between the nuclear energy release rate ε and the net energy flux. Instead, the time rate of change of the inner structure in each mass shell gives an additional (positive or negative) contribution, which, following the 1st Law of thermodynamics, arises from the sum of the change in thermal energy per unit mass, $(3/2)kT/(\mu m_u)$, and the work performed by pressure forces, $-P dv$ ($v = 1/\varrho$ = specific volume), since gravitation does not act directly on a volume element of the gas. We then find:

$$\frac{dL(r)}{dr} = 4\pi r^2 \varrho \left(\varepsilon - \frac{3}{2} \frac{k}{\mu m_u} \frac{dT}{dt} + \frac{P}{\varrho^2} \frac{d\varrho}{dt} \right) . \quad (8.35)$$

8.1.5 Stability of the Stars

Finally, we ask the question, under which circumstances is a star able to maintain its hydrostatic equilibrium (8.1) in the face of disturbances or changes in its structure? Here, the *compressibility* of its matter, i.e. the variation of the pressure P with density ϱ, plays a decisive role. If we take $P \propto \varrho^\Gamma$, then the gravitational force averaged over the whole star varies (due to $\varrho \propto \mathcal{M}/R^3$) with the star's radius R according to:

$$\langle F_G \rangle \propto \varrho \frac{G\mathcal{M}}{R^2} \propto R^{-5} , \quad (8.36)$$

while the pressure gradient varies as

$$\left\langle \frac{dP}{dr} \right\rangle \propto \frac{P}{R} \propto \frac{\varrho^\Gamma}{R} \propto R^{-3\Gamma - 1} \quad (8.37)$$

and thus the ratio of the two terms depends on R according to:

$$\frac{\langle dP/dr \rangle}{\langle F_G \rangle} \propto R^{-3(\Gamma - 4/3)} . \quad (8.38)$$

If $\Gamma > 4/3$, then for a *contraction*, $\langle dP/dr \rangle$ increases more rapidly than $\langle F_G \rangle$, i.e. the star will finally be able to find a new configuration in hydrostatic equilibrium. On the other hand, for $\Gamma \leq 4/3$, the gravitational force increases more and more if the radius is reduced, and the star will not find a stable configuration. We therefore obtain as a *criterion for the stability* of hydrostatic equilibrium that the relation:

$$\Gamma > \tfrac{4}{3} \quad (8.39)$$

must be fulfilled in the interior of stars over a considerable range of masses. This condition is obeyed e.g. by a star consisting of a monatomic ideal gas, since then the adiabatic equation (7.95), (8.10) has the form $P \propto \varrho^\Gamma$ with $\Gamma \equiv \gamma = 5/3$ (γ = specific heat ratio).

If the stability criterion is violated, for example by the excitation of internal degrees of freedom or by a "phase transition" (Sect. 8.2.4), then a stellar *collapse* results, practically in free fall, as long as $\Gamma \leq 4/3$ is maintained. The characteristic time for this collapse is the *free fall time* t_{ff}, which we estimate by extending (8.1) to include an acceleration term $\varrho \, d^2r/dt^2$ or else use the hydrodynamic equation of motion, (7.87):

$$\varrho \frac{d^2 r}{dt^2} \simeq \varrho \frac{R}{t_{ff}^2} \simeq \varrho \frac{G\mathcal{M}}{R^2} \simeq \frac{P}{R} . \quad (8.40)$$

With $\varrho \simeq \mathcal{M}/R^3$, we then find for the dynamic time scale:

$$t_{ff} \simeq \sqrt{\frac{1}{G\varrho}} . \quad (8.41)$$

For the Sun, with $\varrho_\odot = 1400 \text{ kg m}^{-3}$, t_{ff} is of the order of magnitude of 1 h, i.e. considerably shorter than the Helmholtz–Kelvin time (8.34). Conversely, we can conclude from the stability of the Sun over thousands of millions of years that its hydrostatic equilibrium must be maintained with an extremely high precision. Large-scale pressure disturbances are balanced within t_{ff}. As can be seen from the virial theorem (8.32), the free fall time corresponds to the characteristic propagation time R/c_s of an acoustic wave through the star with the velocity of sound $c_s \simeq (kT/\mu m_u)^{-1/2}$ (7.93), or also to the fundamental period of oscillation of the star (7.99).

8.1.6 The System of Fundamental Equations and Their General Results

In order to give a better overview, we summarize the four basic equations of the theory of stellar interiors from the preceding sections. Their numerical solution is carried out exclusively using fast electronic computers.

The System of Fundamental Equations. The structure of stars is described by the following four coupled differential equations, from which the r-dependence of $\mathcal{M}(r)$, the mass within r, the pressure P, the shell luminosity $L(r)$, and the temperature T can be determined.

Conservation of mass (8.2):
$$\frac{d\mathcal{M}(r)}{dr} = 4\pi r^2 \varrho$$

Hydrostatic equilibrium (8.1):
$$\frac{dP}{dr} = -\varrho \frac{G\mathcal{M}(r)}{r^2}$$

Thermal equilibrium (8.26):
$$\frac{dL(r)}{dr} = 4\pi r^2 \varrho \, \varepsilon \qquad (8.42)$$

Energy transport
a) by radiation (8.9):
$$\frac{dT}{dr} = -\frac{3\kappa\varrho}{4acT^3} \frac{L(r)}{4\pi r^2}$$
b) by convection (8.10):
$$\frac{dT}{dr} = \left(1 - \frac{1}{\gamma}\right) \frac{T}{P} \frac{dP}{dr} \, .$$

In addition, we need the *equation of state* and the equations which relate the *material functions*, the nuclear energy release rate ε, the opacity κ and the specific-heat ratio γ, to two of the state variables P, T, or ϱ. All of these relationships depend essentially on the *chemical composition* of the stellar matter – this is important in what follows.

Finally, our problem is completely determined by the *boundary conditions*:

1. At the center $r = 0$ of a star, we must of course have $\mathcal{M}(0) = 0$ and $L(0) = 0$.
2. At the surface, the equations for the stellar interior must make a smooth transition to those discussed previously for the stellar atmosphere. As long as we are interested *only* in the interior of the star, we shall frequently be able to apply the above equations (8.42) out to $T \to 0$ or $T \to T_{\text{eff}}$ for $r = R$. Furthermore, $\mathcal{M}(R)$ is naturally $\mathcal{M}$, the total mass, and $L(R) = L$, the total luminosity of the star.

General Results. We can readily gain an intuitive understanding of some general results from the theory of stellar structure; of course, they can also be derived by formal calculation.

We imagine the total mass $\mathcal{M}$ of gas to be provided with given nuclear energy sources. This object, initially not at all defined in terms of its spatial structure, we then allow in a thought experiment to consolidate itself into a star of mass $\mathcal{M}$ and luminosity L. It will adjust itself to a particular radius R, assuming that a stable configuration is possible. Since on the other hand, R and L are related to the effective temperature T_{eff} by $L = 4\pi R^2 \sigma T_{\text{eff}}^4$, i.e. luminosity = surface area · total radiation flux (6.43), the effective temperature T_{eff} is determined. There must then exist for stars of similar structure and composition (we must not forget that ingredient!), socalled *homologous* stars, a unique relationship between the mass $\mathcal{M}$, the luminosity L, and the radius R or the effective temperature T_{eff}:

$$\varphi(\mathcal{M}, L, T_{\text{eff}}) = 0 \, . \qquad (8.43)$$

Such a relation was indeed discovered in 1924 by A. S. Eddington. According to his calculations, the dependence of the function on T_{eff} was so weak that he referred simply to a *mass–luminosity relation*. Its agreement with observations initially appeared to be rather good; later, several exceptions were found, which however in the light of the general theory are by no means unexpected.

If we also take into account the fact that theory relates the nuclear energy release ε to the state variables (e.g. T and ϱ) for a given element mixture, there is an additional relation between the three quantities $\mathcal{M}$, L and T_{eff}. For stationary stars of similar structure, an equation of the form

$$\Phi(L, T_{\text{eff}}) = 0 \qquad (8.44)$$

thus holds.

These stars must then lie on a particular *line* in the Hertzsprung–Russell diagram or in the color–magnitude diagram. This result is sometimes called the *Russell–Vogt theorem*. We shall indeed meet up with such a line in the Zero Age Main Sequence. On the other hand, the very existence of the red giants and supergiants tells us that at least one other parameter enters the picture. We shall see that it is the *chemical composition* of the stars, which changes along with their *ages* as a result of nuclear reactions. The interior structure of the stars therefore changes and our above considerations do not necessarily remain applicable.

Conversely, in addition to the the mass of a star, its chemical composition must be known before we can calculate its structure from the system of fundamental equations. The composition is however in general known only for very young stars, which have not yet undergone any essential changes in their element mixtures by nuclear reactions since their formation from the interstellar medium. In practice, we must therefore follow the evolution of a star "from the beginning", i.e. starting from a homogeneous composition, in order to learn about its structure in the later stages of evolution.

A direct comparison of the theory of stellar structure and evolution with observations is, to be sure, not possible, due to the extremely long times required for the evolution as a rule; however, investigations of *star clusters* represent an important test for model calculations. In a star cluster, all members were formed nearly *at the same time* and with the *same* chemical composition; on the other hand, in a "snapshot view", they exhibit a wide range of different stages of evolution, as a result of their differing masses.

8.2 Stellar Evolution

We begin in Sect. 8.2.1 with the structure of the main-sequence stars. In Sect. 8.2.2, we then describe our Sun in more detail and consider the possibility of testing the model for this star which is nearest to us by observing its neutrino emissions. The question of how the stars get onto the main sequence will be postponed; we take it up again in connection with star formation from the interstellar medium in Sect. 10.5. In Sect. 8.2.3, we turn to the evolution of stars away from the main sequence into the region of red giants, where helium burning occurs. Sect. 8.2.4 is devoted to the late phases of stellar evolution, and in Sect. 8.2.5, we consider the role of the stars, with their nuclear energy production, in the formation of the chemical elements. Finally, in Sect. 8.2.6 we discuss the special aspects of stellar evolution in a close binary star system, and in Sect. 8.2.7 the physics of accretion disks, which play a major role not only in many binary systems, but also in the central regions of galaxies (Sect. 12.3.6).

8.2.1 Main Sequence Stars: Central Hydrogen Burning

For *homologous* stars, i.e. for stars with the same structure and chemical composition, a relation of the form $\Phi(L, T_{\text{eff}}) = 0$ (8.44) holds. When hydrogen burning is initiated in their centers, the young, still chemically homogeneous stars all lie along a line in the (L, T_{eff}) diagram or in the color–magnitude diagram. Stars with a large mass have a high luminosity and those with a small mass have a lower luminosity. This line is termed (not quite correctly) the Zero Age Main Sequence (ZAMS), from which further nuclear evolution proceeds "to the upper right" in the Hertzsprung–Russell diagram.

Empirically, the ZAMS is defined for Population I stars, to which we shall limit ourselves in this section, by the envelope of all the color–magnitude diagrams of the open star clusters (Sect. 9.1.3).

The stars remain in the immediate neighborhood of the main sequence until a considerable portion of their hydrogen has been consumed. In order to estimate the length of this stage, we first consider the energy balance of the *Sun*. Its central temperature $T_c = 1.5 \cdot 10^7$ K, as calculated by the theory (Table 8.1), has clearly adjusted itself so that the pp-process takes over energy production. Now, the mass $\mathcal{M}_\odot = 1.98 \cdot 10^{30}$ kg of the sun, if we initially consider it to be homogeneous, consists of hydrogen to about 70%. Its complete conversion into He by the pp-process would provide an energy of $8.8 \cdot 10^{44}$ J. At its current luminosity, $L_\odot = 3.85 \cdot 10^{26}$ W, the Sun therefore burns of the order

Table 8.1. The inner structure of the Sun. Age $4.5 \cdot 10^9$ yr. Original element mixture $X : Y : Z = 0.73 : 0.25 : 0.015$

Spectral class	Effective temperature T_{eff} [K]	Mass $\mathcal{M}/\mathcal{M}_\odot$	Luminosity $L/L_\odot$	Evolution time t_E [yr]
O 5 V	44 500	60	$7.9 \cdot 10^5$	$5.5 \cdot 10^5$
B 0 V	30 000	18	$5.2 \cdot 10^4$	$2.4 \cdot 10^6$
B 5 V	15 400	6	$8.3 \cdot 10^2$	$5.2 \cdot 10^7$
A 0 V	9500	3	$5.4 \cdot 10^1$	$3.9 \cdot 10^8$
F 0 V	7200	1.5	6.5	$1.8 \cdot 10^9$
G 0 V	6050	1.1	1.5	$5.1 \cdot 10^9$
K 0 V	5250	0.8	$4.3 \cdot 10^{-1}$	$1.4 \cdot 10^{10}$
M 0 V	3850	0.5	$7.7 \cdot 10^{-2}$	$4.8 \cdot 10^{10}$
M 5 V	3250	0.2	$1.1 \cdot 10^{-2}$	$1.4 \cdot 10^{11}$

of 10% of its hydrogen – this should lead to a noticeable change in its properties – in a time of $7.3 \cdot 10^9$ yr. Since the time when the Earth obtained its solid crust, the Sun has thus hardly changed.

Now, what is the situation for the energy balance of the other *main sequence stars*? Their central temperatures T_c increase from low values at the cool end of the main sequence to around $3.5 \cdot 10^7$ K for the B0 stars, etc. Somewhat above the Sun on the main sequence, according to Fig. 8.3, the CNO cycle takes over the task of energy production, without a serious change in overall efficiency. From the known values of the masses, $\mathcal{M}/\mathcal{M}_\odot$, and the luminosities $L/L_\odot$ of the main sequence stars (of Population I), we can now readily calculate the time after which they will consume 10% of their hydrogen, which for brevity we term their *evolution time* t_E (Table 8.2):

$$t_E \text{ [yr]} = 7.3 \cdot 10^9 \, \frac{\mathcal{M}/\mathcal{M}_\odot}{L/L_\odot} \, . \qquad (8.45)$$

Since their formation, which – as we assume in advance here – cannot have been more than about $2 \cdot 10^{10}$ yr ago, the main sequence stars below G0 can thus have burned only a small portion of their original hydrogen. On the other hand, the hot stars of the early spectral classes burn their hydrogen so rapidly that they must have been formed relatively short times ago, of the order of t_E. The age of the O and B stars is in fact considerably shorter than the rotational period of the Milky Way in our neighborhood ($2.4 \cdot 10^8$ yr); these stars must thus have been formed in the same regions where they are found today.

Table 8.2. The main-sequence stars and their evolution times

$r/R_\odot$	$\mathcal{M}(r)/\mathcal{M}_\odot$	$L(r)/L_\odot$	P [Pa]	ϱ [kg m^{-3}]	T [K]
0.0	0.00	0.00	$2.5 \cdot 10^{16}$	$1.6 \cdot 10^5$	$1.5 \cdot 10^7$
0.1	0.08	0.45	$1.4 \cdot 10^{16}$	$9.2 \cdot 10^4$	$1.3 \cdot 10^7$
0.15	0.20	0.79	$8.4 \cdot 10^{15}$	$5.8 \cdot 10^4$	$1.1 \cdot 10^7$
0.2	0.35	0.94	$4.5 \cdot 10^{15}$	$3.6 \cdot 10^4$	$9.3 \cdot 10^6$
0.3	0.62	1.00	$1.1 \cdot 10^{15}$	$1.2 \cdot 10^4$	$6.6 \cdot 10^6$
0.4	0.80	1.00	$2.6 \cdot 10^{14}$	$3.8 \cdot 10^3$	$5.0 \cdot 10^6$
0.5	0.89	1.00	$7.0 \cdot 10^{13}$	$1.4 \cdot 10^3$	$3.9 \cdot 10^6$
0.8	0.99	1.00	$1.9 \cdot 10^{12}$	$9.0 \cdot 10^1$	$1.7 \cdot 10^6$
1.0	1.00	1.00	$(10^4)^a$	$(3 \cdot 10^{-4})^a$	$(6 \cdot 10^3)^a$

[a] Characteristic values for the photosphere

Before we consider the evolution of stars *away* from the main sequence, we first consider the structure of the Sun in more detail.

8.2.2 The Internal Structure of the Sun. Solar Neutrinos

In order to calculate a model for our Sun in its present state from the system of fundamental equations (8.42), we begin with a star of 1 $\mathcal{M}_\odot$ with a *homogeneous* element mixture (H : He : heavy elements; relative mass abundances) of

$$X : Y : Z = 0.73 : 0.25 : 0.02 \, , \qquad (8.46)$$

which corresponds to the atmosphere of the Sun and the Population I stars as well as to the interstellar medium.

Solar Model. We begin the calculation of the model at the time of ignition of nuclear energy release through hydrogen burning, since the preceding, relatively brief evolutionary phases following the formation of the Sun from the interstellar medium (of the order of the Helmholtz–Kelvin time $t_{HK} \simeq 3 \cdot 10^7$ yr) had only a minimal influence on its present structure. Beginning at the time of ignition of hydrogen burning, we then construct a series of models for the time evolution in such a way that after $4.5 \cdot 10^9$ yr, the age of the Sun and the Solar System (3.26), the currently observed luminosity $L_\odot$ and effective temperature $T_{\text{eff},\odot}$ are obtained for one solar radius $R_\odot$. We thus find the model shown in Table 8.1 for the internal structure of the Sun.

Half of the solar mass is found to be within about $0.25\,R_\odot$. In most of the interior, energy transport occurs via radiation; only outside $\geq 0.7\,R_\odot$ (out to the lowest layers of the photosphere), i.e. for only about 1% of the overall mass, is convection important. From hydrogen burning during $4.5 \cdot 10^9$ yr, about half of the hydrogen originally present at the center (c) of the Sun has been converted to helium; here, we have $X_c : Y_c : Z_c = 0.36 : 0.62 : 0.02$. Outside about $0.2\,R_\odot$ we still find the original chemical composition (8.45). The energy release in the central regions is mainly due to the pp chain (8.18); at $0.2\,R_\odot$, $L(r)$ has already reached 95% of the total solar luminosity $L_\odot$.

Neutrino Astronomy. Several percent of the energy released in the Sun's interior by nuclear reactions escapes as neutrino radiation directly into space. The extraordinarily small interaction cross-sections for neutrinos with all types of matter allows them to pass through a star, indeed even through the whole universe, practically without any collisions. Neutrino astronomy can thus give us direct information about the energy-releasing core of the *Sun*, while the flux of neutrino radiation from distant stars is far below the present limit of detection. An exception was the explosion of the supernova SN 1987A in the "nearby" Large Magellanic Cloud (Sect. 7.4.7).

The *detection* of (electron) neutrinos ν, which is very difficult owing to their weak interactions with matter, can be carried out via radiochemical methods, or also by experiments involving the scattering of neutrinos with electrons.

The *radiochemical* detection is based on inverse β decay ($\nu + n \rightarrow p + e^-$), whereby within a suitable atomic nucleus X a neutron is transformed into a proton:

$$^A_Z X(\nu, e^-)^A_{Z+1} X^* \,. \tag{8.47}$$

The threshold energy for this process is given by the mass difference of the two nuclei. The excited product nucleus decays back to the original nucleus X by capturing an electron, mainly from the *K* electronic shell of the atom. The energy which is released in this process excites Auger electrons from the atomic shells, which can then be detected. Although the neutrino flux arriving at the Earth is relatively high, around $6 \cdot 10^{14}$ neutrinos s^{-1} m^{-2}, due to their extremely small reaction cross-sections, the measured reaction rates are very small and put great demands on the experimental sensitivity. Typically, in current neutrino detectors some 10 particles must be separated from 10^{30} atoms(!) and counted.

The measurements of solar neutrino radiation are quoted in units of

$$1 \text{ SNU } (\textit{Solar Neutrino Unit}) \tag{8.48}$$
$$= 1.0 \cdot 10^{-36} \text{ neutrino events s}^{-1} \text{ per nucleus}$$
$$= 8.6 \cdot 10^{-32} \text{ neutrino events d}^{-1} \text{ per nucleus}$$

where here the interaction cross-section of the target material used for the neutrino reaction enters into the quoted value. This neutrino unit reflects the extremely small reaction probability: in order to measure e.g. 1 SNU with a detector containing 10^{31} nuclei, we would, according to (8.48) have to be content with a counting rate of only *one* neutrino capture per *day*. For such a detector, containing e.g. a medium-heavy element with an atomic weight of about $A = 50$, we would require 830 t of the element!

The first researcher to carry out such experiments, beginning in 1964, was R. Davis Jr., using the reaction of solar neutrinos with ^{37}Cl, which makes up 23% of naturally-occurring chlorine:

$$^{37}\text{Cl}(\nu, e^-)^{37}\text{Ar}^* \,, \quad \text{threshold: } 814 \text{ keV} \,. \tag{8.49}$$

In order to avoid as much as possible disturbances due to cosmic radiation, the apparatus is located 1.5 km deep in the Homestake gold mine in South Dakota (USA). The "receiver" is a tank holding about 615 t of tetrachloroethylene, C$_2$Cl$_4$ (which is normally used as a cleaning fluid), containing $2 \cdot 10^{30}$ nuclei of the isotope ^{37}Cl. The noble gas ^{37}Ar which results from the reaction decays back to ^{37}Cl with a half-life of 35.0 d. After each 2 to 3 half-lives, it is washed out with helium and its decay electrons are counted.

Since about 1990 two radiochemical experiments which are based on the reaction

$$^{71}\text{Ga}(\nu, e^-)^{71}\text{Ge}^* \,, \quad \text{threshold: } 233 \text{ keV} \tag{8.50}$$

have been running. The radioactive ^{71}Ge decays with a half-life of 11.4 d back to ^{71}Ga, where again, – analogously to the chlorine experiment – the resulting Auger electrons are detected.

In the case of GALLEX (measurements from 1991/97), an essentially European cooperation, the detector is located in a tunnel under the Gran Sasso in Italy and contains 30 t of gallium in the form of a GaCl$_3$ solution. In SAGE, the *Soviet–American Gallium Experiment*, in a mine near Baksan in the Northern Caucuses in Russia, 60 t of liquid metallic gallium serve as detector.

While the radiochemical methods – integrated over a time period of several weeks – detect all the neutrinos above a certain threshold energy, elastic *scattering*

$$\nu + e^- \rightarrow \nu + e^- \tag{8.51}$$

of neutrinos by electrons allows the *direct* determination of energy, direction and arrival time of the incoming neutrino from the recoil of the electron. The recoil electrons are detected (e.g. in water) through their Cherenkov radiation (Sect. 5.3.1), resulting from their high velocities.

In Kamiokande, the *Kamioka* Neutrino *D*etection *E*xperiment, which has been running in the Kamioka mine in Japan since 1987, the detector – for neutrinos with energies above 7 MeV – consists of a large tank containing over 2000 t of extremely pure water, surrounded by numerous large-area photomultipliers for the detection of the Cherenkov radiation. *Super*-Kamiokande, 300 km west of Tokyo, is based on the same detection scheme as Kamiokande and since 1996 has been detecting neutrinos of energies above 5 MeV. Its Cherenkov detector consists of 50 000 t of highly purified, normal water with around 14 000 photodetectors.

Detectors of this kind are also used to observe possible proton decays and neutrino oscillations, two fundamental experiments for elementary particle theory and also for cosmology.

Neutrino oscillations, i.e. transformations of various types of neutrinos into one another, are predicted by some elementary particle theories which extend beyond the standard model (Sect. 13.3.2) and are based upon the hypothesis of a nonzero neutrino *mass*. Super-Kamiokande is able to detect both electron and muon neutrinos (Table 13.3). Thus, in 1998, a Japanese–American collaboration, by analyzing the ratio of the two types of neutrinos in the flux which is produced in the atmosphere by incident cosmic radiation, were able to detect neutrino oscillations indirectly ($\nu_\mu \leftrightarrow \nu_\tau$) and therefore to obtain an indication of the neutrino mass.

With Kamiokande – and with the H$_2$O-Cherenkov detector of the Irvine-Michigan-Brookhaven experiment in a salt mine in Ohio (USA) – the neutrino pulse from the *Supernova* SN 1987A in the Large Magellanic Cloud were detected. This type of detector would thus be able to observe future supernova explosions in our Milky Way galaxy.

Solar Neutrinos. According to *theory*, i.e. from the standard model of the Sun (Table 8.1), we expect a *neutrino spectrum* as shown in Fig. 8.5, in which the neutrinos from the reaction ^{1}H(p, e$^+\nu$)^{2}D of the pp chain (8.18) with a continuous energy distribution below 0.42 MeV are strongly dominant. These neutrinos can be detected only by the gallium experiments, which have a lower threshold energy (8.50).

In the chlorine experiment, in contrast, owing to the higher threshold of 814 MeV (8.49), only the continu-

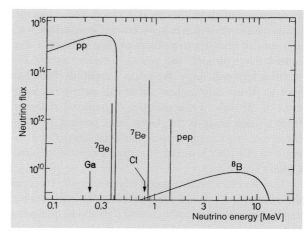

Fig. 8.5. The theoretical neutrino spectrum of the Sun. The neutrino flux at the location of the Earth from the reactions of the pp chain (8.18) and the "pep" reaction, p + e$^-$ + p → ^{2}D + ν + 1.44 MeV, is shown. The contributions from ^{13}N and ^{15}O from the CNO cycle are not included here. The continuous neutrino radiation is given in [Neutrinos m^{-2} s^{-1} MeV^{-1}], the monoenergetic radiation in [Neutrinos m^{-2} s^{-1}]. The thresholds for detection by the ^{37}Cl or ^{71}Ga reactions are indicated by arrows. From T. Kirsten (1983) (with the kind permission of the VCH Verlagsgesellschaft, Weinheim)

ous neutrino emission due to the β$^+$ decay of ^{8}B from the pp III reaction chain can be detected. This reaction is strongly dependent on the temperature at the center of the Sun, and plays only a subordinate role in energy production through hydrogen burning. Kamiokande also detects only the neutrinos from the ^{8}B decay.

The current result of the chlorine experiment, which has been running in the Homestake mine for many years, is (2.55 ± 0.25) SNU. It lies about a factor of 3 *below* the theoretical value of 6 to 10 SNU. The gallium experiments, Kamiokande, and Super-Kamiokande also observe a *smaller* neutrino flux than predicted, of only about 50 to 60% of the theoretical value.

In principle, the current neutrino experiments confirm our expectation that the energy production in the Sun is a result of hydrogen burning. Kamiokande, through its determination of the direction of the incoming neutrinos, has also shown that they do, in fact, originate in the Sun.

The cause of the discrepancy of about a factor of 2 to 3 is not yet clear. It is possible that the experimental accuracy has been overestimated, or else the solar model is still imperfect. A further possibility is that the *electron* neutrinos which are produced in the Sun are in part transformed on their way to the Earth into muon and tau neutrinos, which then are not detected by the experiments performed up to now. Such *neutrino oscillations* (see above) would for the most part occur in the interior of the Sun through interactions with the electrons there.

A solution of the "neutrino problem" is expected from the experiments which have been started in recent years. For example, since 1996, *Super-Kamiokande* has been detecting solar neutrinos at a rate of about 10 per day. We also mention the *Sudbury Neutrino Observatory* (SNO) in a Nickel mine at a depth of 2.1 km in Ontario, Canada, which uses about 1000 t of heavy water, D_2O, and Cherenkov detectors to observe neutrinos of energies above about 5 MeV. The deuterium interacts not only with electron neutrinos, but also with the other types, the muon and tau neutrinos, so that the SNO can also be used to investigate neutrino oscillations.

High-Energy Neutrinos. Aside from the neutrinos from the Sun and the stars, with relatively low energies ($\leq$ a few 10 MeV), it is expected that among other sources, accretion onto neutron stars (Sect. 7.4.6) and the active nuclei of galaxies (Sect. 12.3.6), where high-energy particles and photonen interact with one another, also neutrinos of *high* energies ($\geq 10^6$ MeV) will be produced, which then could be detected by means of the high-energy muons $\mu^{\pm}$ which are generated in the detector. Since their mean free path in water is of the order of 1 km, an underwater observatory for example represents an enormous "detector" of around 10^9 t. After the pioneering project DUMAND (*Deep Underwater Muon And Neutrino Detector*) at a depth of 4.5 km near Hawaii was discontinued due to technical problems in 1996 following a long construction period, at present several underwater or underground observatories are being built; among them is AMANDA (*Antarctic Muon And Neutrino Detector Array*), in the 3 km thick antarctic ice near the American Amundsen–Scott Station.

8.2.3 From Hydrogen to Helium Burning

The course of evolution of the stars away from the main sequence depends essentially on whether the transformed matter in the stellar interior mixes with the remaining matter, or whether it remains where it was formed in the core or the particular convection zone, if one is present. F. Hoyle and M. Schwarzschild were the first to show, in 1955, that only the latter possibility leads to an acceptable theory of stellar evolution.

Massive Stars. We shall first describe what happens in detail, using the example of a star of $5\,\mathcal{M}_\odot$ (Fig. 8.6). It begins its evolution as a completely homogeneous B5 V Population I star with the chemical composition given in (8.46), an effective temperature $T_{\text{eff}} = 17\,500$ K, a radius of $2.6\,R_\odot$, and thus an absolute bolometric magnitude of $M_{\text{bol}} = -2.2$ mag. In its center, the temperature is $T_c = 2.6 \cdot 10^7$ K and the pressure is $P_c = 5.5 \cdot 10^{15}$ Pa. At the very center, there is a hydrogen burning zone, the nuclear fusion reactor, so to speak, where hydrogen is burned to helium by the CNO cycle. Adjacent to this hydrogen burning zone is a convection zone, within which the reaction products are thoroughly mixed. This phase of evolution (A→B→C in Fig. 8.7a)

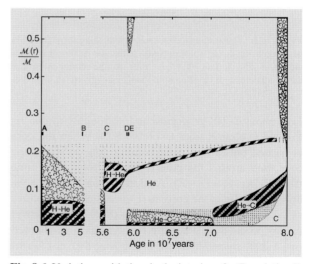

Fig. 8.6. Variations with time in the interior of a (Population I) star of $5\,\mathcal{M}_\odot$. The abcissa gives the age, reckoned in 10^7 yr, since the star left the main sequence. The letters A to E denote the correspondence to the evolutionary paths shown in Fig. 8.7a. The ordinate is $\mathcal{M}(r)/\mathcal{M}$, the fraction of the mass within r. The stippled regions correspond to convection zones, and the barred regions are zones of nuclear energy release; the dotted regions are those in which the H or He content decreases on going inwards. (After R. Kippenhahn, H. C. Thomas and A. Weigert, 1965)

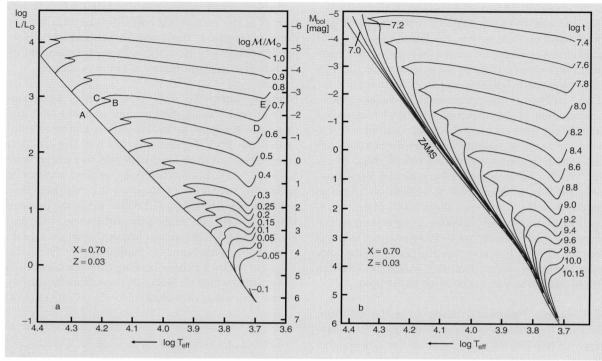

Fig. 8.7a,b. Theoretical color–magnitude diagrams for stars of Population I. **(a)** Evolutionary paths for different masses $\mathcal{M}/\mathcal{M}_\odot$. The letters A to E refer to the changes in the inner structure of a star with $5\,\mathcal{M}_\odot$ as given in Fig. 8.6. **(b)** Isochrones for the evolutionary paths given in (a), with the ages t in years, as indicated. The lower envelope is the initial main sequence (ZAMS = Zero Age Main Sequence)

lasts about $6 \cdot 10^7$ yr, corresponding roughly to our estimated evolution time t_E (Table 8.2). When the core is burned out, a hydrogen burning zone in the shape of a *shell* around the core is formed for a short time ($0.3 \cdot 10^7$ yr, C→D→E). At E, a *helium burning zone* arises, initially in the core; in it, at central temperatures of 1.3 to $1.8 \cdot 10^8$ K, the 3α-process (8.19, 20) takes over energy production. When this He core has in turn burned out, a shell-shaped helium burning zone then also forms around it; but the thin hydrogen burning zone, which is continually moving outwards, still makes a considerable contribution to energy production after E. Finally, the star moves quickly into the stages of red giant and supergiant, where it loses part of its mass to the stellar wind, which increases with increasing luminosity (7.105).

The evolutionary paths of the more massive, luminous stars in the Hertzsprung–Russell diagram depend sensitively on the rate of *mass loss*, $\dot{\mathcal{M}}$, which at present can only be taken into account in theoretical calculations as an empirical, rather uncertain parameter. Stars with $\geq 10\,\mathcal{M}_\odot$ on the initial main sequence lose on the order of magnitude of 1/4 to 1/3 of their mass before reaching the region of red supergiants. Their essentially "horizontal" evolution away from the main sequence at a roughly constant luminosity (Fig. 8.7a) takes place at smaller luminosities for increasing $\dot{\mathcal{M}}$. In the case of extreme mass losses (those which persist for long times, with $\geq 10^{-5}\,\mathcal{M}_\odot\,\text{yr}^{-1}$), the evolutionary path bends downwards towards the main sequence, so that the red giant stage is not reached. (In the hypothetical limiting case of $\dot{\mathcal{M}} \to \infty$, the evolutionary path would coincide with the main sequence.)

All stars with masses $\geq 2.5\,\mathcal{M}_\odot$ have an evolution during the stages of hydrogen and helium burning which is qualitatively similar to what we have described for $5\,\mathcal{M}_\odot$. In particular, helium burning takes place *"hydrostatically"*, i.e. the star remains continuously in hydrostatic equilibrium (8.1). This follows essentially from the equation of state (8.3) of an *ideal* gas, according to which a change in the temperature T is accompanied by a change in pressure. If, for example, T increases, and with it the rate of nuclear energy release, then the accompanying pressure pulse causes an expansion and cooling, which leads to a decrease in the energy production, thus stabilizing the star.

Stars of Low Mass. For stars with masses $\leq 2.5\,\mathcal{M}_\odot$, the helium burning stage, in contrast to the more massive stars, takes place *explosively*. These stars attain such high densities in their interiors in the course of their evolution "straight up" in the Hertzsprung–Russell diagram, on the socalled first giant branch (A→E in Fig. 8.8), that a *Fermi–Dirac degeneracy* (Sect. 8.3.2) of the electron gas becomes established before helium burning can start at $T \simeq 8 \cdot 10^7$ K. The pressure of this

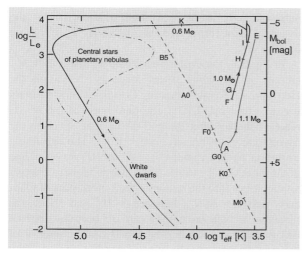

Fig. 8.8. The evolutionary paths in the Hertzsprung–Russell diagram of Population I stars having $1.0\,\mathcal{M}_\odot$ and $1.1\,\mathcal{M}_\odot$, from central hydrogen burning (A) to the helium flash (E), without taking mass losses into account. After A. V. Sweigart and P. G. Gross (1978). The ejection of a mass of $0.1\,\mathcal{M}_\odot$ during the helium flash was assumed. The further evolution of the star of $1.0\,\mathcal{M}_\odot$ was calculated taking the mass loss according to (7.105) into account, after D. Schönberner (1979). F → G: the asymptotic giant branch; only one of the thermal pulses (helium flashes) which occur after I is drawn in, at J. The mass loss becomes important at H and leads to a final mass of $0.6\,\mathcal{M}_\odot$, which is reached at K

Fermi gas, unlike that of an ideal gas, does *not* depend on the temperature T (8.69), so that when the energy release increases, no expansion and cooling can occur. On the contrary, a *helium flash* takes place (E): a strong temperature rise within a very short time, of the order of the free fall time (8.41). This temperature rise stops only when T becomes so high that the electron degeneracy is again lifted. The resulting explosive shock wave in the stellar interior is damped by the massive shell outside the helium zone, as is shown by tedious detailed calculations, so that the star "survives" the central helium flash at the tip of the giant branch (at roughly $2 \cdot 10^3\,L_\odot$). Afterwards, it finds a new equilibrium configuration (F) with central, hydrostatic helium burning and a concentric, hollow hydrogen burning zone.

Both during its evolution on the first giant branch and in the helium flash, the star loses a considerable portion of its mass; e.g. a star of originally $1\,\mathcal{M}_\odot$ loses in the range of around 0.1 to 0.5 $\mathcal{M}_\odot$.

Now how does the stellar evolution continue after the helium core has been consumed? At first, a shell-shaped helium burning zone forms. The star moves – as long as $\mathcal{M} \geq 0.6\,\mathcal{M}_\odot$ – for the second time towards higher luminosities into the region of red giants and supergiants, along the *asymptotic giant branch* or AGB (F→I in Fig. 8.8).

Before we take up the further evolution of stars, we turn briefly to those stars at the lower end of the mass distribution. If the original mass was $\leq 0.4\,\mathcal{M}_\odot$, the star never reaches the ignition temperature for helium burning. If the mass lies below $\leq 0.1\,\mathcal{M}_\odot$, then not even the hydrogen-burning phase is reached, which we considered to be characteristic of a *star*. These "substellar" masses or *brown dwarfs* can – depending on their structures – be considered to be giant planets, similar to Jupiter (Sect. 8.3.1).

8.2.4 Late Phases of Stellar Evolution

As we saw in the preceding section, stellar evolution of both massive and of less massive stars leads to a state in which the star has two *shell-shaped regions*, which are sources of energy production. This structure is characteristic of the asymptotic giant branch: an inner helium burning shell, which surrounds a central region containing ^{12}C and ^{16}O as well as degenerate electron gas, and an outer hydrogen burning zone.

Thermal Pulses. Energy production in the helium burning shell is thermally *unstable*. The fusion takes place as a series of brief pulses (helium-shell flashes) at typical intervals of about 1000 yr. In connection with the pulses, a strong mass loss occurs. Finally, a *planetary nebula* containing several tenths of a solar mass (Fig. 8.9) is ejected. We will discuss its properties in Sect. 10.3.2 in connection with the luminous gas nebulae. The masses of the remnant central stars of the planetary nebulae lie in a relatively narrow range around $0.6\,\mathcal{M}_\odot$.

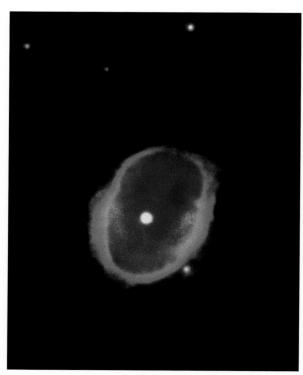

Fig. 8.9. The planetary nebula NGC 3132 in Vela; a color image made by S. Laustsen using the 3.6 m telescope at the ESO. The angular diameter of the nebula, about $50''$, corresponds to 0.24 pc at its distance of around 1 kpc. The inner, hotter portion appears bluish due to the emission lines of ionized oxygen, and the outer portion appears reddish from the $H\alpha$ emission of neutral hydrogen. The nebula is excited to luminosity by an extremely hot central star ($T_{\rm eff} \simeq 150\,000$ K), which is very faint in the visible range and cannot be seen in this picture. It is only $1.65''$ from the bright star (of spectral type A) in the middle of the nebula, which itself is too cool to produce significant excitation. (With the kind permission of the European Southern Observatory)

Evolution Towards a White Dwarf. For the later evolutionary stages of stars with original masses $\leq 8\,\mathcal{M}_\odot$, the *mass loss* in the region of red giants and supergiants is of decisive importance. If the loss is so great that the stellar remnant has a mass lower than Chandrasekhar's limiting mass of $1.4\,\mathcal{M}_\odot$ (8.75), then the star finishes its evolution, after its nuclear energy sources (hydrogen and helium burning layers) have been exhausted, as a stable final configuration in the form of a *white dwarf*, containing ^{12}C and ^{16}O. In such stars, the pressure of the degenerate electron gas holds the gravitational forces in balance (Sect. 8.3.3).

The existence of white dwarfs in open star clusters whose bend-over point from the main sequence in the color–magnitude diagram (Sect. 9.1.3) still corresponds to a mass of $8\,\mathcal{M}_\odot$ demonstrates that, indeed, a large fraction of stars of $\lesssim 8\,\mathcal{M}_\odot$ lose sufficient mass through stellar winds and the casting off of planetary nebulae that they become white dwarfs. On the asymptotic giant branch, these white dwarfs are already well "hidden" within the innermost $10^{-2} R_\odot$ of the red giants, which themselves have radii of about $10^2 R_\odot$; the dwarf takes the form of a dense (10^8 to 10^9 kg m^{-3}) core of $\gtrsim 0.6\,\mathcal{M}_\odot$.

In Fig. 8.8, the evolutionary path of a star of $1\,\mathcal{M}_\odot$ ending at the white dwarf stage is shown on the Hertzsprung–Russell diagram. Corresponding to a mass loss as given by (7.105), the final mass of $0.6\,\mathcal{M}_\odot$ is reached shortly after the end of the time spent on the asymptotic giant branch, some 10^6 yr. The remaining evolution occurs rapidly, in 10^4 to 10^5 yr, on a line going horizontally to the left towards higher effective temperatures into the region of the central stars of planetary nebulae. At $T_{\rm eff}$ between $3 \cdot 10^4$ and 10^5 K, the planetary nebula previously cast off becomes ionized and is excited to luminosity. Finally, in the interior of the star, nuclear energy release ceases to be possible; its evolutionary path bends down and merges into the sequence of the white dwarfs. In some 10^9 yr, the star cools gradually to effective temperatures below 4000 K, the lowest values which have been observed in white dwarfs.

Those stars which have originally $\leq 8\,\mathcal{M}_\odot$ and whose mass exceeds the limiting mass for white dwarfs near the end of their evolutionary phase on the asymptotic giant branch finally attain such high temperatures in their helium cores with degenerate electron gas that, in spite of large energy losses due to the emission of neutrinos (Sect. 8.1.3), *carbon burning* $^{12}C + ^{12}C$ (8.24) is ignited. This process should proceed explosively, similarly to helium burning in degenerate matter. Complex hydrodynamic model calculations for the energy balance and expansion of the combustion front, which are difficult to carry out, do not yet give a satisfactory answer to the question of the further evolution of stars in this mass range. The explosive carbon burning is possibly the cause of some of the observed supernova outbursts (Sect. 7.4.7).

Stars with Masses $\geq 8\,\mathcal{M}_\odot$. In these cases, at temperatures of 5 to $8 \cdot 10^8$ K, the reaction $^{12}C + ^{12}C$ is initiated in *non*degenerate matter, so that the stability of the star is maintained during the phase of *hydrostatic* carbon burning which lasts only the order of 100 yr. After the formation of a concentric-shell combustion zone, a core region consisting of ^{16}O, ^{20}Ne and ^{24}Mg is left at the center of the star. In stars with $\geq 13\,\mathcal{M}_\odot$, a rapid sequence occurs, with increasing temperatures and densities in the core, consisting of neon burning (lasting on the order of 1 yr), oxygen burning (several months), and silicon burning (1 d) (see Sect. 8.1.3). The star forms an *"onion-layer structure"* (Fig. 8.10), with an *"iron" core* of mass in the range 1.3 to 2.5 $\mathcal{M}_\odot$. Depending on the physical conditions, different nuclides are predominant in this core, e.g. ^{56}Fe or the neutron-poorer radioactive ^{56}Ni, which decays finally to ^{56}Fe. The synthesis of the nuclides in the iron group marks the maximum of the nuclear binding energy and thus the end of nuclear energy release in the star.

The nuclear evolution of stars in the mass range from about 8 to 13 $\mathcal{M}_\odot$, in which the densities in the (^{16}O, ^{20}Ne, ^{24}Mg) core of $\leq 1.4\,\mathcal{M}_\odot$ can be so high that electron degeneracy is produced and the following combustion processes are no longer hydrostatic, is exceedingly complex; we cannot treat it further here.

Collapse of the Central Region. After the star no longer has any more nuclear energy sources at its disposal, the central region contracts further; this is accompanied by an increase in temperature, until a "phase transition" takes place, whose nature depends on the range of (ϱ, T). In this transition, the compressibility of the stellar matter becomes so high that the stability con-

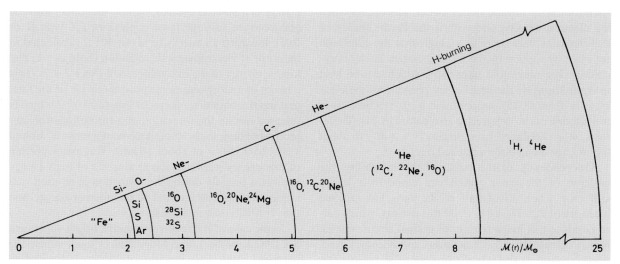

Fig. 8.10. The "onion-layer structure" of an evolved Population I star of $\mathcal{M} = 25\,\mathcal{M}_\odot$ after the conclusion of silicon burning. The zones of maximum energy release and the most abundant product nuclides within each fusion layer are indicated (after J. R. Wilson et al., 1985). Outside of $\mathcal{M}(r) = 8.43\,\mathcal{M}_\odot$ or $r = 0.55\,R_\odot$, the star still has its original H- und He-rich composition. The neutronized Fe-core contains $\mathcal{M}(r) = 2.1\,\mathcal{M}_\odot$, corresponding to $r = 4.2 \cdot 10^{-3}\,R_\odot = 2900$ km; at its center, the temperature is $T_c = 8.2 \cdot 10^9$ K and the density is $\varrho_c = 2.2 \cdot 10^{12}$ kg m^{-3}. In the Hertzsprung-Russell diagram, this star is in the region of red supergiants (effective temperature 4400 K, luminosity $3 \cdot 10^5\,L_\odot$, and radius around $10^3\,R_\odot$)

dition $\Gamma > 4/3$ (8.39) is violated and a *collapse* occurs within the free fall time (8.41).

For $\mathcal{M} \gtrsim 100\,\mathcal{M}_\odot$, the instability is caused by the production of *electron–positron pairs*,

$$\gamma + \gamma \to e^+ + e^- \tag{8.52}$$

as soon as the temperature exceeds a few 10^9 K, i.e. when $kT \gtrsim m_e c^2$ ($m_e c^2 = 0.51$ MeV = rest energy of the electron and positron).

In the mass range of 13 to around $100\,\mathcal{M}_\odot$, above about $5 \cdot 10^9$ to 10^{10} K, the *photodisintegration* of nuclides by energetic thermal γ quanta triggers the collapse: in particular,

$$\gamma + {}^{56}\mathrm{Fe} \to 13\,{}^4\mathrm{He} + 4\,\mathrm{n}\,, \tag{8.53}$$

followed by

$$\gamma + {}^4\mathrm{He} \to 2\,\mathrm{p} + 2\,\mathrm{n}\,.$$

At the beginning of the collapse, for example in a star of $25\,\mathcal{M}_\odot$, the temperature in the core is roughly $8 \cdot 10^9$ K and the density about $4 \cdot 10^{12}$ kg m^{-3}; the corresponding free fall time is around 0.1 s.

At smaller stellar masses (8 to 13 $\mathcal{M}_\odot$), the collapse of the degenerate electron-gas core region is caused by "*neutronization*" of the stellar matter; this occurs when the electron density, and with it the Fermi energy of the electrons, has become so high that the threshold for *electron capture* by the more abundant nuclei is exceeded. The maximum mass $\mathcal{M}_{\mathrm{Ch}}$ which can be held in equilibrium by the degeneracy pressure of the electrons is, from (8.75), proportional to μ_e^{-2}, where μ_e is the "atomic mass" referred to an electron (8.71). Therefore, $\mathcal{M}_{\mathrm{Ch}}$ is reduced from originally $1.4\,\mathcal{M}_\odot$ to about $0.8\,\mathcal{M}_\odot$ for the neutron-enriched matter.

Independently of the cause of the instability, the collapse of the inner region can come to a standstill only when about half of its mass has reached a density of $\gtrsim 2 \cdot 10^7$ kg m^{-3}. At this density, which corresponds to that in the interior of normal atomic nuclei, matter consists mainly of neutrons and becomes practically incompressible. In the interior of a massive star ($\gtrsim 10\,\mathcal{M}_\odot$), a *neutron star* (Sect. 8.3.4) is thus formed within a short time ($\lesssim 1$ s); its surface then brakes the fall of the remaining material from the rest of the star.

The neutron star oscillates like an elastic sphere, and initially springs back somewhat; this creates a shock wave which passes out through the matter which is still falling inwards. Behind the shock wave front, the motion of this matter is reversed, but at the initially high temperatures, the atomic nuclei dissociate into free protons and neutrons and thereby strongly damp the shock wave. Its original kinetic energy of several 10^{44} J is dissipated after it has passed through only a few 100 km or about $0.5\,\mathcal{M}_\odot$ of stellar material. The subsequent fate of the star depends sensitively on its density structure and on energy transport in the layers which are outside the central region: if the shock wave has to pass through only a relatively small amount of matter, or if it acquires sufficient energy, e.g. by absorption of high energy neutrinos from the inner regions, then it can reach the surface of the star and lead to the *casting off of a shell*. The neutron star remains as a remnant. If, on the other hand, the shock wave comes to a standstill within the star, more and more matter will be collected within the standing wavefront until finally the limiting mass for a neutron star, about $1.8\,\mathcal{M}_\odot$, is exceeded. Then a stable configuration no longer exists; the matter will collapse into a *black hole* (Sect. 8.4.4).

Supernova Explosion. The theoretical model of the final phase in the evolution of massive stars, in which a shell of matter is ejected, can be identified with the phenomenon of supernova explosions (of type II or Ib/c), in which an energy of about 10^{44} J is released in the form of electromagnetic radiation (light curve) and kinetic energy of the expanding shell (Sect. 7.4.7). This amount of energy represents only about 1% of the total gravitational binding energy of roughly 10^{46} J which is released during the formation of a (proto) neutron star (corresponding to the compression of a mass of around $1\,\mathcal{M}_\odot$ to a radius of about 10 km). The major portion of this energy goes into the production of high-energy *neutrinos* (mainly during the formation of neutron-rich matter) and into an increase in the internal energy, and thus is finally lost to the star.

In the case of the relatively nearby, bright supernova SN 1987A in the Large Magellanic Cloud, supernova neutrinos, with a total energy of 10^{45} to 10^{46} J and an average particle energy of 10 MeV were detected for the first time; they were emitted several hours before the optical luminosity increase (Sect. 7.4.7).

As we have seen, the shock wave which moves outwards at first has a kinetic energy of only a few times 10^{44} J, so that only very precise and difficult numerical calculations can show whether our theoretical picture of stellar evolution is, in fact, able to explain the observed light curves and the expansion of the gas shell in supernovae. These hydrodynamic calculations require knowledge of the equation of state of stellar matter at high densities, of a complex chain of nuclear reactions, and in particular, of the cross-sections for energy and momentum transfer from neutrinos to the stellar matter, as well as of the internal structure of the pre-supernova. In spite of their extremely small cross-sections, at densities of $\gtrsim 4 \cdot 10^{14}$ kg m^{-3}, the neutrinos are almost completely absorbed in the central region and are thus "stored" for a short time, before they can escape from the star. Even the absorption of a small fraction of the enormous number of neutrinos is sufficient to overcome the damping of the outgoing shock wave.

Our concept that the evolution of a more massive star after the collapse of its interior can lead to a supernova explosion of type II is given strong support by, for one thing, the observation of neutrinos from the supernova SN 1987A. For another, the complicated model calculations which have thus far been carried out fundamentally confirm it. However, at present, a number of important questions remain unanswered. For example, it is not known with certainty how the evolution of stars with masses above 8...10 $\mathcal{M}_\odot$ proceeds, whether a neutron star always remains as a remnant after a supernova explosion, whether neutron stars can also be formed without a supernova explosion, how much mass is required for the evolution to end in a black hole, etc.

While our theoretical picture for supernovae of type II seems to be well established, at least in its basic features, the origin of supernova explosions of *type I* is still not certain. Their spectra indicate that the star has already lost its hydrogen shell before the explosion (Sect. 7.4.7). It is thus reasonable to attribute the supernovae of type Ib/c to explosions of Wolf–Rayet stars with original masses of $\geq 35\, \mathcal{M}_\odot$; they no longer have hydrogen-rich shells and occur as individual stars or in binary star systems (Sects. 6.4.2, 8.2.5).

On the other hand, the shape of the light curves of *type Ia supernovae* may best be explained by the explosion of a compact star. Since the SN Ia also occur in *all* types of galaxies, they must be attributed to relatively old objects. Thus, one model of supernovae Ia starts with a *white dwarf*, which consists mostly of ^{12}C and ^{16}O and, as was shown above, can be formed by the evolution of a star of $\leq 8\, \mathcal{M}_\odot$ as a result of major mass losses. If the white dwarf is in a *close binary star system*, there are two possibilities for a SN Ia explosion. For one thing, it can collect matter from its companion star. If the accretion rate is sufficient ($\geq 10^{-7}\, \mathcal{M}_\odot$ yr^{-1}), the instability which leads to a nova outburst (Sect. 8.2.6) can be avoided and the critical mass of its (^{12}C,^{16}O) core of around 1.39 $\mathcal{M}_\odot$ for the ignition of explosive carbon burning can be attained. The other possibility is that in a very close binary system, *both* components have developed into white dwarfs and their orbital angular momentum is decreased by the emission of gravitational radiation (Sect. 8.4.5), so that they approach each other and finally *merge*; then the critical mass can be exceeded. Since the time scale for this process is determined by the rate of emission of gravitational radiation, it can, depending on the distance of the components, easily be of the order of magnitude of the age of the galaxy, so that the occurrence of SN Ia e.g in elliptical galaxies, as well at its rate, can be understood.

In both cases, the star is probably torn apart by a *carbon detonation*, without leaving a remnant star (neutron star). This model can explain in principle the similar shapes of the light curves from supernovae of type Ia and their exponential brightness decreases. In the expanding combustion front, about 0.6 $\mathcal{M}_\odot$ is converted into nuclides of the iron group, mainly into ^{56}Ni. The principal energy source for these emissions is, as also for supernovae of type II, the radioactive decay of ^{56}Ni to ^{56}Co with a half-life of 6.1 d; the latter nuclide further decays with a half-life of 77.3 d into stable ^{56}Fe.

To summarize, we show in Fig. 8.11 a greatly simplified schematic representation of the possible final evolutionary stages of a star, depending on its mass.

8.2.5 Nucleosynthesis in Stars

We now consider nuclear reactions in stellar evolution from the point ov view of nucleosynthesis, or, as it is also termed, the *formation of the elements* in stars, since each nuclear reaction indeed results in the transmutation of elements. As a rule, these nuclear processes take place in the interiors of the stars; their products can however be seen in the spectra of the stars, if they in some manner

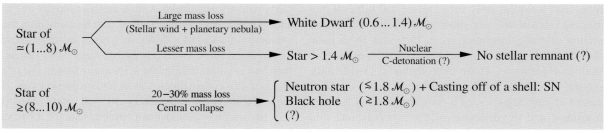

Fig. 8.11. Schematic of the possible evolutionary paths for a star

reach the stellar atmosphere. This can occur through "mixing up" through convection zones, or also through ejection of outer shells, so that the former reaction zones are exposed.

In the 1940's, G. Lemaître and G. Gamow first developed the idea that the mixture of chemical elements, which at that time was regarded as universal, had been formed as a result of the beginning of cosmic evolution, the "Big Bang". In the course of time, the nuclear physicists however pointed out that in that case, the formation of the heavier elements would have had to end at the mass number $A = 5$ (Sect. 13.2.4). On the other hand, it was discovered that stars such as the high-velocity stars or the subdwarfs (Sect. 7.2.7) are "metal poor" in differing degrees and thus that a chemical evolution must have taken place within our Milky Way galaxy.

B²FH Theory. The first attempt to combine the two problem areas – that of the elemental abundance distributions and that of the evolution of the *stars*, and thus also of the galaxies – was made in 1957 by E. M. and G. R. Burbidge, W. A. Fowler, and F. Hoyle (B²FH). Here, for the first time, nuclear physics was applied to astrophysics on a broad scale. In particular, the measurements of nuclear reaction cross-sections at low energies by W. A. Fowler, with regard to energy release in the stars, first placed this area of research on a secure foundation. The theory, from an astronomical point of view, starts with a Universe which as a result of the Big Bang (we will soon take up the newer models for the origin of the Universe) consisted of hydrogen and helium (10 : 1) with only traces of heavier elements (Sect. 13.2.4).

We shall treat aspects of galactic evolution later, in Sect. 12.5.4, while here, we give an outline of the nuclear processes envisaged in the B²FH theory, in particular since their terminology is still in wide use.

In Sect. 8.1.3, we have already mentioned the processes which are important for energy release in the interiors of stars: at $\simeq 10^7$ K, nuclear fusion of hydrogen to helium begins, initially through the pp process, then, at somewhat higher temperatures, in the main through the CNO cycle. Above $\simeq 10^8$ K, helium is then burned to carbon, and at still higher temperatures, the carbon itself burns.

Beginning with ^{12}C, the α-*process* leads to the formation of heavier nuclei with mass numbers which are multiples of four by fusing α particles, up to ^{40}Ca (at $\simeq 10^9$ K).

The *e-process* produces V, Cr, Mn and the elements of the iron group, Fe, Co, and Ni, in thermal *e*quilibrium at around $4 \cdot 10^9$ K.

The heavier atomic nuclei are not accessible to charged particles due to their strong Coulomb repulsion. Their formation can therefore take place only by neutron processes. We distinguish:

The *s-process*, which consists of neutron capture by the elements of the iron group; it takes place *s*lowly in relation to the competing β decays. The s-process produces e.g. Sr, Zr, Ba, Pb,... and in general the *stable* nuclides beyond the iron group. If one plots the binding energies (per nucleon) of the atomic nuclei above a plane with the coordinates $N =$ neutron number and $Z =$ atomic number or proton number (nuclear charge), the resulting energy surface gives an intuitively clear representation of the nuclear reactions including their energetics. The stable nuclei are to be found in the neighborhood of the *valley* in the energy surface. The fact that the s-process plays a role in the formation of e.g. the solar elemental abundance distribution was already clear to G. Gamow, who noted that the product of the abundance times the neutron absorption cross-section at around 25 keV for the nuclides concerned is a smooth function of the mass number A.

The *r-process* refers to neutron captures which take place *r*apidly with respect to the competing β decays. This process produces the neutron-*rich* isotopes of the heavy nuclei; it is in particular responsible for the formation of the radioactive elements, e.g. ^{235}U and ^{238}U (at the expense of the Fe group).

The *p-process* yields the neutron-poor or proton-rich isotopes of the heavy elements from a hydrogen-rich medium (*protons*).

The original theory of B^{2}FH was modified in many aspects as a result of progress in nuclear physics and in the theory of stellar evolution. In particular, it was found that in the later stages of stellar evolution, e.g. in the rapid collapse phase and in the shock waves of supernovae, the hydrodynamic time scales occur in part at times considerably less than 1 s, and it is no longer true that all the nuclear reactions take place in thermodynamic equilibrium, e.g. according to the e-process. The complex calculations of "explosive nucleosynthesis", beginning with the work of J. W. Truran, W. D. Arnett and others (1971), have yielded nuclide distributions which differ clearly from those of the equilibrium processes. The newer results were already quoted in Sects. 8.1.3 and 8.2.4 in connection with stellar evolution. Concerning the late stages of evolution, owing to the complex interactions between nuclear reactions and dynamic processes, many questions remain open, for example regarding the course of the r- and p-processes. The early evolutionary stages, whose reactions are also responsible for energy release, are in contrast well understood.

Stars with He- and CNO-Anomalies. In the first thermonuclear fusion phase, that of hydrogen burning, H is initially converted into ^{4}He, and, in the CNO cycle, C, N and O are converted mainly into ^{14}N, while the isotopic ratio of ^{12}C/^{13}C $\simeq 4$ is produced (Sect. 8.1.3). For a portion of this matter, ^{4}He is later burned in the star, yielding mainly ^{12}C. Mixing processes, together with mass losses, as well as matter exchange between the components in close binary systems, can move the products of nuclear reactions from the stellar interior to the surface and thus make them "visible" in the spectra. The fact that this really does occur in some cases, even in the early evolutionary phases, is shown by the anomalous spectra of particular groups of stars, which exhibit enhanced intensities for the lines of product nuclides from hydrogen or helium burning.

Thus we find among the hotter stars evidence for the CNO cycle in the *helium stars*, whose spectra are dominated by strong helium lines, and to a lesser extent in many OB stars with minor anomalies of the C and N lines. In the *Wolf–Rayet stars* (Sect. 6.4.2), with effective temperatures around 50 000 K and masses of about 10 $\mathcal{M}_\odot$, an extremely high mass loss has apparently laid bare the layers of helium- and nitrogen-rich matter from CNO burning (type WN) or even the carbon-rich matter from helium burning (type WC).

A great variety of anomalous spectra are observed in the cool *red giants*, which, depending on their evolutionary phase, can be either on the first giant branch or on the asymptotic giant branch. For example, low ^{12}C/^{13}C ratios are derived from the molecular bands of otherwise normal G and K giants, as predicted to result from CNO burning. The spectra of the *carbon stars* (spectral type C, CH stars, etc.; cf. Table 6.1), with their strong bands from CN, CH, and C$_2$, indicate that in their atmospheres, in contrast to the normal stars (G, K, M), the ratio C/O is > 1, so that here, matter which has been modified by helium burning is present on the star's surface.

s-Process Elements. Particularly noteworthy are the lines of elements such as Sr, Y, Zr and Ba, which occur with enhanced intensities in the spectra of many types of red giants (C and S stars, CH stars, barium stars), as well as the lines of Tc, which has no stable isotope, in some S stars (Table 6.1). These elements must have been produced in the stars in a few 100 yr during the red giant stage. They must be due to neutron capture by the nuclides of the iron group, in the s-process.

Model calculations for stellar evolution show that on the asymptotic giant branch, as a result of the thermally unstable, pulsed helium-shell burning (Sect. 8.2.4), for one thing, sufficient neutrons are released, and for another, convection zones are formed, which can mix the s-process elements up to the surface. The most important source of these neutrons is probably the reaction ^{22}Ne$(\alpha, n)^{25}$Mg (8.23), which is initiated at the base of the helium-rich zone at sufficiently high temperatures according to (8.22) by reaction of ^{4}He with the ^{14}N produced by the CNO cycle. Another possibility for neutron production is the series

$$^{12}C(p, \gamma)^{13}Ne(e^+\nu)^{13}C(\alpha, n)^{16}O \ , \qquad (8.54)$$

which can occur when fresh, hydrogen-rich matter is mixed downwards in the star and comes into contact with the ^{12}C resulting from helium burning.

Although the complex events in this phase of stellar evolution are not yet understood in detail, the asymptotic giant branch can still be considered with some certainty to be the place where the s-process elements are produced.

FG Sagittae. In the notable case of this variable central star of a planetary nebula, we can, so to speak, watch stellar evolution, in particular the mixing upwards to the surface of nuclear products: following about 1890, the brightness of FG Sge increased roughly monotonically by 4 magnitudes to a maximum in 1970 at $V \simeq 9$ mag, and since then has again gradually decreased. In this astonishingly short time, this star has rushed through the Hertzsprung-Russell diagram, going from spectral class B to G and K, while its bolometric magnitude remained practically unchanged, and its radius increased correspondingly from around 1 $R_\odot$ to nearly 200 $R_\odot$. After the time of the visual brightness maximum, when the spectrum of FG Sge was similar to that of a normal A supergiant such as α Cyg, over a time period of about 10 yr(!), an increase occurred in particular of the intensity of the lines of the s-process elements such as Zr II, Ba II, and some of the rare earths, until an enrichment by a factor of about 25 was reached – behavior such as we would expect from a star at a stage of evolution following the asymptotic giant branch, as the result of a pulse of helium layer burning (Sect. 8.2.4). The visual magnitude of this unusual star by the way decreased strongly within a few weeks in 1992, down to nearly its original value of 100 yr ago, and then in the following years again increased. This is similar to what is observed from an R CrB star (Sect. 7.4.1) when it ejects a dust shell.

Supernovae. Stars with $\geq (8\ldots 10)\,\mathcal{M}_\odot$, which pass through the various phases of nuclear fusion from carbon to silicon burning on a very short time scale and finally explode as supernovae of *type II*, play a decisive role in the nuclear synthesis of the elements between oxygen and the iron group. On the one hand, the major portion of the Fe-Ni core of the "onion-layer structure" from the pre-supernova (Fig. 8.10) remains trapped in the newly-formed neutron star. On the other hand, at the high temperatures and densities behind the expanding shock wave front, nuclear reactions take place that change the nuclear composition of those layers, which then are cast off by the star and mix with the surrounding interstellar medium. For example, model calculations for a star of 25 $\mathcal{M}_\odot$ show that the re-ignition of oxygen and silicon burning increases the abundances of, in particular, the elements O, Si, S, Ca, and Fe by an order of magnitude in comparison to the solar mixture, while retaining about the same relative abundance ratios as in solar matter. The products of these "explosive" nuclear processes behind the shock wave front differ from those of the corresponding "hydrostatic" fusion reaction phases, for which sufficient time is available for equilibrium to be established. In a supernova explosion, presumably also the neutron-rich isotopes of the heavy elements following the iron group can be formed by neutron capture through the r-process.

In contrast to the supernovae II, the supernovae of *type Ia* contribute essentially only to the elements of the iron group.

The quantitative results of calculations of nuclear synthesis in supernova explosions naturally also reflect the large uncertainties which we have already met in the models for the supernovae themselves.

As we have seen, the evolution of stars is inseparably bound up with the question of the formation of the elements. In particular, supernovae release matter to their surroundings which contains heavy nuclides formed by a series of nuclear reactions. From this matter, new generations of stars condense, with a chemical composition which is enriched in the heavy elements. Up to this point, we have discussed only the individual contributions of stars of different masses to nuclear synthesis; the interactions of all the stars, through stellar genesis and mass loss, with the interstellar matter in our Milky Way system and in other galaxies will be treated in Sect. 12.5.4.

8.2.6 Close Binary Star Systems

Nuclear evolution in close binary star systems can follow quite a different course than in the individual stars which we have considered up to now; in such systems, the components can influence each other through their gravitational fields, through stellar winds, and especially through an exchange of gas flows. Following the fundamental observations of O. Struve in the 1940's and 50's, R. Kippenhahn and A. Weigert, B. Paczyński and M. Plavec and their coworkers began in the mid-1960's to apply theory and to develop complex computer programs to describe stellar evolution in close binary systems.

We have already met up with the common gravitational potential of a binary star system, with its Roche surface and the Lagrange points (Fig. 6.15); we have also identified important groups of stars, such as the cataclysmic variables (novae,...) and the X-ray pulsars, as close binary systems (Sects. 7.4.5, 6).

Fundamentals of the Evolution. Here, we describe the basics of the evolution of close systems, building on the results for the evolution of individual stars. What happens in detail depends on the two original masses $\mathcal{M}_1$ and $\mathcal{M}_2$ and on the distance a between them, as well as on the losses of mass and angular momentum which the system sustains. We can give only an outline here of the great variety of possible binary star configurations.

The originally more massive component (star 1) is the *primary* component. This term is retained even when the mass ratio is reversed in the course of the evolution of the system. Corresponding to the evolution times t_E for the different stellar masses (Table 8.2), component 1 first evolves into a giant star, increasing its radius towards the end of the hydrogen burning phase, during hydrogen-shell burning, and then during helium burning (Fig. 8.7a). If it grows larger than its Roche surface during one of these phases, matter will flow through the inner Lagrange point L_1 (Fig. 6.15a) onto component 2. In this way, a binary system can develop in which the more evolved component has the smaller mass and no longer fits on the mass–luminosity curve (Fig. 6.14).

The change in separation a of the two components as a result of mass exchange can be given for the case, which often occurs in fact, that the whole system loses no mass and that the total orbital angular momentum (2.53)

$$L = a_1^2 \mathcal{M}_1 \omega + a_2^2 \mathcal{M}_2 \omega = \text{const} \tag{8.55}$$

is conserved. Here, ω is the circular frequency of the orbital motion ($\omega = 2\pi/$orbital period P) and a_i is the distance of component i from the common center of gravity. Inserting the definition of the center of gravity (2.54) and Kepler's 3rd law, $\omega^2 a^3 = G(\mathcal{M}_1 + \mathcal{M}_2)$ (2.57) into (8.55), we find that the separation is proportional to the following function of the mass ratio $q = \mathcal{M}_1/\mathcal{M}_2$:

$$a \propto \frac{(1+q)^4}{q^2} \tag{8.56}$$

(Fig. 8.12). The two stars have their minimum separation when q equals 1. The radius of the two Roche surfaces depends upon q, and is also directly proportional to the separation a.

While a few binary star systems, the contact systems or W UMa stars (Fig. 6.15b), are so close that the Roche surfaces of the two stars touch each other already during the main sequence stage, in general, a strong interaction

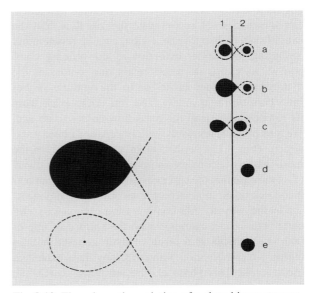

Fig. 8.12. The schematic evolution of a close binary star system with an initial mass ratio of $q = \mathcal{M}_1/\mathcal{M}_2 = 2$ (time a); star 1 = primary component. The axis of rotation around the center of gravity of the system is drawn in as a vertical line, and sections through the Roche surfaces are shown as dashed curves; the volumes of the two components are indicated by the filled areas. Time b: beginning of mass loss from star 1 as it fills its Roche surface. Time c: $q = 1/2$. Time d: the end of mass loss at $q = 1/10$; the last contact of star 1 with its Roche surface. The primary component contracts into a compact star, for example a white dwarf (at time e)

of the two components begins only when the more massive star has evolved away from the main sequence. Even before it has filled its Roche surface, it can heat up its cooler companion by radiation, and, among other things, can initiate surface activity on the companion. The strong X-ray and radiofrequency emissions of the RS CVn variables (Sect. 7.4.3) are probably produced in this way.

Once star 1 has reached its Roche surface, there is at first a phase with a strong mass flow through the inner Lagrange point within a relatively short time, of the order of the Helmholtz-Kelvin time, until the masses of the two components have become equal. Thereafter follows a phase with a slower gas flow, as a rule on a nuclear time scale. The observations of eclipsing binaries (Sect. 6.5.2) show that, for example, β Lyr is in the first phase and Algol (β Per) in the second. The secondary component is often surrounded by a dense accretion disk during the rapid phase. Observations of a common gas shell around both components indicate that star 2 can sometimes not accomodate the gas flow, and the overall system then loses mass and angular momentum.

During the further evolution of the primary component, its radius decreases again due to loss of the hydrogen-rich outer layers or to the ignition of helium burning (Sect. 8.2.3). The star draws back from its Roche surface and the mass flow comes to an end. We thus obtain a relatively widely-separated binary star system (Fig. 8.12), with a main sequence star as the more massive secondary component and a white dwarf or, in the case that the supernova explosion does not tear the system apart, a neutron star as primary component.

On the way to this configuration, anomalous element abundances can appear on the surface of one or both components as a result of hydrogen and helium burning in combination with the strong mass exchange (Sect. 8.2.4). This phase can be attributed to e.g. the helium stars, the OB stars with CNO anomalies, the Wolf–Rayet stars (at least those which belong to binary star systems) and the barium stars.

Owing to the strong dependence of the evolution time t_E along the main sequence on the mass of the star, the primary star usually ends its evolution before the *secondary* star has left the main sequence and evolved into a giant star. Then, the mass transfer is replayed with reversed roles. The compact primary component can, as a result of its deep gravitational potential, at first effectively collect the stellar wind from the more massive star 2, and later its gas flow through the Lagrange point. Accretion onto a neutron star as primary component yields the various forms of X-ray binary stars (Sect. 7.4.6); continued accretion can lead to the formation of a black hole. If, on the other hand, the matter flows over onto a white dwarf star, the result, in connection with nuclear fusion processes, is the great variety of cataclysmic variables (Sect. 7.4.5) such as novae, recurrent novae, etc.

Finally, the evolution of the secondary component also ends with the formation of a white dwarf plus planetary nebula, a neutron star, or a black hole. In the first case, the result is a relatively widely-spaced binary system consisting of two white dwarfs or of a white dwarf and a neutron star (primary component). On the other hand, if the more massive star 2 explodes as a supernova, the binding of the binary system is destroyed and each component flies away as a "runaway star" with a velocity of the order of 100 km s^{-1}. This explains the high spatial velocities observed for many radio pulsars (Sect. 7.4.7). If star 2 has lost a great deal of mass, however, a weaker supernova explosion is possible, leaving a bound system consisting of two neutron stars. The binary pulsar PSR J1915+1606 (Sect. 6.5.6) was perhaps formed in this way. The remaining possible final configurations, pairs consisting of one neutron star and one black hole, or of two black holes, would be rather difficult to observe.

Novae. After becoming acquainted with the observational facts concerning novae, and the fundamentals of their explanation in Sect. 7.4.5, we take up here the *nuclear* processes which cause their outbursts.

In a close binary system, hydrogen-rich matter flows from the secondary component via an accretion disk onto the white dwarf (Fig. 7.35). In a time of about 10^4 to 10^5 yr, it collects on the surface of the latter, causing the temperature T and the density ϱ to increase, until a critical mass is reached and at its base, at several 10^7 K and $\varrho \simeq 10^7$ kg m^{-3}, *hydrogen* burning via the CNO cycle (8.18) ignites. Since here the electrons are *degenerate*, following Fermi-Dirac statistics, the fusion reaction occurs *explosively*. Analogously to the helium flash in degenerate matter (Sect. 8.2.3), a *"thermonuclear runaway"* process occurs, in which the temperature increases to over 10^8 K within a few s, until the degeneracy of the gas is lifted, the accreted matter expands, and T and ϱ again decrease.

During the explosive hydrogen burning, in contrast to the hydrostatic burning which occurs in the course of the "normal" evolution of an individual star, due to the high temperatures (p, γ) reactions become important, so that the proton-rich radioactive nuclides ^{13}N, ^{14}O, ^{15}O, and ^{17}F are formed; their half-lives for β^+-decay lie between 70 and 600 s. Since the explosive H burning is accompanied by a convective mixing of the accreted matter on a typical time scale of 100 s, these nuclides reach the surface before their decay and form an important additional energy source for ejection of the nova shell at a velocity ≥ 1000 km s^{-1}.

The explosion is all the more violent, the higher the concentration of C, N and O, so that the abundances of these elements determine practically all the phenomena connected with the nova outburst, in particular the brightness increase and the expansion velocity. In agreement with theory, increased abundances of C, N and O relative to the solar mixture are observed in the spectra of novae; these cannot result from the explosion itself, but rather are formed in a previous process, which is still not completely understood, probably in the interior of the white dwarf from whence they were mixed into the accreted hydrogen layer.

8.2.7 The Physics of Accretion Disks

Using the novae as an example, we have seen in Sect. 7.4.5 that the matter which flows over from the companion star forms an accretion disk as a result of angular momentum conservation. It often gives the main contribution to the luminosity of the system (together with the hot spot, Fig. 7.35). Accretion disks also play a decisive role in the centers of galaxies, as we shall see in Sect. 12.3.6. It thus seems appropriate at this point to discuss the physics which lies behind an accretion disk.

In a disk which consists of point masses, the particles move around a central object of mass $\mathcal{M}$ on stable circular orbits, according to Kepler's 3rd law (2.57); their angular velocity decreases on going outwards, as we know from e.g. our Solar System. Here, however, we are dealing with a diek of *gas*, in which the differential rotation leads to frictional forces, that in turn cause loss of kinetic energy and modify the distribution of angular momentum, thus giving rise to deviations from Keplerian motion. While stellar accretion disks are often of low mass, the mass of the disks in the centers of galaxies is frequently not negligible in comparison to $\mathcal{M}$, so that (2.57) does not hold.

If now matter is added to the outer edge of the disk (around the companion star) at a rate of $\dot{m}$, it does not directly reach the surface of the central object (at R). The gas particles instead lose a portion of their kinetic energy through friction (viscosity) within the disk and move inwards on spiral orbits; at the same time, their angular momentum is transported outwards in the disk. The energy dissipation as a result of friction leads to heating of the disk and emission of radiant energy. Gravitational energy can thus be converted in an efficient manner into radiation by an accretion disk, especially one around a compact object. The parameters which control this process are, along with the density of matter in the disk, the *accretion rate* $\dot{m}$ and the *viscosity* ν.

Keplerian Disks. Here, we limit ourselves for simplicity to *thin* disks ($z \ll r$), whose mass is small relative to the central mass $\mathcal{M}$, so that their own gravitational forces play no significant role; r is the radial distance from $\mathcal{M}$ in the symmetry plane of the disk and z is the distance above or below the plane of the disk. In these Keplerian disks, the gas particles move on nearly circular orbits according to (2.57) around the mass $\mathcal{M}$ with the angular velocity $\omega(r)$, or the orbital velocity

$$v_\varphi(r) = r\,\omega(r) = \sqrt{\frac{G\mathcal{M}}{r}} \qquad (8.57)$$

(φ = azimuthal angle). In a Keplerian disk, we have for both the radial and for the perpendicular velocity components $|v_r|, |v_z| \ll v_\varphi$.

The vertical structure of the disk is found at any distance r from the hydrostatic equilibrium, (8.1):

$$\frac{\partial P}{\partial z} = -\varrho g_z(r) = -\varrho \frac{G\mathcal{M}z}{(r^2+z^2)^{3/2}} \simeq -\varrho \frac{G\mathcal{M}z}{r^3}, \qquad (8.58)$$

where P is the pressure, ϱ the mass density and g_z the vertical component of the gravitational acceleration from the star. If we take $z \simeq H$ for a thin disk, and define the vertical scale height H using $dP/P = -dz/H$

(3.16), it follows from (8.58) that

$$H \simeq \frac{c_s}{\omega}, \quad \frac{H}{r} \simeq \frac{c_s}{v_\varphi}, \qquad (8.59)$$

where $c_s \simeq \sqrt{P/\varrho}$ is the velocity of sound (7.93). The condition for a thin disk ($H \ll r$) is thus equivalent to $c_s \ll v_\varphi$, i.e. the disk rotates with a highly supersonic velocity.

In the following, we neglect the vertical structure and use only the "matter content" of the disk described by its *surface density*

$$\Sigma(r) = \int_{-\infty}^{\infty} \varrho(r, z) \mathrm{d}z. \qquad (8.60)$$

Accretion Luminosity. The maximum energy which can be obtained at the cost of gravitational energy, when a mass m falls from a large distance into the disk and onto the surface R of the central star, is given by (2.44) as the binding energy of the innermost orbit of the disk:

$$E_a(R) = \frac{G \mathcal{M} m}{R}. \qquad (8.61)$$

Correspondingly, if a mass flow is directed inwards, the rate $\dot{m}$ is available as accretion luminosity

$$L_a(R) = \frac{G \mathcal{M} \dot{m}}{R}. \qquad (8.62)$$

Since the frictional forces and thereby the energy dissipation are distributed over the entire disk, the emission of radiation takes place over the surface of the *whole* disk and does not occur only when the accreted mass strikes $\mathcal{M}$. However, according to (2.59) and (2.60), the total energy of a Keplerian orbit close to the star's surface is given by

$$E(R) = \frac{1}{2} \frac{G \mathcal{M}}{R} = \frac{1}{2} E_a, \qquad (8.63)$$

so that *within* the disk, only *half* of the accretion energy is converted into radiation. The other half remains as kinetic energy of the rotational motion of the disk and has to be "braked" within an extremely thin *boundary layer* directly at the surface e.g. of the white dwarf. The structure of this boundary layer, which is hard to calculate in detail, thus determines a considerable part of the radiation emission from an accretion disk.

Viscosity. The frictional force which acts on a surface A is given by $F = \varrho v \cdot A (\mathrm{d}v/\mathrm{d}x)$, where $\mathrm{d}v/\mathrm{d}x$ is the velocity perpendicular to A and v the (kinematic) viscosity with dimensions of velocity times length, $[\mathrm{m}^2\,\mathrm{s}^{-1}]$.

The viscosity determines in particular the radial velocity v_r or the mass flow $\dot{m} = 2\pi r \Sigma(r) v_r$. As an order of magnitude, we have $r v_r \simeq v$ and thus, in the *stationary* case, $\dot{m} \propto v \Sigma$.

In cosmic accretion disks, the friction is caused by the *turbulence* of the gas; magnetic forces can also contribute to it. In contrast, the molecular friction of the gas plays no role. In the absence of an exact theory for the calculation of the turbulent viscosity, usually the α approach of N. I. Shakura and R. A. Sunyaev (1973):

$$v = \alpha c_s H \simeq \alpha \frac{c_s^2}{\omega} \qquad (8.64)$$

is used; here, $\alpha \leq 1$ is a free parameter, almost always taken to be constant.

Emission of Radiation. We consider a ring of radius r and the radial thickness $\mathrm{d}r$ within the accretion disk; then the viscosity finally leads to the emission of the energy $4\pi r L(r) \mathrm{d}r$ per unit time from its (upper and lower) surfaces (we simply give this result here without proof). The resulting "ring luminosity" is given by:

$$L(r) = \frac{1}{2} v \Sigma \cdot \left(r \frac{\mathrm{d}\omega}{\mathrm{d}r} \right)^2. \qquad (8.65)$$

For Keplerian orbits, it follows from (8.57) that $\mathrm{d}\omega/\mathrm{d}r = -(3/2)(\omega/r)$ and thus $(r \mathrm{d}\omega/\mathrm{d}r)^2 = (9/4) G \mathcal{M} / r^3$.

In the limiting case of *stationary* disks, owing to $\dot{m} \propto v \Sigma$ (see above), the overall emission of a ring, and therefore of the whole disk, does not depend explicitly on the viscosity v. However, the spectral distribution of the radiation depends sensitively on v.

From $L(r)$, we can also estimate the order of magnitude of the temperature $T(r)$ at the surface of the disk using the Stefan–Boltzmann law, (4.66).

8.3 The Final Stages of Stellar Evolution

In this subchapter, we describe the structures of the *stable*, compact final configurations of stellar evolution: brown and white dwarfs and the neutron stars. An additional possible end product of stellar evolution, the collapse of matter into a black hole, will be treated in a later section, 8.4.4, together with the Theory of General Relativity.

We begin in Sect. 8.3.1 with the brown dwarfs, whose masses lie below $0.08\, \mathcal{M}_\odot$ and are therefore too small for them to be "real" stars with hydrogen burning in their centers (Sect. 8.2.3). The structures of white (and also brown) dwarfs are determined in the main by the Fermi–Dirac degeneracy of their electrons, so that we first concern ourselves in Sect. 8.3.2 with the equation of state for electron-degenerate matter, before taking up the white dwarfs in Sect. 8.3.3. They are the end products of the evolution of stars with masses $\leq 8\, \mathcal{M}_\odot$ (Sect. 8.2.4). Finally, in Sect. 8.3.4, we treat the neutron stars, with their extremely high densities; they are formed as remnant stars by supernova explosions (Sect. 8.2.4) and are usually observed as radio pulsars (Sect. 7.4.7) or as X-ray binary stars (Sect. 7.4.6).

8.3.1 Brown Dwarfs

Objects with the solar elemental mixture, whose masses are less than $0.08\, \mathcal{M}_\odot$ or $80\, \mathcal{M}_J$ (Jupiter masses), never attain the stage of hydrogen burning in equilibrium (S. Kumar, 1963). These brown dwarfs are transitional objects between the "actual" stars and Jupiter-like planets. Their maximum interior temperatures of $T_c \leq 10^6$ K permit, for masses of at least $13\, \mathcal{M}_J$, the ignition of *deuterium burning* $^2\mathrm{D}(\mathrm{p},\gamma)^3\mathrm{He}$, in which the deuterium formed in the Big Bang (Sect. 13.2.4) is consumed. (Some authors *define* a brown dwarf as an object in which deuterium fusion can take place.) This nuclear energy release is not sufficient to compensate the rediative energy loss of the brown dwarfs, so that they cool down gradually over very long periods of time.

The interior of a brown dwarf consists, like large parts of that of Jupiter (Sect. 3.4.1), mainly of liquid metallic hydrogen and of helium. The pressure there is generated principally by the degenerate electrons, which obey Fermi–Dirac statistics (Sect. 8.3.2); it is of the order of 10^{16} Pa $= 10^5$ Mbar. The densities lie in the range of 10^4 to 10^7 kg m^{-3}, and are therefore, like the pressures, higher than those found on Jupiter. The radii of the brown dwarfs, on the other hand, are not strongly dependent on their masses and are in the neighborhood of 1 R_J (Jupiter radius) $\simeq 0.1\, R_\odot$.

The extremely thin atmospheres of the brown dwarfs contain numerous molecules and dust. An important spectroscopic criterion for identifying substellar masses is a high abundance of *lithium*. In cool stars with masses $\geq 70\, \mathcal{M}_J$, ^{7}Li from the atmosphere is mixed into the interior by powerful convection zones and destroyed there by reaction with thermal protons. In the brown dwarfs, in contrast, the temperatures are too low for this process to occur, so that the original Li remains for the most part intact.

Owing to their weak luminosities, brown dwarfs are very difficult to observe. By applying the lithium criterion, however, a number of brown dwarfs have been identified since the 1990's among the *field* stars of later spectral classes, $\geq$ M7, L, and T (Sect. 6.4.3). Several of these objects have also been found since 1995 as *companions* of nearby stars (Sect. 6.5.7). The problem of distinguishing them from Jupiter-like planets will be discussed in Sect. 14.1.5, together with the question of the existence of planetary systems around other stars.

8.3.2 The Equation of State of Matter with Degenerate Electrons

The average densities of white dwarf stars (Sect. 6.4.2), whose radii are comparable to the that of the Earth, while their masses are of the order of the Sun's mass, are in the range of 10^8 to 10^9 kg m^{-3}; they are thus about 10^6 times greater than the mean density of the Sun. At such densities, the equation of state for an ideal gas (8.3) is no longer applicable. As R. H. Fowler recognized in 1926, the matter in these stars (more precisely, the electron component of the matter) is degenerate in the sense of Fermi–Dirac statistics, with the exception of a quite thin atmospheric layer.

Fermi–Dirac Degeneracy. In classical or Maxwell–Boltzmann statistics, the velocity distribution of electrons or the equation of state of gases, for example, are derived by determining the distribution of the particle states in *phase space*. A 6-dimensional volume element $\Delta \Omega$ in phase space consists of the usual volume element $\Delta V = \Delta x \Delta y \Delta z$ in position (configuration) space, multiplied by a volume element in momentum space, $\Delta p_x \Delta p_y \Delta p_z$, following the rules of probability theory. The applicability of this method is however limited for an electron gas by the *Pauli principle*, which requires that one quantum state or one quantum cell of phase-space volume $\Delta\Omega = h^3$ (h = Planck's constant) can be occupied by at most one electron of each spin direction, i.e. all together by 2 electrons. This fact is taken into account in Fermi–Dirac statistics. In the interior of a white dwarf star, the matter is so strongly compressed due to the enormous pressures and relatively low temperatures that all phase-space cells h^3 are completely occupied up to a particular limiting energy E_0 or the corresponding momentum p_0, i.e. each cell contains two electrons. This state is referred to as *complete degeneracy* of the Fermi gas.

The equation of state of the degenerate electron gas is readily derived (the associated proton gas becomes degenerate only at much higher densities; its pressure can be neglected):

In a volume V, with a particle density n_e, there are a total of $V \cdot n_e$ electrons. In momentum space, $\{p_x p_y p_z\}$ (Fig. 8.13), these electrons fill a homogeneous sphere up to a maximum limiting momentum p_0, corresponding to the maximum energy (the *Fermi energy*) E_0. In phase space, we thus have a volume $(4/3)\pi p_0^3 V$, and with 2 electrons per phase-space cell of volume h^3, we obtain the relations:

$$n_e V = \frac{2}{h^3} \cdot \frac{4\pi}{3} p_0^3 \cdot V,$$

$$p_0 = \left(\frac{3h^3}{8\pi}\right)^{1/3} n_e^{1/3}. \quad (8.66)$$

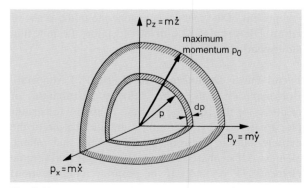

Fig. 8.13. A degenerate electron gas (Fermi–Dirac statistics). In momentum space, of which only one octant is shown here, the electrons fill uniformly the Fermi sphere, of radius p_0

Nonrelativistic Electron Gas. For particle velocities $v \ll c$ or energies $E \ll m_e c^2$ (m_e: rest mass of the electron), the relation $E = p^2/(2m_e)$ holds between the kinetic energy $E = m_e v^2/2$ and the momentum $p = m_e v$; thus the Fermi energy is given by $E_0 = p_0^2/(2m_e)$. The pressure of the electron component is given by (from the well-known relation of kinetic gas theory, as for an ideal gas)

$$P_e = \frac{2}{3} n_e \overline{E}, \quad (8.67)$$

where $\overline{E}$ is the mean energy per electron. The relationship between $\overline{E}$ and E_0 can be calculated by referring to Fig. 8.13; we then find:

$$\overline{E} = \frac{\int_0^{p_0} E \cdot 4\pi p^2 dp}{\int_0^{p_0} 4\pi p^2 dp} = \frac{3}{5} \frac{p_0^2}{2m_e} = \frac{3}{5} E_0, \quad (8.68)$$

and we thus obtain the equation of state of a completely degenerate electron gas:

$$P_e = \frac{1}{5m_e} \left(\frac{3h^3}{8\pi}\right)^{2/3} n_e^{5/3}. \quad (8.69)$$

The temperature does not enter this equation at all; this is a characteristic of complete degeneracy. It can readily be verified that the pressure of the equally dense proton gas is much smaller than that of the degenerate electron gas, so that to a good approximation the total pressure is

$$P \simeq P_e. \quad (8.70)$$

The relation between n and the mass density ϱ is most simply expressed in terms of the *molecular weight* μ_e *per free electron* (in atomic mass units), i.e. the mass which corresponds to *one* electron. We then have:

$$\varrho = \mu_e m_H n_e. \quad (8.71)$$

Thus for completely ionized matter, in the case of hydrogen, $\mu_e = 1$, and for helium and heavier elements, $\mu_e \simeq 2$.

The Relativistic Electron Gas. The energy of the electrons increases with increasing density ϱ initially $\propto \varrho^{2/3}$, until at $E \geq m_e c^2$ the relativistic mass variation of the electrons becomes important. In the case of *complete* relativistic degeneracy, $\overline{E} \gg m_e c^2$, we have $E = pc$ (4.26) and $P_e = (1/3) n_e \overline{E}$ (4.47). A recalculation of the above formulas using these results of the Theory of Special Relativity yields the equation of state:

$$P_e = \frac{c}{4}\left(\frac{3h^3}{8\pi}\right)^{1/3} n_e^{4/3}, \quad (8.72)$$

i.e. with $P \propto \varrho^{4/3}$, the matter is less compressible.

8.3.3 The Structure of the White Dwarfs

We first try to clarify the relationship between ϱ, n_e, and the overall pressure P in the nonrelativistic case by means of an estimate of the density of a white dwarf, $\varrho = 10^9$ kg m^{-3} (with $\mu_e = 2$). From (8.71), we then have $n_e \simeq 3 \cdot 10^{35}$ electrons per m^3, and, from (8.69) and (8.70), a pressure of the degenerate gas equal to $P \simeq 3 \cdot 10^{21}$ Pa, i.e. about 10^5 times higher than e.g. in the center of the Sun. An ideal gas at this pressure would have to be at a temperature of about 10^9 K.

Mass–Radius Relation. The equation of state $P \simeq P_e \propto \varrho^{5/3}$ can be used together with our earlier estimate of the pressure in the interior of a star with mass $\mathcal{M}$ and radius R from (8.1), i.e. $P \propto \varrho \, G\mathcal{M}/R$, and the trivial relation $\varrho \propto \mathcal{M}/R^3$, to obtain an approximate mass–radius relation for white dwarf stars, namely:

$$R \propto \mathcal{M}^{-1/3}. \quad (8.73)$$

That is, with increasing mass, the radius of the star decreases. We have here neglected the fact that the outermost part of the star is *not* degenerate and that near its center, it is relativistically degenerate.

With (8.73), we can also write down the luminosity $L = 4\pi R^2 \sigma T_{\text{eff}}^4$. White dwarf stars of mass $\mathcal{M}$ should thus be ordered according to their mass in the Hertzsprung–Russell diagram on lines:

$$L \propto \mathcal{M}^{-2/3} T_{\text{eff}}^4. \quad (8.74)$$

Comparison with observations yields the initially surprising result that *nearly all* white dwarfs have masses between 0.5 and $0.6\mathcal{M}_\odot$.

The Limiting Mass. As was recognized by S. Chandrasekhar in 1931, the radius of a white dwarf star tends with increasing relativistic degeneracy to the limiting value $R \to 0$ even at a finite limiting mass:

$$\mathcal{M}_{\text{Ch}} \simeq \frac{5.8}{\mu_e^2} \mathcal{M}_\odot. \quad (8.75)$$

Since $\mu_e \simeq 2$ as a result of the evolutionary history of the white dwarfs, these stars are stable or able to exist only for masses $\mathcal{M} \le 1.44\mathcal{M}_\odot$. The appearance of *Chandrasekhar's limiting mass* is caused by the violation of the stability condition for hydrostatic equilibrium (8.39) as a result of the pressure dependence $P \propto \varrho^{4/3}$ of the relativistically degenerate electron gas.

The formation and evolution of white dwarfs has already been discussed in Sect. 8.2.4. Stars, or the central portions of stars, whose mass at the end of their nuclear evolution is greater than Chandrasekhar's limiting mass of about $1.4\,\mathcal{M}_\odot$, have only one way to form a stable structure: through a collapse to much higher densities. The electrons and protons will finally be pressed together so strongly that, in a reversal of β-decay, they fuse to form neutrons and thus give rise to a neutron star.

8.3.4 Neutron Stars

In the year of the discovery of the neutron, 1932, L. Landau already discussed the possibility of the existence of stable neutron stars. W. Baade and F. Zwicky predicted in 1934 that neutron stars are formed in supernova explosions (Sect. 7.4.7), in which the energy for the explosion is provided by the gravitational energy released (Sect. 8.1.4) in the collapse to higher densities. The first models for stars consisting of degenerate neutron matter were calculated by J. R. Oppenheimer and G. M. Volkoff in 1939; neutron stars were finally discovered in 1967 as pulsars (Sect. 7.4.7) by A. Hewish and his coworkers.

The Matter in Neutron Stars. Neutron matter has great similarities to the nuclear matter of heavy atoms, since the interaction forces between protons and neutrons are equally strong (aside from the electrostatic Coulomb force, which is unimportant here). On this basis, we first make an elementary estimate, by considering the neutron star as a giant atomic nucleus, so to speak:

A nucleus of e.g. ^{56}Fe has a mass of $m_{\text{Fe}} = 56 \cdot 1.67 \cdot 10^{-27}$ kg $= 9.4 \cdot 10^{-26}$ kg and a radius of about

$5.6 \cdot 10^{-15}$ m, and thus a mean density of $\varrho \simeq 10^{17}$ kg m^{-3}. If a star of, for example, 1 $\mathcal{M}_\odot$ is compressed to have a density equal to that of nuclear matter, it must shrink its radius to about $5.6 \cdot 10^{-15} \, (\mathcal{M}_\odot/m_{\text{Fe}})^{1/3}$ [m] or about 16 km! The comparison between atomic nuclei and neutron stars should, however, not be taken too literally, since the latter cannot have an electric charge. While roughly equal numbers of protons and neutrons are present in an atomic nucleus, neutrons are strongly predominant in the star, where each proton has been "squeezed together" with an electron to form a neutron: $p^+ + e^- \rightarrow n + \nu$. A free neutron decays with a half-life of 617 s ($n \rightarrow p^+ + e^- + \bar{\nu}$), but in the star, this decay is prevented by the high electron density. If the Fermi energy E_0 of the electrons namely exceeds 0.78 MeV, the equivalent of the mass difference between neutron and proton (1.29 MeV) minus that of the electron's rest mass (0.51 MeV), then the decay electrons would 'find no more room' in phase space. This value of $E_0 = pc$ is attained according to (8.72), and using $P_e = (1/3)n_e E$, at about $3 \cdot 10^{36}$ electrons m^{-3}, or a mass density of $1 \cdot 10^{10}$ kg m^{-3}, i.e. somewhat above the density characteristic of white dwarfs. (Above 0.51 MeV, the electrons are relativistically degenerate.)

Structure. The transition from the low densities at the surfaces of neutron stars to a neutron fluid with $\geq 10^{17}$ kg m^{-3} in their interiors poses a number of very interesting problems for the theoretician. For a neutron star of 1 $\mathcal{M}_\odot$ with a radius of $R \simeq 16$ km, the density initially increases in the *outer* crust from about 10^7 kg m^{-3} to about $4 \cdot 10^{14}$ kg m^{-3} at a depth of 1 km. Here, the matter, similar to the interiors of white dwarfs, consists of a degenerate electron gas and atomic nuclei, which form a crystal lattice. While in the outer layers ^{56}Fe nuclei are most abundant, further inside we find, with increasing density, more and more neutron-rich nuclei. Above $4 \cdot 10^{14}$ kg m^{-3}, in the *inner* crust, the nuclei begin gradually to dissociate and free neutrons are present. Finally, above about $2 \cdot 10^{17}$ kg m^{-3}, at radii of $R \leq 11$ km, the nearly incompressible neutron fluid is formed; it still contains a small percentage of protons and electrons. The central density of the neutron star is in the range $\geq 4 \cdot 10^{17}$ kg m^{-3}. In this region, the equation of state of matter is only imprecisely known; hyperons, pions, or quarks may be present.

This neutron matter exhibits interesting physical properties: the neutrons can to some extent interact pairwise with each other and form a *superfluid* constituent, while the proton constituent is *superconducting*. We cannot go here into the details of the influence of the superfluidity on the rotation of neutron stars in connection with their extremely strong magnetic fields.

Thermal Radiation. Following its origins in a supernova explosion (Sect. 7.4.7), the interior of a neutron star is initially very hot ($\simeq 10^7$ K), and cools in the course of several thousand years to a few 10^6 K. During this phase, the radiation emitted is mainly in the soft X-ray range (from around 0.1 to 1 keV). In a few cases, the X-ray satellite ROSAT was able to detect directly the (not pulsed) thermal emission of cooling neutron stars or pulsars with X-ray luminosities of $L_x \geq 10^{26}$ W.

Gravitational Energy. The gravitational energy $G\mathcal{M}^2/R$ (8.31) of a neutron star of mass 1 $\mathcal{M}_\odot$ and radius $R = 16$ km is about $2 \cdot 10^{46}$ J, i.e. about one-tenth of the energy equivalent of its rest mass, $\mathcal{M}c^2$. In such strong gravitational fields, Newton's Theory of Gravitation is no longer sufficient for the calculation of its structure; instead, Einstein's General Relativity theory (Sect. 8.4.1) must be applied. The deviations from Newton's theory are of the order of $(G\mathcal{M}^2/R)/(\mathcal{M}c^2) = G\mathcal{M}/(Rc^2) \simeq 0.1$. In particular, the *gravitational* mass $\mathcal{M}$ which an external observer would derive e.g. from the motion of a neutron star in a binary star system, is of the order of 10% *smaller* than the mass which is required to form the star from an originally extended configuration. The difference (mass defect) is for the most part due to the negative gravitational binding energy.

Limiting Mass. As was first shown by J. R. Oppenheimer and G. M. Volkoff in 1939, an upper limiting mass exists for neutron stars, just as for the white dwarfs. Owing to the uncertainties in the equation of state of material at densities above those of nuclear matter, the limiting mass is not very precisely known; it probably lies near 1.8 $\mathcal{M}_\odot$. If this mass is exceeded, a stable final configuration is no longer possible; the matter of the star then collapses to form a black hole (Sect. 8.4.4).

8.4 Strong Gravitational Fields

If massive bodies or particles move at velocities near to that of light, Newtonian mechanics and gravitational theory (Sect. 2.3) can no longer be applied. We begin by making a rough estimate of the strength which the gravitational field or the gravitational potential $G\mathcal{M}/R$ of a sphere of mass $\mathcal{M}$ and radius R must have, in order that the escape velocity v from its field region becomes of the order of the velocity of light, c. If we naively set $v = c$ in the relation $v^2 = 2G\mathcal{M}/R$ (2.61), we obtain:

$$\frac{2G\mathcal{M}}{Rc^2} = 1 \ . \tag{8.76}$$

Thus, if the gravitational potential becomes comparable to c^2 or the gravitational energy $G\mathcal{M}^2/R$ becomes comparable to the relativistic rest energy $\mathcal{M}c^2$, then a relativistic theory of gravitation must replace the Newtonian theory: this is the Theory of General Relativity formulated by A. Einstein in 1916.

While on the surface of the Sun, $2G\mathcal{M}_\odot/(R_\odot c^2) \simeq 4 \cdot 10^{-6}$, so that we can expect only very small deviations from Newtonian theory in the Solar System and for ordinary stars, we have learned that neutron stars (Sect. 8.3.4) represent a mass concentration with $G\mathcal{M}/(Rc^2)$ of the order of 0.1. In the calculation of their structure, especially of their limiting mass, the effects of General Relativity are important. The strong gravitational fields in the sense of (8.76), characteristic of *relativistic astrophysics*, and the highly energetic particles and radiation fields which occur in connection with them, will be found again for one thing in galactic centers (Sects. 11.3.4 and 12.3.6). For another, cosmology, which treats the structures and evolution of the Universe as a whole (Sect. 13.1.3), is based on General Relativity as the theory of gravitation.

We begin in Sect. 8.4.1 with an overview of the theory of General Relativity. This section is intended merely to give an impression of the structure of General Relativity theory; it cannot take the place of a proper introductory study of the theory. In Sect. 8.4.2, we then give selected results of the theory of General Relativity for a spherically symmetrical gravitational field, in particular the Schwarzschild metric and the Schwarzschild radius, the precession of the perihelion of planetary orbits, and the gravitational red shift. In Sect. 8.4.3, we treat the deflection of light rays and the action of "gravitational lenses", and in Sect. 8.4.4 we describe black holes, which follow from General Relativity theory, and briefly discuss their properties. Finally, in Sect. 8.4.5, we introduce gravitational waves, their sources and the possibilities for their observation.

8.4.1 The Theory of General Relativity

The precursor of General Relativity was the Theory of Special Relativity, developed by A. Einstein in 1905. This starts with the postulate that no material motion and no signal propagation can exceed the velocity c of light in a vacuum, and further requires that all natural laws be invariant with respect to Lorentz transformations (Sect. 4.2).

Should it not be possible to formulate the natural laws in such a way that they are invariant with respect to even *arbitrary* coordinate transformations? In 1916, Einstein had the ingenious idea of combining this appealing requirement with the theory of gravitation. The identity of *gravitational* and *inertial* mass, independent of the type of matter, was so to speak a "miracle" in the framework of Newtonian mechanics, but what did it mean? Einstein raised the experimental observation that in a freely falling coordinate system (elevator), the gravitational force (mg) appears to be exactly compensated by the inertial force ($m\ddot{z}$) to the status of a *basic postulate*. This means that gravitational force and inertial force are, in the end, *the same*.

Isaac Newton himself, then F. W. Bessel and later R. v. Eötvös had verified the equality of gravitational and inertial mass experimentally with increasing accuracy. In his classical experiment, Eötvös used the gravitational field of the Earth and the centrifugal force of the Earth's rotation, and as early as 1922 obtained an accuracy of 10^{-9}. Modern measurements have improved this value to about $5 \cdot 10^{-13}$.

Field Equations. The forces of gravity and inertia can be transformed away by local transformations on a four-dimensional Cartesian coordinate system with the "Euclidian" Minkowski metric (4.12). Conversely, as is found, the coefficients g_{ik} of the Riemannian metric

$$ds^2 = \sum_{i,k} g_{ik} dx^i dx^k \tag{8.77}$$

of an arbitrary coordinate system, and its spatial connectivity, determine the gravitation and inertial fields acting there.

This connection is formulated in Einstein's field equations

$$G_{ik} = -\frac{8\pi G}{c^4} T_{ik} \qquad (8.78)$$

in which $G = 6.67 \cdot 10^{-11}$ m^{-3} s^{-2} kg^{-1} is the usual Newtonian gravitational constant, G_{ik} is the Einstein tensor, and T_{ik} is the energy-momentum tensor ($i = 1, 2, 3, 4$). G_{ik} contains the geometrical properties of the four-dimensional space and depends only on the metric g_{ik} and its first and second derivatives ($\partial g_{ik}/\partial x^\ell$, $\partial^2 g_{ik}/\partial x^\ell \partial x^m$), while T_{ik} contains the energy and momentum and thus affects the properties of matter, the electromagnetic field, etc. Since the tensors G_{ik} and T_{ik} are symmetric ($G_{ik} = G_{ki}$, $T_{ik} = T_{ki}$), (8.78) represents a system of 10 coupled, nonlinear differential equations.

For applications in cosmology (Sect. 13.1.3), Einstein later added another term to the field equations, containing the *cosmological constant* Λ, which becomes important only at "cosmological" distances:

$$G_{ik} + \Lambda\, g_{ik} = -\frac{8\pi G}{c^4} T_{ik} \ . \qquad (8.79)$$

The field equations (8.78) and (8.79) of course contain the Newtonian theory of gravitation as a limiting case for weak gravitational fields.

Geodesics. The motion of a point mass under the influence of gravitation alone follows geodesic (i.e. shortest) curves, in a generalization of the Galilean law of inertia.

The propagation of light is along the socalled *null* geodesic

$$ds^2 = 0 \ , \qquad (8.80)$$

analogously to the expression (4.11) from Special Relativity.

Testing the Theory. Within the Solar System, the predictions of General Relativity differ only slightly from those of Newtonian mechanics and gravitational theory. The tests which can be used to distinguish between the two require great metrological precision.

Following the first important successes of Einstein's theory, the explanation of the precession of the perihelion of Mercury and of the deflection of light by the Sun (Sect. 8.4.3), a series of complex experiments, especially in the years from 1960 to 1980, have reached precisions of better than 1% for all of the effects predicted by the General Relativity Theory, so that we can on the whole speak of an excellent verification of this theory. Considerably stronger gravitational effects *outside* our Solar System, in particular in the binary star pulsar PSR J1915+1606 (Sect. 6.5.6), have likewise shown a very good agreement with the theory. The newer tests have essentially eliminated all competing theories of gravitation.

8.4.2 Spherically Symmetric Fields in Vacuum

The gravitational field ($T_{ik} = 0$) outside a mass $\mathcal{M}$, e.g. that of the Sun, can in the simple spherically symmetric case, as K. Schwarzschild showed in 1916, be represented using the metric:

$$ds^2 = \left(1 - \frac{R_S}{r}\right) c^2 dt^2 \qquad (8.81)$$
$$- \frac{dr^2}{1 - (R_S/r)} - r^2(d\theta^2 + \sin^2\theta\, d\varphi^2) \ ,$$

where r, θ, and φ are the spatial spherical coordinates and t is the time. The theory of planetary motions can thus be based on this metric. The constant of integration

$$R_S = \frac{2G\mathcal{M}}{c^2} \qquad (8.82)$$

is called the *Schwarzschild radius* or gravitational radius of the mass $\mathcal{M}$. Its value

$$R_S = 2.9\, \frac{\mathcal{M}}{\mathcal{M}_\odot} \quad [\text{km}] \qquad (8.83)$$

relative to the radius R of the mass $\mathcal{M}$ determines the deviations from the Euclidian metric of empty space (cf. (8.76)).

R. P. Kerr in 1963 found a considerably more complicated metric which represents a mass with angular momentum (and with an electric charge).

Planetary Orbits. Using the Schwarzschild metric, it is found that General Relativity also predicts planar orbits for a point mass orbiting a central mass $\mathcal{M}$ and that the law of areas remains valid. The *gravitational potential* however differs from the Newtonian case by an additional term $\propto 1/r^3$.

8.4 Strong Gravitational Fields

This term has on the one hand the effect that there are no stable closed orbits very close to $\mathcal{M}$. The innermost circular orbit which is still just stable has a radius of $r = 3R_S$.

On the other hand, a bound orbit is only *approximately* given by an ellipse, since its semimajor axis, and with it the perihelion, rotate during each orbit of the planet in the same direction by an amount

$$\Delta \omega = 6\pi \frac{G\mathcal{M}}{a(1-e^2)c^2}, \tag{8.84}$$

where ω is the angular distance (in radians) of the perihelion from the node, and a is the semimajor axis and e the eccentricity of the orbit (Sect. 2.2.1). As a result of this *precession of the perihelion*, the planet follows a *rosette orbit*.

Observations of the innermost planet of our Solar System, *Mercury*, over many years yielded a larger precession of its perihelion than could be derived from perturbations due to the other planets. This deviation was first explained by Einstein with his General Relativity theory. Taking the orbital elements of Mercury from Table 2.2, we obtain with (8.84) the precession of the perihelion $\Delta\omega_M = 0.1038''$ per orbit or $43.03''$ in one century.

This result is an important verification of General Relativity, since the precession of the perihelion is a 2nd order effect, while the deflection of light, the delay of radar signals and the gravitational red shift (see below) are only 1st order effects.

The calculated values for the precession of the perihelia per 100 yr for the three innermost planets are, in very good agreement with the observed values:

$$\begin{array}{lll} \text{Mercury} & \text{Venus} & \text{Earth} \\ 43.0'' & 8.6'' & 3.8''. \end{array} \tag{8.85}$$

Expression (8.84) for the precession of the perihelion can also be applied to the precession of the *periastron* in a binary star system, if, corresponding to the two-body problem (Sect. 2.4.2), the reduced mass and the relative orbital dimensions are introduced. The observations of the binary-star pulsar PSR J1915 + 1606, which belongs to an unusually close binary system (Sect. 6.5.6), show a considerably larger effect than can be observed in our Solar System, since, for comparable masses, the size of this system, of about 1.3 $R_\odot$, is much smaller. Here, a precession of the periastron of 4.23° per year(!) has been measured.

Gravitational Red Shift. A light quantum $h\nu$ which passes through a gravitational potential difference in the field of a mass $\mathcal{M}$ will experience a red shift relative to a light source in the laboratory frame:

$$\frac{\Delta \nu}{\nu} = 1 - \frac{1}{\sqrt{1-R_S/R}} \simeq -\frac{G\mathcal{M}}{Rc^2}. \tag{8.86}$$

The corresponding energy difference is given by $\Delta(h\nu) = -G\mathcal{M}m_\nu/R$, where $m_\nu = h\nu/c^2$ is the "mass" of the photon.

If e.g. the photon passes through the potential difference $G\mathcal{M}_\odot/R_\odot$ from the Sun to the Earth, then a frequency shift occurs which is formally equivalent to that due to the Doppler effect, $c\Delta\nu/\nu = 0.64$ km s^{-1}. Measurements verify this effect within about 1%, but the separation of the relativistic red shift from the Doppler effect due to turbulence in the solar atmosphere introduces uncertainties. Experiments with the extremely sharp γ-ray lines observed using the Mössbauer effect, e.g. in the gravitational field of the Earth, permit a much higher precision and also verify the calculated frequency shifts.

By means of experiments with extremely stable oscillators on space probes, the red shifts in the gravitational fields of some other planets (Voyager) and of the Sun (Galileo) have been determined in recent times.

8.4.3 Light Deflection and Gravitational Lenses

The gravitational field of a mass $\mathcal{M}$ deflects the light from a distant point source, which passes within a distance a from $\mathcal{M}$, by an angle (in radians) of

$$\theta = 2\frac{R_S}{a} = 4\frac{G\mathcal{M}}{ac^2} \tag{8.87}$$

(A. Einstein, 1915). If we calculate the gravitational deflection of a photon of "mass" $h\nu/c^2$ in classical mechanics, which does not take into account the curvature of space according to the theory of General Relativity, then we obtain just half the value predicted by (8.87).

Deflection of Light by the Sun. In the gravitational field of the Sun, the light beam from a star passing at a distance a experiences a deflection which we can relate to the radius of the Sun, $R_\odot$; from the center of the Sun, according to (8.87), there is a deflection (in seconds of arc) of

$$\theta''_\odot = \frac{1.75''}{a/R_\odot}, \qquad (8.88)$$

i.e. a maximum of $1.75''$ for a beam which just passes the edge of the solar disk. During a total eclipse of the Sun in 1919, the corresponding extremely difficult measurements were made with an accuracy of only about 20%, and led to the first experimental verification of the predictions of Einstein's General Relativity Theory.

Since 1969, there have been new possibilities for making such tests with long-baseline interferometers in the cm-wavelength range, which have an angular precision of less than $0.001''$, using e.g. the two bright quasars 3C 273 and 3C 279, which each year are completely or partially eclipsed by the Sun.

Together with the deflection of light, General Relativity predicts a *delay* in signals. I. Shapiro found in 1964 that a radar signal that passes near to the Sun is delayed by a time of the order of $2 \cdot 10^{-4}$ s. Radar reflections were obtained first from Mercury and Venus, then later from space probes; they confirmed very well the predictions of the theory.

Gravitational Lenses. Closely related to the deflection of a light beam by a mass $\mathcal{M}$ is the possibility of focusing light beams and, in analogy to an optical lens, of generating *images* of a light source. After the formulation of the theory of General Relativity, among others A. S. Eddington (in 1920) and A. Einstein (in 1936) noted that stars could act as gravitational lenses. F. Zwicky then pointed out in 1937 that a whole stellar system, such as a *galaxy*, would be more effective for this purpose.

From (8.87), we can see that the effect of a lens is stronger the larger the mass $\mathcal{M}$ or the Schwarzschild radius R_S, and the more precisely the light source Q, the gravitational lens L, and the observer O are all arrayed along a line. If we initially consider Q and L to be point sources, then we find as the condition for gravitational image formation in the case of weak gravitational fields and small deflection angles the *lens formula*:

$$\delta = \phi - \phi_Q = \theta \cdot \frac{r_Q - r_L}{r_Q}$$
$$= \frac{2R_S}{a} \cdot \frac{r_Q - r_L}{r_Q}, \qquad (8.89)$$

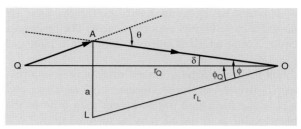

Fig. 8.14. The deflection of the light from a source Q at a distance r_Q by a gravitational lens L at a distance r_L from an observer at O. For small angles of deflection, $\theta = 2R_S/a$ (8.87), and $\delta = \phi - \phi_Q$ follows the lens formula (8.89) which results from the law of sines applied to the triangle OQA with $|QA| \simeq r_Q - r_L$ and $|OA| \simeq r_L$

where r_Q is the distance to the source and r_L that to the lens from the observer. ϕ_Q is the true angular distance of the source from the lens L and ϕ is the angular distance (of L), at which the source, after deflection, at a distance of a from L, is observed under an angle θ (8.87) (cf. Fig. 8.14).

The lens formula (8.89) sets up a relation between the distances of the source and of the observer from the gravitational lens. A suitable choice of a always allows us to find a deflection angle θ such that the light beam is focused onto O. The most favorable situation for a gravitational lens is found when $r_L \simeq r_Q/2$ is nearly fulfilled.

If the gravitational lens is not a point mass, but instead for example an extended galaxy, we can obtain its effect simply by summing over all the mass elements. For this sum, it is sufficient to consider the area density of the masses projected onto a plane through L and perpendicular to the observer. Corresponding to the lens formula, a *minimum* surface density of the gravitational lens is required to produce an image at the position of the observer.

Due to the deflection of the light rays or to the bending of the wavefronts, *several* images of a single source can be projected. The intensity or surface brightness of the source is in fact conserved in an image formed by a gravitational lens (Sect. 4.3.1), but the solid angle Ω_i under which the image i appears can be changed relative to the "direct" solid angle Ω_Q subtended by the source. A gravitational lens can thus amplify or attenuate the brightness, i.e. the flux (4.40) of an image by the geometrical factor

$$\gamma_i = \Omega_i / \Omega_Q \,. \tag{8.90}$$

The factor for all images together is $\gamma = \sum_i \gamma_i$. One can show that at least one of the images must be *amplified*. The multiplicity of the images depends on the often quite complex shape of the focal surface (caustic).

The discovery in 1979 of the first gravitational lens by means of the *"twin quasar"* 0957+561 A and B, which consists of two quasars with identical spectra at a small angular distance from one another, opened up interesting possibilities for applications in the study of distant galaxies. We observe gravitational-lens effects in particular for the most distant and brightest galaxies, the quasars (Sect. 12.3.4), as well as also for distant galactic clusters (Sect. 12.4.2), since in these cases, the probability is greatest that a massive galaxy is to be found at roughly half the distance to the quasar near to the line of sight.

Image formation by a lens is independent of the energy or frequency of the light, so that e.g. optical images and radio images are similar to one another.

In connection with the deflection of light, there are also differences in the *propagation times* of the light signals, which can be used for the determination of cosmic distances (Sects. 12.3.4 and 13.1.1).

Einstein's Rings. If source, lens and observer are *exactly* aligned along a line ($\phi_Q = 0$), then from symmetry grounds, a *ring-shaped* image is formed. From (8.89) and using $\phi \simeq a/r_L$ (Fig. 8.14), we immediately obtain the angular radius of such an Einstein ring:

$$\phi_E \simeq \sqrt{2R_S \cdot \frac{r_Q - r_L}{r_Q r_L}} \,. \tag{8.91}$$

In the case of an *extended* source, geometric distortions occur, so that only parts of the ring can be observed, i.e. as *arcs*.

Along with multiple lens images of quasars, in recent times also several Einstein rings and arcs have been discovered (Sects. 12.3.4 and 12.4.2).

The angular radius ϕ_E of the Einstein ring also gives the order of magnitude of the angular separations of the images in the *general* case of a gravitational lens. With $r_L = r_Q/2$, using (8.91) we can estimate the most favorable case for the extent of the lens effect on the celestial sphere: $\tilde{\phi}_E \simeq \sqrt{R_S/r_L}$, or, expressed in seconds of arc for easier application,

$$\tilde{\phi}_E'' \simeq 6.3 \cdot 10^{-5} \cdot \sqrt{\frac{\mathcal{M}}{\mathcal{M}_\odot} \cdot \frac{1}{r_L \,[\text{Mpc}]}} \,. \tag{8.92}$$

A massive galaxy of $10^{12} \, \mathcal{M}_\odot$ at a distance of $r_L = 1000$ Mpc thus produces an image pattern with a typical angular splitting of $2''$.

Gravitational Microlenses. Within a galaxy which forms a gravitational image, the *individual stars* can also act as lenses, as was pointed out by K. Chang and R. Refsdal in 1979. The image splitting from these "microlenses" is, to be sure, in the range of micro- to milliseconds of arc (8.92) and thus much too small to be directly observed, but the motion of a star perpendicular to the line of sight can lead to an observable *increase in brightness* for a short time (Fig. 12.25). Since this effect is independent of frequency or wavelength, it can be distinguished from the many other brightness fluctuations of the stars.

The effects of microlenses are suitable for the detection of extremely faint *"dark" objects* such as white and brown dwarfs, neutron stars, planets and black holes. In an organized search for the corresponding brightness fluctuations, several international cooperations have been continuously observing an enormous number (all together about 20 million) stars as background sources in the rich starfields of the galactic bulge (Sect. 11.1.4) in Sagittarius and in the Large and Small Magellanic Clouds (Sect. 12.1.2) since the early 1990's. The goal of e.g. the Macho project (*M*assive *C*ompact *H*alo *O*bjects) is to locate faint, low-mass objects in the extended halo of the Milky Way (Sect. 11.1.4).

The evaluation of the enormous collection of data (several billion measurements!) has yielded, along with the discovery of several 10 000 new variable stars, a few 100 events which are due to gravitational microlenses. These were mainly observed in the dense regions of the Milky Way galaxy, while in the direction of the Magellanic Clouds, i.e. towards the galactic halo, so far only of the order of 10 objects could be found. The presently available observations are not yet sufficient to draw conclusions about the mass distributions of dark objects.

8.4.4 Black Holes

The Schwarzschild metric (8.81), besides the very small effects described above, predicts also a much more spectacular possibility, whose significance has already been mentioned in connection with stellar evolution (Sect. 8.2.4) and which we shall meet again when we consider galaxies in Sect. 12.3.6: the existence of black holes.

If, namely, the radius of our mass $\mathcal{M}$ becomes smaller than the gravitational radius R_S (8.82), the coefficients of the spacelike and timelike elements dr^2 and $-c^2 dt^2$ in (8.81) reverse their signs within the sphere $r < R_S$. The result, which we cannot derive in detail here, is that neither matter nor radiation quanta, i.e. no signals at all, can escape from the region within $r < R_S$ to the outside world ($r > R_S$); this region is thus termed a "black hole". A black hole, more precisely a Schwarzschild-type of black hole, is observable *only* through its gravitational field.

Within an astronomical object, a black hole can therefore be detected with certainty when other compact masses can be ruled out as sources of an observed gravitational effect. In the case of an object with stellar dimensions, for example, neutron stars, whose upper limiting masses are not very precisely known, must be eliminated; in the case of galaxies, we must be sure that the effect is not due to a dense, massive star cluster.

The result – that a black hole can be observed only through its gravitational effects – must be modified when gravitational theory is combined with thermodynamics and quantum mechanics, as shown by J. D. Bekenstein, S. W. Hawking and others in 1972/75: a black hole of mass $\mathcal{M}$ is associated with a nonvanishing entropy or temperature

$$T = \frac{hc^3}{16\pi^2 kG\mathcal{M}} \simeq 10^{-7} \frac{\mathcal{M}_\odot}{\mathcal{M}} \text{ [K]} . \quad (8.93)$$

Corresponding to this *Hawking temperature*, which is exceedingly low for stellar or larger masses, a black hole "evaporates" extremely slowly by emission of particles (with a thermal energy spectrum) owing to quantum effects.

If we write (8.82) in the form $G\mathcal{M}^2/R_S = \mathcal{M}c^2/2$, it can be seen that the gravitational collapse of a mass $\mathcal{M}$ to a black hole or a related configuration offers the only possibility for releasing a major portion of its relativistic rest energy $\mathcal{M}c^2$. Even with the highest-yield nuclear process, $4\,^1\text{H} \rightarrow\,^4\text{He}$, according to (8.18) only 0.7% of this energy would be available.

8.4.5 Gravity Waves

To conclude this chapter, we discuss briefly the interesting phenomenon of gravity waves. Soon after his discovery of the gravitational equations, Einstein derived from them the result that a system of moving masses emits gravity waves: transverse waves with two polarization states which propagate with the velocity of light c. The emission of gravity waves is caused by time variations of the *quadrupole* and higher moments of a mass distribution; in contrast to electromagnetic waves, there is no dipole radiation in the case of gravity waves.

If a (polarized) gravity wave interacts with matter, the latter will be compressed in a direction perpendicular to the wave propagation vector (at a particular time), and expanded in the direction perpendicular to both; after a half-oscillation, the reverse deformation occurs. The oscillation pattern for the "other" polarization is found by performing a 45° rotation about the propagation vector.

For the description of the effects of a gravity wave at a great distance from its source, we can limit ourselves to considering small deformations of the Riemann metric

$$g_{ik} = \eta_{ik} + h_{ik}, \qquad |h_{ik}| \ll 1 \quad (8.94)$$

relative to the "flat" Minkowski metric with $|\eta_{ik}| = 1$ (4.12). In order to describe the *order of magnitude* of the effect, a suitable mean value h (which will not define in detail here) of the h_{ik} will be sufficient.

The deformation produced by a gravity wave corresponds to that of a tidal force, i.e. $\delta\ell$ is proportional to the dimension ℓ of the system: $\delta\ell = h\,\ell$. The dimensionless *relative* deformation h we will denote as the *amplitude* of the gravity wave.

Cosmic Sources. Possible sources of gravity waves which are sufficiently strong that the waves could be detected are (among others) close binary star systems and the gravitational collapse to compact configurations.

The orbital motions in a (close) *binary star system* with neutron stars or black holes as components can cause the emission of gravity waves at twice the frequency of the orbital rotation. This emission leads in the course of time to a reduction of the orbital size and eccentricity. From a system at a distance of 100 pc, we would expect an amplitude for the waves of $h \simeq 10^{-20}$. When two components fuse together, there would be a short, very intense burst of gravitational radiation; the rate of such events is, to be sure, completely unknown.

A further source of gravitational radiation would be the *collapse of a star* to give a neutron star or a black hole, which is accompanied by a supernova explosion (Sect. 7.4.7). Here, a burst of duration in the range of 1 ms to 1 s would be expected, whose intensity is however rather difficult to estimate, since only the time variations of the deviations from a spherical or dipole-shaped mass distribution give rise to gravity waves. A galactic supernova, which on the average should occur each 50 yr, would produce a typical amplitude of $h \simeq 10^{-17}$; a supernova in another, not too distant galaxy (within the Virgo cluster, cf. Sect. 12.4.2) in contrast would give at most $h \simeq 10^{-20}$, but with a rate of perhaps one event per month.

In the centers of galaxies, also, the collapse of matter into a *massive* black hole would give rise to a burst of gravitational radiation.

Detectors. The detection of the extremely small amplitudes $h \leq 10^{-20}$, together with the short duration of the bursts, places heavy demands on the sensitivity of detectors for gravity waves.

Beginning in about 1958, J. Weber was the first to carry out intensive efforts to detect gravity waves from outer space, and he continued them for more than a decade. As detectors, he used massive, large (about 1.5 m) aluminum cylinders, which were suspended horizontally and protected as much as possible from (terrestrial) vibrations; their sensitivity was about $h \simeq 3 \cdot 10^{-17}$, corresponding to detectable deformations of $\delta \ell \simeq 10^{-17}$ m. The results published by Weber in 1968, which consisted of short pulses at a high repetition rate, could however not be verified in independent series of observations.

The *direct* detection of gravity waves from space will probably be attained only by future, considerably more sensitive detectors. At present (2001), improved low-temperature Weber detectors are under development; they should attain a sensitivity of $h \simeq 10^{-19}$.

In addition, *laser systems* with suspended mirrors, based on a Michelson interferometer, are under construction. If the positions of the mirrors are varied by a gravity wave, e.g. from a burst of some ms duration, then the corresponding variations in the light paths can be detected through changes in the interference pattern. The attainable sensitivity of around $h \simeq 10^{-21}$ at interferometer arm lengths of several 100 m up to 4 km is limited by background vibrations and by the stability of the lasers.

Indirect Detection. The observations of the binary star pulsar PSR J1915+ 1606, which belongs to an unusual, very close binary star system (Sect. 6.5.6), indicate a small decrease in the orbital period, and thus of the energy of the system, over time. The rate of the observed decrease agrees well with the theoretical value for gravity-wave emission by the moving masses, and thus gives a further, indirect confirmation of the theory of General Relativity.

IV. Stellar Systems. Cosmology and Cosmogony

The Advance into the Universe
Historical Introduction to the Astronomy of the 20th Century

Around 1900, H. v. Seeliger, J. Kapteyn and others attempted (following the "star gauging" of W. and J. Herschel) to apply statistical methods to the investigation of the structure of the Milky Way Galaxy. Although they did not attain their goal, the enormous amount of work expended has proved to be extremely valuable in other respects.

The decisive step forward was made in 1918 with H. Shapley's method of photometric distance determinations using Cepheid variables (cluster variables). The *period–luminosity relation*, i.e. the relationship between the period of their brightness variations and their absolute magnitudes, made it possible to measure the distance to any cosmic object in which Cepheids could be found.

More precise investigations later revealed two difficulties in the basic assumptions of this method: (a) the lack of a correction for interstellar absorption, and (b) the question of the applicability of a single period–luminosity relation to all types of Cepheids. These later made it necessary to apply significant corrections: in 1930, R. J. Trumpler discovered the general interstellar absorption and reddening, and in 1952, W. Baade recognized that the period–luminosity relationships of the classical Cepheids and of the W Virginis stars, i.e. the pulsation variables of stellar populations I and II (see below), differ by 1 to 2 magnitudes. In the following chapters, we shall consistently include these corrections where relevant in all numerical values quoted, and thus will not adhere strictly to the purely historical standpoint.

The distances to the globular clusters determined by H. Shapley made it clear that they form a somewhat flattened system whose center lies about 10 kpc or 30 000 light years from the Earth, in the direction of Sagittarius.

From this beginning, our present view of the Milky Way Galaxy rapidly developed: the main portion of its stars form a flat disk about 30 kpc in diameter, containing the spiral arms. We can view the outer parts of its central regions as bright swarms of stars in the direction of Scorpio and Sagittarius; the galactic center itself is hidden from us by thick, dark interstellar clouds. Only after the inception of radio astronomy and then later with infrared astronomy did it become accessible to direct observation. Our own Solar System is in the outer portion of the disk, about 10 kpc from its center. The disk is surrounded by a considerably less flattened halo, which contains the globular clusters and certain types of individual stars.

As early as 1926/27, B. Lindblad and J. Oort were able to elucidate the kinematics and dynamics of the Milky Way to a considerable extent. The stars in the disk circle the galactic center under the influence of the gravitational force of the large mass which is concentrated there. In particular, the Sun completes its circular orbit with a radius of 10 kpc at a velocity of 220 km s^{-1} in around 250 million years. We initially detect only the *differential* rotation, however: away from the galactic center, the stars move somewhat more slowly (like the planets around the Sun), and towards the center, they move somewhat more rapidly than we do. From this motion, an estimate of the mass can be obtained; after applying various corrections, the mass of the whole system is found to be about $2 \cdot 10^{11}$ solar masses.

While the stars of the galactic disk move on circular orbits, the globular clusters and the stars of the halo describe extended elliptical orbits about the galactic center: their velocities relative to the Sun are therefore of the order of 100 to 300 km s^{-1}. This is J. H. Oort's explanation of the socalled "high-velocity stars".

In addition to the stars, the *interstellar matter* in the Milky Way system plays an important role, although it makes up only a small percentage of the mass. Precise distance measurements became possible only after the interstellar absorption and reddening of starlight by cosmic dust were quantitatively described by R. J. Trumpler in 1930. Somewhat earlier, in 1926, A. S. Eddington had clarified the physics of interstellar gas and the interstellar absorption lines, while H. Zanstra and I. S. Bowen (1927/28) had dealt with the luminosity of galactic and planetary *nebulae*. The surprising discovery of the polarization of starlight by W. A. Hiltner und J. G. Hall (1949) finally made it clear that there is a galactic magnetic field in the disk of about 10^{-9} Tesla.

In 1924, E. Hubble, using the 2.5 m telescope of the Mt. Wilson Observatory, succeeded after major refinements of photographic techniques in resolving the outer regions of the Andromeda nebula (and other spiral nebulas which are not too distant) to a considerable extent into individual stars, and in finding there (classical) Cepheids, novae, bright blue O and B stars, etc. These in turn made it possible to determine the distance by photometric methods; it was found to be about 700 kpc or 2 million light years. It was thus settled, after long and tedious controversies, that the Andromeda nebula (M 31) and our own Milky Way system are basically similar cosmic objects. Following Hubble's investigations, it has become customary to call the "relatives" of our Milky Way *galaxies*, and to reserve the term "nebula" for gas or dust masses *within* the galaxies.

In 1929, Hubble made a second discovery of great import: the spectra of the galaxies show a red shift proportional to their distances. We interpret this as due to a uniform *expansion* of the cosmos; one can readily imagine that dwellers in other galaxies would observe exactly the same effect as we have. If the spreading of the galaxies is extrapolated (somewhat schematically) backwards, it is found that the whole cosmos would have been concentrated closely together at a time around $\tau_0 \simeq 10^{10}$ yrs. ago. We call τ_0 the Hubble time; it gives a first indication of the age of the Universe. What might have occurred further back lies outside the grasp of our research, and in any case, the Universe at a time τ_0 ago must have been "very different" from its present state. The forerunners of Hubble's discovery, V. M. Slipher, C. Wirtz et al., as well as M. Humason's collaboration, can only be mentioned briefly here. These observations made the cosmos as a whole the object of exact science for the first time. Theoretical approaches to *cosmology* had been developed within the framework of General Relativity theory, which initially aimed at an explanation of gravitation and inertial forces, as early as 1916 by A. Einstein, then by W. de Sitter, A. Friedmann, G. Lemaître and others. On the other hand, for the investigation of distant galaxies, it is of fundamental importance that their distances, and thus their absolute magnitudes, their true dimensions, etc. can be determined from the red shift of their spectral lines.

The recognition of an age of the cosmos, which is of the order of 10^{10} yr, i.e. not very much more than the age of the Earth $(4.5 \cdot 10^9$ yr), created a strong impulse for the study of the evolution of stars and stellar systems.

After H. Bethe and C. F. von Weizsäcker in 1938 described the thermonuclear reactions which allow hydrogen to undergo fusion to helium in the interiors of main sequence stars, Bethe and then A. Unsöld in 1944 pointed out that the nuclear energy sources would have lasted for a time of the order of the age of the Universe only in the cooler main sequence stars. For the hotter stars, shorter lifetimes were obtained, going down to only about 10^6 yr in the case of the O and B stars, with their high luminosities. In such a short time, the stars cannot have moved far from the places where they were formed; they must therefore have arisen at nearly the same locations where we observe them today. In reply to numerous speculative hypotheses, W. Baade has pointed out that the close spatial connection of the blue OB stars with dark clouds, e.g. in the Andromeda galaxy, indicates that the stars were formed from interstellar matter.

The investigations by A. R. Sandage, H. C. Arp, H. L. Johnson and others, likewise mostly based on suggestions by W. Baade, of the color–magnitude diagrams (CMD) of the star clusters, have led even further. Together with the theory of stellar structure, they reveal the following picture: a star which has been formed from interstellar matter at first passes through a relatively short contraction phase. Then, on the main sequence, the star remains for a long period in its hydrogen-burning phase. Finally, it moves to the right on the CMD (M. Schönberg and S. Chandrasekhar, 1942) and becomes a red giant star. The place where the main sequence deviates to the right in the CMD of a star cluster (Fig. 9.3) indicates which stars have used up about 10% of their hydrogen since the formation of the cluster. The evolutionary time necessary for this process likewise indicates the age of the cluster. Clusters with bright blue OB stars such as h and χ Persei are thus very young, while in older clusters, only main-sequence stars earlier than G0 still remain.

The study of numerous color-magnitude diagrams led to the fundamental result that most globular clusters have the same age; recent determinations yield a value of 12 to $15 \cdot 10^9$ yr. In the case of open star clusters, in contrast, there are young and old objects. While the youngest are hardly a million years old, the oldest open star clusters have ages of around $10 \cdot 10^9$ yr, and are thus still younger than the globular clusters.

Turning from stellar evolution, we now consider the distribution of various types of stars in our Milky Way and in other galaxies. It was pointed out in 1944 by W. Baade that different portions of the Milky Way differ not only in their dynamics, but also in their color–magnitude diagrams. Thus, the concept of *star populations* was developed. It soon became apparent that these populations differ essentially in their ages and in their abundances of heavy elements relative to hydrogen (i.e. elements heavier than helium; one usually refers for short to metal abundances). Leaving aside fine distinctions and transitional groups, we define:

1. *Halo population II*: globular clusters, the metal-poor subdwarfs, etc.; they follow elongated galactic orbits and form a structure which is slightly flattened, but with a higher concentration towards the center. Their metal abundance corresponds to about 10^{-3} to 1/5 of the "normal" values; their ages may be roughly equated with that of the Milky Way itself.
2. The *disk population*: most of the stars in our neighborhood belong to this group; they form a strongly flattened structure with an increasing concentration towards the galactic center. These stars have "normal" metal abundances, similar to those found in the Sun.
3. The *spiral-arm population I*: it is characterized by young blue stars of high luminosities. Within the disk, the interstellar matter is concentrated in the spiral arms; from it, associations and clusters of young stars are formed.

The *galactic classification scheme* developed in 1926 by E. Hubble, based on the shapes of galaxies, later proved to be basically a classification according to the predominance of population II or population I characteristics: the elliptical galaxies contain in the main old stars, similar to the halo population or the disk population in the Milky Way. At the other end of the scale, the Sc and Irr I galaxies, which have quite strongly differentiated shapes, contain much gas and dust, well-developed spiral arms or other structures, and bright blue O and B stars.

Observations made since the 1970's have shown that the rotational velocities of stars and gas in the outer portions of the Milky Way (and many other spiral galaxies) do not decrease on going outwards (as would be expected for large distances from the main concentration of mass), but instead are nearly constant. This means that an important, probably even the major portion of the mass of the galaxy is "invisible" and forms a roughly spherical system of about 50 to 100 kpc diameter. The nature of the matter in this dark halo is still a complete riddle.

From our own Milky Way galaxy, we now turn to the current ideas about galactic evolution, and especially the formation of the chemical elements and their abundance distributions. A first venture into this area was made by G. Lemaître and G. Gamow in 1939, with their proposal that the expansion of the Universe began with a promordial explosion ("Big Bang"), during whose initial stages the "cosmic" (i.e. roughly the solar) abundance distribution of the elements already became established. The discovery of metal-poor stars and especially the difficulty of explaining the continued formation of the elements above mass number $A = 5$ at first seemed to discredit this theory. Only after the detection of the cosmic 3 K radiation in the microwave region by A. A. Penzias and R. W. Wilson in 1965, which has been explained as a remnant of the Big Bang, did the theory return in a more modest form: in an initial "Big Fireball", essentially only hydrogen and helium atoms are produced.

The explanation of nucleosynthesis according to the theory developed by E. M. and G. R. Burbidge, W. Fowler and F. Hoyle in 1957 is based on the idea that the elements and their relative abundances are for the most part (except for H and He) produced in the *stars* as a result of their nuclear evolution. It begins with a protogalaxy composed of hydrogen plus 10% helium and possibly also traces of heavier elements. Then, the first halo stars are formed; they produce heavy elements, decay (by supernova explosions), and thus enrich the interstellar matter in heavier elements. From this matter, a new generation of metal-richer stars is formed, etc. Before the inventory of heavy elements has reached that in current "normal" stars, the galactic disk with its spiral arms is formed by collapse of the remaining halo material, which has not yet condensed into stars.

A completely new era in the investigation of our Milky Way and the more distant galaxies was ushered in by *radio astronomy*, beginning with K. G. Jansky's discovery of the meter-wave radiation from the Milky Way in 1932.

The thermal free–free radiation from plasmas at 10^4 K in the interstellar gas, in H II regions, planetary nebulae etc. was soon detected. In 1951, the first observations were made of the 21 cm line of atomic hydrogen, predicted in 1944 by H. C. van de Hulst; its Doppler effect gave completely new insights into the structure and dynamics of interstellar hydrogen and thus into those of whole galaxies. Additional lines in the mm to dm ranges could be attributed to transitions between states of very high quantum number in hydrogen and helium atoms and in part to diatomic and polyatomic molecules, which are in some cases surprisingly complex. The 2.6 mm line of the abundant CO molecule, discovered in 1970, has particular significance: it has been used to investigate the distribution of cold (≥ 10 K) constituents of interstellar matter which are concentrated in massive *molecular clouds* in the Milky Way. These concentrations, which are mainly composed of hydrogen molecules, have been found to be the real locations of *star formation*; only after the formation of highly luminous O and B stars do they become visible in the optical wavelength region as glowing nebulae (H II regions).

The radio continuum originally observed by Jansky is nonthermal synchrotron radiation. This was postulated in 1950 by H. Alfvén and N. Herlofson; soon thereafter, I. S. Shklovsky, V. L. Ginzburg and others showed that this radiation is produced when high-energy electrons move on spiral orbits around the force lines of cosmic magnetic fields.

Along with the elucidation of the mechanisms which can lead to the emission of radiofrequency radiation, the experimental development of better and better angular resolution and more precise radio position determinations was of equal importance. We have already described the construction of continually improved and larger radiotelescopes and of radiointerferometers up to the intercontinental long-baseline interferometers in Sect. 5.2.1; here, we add some historic dates in radioastronomical research: in 1946, J. S. Hey employed the fluctuations of radio emissions (later recognized as scintillation originating in the ionosphere) to find the first cosmic *radio source*, Cygnus A. In 1949, J. G. Bolton, G. J. Stanley and O. B. Slee identified the radio source Taurus A with the Crab nebula, and in 1952, W. Baade and R. Minkowski were able to demonstrate that the radio source Cassiopeia A is, like Taurus A, the remnant of an earlier supernova.

Cygnus A, in contrast, could be identified with a peculiar galaxy having emission lines from highly excited states: the first *radio galaxy*. This opened the door to the rapid development of the field of *extragalactic* radio astronomy, which is unparalleled in the whole history of astrophysics. In 1962/63, M. Schmidt was able to identify the quasars, whose optical images are hardly distinguishable from those of stars, as very distant galaxies with extremely high optical and radio luminosities. Their red shifts, which are much larger than those known previously, opened unsuspected perspectives for cosmology. It was further found that in the centers of the quasars, radio galaxies, Seyfert galaxies, and, to a lesser extent, even in normal galaxies like our own, a characteristic *"activity"* occurs in the form of non-thermal optical and radiofrequency emissions and emission lines with non-thermal excitation. Often, two narrow beams of relativistic particles or plasma (jets) are found to emerge from the galactic nucleus in opposite directions along the axis of rotation of the galaxy and to extend out over enormous distances of several 100 kpc; their synchrotron radiation covers the spectral regions from the radio to the X-ray range. Structural changes in the central emission regions, which apparently correspond to velocities exceeding that of light, can be explained in terms of relativistic effects in jets which are moving with near-light velocity close to the direction towards the observer. The compact nuclei of galaxies thus contain energy sources of unimaginable strength, whose physical origin is probably the release of gravitational energy in the neighborhood of an extremely massive black hole. The theory of the activity of galactic nuclei, which is still not clear in all its details, will depend decisively on the results from high-energy astronomy.

As mentioned in the introduction to Part II, an important complement to the radio astronomy of synchrotron radiation, which arises from relativistic electrons, has been developed since the 1960's in *X-ray* astronomy, and since the 1970's in *gamma-ray* astronomy. Parallel to the rapid progress of astronomy in these "high energy" spectral ranges, a turbulent development in *infrared* astronomy has also taken place. In these new spectral regions, as well as in the far ultraviolet, observations must necessarily be carried out from rockets and space vehicles, or at least from stratospheric balloons. The opening of possibilities for observations from space has been accompanied since about 1970 by an enormous extension of the range of earthbound *optical* astronomy, to fainter and fainter light sources; this has been made possible in particular by the development of highly sensitive photodetectors, the use of active and adaptive optics, and, since the 1990's, by the construction of very large telescopes with diameters up to 10 m. Since 1990 (or rather, since 1993, following the correction of the imaging error in its mirror system), optical astronomy has had the benefit of the 2.4 m Hubble Space Telescope. It has produced images with excellent angular resolution, without disturbances by the Earth's atmosphere, of objects ranging from within the Solar System out to the most distant galaxies.

These improvements in astronomical observation techniques have led to a great increase in our knowledge and to quite new ideas in many areas of astronomy. We can describe here only a limited selection of the new observations made in recent decades.

The observations in the ultraviolet and X-ray regions have demonstrated the existence within the *Milky Way Galaxy* of hot interstellar gas which extends far out into the halo.

Our view in the optical region into the molecular clouds, with their regions of *star formation*, is for the most part blocked by the extinction due to interstellar dust, and the central star is often covered by a dust-filled accretion disk. Thus, the important phenomena here could be observed only in the infrared and radiofrequency ranges. Along with inflowing matter, young stars often show surprisingly energetic *out*flows of material which is frequently strongly focused into jets. In the nearest large region of star formation, the Orion complex, the images made by the Hubble telescope show a fantastic variety of structures such as disks and ionized "tails" around the newly forming stars.

Also in the infrared and radiofrequency ranges, the finer structures and movements in the region of the galactic *center*, with its rich star clusters and the strong nonthermal "pointlike" radio source Sagittarius A*, become accessible; the latter very probably indicates the existence of a black hole of about 10^6 solar masses.

In the sky surveys carried out by the IRAS and ISO satellites, many *infrared galaxies* were discovered, which emit their radiation almost exclusively in the infrared range and, since the dust they contain has been heated by the radiation from many hot, young stars, also must exhibit a particularly high rate of star formation. Interactions during near collisions with other galaxies are responsible for this phenomenon.

The *galaxies* of all types have recently been observed with steadily improving resolution and sensitivity in all regions of the electromagnetic spectrum. In the elliptical galaxies, structures which deviate from rotational symmetry were thus discovered; they can be attributed to interactions with other systems. On the basis of observations of many other objects, the realization has gradually taken root that the galaxies are not isolated systems, but rather that they *interact* with one another, and that galactic fusions are widespread occurrences within the cosmos.

The *active nuclei* of the galaxies, with their complex structures and variability, have been the object of particular attention. The highest degree of activity has been found in the quasars. Based on the numerous observations, there can today be no more doubt that these activity phenomena are due to interactions of a massive black hole of 10^8 or more solar masses with its surroundings in the galactic centers, serving as energy source.

Among the strongest extragalactic sources in the X-ray region are the *galaxy clusters*. Their radiation, discovered by the Einstein satellite in 1979, originates from extremely hot gases (up to 10^8 K) in the central regions of the dense regular clusters; they were lost by the galaxies in the course of collisions with each other. The interactions between the galaxies within groups and clusters influence to a considerable extent their development, and thereby the morphological type of the galaxies.

Optical spectroscopy of the faintest and most distant galaxies with sensitive solid-state detectors, beginning in the mid-1970's, was able to confirm the organization of the galaxy clusters into superclusters, which had earlier been postulated on the basis of their distribution on the celestial sphere. The unique large-scale structure of the cosmos, consisting of connected filaments and disks of the superclusters surrounding large empty regions, is likely to be a key to theories of the formation of the galaxies and galaxy clusters.

The light of distant galaxies is sometimes deflected in its path over cosmological distances by the gravitational fields of strong mass concentrations which lie closer to the observer. The images from these gravitational lenses, e.g. of "twin quasars", give us valuable information about the distribution of matter in the cosmos.

In the spectra of the quasars with the largest red shifts, one can observe numerous absorption components of Lyman α, the resonance line of hydrogen, which are shifted from the rest-frame wavelength of $\lambda = 121.6$ nm into the optical spectral region. These "clouds" of neutral hydrogen along the line of sight are in many cases *protogalaxies*. Information about the early phases of the galaxies is also offered to us by long exposures using the Hubble telescope of regions previously found to be completely free of galaxies. We see an impressive variety of faint galaxies, some with unusual shapes which no longer occur in present-day systems, and we can thus practically "watch" the formation of these most distant galaxies.

At the conclusion of our historical introduction, we return to the evolution of the cosmos as a whole. While *cosmology* was dominated in the first half of the 20th century by models of the Universe based on Einstein's theory of General Relativity and Hubble's discovery of the expanding Universe, the observation of the 3 K radiation in the microwave region by A. A. Penzias and R. W. Wilson in 1965 made it possible to establish physical models of cosmic evolution. Progress in elementary-particle physics and the fundamental interactions since the 1960's has allowed a physical description of the earliest stages of the Universe, with their extremely high energies and temperatures. Finally, beginning about the year 2000, precise measurements of the fluctuations of the 3 K radiation have permitted far-reaching conclusions to be drawn about the parameters and the evolution of the Universe. As a result of this progress, a fascinating development in modern cosmology, which is at present by no means completed, has been set in motion.

The most recent period has brought important new knowledge not only "at the frontiers" of the cosmos, but also in our immediate neighborhood within the Milky Way: the enormous increases in the precision of spectroscopic methods led M. Mayor and D. Queloz in 1995 to the discovery of a *planet* around another star, which was based on the tiny fluctuations of its radial velocity. In a rapid series of discoveries, there followed numerous other planets and candidates for planets, most of them with a mass near that of Jupiter. The discovery of *earthlike* planets still lies just beyond the threshold of today's observational capabilities. Infrared astronomy has also contributed in a major way to understanding of the formation of planetary systems through the observation of numerous cold, dust-filled accretion disks around stars, which we can consider to be the precursors of possible planetary systems.

The rapid increase in the mass of observational data in the past decades, which has accompanied progress in the development of observational techniques, appears to be by no means at an end. On the threshold of the third millenium, we may expect many new insights, both concerning the most distant objects in the cosmos and also concerning our immediate neighborhood within the Milky Way and other planetary systems, especially from the new generation of large optical telescopes and interferometers.

9. Star Clusters

In this chapter, we concern ourselves with concentrations of stars whose structure can be recognized in part with the naked eye, in part only on suitable telescopic images: the globular star clusters, in which the stars appear to be clustered together like the bees in a swarm, the less "concentrated" open star clusters, and the stellar associations. As we have pointed out in the historical introduction to Part IV, investigations of the star clusters have contributed in important ways both to our knowledge of the structure of our Milky Way Galaxy as well as to understanding of stellar structure and evolution, the latter through study of their color–magnitude diagrams.

The classical position catalogues for all of the "nebulae", which we recognize today as galaxies on the one hand, and as galactic and planetary gaseous nebulae on the other, are those of Ch. Messier (M), dating from 1784, as well as the New General Catalogue (NGC) by J. L. E. Dreyer (1890) with its continuation in the Index Catalogue (IC) of 1895 and 1910. Along with the classical treatise by H. Shapley, "Star Clusters" (1930), we should also mention the "Catalogue of Clusters and Associations" by G. Alter, J. Ruprecht and V. Vanysek (1970).

We begin in Sect. 9.1 with a discussion of the open or galactic star clusters and the stellar associations, and then in Sect. 9.2, we treat the globular (star) clusters. Here, the discussion of the clusters as *stellar systems* occupies the foreground; their spatial distribution and their significance for the Milky Way and other galaxies will be treated in connection with these only in Chaps. 11 and 12.

9.1 Open Star Clusters and Stellar Associations

We begin in Sect. 9.1.1 with a brief description of the properties of the open star clusters, and treat stellar associations in Sect. 9.1.2. In Sect. 9.1.3, we consider in detail the color–magnitude diagrams of the open clusters, which we can understand by applying the theory of stellar evolution from Chap. 8, and which permit the determination of the ages of the star clusters.

9.1.1 Open (Galactic) Star Clusters

The open star clusters (Fig. 9.1), whose individual stars can be readily resolved, are found in the sky along the whole bright band of the Milky Way. Some of them have a particularly large number of stars, many hundreds (but still many fewer than the globular clusters); others are poor in stars and contain only a few dozen. Their masses are for the most part in the range of 10^2 to $10^3 \mathcal{M}_\odot$, and their diameters are of the order of 1 to 10 pc. The concentration of the stars towards the center of the clusters, i.e. the compactness of the cluster, is quite variable.

Among the best-known are the Pleiades and the Hyades in Taurus and the double cluster h and χ Persei.

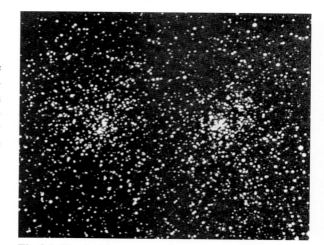

Fig. 9.1. The double open star cluster h and χ Persei

In our Milky Way, about 1000 open clusters are known; taking the more distant regions of the galaxy and those hidden by dark clouds into consideration, the total number is estimated to be around 20 000.

The open star clusters are closely related to the loose moving clusters which were already mentioned in Sect. 6.2.3, as well as to the OB and the T associations, which we shall treat in the following section.

9.1.2 Stellar Associations

Associations are relatively loose collections of stars of a particular type, such as OB stars or T Tau stars, with a lower density than that of the surrounding star field. However, the concentration of the stars of the particular type is noticeably higher than that of the same type in the surroundings. We know of about 100 stellar associations in the Milky Way.

OB Associations, for example Cyg OB2, are loose groups of roughly 50 bright O and B stars of early spectral classes ($\leq$ B2) within a region of 40 to 200 pc diameter. One can observe an expansional motion of the OB stars directed outwards from the center of the association, from which, by extrapolating their velocities backwards, a time of formation of the order of 10^6 to 10^7 yr ago can be derived, comparable to the nuclear evolution time of these stars (Table 8.2).

The OB associations often surround an open star cluster, as for example the ζ Per association around the double cluster h and χ Per.

T Associations, such as Tau T2, are collections of young pre-main-sequence stars (Sect. 10.5.1), in particular of the T Tauri or RW Aurigae variables (Sect. 7.4.3) and other stars ($\geq$ G) which are near the lower part of the main sequence. They are often found spatially close to OB associations.

The OB and T associations are clearly regions of *star formation* or stellar genesis, whose importance to cosmogeny, as very young formations, was pointed out in 1947 by V. A. Ambarzumian. They occur in close conjunction with luminous nebulae (H II regions), which are excited by the ultraviolet radiation of the OB stars, and with molecular clouds (Sect. 10.2.3). Presumably, all OB stars were formed at some time in associations (or in star clusters).

9.1.3 Color–Magnitude Diagrams and the Age of Open Clusters

Our present concepts of the formation, evolution, and final fate of stars were derived from investigations of the color–magnitude diagrams of star clusters. These clusters represent groups of stars which are all at the *same distance* from us. From their magnitudes and colors, which can be determined with high precision, subtraction of the common distance modulus and a uniform correction for interstellar absorption and reddening (Sect. 10.1.2) leads to the (generally used) values of the true

Absolute magnitudes $M_{V,0}$ and

Color indices $(B - V)_0$.

As we have seen, the Hertzsprung–Russell diagram is fundamentally equivalent to the color–magnitude diagram. However, the color indices can still be measured even for extremely faint stars, for which classifiable spectra can no longer be obtained.

Observations. The pioneering works of R. Trumpler in the 1930's dealing with the Hertzsprung–Russell diagrams of open star clusters can be only briefly mentioned here. In the following, we turn directly to the investigations which were first carried out (using photoelectric magnitude scales) by H. L. Johnson, W. W. Morgan, A. R. Sandage, O. J. Eggen, M. Walker and others. For a definitive decision about whether this or that star belongs to a particular cluster, one must depend on proper motions and possibly radial velocity measurements; here, progress has naturally not been so rapid.

In Fig. 9.2, we first show the raw observational data for Praesepe and for NGC 188. Figure 9.3 then gives a schematic summary of the color–magnitude diagrams, $M_{V,0}$ vs. $(B-V)_0$, of the open clusters NGC 2362, h and χ Persei, the Pleiades, M 11, NGC 7789, the Hyades, NGC 3680, M 67 and NGC 188. The lower parts of the main sequence (up to about the "Sun", G2) can be readily brought onto one curve. Further up, in contrast, the main sequence bends over to the right: earlier, in h and χ Per, already at the O and B stars; or later, in Praesepe, near the A stars. Close to the absolute magnitude at this bend, the socalled "knee", some red giant stars are found to the right, at larger (positive) values of $(B - V)_0$. In NGC 188, M 67..., the transition from the main sequence to the red-giant branch is continuous.

Theoretical Elucidation. We now turn to a comparison of the observations with the results of the theory of stellar evolution. In Fig. 8.7a, the initial evolutionary paths of Population I stars of different masses are

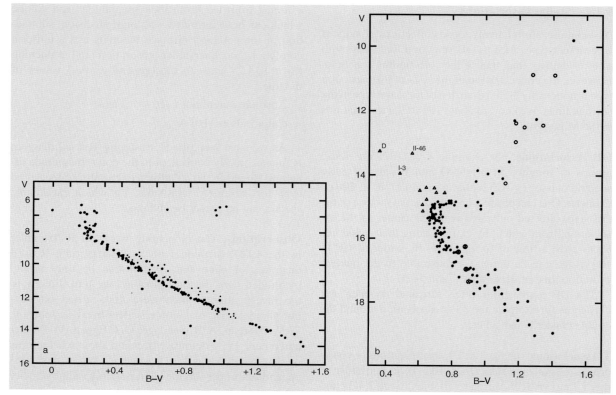

Fig. 9.2. (a) The color–magnitude diagram of Praesepe (after H. L. Johnson, 1952). The apparent magnitudes V are plotted against (B−V); the module of distance is (6.2 ± 0.1) mag. The stars at about 1 mag above the main sequence are most probably binary systems. The cosmic variation of the magnitudes along the main sequence is ±0.03 mag. **(b)** The color–magnitude diagram of NGC 188, one of the oldest open star clusters (after O. J. Eggen and A. R. Sandage, 1969): uncorrected observational data for the apparent magnitudes V and the color indices (B−V). From the two-color diagram (Fig. 6.12) and the position of the main sequence, the interstellar reddening, extinction, and the true distance modulus, $(m − M)_0 = 10.85$ mag, can be computed. • and ○ represent newer and older data, respectively; △ are "blue stragglers" which were formed by the evolution of close binary star systems. ⊗ are four eclipsing binary systems (each plotted as one star)

shown in a theoretical color–magnitude diagram, with the luminosity L or the absolute bolometric magnitude $M_{V,0}$ plotted against the effective temperature T_{eff}; for 5 $\mathcal{M}_\odot$, the evolutionary phases (A to E) are taken from Fig. 8.6.

We first summarize here the most important features of nuclear evolution from Sect. 8.2:

The still homogeneous young stars are ordered in the color–magnitude diagram on the initial main sequence. In their interiors, they form a hydrogen fusion zone, in which energy release takes place at higher central temperatures (larger masses) through the CNO cycle, and at lower central temperatures (lower masses) through the pp process. These stars remain near the main sequence until they have consumed about 10% of their hydrogen, i.e. during a time period of the order of t_E (8.45). Then their evolution leads within much shorter times first to the right and upwards into the region or red giants, where helium fusion begins. The evolutionary phases which follow helium burning are passed through more rapidly and can practically no longer be followed in the color–magnitude diagram.

The *zero-age main sequence* for the stars of Population I (Table 9.1) is defined by the envelope of all the

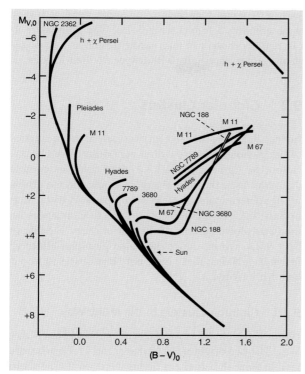

Fig. 9.3. The color–magnitude diagrams of open star clusters (after A. R. Sandage and O. J. Eggen, 1969; A. R. Sandage, 1957). The absolute magnitudes $M_{V,0}$ corrected for interstellar absorption and reddening are plotted against the color indices $(B-V)_0$. The bending of the main sequence to the right ("knee") reveals the age of the star cluster. While the youngest clusters, NGC 2362 and h and χ Persei, are only a few million years old, the oldest cluster, NGC 188, has an age of about $6 \cdot 10^9$ yr. The Sun itself is still on the (nearly) "unevolved" main sequence

Table 9.1. The Zero-Age Main Sequence (ZAMS) of Population I, after T. Schmidt-Kaler (1982). $(B-V)$ and M_V are in [mag]

B − V	M_V	B − V	M_V	B − V	M_V
−0.30	−3.3	+0.30	+2.8	+1.00	+6.7
−0.20	−1.1	+0.40	+3.4	+1.20	+7.5
−0.10	+0.6	+0.50	+4.1	+1.40	+8.8
0.00	+1.5	+0.60	+4.7	+1.60	+12.0
+0.10	+1.9	+0.70	+5.2	+1.80	+14.2
+0.20	+2.4	+0.80	+5.8	+2.00	+16.7

Using the evolution times which correspond to the evolutionary paths shown in Fig. 8.7a, we can draw curves in the color–magnitude diagram onto which a group of stars which started on the initial main sequence at time $t = 0$ will arrive at time t. These curves are termed *isochrones* (Fig. 8.7b) and allow us to interpret the color–magnitude diagrams of the open star clusters (Fig. 9.3) as an *age sequence*, thereby refining our previous age estimates (t_E).

color–magnitude diagrams of the open star clusters in Fig. 9.3. In the upper part, it lies a bit below the main sequence of luminosity class V (Fig. 6.11), and merges with the main sequence for the G stars.

The position of the zero-age main sequence in the theoretical color–magnitude diagram (Figs. 8.7b and 9.4) is determined by the chemical composition of the stars. The curve for Population I stars with a normal (solar) element mixture is given by a mass ratio of hydrogen X to helium Y to heavier elements Z of $0.73 : 0.25 : 0.02$ (8.46). This corresponds to a helium abundance (relative numbers of atoms) of He/H = 0.09.

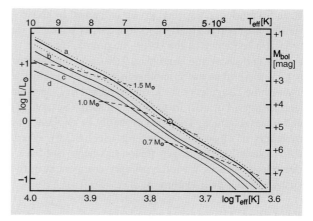

Fig. 9.4. The zero-age main sequence in the theoretical color–magnitude diagram for various chemical compositions, after P. M. Hejlesen (1980). *a*: $(X, Y, Z) = (0.70, 0.27, 0.03)$, characteristic curve for Population I; the adjacent dotted curves differ by $\Delta Z = \pm 0.01$ from the main curve; *b*: $(0.70, 0.29, 0.01)$; *c*: $(0.70, 0.296, 0.004)$; *d*: $(0.70, 0.30, 0.0004)$, corresponding to $\log \varepsilon \simeq -1.9$. This is typical of extremely metal-poor halo stars. Curves of constant mass for 1.5, 1.0 and 0.7 $\mathcal{M}_\odot$ are also drawn. The helium-rich mixture $(0.60, 0.38, 0.02)$ yields a zero-age main sequence which is practically identical to the metal-poor sequence *b*

The star cluster h and χ Persei with its extremely bright blue supergiants, which burn their hydrogen wastefully, is a very young cluster. The knee at $M_V = -6$ mag indicates an *age* of a few million years. The few red supergiants to the right of the upper end of the main sequence are separated from the others by the Hertzsprung gap which has long been known empirically; it continues down to about the F0 III stars, becoming narrower. This can be explained by the fact that e.g. in Fig. 8.7a, for a star of mass $5\,\mathcal{M}_\odot$, the path CD takes only $3 \cdot 10^6$ yr, compared to $2 \cdot 10^7$ yr for the following red-giant stage or $6 \cdot 10^7$ yr for the hydrogen-burning time on the main sequence.

The color–magnitude diagrams e.g. of the Pleiades, ... Praesepe, ... up to NGC 188, whose main sequences bend further and further "down" towards the giant branch, indicate an increasing age. Without doubt there are no diagrams which bend much further down than that of NGC 188, i.e. for the open clusters, there is a *maximum* age, of about $6 \cdot 10^9$, perhaps up to $10 \cdot 10^9$ yr. (The absolute age values are unfortunately accurate only up to a few 10^9 yr.)

Blue Stragglers. A special group is formed by those – for the most part normal – stars in a cluster which lie on the main sequence in the color–magnitude diagram to the left and *above* the bend (Fig. 9.2b), and thus appear to be *younger* than the remaining members of the cluster. These blue stragglers can be observed not only in open star clusters, but also in globular clusters (Sect. 9.2.3) and in associations. They can probably be explained as members of close *binary* systems (Sect. 8.2.6), where their "clock" for hydrogen fusion has been "set back" due to a mass flow from the primary component of the system during their evolution.

Field Stars. The well-known color–magnitude diagram of the field stars in our neighborhood can be interpreted as that of a mixture of stars from many associations and open clusters which have drifted apart with time. The calculated evolution times make it immediately clear why the main sequence closely parallels the initial main sequence. The concentration of yellow and red giant stars in the giant branch can be explained by the fact that in this region, the evolutionary curves of the more massive and brighter stars run together from left to right, and those of the less massive and fainter stars from the lower part of the main sequence to the upper right, as if they were entering a funnel. (Fig. 8.7a). The giant stars in our neighborhood can thus *not* be treated as a homogeneous group, especially with respect to their masses.

9.2 Globular Clusters

In Sect. 9.2.1, we first give an overview of the properties of the globular clusters within our Milky Way Galaxy. Before we turn in Sect. 9.2.3 to their color–magnitude diagrams, their theoretical explanations, and the resulting age determinations, we give a preparatory discussion in Sect. 9.2.2 of the influence of metal abundances on the two-color diagrams of the star clusters. Finally, in Sect. 9.2.4, we introduce the "blue globlular clusters" in *other* galaxies; these have not been observed in our own Milky Way.

9.2.1 Globular Clusters in the Milky Way

About 150 globular star clusters are known in our Milky Way. The two brightest, ω Centauri and 47 Tucanae, are visible from the southern hemisphere. In the northern sky, M 13 = NGC 6205 in Hercules (Fig. 9.5) can still be seen with the naked eye; its brightest stars are of about 13.5 mag. The globular clusters show a strong concentration in the sky in the direction of Scorpius and Sagittarius. Images made with large telescopes resolve more than 50 000 stars in the brighter globular clusters, whereby – from the ground – the individual stars in the centers of the clusters cannot be separated. The Hubble Space Telescope was first able to resolve the stars in the centers of the globular clusters (Fig. 9.6).

A typical globular cluster contains within a region of 40 pc diameter several hundred thousand stars, so that the average density of stars is around ten times greater than in the open star clusters. Towards the center of the cluster, the star density increases so strongly that the night sky there would be rather bright! The absolute visual magnitude of an average globular cluster is around -7.3 mag. Their total masses can be estimated from the scatter in the radial velocities of the stars, using well-known methods from the kinetic theory of gases; the resulting values are several $10^5\,\mathcal{M}_\odot$.

The globular clusters exhibit clear individual differences: their masses extend over a range from only 10^3 up

Fig. 9.5. The globular cluster M 13 = NGC 205 in the constellation Hercules. Distance 6.4 kpc

Fig. 9.6. The globular cluster M 15 = NGC 7078 at a distance of 11.5 kpc: an image of the central region, $9'' \cdot 9''$ or $0.5\,\text{pc} \cdot 0.5\,\text{pc}$, taken by P. Guhathakurta et al. in 1996 with the Wide Field Planetary Camera 2 of the Hubble Space Telescope. In this dense cluster, within a distance of $2'$ (6.7 pc) from the center, 30 000 stars can be resolved; the star density increases towards the center at least to within a distance of $0.3''$. (© Association of Universities for Research in Astronomy AURA/Space Telescope Science Institute STScI)

to several $10^6\,\mathcal{M}_\odot$, their diameters cover the range from 20 to 150 pc, and their absolute visual magnitudes lie between $M_V = -1.7$ mag and (for ω Cen) -10.1 mag.

In about 20% of the clusters, we find in the center an enormously steep increase of the star density, which, as shown by N-body simulations, develops in the course of time through gravitational interactions of the stars. Globular clusters furthermore contain a high proportion of *binary stars*, which can be observed in part as X-ray sources, such as the X-ray burster MXB 1730-335 (Sect. 7.4.6), and also as millisecond pulsars (Sect. 7.4.7).

We cannot treat here the dynamic *evolution* of the globular clusters as a result of the mutual interactions of their stars and with the other stars in their mother galaxies.

9.2.2 Metal Abundances and Two-Color Diagrams

Spectroscopic analyses (Sect. 7.2.7) show that in the stars of the halo population (both field stars and those in globular clusters), the abundances of all the heavier elements (of nuclear charge $Z > 6$; usually referred to as "metals" M) relative to hydrogen are reduced in comparison to the Sun by factors up to $\geq 10^3$, although the abundance *ratios* of the heavier elements among themselves are more or less the same as in the Sun, except for a few elements such as O, Mg and some others. Helium ($Z = 2$) is in general not reduced like the metals; both in Population I and in Population II stars, we find essentially the original atomic ratio of helium to hydrogen, resulting from the Big Bang, of about 1/10 (Sect. 13.2.4).

For many purposes it is sufficient to employ the *metal abundance* $\varepsilon(M)$, where we relate the abundances of the heavier elements M according to their particle density n to that of hydrogen and compare the resulting values with the solar ones (7.67). We use the abbreviation (cf. (7.68))

$$[M/H] = \log \varepsilon(M) = \log \frac{\bigl(n(M)/n(H)\bigr)_*}{\bigl(n(M)/n(H)\bigr)_\odot}. \qquad (9.1)$$

If the metal abundance is quoted as a *mass* ratio, then usually the notation Z/X is used, cf. (8.46).

Since detailed spectroscopic analyses can, for practical reasons, be carried out only for relatively few, usually brighter objects, it is important to be able to determine the metal abundances of the stars at least globally by referring to the two-color diagram. In the cooler *metal-poor* halo stars, in comparison to normal metal-rich stars, the metal lines, which are crowded increasingly closely together towards the short-wavelength end of the spectrum, are so much weaker that the color index $(U-B)$ is rather strongly shifted to smaller values, while the index $(B-V)$ is only slightly shifted. Because of the large number of iron lines in the optical spectra of cooler stars, the photometrically derived metal abundance is practically equal to that of iron, $[M/H] \simeq [Fe/H]$.

In the two-color diagram (Fig. 6.12), the curve derived from normal stars is shifted by this effect upwards in the region of $(B-V) \geq 0.35$ mag by at most (i.e. for extremely metal-poor stars) $\delta(U-B) \simeq 0.25$ mag; cf. the two-color diagram of the extremely metal-poor globular cluster M 92 in Fig. 9.7. This of course has to be taken into consideration in determining the interstellar reddening from the two-color diagram.

The *UV excess* $\delta(U-B)$, which is more precisely defined as the difference in $(U-B)$ compared to main-sequence stars of the *Hyades* with the same $(B-V)$ values, also gives a measure of the metal abundances of fainter stars. Still more precise results are obtained from narrow-band photometry by the method of B. Strömgren, which is however limited to somewhat brighter stars.

Applying the theory of stellar atmospheres, a calibration of the dependence of $\delta(U-B)$ and other corresponding indices $[M/H]$ on the metal abundance can be carried out for various effective temperatures and surface gravities of the stars.

9.2.3 Color–Magnitude Diagrams and the Ages of the Globular Clusters

The structure of the color–magnitude diagrams of the globular clusters remained unclear until in 1952, under the leadership of W. Baade, a group of young astronomers at the Mt. Wilson and Mt. Palomar observatories, including A. R. Sandage, H. C. Arp, W. A. Baum and others, took up the task of determining their main sequence, for favorable objects in the range from 19 to 21 mag. Only as a result of this work has it been possible to make comparisons with the stars in our neighborhood and with the open star clusters.

Observed Diagrams. Figure 9.7 shows, as an example, the color–magnitude diagram of M 92, and in Fig. 9.8 we show a composite diagram of some globular clusters which were investigated with great precision by A. Sandage and his coworkers. Making use of a careful determination of the distance modulus and the interstellar reddening (Sect. 10.1.2), the data have been reduced to absolute magnitudes $M_{V,0}$ and true color indices $(B-V)_0$.

The main sequence extends from the faintest (then) observable stars up to the "knee" and is continued upwards by the subgiants B and then the *red giant branch* A. From the tip of this sequence, at somewhat smaller values of $(B-V)$ or effective temperatures, the *asymptotic* giant branch C leads down and to the left; the *horizontal branch* D continues up to the B stars of absolute magnitudes $M_V \simeq +2$ mag. Embedded in the horizontal branch is the well-defined "gap" of the pulsating cluster variables or *RR Lyrae stars* (Sect. 7.4.1).

Through the development of sensitive solid-state detectors (CCD's) and the use of the Hubble Space Telescope, in recent years not only have the centers of the globular clusters been resolved into their individual stars, but also their main sequences in the color–magnitude diagram have been measured down to considerably fainter magnitudes.

The shape of the main-sequence curve and the sequence of subgiants and giants is similar for all the globular clusters; however, they differ markedly – corresponding to their metal abundances – in terms of their positions in the color–magnitude diagram. Noticeable differences, also caused by differing metal abundances, also exist in the occupation of the horizontal branch among different globular clusters. Thus, in M 92 and other metal-poor clusters, the red part of the horizontal branch to the right of the region of the variables is missing.

9.2 Globular Clusters

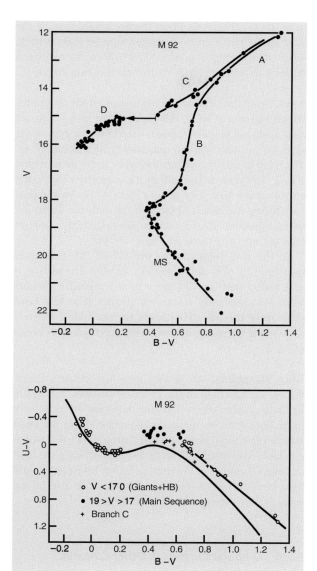

Fig. 9.7. The color–magnitude diagram of the globular cluster M 92, after A. R. Sandage (1970). The apparent magnitudes V are plotted against the color indices (B − V) without the (very small) corrections for interstellar absorption and reddening. A: giant branch; B: subgiant branch; C: asymptotic giant branch; D: blue horizontal branch; MS: main sequence. The cluster variables occur in the gap in the horizontal branch at V = 15 mag. The apparent visible modulus of distance is $m - M = 14.4$ mag. The two-color diagram is also shown below. The heavy curve indicates the two-color diagram of the Hyades; the curve at the right shows its increase for the extremely metal-poor stars

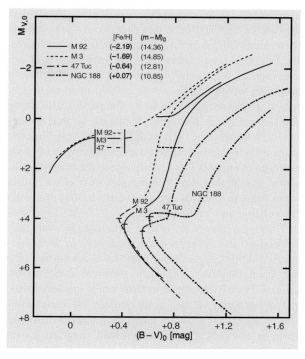

Fig. 9.8. Color–magnitude diagrams for the three globular clusters M 92, M 3, and 47 Tuc, and the old open cluster, NGC 188. The absolute magnitudes $M_{V,0}$, corrected for interstellar extinction, are plotted against the true color indices $(B - V)_0$; also, the metal abundance [Fe/H] and the true distance modulus, $(m - M)_0$, are given for each object. Although the globular clusters are considerably older (about $12 \cdot 10^9$ yr) than NCG 188 (about $6 \cdot 10^9$ yr), their giant branches lie to the left of the giant branch of the metal-richer open star cluster; this results from the different metal abundances. From A. Sandage, Astrophys. J. **252**, 574 (1982). (Reprinted courtesy of the University of Chicago Press; (c) 1982, The American Astronomical Society)

Theoretical Elucidation. We now turn to the explanation of the color–magnitude diagrams of the globular clusters (Figs. 9.7 and 9.8), which were in fact the starting point for the modern theory of stellar evolution.

The zero-age main sequence of the metal-poor halo stars, also denoted as subdwarfs (sd), is below that of Population I for the same helium abundance (Fig. 9.4). Since the metals make the main contribution to the absorption coefficient κ at the cooler end of the sequence, the luminosity L increases for a given mass $\mathcal{M}$ when the metal abundance Z decreases (and thus κ

also decreases), following the mass–luminosity relation $L \propto \mathcal{M}^3/\kappa$ (8.43). Since, however, from the equations governing stellar structure (whose details we omit here), the effective temperature T_{eff} also increases with decreasing κ, the overall result is a metal-poor main sequence which lies *below* the normal main sequence, but approximately parallel to it. The mass scales along the sequences are relatively displaced, so that for example, a Population I star with a mass of $1\,\mathcal{M}_\odot$ has the same effective temperature as an extremely metal-poor halo star of about $0.7\,\mathcal{M}_\odot$. A change in the helium abundance would also lead to a parallel shift of the initial main sequence, in the opposite sense to that produced by a change in metal abundance.

From the theory of stellar structure, we can thus understand the different main sequences in the color–magnitude diagrams of the globular clusters (and also the open clusters) with their different metal abundances. For a quantitative calibration of the sequences, the dependence of the color indices on metal content (Sect. 9.2.2) must be taken into account in going from a theoretical color–magnitude diagram to an $(M_V, B - V)$ diagram.

In the region of the "knee", the color–magnitude diagrams of the globular clusters are to a great extent similar to those of the older open star clusters; the difference is that their giant branch is steeper. This results from the lower metal abundances in the globular clusters, as shown by calculations of stellar evolution. The spectra of the red giants reveal that they have for the most part metal abundances in the range [M/H] $\simeq -1$ to -2; some clusters have however nearly solar element abundances. As a result of lower metal abundances, the opacity and energy production are radically changed.

As we saw in Sect. 8.2.3, the end of the evolutionary path at the point of the first giant branch is ascribed to the sudden ignition of central *helium* burning (helium flash). After the helium flash, and following a considerable mass loss in the red-giant stage, the metal-poor stars of Population II occupy the horizontal branch. They produce their energy through central helium burning and hydrogen shell burning; their masses at the blue end of the horizontal branch are around $0.5\,\mathcal{M}_\odot$, and at the red end, they are about $0.9\,\mathcal{M}_\odot$. In Population I with normal metal abundance, the horizontal branch corresponds to a not very noticeable clump of stars in the color–magnitude diagram close to the giant branch near $10^2\,L_\odot$ (Fig. 8.8). From the horizontal branch, the stars finally move up along the asymptotic giant branch, where their energy is produced by helium and hydrogen burning in separate shell-shaped combustion zones.

Ages. From the bend in the color–magnitude diagrams, which like that of NGC 188 occur at $M_V \simeq +4$ mag, we can (using model calculations) determine the ages of the globular clusters with an uncertainty of several 10^9 yr. The relative ages, however, should be accurate to within 1 to $2 \cdot 10^9$ yr. It is found that the majority of globular clusters have an age of about $(15 \pm 3) \cdot 10^9$ yr and thus are among the oldest objects in our galaxy. There are several globular clusters which are noticeably younger than the majority (around 3 to $4 \cdot 10^9$ yr).

In the late 1990's, with the aid of the parallaxes measured by the astrometric satellite Hipparcos, it was found that the distances to the globular clusters were systematically somewhat greater than had been previously thought. This leads for the oldest clusters to a small decrease in their estimated ages, to about $(12 \pm 2) \cdot 10^9$ yr.

Blue Stragglers. As in the open star clusters (Sect. 9.1.3), we find also in the globular clusters, especially in their centers where the star densities are relatively high, blue stars on the main sequence above the bend which are "too young". Such "stragglers", of which e.g. in 47 Tuc around 20 were discovered using the high angular resolution of the Hubble Space Telescope, could have been formed in binary star systems or also by the fusion of two stars following a collision.

9.2.4 Globular Clusters in Other Galaxies

Aside from the clusters which we know within our Milky Way galaxy, we also find globular star clusters in other galaxies which do *not* correspond to any observed in the Milky Way.

It has long been known that in our neighboring irregular galaxies, the Large and Small Magellanic Clouds (Sect. 12.1.2), there are star clusters whose masses and compact structures are to be sure quite similar to those of globular clusters in our own galaxy, but whose light is dominated by *young, blue* (metal-poor) stars and not, as in the Milky Way, by red giant stars. The brightest cluster in the Large Magellanic Cloud is the Blue Globular Cluster NGC 1866 with a mass of about $10^5\,\mathcal{M}_\odot$ and an age of only around 10^8 yr. In addition, in these galaxies there are also old globular clusters similar to those in our Milky Way, to be sure only a very few.

Numerous blue globular clusters have also been discovered in recent years with the Hubble Space Telescope in elliptical galaxies and in interacting galaxies (Sects. 12.1.2 and 12.5.3). For example, the giant elliptical galaxy M 87 is surrounded by a system of roughly 100 000 bright blue globular clusters, which, due to their great distance, cannot be resolved into individual stars.

We thus should not generalize the observations made within our Milky Way galaxy that the globular clusters are among the oldest objects; instead, we must ask which conditions in our galaxy led to the result that it contains no younger globular clusters.

10. Interstellar Matter and Star Formation

The matter which is finely distributed between the stars of the Milky Way at first came to the attention of astronomers in the form of *dark clouds*, which weaken and redden the light of those stars which are behind them, due to absorption and scattering. But it was only in 1930 that R. J. Trumpler was able to show that even outside the recognizable dark clouds, *interstellar* extinction and reddening are by no means negligible in the photometric determination of distances of a few hundred parsec throughout the Milky Way Galaxy. Already in 1922, E. Hubble had recognized that the galactic (diffuse) *reflection* nebulae (like the one which surrounds the Pleiades, for example) are due to scattering of the light from relatively cool stars in cosmic dust clouds, while in the galactic (diffuse) *emission nebulae*, interstellar gas is excited by the radiation from hot stars and therefore emits line spectra. Following his observations, the investigation of the *interstellar gas* quickly gained momentum in the years 1926/27. The "stationary" Ca II lines had already been discovered in 1904 by J. Hartmann; they occur in the spectra of binary stars but do not show Doppler shifts corresponding to the orbital motion. Only in 1926 was an explanation developed, theoretically by A. S. Eddington, and based on observations by O. Struve, J. S. Plaskett, and others: the interstellar Ca II, Na I,... lines are produced in a gas layer which is partially ionized by the stellar radiation. This gas layer fills the entire disk of the Milky Way and participates in its rotation. About the same time, in 1927, I. S. Bowen succeeded in making the long-sought identification of the "nebulium lines" in the spectra of gaseous nebulae, finding that they are due to forbidden transitions in the spectra of [O II], [O III], [N II],...; and H. Zanstra developed the theory of nebular luminescence. Only about ten years later was it recognized that in the interstellar gas, as in stellar atmospheres, hydrogen is the strongly predominant constituent. O. Struve and his coworkers discovered with the aid of their nebula spectrograph, which had great light-gathering power, that many O and B stars are surrounded by well-defined regions which fluoresce in the red hydrogen recombination line, Hα. Here, the interstellar hydrogen must thus be *ionized*. The theory of these H II regions was formulated in 1938 by B. Strömgren.

Neutral hydrogen (one speaks of H I regions) at first seemed not to be directly observable, until in 1944, H. C. van de Hulst calculated that the transition between the two hyperfine levels of its ground state must lead to a radiofrequency emission of measurable intensity at $\lambda = 21$ cm. This line was first observed in 1951, almost simultaneously at the Harvard Institute, in Leiden, and in Sydney, and led to completely new insights into the structure and dynamics of interstellar hydrogen and thus of the galaxies. Thanks to progress in amplifier technology in the mm to dm range, numerous other lines in the radiofrequency region have been detected, for example transitions between energy states with very large quantum numbers in hydrogen and helium atoms, lines of the OH radical at $\lambda = 18$ cm with unusual intensities produced by the maser amplification principle, and the rotational transition of the abundant CO molecule at $\lambda = 2.6$ mm.

The surprising discovery of the first *polyatomic* molecule, NH_3, in interstellar space by C. H. Townes and coworkers in 1968 has been followed by the detection of over a hundred species of di- and polyatomic molecules up to the present.

Progress in radio and infrared astronomy has allowed the recognition that large, dense *molecular clouds* are, in fact, the locations of *star formation*. The luminous nebulae, very noticeable in the optical region, appear in the border areas of these molecular clouds in connection with the formation of OB stars.

At the other end of the spectrum, observations from satellites in the ultraviolet and X-ray regions show the existence of very hot interstellar gases (10^4 to 10^6 K). The most abundant interstellar molecule, H_2, could be observed with the aid of its emission bands near $\lambda \simeq 100$ nm. Finally, using the tools of gamma-ray astronomy, the interaction of cosmic radiation with the interstellar gas in the Milky Way has been investigated. This *cosmic radiation* itself consists for the most part of highly energetic protons (and some heavier atomic nuclei); it was discovered in 1912 by V. Hess on the basis of its ionizing effect in the Earth's upper atmosphere.

In this chapter, we discuss the results of observations and the properties of the various components of the interstellar medium in our Milky Way galaxy in sequence, beginning in Sect. 10.1 with the dust and continuing in Sect. 10.2 with neutral gas, which contains hydrogen in atomic or molecular form and which is concentrated in the H I clouds and molecular clouds with a varying density, mixed with dust. In Sect. 10.3, we then take up the ionized gas component, which appears in the main as luminous nebulae, and in Sect. 10.4 we treat the magnetic fields, the high-energy particles, and the gamma radiation in the Milky Way. The distribution of all these components within the galaxy and their significance for its structure and evolution will be treated in the following Chap. 11. We conclude this chapter in Sect. 10.5 with the process of star formation, which is closely connected with the physics of the interstellar medium and with the early evolutionary phases of the stars.

Fig. 10.1. The southern Milky Way in the region of the constellations Centaurus and Crux; this picture was taken by C. Madsen, European Southern Observatory (ESO). At the far right is the Southern Cross, and to its left the "Coal Sack", a dark cloud which is relatively near to us, with an angular size of $5° \cdot 8°$ at a distance of 170 pc. The bright star to the right near the Coal Sack is α Cru; the two bright stars in the left half of the picture are α Cen (*left*) and β Cen. The galactic equator is a horizontal line near the center of the picture, passing somewhat to the north of α Cen and α Cru, and through the center of the Coal Sack. (With the kind permission of the European Southern Observatory)

10.1 Interstellar Dust

The interstellar dust makes itself known on the one hand through the readily-seen dark clouds (Sect. 10.1.1), and on the other through the "general" extinction and reddening which it causes (Sect. 10.1.2), along with the polarization (Sect. 10.1.3) of starlight. In Sect. 10.1.4, we derive the properties of the dust particles, such as their sizes, compositions, and mean density within the Milky Way, and discuss the thermal radiation which they emit in the infrared. The diffuse absorption bands in the spectra of distant stars (Sect. 10.1.5) are in part also due to the interstellar dust.

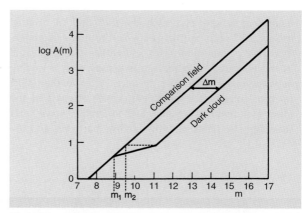

Fig. 10.2. A Wolf diagram for determining the distance to galactic dark clouds. The number $A(m)$ of stars per square degree in the magnitude interval $m - 1/2$ to $m + 1/2$ is plotted as a function of m. The front and rear limits of the cloud correspond to the mean stellar magnitudes m_1 and m_2; their extinction is equal to Δm magnitudes

10.1.1 Dark Clouds

Even with the unaided eye, we can see dark clouds against the background of the bright starfields, especially in the southern Milky Way, such as the well-known "Coal Sack" in the Southern Cross (Crux; Fig. 10.1) or the dark cloud in Ophiuchus. E. E. Barnard, F. Ross, M. Wolf and others made early on quite beautiful photographs with relatively small cameras of high light-gathering power.

There is a strong concentration of dark clouds towards the plane of the Milky Way. The well-known "division of the Milky Way" is clearly caused by a long, extended dark cloud. Pictures of distant galaxies give an even clearer image of the relationship between these dark nebulae and the spiral arms. M. Wolf was the first to estimate the distances to some dark nebulae using the diagram which bears his name: the number of stars $A(m)$ in the magnitude range from $m - 1/2$ to $m + 1/2$ per square degree is counted in the region of the dark cloud and in one or more comparison fields. If all stars had the same absolute magnitude M, a dark cloud in the distance range r_1 to r_2 or with the reduced (i.e. "absorption-free") distance modulus $m_1 - M$ to $m_2 - M$ (6.32) would produce an extinction of Δm magnitudes and reduce the number of stars $A(m)$ in a manner readily seen from the schematic drawing in Fig. 10.2. Because of the scatter in the absolute magnitudes which is in reality present, the precision of the method is not very great, but it is sufficient to show that many of the noticeable dark clouds are no more than a few hundred parsecs from us. Probably even the large complexes in Taurus and Ophiuchus are connected together past the direction of the Sun.

Bright, diffuse nebulae with continuous spectra, the *reflection nebulae*, such as for example the one surrounding the Pleiades, occur where a dust cloud is illuminated by bright stars having temperatures below 30 000 K. Often, the transition from a dark to a bright nebula can be seen directly in photographs.

Both in our Milky Way Galaxy and in more distant galaxies, the form of the dark clouds gives the impression that structures of only a few parsecs in cross-section are stretched out to a length of a hundred and more parsecs.

10.1.2 Interstellar Extinction and Reddening

Although the extinction of starlight in the extended and often not sharply bounded regions of the dark nebulae can readily be detected, it was only in 1930 that the idea of a *general* interstellar extinction (and reddening) became accepted; it plays a decisive role in the photometric determination of larger distances.

If a star of absolute magnitude M is located at a distance r [pc], without interstellar extinction its apparent magnitude m would be given by (6.32) in terms of the *true* distance modulus, as we now more precisely term it:

$$(m - M)_0 = 5 \log r \, [\text{pc}] - 5 \quad [\text{mag}], \quad (10.1)$$

which is thus simply a measure of its distance. If, however, the starlight is subject to an extinction of γ [mag pc^{-1}] on its way from the star, so that overall $A = \gamma r$ [mag], then we obtain, as the difference between the actually measured apparent magnitude and the absolute magnitude, the *apparent* distance modulus (6.34):

$$m - M = 5 \log r \, [\text{pc}] - 5 + \gamma r \quad [\text{mag}]. \quad (10.2)$$

In Fig. 10.3, we have plotted the relation between $m - M$ and r, on the one hand, and between $(m - M)_0$ and r on the other, for $\gamma = 0$ (no extinction) and for varying degrees of extinction γ. It is evident that at extinctions of 1 to 2 mag kpc^{-1}, our view out to distances of more than a few thousand parsecs is practically cut off.

The contribution γ to the mean interstellar extinction in the plane of the Milky Way was first estimated quantitatively in 1930 by R. Trumpler using a comparison of the angular diameters and the magnitudes of open star clusters of similar structure as functions of their distances. He was able to find an elementary relation between geometric and photometric distance determinations. Trumpler's discovery that the extinction is always accompanied by a *reddening* of the starlight is equally important.

On the average, one can expect a (visual) extinction *within* the plane of the Milky Way but outside the directly recognizable dark clouds of $\gamma \simeq 0.3$ mag kpc^{-1}; if we do not exclude dark clouds, this value rises to 1 to 2 mag kpc^{-1}.

The interstellar reddening is described in the framework of spectrophotometry by the color excesses:

$$E_{\text{X-Y}} = (X - Y) - (X - Y)_0 \quad (10.3)$$

which indicate the increase of a color index $(X - Y)$ (6.24) relative to its extinction-free value (X and Y are the measured magnitudes in two wavelength regions for some color index system).

Spectrophotometric measurements have shown that in the optical region, the dependence of the interstellar

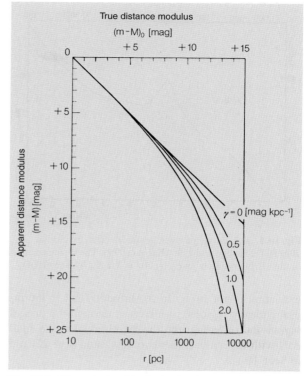

Fig. 10.3. The relationship between the apparent distance modulus $m - M$ and the distance r [pc] of stars without interstellar extinction ($\gamma = 0$), and with interstellar extinctions, taken to be uniform, of $\gamma = 0.5$, 1 and 2 mag kpc^{-1}

extinction $A(\lambda)$ on the wavelength λ is given to a good approximation by a proportionality to $1/\lambda$. From this relation, as well as from the photometry of objects of known color, we obtain as an *average* relation between, for example, the decrease of visual magnitude, A_V, and the color excess $E_{\text{B-V}}$:

$$A_V = (3.1 \pm 0.1) \, E_{\text{B-V}}. \quad (10.4)$$

With knowledge of the dependence of the interstellar extinction, *reddening independent* indices can be empirically defined, as for example for the UBV system of S. van den Bergh

$$Q = (U - B) - 0.72 \, (B - V) \quad (10.5)$$

with a corresponding index Δm_1 for narrow-band photometry by Strömgren's method.

The dependence of A_λ on wavelength over a large range of λ is shown in Fig. 10.4. While the interstellar

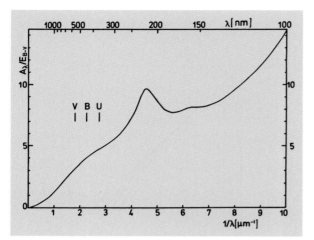

Fig. 10.4. The curve of average interstellar extinction, A_λ, after B. D. Savage and J. S. Mathis (1979). The normalization in the visible range is given by $A_V = 3.1\, E_{B-V}$, see (10.4)

extinction is not strong in the infrared (and in the radiofrequency regions), it increases through the optical region and on going into the far ultraviolet. We shall deal with the noticeable broad maximum near 220 nm in Sect. 10.1.5.

Since the color excesses, e.g. for the color indices $(U-B)$ and $(B-V)$, are proportional to one another $(\propto 1/\lambda_{\rm eff})$, the interstellar reddening displaces a star in the two-color diagram (Fig. 6.12) along a straight interstellar reddening line, whose direction is indicated in the figure. If, for example, we know of a star that it belongs to the main sequence, we can extrapolate from the measured color indices $(U-B)$ and $(B-V)$ along an interstellar-reddening line of the slope indicated towards the "main-sequence line" and read off the two color excesses and the unshifted color indices. From the color excess E_{B-V}, according to (10.4) we immediately obtain the magnitude of the (visual) interstellar extinction. This technique, which can of course be modified in several ways, is one of the most important methods of stellar astronomy.

10.1.3 Polarization of Starlight by Interstellar Dust

In 1949, W. A. Hiltner and J. S. Hall made the startling observation that the light of distant stars is partially *linearly polarized* and that the degree of polarization increases roughly proportionally to the interstellar reddening E_{B-V} or the interstellar extinction A_V. The electric vector of the light waves (perpendicular to the conventional plane of polarization) oscillates preferentially parallel to the galactic plane.

If we denote the intensity of the light oscillating parallel to the plane of polarization by $I_\parallel$ and that oscillating perpendicular to it by $I_\perp$, then the degree of polarization is defined as:

$$P = \frac{I_\parallel - I_\perp}{I_\parallel + I_\perp}\,. \tag{10.6}$$

The polarization is often quoted in terms of magnitudes,

$$\Delta m_P = 2.5 \log \frac{I_\parallel}{I_\perp} \quad \text{oder}$$
$$\Delta m_P \simeq 2.17 P \quad \text{for} \quad P \ll 1\,. \tag{10.7}$$

The largest values of the degree of polarization P are of the order of a few percent; as a function of wavelength, it shows a flat maximum at around 550 nm. The interstellar polarization (in the visual range) is correlated with the interstellar reddening E_{B-V} and the extinction A_V. We find:

$$\Delta m_P \leq 0.065\, A_V\,, \tag{10.8}$$

where, on the average, $\Delta m_P \simeq 0.03\, A_V$. The interstellar polarization indicates that the particles which cause the extinction and reddening are anisotropic, i.e. that they are needle-shaped or disk-shaped and partially oriented. The orientation of the particles has been attributed by L. Davis and J. L. Greenstein to a galactic magnetic field of at least a few 10^{-10} Tesla. In this field, the particles precess, so that the axes of their largest moments of inertia stay parallel to the magnetic lines of force, while the remaining components of their motion are damped.

Figure 10.5 shows the preferred directions of oscillation of the electric vector, and the degree of polarization, for light from about 7000 stars. The strongest polarization is apparently observed under otherwise similar conditions when the lines of force are perpendicular to the line of sight.

10.1.4 Properties of the Dust Grains

In order to obtain a picture of the composition and structure of the interstellar dust which will explain the

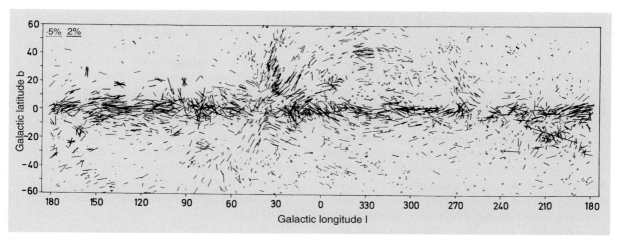

Fig. 10.5. Interstellar polarization, from D. S. Mathewson and V. L. Ford (1970), plotted in galactic coordinates (Sect. 11.1.1). The lines, each of which represents a star at its midpoint, denote the direction of the electric field vector of the optical polarization, and their length gives the degree of polarization P; small circles denote stars with $P < 0.08\%$. The scales for the degree of polarization (*upper left corner*) are to be understood as follows: the left scale is for stars with $P < 0.6\%$ (*light lines*), the right scale for stars with $P \geq 0.6\%$ (*heavy lines*). Roughly speaking, this picture can be regarded as the analog of the familiar experiments with iron filings sprinkled onto a sheet of paper above a magnet

observations, we apply the theory of scattering and absorption of light from colloidal particles, formulated by G. Mie, H. C. van de Hulst, and others.

Grain Sizes. Large particles ("sand"), with radii $a \gg \lambda$, absorb and scatter independently of the wavelength λ, in a manner corresponding approximately to their geometrical cross-section πa^2; very small particles, of $a \ll \lambda$, have (according to Rayleigh) cross-sections proportional to λ^{-4}. The dust particles which give rise to the general interstellar extinction and reddening must therefore have *radii* of the order of the wavelength of visible to ultraviolet light, i.e. $a \simeq 0.3\,\mu\text{m}$. Assuming a density of $3000\,\text{kg m}^{-3}$, their average mass is then about $3 \cdot 10^{-16}\,\text{kg}$.

Density of the Dust. The average particle density of the interstellar dust in the Milky Way can now be readily estimated from the observed extinction, e.g. in the visual range. Its order of magnitude of $1\,\text{mag kpc}^{-1}$ corresponds over a distance of $L \simeq 1\,\text{kpc}$ to an optical thickness of $\tau_V \simeq 1$ (Sect. 6.3.3). On the other hand, from (4.55), we have:

$$\tau_V = Q_{\text{ext},V} \pi a^2 n_D L \,, \tag{10.9}$$

where n_D is the particle density of the dust particles. The extinction coefficient per particle (4.51) has been replaced here by the extinction factor Q_{ext} expressed in units of the geometric cross-section:

$$k_{\nu,D} = Q_{\text{ext},\nu} \pi a^2 \,. \tag{10.10}$$

For particles with $a = 0.3\,\mu\text{m}$, in the visual range, $Q_{\text{ext},V} \simeq 1$, so that we obtain a mean dust particle density of $n_D \simeq 10^{-7}\,\text{m}^{-3}$, or several $10^{-23}\,\text{kg m}^{-3}$ in terms of mass density. The mass due to dust is thus only about 1% of the total mass of interstellar matter; only about *one* dust particle occurs in interstellar space for every 10^{13} hydrogen atoms (Sect. 10.2.2). In spite of this low particle density, in the optical region the extinction due to dust particles is predominant, owing to their enormous interaction cross-sections ($\pi a^2 \simeq 3 \cdot 10^{-13}\,\text{m}^2$) as compared to atomic or molecular scattering cross-sections.

Composition of the Dust Particles. According to the Mie Theory, the frequency dependence of the extinction factor can be calculated for various materials (different indices of refraction of dielectric or conducting materials) and for simple geometric forms (spheres, long needles, etc.). It is composed of an absorption part and a scattering part:

$$Q_{\text{ext}} = Q_{\text{abs}} + Q_{\text{sca}} \,. \tag{10.11}$$

However, these idealized assumptions do not lead to a unique interpretation of the interstellar extinction curve, in particular since it must be taken into account that the dust particles have a broad distribution of "radii" a. The interstellar polarization indicates that the dust particles are anisotropic and partially oriented; the absorption bands in the ultraviolet and infrared (Sect. 10.1.5) show that the dust consists of a mixture of various kinds of particles (graphite, silicates, ices).

We can obtain indirect information about the chemical composition of the dust from an analysis of the atomic absorption lines of the interstellar *gas*: its deviations from the normal (solar) element mixture can be explained by the fact that some elements are missing in the gas phase because they are bound up in the dust component (Sect. 10.2.1).

Temperature and Emitted Radiation. The interstellar dust is subjected to the radiation field of the stars in the Milky Way and is "heated" by it to a temperature T_D; this temperature corresponds to an equilibrium between irradiation and re-radiation. Analogously to our considerations of the global thermal balance of a planet (3.2), we obtain for a (spherical) dust particle of radius a, which is struck by a radiation flux $f_\nu(r)$ at a distance r from a star,

$$\int_0^\infty \pi a^2 Q_{\text{abs},\nu} f_\nu(r) \, d\nu = \int_0^\infty \pi a^2 Q_{\text{abs},\nu} B_\nu(T_D) \, d\nu \,, \tag{10.12}$$

where the re-radiation is assumed to follow Kirchhoff's law (4.67). $B_\nu(T_D)$ is the Kirchhoff–Planck function for a dust temperature T_D. Only the absorption part of the extinction factor (10.11) enters here; a fraction $Q_{\text{sca}}/Q_{\text{ext}}$ of the incident radiation is reflected. Since the absorption of stellar radiation is predominantly in the ultraviolet, but re-radiation predominates in the infrared, we carry out suitable averaging over the corresponding spectral regions for simplicity:

$$\overline{Q}_{\text{abs,UV}} \overline{f}_{\text{UV}}(r) = 4 \overline{Q}_{\text{abs,IR}} \sigma T_D^4 \tag{10.13}$$

with the Stefan–Boltzmann radiation constant σ (4.67).

Depending on the size a and on the composition of the dust particles, we find from (10.13) for dust in the vicinity of *hot stars*, e.g. in H II regions (Sect. 10.3.1), temperatures of up to several 100 K. According to Wien's displacement law, (4.65), these dust particles radiate mainly at wavelengths of $\lambda \leq 30\,\mu\text{m}$. This infrared thermal self-radiation of the "warm" dust is observed from the H II regions as an excess over the free–free emissions of the ionized gas (Fig. 10.9).

Dust which is subjected only to the *general* stellar ultraviolet radiation field of the Milky Way attains temperatures between 15 and 50 K and thus emits only in the long-wavelength portion of the infrared spectrum at $\lambda \geq 100\,\mu\text{m}$.

Depending on their material, dust particles evaporate at temperatures above about 1500 to 1800 K. Dust formation can take place through condensation ("smoke") of the gas in the cooler outer atmospheres of red giant stars, in the expanding shells of novae, and in connection with the formation of stars and planets. The relative importance of the different processes for the global balance of dust formation and removal in the Milky Way is still unclear.

10.1.5 Diffuse Interstellar Absorption Bands

In addition to the known interstellar lines in the spectra of stars, which are notable for their sharpness (Sect. 10.2.1), P. W. Merrill discovered in 1934 several *broad* interstellar absorption bands. The strongest lies at $\lambda = 443$ nm, with a full width at half-maximum of 3 nm.

In the *ultraviolet*, we observe a strong, broad interstellar absorption band at $\lambda = 220$ nm with a half-width of around 40 nm (Fig. 10.4). On the basis of model calculations and laboratory experiments, this band can probably be attributed to small *graphite particles*.

In the *infrared*, where the interstellar extinction is in general weaker, a series of interstellar bands from dense clouds can be observed in absorption and to some extent also in emission. The strongest are a band at $\lambda = 9.7\,\mu\text{m}$, which occurs together with a weaker structure at 18 μm, as well as a band at $\lambda = 3.1\,\mu\text{m}$, whose strength is not correlated with that of the 9.7 μm band. The extinction at 9.7 and 18 μm is probably due to vibrations of the SiO_4 tetrahedron, which occurs as a functional group in *silicates* such as e.g. olivine $(Mg, Fe)_2SiO_4$. Laboratory experiments show that, due to the lack of structures within the bands, we are not dealing here with crystalline silicate particles, but rather with amorphous silicates. Water or ammonia *ice* is the probable source of the 3.1 μm band.

All together, in the optical and infrared regions there are about 200 diffuse bands known today, whose *identification* presents a tedious problem, not yet satisfactorily solved in most cases. Considering their broadness, they can be due only to solid particles or large molecules. A recent suggestion seems promising, according to which these diffuse bands can be attributed to *polycyclic aromatic hydrocarbons* (PAH) with perhaps 10 to 100 carbon atoms, or, in the case of optical transitions, to their (positively charged) cations such as that of coronene, $C_{24}H_{12}$.

Coronene

The identification of the diffuse bands in terms of the relatively stable polycyclic aromatic hydrocarbons, which consist of several benzene rings in a plane, is supported by laboratory experiments, although it cannot be considered to have been conclusively proven.

It was recently suggested that two diffuse bands in the infrared can be attributed to *Fullerene* cations, C_{60}^+, i.e. cations of the very large, stable "soccer-ball molecule".

10.2 Neutral Interstellar Gas

Neutral atomic gas (mixed with interstellar dust) occurs in our Milky Way galaxy in irregular accumulations of widely differing concentrations (H I regions). We can observe it – at distances which are not too great – by means of many absorption lines from the more abundant elements, mostly in the ultraviolet (Sect. 10.2.1), and also, throughout the whole galaxy, by means of the 21 cm line of neutral hydrogen (Sect. 10.2.2). The cooler, still denser molecular clouds are remarkable for their great variety of interstellar molecular lines, mostly observed in the radiofrequency region (Sect. 10.2.3), of which the strongest is at $\lambda = 2.6$ mm and is due to carbon monoxide.

10.2.1 Atomic Interstellar Absorption Lines

The discovery of the interstellar calcium lines by J. Hartmann in 1904 gave the first indication at all of the existence of interstellar atoms or ions. It was recognized that the Ca II H and K lines from the binary star δ Ori show no Doppler shifts corresponding to its orbital motion; they were therefore initially called "stationary lines". In the course of time, interstellar lines of the following atoms, ions, molecules, and molecule-ions were discovered in the *optical* region:

Na I, K I, Ca I, Ca II, Ti II, Fe I; CH, CH^+, CN .

By far the most intense are the H and K lines of Ca II at $\lambda = 393.3/396.8$ nm and the D lines of Na I at $\lambda = 589.0/589.6$ nm.

Using the Copernicus satellite, in 1972/3 a group of researchers from Princeton discovered interstellar lines in the *ultraviolet* from numerous, in part multiply-ionized atoms and molecules. In the spectral region from 95 to 300 nm, nearly 400 lines are known. By far the strongest line is the Lyman-α line from H I at $\lambda = 121.6$ nm (Fig. 10.6). Additional strong lines are due to:

H I, C I, C II, N I, N II, O I, Mg I, Mg II, Al II,
Al III, Si II, Si III, P II, S II, S III, Ar I, Mn II,
Fe II, Zn II .

A major portion of the ultraviolet interstellar lines belongs to the Lyman and Werner bands of the molecules H_2 and HD around $\lambda \simeq 100$ nm; furthermore, lines of CO are observed.

Using the high-resolution spectrograph of the Hubble Space Telescope, (Fig. 5.19), in the 1990's weaker lines were also detected and thereby also some less abundant elements (Ga, Ge, As, Kr, Sn, Tl and Pb).

The Degree of Ionization of the Gas. In all the interstellar lines, the corresponding transitions originate from the ground state term, i.e. they are *resonance lines*.

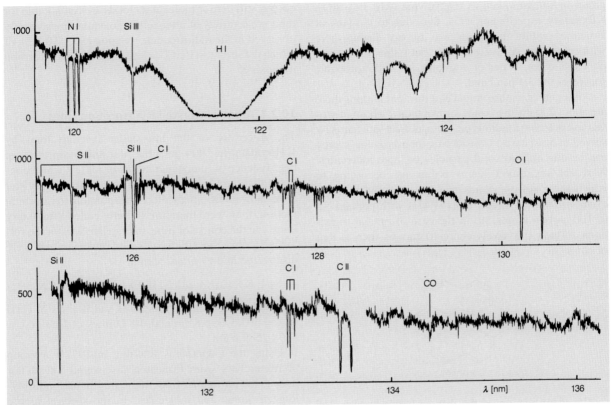

Fig. 10.6. Interstellar absorption lines in the ultraviolet spectrum (119.6–135.6 nm) of the O9 V star ζ Oph, from observations with the Copernicus satellite. The intensities are not corrected for sensitivity variations of the spectrometer, etc. The strongest interstellar line is the broad, saturated Lα line at $\lambda = 121.6$ nm from H I (in the middle of which the weak emission of the geocorona can be seen). Some of the strongest interstellar lines, which can be distinguished from stellar lines by the sharpness of the latter, are identified in the figure according to the absorbing ion. From D. C. Morton, Astrophys. J. **197**, 85 (1975); reprinted courtesy of the author and the University of Chicago Press, and The American Astronomical Society

The *ionization* of interstellar matter so far from thermodynamic equilibrium takes place at the low pressures present only through photo processes starting from the ground state; thus, for example, to ionize neutral calcium, with an ionization potential of 6.1 eV, requires radiation of $\lambda \leq 204$ nm. At a distance r from a star of radius R, its radiation acts with a *dilution factor*:

$$W = \frac{\pi R^2}{4\pi r^2} = \frac{1}{4}\left(\frac{R}{r}\right)^2 . \tag{10.14}$$

Recombination, on the other hand, takes place only by means of two-particle collisions between ions and electrons. The degree of ionization reaches a dynamic equilibrium value for which the rates of ionization and recombination are equal (Sect. 4.5.3).

If we now compare the interstellar lines, for example from Ca II and Ca I or Na I, we find, surprisingly, that their intensity ratios are not too different from those found for an F star. How can we understand this? In the interstellar gas, as well as in the stellar atmosphere, the above-mentioned processes must be in equilibrium. However, in interstellar matter, the electron density is, on the one hand, about 10^{15} times smaller than in the stellar atmosphere (10^5 compared to 10^{20} m^{-3}), and the recombination rate is correspondingly smaller. On the other hand, the radiation field in interstellar space is diluted relative to that in the stellar atmosphere by about the same factor, $W \simeq 10^{-15}$, if we assume as orders of magnitude $R \simeq 2R_\odot$ and $r \simeq 1$ pc, so that the ionization rate is also reduced by this factor. Thus, in interstellar space, both processes occur with rates about 10^{15} times slower than in the stellar atmosphere, but the degree of ionization remains about the same!

The unexpected observations of ultraviolet absorption lines of more *highly ionized* ions such as C IV, N V and O VI, which indicate very much hotter gas in the interstellar medium, will be discussed in Sect. 10.3.4.

Distance Determinations. The equivalent widths of the interstellar lines increase with increasing distance of the star in whose spectrum they are observed, in a fairly regular manner. Conversely, they can thus be used to estimate the star's distance. Observations with high spectral resolution (corresponding to 0.5 to 1 km s^{-1}) show that, for example, interstellar calcium lines usually consist of several *components*, which can be attributed to several regions of concentration of interstellar matter, or interstellar clouds, through which the light has passed. On the average, about 5 to 10 such clouds occur within a distance of 1 kpc. The radial velocities of the stronger components correspond to those of the spiral arms of the Milky Way.

Chemical Composition. Taking into account the component structure as well as local differences in the ionization and excitation, the chemical composition of the interstellar gas can be derived from the measured equivalent widths of the absorption lines.

It is found that along lines of observation with little interstellar extinction or reddening ($E_{B-V} \leq 0.05$ mag), i.e. when the light has not passed through any large concentration of interstellar matter, the abundance distribution of the chemical elements in the gas is roughly the same as that on the Sun. On the other hand, in the *denser* interstellar clouds, many elements show *under*abundances of varying degrees compared to the solar composition: for example, Ca, Al and Ti are up to several 1000 times and Fe up to 100 times less abundant than on the Sun, while C, N, O, S, Ar, Sn and Tl are only slightly underabundant. Clearly, many elements are not to be found in the gas phase in the cooler, more dense areas; rather, they are bound up in the dust components (Sect. 10.1).

10.2.2 The 21 cm Line of Neutral Hydrogen. H I Clouds

The 21 cm line, which is important for the investigation of the large-scale structure and dynamics of the interstellar gas in the Milky Way, results from the transition within the ground state $1s\,^2S_{1/2}$ of H I between its two hyperfine-structure levels with total spin $F = 1$ (nuclear and electronic spins parallel) and $F = 0$ (nuclear and electronic spins antiparallel). The small energy difference of $6 \cdot 10^{-6}$ eV corresponds to a line in the radiofrequency range at

$$\lambda_0 = 21.1 \text{ cm} \quad \text{or} \quad \nu_0 = 1420.4 \text{ MHz}.$$

This transition is "forbidden" (magnetic dipole radiation) and has an extremely small transition probability (Einstein coefficient) of $A = 2.87 \cdot 10^{-15}$ s^{-1} or, from (4.125), an oscillator strength of $f = 5.8 \cdot 10^{-12}$. The average lifetime of the upper level with respect to the emission of a 21 cm photon is $A^{-1} = 1.1 \cdot 10^7$ yr.

Level Populations. Under the conditions in the interstellar medium, an equilibrium distribution of hydrogen atoms in the two hyperfine levels 0 and 1 is established by (electron-exchange) collisions of the atoms among themselves, with an average time between collisions of "only" 400 yr. From the Boltzmann formula, (4.88), the ratio of the populations n is given by the statistical weights $g = 2F + 1$, since the exponential function is practically equal to one because of the small value of $h\nu_0$. We thus have $n_1/n_0 \simeq 3$ or, since the overall density of H atoms is $n(\text{H}) \simeq n_1 + n_2$, $n_1 = (3/4)n(\text{H})$ and $n_0 = (1/4)n(\text{H})$.

Radiation Intensity. The intensity emitted from an optically *thin* layer of thickness ℓ (optical thickness $\tau_\nu = \kappa_\nu \ell \ll 1$), in which we here assume a constant temperature, pressure, etc. for simplicity, is given by (4.60):

$$I_\nu = \kappa_\nu \frac{2\nu^2 kT}{c^2} \ell. \tag{10.15}$$

For the Kirchhoff–Planck function, owing to $h\nu/(kT) \ll 1$ we can use the Rayleigh–Jeans approximation, (4.63). T is the temperature which corresponds to the thermal velocity distribution of the H atoms. The absorption coefficient of the 21 cm line is found from

(4.133) using $n_0 = n(\text{H})/4$ to be:

$$\kappa_\nu = \frac{1}{4\pi\varepsilon_0} \frac{\pi e^2}{m_e c} f \frac{h\nu_0}{kT} \phi(\nu) \frac{1}{4} n(\text{H}), \qquad (10.16)$$

where, in the Rayleigh–Jeans approximation, stimulated emission has to be taken into account by the factor $[1 - \exp(-h\nu_0/kT)] \simeq h\nu_0/(kT) \ll 1$.

The case of optically *thick* layers occurs in only a very few places in the Milky Way. These can be recognized by the fact that the intensity takes on a constant maximum value over a wide range of frequencies:

$$I_\nu = B_\nu(T) = \frac{2\nu^2 kT}{c^2} \qquad (10.17)$$

from which the *temperature* of the interstellar hydrogen can be immediately calculated to be $T \simeq 125\,\text{K}$. The precise value is fortunately unimportant for the emission from the optically thin layers, since T then drops out of (10.15) due to (10.16).

Velocity Distribution. The frequency dependence of the absorption coefficient or the profile function $\phi(\nu)$ is determined entirely by the *Doppler effect* due to the motion of the interstellar hydrogen. If there are $n(V_r)\,\mathrm{d}V_r$ hydrogen atoms per unit volume in the radial velocity interval V_r to $V_r + \mathrm{d}V_r$, then we find

$$\frac{n(V_r)\,\mathrm{d}V_r}{n(\text{H})} = \phi(\nu)\,\mathrm{d}\nu = \phi(\nu)\frac{\nu_0}{c}\mathrm{d}V_r. \qquad (10.18)$$

Measurement of the line profiles thus initially yields the number of H atoms along the direction of observation in a column of cross-sectional area $1\,\text{m}^2$, whose radial velocities V_r or frequencies ν lie in the given interval

$$\begin{bmatrix} V_r, V_r + \mathrm{d}V_r \end{bmatrix} \text{ or }$$
$$\begin{bmatrix} \nu_0 - \frac{V_r}{c}\nu_0, \nu_0 - \frac{V_r + \mathrm{d}V_r}{c}\nu_0 \end{bmatrix}. \qquad (10.19)$$

The velocity distribution of the interstellar hydrogen is composed of two parts: on the one hand, there are statistical velocities, a kind of turbulence, whose distribution function is similar to a Maxwell–Boltzmann distribution. Their mean value was known from the interstellar Ca II lines to be about $6\,\text{km s}^{-1}$. More important, however, is the contribution from the differential *galactic rotation*, which we shall discuss in Sect. 11.2.1.

Optical Depths. We can estimate here the optical depth for the 21 cm radiation in our galaxy. We assume a Gaussian distribution (4.28) for the statistical velocities, with a full width at half-maximum of ΔV_r; its value at maximum is then $c/(\nu_0 \Delta V_r)$ and the optical thickness at the center of the 21 cm line (index 0) is found from (10.16) and (10.18) to be

$$\tau_{\nu_0} \simeq \frac{1}{4\pi\varepsilon_0} \frac{\pi e^2}{m_e c} f \frac{h\nu_0}{kT} \frac{c}{\nu_0 \Delta V_r} \frac{1}{4} n(\text{H}) \ell \qquad (10.20)$$

or numerically, using convenient units,

$$\tau_{\nu_0} \simeq 2 \cdot 10^{-3} \frac{n_\text{H}\,[\text{m}^{-3}] \cdot \ell\,[\text{kpc}]}{T\,[\text{K}] \cdot \Delta V_r\,[\text{km s}^{-1}]}. \qquad (10.21)$$

For average values in the Milky Way of $n(\text{H}) = 4 \cdot 10^5\,\text{m}^{-3}$ (Sect. 11.2.2), $T = 100\,\text{K}$ and $\Delta V_r = 10\,\text{km s}^{-1}$, the 21 cm line, in spite of its extremely small transition probability, would therefore become optically dense even for a thickness of only about 1 kpc. Only the "spreading out" of the velocities owing to the differential rotation, of the order of $\Delta V_r = 100\,\text{km s}^{-1}$, makes it possible to observe the *whole* Milky Way in the 21 cm line. We will treat the extended, spiral-shaped distribution of neutral hydrogen in the Milky Way galaxy in Sect. 11.2.2; here, we describe the results of observations of smaller structures.

H I Clouds. Surveys detecting the 21 cm line with a high angular and velocity resolution ($\delta\phi \simeq 10'$, $\delta V_r \simeq 1\,\text{km s}^{-1}$) have shown that the spiral arms are composed of numerous concentrations or diffuse H I clouds, which have a wide range of sizes. Typical values for H I clouds are: diameter of around 5 pc, average density of $2 \cdot 10^7\,\text{m}^{-3}$, and mean temperature of about 80 K, and thus a mass of around $30\,\mathcal{M}_\odot$.

The concentration of hydrogen in the diffuse H I clouds is correlated with the concentration of interstellar dust clouds, as long as the extinction of the latter is not too strong. In the denser regions with higher extinctions due to dust ($E_{B-V} \geq 0.3$ mag), i.e. within the dark clouds themselves, the hydrogen is present mainly in molecular form and can thus not be observed by means of the 21 cm line.

Warm H I Gas. In the line profiles of the 21 cm line, often a very faint, much *broader* component can be distinguished as a background to the contributions from individual H I clouds. This emission originates from "warm" gas at temperatures around 6000 K and low densities of a few 10^5 m^{-3}, in which the hydrogen is partially (10 to 20%) ionized. The diffuse H I clouds are presumably embedded in this socalled "intercloud gas".

10.2.3 Interstellar Molecular Lines. Molecular Clouds

In the diffuse H I clouds, where hydrogen occurs mainly in atomic form at densities of $\leq 10^8$ m^{-3}, we find only a few simple molecules such as CH, CH$^+$ and CN, which have long been known from their optical absorption lines; the hydroxyl radical OH, discovered in 1963 at $\lambda = 18$ cm; and the molecules H$_2$, HD and CO, which can be observed by means of their absorption lines in the ultraviolet (Sect. 10.2.1). Through observations of their radiofrequency lines in 1968/69, the first *polyatomic* interstellar molecules in interstellar space were discovered: ammonia, NH$_3$, water, H$_2$O, and formaldehyde, H$_2$CO. Continuing up to the present (2001), about 120 different molecules have been observed, including some surprisingly complex ones, in the more dense and cooler regions of the Milky Way. In these *molecular clouds*, hydrogen is present mainly in molecular form, H$_2$.

We have already summarized the fundamentals of molecular spectroscopy in Sect. 7.1.5.

The Hydrogen Molecule, H$_2$. Hydrogen, by far the most abundant interstellar molecule, has no spectral lines in either the radiofrequency or the infrared range which could be used to determine e.g. the large-scale distribution of cool, molecular hydrogen gas in the Milky Way. The Lyman and Werner bands in the ultraviolet can give information only from our immediate galactic neighborhood, owing to the strong interstellar extinction at these wavelengths; the rotational–vibrational spectra and the quadrupolar transitions in the infrared are sensitive to considerably hotter H$_2$ gas, which is present only near regions where star formation takes place.

Carbon Monoxide, CO, the second-most abundant molecule, has an abundance relative to the hydrogen molecule of CO/H$_2 \simeq 10^{-4}$ (in particle numbers). The transition from the first excited rotational level ($J = 1$) to the ground state ($J = 0$) occurs for the most abundant isotopes, ^{12}C^{16}O (^{12}C/^{13}C $\simeq 40\ldots 80$), at

$$\lambda_0 = 2.60 \text{ mm} \quad \text{or} \quad \nu_0 = 115.27 \text{ GHz } (^{12}\text{CO}) .$$

Since the emission in this line is often optically thick, the weaker line corresponding to a transition in the molecule containing the isotopes ^{13}C^{16}O, at

$$\lambda_0 = 2.72 \text{ mm} \quad \text{or} \quad \nu_0 = 110.20 \text{ GHz } (^{13}\text{CO}) ,$$

may also be used for the observation of molecular clouds.

The abundance of the stable carbon monoxide molecule, CO, varies only slightly from cloud to cloud, in contrast to those of other molecules. The galactic distribution of H$_2$ can thus be determined *indirectly* by observations of CO.

Molecular Clouds. Like the interstellar dust and atomic hydrogen, molecular hydrogen occurs in non-uniform concentrations, in the form of clouds. These molecular clouds ("CO clouds") occur in a great variety of sizes (roughly 1 to 200 pc), densities (about 10^9 to 10^{12} m^{-3}), and masses (from 10 to 10^6 $\mathcal{M}_\odot$); they are correlated with dark clouds (Sect. 10.1.1). Their extinction in the visible region, A_V, ranges from about 1 mag to over 25 mag. The *giant molecular clouds* with masses $\geq 10^5$ $\mathcal{M}_\odot$ are, along with the globular star clusters, the most massive objects in the galaxy; their number is estimated to be around 4000. Near the center of the galaxy, we find particularly massive giant molecular clouds (Sect. 11.3.2); Sgr B2 is notable among all the molecular clouds for the large variety of molecules that it has been observed to contain.

The temperatures of the molecular clouds are in the range 10 to 30 K. Often, dense *compact* condensates with diameters of the order of 1 pc are observed in the far infrared ($\lambda \geq 20$ µm) in the larger molecular clouds; in them, and in their immediate vicinities, the temperatures are often considerably higher (10^2 to 10^3 K). These condensates are *protostars* which are surrounded by a dense shell of dust and can only be recognized by their radiation in the infrared. We shall return in Sects. 10.5.2 and 10.5.3 to the subject of star formation, which takes place in the most dense parts of the molecular clouds.

Interstellar Molecules. We will now have a look at the great variety of interstellar molecules which can be observed by molecular spectroscopic methods in the radiofrequency and infrared regions. Favorable observation conditions, with a large number of molecular species, are offered for example by the giant molecular cloud Sgr B2 in the region of the galactic center, the nearby molecular cloud Ori MC1 (OMC 1, *Orion Molecular Cloud*) in Orion (Fig. 10.24.), the dark cloud Tau MC1 (TMC 1) in the Taurus complex, or the cold gas shell around the carbon star IRC $+10°216$.

Nearly all interstellar molecules are composed of the abundant elements H, C, N, O, S and Si. Along with the simple molecules and radicals, such as CH and OH, the following polyatomic molecules are relatively abundant and thus observable also in other galaxies: ammonia, NH_3; water, H_2O; hydrocyanic acid (hydrogen cyanide), $HC\equiv N$ and its isomer, hydrogen isocyanide, $H=N=C$; formaldehyde (methanal), $H_2C=O$; as well as the oxomethylium ion, HCO^+.

Further inorganic molecules which are observed are mainly oxides and sulfides such as NO, NS, SiO, SiS, SO, SO_2 and O_3, also H_2S, HNO, N_2H^+, SiN, and SiH_4.

Because of carbon's ability to form complicated chain compounds, the organic molecules also exhibit great variety in interstellar space. We find (in addition to those already named) simple molecules, ions, and radicals such as C_2, CS, HCO; C_2H, C_3H up to C_7H, C_8H; C_3N, HCS^+; and in addition to HCO^+, also HOC^+.

We also observe among others the hydrocarbons methane, CH_4; ethyne (acetylene), $HC\equiv CH$ and propyne, $H_3C-C\equiv CH$; the alcohols methanol, CH_3OH, methanethiol, CH_3SH and ethanol, CH_3CH_2OH; formic acid, HCOOH, acetic acid, CH_3COOH, and ethanal (acetaldehyde), H_3C-CHO. In addition, numerous nitrogen-containing compounds occur, such as methylamine, H_3C-NH_2, methylimine, $H_2C=NH$, cyanamid, $H_2N-C\equiv N$, isocyanic acid, $O=C=NH$, and isothiocyanic acid, $S=C=NH$, as well as several nitriles. The series of long, chain-like polyenenitriles (cyanopolyenes) which are observed is astounding: they range from hydrocyanic acid, $HC\equiv N$, up to undecapentaene nitrile,

$$HC\equiv C-C\equiv C-C\equiv C-C\equiv C-C\equiv C-C\equiv N,$$

and some of them have yet to be synthesized under terrestrial conditions. Further "unusual" molecules include tri-carbon monoxide, $^\ominus:C\equiv C-C\equiv O:^\oplus$; 1,3-pentadiene, $H_3C-C\equiv C-C\equiv CH$; and the deuterated ethyne radical, $\cdot C\equiv C-D$.

In contrast to terrestrial chemistry, at first glance under the conditions of the interstellar medium, organic ring compounds and branched chains would seem to be less favored than the linear chain molecules. Only a few ring compounds have been discovered through their radiofrequency lines, such as silicon carbide, SiC_2 and cyclopropenylidene, C_3H_2, its radical, C_3H

$$\begin{array}{ccc} \text{Si} & \ddot{\text{C}} & \ddot{\text{C}} \\ /\ \backslash & /\ \backslash & /\ \backslash \\ \text{C}\equiv\text{C} & \text{HC}=\text{CH} & \text{HC}=\text{C}\cdot \end{array}$$

and its derivatives, $H_3C(C\equiv C-C\equiv C-H)$ and $H_3C(C\equiv C-H)$ with a carbon chain in place of an H atom. Finally, – as we saw in Sect. 10.1.5 – polycyclic aromatic hydrocarbons have been suggested to explain the diffuse interstellar bands.

Molecular Genesis. Even in the most dense molecular clouds, the densities ($\geq 10^{12}$ m^{-3}) and the temperatures are so low that molecular genesis takes place far from thermodynamic equilibrium. Dust, which nearly always accompanies the gas, shields it from ultraviolet radiation and prevents the destruction (photodissociation) of the molecules once they are formed. On the other hand, the more energetic protons of *cosmic radiation* can also lead to partial *ionization* in molecular clouds. The (positive) ions (e.g. H_2^+) which are produced in this manner can react efficiently with H_2 and other molecules and lead to the formation of still more complex molecular species. For these usually exothermic ion–molecule reactions in the *gas* phase (E. Herbst, W. Klemperer, 1973), the H_3^+ ion plays a key role. This simplest, stable polyatomic molecule–ion is formed from H_2^+ in the reaction

$$H_2^+ + H_2 \longrightarrow H_3^+ + H.$$

More complex molecules are then formed through reaction chains which begin with proton transfer reactions of the type

$$H_3^+ + X \longrightarrow XH^+ + H_2.$$

We thus obtain for example (X := O) from OH^+ by reaction with H_2 the H_2O^+ ion, and by a further reaction with H_2, H_3O^+, from which, finally, by electron attachment, H_2O and OH are produced.

H_3^+, whose existence was at first only predicted theoretically, was finally also detected in 1996 by T. R. Geballe and T. Oka in interstellar molecular clouds.

H_2 and the complex chain molecules are probably formed through catalytic reactions on the interstellar *dust particles* (D. J. Hollenbach and E. E. Salpeter, 1971). Gas atoms or molecules which strike a dust particle can remain adsorbed there and can find reaction partners by diffusion on its surface. Under favorable conditions, the resulting molecule may then desorb from the surface.

Element Abundances. In order to derive element abundances quantitatively from the observed intensities of the molecular lines, the individual processes of excitation, dissociation, ionization, charge exchange, etc., must be considered separately under the conditions in the interstellar gas, which deviate strongly from thermodynamic equilibrium; this is analogous to the case of atoms (Sect. 4.5.3). However, the number and variety of relevant processes for molecules is considerably greater than for atoms.

If several isotopes may be present in a particular molecule, the excitations of its energy levels hardly depend in general on the difference in isotopic masses. We can thus determine *relative* isotopic abundances rather precisely by observing the corresponding transitions. Using the molecules HCO^+, H_2CO, HCN, CS and SiO, it is possible to gain information about the less abundant isotopes

$$^2D, \ ^{13}C, \ ^{15}N, \ ^{18}O, \ ^{33,34}S \quad \text{and} \quad ^{29,30}Si \ .$$

The isotopic ratios in interstellar space, in turn, can help elucidate the nuclear reactions which take place in the course of stellar evolution (see Sect. 8.2.3).

OH Masers. Quite unusually high intensities are found for the lines of the OH radical at $\lambda = 18$ cm; expressed formally in terms of radiation temperatures (Sect. 6.1.3), they correspond to 10^{12} to 10^{15} K! The lines are exceedingly sharp and are emitted by very compact sources (≤ 10 AU), as was shown by angular measurements using very long-baseline interferometry; in many cases, circular polarization is also observed. These extreme non-equilibrium properties can be explained only by the maser amplification mechanism (Maser = *m*icrowave *a*mplification by *s*timulated *e*mission of *r*adiation): the upper levels corresponding to the transitions are strongly overpopulated by a "pump process" (which is not understood in detail), so that stimulated emission (Sect. 4.5.4), produced coherently by a radiation field of the same frequency, predominates over spontaneous emission. This "natural" OH maser functions in the microwave range in a manner similar to a laser, which can produce concentrated, intense beams of radiation in the optical or infrared regions.

The OH masers are classified into two types, depending on their relative line intensities. Type I usually occurs in groups in the neighborhood of strong infrared sources, i.e. in regions where star formation is taking place (Sect. 10.5.2); the other type (II) is found, in contrast, to be correlated with late phases of stellar evolution (supergiants, Myra variables; see Sect. 7.4.1). Maser lines from other molecules are also observed, e.g. from H_2O at $\lambda = 1.35$ cm, and, less strongly, from SiO, H_2CO and other species.

10.3 Ionized Gas: Luminous Gaseous Nebulae

In the neighborhoods of bright O and B stars, the interstellar gas, and in particular the hydrogen, becomes ionized and is excited to luminosity; we then observe a diffuse nebula or an H II region. We begin our treatment in Sect. 10.3.1 with a discussion of these H II regions and of the physical processes which are important in them. Then, in Sect. 10.3.2, we describe the planetary nebulae, so called because of their appearance, which are related to the diffuse gaseous nebulae in terms of the physics of their luminosity, however not in terms of their importance in the cosmos. Another group of luminous nebulae are the supernova remnants (Sect. 10.3.3), which in fact are not excited to luminosity by the radiation from a central star, but instead by the high temperatures in shock waves from the outgoing shell of the supernova. In Sect. 10.3.4, finally, we treat the hot gas component of the interstellar medium, which makes itself known through spectral lines from multiply ionized atoms and through emission in the soft X-ray range, and whose temperatures are considerably higher than those in the ionized H II regions.

10.3.1 H II Regions

The excitation of the H II regions (Figs. 10.7, 8) occurs, as was recognized by H. Zanstra in 1927, in the following way: when a neutral hydrogen atom absorbs the light from a star in the Lyman region at $\lambda < 91.1$ nm, it will be ionized (cf. Fig. 4.7). The resulting photoelectron will later be recaptured by a positive ion (proton). This recombination leads only rarely directly to the ground state; usually, there are cascade transitions via several energy levels, accompanied by the emission of lower-energy light quanta. As an estimate, we can say that for each Lyman quantum absorbed, with $h\nu > 13.6$ eV or $\lambda < 91.1$ nm, among other quanta, about one $H\alpha$ quantum will be emitted. If it can further be assumed that the nebula absorbs practically all the Lyman radiation from the star, then following H. Zanstra we can use the $H\alpha$ radiation of the nebula to estimate the Lyman radiation of the star. Comparing it with the optical radiation, we can arrive at an estimate of the temperature of the star by applying the theory of stellar atmospheres. For the O and B stars, values are obtained which lie within the range of temperatures determined spectroscopically.

Strömgren Spheres. The size of a H II region, which we for simplicity consider to be a sphere of radius $R_{\mathrm{H\,II}}$, can be estimated as follows: in equilibrium, the total number $\mathcal{N}_{\mathrm{UV}}$ of UV photons of wavelength $\lambda \leq 91.1$ nm emitted per unit time by the star must be equal to the number of recombination processes per unit time within the whole nebula. Since the number of recombination processes per unit volume is proportional to the number of electrons, n_{e}, times the number of recombining ions n_{ion} in the unit volume, we obtain

$$\mathcal{N}_{\mathrm{UV}} = \frac{4\pi}{3} R_{\mathrm{H\,II}}^3 \cdot \alpha \, n_{\mathrm{e}} \, n_{\mathrm{ion}} \,, \qquad (10.22)$$

where $\alpha \, [\mathrm{m}^3 \, \mathrm{s}^{-1}] \simeq 2 \cdot 10^{-16} \, (T_{\mathrm{e}} \, [\mathrm{K}])^{-3/4}$ is the recombination coefficient (and T_{e} is the electron temperature, see below).

Since furthermore, each H atom in a completely ionized hydrogen plasma releases *one* electron on ionization, the ion density is equal to n_{e}, so that we obtain for the *Strömgren radius*

$$R_{\mathrm{H\,II}} = \left(\frac{3}{4\pi\alpha}\right)^{1/3} \mathcal{N}_{\mathrm{UV}}^{1/3} \, n_{\mathrm{e}}^{-2/3} \,. \qquad (10.23)$$

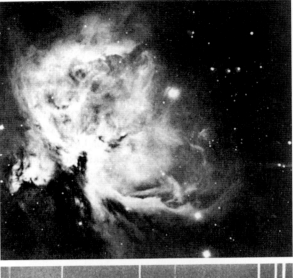

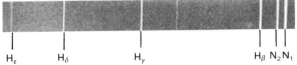

Fig. 10.7. The Orion Nebula (M 42 = NGC 1976), a diffuse or galactic nebula, and its emission spectrum in the blue. N2 and N1 are the "nebulium" lines of [O III] at $\lambda = 495.89$ and 500.68 nm

Fig. 10.8. The central portion of the η Carinae nebula NGC 3372. This color picture is a composite image (made using blue and red filters and narrow-band filters in the region of the green [O III] line) by S. Laustsen and J. Surdej with the 3.6 m telescope of the ESO. The whole nebula has a diameter of about 30 or 130 pc at a distance of 2.5 kpc; the section shown has angular dimensions of $10' \cdot 13'$. The bright star in the center of the nebula (to the left of the center of the picture) is η Car, one of the absolutely brightest stars of the "luminous blue variable" type. η Car, which itself is surrounded by a small nebula, exhibits strong brightness fluctuations and a substantial mass loss due to its stellar wind. Although η Car today has a magnitude of about 6, in 1843 it was the second-brightest star in the sky, with a magnitude of -0.8 mag. (With the kind permission of the European Southern Observatory)

10.3 Ionized Gas: Luminous Gaseous Nebulae

$R_{\mathrm{H\,II}}$ can also be interpreted as the distance within which the ionizing photons are all "used up".

For an O star emitting $\mathcal{N}_{\mathrm{UV}} \simeq 10^{49}$ photons s^{-1} in a nebula with a density of e.g. $n_\mathrm{e} \simeq 10^8$ m^{-3} (and $T_\mathrm{e} \simeq 10^4$ K), the Strömgren radius is found to be equal to $R_{\mathrm{H\,II}} \simeq 3$ pc; or, with $n_\mathrm{e} \simeq 10^6$ m^{-3}, it would be about 65 pc.

The Emission Measure. The brightness of a nebula in Hα and all other recombination lines is proportional to the number of recombination events and thus proportional to its emission measure as introduced by B. Strömgren:

$$EM = \int n_\mathrm{e}^2 \, dr \,. \qquad (10.24)$$

The integration is to be carried out along the line of observation; r is usually measured in [pc], and the electron density n_e in [cm^{-3}] or in [m^{-3}]. The emission measure EM of a H II region is of the order of $n_\mathrm{e}^2 R_{\mathrm{H\,II}} \propto n_\mathrm{e}^{4/3}$ (10.23) and has a value on the average of $\geq 10^{15}$ m^{-6} pc. Estimating n_e by assuming the longitudinal and transverse dimensions of the nebula (which are in the range of about 1 to 100 pc) to be equal, we find that the electron density in the H II regions is of the same order of magnitude as the density of neutral atoms in the H I clouds, i.e. 10^7 to 10^8 m^{-3}.

In the large diffuse nebulae, such as for example the Orion nebula (Fig. 10.7), values of $n_\mathrm{e} \simeq 5 \cdot 10^9$ m^{-3} are attained. Radio-astronomical surveys however indicate a wide range of H II regions in terms of size and electron density. At one end of the scale, we find *giant H II regions*, whose radio emissions surpass those of the Orion nebula by a large factor. Of these, about 80 are known within the Milky Way; the giant H II regions in the neighborhood of the center of the galaxy will be discussed in more detail in Sect. 11.3.2. At the other end of the scale, we see (at high angular resolution) *compact* H II regions with diameters of only 0.05 to 1 pc or less, considerably higher densities, of $n_\mathrm{e} \geq 10^9$ m^{-3}, and emission measures of $EM \geq 10^{19}$ m^{-6} pc, which are embedded in more extended, less dense regions of ionized hydrogen and are for the most part not recognizable in the optical region. The most dense of these are often in the same location as sources of emission of molecular lines (Sect. 10.2.3) and are regions of star formation (Sect. 10.5.2).

Recombination Lines. The optical and ultraviolet spectra of gaseous nebulae are produced under conditions which are far from thermal equilibrium, so that a treatment of the individual elementary processes is necessary to describe them (Sects. 4.5.3, 4). The effect of radiation from the star which excites them is reduced by a dilution factor, $W \simeq 10^{-16}$ to 10^{-14} (10.14). Therefore, in their atoms and ions, only the ground states as well as long-lived metastable states are occupied to any extent, the latter in lower terms with very small transition probabilities.

Hydrogen and helium, the two most abundant elements, are, as we have seen, ionized by the diluted stellar radiation ($\lambda \leq 91.1$ or ≤ 50.4 nm, respectively). Recombination of the ions and electrons takes place into all possible quantum states. From these initial recombination states, the electrons drop back to the ground state, for the most part via cascade transitions. We thus obtain the entire spectra of H, He I and possibly He II and some other ions as recombination lines.

In the range of mm to dm waves, recombination lines of neutral H, He and C occur, corresponding to transitions between neighboring levels with *high* quantum numbers, $n \simeq 100$ to 200. Such excited atoms are "blown up" to a radius of about $a_0 n^2$ (Sect. 7.1.1), i.e. about 1 μm! From the Doppler effect of these radio lines, which are not attenuated by interstellar extinction, the kinematics of ionized hydrogen in the Milky Way and the velocity fields in the H II regions can be determined.

Furthermore, by comparing the intensities of corresponding transitions, the *abundance ratio He/H* can be measured. Suitable lines for this purpose are pairs which lie around e.g. $\lambda \simeq 6$ cm

H 109α (5.009 GHz), He 109α (5.011 GHz) and
H 137β (5.005 GHz), He 137β (5.007 GHz) .

(The transition $n+1 \rightarrow n$ is denoted as $n\alpha$, and that from $n+2 \rightarrow n$ as $n\beta$, etc.)

An order of magnitude of He/H $\simeq 0.1$ is found, in sufficiently good agreement with the analyses of OB stars, which were formed from the interstellar gas only 10^6 to 10^8 yr ago. The He/H ratio increases from the outer regions of the Milky Way towards the center from about 0.07 to 0.11 (Sect. 12.1.6).

Bowen Fluorescence. In a few special cases, certain transitions are excited by fluoresence (I. S. Bowen). In particular, in the recombination of He^+, its resonance line at $\lambda = 30.38$ nm is emitted; it can, by coincidence, be absorbed by the ground state of O^{++}, exciting its 3d $^3P_2$ term. This, in turn, emits a number of O III lines, which can be observed in the near ultraviolet.

Forbidden Transitions. In 1927, I. S. Bowen made a discovery which created a certain sensation among astronomers: he found that the strong lines at $\lambda = 495.89$ and 500.68 nm, observed in all nebular spectra, and which had long been attributed to a mysterious element called "nebulium" (Fig. 10.7), could be interpreted as *forbidden* lines in the O III spectrum, originating with a low-lying metastable level in the ground state term of the ion. While the "usual" allowed lines have transition probabilities of the order of $A \simeq 10^8 \text{ s}^{-1}$ (electric dipole radiation, Sect. 7.1.1), these are e.g. for the the [O III] nebula lines only 0.007 or 0.021 s^{-1}, respectively. (Forbidden transitions are denoted by setting the symbol for the spectrum in square brackets.) They represent examples of magnetic dipole radiation; in other cases, electric quadrupole emissions from ions are observed. The excitation of the metastable levels takes place through electronic collisions with photoelectrons which are produced by the photoionization of H and He.

Of significance, especially for the energy balance of the H II regions, are in addition forbidden transitions between fine-structure levels *within* the ground-state terms of the more abundant ions, which lie mostly in the infrared; e.g. the lines [O I], $\lambda = 63.1$ μm; [O III], $\lambda = 88.3$ and 51.8 μm; [Ne II], $\lambda = 12.8$ μm; [S III], $\lambda = 18.7$ and 33.5 μm; as well as [C II], $\lambda = 157$ μm occur in H II regions.

In contrast to the lines in the optical region, these lines, like the radiofrequency recombination lines of H and He, are not attenuated by the interstellar dust and thus can be observed from throughout the Milky Way.

Now how does it happen that in nebulae, forbidden and allowed transitions occur with comparable intensities? We consider a line at the frequency v_0 corresponding to a transition from an excited state 1 into the ground state 0 (Fig. 7.4). Under the conditions which apply in a nebula, the excitation takes place only by electron impact (rate C_{01}), while the depopulation of level 1 can also occur through collisions (rate C_{10}), and through spontaneous emission (rate A_{10}). From (7.14), the population ratio is then given by

$$\frac{n_1}{n_0} = \frac{C_{01}}{A_{10} + C_{10}}. \quad (10.25)$$

Since C_{01} and C_{10} are both proportional to the electron density n_e, for a given A_{10} and with sufficiently low densities, $C_{10} \ll A_{10}$ and therefore $n_1/n_0 = C_{01}/A_{10}$ ($\ll 1$). This occurs e.g. for the [O III] lines, corresponding to their value of $A_{10} \geq 0.01$ s^{-1}, for $n_e \ll 10^{11}$ m^{-3}. The line intensity, which in the optically thin case is given by the integral over emission coefficients (4.49), then becomes

$$\int j_\nu d\nu = \frac{h\nu_0}{4\pi} A_{10} n_1 = \frac{h\nu_0}{4\pi} C_{01} n_0. \quad (10.26)$$

It is therefore *independent* of the transition probability A_{10} and is proportional to the electron density times the density of the atom in question, $n_0 + n_1 \simeq n_0$. For very small n_e, the electronic collisions are so rare that even with the small probability of a forbidden transition (and quite certainly for allowed transitions!), the excited state is nearly always depopulated by spontaneous emission.

Now that we have understood how in nebulae the forbidden and allowed lines can have similar intensities, other conditions being the same, e.g. the densities n_0 and the elemental abundances, it remains to be explained how the [O III] lines can be even as strong as the allowed H β line from the much more abundant element hydrogen (Fig. 10.7). The difference in abundances is compensated here by the fact that electron impact excitation is much more effective for O^{++} than is excitation by recombination of H.

Thermal Radio Emissions. In the radiofrequency range, we can observe the thermal radiation of the H II regions as a free–free continuum. This radiation is produced (Fig. 4.7) when an electron is not actually captured by a proton in a near collision, but instead is only deflected. Its intensity is, like that of the recombination lines, proportional to the emission measure EM (10.24). The free–free radio emissions, like infrared radiation, pass through the interstellar dark clouds practically without hindrance and thus allow the observation of the H II regions in the entire Milky Way.

The free–free radiation can be recognized by the fact that its intensity I_ν in emission from an optically *thin* layer is practically independent of the frequency ν

[C II], [Ne II], ... as well as from the heated dust. The *electron temperature* T_e of a H II region is therefore lower than would be expected from the temperatures of the stars, about 7000 to 10 000 K.

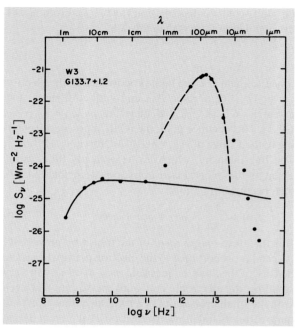

Fig. 10.9. Radio and infrared spectra of the H II region W3 = IC 1795 = G133.7 + 1.2 (W denotes the number in the sky survey by G. Westerhout, 1958; G: galactic coordinates, $l = 133.7°$, $b = +1.2°$, see Sect. 11.1.1). The observations (●) of the (spatially integrated) H II region show the thermal free-free radiation of the ionized gas (theoretical model), which is optically dense for $\lambda \geq 20$ cm, as well as the thermal emission from "warm" dust for $\lambda \leq 1$ mm (– – – black body radiation for $T = 70$ K). With a higher angular resolution, several compact H II regions or infrared sources can be seen in W3; they have qualitatively similar spectra. (Courtesy of the Publications of the Astronomical Society of the Pacific)

(Fig. 10.9). According to theory, the absorption coefficient κ_ν or the optical thickness τ_ν is approximately proportional to $n_e^2 \nu^{-2}$, so that from (4.60), the frequency dependence in the Rayleigh–Jeans region (4.63) cancels out. At lower frequencies (or higher electron densities), $\tau_\nu \gg 1$ and the intensity is frequency-dependent, $I_\nu = B_\nu \propto \nu^2$. In the infrared, thermal radiation from dust in the H II regions which is heated to about 100 K predominates over the free–free continuum.

Energy Balance. The electrons of course lose energy in the course of the collisions which excite the metastable states; it is then radiated away from the nebula, mainly in the forbidden lines in the far infrared from [O I], [O III],

10.3.2 Planetary Nebulae

We know of over 1100 planetary nebulae (Fig. 8.9) in the Milky Way. In contrast to the H II regions, they are not excited to luminosity by a young, bright O or B star, but rather by a smaller, very hot star, which ejected the nebula in a late phase of its evolution within a relatively short time (Sect. 8.2.4) and is located on the Hertzsprung–Russell diagram on the path from the red giant stage (on the asymptotic giant branch) to the white dwarf stage (Fig. 8.8). Since the optical and ultraviolet spectra of the planetary nebulae (Figs. 10.10 and 5.20) are very similar to those of the diffuse gas nebulae, we can assume that in spite of the different stars which provide their excitations, the physics of their luminosity is related to that of the H II regions, and we can expect the same basic ionization processes to apply here as in Sect. 10.3.1.

Zanstra's method, in agreement with the theory of stellar evolution, yields for the central stars of the planetary nebulae temperatures of 30 000 up to 150 000 K. The luminous shells of the planetary nebulae, whose apparent sizes are in the range of a few minutes of arc for the objects nearest to us, have radii of the order of 0.2 pc or of several 10^4 AU, electron densities of around 10^9 to 10^{10} m^{-3}, electron temperatures of 10^4 K, and masses over a wide range up to about 0.2 $\mathcal{M}_\odot$.

Expansion. The observed splitting of the emission lines indicates an expansion of the nebula's shell at a velocity of about 25 km s^{-1}; the front side is approaching us and the rear side is moving away. This expansion corresponds to a dynamic lifetime of the nebulae of only about 10^4 yr. In order for us to see the luminous shell at all, the evolution time of the central star, with its ionizing UV photons, must also "fit" the lifetime of the shell.

The rapid evolution of the central star not only causes a correspondingly rapid change in its ultraviolet radiation, but also has an important influence on the dynamics of the planetary nebula. As a red giant, the central star had been losing mass at a rate of from 10^{-6} to

10.3 Ionized Gas: Luminous Gaseous Nebulae

and Si IV in the ultraviolet with the IUE satellite, the small central star also loses mass at a relatively low rate of around 10^{-9} to $10^{-7}\,\mathcal{M}_\odot\,\mathrm{yr}^{-1}$ as a hot, fast wind, which streams out at velocities of up to $4000\,\mathrm{km\,s^{-1}}$ and produces a second shock wave front at the inner boundary of the ejected shell.

Condensations. In the planetary nebula which is nearest to us, NGC 7293, the "Helix Nebula" in Aquarius at a distance of about 150 pc, density condensations have been observed near to the resolution limit of earthbound telescopes; these show motion relative to the nebula shell. Only with the Hubble Space Telescope did it become possible to resolve their structures. In the impressive images made by C. R. O'Dell and K. P. Handron in 1996, one can make out several hundred extended "cometary clumps", whose "tails" are pointing away from the central star and whose heads, pointing inwards, are ionized by the ultraviolet radiation of the white dwarf (Fig. 10.11). The typical size of a node ranges from 100 to 300 AU, its mass is around $10^{-5}\,\mathcal{M}_\odot$ or a few Earth masses. These concentrations move outwards at about $10\,\mathrm{km\,s^{-1}}$, more slowly than the nebula shell, and are probably remnants of the wind from the red giant star which have remained due to instabilities.

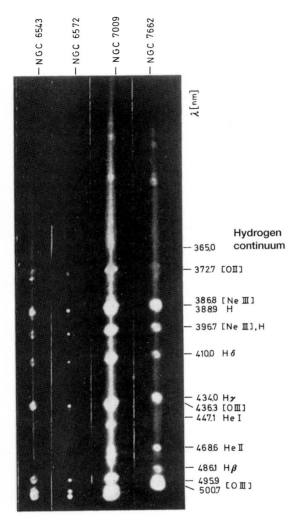

Fig. 10.10. Optical spectra of planetary nebulae, taken by a spectrograph without an entrance slit. This allows the distribution of emission of different lines in the nebular shell to be seen. NGC 6543 also clearly shows the continuous spectrum of the central star

$10^{-4}\,\mathcal{M}_\odot\,\mathrm{yr}^{-1}$ over a period of roughly 10^6 yr owing to its stellar wind which streamed out from the star at velocities around $10\,\mathrm{km\,s^{-1}}$ (Sect. 7.4.3). This matter collected in the neighborhood of the star. The ejected shell collides with this slow wind, so that at its outer boundary, a shock wave front is formed.

As shown by observations of the P Cyg profiles (Sect. 7.4.4) of the resonance lines of C IV, N V, O V

10.3.3 Supernova Remnants

In Sect. 7.4.7, we have already met up with the supernovae, their various types and the neutron stars or pulsars which remain as remnant stars after the supernova explosions. We now want to consider in more detail the shells ejected in the explosions, denoted as SNR (*Supernova Remnants*). They move out into the interstellar medium with initial velocities of the order of $10\,000\,\mathrm{km\,s^{-1}}$, gradually slow down, and finally, after about 10^6 yr, they disperse.

Excitation. The supernova remnants, in contrast to the H II regions, are not excited by the ionizing radiation from a central star. In the cases of most of the remnants, shock wave fronts instead form at the collision boundary of the expanding shell with the interstellar medium; here, the temperatures are so high that the thermal radiation is emitted mainly in the X-ray spectral region.

Fig. 10.11. The Helix Nebula NGC 7293, a ring-shaped planetary nebula at a distance of 150 pc from the Earth, with a diameter of about 12′ (0.5 pc). The picture shows an area of roughly 2.5′ · 2.5′, taken by C. R. O'Dell and K. P. Handron in 1996 with the Wide Field Planetary Camera 2 of the Hubble Space Telescope. The color scale corresponds to the ionization energies of the emitting atoms: [O III] lines are blue, Hα green, [N II] lines red. Between about 115″ and 180″ distance from the central star which provides the excitation energy (it is above and to the right, outside the picture), numerous comet-like, radially oriented *condensations* can be seen; they move outwards at only about 10 km s^{-1}, while the ring nebula itself expands at around 20 km s^{-1}. *Insert:* An image of the whole Helix Nebula with the enlarged area marked. (© Association of Universities for Research in Astronomy AURA/Space Telescope Science Institute STScI)

The predominant emission lines in the optical spectra are qualitatively similar to those of the diffuse gas nebulae, but can be distinguished from the latter in particular by their noticeably higher ratio of the intensities of the forbidden sulfur lines [S II], $\lambda = 671.6/673.1$ nm to that of the neighboring Hα line.

In a few young supernova remnants, such as e.g. the Crab nebula, the central pulsar is in fact the main source of the emitted energy. The rapidly rotating neutron star produces high-energy particles which emit nonthermal synchrotron radiation through interaction with the magnetic field.

We cannot go further here into the details of excitation and ionization in shock wave fronts or by means of energetic particles.

Expansion. At first, the shell ejected by a supernova explosion moves out at about 10^4 km s^{-1} almost without hindrance, in a free expansion. After around 200 yr, when its diameter has increased to about 2 pc, it slows down noticeably due to the interaction with interstellar matter. Most of the observed supernova remnants are in this latter phase, which lasts around 10^5 yr, with diameters of ≤ 30 pc and expansion velocities of several 100 km s^{-1}, before they finally "dissolve" in the interstellar medium after they have reached a size of roughly 100 pc and their velocities have decreased to around 10 km s^{-1}, i.e. the same order as those of the matter in interstellar space.

In detail, however, a supernova remnant develops a very complex spatial and kinetic structure through its interactions with the interstellar gas. In optical images, e.g. of the Vela remnant (Fig. 10.12), one can see numerous arches, strings, nodes and "flakes", which can have a variety of different velocities and also different chemical compositions. As shown by high-resolution X-ray images, e.g. of Cas A made with the Einstein satellite (Fig. 10.13), or of the Vela remnant taken by ROSAT (Fig. 10.14), the hot plasma also exhibits a similarly complex structure.

In the following paragraphs, we describe several supernova remnants in more detail, beginning with

10.3 Ionized Gas: Luminous Gaseous Nebulae

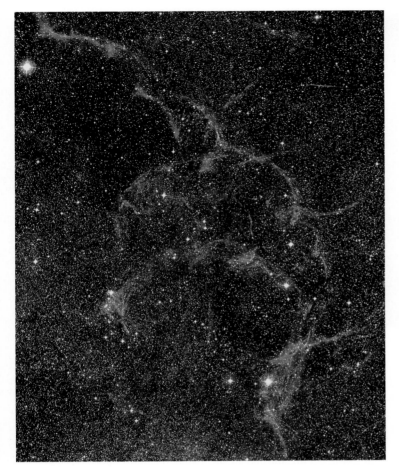

Fig. 10.12. The Vela supernova remnant (a section of angular size $2° \cdot 2.5°$). This is a color photograph taken by H.-E. Schuster with the 1 m Schmidt camera at the ESO. (With the kind permission of the European Southern Observatory)

the most typical example, the Crab nebula; it has already played a decisive role in the development of astrophysics several times in the past.

The Crab Nebula. As early as 1942, W. Baade and R. Minkowski had begun detailed investigations of this interesting object, M1 = NGC 1952, which has a photographic magnitude of 9.0 mag (Fig. 7.41). It consists of an inner, almost amorphous region of $3.2' \cdot 5.9'$, emitting a continuous spectrum without lines, and a shell, whose bizarre filaments, the "legs of the Crab", emit mainly the Hα line. All together, its mass is 1 to 2 $\mathcal{M}_\odot$, and its gas has a higher ratio of He/H than the solar element mixture.

As could be determined from East Asian records, the Crab nebula is the shell of a supernova from the year 1054. A comparison of its radial expansion velocities of about 1000 to 1500 km s^{-1} with the corresponding proper motion outwards confirms the date of the explosion and in particular allows an estimate of the distance to the Crab nebula, giving roughly 2 kpc.

The Crab nebula is a noticeable object not only in the optical spectral range, but is also observed as a strong radio source (Tau A = 3C 144) and an intense source of X-rays (Tau X-1).

Following I. S. Shklovsky's suggestion of 1953, we can assume that the continuum radiation from the Crab nebula from the radiofrequency region up to the X-ray region is *synchrotron radiation* (Sect. 12.3.1) produced by energetic electrons in a magnetic field of the order of 10^{-8} Tesla. This hypothesis is supported by the following arguments: if the radiation were thermal, the

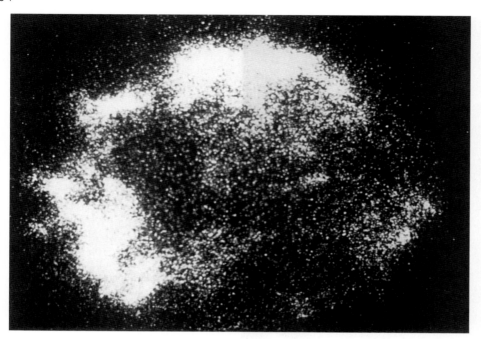

Fig. 10.13. X-ray image of the supernova remnant Cassiopeia A. The image was made with the Einstein satellite at an angular resolution of about 4″, corresponding to roughly 0.05 pc at a distance to Cas A of 2.8 kpc. The diameter of Cas A is about 5′ or 4 pc. (Image taken by Stephen S. Murray, Center for Astrophysics, Cambridge, Mass.)

electron temperature T_e would have to be higher than the highest radio brightness temperatures, i.e. $\geq 10^9$ K, while all gaseous nebulae have $T_e \leq 10^4$ K. Secondly, the amorphous core of the nebula emits only a continuum in the optical region, and no spectral lines at all. A conclusive argument in favor of the synchrotron theory was however given by the detection of the *polarization* (Fig. 12.16) of the continuum, as expected for synchrotron radiation, initially in the optical region by Dombrowsky and Vashakidze (1953/54). Later, in spite of the difficulties presented by the Faraday effect, polarization measurements were successfully carried out also in the cm- and dm-wavelength ranges. A thorough consideration of the lifetime of the relativistic electrons in the Crab nebula led J. H. Oort to the conclusion that they are still being emitted even today, nearly a thousand years after the supernova explosion.

The X-ray emissions of the Crab nebula also result from an extended source of around 1′ in size, as shown from the first observations, spatially resolved with the aid of lunar occultations. The high angular and time resolution of the Einstein satellite later made it possible to resolve the pulsar from the extended source, which is located northwest of it (Fig. 7.42). The average contribution of the pulsar to the emissions in the X-ray range is only about 4% of the total.

Soon after the discovery of the Crab pulsar, it became clear that its *rotation* was, in fact, the energy source for the Crab nebula and that in connection with the rotation and its braking by the surrounding plasma in the magnetic field, enormous numbers of high-energy "superthermal" particles are released. The production of synchrotron radiation in the radiofrequency range requires electrons of about 10^9 eV, and in the X-ray region, it requires about 10^{14} eV. In connection with these electrons, doubtless also a corresponding number of positively-charged atomic nuclei, i.e. cosmic-ray particles (Sect. 10.4.2) are accelerated. Using the high resolution of the Hubble Space Telescope, the time variation of the synchrotron emissions could be clearly detected; the emissions were found to emerge as a jet from one of the poles (and probably from the other, also – this is not visible from the Earth for geometrical reasons), in the direction of the axis of rotation. In addition, wavelike motions can be seen in the equatorial plane, moving outwards with about half the velocity of light. These jets, which could also be observed with the ROSAT satellite in

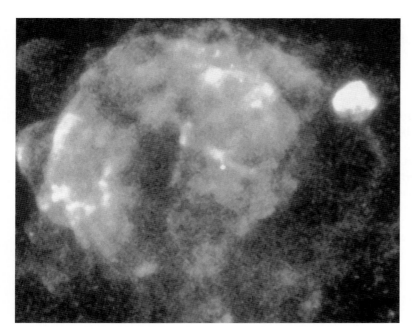

Fig. 10.14. An X-ray image of the supernova remnant Vela, with a diameter of 8.3° or 65 pc at a distance of about 450 pc. The image was made using the multiwire proportional counter on the X-ray satellite ROSAT in the energy range 0.1–2.4 keV by B. Aschenbach, R. Egger, and J. Trümper (1995). The strong source (*upper right*) is the supernova remnant Puppis A, which is roughly 4 times more distant than Vela and emits harder X-radiation. (Image: ROSAT, Max-Planck-Institute for extraterrestrial Physics, MPE Garching and NASA)

the X-ray region, certainly contribute to the transport of high-energy particles to the outer regions of the nebula.

The Crab nebula occupies a unique position among the supernova remnants. On the one hand, we know the time of the supernova explosion in its case; on the other, we can observe the pulsar (neutron star) within the nebula and can investigate in detail the ejected gas shell and the interactions between the pulsar and its environment over the whole of the electromagnetic spectrum.

Vela SNR. This nebula, at a distance of about 450 pc from the Earth, was formed by a supernova explosion around 14 000 yr ago. The remnant star is denoted as the Vela pulsar, PSR 0833-45, with a period of 89 ms, observed from the radiofrequency up to the gamma-ray range (cf. Sect. 7.4.7). Its image in the optical region (Fig. 10.12) shows clearly a filamentary structure, formed by the interaction of the ejected gas shell with the surrounding interstellar medium. Collisional heating is strongest at the front edges of the filament arches, where the lines of ionized oxygen give rise to the blue luminosity. The inner, cooler parts appear reddish due to the Hα emission of neutral hydrogen. In the X-ray region, we see a similarly complex structure in images made with the ROSAT satellite (Fig. 10.14).

As a young supernova remnant, the Vela nebula, like the Crab nebula, shows jets leading out from the pulsar; they can be observed in optical and in X-radiation. The strong thermal emission from the filaments of the Vela remnant, however, is not seen in the case of the Crab nebula.

Optical SNR. As a result of the interstellar extinction, only a few supernova remnants within the Milky Way galaxy are observable via their optical emissions. Aside from the Crab nebula and the Vela remnant, we can see e.g. the remnants of the supernovae of 1572 (Tycho Brahe) and 1604 (Kepler).

Cassiopeia A, the strongest radio source in the northern sky, is the youngest galactic remnant of a supernova which exploded between 1650 and 1700. It is found at the same position as an optical nebula with a diameter of 4 pc, which is expanding at a velocity of 7400 km s^{-1}. Cas A was observed by ROSAT in the gamma-ray region using the line of radioactive ^{44}Ti; this isotope was produced in the supernova explosion and decays with a half-life of 60 yr via the scheme ^{44}Ti $\rightarrow$ ^{44}Sc $\rightarrow$ ^{44}Ca.

Images made with Hα radiation also show some ring-shaped or circular nebulae such as the well-known Cygnus arc (veil nebula).

Radio Emissions. In the radiofrequency region, undisturbed by interstellar extinction, we can observe in the Milky Way galaxy about 200 supernova remnants. Their radio spectra are characterized by *non-thermal* radiation, whose intensity decreases with increasing frequency according to a power law (Sect. 12.3.2). They can thus be distinguished from the (thermal) spectra of other nebulae, such as e.g. the luminous gas nebulae (H II regions), and thereby be identified as supernova remnants.

The enigmatic "radio spur", which stretches across the sky from the neighborhood of the galactic center to the galactic north pole (and likewise several other similar formations) seems to be part of a gigantic ring. All of these radio sources are no doubt related to the remnants of old supernovae.

The younger remnants can be classified according to their radio images into two *types*:

1. The "shell" or "ring-shaped" remnants like the remnant of Tycho Brahe's supernova (Fig. 10.15).

2. The "filled" remnants. These *synchrotron nebulae* or plerions, whose prototype is the Crab Nebula, are filled with highly energetic particles as a result of the "activity" of a rapidly rotating pulsar, and the particles can emit synchrotron radiation when they move in magnetic fields. The supernova remnants of this type are associated with the type II supernovae (Sect. 7.4.7).

Since the emissions of the synchrotron nebulae decay more rapidly than those of the shell remnants, the latter, usually broken up into many segments and filaments, represent the end phase of *all* SNR.

X-ray Emission. Most supernova remnants have been detected as X-ray sources. With the exception of the Crab Nebula, and to some extent the Vela remnant, the X-ray spectra of the supernova remnants observed up to now can be explained in terms of *thermal* emission from a very hot plasma source at temperatures above 10^6 K. Theoretically, we would expect essentially a continuum at temperatures $> 10^7$ K, with only a few lines from highly ionized, abundant elements such as Fe superposed on it; at low temperatures, numerous emission lines dominate the continuum.

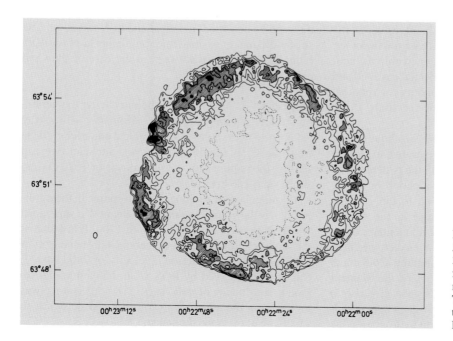

Fig. 10.15. Radio contour lines at $\lambda = 6$ cm (5 GHz) of the remnant 3C 10 of Tycho Brahe's supernova from the year 1572. Observations made with the Westerbork Synthesis Telescope at an angular resolution of $7'' \cdot 8''$ by R. M. Duin and R. G. Strom (1975)

10.3 Ionized Gas: Luminous Gaseous Nebulae

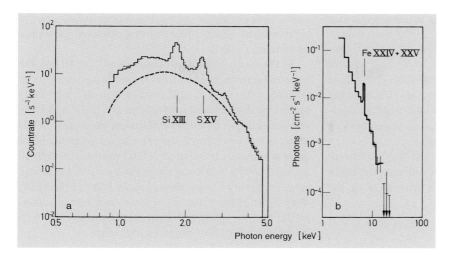

Fig. 10.16a,b. Spectral lines in the X-ray spectrum of the supernova remnant Cas A. (**a**) Observations with the solid-state spectrometer on the Einstein Satellite (HEAO-2) by R. H. Becker et al. (1979). The main contribution to the soft X-ray emissions is from unresolved lines; the dashed curve gives an estimate of the continuum. (**b**) Proportional counter observations from the OSO-8 satellite by S. H. Pravdo et al. (1976)

The (spatially unresolved) X-ray spectrum of the strong radio and X-ray source Cas A fits well into this picture, if we assume the existence of two components at different temperatures, one at $\simeq 4.5 \cdot 10^7$ K, which emits the continuum in the range > 5 keV and the Fe XXIV+XXV line at 6.7 keV; the other at $\simeq 7 \cdot 10^6$ K, represented by many lines in the soft X-ray region, which are unresolved except for those of Si XIII at 1.9 keV and S XV at 2.45 keV (Fig. 10.16).

RX J0852.0-4622. This circular object, recently discovered by observing its hard X-ray emissions above 1.3 keV (B. Aschenbach, 1996), has a diameter of around 2° and, at a distance of about 200 pc, must be the young supernova remnant nearest to the Earth, with an estimated age of 680 yr. In the soft X-ray region, it is outshined by the bright Vela remnant at a distance of around 450 pc (Fig. 10.14); it lies in front of the southwest part of Vela. The discovery of the gamma lines resulting from the decay of ^{44}Ti, similarly to the case of Cas A (see above), confirms that RX J0852.0-4622 is a very young supernova remnant.

SN 1987A. The radiation flash in the ultraviolet which accompanied the explosion of this supernova in the Large Magellanic Cloud ionized the surrounding matter, so that it could be seen for several years in the light of its recombination radiation. Corresponding to the bipolar arrangement of the mass distribution, a "dumbbell" shape, we observe with the Hubble Space Telescope a projection onto the celestial sphere in the form of three bright *rings* (Fig. 10.17): a smaller, inner ring and two larger outer ones.

Fig. 10.17. Rings around the supernova SN 1987A in the Large Magellanic Cloud. An image made with the Wide Field Planetary Camera 2 of the Hubble Space Telescope seven years after the explosion by C. J. Burrows et al. (1994). The size of the inner ring is 1.7″ or 0.4 pc, while that of the two outer rings is around 3.5″ or 0.8 pc. (© Association of Universities for Research in Astronomy AURA/Space Telescope Science Institute STScI)

The supernova shell is expanding at present at a rate of about $0.01''\,\mathrm{yr}^{-1}$ or $2700\,\mathrm{km\,s^{-1}}$. The first indication of the supernova *remnant*, around 10 years after the explosion, was the observation of X-ray emission and non-thermal synchrotron radiation. It is estimated that in the year 2005 (with an uncertainty of ± 3 yr) the ejected shell will collide with the neighboring interstellar matter and that a strong increase in brightness as well as emission of X-rays will result.

From continuing observations of the supernova SN 1987A, we expect important new information about the evolution and the chemical composition of the supernova remnant.

10.3.4 Hot Interstellar Gas

A major modification of our understanding of the interstellar medium was brought about by the discovery of a hot gas component, by means of satellite observations in the ultraviolet and X-ray regions. The temperatures in this hot gas are much higher than those of the partially ionized hydrogen gas which can be observed via its 21 cm emission line, or those of the ionized gas in the H II regions.

Interstellar Absorption Lines. The occurrence of interstellar absorption lines from multiply ionized atoms (Sect. 10.2.1) can be explained only by correspondingly high temperatures, due to the high ionization energies required. Thus, the lines of C II and S III indicate temperatures around $5 \cdot 10^4$ K, while the lines of C IV at $\lambda = 154.8/155.1$, Si IV at $\lambda = 139.4/140.3$, and N V at $\lambda = 123.8/124.2$ nm indicate temperatures $\leq 10^5$ K. This hot matter presumably occurs in "cloudlike" structures which are moving at relatively high velocities of ≥ 20 to $100\,\mathrm{km\,s^{-1}}$ relative to the colder gas. Finally, observations of the O VI resonance doublet at $\lambda = 103.2/103.8$ nm show a component at 2 to $8 \cdot 10^5$ K (with radial velocities $\leq 2000\,\mathrm{km\,s^{-1}}$).

The Extreme Ultraviolet and X-ray Regions. The spectral region below the Lyman edge at 91.1 nm has not yet been investigated in detail, since here, the strong absorption in the Lyman continuum of neutral hydrogen limits the observations to our immediate vicinity in the Milky Way. However, as a result of the inhomogeneous distribution of interstellar matter, objects out to a distance of about 100 pc can be observed in certain directions.

In the X-ray region at below about $\lambda = 10$ nm, the interstellar gas again becomes increasingly transparent. Although the view is still restricted in the soft X-ray region by absorption due to helium and the K and L absorption edges of the more abundant elements, below about 1 nm wavelength the interstellar medium is almost completely transparent. Observations of a thermal radiation in the soft X-ray range (about $\lambda = 2$ to 12 nm), whose intensity is concentrated in the plane of the Milky Way, indicate the existence of a very hot, tenuous interstellar gas at $\geq 10^6$ K with electron densities $\leq 10^4\,\mathrm{m^{-3}}$, which fills about half of interstellar space.

The Galactic Halo. Observations of "interstellar" lines in the ultraviolet spectra of bright stars in the Magellanic Clouds by the IUE satellites showed, surprisingly, several line components with radial velocities which lie between those of our local galactic vicinity and those of the two neighboring galaxies. We can conclude the existence of a gaseous galactic halo, whose nonuniformly distributed matter stretches out to several kpc from the plane of the Milky Way. Since the corresponding absorption lines from Fe II, S II and Si II are observed, as well as from C IV and S IV, and since later also O VI lines were discovered, the halo must contain several regions at different temperatures (up to a few 10^5 K).

The high temperatures and velocities of the hot gas require a considerable energy input to the interstellar medium. Supernova explosions and the stellar winds from OB stars are probably the sources of this energy.

10.4 High-Energy Components

Keeping in mind our later considerations of the production, motion, and storage of cosmic-ray particles and of synchrotron electrons in the Milky Way and other galaxies, we first deal briefly in Sect. 10.4.1 with the methods for measuring interstellar magnetic fields. Then we turn to a discussion of the high-energy components in the interstellar medium and the associated physical processes, treating cosmic radiation in Sect. 10.4.2 and gamma radiation from the Milky Way in Sect. 10.4.3.

10.4.1 Interstellar Magnetic Fields

The *polarization* of starlight by dust particles which are partially oriented through magnetic fields was already mentioned in Sect. 10.1.3. Another possibility for determining the interstellar magnetic fields is based on the *Zeeman effect* (7.72) of the 21 cm line (Sect. 10.2.2).

Later, we shall see (in Sect. 12.3.1) that the non-thermal *synchrotron radiation* from many radio sources is also *polarized*. On the one hand, measurements of this polarization give information about the magnetic field at the place where the radiation originated. On the other hand, when the radiation traverses an interstellar plasma with a magnetic field, the plane of polarization of the radiofrequency radiation will be rotated; this is none other than the well-known *Faraday effect*. The theory of this effect shows that the rotation of the plane of polarization through an angle ψ is proportional to the square of the wavelength λ times a wavelength-independent *rotation measure RM*:

$$\psi = \lambda^2 \cdot RM \qquad (10.27)$$

with

$$RM = 8.1 \cdot 10^3 \cdot \int_0^L n_e B_\parallel \, d\ell \quad [\text{rad m}^{-2}] \, . \qquad (10.28)$$

Here, we give ψ in units of arc [rad], λ in [m], the electron density n_e in [m^{-3}], the longitudinal component of the magnetic induction, $B_\parallel$, in [T = Tesla], and the distance L to the radio source (as well as dℓ) in [pc]. From measurements at several wavelengths, (10.27) can be used to determine the original position of the plane of polarization and the rotation measure. The largest observed values of RM for extragalactic sources lie in the range of 300 rad m^{-2}. If the electron density "on the way" is known, then (10.28) yields a correspondingly averaged value of $B_\parallel$.

To summarize, the following methods are available for the investigation of interstellar magnetic fields:

	Method	Indicating medium	Information about
a	Polarization of starlight	Dust	$B_\perp$
b	Zeeman effect of the 21 cm Line	Neutral hydrogen	$B_\parallel$
c	Synchrotron radiation	Relativistic electrons	$B_\perp$
d	Faraday rotation	Thermal electrons	$B_\parallel$

Corresponding to the different spatial distributions of the contributing media, methods (a–d) yield different average values of the fields, which are not readily comparable. It is therefore not surprising that attempts to construct a model of the galactic magnetic field have not yet converged to a final result. From all the methods, the order of magnitude of the general galactic magnetic field is about $2 \cdot 10^{-10}$ T, with large-scale variations of the same order of magnitude superposed over regions of about 50 pc in size. In the more dense H I clouds and H II regions, we find fields about 10 times stronger; in molecular clouds, the fields are up to 100 times stronger. Measurements of the Zeeman splitting of the lines at $\lambda = 18$ cm of the OH maser (Sect. 10.2.3) indicate fields of up to 10^{-7} T in spatially very limited regions.

10.4.2 Cosmic Radiation

In 1912, during a balloon flight, V. Hess discovered that the electrical conductivity of the atmosphere increases with increasing altitude, and explained this as being due to the ionizing effect of energetic *cosmic radiation* from outer space. The nature of these cosmic rays, as they are now called, i.e. the fact that they consist mainly of positively-charged particles, was established after about 1930, first on the basis of their dependence on the Earth's magnetic field and then by direct observation. Their penetrating power (into deep mine shafts) shows that we are dealing with particle energies which by far exceed those encountered in everyday phenomena.

We first consider the *primary* cosmic rays, which have *not* yet interacted with the matter in the Earth's atmosphere and can be investigated from balloons and rockets; or, before they are influenced even by the geomagnetic field, from satellites and space probes. We have already gained an overview of the detectors and telescopes used for investigating cosmic radiation in Sect. 5.3.2.

In the following, we first consider briefly the cosmic radiation generated by the Sun, and then discuss the galactic component of the primary cosmic rays, which extends to considerably higher particle energies.

Solar Components. We have been able to observe the production of cosmic-ray particles in at least one case from "close up", so to speak, since S. E. Forbush was able to demonstrate in 1946 that the Sun emits particles having energies up to several GeV during major eruptions (flares, cf. Sect. 7.3.6). Their chemical composition is, on the whole, that of solar matter. We cannot describe these and further investigations by A. Ehmert, J. A. Simpson, P. Meyer and others in more detail here; they give information in particular on the propagation of solar cosmic rays in the interplanetary plasma and magnetic field. The physical mechanisms responsible for the acceleration of charged particles in the magnetic plasma of the solar atmosphere or corona up to energies of 10^8 to 10^{10} eV are still not completely understood. Whether the principal mechanism is an induction effect, as in a betatron, or whether the particles are trapped between two magnetic "mirrors" (shock-wave fronts?) and thereby accelerated (E. Fermi) is not certain at present. Empirically, however, it is clear and should be remembered that the acceleration of particles to high energies in the Sun as well as in the giant radio sources (Sect. 12.3.3) always occurs in a highly turbulent plasma with a magnetic field.

We now turn to a discussion of the *galactic* component of the cosmic radiation.

The Energy Spectrum. We first consider the *nucleon* components, consisting of protons, α particles (He^{2+}) and heavy nuclei (i.e. completely ionized heavier atoms). Their energies can be determined, when they are not too high, by experiments involving deflection in electric and magnetic fields. At higher energies ($\geq 10^{14}$ eV), one measures the ionization energy which is deposited in the atmosphere by the large air showers. The results of these measurements over the enormous energy range of $10^8 < E < 10^{20}$ eV are shown in Fig. 10.18. Above $E > 10^{10}$ eV, the influence of the interplanetary magnetic field is unimportant. At higher energies, the *integral* energy spectrum falls off roughly according to a *power law* up to the highest measured energies of $\geq 10^{20}$ eV:

$$J(>E) = J_0 \cdot \left(\frac{E}{E_0}\right)^{-q} . \quad (10.29)$$

Here, $J(>E)$ is the particle flux at energy $>E$, which is usually quoted in units of $[cm^{-2} \, s^{-1} \, sr^{-1}]$. At around 10^{15} eV, the energy distribution shows a change of slope; up to this "knee", the exponent in the power law is $q \simeq 1.7$; above it, the spectrum becomes steeper

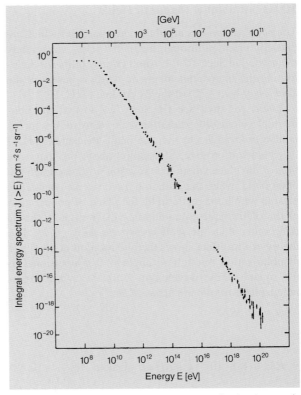

Fig. 10.18. The integral energy spectrum of galactic cosmic radiation. The number $J(>E)$ of particles with kinetic energies of $>E$ which pass through an area of 1 cm² per s and unit solid angle is plotted against the energy E in [eV] or [GeV = 10^9 eV]. In recent times, some particles of energies up to $3 \cdot 10^{20}$ eV have been observed

($q \simeq 2.1$); and finally, above the "ankle" at $\simeq 3 \cdot 10^{18}$ eV, it flattens off again. Owing to the rareness of the large air-shower events, the measurements are somewhat uncertain above 10^{19} eV. The particle flux above 10^{20} eV is thus only about *one* nucleon per km² and year! The highest energies which have as yet been observed for cosmic-ray particles are $3 \cdot 10^{20}$ eV = 48 J.

The *electron* component of cosmic radiation was discovered only in 1961, by P. Meyer et al. Its energy distribution ranges from about 2 MeV up to over 1000 GeV; the flux is only about 1/100 that of the protons. The energy spectrum also obeys a power law, $J_e(>E) \propto E^{-q}$, with $q \simeq 2.0$ in the range from about 1 to 25 GeV. At higher energies, the spectrum decreases more steeply.

The integral spectrum $J(>E) \propto E^{-q}$ corresponds to a *differential* flux $J(E)$, given e.g. in [cm^{-2} s^{-1} sr^{-1} GeV^{-1}], or also to a particle density $n(E)$ for energies in the range E to $E+\mathrm{d}E$; these likewise obey a power law, $J(E), n(E) \propto E^{-p}$, with $p = q+1$.

Here, we cannot go into the complicated processes which accompany the penetration of primary cosmic rays into the atmosphere. We shall, however, discuss in more detail the deflection of the charged cosmic-ray particles by magnetic fields, in particular the geomagnetic field, the interplanetary magnetic field from the Sun, and the interstellar magnetic field in the Milky Way galaxy.

Motion in a Magnetic Field. In a (homogeneous) magnetic field of flux density $\boldsymbol{B}$, a particle of charge e, rest mass m, and velocity $\boldsymbol{v}$ moves, in a plane perpendicular to $\boldsymbol{B}$, on a circular orbit with the cyclotron frequency $\omega_C = 2\pi\nu_C$ and radius $r_C = v/\omega_C$.

Cosmic-ray particles are extremely *relativistic* ($v \simeq c$), since their energies are much larger than the rest energy mc^2 (0.51 MeV for electrons, 938 MeV for protons). As we saw in Sect. 4.2.2, the relativistic mass of the (moving) particle is γm, its momentum is $p = \gamma m v$ (4.20) and its energy is $E = \gamma mc^2$ (4.21), where $\gamma = 1/\sqrt{1-(v/c)^2}$ is the Lorentz factor (4.14).

For the motion perpendicular to $\boldsymbol{B}$, the centrifugal force $\gamma m v^2/r_C$ equals the Lorentz force evB, so that the *cyclotron radius* is given by

$$p = eBr_C \qquad (10.30)$$

and the *cyclotron frequency* by

$$\omega_C = \frac{1}{\gamma}\frac{eB}{m} \qquad (10.31)$$

[B in Tesla]. The magnetic rigidity Br_C of the particle is thus directly related to its momentum p. The cyclotron frequency ω_C is, by the way, twice as large as the Larmor frequency ω_L, which plays a role in the Zeeman effect (7.71).

In the Gaussian system of units, the Lorentz force is evB/c, where B is expressed in [Gauss]; correspondingly, the cyclotron frequency is $\omega_C = eB/(\gamma mc)$ (10.31). (1 G corresponds to 10^{-4} T.)

Since for relativistic particles, $p = E/c$ (4.26), it follows from (10.30) that

$$r_C = \frac{E}{ceB}. \qquad (10.32)$$

If we compute E in [GeV = 10^9 eV] and r_C in [pc], then we obtain

$$r_C \,[\mathrm{pc}] \simeq 1.08 \cdot 10^{-16} \frac{E\,[\mathrm{GeV}]}{B\,[\mathrm{T}]}. \qquad (10.33)$$

The general shape of the orbits of charged particles in a magnetic field results from the superposition of this circular motion with the component of translational motion parallel to the magnetic force lines, which is not affected by the field. The particles thus move on helical orbits along the lines of magnetic force.

The *geomagnetic* field has the effect that relatively low-energy charged particles can reach the Earth's surface only in limited zones around the geomagnetic poles. This latitude effect becomes negligible, as can be shown by setting r_C = Earth's radius and $B = 10^{-4}$ T, at energies above about 100 GeV.

When the cosmic rays enter the Solar System, they interact with the solar wind and with the interplanetary magnetic field, of order 10^{-8} T, which is swept out along with its plasma. It is therefore not surprising that the intensity of cosmic radiation within the Solar System shows a dependence on the 27-day cycle of the synodic solar rotation and on the 11-year cycle of solar activity.

In the *galactic* magnetic field of $2 \cdot 10^{-10}$ T, the cyclotron radius r_C for singly-charged particles is

$$
\begin{array}{llllll}
E = & 1 & 10^3 & 10^6 & 10^9 & [\mathrm{GeV}] \\
r_C = & 5\cdot 10^{-7} & 5\cdot 10^{-4} & 0.5 & 500 & [\mathrm{pc}]. \\
& (0.1\,\mathrm{AU}) & & & &
\end{array}
\qquad (10.34)
$$

Due to the deflection of the particles by the galactic (and for lower energies also by the interplanetary and the terrestrial) magnetic fields, their directional distribution is *isotropic* within the accuracy of the measurements. Therefore, the possibilities of learning something about their origins through radio astronomy and gamma-ray astronomy (Sect. 10.4.3) are all the more important. Only at the highest energies ($\geq 10^{20}$ eV) is it possible to determine the position of the source directly from the air shower.

Element Abundances. The nuclear charges Z of the cosmic-ray nuclei can be determined from the tracks they leave in nuclear emulsion plates, by using suitable arrangements of counters, or from their tracks in solid media, and thus their abundance distributions can be determined (Fig. 10.19). These are, on the one hand, relatively similar to the solar abundance distribution (Table 7.5); but on the other hand, they exhibit some noticeable differences: the light elements Li, Be, and B, which are exceedingly rare in stars, have nearly the same abundances in cosmic radiation as the following heavier elements. Also, the abundance minimum from Sc to Mn is "filled in" in the cosmic radiation. These overabundances are caused by *spallation*, i.e. as a result of the splitting of heavier nuclei, especially C and Fe, by collisions with protons and α-particles in the interstellar medium; these are highly energetic in the rest system of a cosmic-ray particle. The interaction cross-sections for these spallation reactions can be measured with large particle accelerators up to energies of about 10^3 GeV; they are of the order of the geometrical nuclear cross-sections. From the ratio of the numbers of the lightest and heaviest nuclei in the cosmic radiation, it can thus be calculated that the particles have passed through an amount of matter corresponding to about 50 kg m^{-2} or 5 g cm^{-2}. This number should however be regarded as an upper limit, since nuclei have been detected up to the end of the Periodic Table. Such "large" nuclei, and furthermore all the less energetic nuclei, must have passed through only a much smaller amount of matter.

On the basis of suitable assumptions about the distribution of the layers of interstellar matter through which the cosmic rays have passed, and using measured or calculated interaction cross-sections for the production and annihilation of the energetic nuclei, M. M. Shapiro and coworkers (1972) were the first to draw some conclusions about the chemical composition of cosmic rays at their points of origin. The abundance distribution of the elements in cosmic radiation at their point of origin is, within plausible error limits, the same as the abundance distribution in the Sun and other normal stars (Table 7.5), perhaps with the exception of the neon isotopes and "ultraheavy" nuclei of $Z > 26$ (overabundance in the Pt region). Before we follow up this important conclusion, we first consider the energy density of cosmic rays.

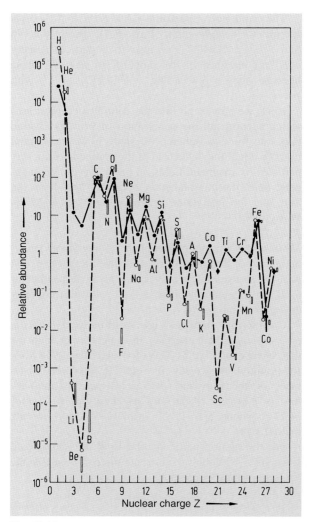

Fig. 10.19. The abundance distribution of nuclei in the cosmic radiation (in the energy range from about 0.1 to 0.3 GeV per nucleon) from observations (●) made with the IMP-8 satellite by the group at the University of Chicago (P. Meyer, 1981). For comparison, the relative element abundances in the Solar System (○) and in the stars and nebulae in our vicinity in the Milky Way (□) are indicated. The abundance distributions are set equal for C ($Z = 6$), and are plotted against the atomic number

Energy Density and Pressure. We start by comparing the energy densities u in our galactic vicinity for the most important cosmic-ray constituents in Table 10.1:

It would seem to be ruled out even by the 2nd Law of thermodynamics that in the Milky Way, roughly the

10.4 High-Energy Components

Table 10.1. Energy densities in the interstellar medium

	u [J m^{-3}]	u [eV cm^{-3}]
Cosmic radiation	$1 \cdot 10^{-13}$	0.7
"Thermal radiation", i.e. overall starlight	$5 \cdot 10^{-14}$	0.3
Kinetic energy $\varrho v^2/2$ of the interstellar matter (10^6 protons m^{-3} at 7 km s^{-1})	$4 \cdot 10^{-14}$	0.2
Galactic magnetic field $B^2/(2\mu_0)$ (with $B = 2 \cdot 10^{-10}$ T)	$2 \cdot 10^{-14}$	0.1

same amount of energy would be released in the form of extremely nonthermal cosmic radiation as in the form of thermal radiation. Indeed, the orbits of the charged cosmic-ray particles are "wound up" by the galactic magnetic field $\boldsymbol{B}$; i.e. they move $\perp \boldsymbol{B}$ on cyclotron orbits whose midpoints carry out a translational motion $\parallel \boldsymbol{B}$. The particles are thus *stored* in the Milky Way on these helical orbits. According to (10.34), we would expect that this mechanism would be effective up to energies of about 10^8 GeV. From the amount of matter the cosmic rays have passed through, about 50 kg m^{-2}, we can estimate their path length, using an average density of $2 \cdot 10^{-21}$ kg m^{-3} for the interstellar matter, and find 800 kpc or $2 \cdot 10^6$ light years, corresponding to a lifetime of roughly $2 \cdot 10^6$ yr. Compared to particles which are not deflected, e.g. photons, which have paths in the galactic disk of the order of 300 pc, this means an average increase in the path length by a factor of $2 \cdot 10^3$. Recent determinations of the abundances of long-lived isotopes in the cosmic radiation have however indicated a longer storage time of about $2 \cdot 10^7$ yr, meaning that the particles could have spent most of their time outside the galactic disk. The end of the "history" of a cosmic-ray particle is in general not determined by nuclear collisions, but instead by its eventual escape from the galaxy.

The energy densities given in Table 10.1 correspond in each case to about the same *pressure* ([J m^{-3}] = [N m^{-2}]). The order-of-magnitude equality of the magnetic pressure and the turbulence pressure in interstellar matter seems plausible in the light of magnetohydrodynamics. The approximate equality of the pressure of the cosmic rays to the magnetic and turbulence pressures can be understood by considering that cosmic radiation collects, "frozen in" by the galactic magnetic field, until its pressure is sufficient to allow it to escape from the disk of the Milky Way, probably taking with it a certain amount of interstellar matter and its associated magnetic field.

Origins. From the abundances of certain isotopes, which are produced in meteorites by spallation processes, we know that the intensity of cosmic rays has remained practically constant over a period of time of at least 10^8 yr. Since on the other hand the average dwell time of a cosmic-ray particle in the disk of the Milky Way is only $2 \cdot 10^6$ yr, the cosmic radiation must be continually "regenerated". Its sources are to be searched for in highly turbulent plasmas with magnetic fields, whose energy densities $B^2/(2\mu_0)$ adjust themselves to correspond roughly to the kinetic energy density $\varrho v^2/2$. These locations correspond precisely to the strong, nonthermal radio emission regions (A. Unsöld, 1949).

This connection is emphasized by the electron component of cosmic radiation; its energy distribution obeys a power law, $n(E) \propto E^{-p}$ (see above). As will be shown in Sect. 12.3.1, relativistic electrons produce *synchrotron radiation* as a result of their motion in magnetic fields: nonthermal radiation with a spectrum of the form $I_\nu \propto \nu^{-\alpha}$. The two exponents p and α are related by the equation $p = 2\alpha + 1$ (12.27). The exponent $p \simeq 3.0$ (or $q \simeq 2.0$) of the cosmic electron component would thus lead to a radio spectrum with $\alpha \simeq 1$, in sufficiently good agreement with the values found for the galactic synchrotron radiation ($\alpha \simeq 0.7 \ldots 1.0$).

It was uncertain for a long time whether the electron component of cosmic radiation was a secondary product from the interaction of protons with the interstellar medium, through the decay chain π meson $\rightarrow$ muon μ (penetrating component) $\rightarrow$ electron; or whether it is produced by *direct* acceleration of electrons *together* with the nucleons of the plasma. In the former case, somewhat more positrons, e$^+$, would be produced than electrons, e$^-$; while in the latter case, the electrons would strongly predominate. The measured ratio e$^-$/e$^+ \simeq 10$ (in the energy range ≥ 10 GeV) indicates that the electron component of the cosmic radiation in the Milky Way is produced by the same sources as the nucleon components.

Which celestial objects are these sources? The abundance distribution of the elements in the sources of cosmic radiation indicates that its matter was initially formed in the interiors of stars by the nuclear processes that take place there. We have already seen that the Sun produces cosmic rays and synchrotron radiation. But all together, the solar flares and the flare stars make only a tiny contribution to the galactic cosmic radiation.

I. S. Shklovsky, V. L. Ginzburg and others have therefore pointed out the importance of the supernovae. In particular, their expanding remnants (Sect. 10.3.3), as well as their remnant stars, the pulsars (Sect. 7.4.7), can accelerate considerable amounts of matter to high energies, as has been demonstrated by the observed synchrotron radiation.

According to our current ideas, the cosmic radiation up to energies of about 10^{18} eV has *galactic* origins. Up to 10^{15} eV, i.e. up to the "knee", it is accelerated by the Fermi mechanism, especially in the shock-wave fronts of the *supernova remnants* of our galaxy; the pulsars give a contribution mainly to the lower-energy particles. That in fact the supernova remnants can be considered as possible sources for such energetic particles has been supported by the observation of non-thermal X-radiation, requiring synchrotron electrons with energies of 10^{14} eV, in the remnant of the supernova from the year 1006 by the ASCA satellite.

Furthermore, the extremely hard gamma radiation from some X-ray binary star systems (Sect. 7.4.6), such as Cyg X-3, indicates that here, we are probably observing a source of highly energetic particles. The contribution of all these objects to cosmic rays, about which only rather uncertain estimates can be made, is in any case of the correct order of magnitude.

The cosmic radiation above about 10^{15} eV is probably also from the Milky Way galaxy, but here we know little about its source. The acceleration of these particles may occur in the multiple, mutually interacting shock-wave fronts of young supernova remnants in the neighborhood of the OB stellar associations, or in extended shock-wave fronts in the galactic halo, or, distributed over a large area, in the magnetic fields of the general interstellar medium. A contribution from the region of the galactic nucleus of the Milky Way, by contrast, probably does not play an important role, since observations of other, similarly structured galaxies show no correlation between the activity of their nuclei and the strength of the non-thermal radio emissions from their disks.

A further source of synchrotron electrons and cosmic rays has been found, in particular by radio-astronomical observations, in the *activity* of galactic nuclei and of quasars (Sect. 12.3.6). This might offer the solution to the riddle of the *most energetic particles* of the cosmic radiation, with energies up to 10^{20} eV. Since the cyclotron radius at 10^{18} eV is already comparable to the thickness of the galactic disk, and at 10^{20} eV, it corresponds to the dimensions of the whole galactic system, such particles can no longer be stored. They are therefore reduced in number relative to the softer particles by a factor of $2 \cdot 10^3$ in the Milky Way system. This suggests that in the energy spectrum of cosmic rays, the *extragalactic* contributions of distant galaxies (galactic nuclei) become increasingly important at the highest energies. This idea gains support from the observed isotropy of even the most energetic particles, although they are deflected only slightly by magnetic fields. Above the "ankle" at around $3 \cdot 10^{18}$ eV, not only does the slope of the spectrum change, but also the composition of the cosmic radiation: we find here practically only protons and helium nuclei.

The extragalactic sources have not been individually identified. Possibilities are the acceleration of the particles in "hot spots" at the ends of the relativistic jets from active galaxies (Sects. 12.3.3, 4), or also acceleration in shock-wave fronts which occur when matter falls into galaxy clusters (Sect. 12.4.3).

In the case of extragalactic particles with energies above 10^{19} to 10^{20} eV, we expect a noticeable reduction of their flux due to interactions with the cosmic microwave background radiation (Sect. 13.2.2), so that these particles can reach us only when they originate from sources within a few 10 Mpc away.

Significant contributions to the area of the origin of cosmic radiation are also made by the investigation of gamma radiation, which is closely connected with the observation of cosmic radiation; it has the advantage, compared to the latter, that it can give us information about the direction of the sources of even the low-energy cosmic radiation.

10.4.3 Galactic Gamma Radiation

Radiation in the gamma-ray range from the whole of the Milky Way can reach us without interstellar extinction and can be observed from satellites. The physical processes which generate gamma radiation were already introduced in Sect. 4.5.6.

10.4 High-Energy Components

Sky Surveys. In the survey charts of the gamma-ray region, particularly at energies ≥ 100 MeV, the emissions of the *Milky Way* predominate as a narrow band, only a few degrees wide, around the galactic equator (Fig. 10.20). This first became clear in the period from 1975–1982 through observations with the COS-B satellite. The intensity exhibits several maxima which seem to be correlated with spiral arm structures; within $|l| \leq 40°$ from the galactic center, we find especially strong emissions.

At the beginning of the 1990's, surveys were then carried out with the greatly improved sensitivity, positional accuracy and angular resolution of the instruments on board the Compton Observatory, CGRO. With the Compton telescope, first studies were made in the low-energy range from 1 to 30 MeV, while the high-energy telescope EGRET (*E*nergetic *G*amma-*R*ay *E*xperiment *T*elescope) carried out observations in the energy range above 100 MeV (Fig. 10.21).

Point Sources. At first view, a number of "point-like" galactic gamma-ray sources stand out above the background. While COS-B, owing to its limited angular resolution, could resolve only about 30 such sources, the chart obtained with EGRET permits us to distinguish more than 130 point sources, even at higher galactic latitudes. However, since even the instruments of the Compton Observatory permit – in the best cases (very bright sources) – a positional accuracy of only around $10'$, a definite identification with known objects is in general difficult and is possible for only a portion of the sources.

In a first survey, over 60 of the point sources, mostly outside the plane of the galaxy, could be identified as

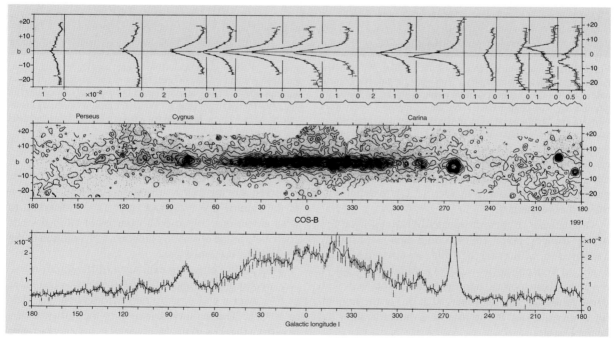

Fig. 10.20. The first picture of the Milky Way in the gamma-ray spectral region (70–5000 MeV), from observations with the gamma-ray telescope on the COS-B satellite by H. A. Mayer-Haßelwander et al. (1982). The measured countrates are shown; they are proportional to the gamma intensity. The error bars indicate the measured points and the smooth curve shows a fit to the data. *Center*: contour lines of constant gamma-ray emission intensity in galactic coordinates (l, b) (cf. Sect. 11.1.1). *Upper part*: Dependence of the intensity on the galactic latitude b, averaged over the range of galactic longitudes indicated. *Lower part*: The intensity along the galactic equator $b = 0°$, averaged over $|b| \leq 5°$. The strongest gamma source at $l \simeq 264°$ is the Vela pulsar PSR 0833-45 (Sect. 7.4.7)

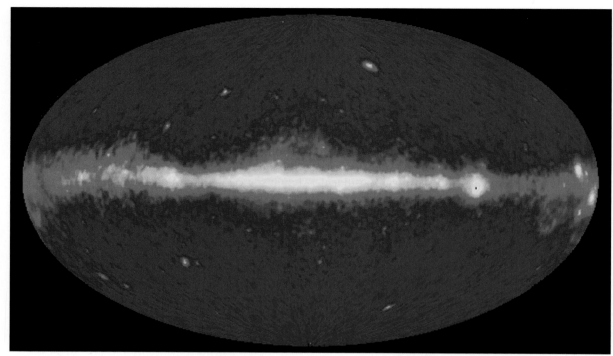

Fig. 10.21. A false-color image of the Milky Way in the gamma-ray region for energies ≥ 100 MeV, in galactic coordinates (Sect. 11.1.1); observations made with the high-energy gamma-ray telescope EGRET on the Compton Observatory (1993). The brightest sources in the galactic disk are *pulsars* such as Vela, Geminga and the Crab pulsar, while the point sources at high galactic latitudes are mainly extragalactic, variable *gamma blasars* (Sect. 12.3.4). The brightest of these gamma sources (*top*, somewhat to the right of the center) is 3C 279. (Image from: EGRET on the Compton Gamma Observatory, Max-Planck-Institute for extraterrestrial Physics/MPE Garching and NASA)

extragalactic objects (quasars). We will discuss their gamma emissions later in in Sect. 12.3.4, together with their other properties.

On the basis of their characteristic variability, a few of the strongest point sources could then be positively attributed to *pulsars* (as of 1998: seven), such as the Crab and Vela pulsars as well as Geminga (Sect. 7.4.7). Several other sources appear to be associated with dense *molecular clouds* (Sect. 10.2.3). For the remaining sources, about 70, up to now no identification has been possible.

In Sect. 7.4.8, we have already described the gamma emissions of some galactic binary star systems, and in Sect. 7.4.9, we treated the gamma bursters.

Diffuse Gamma Radiation. In addition to the point sources, a *diffuse* emission, concentrated towards the galactic disk, with a continuous intensity spectrum is observed. Owing to the still modest angular resolution of the gamma-ray telescopes, it is as yet uncertain to what extent unresolved point sources of various kinds contribute to the observed component, in addition to the genuine diffuse emissions.

This diffuse emission exhibits a (differential) intensity spectrum which approximately follows a power law, $\propto E^{-2}$, over a wide energy range from about 0.1 MeV to 10 GeV. The main contribution to the diffuse radiation from the disk of the Milky Way in the energy range from 1 MeV to 100 MeV is made by bremsstrahlung from relativistic cosmic-ray electrons in the fields of interstellar atomic nuclei; in contrast, above 100 MeV, gamma quanta accompanying the decays of π^0 mesons are predominant. The mesons are produced through inelastic collisions of cosmic-ray protons with protons of the interstellar medium. Inverse Compton scattering of cosmic-ray electrons by starlight in the Milky Way contributes only to the order of 10% within the disk, but its relative importance increases steadily with decreasing energy below about 1 MeV.

At higher galactic latitudes, on the other hand, the gamma radiation produced by inverse Compton scattering dominates that from the other two mechanisms, since the galactic radiation field is much less strongly concentrated in the disk than is the interstellar gas component.

Gamma-Ray Lines. Nuclear spectroscopy of gamma-ray lines which occur in connection with *nucleosynthesis* of the elements in the stars (Sect. 8.2.5) allows the direct detection of their production.

The first gamma-ray line from a radioactive product of nucleosynthesis in the Milky Way was detected in 1982: the 1.809 MeV line of ^{26}Al. This isotope, with a half-life of $7.4 \cdot 10^5$ yr, decays predominantly by emission of a positron to the first excited state of ^{26}Mg* at an energy of 1.809 MeV above the ground state. Its rather short lifetime, in comparison to the age of the Milky Way, gives direct evidence that even "today", nucleosynthesis processes are occurring within our galaxy.

With the imaging Compton telescope of the CGRO satellite it was possible for the first time in 1996 to make a chart of the entire Milky Way in the radiation of the 1.809 MeV line, with an angular resolution of around 2°. The emissions of this line occur mainly in the galactic disk and are strongly concentrated around the center of the galaxy as well as in in Cygnus, Carina and Vela. Their intensity distribution reflects in detail the projection of the distribution of young *massive* stars. With these recent observations, older stars, novae or other point sources, which had earlier been under discussion as possible sources of the 1.809 MeV emissions, can be eliminated. Instead, it is clear that the radioactive aluminum is produced in massive stars ($\geq 25 \mathcal{M}_\odot$), in the main as a result of proton capture reactions which occur during CNO burning at temperatures above $4 \cdot 10^8$ K: ^{25}Mg(p, γ)^{26}Al. It is then finally dispersed in the interstellar medium by stellar winds or ejected gas shells. Candidates for such stars are the massive stars on the asymptotic giant branch, Wolf-Rayet stars, or supernovae; their relative contributions are not yet known precisely.

In the case of *supernovae* (Sect. 7.4.7), a few gamma-ray lines have thus far been discovered using the CGRO, such as the 0.847 MeV line of ^{56}Co, which is produced in the decay of ^{56}Ni, from SN 1987A in the Large Magellanic Cloud, as well as lines of ^{57}Co at 0.122 and 0.136 MeV. From the *supernova remnant* Cas A (Sect. 10.3.3), the 1.156 MeV line of ^{44}Ti was detected; this isotope originated directly in the supernova.

Finally, we mention the emission of gamma-ray lines from the *molecular clouds* in Orion at 4.43 and 6.13 MeV, which can be attributed to the nuclides ^{12}C* and ^{16}O*; they can be raised to excited nuclear states by collisions with cosmic-ray particles.

10.5 Early Evolution and Formation of the Stars

After having considered in Sect. 8.2 the evolution of the stars from the main sequence through their final stages, and having described the properties of the interstellar medium in the earlier sections of the present chapter, we can now return to the starting point of our considerations of star formation or stellar genesis and ask the questions, "How do the stars evolve onto the standard main sequence?", and then: "How and from what are the stars formed?"

The close spatial conjunction of the young O and B stars of high absolute magnitudes with gas and dust clouds in the spiral arms of our own galaxy, and of others such as the Andromeda galaxy, leads us to the conclusion that, quite generally, stars are formed from – or in – cosmic clouds of diffuse matter. The only energy source which is initially available to a protostar, i.e. before the ignition of some sort of thermonuclear reaction, is the *gravitational energy* $E_G \simeq G \mathcal{M}^2/R$ (8.28) which is released by the contraction of the stellar mass $\mathcal{M}$ from originally widely-distributed matter down to a radius R.

Time Scales. The characteristic evolution time for the gravitational contraction phase is the *Helmholtz–Kelvin time* $t_{HK} \simeq G\mathcal{M}^2/(RL)$ (8.34); hydrostatic equilibrium (8.1) can be established in the (whole) star, of mean density $\bar{\varrho}$, when the dynamic time scale or *free fall time* is much shorter than the Helmholtz–Kelvin time:

$$t_{\text{ff}} \simeq \frac{1}{\sqrt{G\varrho}} \ll t_{\text{HK}} \qquad (10.35)$$

(8.41). Near the main sequence, this condition is very well fulfilled; e.g. for a mass of $\mathcal{M} = 1\,\mathcal{M}_\odot$, a luminosity $L = 1\,L_\odot$ and a mean density $\bar{\varrho} = 1400$ kg m^{-3}, we find $t_{\text{ff}} \simeq 3 \cdot 10^3$ s $\simeq 10^{-4}$ yr and $t_{\text{HK}} \simeq 2 \cdot 10^7$ yr. However, with increasing radius, $t_{\text{ff}} \propto R^{3/2}$ increases, while $t_{\text{HK}} \propto R^{-3}$ decreases (due to $L = 4\pi R^2 \sigma T_{\text{eff}}^4 \propto R^2$ (6.43)), if we keep the effective temperature constant for simplicity. At a radius of $R \geq 300\,R_\odot$, we thus see that $t_{\text{ff}} \geq t_{\text{HK}}$, and the hydrostatic equilibrium condition is no longer met.

In the evolution of *proto*stars whose radii are still very large, one must therefore apply *dynamic* model calculations, such as were first carried out by R. B. Larson in 1969, in which the hydrostatic equation (8.1) is extended by the addition of an acceleration term $\varrho\, dv/dt = \varrho\, d^2 r/dt^2$ to the Euler equation (7.87). Only when its evolution has reached the neighborhood of the main sequence can a *global* hydrostatic equilibrium be established in the pre-main-sequence star.

We begin in Sect. 10.5.1 with a discussion of the evolution and the properties of the pre-main-sequence stars. Before treating their earliest evolutionary phases, i.e. the protostars and stellar genesis, we give in Sect. 10.5.2 a summary of the observational data from the regions of star formation, and investigate in Sect. 10.5.3 the conditions under which a mass of interstellar gas can become gravitationally unstable and begin to form a star. In Sect. 10.5.4, we then consider the theory of evolution of protostars, which occurs on a short, "dynamic" time scale and depends sensitively on the amount of angular momentum originally present in the system. Sect. 10.5.5 deals with the various forms of matter flows, such as bipolar winds, jets and accretion disks, which are closely related to the evolution of protostars but are theoretically only poorly understood. Finally, in Sect. 10.5.6, we discuss the statistical distribution of luminosities and of masses of the stars, as well as their rate of formation.

10.5.1 Pre-Main-Sequence Stars

Stars which are in complete *hydrostatic equilibrium*, and whose only energy source is their gravitational energy, are termed pre-main-sequence stars. We first want to consider the path in the color–magnitude diagram which a star must follow as it is formed from matter which was originally very widely distributed in space, while always remaining in hydrostatic equilibrium. We then compare these results with observations of the T Tauri stars, which have been identified as pre-main-sequence stars.

Evolutionary Paths. Giant stars with effective temperatures below 3000 to 4000 K have essentially *convective* structures (Sect. 8.1.2); their interiors are occupied for the most part by extended hydrogen convection zones. C. Hayashi showed in 1961 that no hydrostatic equilibrium can be established in these stars. On the (theoretical) color–magnitude diagram there thus exists, for each mass, a line: a *Hayashi line* which is nearly vertical, i.e. it remains at constant effective temperature, and which separates the "forbidden zone" of the unstable stars (to the right) from the region of the stable stars (left).

A pre-main-sequence star in hydrostatic equilibrium forms an extended *convective* shell as soon as its hydrogen is partially ionized. It moves downwards in the color–magnitude diagram (Fig. 10.22) from higher luminosities, with a slightly increasing temperature, remaining just to the left of the Hayashi line, until (after 10^7 yr, for a mass of $1\,\mathcal{M}_\odot$) it reaches the nearly horizontal line representing radiation equilibrium (calculated earlier by L. G. Henyey and coworkers). On this line, it continues to fill its energy requirements through release of gravitational energy. Ignition of the well-known thermonuclear reactions takes place only shortly before it reaches the initial main sequence.

It is notable that stars can also not cross over the Hayashi line "in the opposite direction". Their evolutionary lines in the region of giant stars therefore lead steeply upwards in Figs. 8.7 and 8.8.

Towards the upper part of the main sequence, the contraction time t_{HK} becomes shorter and shorter, according to (8.34); for a B0 star, it is only about 10^5 yr. At the other end of the scale, stars of smaller mass are formed more slowly; an M star of mass $0.5\,\mathcal{M}_\odot$ requires about $1.5 \cdot 10^8$ yr. Contracting masses of less than about $0.1\,\mathcal{M}_\odot$, as we saw in Sect. 8.3.1, never reach the temperature (in equilibrium) which would be required for the ignition of hydrogen burning and simply form cool, completely degenerate objects, the brown dwarfs.

10.5 Early Evolution and Formation of the Stars

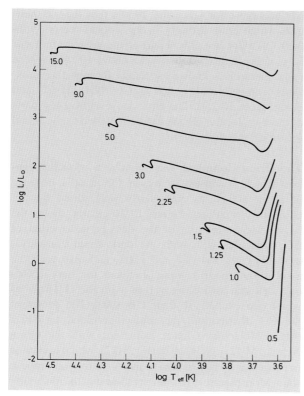

Fig. 10.22. The evolution of hydrostatic pre-main-sequence stars of various masses (0.5 to 15 $\mathcal{M}_\odot$) after I. Iben (1965). Stars of mass $\geq 1 \mathcal{M}_\odot$ require less than a few 10^7 yr to reach the main sequence

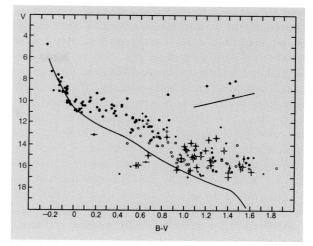

Fig. 10.23. The color–magnitude diagram of the very young open star cluster NGC 2264, after M. Walker (1956). • Photoelectric measurements; ○ photographic measurements; | variable stars; — stars with Hα emission. The apparent magnitudes V are plotted against $(B-V)$. The curves indicate the standard main sequence and the giant branch, corrected for uniform interstellar reddening of the cluster. The apparent distance modulus is 9.7 mag, and the distance is 800 pc

Young Star Clusters. In young open star clusters, which can be recognized by their bright blue stars, M. Walker indeed found stars of middle and later spectral types to the right of the lower part of the main sequence, whose T Tauri variability, Hα emission lines, and in part rapid rotation etc. identify them as young stars. Figure 10.23 shows, as an example, the color–magnitude diagram of the cluster NGC 2264. Calculating on the basis of its brightest star, this cluster has an age of only $3 \cdot 10^6$ yr. In agreement with theory, stars before spectral type A0 ($\leq 3 \mathcal{M}_\odot$) have not yet reached the standard main sequence.

T Tauri Stars. Above the main sequence, in the region of the F to M stars, we find the variable T Tau stars, which can be identified as pre-main-sequence stars with masses below about $3 \mathcal{M}_\odot$. Their spectroscopic properties and their remarkably strong emissions of Hα, Ca II H and K lines, their activity and their X-ray emissions were already described in Sect. 7.4.3. While in the case of the protostars the dense, "dusty" surroundings prevent observations in the optical region, the pre-main-sequence stars become optically visible for the first time in the course of their evolution as *classical* T Tau stars. For the most part, they are surrounded by a dust-filled accretion disk (Sect. 8.2.7), which is "left over" from their formation process and, depending on the direction of the observer, hinders more or less strongly the passage of radiation in the optical range. Their thermal emissions can be seen in the form of an infrared excess relative to the stellar radiation.

Most of the T Tau stars belong to binary star systems; some of them exhibit stellar winds with velocities of up to 100 km s^{-1} and mass loss rates around $10^{-7} \mathcal{M}_\odot \text{ yr}^{-1}$.

During their continuing evolution, their activity decreases and the accretion disk gradually disperses; this can possibly lead to the formation of a planetary system. These "bare" T Tau stars have fainter emission lines than the classical T Tau stars; the equivalent width, e.g. of the Hα line, is ≤ 1 nm. Their X-ray emissions are, in contrast, stronger than those of the classical objects and are correlated with their rotational velocities. They probably originate from arches of magnetic field lines, similar to those we have seen on the Sun, that are formed as a result of "winding up" by the differential rotation.

It is often difficult to distinguish the T Tau stars with faint lines from main-sequence stars. A criterion indicating these young stars is a strong absorption line in the resonance transition of Li I at $\lambda = 670.8$ nm, since lithium is mixed into the core of older stars in the course of their evolution and there, already at temperatures of $2.5 \cdot 10^6$ K, is destroyed by proton capture reactions.

The *Herbig Ae and Be stars* (A and B stars with emission lines), which are related to the T Tauri stars, correspond to the somewhat more massive pre-main-sequence stars of 3 to 8 $\mathcal{M}_\odot$. Pre-main-sequence stars of masses above about $8\mathcal{M}_\odot$, in contrast, cannot be observed in the optical region (Sect. 10.5.4).

10.5.2 Regions of Star Formation

Stability considerations, as we shall see in Sect. 10.5.3, indicate that initially, only relatively large masses of 10^2 to $10^5\,\mathcal{M}_\odot$, which are much greater than observed stellar masses, can condense from the interstellar medium. Indeed, on the basis of observations, V. A. Ambarzumian in 1947 had already arrived at the important conclusion that the stars (at least for the most part) are formed over periods of time of the order of 10^7 yr (Sect. 9.1.2) in groups, with overall masses of $10^3\,\mathcal{M}_\odot$. These are the OB associations, with bright blue stars, and the T associations, with cool stars, especially T Tauri variables of low absolute magnitudes (frequently, both kinds occur together). An OB association, such as those in Orion or Monoceros, with enormous masses of gas ionized by the short wavelength radiation of the O and B stars embedded in it, appears to the observer in the optical region at first as an H II region, dominated by emission of strong Hα radiation. The stars of an association move away from its center with velocities of the order of 10 km s^{-1}; associations are thus not stable. The *expansion age*, obtained by extrapolation backwards of the stellar motions, agrees in general with the evolutionary age of the brightest stars.

As a result of progress in radio astronomy in the millimeter-wave region, and in infrared astronomy, it has become clear since the 1960's that the gas nebulae which emit strongly in the optical region, with their OB stars, are only a part of the gigantic molecular cloud complexes with masses above $10^5\,\mathcal{M}_\odot$. These make themselves known especially through line emissions from molecules such as CO, NH$_3$, H$_2$CO and others (Sect. 10.2.3). The most dense and coolest parts of these molecular clouds can be regarded as the actual *locations of stellar genesis*.

The Orion Complex. The major area of stellar genesis which is closest to us, in Orion at a distance of about 500 pc, is particularly well suited for detailed investigations. Figure 10.24 clearly shows the large-scale correlation of the density distribution of CO molecules with both the dark cloud complexes around Lynds 1640/41 and 1603, and also with the association Ori OB 1, which includes the stars in Orion's belt and sword, as well as with the H II regions which are excited by their OB stars. The center of the CO distribution, the molecular cloud Ori MC 1 (MC = Molecular Cloud), coincides with the Orion Nebula, M 42 = NGC 1976 (Fig. 10.7), which is excited in particular by two (θ^1 Ori C and A) of the four trapeze stars. More precise observations show that the molecular cloud as seen from our position lies directly behind the Orion Nebula. In it, several "point-like" infrared sources can be seen, among them the Becklin–Neugebauer object and the Kleinmann–Low object, named for their discoverers; the luminosities of these objects are several $10^3\,L_\odot$ and their surface temperatures are a few 100 K. These infrared sources form a cluster of very young protostars (see Sect. 10.5.4), with altogether about $10^5\,L_\odot$, which is however not observable in the optical region due to dense dust clouds.

Closely related to the formation of young stars in the Orion complex are a variety of interesting phenomena: we thus find emission lines of H$_2$ from excited vibrational states, which originate in layers at about 2000 K and are probably excited in shock wave fronts. We also observe point sources of intense microwave radiation with extreme excitation conditions, i.e. masers based on transitions in OH as well as in H$_2$O and SiO (Sect. 10.2.3). Furthermore, the line profiles in the optical as well as the infrared and radiofrequency regions indicate surprisingly high velocities up to several 100 km s^{-1} for the gas in the Ori MC 1 region. All

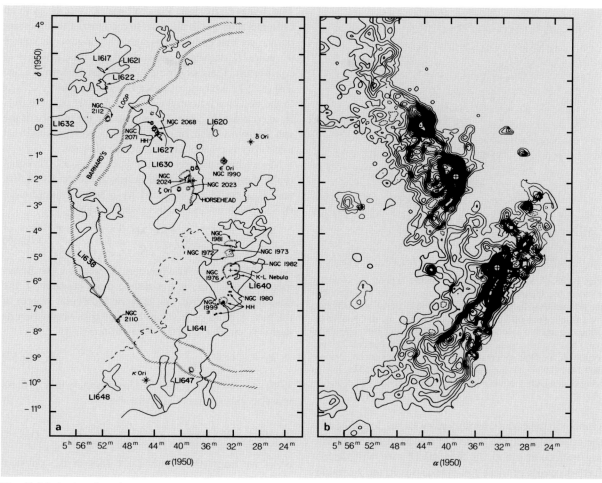

Fig. 10.24a,b. A region of star formation in Orion. (**a**) A sketch of the most noticeable structures in the visible region. *Dashed line*: The boundaries of the emission and reflection nebulae (NGC 1976 = M 42 = Orion Nebula); *Full curves*: Boundaries of the dark dust clouds with the notation of B. T. Lynds (L); HH denotes Harbig–Haro objects; δ, ε and ζ Ori are stars in Orion's belt. The position of the protostar cluster around the Becklin-Neugebauer and Kleinmann–Low objects is indicated as "K-L Nebula". (From M. L. Kutner et al., Astrophys. J. **215**, 521 (1977); reprinted courtesy of the University of Chicago Press, the American Astronomical Society and the authors). (**b**) Intensity contours of the integrated line emissions from the rotational transition $J = 1 \rightarrow 0$ at $\lambda = 2.6$ mm in the CO molecule. From the Goddard-Columbia Sky Survey with a 1.2 m telescope of antenna acceptance angle $8'$. P. Thaddeus, Ann. New York Academy of Sciences **395**, 9 (1982). (With the kind permission of the publisher)

these observations together can be explained in terms of a massive "protostellar wind", which originates in the infrared sources around the Becklin-Neugebauer object and streams away in two oppositely-directed cones. The collimation of this gas flow is probably effected by a thick, dust-filled gaseous disk in the vicinity of the protostars, which also prevents observations in the optical region (Fig. 10.25).

Such bipolar gas flows, with differing degrees of collimation, as well as gas/dust disks, are observed not only in Orion, but also near to almost all young stars; they occur as a characteristic accompanying phenomenon, be-

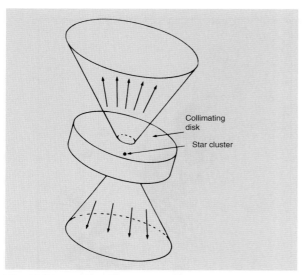

Fig. 10.25. A schematic drawing of the bipolar protostellar wind in the Orion Nebula, which originates in the star cluster around the Becklin–Neugebauer object and is collimated by a thick, dust-filled disk of gas. The axis of the disk is inclined in such a way that the flow in the upper cone (NW part) has a component directed towards the observer, giving a blue shift of the spectral lines. From B. Zuckerman, Nature **309**, 403 (1984). (Reprinted by kind permission of Macmillan Magazines Ltd., London, and of the author)

ginning in the earliest evolutionary stages, together with the collapse of matter onto the star from the surrounding molecular cloud. In Sect. 10.5.5, we will discuss further details of these flow processes in the surroundings of young stars in connection with their evolution. First, however, we describe the theoretical approaches to stellar genesis and to the evolution of protostars.

10.5.3 Gravitational Instability and Fragmentation

We now investigate the far-reaching question, not limited to star formation, of the conditions under which a mass of gas distributed over a region of space can become unstable, so that it is compressed under the influence of its own gravitational field. This process is followed, as we shall see, by a further splitting up of the matter, resulting in the formation of individual stars. Our question can be answered by the criterion of *gravitational instability* discovered by J. Jeans (1902, 1928).

The Jeans Criterion. We shall restrict ourselves to an estimate, which will allow us to recognize the important points; we first consider a roughly homogeneous sphere of radius R, density ϱ and mass $\mathcal{M}$. If this sphere is in equilibrium, then according to the virial theorem ((2.36) and (8.29)), the ratio of twice the kinetic energy, $2E_{\text{kin}}$, to the negative potential energy, $-E_{\text{pot}}$, is equal to *one*. If, in contrast, E_{kin} (i.e. the pressure in the interior of the sphere) is too small, or $-E_{\text{pot}}$ is too large, a gravitational instability is present and the mass collapses.

As we have already seen in Sect. 8.1.4, E_{kin} is equal to the thermal energy of the atoms or molecules (8.30), and E_{pot} could be obtained by a simple integration, (8.31). We thus find for the *stability limit*:

$$\frac{2E_{\text{kin}}}{-E_{\text{pot}}} = \frac{3kT}{\mu m_{\text{u}}} \mathcal{M} \bigg/ \frac{3}{5} \frac{G\mathcal{M}^2}{R} = 1 \ . \tag{10.36}$$

As usual, G is the gravitational constant, k the Boltzmann constant, m_{u} is the atomic mass constant, T the absolute temperature, and μ is the molecular weight of the gas. A mass of gas can thus collapse only when its radius is *smaller* than the *Jeans radius*:

$$R_{\text{J}} = \frac{1}{5} G \mathcal{M} \frac{\mu m_{\text{u}}}{kT} \ . \tag{10.37}$$

With $\mathcal{M} = (4\pi/3)\varrho R^3$ we find immediately that its mass must be *greater* than the *Jeans mass*

$$\mathcal{M}_{\text{J}} = 5.46 \left(\frac{kT}{\mu m_{\text{u}} G} \right)^{3/2} \varrho^{-1/2} \ , \tag{10.38}$$

or, if we insert the numerical values of the physical constants,

$$\frac{\mathcal{M}_{\text{J}}}{\mathcal{M}_\odot} = 3.82 \cdot 10^{-9} \left(\frac{T\,[\text{K}]}{\mu} \right)^{3/2} \\ \times \left(\varrho\,[\text{kg m}^{-3}] \right)^{-1/2} \ . \tag{10.39}$$

If the gas is also subject to turbulent flow, we could insert simply the mean squared velocity $\langle v^2 \rangle$, obtained, for example, from the Doppler effect of a spectral line, in place of $3kT/(\mu m_{\text{u}})$.

Our approach is still not very satisfying, because in reality we cannot assume an initially spherical distribution, but instead must consider a more or less inhomogeneous large mass of gas. This can be taken into account by exerting an (imaginary) pressure on the surface of the sphere. That will favor its collapse and in many cases cause it to begin sooner. Equations (10.38) and (10.39) then are slightly modified, the values of the constants being somewhat changed.

As an important application, we investigate instabilities of the *interstellar gas*. We choose its density ϱ to be in the range 10^{-21} to 10^{-18} kg m^{-3} and its temperature in the range 100 to 20 K (Sect. 10.2), and obtain from (10.39) a Jeans mass between about 10^5 and 10^2 $\mathcal{M}_\odot$. This means that the gas in the disk of a galaxy can initially form an object only of the order of the mass of a *star cluster*. Only when the matter is further compressed can the formation of individual stars begin. The collapse of a mass of gas is favored by an external pressure, as is produced for example by the shock wave projected out from a supernova, or by a density wave in a spiral arm (Sect. 11.2.5).

Regarding many other problems, such as the formation of galaxies, of planetary systems, etc., we mention the following point: the *higher* the density ϱ of the original matter, the *smaller* will be the objects formed from it.

Fragmentation. We know only very little about the process of fragmentation, i.e. how within a few 10^5 yr, the observed stellar masses of 0.1 to 10^2 $\mathcal{M}_\odot$ are separated out of the molecular clouds which condense from the interstellar medium with large masses of 10^2 to 10^4 $\mathcal{M}_\odot$. Observations of the protostars show that as end products of fragmentation, mainly (around 60%) binary or multiple systems are formed. Along with the direct fragmentation of a diffuse molecular cloud into stellar masses, this process also probably – in the case of rapidly rotating clouds – proceeds via the formation of an accretion disk, which later breaks up into individual masses.

Fragmentation also involves the conversion of a considerable part of the angular momentum of the collapsing molecular cloud into *orbital* angular momentum of the fragments, and thus favors star formation.

10.5.4 The Evolution of Protostars

We now ask the question of how a star, following its formation from compressed regions of the interstellar medium, can evolve into the (hydrostatic) pre-main-sequence stage. We have already estimated that for these earliest evolutionary phases, dynamic time scales (the free fall time) are important. We first consider the simple spherically symmetrical case, which already exhibits essential features of protostar evolution. For example, the evolution of low-mass stars is clearly different from that of the more massive stars. We then discuss the role of angular momentum, which cannot be neglected in a realistic discussion of the early phases of stellar evolution.

Spherical Collapse. Since we cannot calculate the fragmentation process of larger masses into stellar masses, we follow R. B. Larson and begin the evolutionary calculations by assuming a homogeneous sphere of molecular hydrogen and dust to have a stellar mass, and choose its density and temperature to keep it just gravitationally unstable according to the Jeans criterion (10.39). In this process, we orient our considerations to the physical properties of the denser parts of cool molecular clouds (Sect. 10.2.3). For $T \simeq 10$ K and $\varrho = 10^{-16}$ kg m^{-3} or $n = 10^{11}$ m^{-3}, masses of ≥ 1 $\mathcal{M}_\odot$ are unstable. The radius of a sphere of 1 $\mathcal{M}_\odot$ with $\varrho = 10^{-16}$ kg m^{-3} is $R = 1.7 \cdot 10^{15}$ m $\simeq 10^4$ AU $\simeq 0.05$ pc, and the corresponding free fall time (10.35) is $t_{\text{ff}} = 4 \cdot 10^5$ yr.

Already within a time of the order of t_{ff} after the beginning of the collapse, within the originally homogeneous protostellar cloud of mass $\mathcal{M}$, a *core region* of mass ≤ 0.01 $\mathcal{M}$ has formed which is many orders of magnitude denser. The collapse of the core is halted when it becomes optically dense as a result of the efficient absorption by dust particles of the radiation which it emits, mainly in the infrared. The impact of matter which continues to fall from the outer regions of the cloud onto the surface of the core produces a shock wave front, which, together with the slow gravitational contraction of the core, causes a continual rise in its temperature. When $T \geq 2000$ K, the H$_2$ dissociates and the effective adiabatic exponent Γ drops below 4/3. According to (8.39), in the innermost part a (second) collapse then takes place, lasting only a few months or years, owing to the high densities; it stops when the density has reached 1 to 10 kg m^{-3} and the temperature is about 10^4 K. This leads to the formation of a small core of mass $\leq 10^{-3}$ $\mathcal{M}$. After some time, when the remaining material from the first core has fallen onto the second, we have a protostar with the following *characteristic structure*: onto a dense core region of mass $\mathcal{M}_c$

and radius R_c in hydrostatic equilibrium, the matter of the extended, much less dense gas shell drops practically in free fall; it is braked at the surface of the core by a shock wave front. The kinetic energy released in this process, $\dot{M} G \mathcal{M}_c / R$ ($\dot{M}$ = rate of mass fall onto the core) supplies nearly the total luminosity of the protostar. In the course of time, $\mathcal{M}_c$ increases continually, while $\dot{M}$ decreases; for example, for 1 $\mathcal{M}_\odot$, after about 10^5 yr, the mass of the core is $\mathcal{M}_c \simeq 0.6 \, \mathcal{M}_\odot$ and $\dot{M}$ has dropped from an initial value of some 10^{-3} to a few $10^{-6} \mathcal{M}_\odot$ yr^{-1}.

Low-Mass Stars. In the less massive stars ($\mathcal{M} \leq \mathcal{M}_\odot$), the entire shell falls within a relatively short time onto the core; e.g. for 1 $\mathcal{M}_\odot$, after 10^6 yr practically the whole mass of the protostar is contained in the core. With decreasing $\dot{M}$, the contribution of the kinetic energy to the luminosity also decreases. When it becomes negligible compared to the contraction energy $G \mathcal{M}_c^2 / R_c$ of the core, the protostar has become a pre-main-sequence star, and its evolutionary path on the Hertzsprung–Russell diagram merges into a "hydrostatic curve" (Fig. 10.26).

In their early evolutionary stages, the protostars are unobservable in the optical region due to their dust-filled shells; however, they can be detected as infrared sources or in the adjacent submillimeter and millimeter wavelength ranges. Only in the transition stage between the dynamic and the hydrostatic phases can we see the young stars in the optical region as T Tauri stars (Sect. 10.5.1). Some of them, the YY Ori stars, evidence the falling matter directly in their spectra as absorption components shifted towards longer wavelengths (inverse P Cygni profiles).

Massive Stars. The Helmholtz–Kelvin time for the contraction of the hydrostatic core of the more massive stars of $\mathcal{M} \geq 3 \, \mathcal{M}_\odot$ is *shorter* than the free-fall time of the shell, in contrast to the less massive stars. Therefore, in these protostars, hydrogen burning begins already while a considerable portion of the protostar mass is still falling onto the core; they thus never pass through the hydrostatic phase of a pre-main-sequence star (Fig. 10.22). Hydrogen burning causes a strong increase in the luminosity of the core; the resulting radiation pressure, which acts in particular on the dust particles in the shell, slows the falling matter until finally its motion is reversed. The star casts off its shell, and with it, a major portion of its original mass; for example, a protostar of originally 60 $\mathcal{M}_\odot$ becomes a main sequence star of only 17 $\mathcal{M}_\odot$.

Young massive stars are thus observable before they enter the main sequence as relatively cool infrared sources of very high luminosities, whose spectra are typified by their dense dust shells (cocoons). Characteristic temperatures (of the dust) are in the range 100 to 600 K, and the luminosities are 10^3 to 10^6 $L_\odot$. An example of a massive protostar is the Becklin–Neugebauer object in Orion, which radiates mainly in the region from 3 to 10 μm and shows strong absorption structures at 3.1 μm (ice) and at 9.7 μm (silicates).

Angular Momentum. Along with the mass, the initial angular momentum J_0 of a collapsing cloud is decisive for its evolution. In fact, we almost never observe a spherically symmetric distribution of the matter in the near vicinity of young stars.

The net angular momentum, which is more or less randomly determined for a particular molecular cloud during its formation, is typically about two orders of

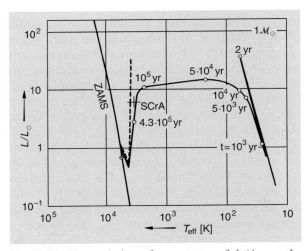

Fig. 10.26. The evolution of a protostar of 1 $\mathcal{M}_\odot$ on the Hertzsprung–Russell diagram, with times given in years since the formation of the hydrostatic core. Shortly before the zero-age main sequence (ZAMS), the evolutionary path merges into a Hayashi line (*dashed curve*) for fully convective stars in hydrostatic equilibrium. The position of the T Tauri star S CrA is indicated by a *cross*

magnitude larger than the value which a compact stellar mass can have without flying apart from too-rapid rotation. Conversely, a star can form from an extended cloud only if it succeeds in getting rid of 99% of the original angular momentum during the collapse, either by transport to the outer regions of an accretion disk, or by conversion into orbital angular momentum of a companion.

For small values of J_0, the spherically symmetrical models described above represent good approximations, but numerical calculations for large J_0 show that instead of a dense core, a rotating ring-shaped density distribution is at at first formed, with the angular momentum retained in that part of the cloud. Presumably, the ring later breaks up into several fragments, so that in this way, binary stars and multiple systems can be formed. However, if the angular momentum is transferred from the inner part of the protostellar cloud to the outer parts, e.g. by viscosity in turbulent flow patterns, then a starlike core is formed, surrounded by a rotating disk.

This kind of model can also describe the fundamental aspects of the formation of our Solar System. In the Solar System, over 99% of the total angular momentum is stored in the orbital motions of Saturn and Jupiter, while the mass is nearly completely concentrated in the Sun (Sect. 14.1).

The theory of the evolution of protostars, taking their angular momentum into account, is still in its early stages and cannot yet quantitatively describe the great variety of matter flows (Sect. 10.5.2) observed in star-formation regions.

10.5.5 Matter Flows in the Vicinity of Protostars

An indirect indication of disks in the immediate neighborhood of protostars or young stars is given by the outward flows of gas with bipolar characteristics observed in many cases (Figs. 10.25 and 10.27). Observations with improved sensitivity and angular resolution since the 1980's have shown ever more clearly that star formation and the early evolutionary stages are closely connected not only with *in*flows of matter, but at the same time with marked *out*flows.

Before we describe the circumstellar accretion disks, from which matter can fall into the star, we turn first to some of the more remarkable phenomena connected to the *out*flow of matter from the protostars.

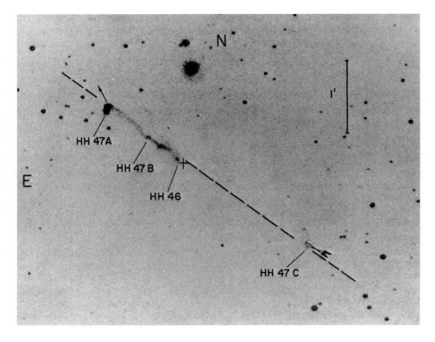

Fig. 10.27. The bipolar mass outflow from an infrared object of $12\,L_\odot$: Herbig–Haro objects HH 46/47 and jets in the globule ESO 210-6A, of diameter 0.4 pc and mass $25\,\mathcal{M}_\odot$. This image was made using the light of the forbidden [S II] transitions at $\lambda = 671.6$ and 673.1 nm by J. A. Graham und J. R. Elias in 1983. The distance to the object is 400 pc, and $1'$ corresponds to 0.12 pc. The central infrared source (+) is obscured in the optical region by dust absorption of $A_V \simeq 20$ mag. Measurements of the proper motions of the HH objects were made by R. D. Schwartz et al. (1984). The arrows show the distance covered in 500 yr

Bipolar Molecular Outflows. Observations of CO and other molecules in the millimeter wavelength range show – not only for massive, luminous objects, but also for less massive stars with luminosities only a few times that of the Sun – energetic protostellar or stellar winds, which flow out into the gas surrounding the star and "collect" it. The matter flows in two opposite directions away from the central object, which itself is rendered invisible in the optical region by a dust-filled accretion disk, but can be observed in the infrared; the axis of the double cone is perpendicular to this disk. The mass loss typically lies between 10^{-8} and $10^{-5}\,\mathcal{M}_\odot\,\mathrm{yr}^{-1}$.

Often, weakly focused flows with velocities up to $20\,\mathrm{km\,s}^{-1}$ occur together with strongly collimated, faster flows with velocities of the order of $100\,\mathrm{km\,s}^{-1}$. The faster component, in particular, often shows an irregular, lumped density distribution. These flows involve a variety of masses in the range of less than $1/100$ up to about $100\,\mathcal{M}_\odot$; their range is up to several pc.

Jets and Herbig–Haro Objects. Frequently, a portion of the bipolar outflow is very strongly collimated to yield jets, which stream out from the protostar in opposite directions into the surrounding denser medium at several $100\,\mathrm{km\,s}^{-1}$, i.e. with velocities well above that of sound, and "bore into" it. In this process, shock waves are formed, which ionize the gas in certain regions and excite it to luminosity. In 1951/52, G. Herbig and G. Haro discovered luminous, small nebula patches that accumulate in the near neighborhood of the young stars. The emission line spectra of these Herbig–Haro objects (HH), of which currently several hundred are known, show (especially through the high intensities of their [S II] lines at $\lambda = 671.6/673.1$ nm) excitations corresponding to around 10^4 K from shock wave fronts. These remarkable objects are thus "nodes" of particularly high energy dissipation in the gas which is flowing out at supersonic speeds. Frequently, an especially bright Herbig–Haro object, such as HH 47A (Fig. 10.27) is found at the end of a jet; it probably indicates the "bow shockwave" resulting from the collision of the supersonic jet flow with the matter of the surrounding molecular cloud.

In comparison with the Herbig–Haro objects, the long (several 0.1 pc up to 1 pc), thin jets are themselves not very noticeable. They can be observed in the optical and to some extent also in the radiofrequency regions and exhibit lumpy structures along their flow directions, which indicate hydrodynamic instabilities and irregular matter ejection from the star. Often, jets are connected with broadly fanned-out, bipolar CO outflows, which have the same orientation of the axis of their flow fields as the jets.

The central protostar which emits the jets is invisible in the optical region; sometimes the absorption by dust is so strong that it cannot be observed even in the infrared, but only in the millimeter wave range.

In favorable cases, such as the spherical dust cloud ESO 210-6A (Fig. 10.27), the proper motions of the Herbig–Haro objects can also be determined; they are of the order of $100\,\mathrm{km\,s}^{-1}$ and are directed away from the central star.

A precise explanation of how protostars with luminosities of only 1 to 100 $L_\odot$ can produce these energetic, more or less strongly collimated gas flows has yet to be offered. The only possible energy source is gravitational energy which is released during the *accretion* of matter when it flows inwards on spiral orbits in the disk. This model is supported by the observation that the mass loss from outflows is closely connected with the accretion rate, and that it stops in the course of the star's evolution, when the disk has dispersed. The jets and the bipolar winds probably also carry off angular momentum (and also magnetic fields) from the protostar and the accretion disk and transport them outwards to the surrounding material, and thus modify the rotation of the star.

Circumstellar Disks. The accretion disks around protostars can be observed directly, with the Hubble Space Telescope, but only in the nearest regions of star formation such as the Orion complex (see below). However, due to their large diameters of 100 to 1000 AU, they can be indirectly detected with certainty in the infrared through the thermal emissions of their dust, along with the radiation of the much smaller star, even when they cannot be spatially resolved. To some extent, the outer regions of protostellar disks can also be recognized on the basis of their molecular lines using interferometers in the millimeter spectral range.

Accretion disks occur in a high proportion of the protostars, perhaps even in all of them. Many disks also have properties which should be favorable for the formation of planetary systems.

The observations of protostars clearly show that the occurrence of disks is closely connected with the bipolar outflows and jets, whose direction is determined by the axis of rotation of the disk. The matter *inflow* from the molecular cloud to the star takes place essentially via the accretion disk, and the gravitational energy released in the process delivers the energy for the *outwardly* directed gas motions. The disk also plays an important role in the redistribution of the angular momentum, which is a "hindrance" to the contraction of the star; it is transported from within the disk outwards and finally given up to the surrounding molecular cloud. In the accretion disk, there are also probably *magnetic fields* which were "frozen in" from the original cloud and which become twisted by the differential rotation, so that magnetic energy can be stored for a time and then again released due to instabilities. Presumably, magnetic fields which are anchored in the disk also contribute to the collimation of the jets. Through interactions of the magnetic fields in the boundary layer at the inner edge of the disk with those of the protostar, the rotation of the latter is probably influenced.

Although we still do not understand many details of the formation and early evolution of the stars, it seems to be certain that the close interaction of accretion disks, jets and the protostar itself with their magnetic fields and angular momenta are of decisive importance in the process. Finally, the outflow of matter also influences stellar genesis in the surrounding molecular cloud, since a sufficiently large mass loss can prevent any further star formation in it for a while.

Protostars in Orion. The nearest large region of star formation lies in Orion at a distance to the Earth of around 500 pc (Sect. 10.5.2). At the nearest edge of the molecular cloud, the Orion Nebula was formed by the ionization of several solar masses of gas by the O stars of the Trapezium, a branch of the molecular cloud with weak extinction lies in front of it. The Trapezium stars are part of a massive star cluster with nearly 1000 stars, the *Trapezium cluster*. The nebula and the stars all have comparable ages of somewhat less than 10^6 yr.

Since the angular resolution of the Hubble Space Telescope of about $0.1''$ corresponds to dimensions of 50 AU ($\simeq 2.4 \cdot 10^{-4}$ pc) at a distance of 500 pc, this instrument can be used to observe a large number of details which give direct information about processes related to the formation and evolution of the protostars. The images from the Hubble Space Telescope (Fig. 10.28) show, for one thing, near roughly half the stars, several flattened, long formations, which appear dark in front of the luminous background of the Orion Nebula. Their sizes are several 100 AU and they are thus not much larger than the Solar System, where the orbit of Pluto has a diameter of about 80 AU (Table 2.2). We have here direct pictures of protostellar *accretion disks* which we can view from a range of different angles.

The Orion complex furthermore offers the unique possibility of investigating the *influence of the surroundings* on the evolution of protostars: on the one hand, the density of the stars in the Trapezium cluster is relatively high, so that tidal forces occur due to stars passing nearby; on the other hand, the ultraviolet radiation and the stellar wind of in particular the hot, bright ($10^5 L_\odot$) star θ^1 Ori C in the Trapezium have a strong effect on the outer portions of the protostars. The images made with the Hubble Space Telescope show a variety of types of interactions, especially ionization and partial vaporization of the gas shells and accretion disks on the side turned towards θ^1 Ori C, and "tails" on the opposite side, consisting of ionized matter which was blown off by the stellar wind. On the other hand, the disks must have considerable resistance to effects of this kind, since they persist for several 10^6 yr.

10.5.6 Stellar Statistics and the Star Formation Rate

In the 1920's, J. Kapteyn, P. J. van Rhijn and others carried out a statistical analysis of stars of known parallax and obtained the *luminosity function* for stars in our vicinity; it provides the empirical basis for the mass function and for the rate of star formation.

The Luminosity Function. We shall limit ourselves here to the stars on the main sequence in early stages of their evolution, and define the luminosity function $\Phi(M_V)$ as the number of main sequence stars with absolute magnitudes M_V per pc^3 and in the magnitude interval $\Delta M_V = 1$ mag (sometimes $= 0.5$ mag is used instead). In the vicinity of the Sun, $\Phi(M_V)$ decreases rapidly from $M_V \simeq 3.5$ mag towards brighter magnitudes (Fig. 10.29). This was explained by E. E. Salpeter

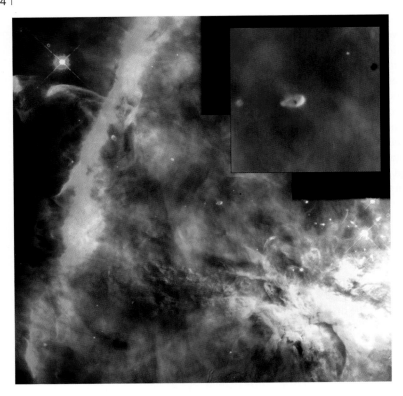

Fig. 10.28. The Orion Nebula (M 42), at a distance of about 500 pc. This high-resolution image taken by C. R. O'Dell et al. in 1993 with the Wide Field Planetary Camera 2 of the Hubble Space Telescope shows numerous circumstellar disks around young ($\leq 10^6$ yr) stars. The brightest star on the *right* edge of the picture is the luminous, hot Trapezium star θ^1 Ori C. *Inset:* An enlarged section from the center of the image, showing several circumstellar disks whose outer portions on the side facing θ^1 Ori C are ionized by its ultraviolet radiation. The extent of the largest disk is roughly 1100 AU, corresponding to 13 times the diameter of Pluto's orbit. (© Association of Universities for Research in Astronomy AURA/Space Telescope Science Institute STScI)

in 1955 with the hypothesis that stars fainter than 3.5 mag have collected since the formation of the galaxy roughly $T_0 \simeq 2 \cdot 10^{10}$ yr ago, without essential changes, while the brighter stars depart from the main sequence after about one evolution time t_E (Table 8.2), reckoned from the time of their formation, and finally, after a time which is in any case $\ll T_0$, become white dwarfs, neutron stars, etc. Thus, we can readily calculate the *initial* luminosity function $\Psi(M_V)$, which tells us how many stars of magnitude M_V in the interval ΔM_V have been formed in the Milky Way per pc^3 during the time of its existence (the external conditions are assumed to have remained constant). For the brighter stars, we find:

$$\Psi(M_V) = \Phi(M_V) \frac{T_0}{t_E(M_V)} \ . \tag{10.40}$$

For the fainter stars, $\Psi(M_V)$ merges continuously into $\Phi(M_V)$. In Fig. 10.29, we have included the initial luminosity function $\Psi(M_V)$ as calculated by A. Sandage, who continued the computations of E. E. Salpeter.

If our ideas are correct, then the luminosity function of young open star clusters should correspond to the initial luminosity function Ψ and not to the luminosity function Φ of our vicinity. Indeed, the investigation of various clusters confirms the concept that the division of an originally-present mass of gas into stars proceeds everywhere according to the same initial luminosity function $\Psi(M_V)$.

The difference $\Psi(M_V) - \Phi(M_V)$, summed over all M_V, corresponds to those stars which have evolved *away* from the main sequence since the formation of the galaxy. By far the major portion of these stars must at present be white dwarfs. In fact, the spatial density of white dwarfs calculated from Fig. 10.29 agrees with the observed density as well as can be expected within the uncertainties in the data.

For the faint stars at the *lower end* of the main sequence, a relatively complete stellar census can be obtained only if we limit ourselves to the immediate vicinity of the Sun in the Milky Way. In Fig. 10.30, the luminosity function of the stars at a distance of ≤ 20 pc

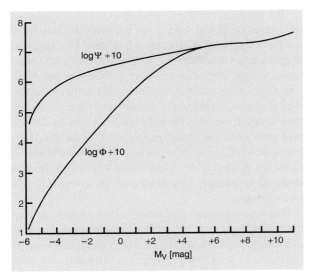

Fig. 10.29. The luminosity function $\Phi(M_V)$ and the initial luminosity function $\Psi(M_V)$ of the main sequence stars in the vicinity of the Sun. Φ and Ψ give the number of stars per pc^3 in the magnitude range $M_V - \frac{1}{4}$ to $M_V + \frac{1}{4}$ which are present or have been formed during the existence of the Milky Way

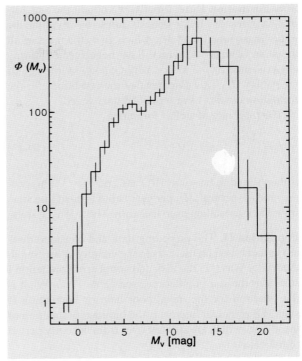

Fig. 10.30. The luminosity function $\Phi(M_V)$ of the stars within 20 pc distance from the Sun, after R. Wielen, H. Jahreiss and R. Krüger (1983). Φ is the number of stars in a sphere of radius 20 pc with absolute magnitudes M_V per $\Delta M_V = 1$ mag

from the Sun according to R. Wielen et al. (1983) is drawn on the basis of the catalog of W. Gliese (1969), with the inclusion of some more recently discovered faint stars. The brighter stars up to about $M_V \simeq 8$ mag and out to that distance are all included; for the fainter stars, to about 13.5 mag, the number can be extrapolated from a smaller volume around the Sun. For still fainter stars, Φ represents only a lower limit for the number of stars. The maximum of the luminosity function at $M_V \simeq 13$ mag is, however, probably genuine.

The Mass Function. Equivalent to the luminosity function is the mass function $\varphi(\mathcal{M})$, which gives the number of main sequence stars of mass $\mathcal{M}$ per pc^3 and per unit mass. It is obtained from the luminosity function $\Phi(M_V)$ by using the mass–luminosity relation (Fig. 6.14), whereby $\varphi(\mathcal{M}) \, d\mathcal{M} = \Phi(M_V) \, dM_V$. Particularly important for the theory of stellar genesis is the *initial mass function* (or IMF) $\psi(\mathcal{M})$, which is derived from the results of stellar evolution theory analogously to $\Psi(M_V)$ (cf. (10.40)).

The initial mass function is also usually expressed in terms of $\xi(\mathcal{M}) = \mathcal{M} \psi(\mathcal{M})$, the *total mass* of all stars with individual masses $\mathcal{M}$ per pc^3 and per unit mass,

where

$$\xi \, d\mathcal{M} = \mathcal{M}\psi(\mathcal{M}) \, d\mathcal{M} = \mathcal{M}\xi \, d\ln\mathcal{M} \, . \quad (10.41)$$

ξ can therefore also be regarded as the number of stars with $\mathcal{M}$ per pc^3 and per *logarithmic* mass interval, $\Delta \ln \mathcal{M} = 2.303 \, \Delta \log \mathcal{M} = 1$.

According to E. E. Salpeter (1955), the initial mass function for the more massive stars ($\geq 1 \mathcal{M}_\odot$) can be approximated by a power law:

$$\psi(\mathcal{M}) \propto \mathcal{M}^{-\beta-1} \quad \text{or} \quad \xi(\mathcal{M}) \propto \mathcal{M}^{-\beta} \quad (10.42)$$

with $\beta = 1.35$. More recent investigations have indicated a steeper decrease at the more massive end, although the results of different determinations show considerable scatter ($1.4 \leq \beta \leq 2$). Towards smaller masses ($\leq 1 \mathcal{M}_\odot$), the mass function flattens out sharply ($\beta \simeq 0.3$).

Star Formation Rate. Finally, we obtain the star formation rates or stellar genesis rates, by dividing the initial mass function $\xi(\mathcal{M})$, which gives the mass of all the stars of mass $\mathcal{M}$ which have been formed during the existence of the Milky Way Galaxy, by the age of the galaxy, T_0. This assumes that the conditions for the formation of stars have remained unchanged.

The *total* rate of stellar formation

$$r = \frac{1}{T_0} \int_0^\infty \xi(\mathcal{M}) \, d\mathcal{M} \tag{10.43}$$

is then found to be some $10^{-12} \mathcal{M}_\odot \, \text{pc}^{-3} \text{yr}^{-1}$. This corresponds to several $\mathcal{M}_\odot$ per year being formed into stars from the interstellar gas in the *entire* Milky Way system.

Population II. The preceding data and considerations are in fact based on the stars in the vicinity of the Sun in the Milky Way, i.e. the disk and spiral arm Population I stars (for the star populations, see Table 11.1). What is the situation for the metal-poor halo or Population II stars? Of principal interest is the luminosity function of those stars which are still left over from the early period of the galaxy.

The luminosity function of a globular cluster as a typical representative of Population II was first investigated, for the case of M3, by A. Sandage in 1957. As expected, brighter stars ($M_V \leq +4$ mag) are lacking. Apart from a maximum at $M_V \simeq 0$ mag, which is produced by the cluster variables, the observed luminosity function of the globular cluster in the range $M_V \leq +6$ mag differs little, according to Sandage, from that of the stars of our neighborhood.

The extension of the observations to the fainter stars of Population II which have not yet evolved away from the main sequence is difficult due to their low apparent magnitudes. Only recently, with the use of CCD photometry, has it become possible to measure the main sequence of the globular clusters with sufficient accuracy down to considerably fainter magnitudes; for the later spectral classes, the infrared brightnesses M_I are preferred over the visual magnitudes. With the Hubble Space Telescope, it is planned in the near future to carry out extensive observations of Population II stars down to (apparent) magnitudes of about $m_V \simeq 28$ or $m_I \simeq 21$ mag.

The measurements performed up to now indicate that the luminosity function, and thus also the mass function, of the halo population exhibits a maximum in the range of $M_V \simeq 11$ to 13 mag ($M_I \simeq 9$ to 10 mag) and then decreases towards fainter stars. Furthermore, the luminosity function seems to become steeper, the lower the metal abundance of the stars.

Thus far, we have considered only the "local" aspects of star formation. Concerning the question of whether the rate of star formation was the same everywhere within the Milky Way and at all times, and concerning "global" points of view, e.g. why the stars are preferably formed in the spiral arms of the galaxy, we shall continue the discussion in Sect. 11.2.5, together with the problems of the formation and evolution of the Milky Way galaxy, its star populations and the distribution of the chemical elements. Before we can take up these topics, we need to become more familiar with the Milky Way and the other galaxies.

11. The Structure and Dynamics of the Milky Way Galaxy

Now that we have become familiar with the stars, their properties and their motions, and also with the various components of the interstellar medium, we have collected all the "parts" we need in order to consider our Milky Way galaxy. We begin in Sect. 11.1 with the information about the structure of the Milky Way which can be gained from observations of the distribution and motions of the stars within it. In Sect. 11.2, we treat the rotation of the Milky Way, its distribution of various kinds of matter, and the star populations, as well as the phenomenon of the spiral arms. Finally, in Sect. 11.3, we describe the central core region of the Milky Way, which cannot be observed optically and must be investigated by observations in the radiofrequency and infrared spectral regions.

11.1 Stars and the Structure of the Milky Way

First of all, in Sect. 11.1.1 we introduce the galactic coordinate system appropriate to the investigation of the Milky Way galaxy. We can get a first idea of the structure of our galaxy by means of "star gauging" (Sect. 11.1.2). This method is, however, severely limited by interstellar absorption. We then discuss the distribution and motions of the stars and star clusters in the Milky Way galaxy. Sect. 11.1.3 deals with stellar kinematics. In Sect. 11.1.4, we consider the star clusters: how with the aid of the open star clusters and the globular clusters, distances in the whole galaxy can be determined, so that we can obtain an initial overview of its structure.

11.1.1 Galactic Coordinates

It is appropriate at this point to introduce a system of galactic coordinates which we can use to describe the Milky Way galaxy: these are the *galactic longitude l* in the plane of the galaxy, and the *galactic latitude b* in the perpendicular direction, which is positive towards the north and negative towards the south. In 1958, the system of galactic coordinates (l, b) in use today was introduced; it has been improved by the addition of information from, in particular, radio astronomical observations. These coordinates are defined by the equatorial coordinates 1950-0 of the galactic *north pole*:

$$\alpha = 12 \text{ h } 49 \text{ min}, \quad \delta = +27.40°. \quad (11.1)$$

Galactic longitude is measured starting from the galactic *center*:

$$\alpha = 17 \text{ h } 42.4 \text{ min}, \quad \delta = -28.92° \quad (11.2)$$

the plane of the galaxy is inclined to the celestial equator by an angle of 62.6°.

Instead of the formulae from spherical trigonometry, we give here in Fig. 11.1 the *charts* for converting between the equatorial (α, δ) and galactic (l, b) coordinates.

11.1.2 Star Gauging

W. Herschel (1738–1822) was the first to attempt to penetrate the secrets of galactic structure with his *star gauging*, in which he counted the number of stars which he could see in different directions down to a particular limiting magnitude.

What should we expect if space were completely transparent and uniformly filled with stars? For stars of a given absolute magnitude, the apparent brightness decreases as a function of the distance r proportional to $1/r^2$ (6.32). The stars brighter than m thus fill a sphere with $\log r = 0.2$ m + const. Their *number* $N(m)$ is $\propto r^3$; we thus find:

$$\log N(m) = 0.6 \text{ m} + \text{const}. \quad (11.3)$$

In Fig. 11.2, we compare the number of stars per square degree according to F. H. Seares (1928) for the direction of the galactic plane and for the direction towards the galactic pole (i.e. $b = 0°$ and $90°$) with what is expected from (11.3). The much slower increase of $N(m)$ observed at fainter magnitudes can

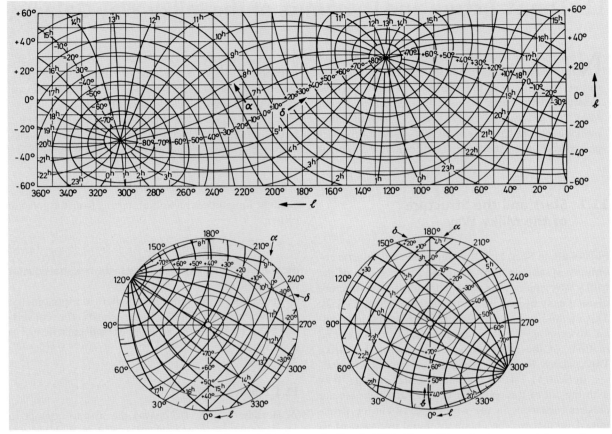

Fig. 11.1. Charts for converting galactic coordinates l, b (plotted above as abcissa and ordinate, respectively; below, around the circumference or as the radius of the circles, respectively) into right ascension α and declination δ, and *vice versa*, for the Epoch 1950. *Above*: The galactic equatorial zone. *Lower left*: The galactic north pole. *Lower right*: the galactic south pole. After G. Westerhout

have only two causes: (i) a decrease in the density of stars at larger distances, or (ii) interstellar absorption; or both. H. v. Seeliger and J. C. Kapteyn took the scatter of the absolute magnitudes of the stars into account and showed how their numbers $N(m)$ can be correctly represented by a superposition (mathematically speaking, by a convolution) of (1) the *density function* $D(r)$ = the number of stars per pc^3 at the distance r and a given direction with (2) the *luminosity function* $\Phi(M)$ = the number of stars per pc^3 in the interval of absolute magnitudes between $M - 1/2$ and $M + 1/2$, possibly also taking into account (3) an *interstellar extinction* of $\gamma(r)$ magnitudes per pc. Even taking the luminosity function $\Phi(M)$ to be everywhere constant and determining it by using stars of known parallax in a region of 5 or 10 pc, one could not separate the functions $D(r)$ and $\gamma(r)$. We can therefore discard the results of older stellar statistics. The concepts just introduced remain important, as does the great sampling survey of the entire sky (magnitudes, color indices, spectral types) in the Kapteyn fields.

11.1.3 Spatial Velocities of the Stars

The study of the motions of stars initially proved to be more fruitful than did stellar statistics. We have already mentioned the spatial velocity v of the stars (6.15), motions. Later, it was found that, at a higher precision, the motion of the Sun depends upon *which* stars are used to determine it; this was the first indication of the systematic effect of the stellar motions.

The socalled standard solar motion which is generally used to reduce stellar motions to the *Local Standard of Rest* is obtained from the proper motions and the radial velocities of the stars in our neighborhood as:

Solar Motion: $v_\odot = 20 \, \text{km s}^{-1}$

to the Apex: $\alpha = 18\,\text{h}, \; \delta = +30°$. (11.4)

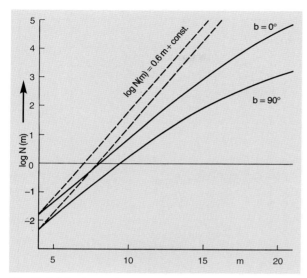

Fig. 11.2. Numbers of stars $N(m)$, i.e. the number of stars brighter than m per square degree, from the stellar census by F. H. Seares in 1928, at the galactic equator ($b = 0°$) and at the galactic pole ($b = 90°$) (*full curves*). Calculated curves (*dashed*): $\log N(m) = 0.6\,m + \text{const}$ for constant stellar density, without galactic absorption. (The constant was adjusted to the observational data for $m = 4$ mag)

Galactic Components. The galactic components of the spatial velocities of stars are defined by (positive signs)

U towards center of the galaxy ($l = 0°, b = 0°$),
V in the direction of galactic rotation ($l = 90°$),
W perpendicular to the galactic plane, towards
the galactic north pole ($b = 90°$).

Here, it must be noted whether or not the *solar* motion was subtracted.

The *standard solar motion* relative to the average motions of neighboring stars (11.4) is:

$U_\odot = +9 \, \text{km s}^{-1}$,
$V_\odot = +12 \, \text{km s}^{-1}$,
$W_\odot = +7 \, \text{km s}^{-1}$ (11.5)

towards the apex $l \simeq 56°, b \simeq +23°$.

which is composed of the spectroscopically determined radial velocity from the Doppler effect, V_r (6.13), and the proper motions μ (PM) or the tangential component V_t (6.14).

Since, for example, a star of 6th magnitude has on the average a parallax of 0.012″, but a proper motion of 0.06″ yr^{-1}, we can delve ca. 100 times further out into interstellar space by studying the proper motions than with parallax measurements.

The Apex. Let us first make the simplifying assumption that the stars are at rest and that only the Sun moves relative to them with a velocity $v_\odot$ towards the *apex*; then we expect the distribution of radial velocities V_r and tangential velocities V_t or proper motions μ shown in Fig. 11.3 as a function of the angular distance χ of the star from the apex.

If the motions of the stars are distributed randomly in space, we can clearly still use this model, by *averaging* over many stars. The first apex determination was made in 1783 by W. Herschel, using only a few proper

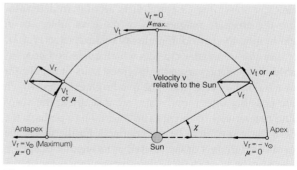

Fig. 11.3. Motion of the Sun, relative to the surrounding stars, with velocity $v_\odot$ towards the apex. The observed parallactic motions of the stars are the reflection of this solar velocity. The figure explains the dependence of the radial velocity V_r and the tangential velocity V_t (or the proper motions μ) of the stars on their angular distance χ from the apex

11.1.4 Star Clusters: Distance Determinations and the Structure of the Milky Way

All of our knowledge about the structure and size of the Milky Way Galaxy is based essentially on the method of *spectrophotometric* distance measurement, since trigonometric parallaxes and stream parallaxes (Sect. 6.2.1) are by far too inaccurate for galactic distance determinations. From the $1/r^2$ law of photometry, we see a star of absolute magnitude M and parallax p or distance $r = 1/p$ with the apparent magnitude m, so that the distance modulus is given by:

$$m - M = 5 \log r \, [\text{pc}] - 5 + A \, [\text{mag}] \qquad (11.6)$$

(6.34). Here, A [mag] is the interstellar extinction (Sect. 10.1.2). In the final analysis, we must always refer to the absolute magnitudes of certain objects which are known from trigonometric parallaxes, stream parallaxes etc.

In order to gain an insight into the structure of our Milky Way, it is reasonable to begin with groups of stars whose structures can be readily recognized, in part by the unaided eye, in part in suitable images: the globular star clusters and the less concentrated open (galactic) star clusters, and the stellar associations (Chap. 9).

Globular Clusters. The decisive turning towards modern astronomy was taken by H. Shapley in 1917, when he measured the distances to numerous globular clusters by means of the cluster variables contained in them (RR Lyr stars); the essential difficulty lay in determining the absolute magnitudes of these variables, or in calibrating the period–luminosity relation (Sect. 7.4.1).

For those globular clusters which contained no variable stars, Shapley used the brightest stars of the cluster as a secondary criterion. The five brightest stars, which are possibly foreground stars, are discarded; the absolute magnitudes of the next brightest, up to perhaps the 30th, have proven to be well defined. Furthermore, the overall brightness of the cluster, or its angular diameter, can also be used (with certain precautions) as criteria. From his observational data of (at that time) 69 clusters, Shapley was able to draw the conclusion that they form a system which is *barely* flattened in the plane of the Milky Way, the halo (Fig. 11.4); its center lies about 9 kpc from us (the value quoted by Shapley was 13 kpc)

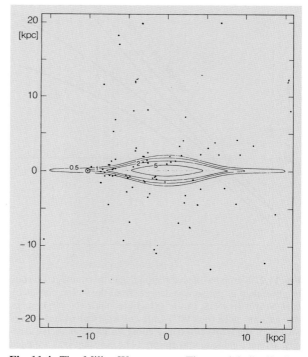

Fig. 11.4. The Milky Way system. The spatial distribution (halo) of the globular clusters is projected onto a plane perpendicular to the galactic plane and passing through the Sun ⊙; the contours show surfaces of constant mass density (relative to the vicinity of the Sun). In the galactic plane, the thin layer of interstellar matter and the extreme Population I are indicated by dots (after J. H. Oort, 1965). The disk of the galaxy, the central bulge and the halo of globular clusters are, as confirmed by dynamic models, embedded in a considerably larger system with a radius of about 60 to 100 kpc, the invisible *"dark" outer halo*, which contains a major part of the mass of the Milky Way system; its composition is currently unknown (Sect. 11.2.4)

in Sagittarius. This work laid the foundations for further investigations of the Milky Way.

Open Star Clusters. The open clusters and associations belong to Population I and are concentrated towards the galactic plane. If we may assume that the main sequence is the same in the color–magnitude diagrams for all systems, as well as for the field stars in our neighborhood which are not recognizably a part of any system, then the vertical distance of the main sequence,

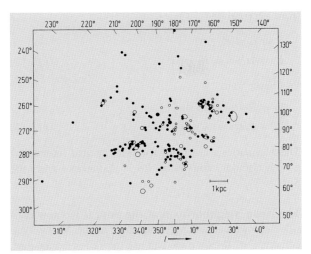

Fig. 11.5. The distribution of young open clusters (•) (which contain early spectral types, O to B3 stars) and the OB associations (○) in the plane of the Milky Way, after R. M. Humphreys (1979). The Sun ⊙ is at the origin of the coordinate system, and the galactic longitude $l = 0°$ (Sect. 11.1.1) points towards the center of the galaxy. The objects of extreme Population I are distributed along the spiral arms (from the center outwards: Sagittarius arm, local or Orion–Cygnus arm, Perseus arm)

e.g. in the $(B-V, m_V)$ diagram of a cluster and in the $(B-V, M_V)$ diagram of our immediate neighborhood, yields directly the distance modulus $m_V - M_V$ and thus the distance to the cluster; the influence of the interstellar extinction must still be taken into account.

In principle, the same method can also be used to determine the distances to the globular clusters. However, they are in the main further away from us than the open star clusters, so that it is difficult to obtain precise magnitudes for their main sequence stars; furthermore, the influence of their different metal abundances must be taken into account (Sect. 9.2.3).

In Fig. 11.5 we show the results for the distribution of the young open star clusters and the OB associations. Their arrangement into long, stretched-out regions is obvious at first glance; these are the neighboring portions of the galactic spiral arms. Like the OB stars, the H II regions and the absolutely brightest Cepheids (with periods ≥ 11 d) are distributed along the spiral arms.

The Structure of the Milky Way. We now summarize what we have already learned about the structure of our Milky Way galaxy from the observations of stars and star clusters. We must take into account the fact that the observations within the galactic disk in the optical spectral region are limited, even for very bright objects, to distances of ≤ 4 kpc. In contrast, in the near infrared ($\lambda \geq 2$ μm), star clusters and individual K and M supergiants can still be seen even in the vicinity of the center of the galaxy.

In Fig. 11.4, we can distinguish three components in the distribution of stars:

1. The *disk* in the plane of the Milky Way, in which (among others) OB stars and open star clusters are arranged in spiral arm structures; the later spectral types, in contrast, are not noticeably concentrated in the arms;
2. The *bulge*, the central distension or central "lens", which appears e.g. in surveys in the near infrared; and
3. The roughly spherical *halo*, with the globular clusters and the Population II stars (high-velocity stars).

In Sects. 11.2.2–6, we will refine and complete this first view of our Milky Way galaxy.

11.2 The Dynamics and Distribution of Matter

In Sect. 11.2.1, we first consider the rotation of the Milky Way. It was recognized as early as 1926/27 by B. Lindblad and J. H. Oort that the stars in the vicinity of the Sun are carrying out a differential rotation around the center of the galaxy; only later did radio astronomical observations of the 21 cm line of neutral hydrogen clarify the rotational velocities in the whole Milky Way. After a brief discussion of the distribution of interstellar matter in the galaxy in Sect. 11.2.2, and of the orbits of stars and the density of matter in the immediate neighborhood of the Sun in Sect. 11.2.3, we turn to the overall mass distribution within the Milky Way in Sect. 11.2.4. In the following Sect. 11.2.5 we discuss the explanation of the spiral arm structure of our Milky Way and the other spiral galaxies. Finally, in Sect. 11.2.6, we consider the star populations introduced by W. Baade in 1944, together with the abundance distributions of the chemical elements in the Milky Way galaxy.

11.2.1 The Rotation of the Galactic Disk

We now turn to the kinematics and dynamics of the Milky Way Galaxy, as developed by B. Lindblad und J. H. Oort in 1926/27 in their theory of the *differential rotation* of the galactic disk. We first assume that all motions (Fig. 11.6) take place on planar circular orbits around the galactic center (Z).

Let the angular velocity ω of a star P at the galactic longitude l as a function of its distance R from the center be $\omega = \omega(R)$; thus $V = \omega R$ is its linear orbital velocity on the circular galactic orbit. Conversely, for the angular velocity and its derivative, we have

$$\omega = \frac{V(R)}{R}, \quad \frac{d\omega}{dR} = \frac{1}{R}\left(\frac{dV}{dR} - \frac{V}{R}\right). \quad (11.7)$$

For the Sun ($\odot$), let $R = R_0$, $\omega(R_0) = \omega_0$ and $V_0 = \omega_0 R_0$. Precisely stated, we always relate the motion here and in the following sections to an average in the neighborhood of the Sun, the *Local Standard of Rest* (Sect. 11.1.3), by subtracting the solar motion (11.4) from all observed coordinates.

We now decompose the velocity vector $\boldsymbol{V} - \boldsymbol{V_0}$ of the star P *relative* to the Sun into its components in the direction $\odot$P and perpendicular to it, in order to obtain the radial velocity

$$V_r = V \sin\alpha - V_0 \sin l \quad (11.8)$$

and the proper motion μ or the tangential velocity

$$V_t = V \cos\alpha - V_0 \cos l = \mu r, \quad (11.9)$$

where r is the distance of the star from the Sun. We can eliminate the auxiliary angle α by using $|ZQ| = R\sin\alpha$ and $|PQ| = R\cos\alpha$, which we can read off from the triangle $\odot$ZQ:

$$R \sin\alpha = R_0 \sin l,$$
$$R \cos\alpha + r = R_0 \cos l. \quad (11.10)$$

We then obtain

$$V_r = R_0(\omega - \omega_0)\sin l,$$
$$V_t = R_0(\omega - \omega_0)\cos l - \omega r. \quad (11.11)$$

These equations are valid for stars or also interstellar gas on circular orbits at *arbitrary* distances r from the Sun.

The Near Neighborhood of the Sun. If we now consider our immediate neighborhood in the Milky Way Galaxy, $r \ll R_0$, then we can use the series expansion:

$$\omega - \omega_0 \simeq \left(\frac{d\omega}{dR}\right)_0 (R - R_0)$$
$$\simeq -\left(\frac{d\omega}{dR}\right)_0 r \cos l. \quad (11.12)$$

We next introduce *Oort's constants* for the differential galactic rotation:

$$A = -\frac{R_0}{2}\left(\frac{d\omega}{dR}\right)_0 = \frac{1}{2}\left[\frac{V_0}{R_0} - \left(\frac{dV}{dR}\right)_0\right],$$

$$B = -\frac{R_0}{2}\left(\frac{d\omega}{dR}\right)_0 - \omega_0$$
$$= -\frac{1}{2}\left[\frac{V_0}{R_0} + \left(\frac{dV}{dR}\right)_0\right] \quad (11.13)$$

or also

$$A + B = -\left(\frac{dV}{dR}\right)_0,$$
$$A - B = \omega_0 = \frac{V_0}{R_0}, \quad (11.14)$$

where we have expressed them using (11.7) in terms of the orbital velocities $V(R)$.

With Oort's constants A and B, the approximation (11.12), and the identity $2\cos^2 l = 1 + \cos 2l$, the radial and tangential velocities for our neighborhood as functions of the galactic longitude l finally take on the simple form:

$$V_r = Ar \sin 2l,$$
$$V_t = Ar \cos 2l + Br. \quad (11.15)$$

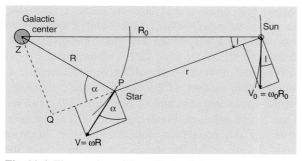

Fig. 11.6. The rotation of the galactic disk

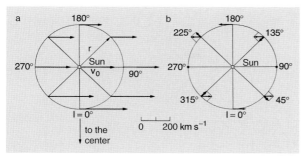

Fig. 11.7a,b. Differential galactic rotation. (**a**) Absolute velocities of the stars at a distance r from the Sun. In our drawing, $r = 3$ kpc. The length of the velocity vectors corresponds to the distance traveled by the stars in 10^7 yr. (**b**) Velocities of the same stars relative to the Sun and their radial components (*heavy arrows*), illustrating the double-wave dependence of the radial velocities from (11.15)

After averaging out of the peculiar motions, observations verify this "double wave" ($\sin 2l$!) of the two velocity components very well (Fig. 11.7). While the amplitudes of V_r and V_t increase proportionally to the distance r, the amplitude of the proper motion $\mu = V_t/r$ is independent of r. The numerical values of Oort's constants are found to be:

$$A = +14 \text{ km s}^{-1} \text{ kpc}^{-1},$$
$$B = -12 \text{ km s}^{-1} \text{ kpc}^{-1}. \quad (11.16)$$

If we had considered rigid rotation ($\omega = \omega_0$), then we would have found $A = 0$ ($V_r = 0$), and $B = -\omega_0$ ($V_t = -\omega_0 r \neq 0$).

The Distance of the Sun from the Galactic Center. The distance of the Sun from the galactic center was initially taken to be the same as that to the center of the system of globular clusters. However, the period–luminosity relation can also be applied directly to the RR Lyr stars in the regions of the Milky Way which are not too strongly covered by dark clouds of cosmic matter. With an uncertainty of about 15%, the result is

$$R_0 = 8.5 \text{ kpc}. \quad (11.17)$$

Thus, with (11.14), we obtain the circular velocity of the Sun:

$$V_0 = 220 \text{ km s}^{-1} \quad (11.18)$$

or $\omega_0 = 26$ km s^{-1} kpc^{-1}, corresponding to a rotational period of $2.4 \cdot 10^8$ yr. Since the beginning of the middle age of the Earth in the Triassic (Table 3.3), we have therefore made one trip around the center of the galaxy.

The numerical values for R_0 and V_0 (and for Oort's constants) are based on the evaluation of a large amount of observational data, which we cannot discuss further here. They were adopted in 1985 by the International Astronomical Union and replaced the older, somewhat higher $R_0 = 10$ kpc and $V_0 = 250$ km s^{-1}.

In the immediate neighborhood of the Sun, according to (11.13) and (11.14), and using the values of Oort's constants, (11.16), both the angular velocity and the linear orbital velocity decrease with increasing R; for example, $V(R)$ changes by -2 km s^{-1} kpc^{-1}.

21 cm Observations. Outside the range of optical observations in the plane of the Milky Way, we must rely on *radio astronomical* measurements for the derivation of the rotation curve, in particular those of the 21 cm line of neutral hydrogen (Sect. 10.2.2). In contrast to the stars and the star clusters, however, the distance to the interstellar gas cannot be directly determined.

The directional dependence of the radial velocity V_r of the 21 cm line shows the characteristic double wave form (11.15), $\propto \sin 2l$. For larger distances from the Sun, we must use the exact relation, (11.11). It can be seen from Fig. 11.8 that V_r attains its maximum value along an observation direction in the direction of the galactic longitude l ($|l| < 90°$)

$$V_m = R_0 \left[\omega(R_m) - \omega_0 \right] \sin l \quad (11.19)$$

at that point where the line of sight touches the galactic circular orbit of "minimal" radius $R_m = R_0 \sin l$, i.e. at the point D. The line profile at V_m therefore shows a steep decrease towards larger radial velocities. By combining the V_m for various galactic longitudes l, one can find the *rotational velocity* $V(R)$ as a function of the distance R from the center of the galaxy (Fig. 11.9).

11.2.2 The Distribution of the Interstellar Matter

In Sects. 10.1–4, we introduced the individual components of the interstellar matter in our Milky Way galaxy, from the dust particles through the gas at varying temperatures to the high-energy cosmic radiation. The gas component includes on the one hand very hot ($T \simeq 10^6$ K), tenuous gas, ranging at the other extreme out to the dense, cooler ($T \simeq 10$ K) molecular clouds.

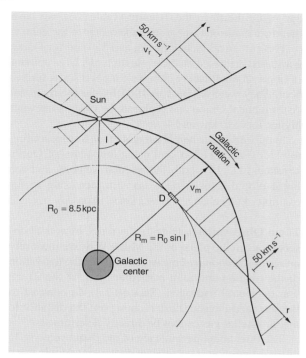

Fig. 11.8. The differential galactic rotation of interstellar hydrogen. The radial velocity V_r of the interstellar hydrogen relative to the vicinity of the Sun is plotted along a line of sight of galactic longitude l as a function of the distance r. It has a maximum V_m at D, where the line of sight is tangential to a circular orbit. Compare this plot with the approximation for $r \ll R_0$ in Fig. 11.7

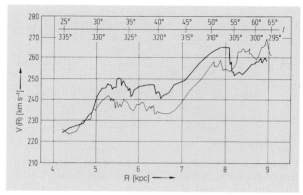

Fig. 11.9. Rotation curve $V(R)$ of the Milky Way, from observations of the 21 cm line of neutral hydrogen for $4 \text{ kpc} \leq R \leq 10 \text{ kpc}$, after F. J. Kerr (1964). The observations from the northern hemisphere ($l < 90°$, *heavy curve*), and from the southern hemisphere ($l > 270°$, *light curve*) deviate systematically from each other (large-scale asymmetry of the galaxy or slight expansion?). The irregularities in both curves ($\leq 10 \text{ km s}^{-1}$) are probably related to the spiral structure of the Milky Way. Within $R < 4$ kpc, there are strong deviations of the motion of H I from circular orbits. (The numerical data in this figure are based on older values of R_0 and V_0 which are somewhat too high; compare (11.17) and (11.18))

In this section, we wish to get an overview of the arrangement of this interstellar medium within the Milky Way, however leaving the central region of the galaxy ($R < 3$ kpc) out of the picture for the moment.

Dust. Observations in the optical range indicated that the interstellar dust in our vicinity is strongly concentrated towards the galactic plane and is very nonuniformly distributed (Fig. 10.1). On the average, the dust extinction γ in the optical region – as we saw in Sect. 10.1.2 – has values within the galactic *disk* of 1 to 2 mag kpc^{-1} and therefore permits observations only out to a distance of a few kpc.

E. Hubble's discovery of the *"zone of avoidance"* (in 1934, while he was investigating the distribution in the sky of galaxies above a certain limiting magnitude) gave us a picture of the distribution of absorbing matter *perpendicular* to the plane of the Milky Way, or, equivalently, of the dependence of γ on the galactic latitude b. The number of such galaxies per square degree is nearly constant at the galactic poles. From 30° down to 40° in galactic latitude, it decreases rapidly going towards the galactic equator, so that in its vicinity, a nearly galaxy-free zone is found. Hubble concluded from this that the absorbing matter in the Milky Way forms a flat disk, within which we are located, so that extragalactic objects are subject to a visual extinction of around $0.2 \csc b$ [mag]. Observations of stars in our galactic neighborhood then further showed that the (full) half-maximum width of the absorbing layer is about 200 pc, roughly corresponding to that of the hydrogen (as was later determined).

Information concerning the *large-scale* distribution of dust in the Milky Way, and in other galaxies, can be obtained from observations in the *far infrared* at $\lambda \geq 100\,\mu\text{m}$. In this spectral region, we can see the thermal radiation from the dust, without hindrance due to the interstellar extinction; the dust is heated by the general galactic radiation to temperatures in the range of 15 to 50 K. As was first shown by investigations with infrared satellites, the dust in the whole galaxy is clearly concentrated towards the galactic disk. IRAS observed in the far infrared in addition some filamentary emission regions at higher galactic latitudes, the "galactic cirrus".

Neutral Gas. From the intensity of the H I and H$_2$ lines in the ultraviolet (as well as from radio astronomical observations), we can derive the fact that in our vicinity in the Milky Way, about half of the interstellar hydrogen is in the atomic and half in the molecular state. The ultraviolet and optical spectral regions can, however, give us no information about the global properties of the interstellar matter in the Milky Way owing to the strong extinction in those wavelength ranges. Here, observations in the infrared and especially in the radiofrequency region must take over this task.

From the measured intensity profiles of the 21 cm line of neutral hydrogen, I_ν or $I(V_r)$ ((10.15) and (10.18)), which in general show a complex structure consisting of several components, we can determine the density distribution of neutral hydrogen, since the known shape of the rotation curve $V(R)$ of the galaxy (Fig. 11.9) permits the calculation of the distribution of radial velocities V_r along any line of sight. The ambiguity which occurs at $|l| < 90°$, i.e. whether a particular V_r belongs to the corresponding point in front of or behind the tangent point to the circle of radius R_m (point D in Fig. 11.8), can often be resolved by applying the consideration that the more distant object in general shows a smaller extent perpendicular to the Milky Way, i.e. along b. Furthermore, in analyzing the 21 cm observations, one must always keep in mind the consistence of the resulting picture.

The distribution of neutral hydrogen within the galactic plane, first obtained by Dutch and Australian radio astronomers, is shown in Fig. 11.10. One can clearly recognize its concentration in the *segments of the spiral arms*; however, the large-scale structure of the spiral arms, – which we can easily see as external observers in the case of other galaxies (Figs. 12.1, 5, 11) – is not

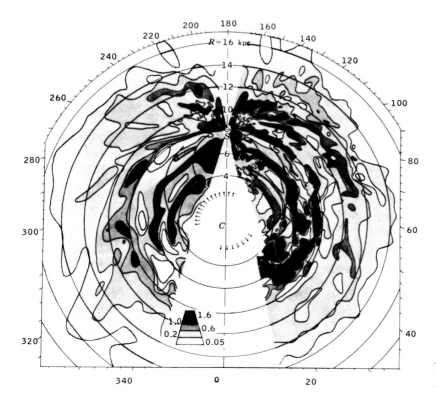

Fig. 11.10. An historic chart of the distribution of neutral hydrogen in the galactic plane (maximum densities projected onto the plane). The density scale is in number of atoms per cm^3. *Outer scale*: galactic longitude l

clearly visible here. Our view of the distribution of interstellar hydrogen rests essentially on the assumption that the gas is moving on *circular orbits* around the galactic center (distance determination!). If serious deviations from this assumption occur, then the galactic distribution can no longer be determined simply from the 21 cm line profiles.

While outside of $R \geq 4$ kpc in the plane of the Milky Way no systematic differences of more than 10 km s^{-1} relative to the circular-orbit velocities are seen, we find major deviations in the inner part of the galaxy. We thus see the 21 cm line in absorption against the strong continuous background of radio emissions from the center of the galaxy with a radial velocity of $V_r = -53 \text{ km s}^{-1}$. A more thorough investigation of the dependence of the line profiles on the galactic longitude indicates a spiral arm at $R \simeq 3.7$ kpc, which also participates in the galactic rotation, but at the same time is *expanding away* from the center at 50 km s^{-1}. Behind the galactic center, there is a counterpart to this "3 kpc arm" with a velocity of $V_r = +82 \text{ km s}^{-1}$, and another part of an arm with $V_r = +135 \text{ km s}^{-1}$. Additional unusual gas flows in the central region of the Milky Way will be discussed in Sect. 11.3.

The observations of the 21 cm emissions *perpendicular* to the plane of the Milky Way indicate that the neutral hydrogen on the average forms a *flat disk*. The distance between the two surfaces at which the density has decreased to one-half of its average value in the galactic plane is – for $4 \leq R \leq 10$ kpc – about 240 pc. Further out, the disk becomes increasingly broad and deviates systematically from the center plane (cf. also Fig. 12.12), possibly under the influence of tidal forces from the Magellanic Clouds (Sect. 12.1.2).

The mean *particle density* of H I in the plane of the Milky Way at a distance between 4 and 14 kpc from the galactic center is around $4 \cdot 10^5 \text{ m}^{-3}$ or 0.4 cm^{-3}; it decreases on going further inwards or outwards (Fig. 11.11). The overall mass of interstellar H I in the Milky Way is about $2.5 \cdot 10^9 \, \mathcal{M}_\odot$.

The distribution of *molecular hydrogen* in the Milky Way can be found from the CO lines (Sect. 10.2.3). In contrast to atomic hydrogen, the H_2 molecules are concentrated on a large scale in a flat, broad ring between about 4 and 8 kpc from the galactic center (Fig. 11.11), which is only around 100 pc thick in the direction perpendicular to the plane of the galaxy. The overall mass

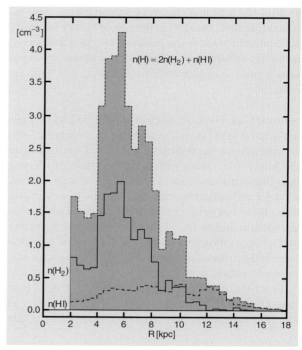

Fig. 11.11. The density distribution of atomic and molecular hydrogen in the plane of the Milky Way as a function of the distance to the center of the galaxy, after M. A. Gordon and W. B. Burton (1976). The distribution of H_2 is derived from observations of the $\lambda = 2.6$ mm line of the CO molecule

of H_2 in the Milky Way is about $2 \cdot 10^9 \, \mathcal{M}_\odot$ and is thus comparable to that of neutral atomic hydrogen.

In summary, we find for the distribution of neutral gas and dust the following picture: the gas (at temperatures below about 10^4 K) *and* the dust are concentrated in a flat disk, whose thickness – at a distance of 4 to 10 kpc from the galactic center – is around 200 pc. Within this disk, the gas and dust are concentrated in the spiral arms; along the arms, density condensates (clouds) are formed and are themselves divided up into finer structures. In the denser clouds, hydrogen is found also in molecular form.

Ionized Gas. The close connection between the more dense molecular clouds as the locations of star formation, and the H II regions, which are ionized and excited by the ultraviolet radiation of the young OB stars, can also be seen in the similarities of the large-scale density distributions of ionized and molecular hydrogen in the

Milky Way (Fig. 11.11). The *ionized* hydrogen also is found to be concentrated in a flat ring between about 4 and 8 kpc distance from the galactic center; on the other hand, it is also found in the central region of the Milky Way. The overall mass of H$^+$ in the Milky Way is only about $2 \cdot 10^8 \mathcal{M}_\odot$.

The *hot* ($T > 10^4 \ldots 10^6$ K) gas component of the interstellar medium (Sect. 10.3.4) occupies about one-half the volume of the disk and and stretches out to considerably greater distances (several kpc) from the plane of the Milky Way than do the cooler gas and dust.

Dynamics. The gas pressure $P = nkT$, both in the cool diffuse H I clouds and in the partially ionized "warm" gas, as well as in the hot gas component, is of the same order of magnitude (10^{-14} to 10^{-13} Pa = J m^{-3}), so that among these components in the disk, there is an approximate *pressure equilibrium*. According to Table 10.1, the corresponding energy densities are also comparable with those of the interstellar magnetic fields and of the cosmic radiation.

On the other hand, the extremely complex mutual interactions of the different components of the interstellar medium and their energy budgets are influenced in essential ways by *dynamic* processes. Among these are in particular the supernova outbursts with their ejected gas shells and emissions of X-rays, but also stellar winds and the ultraviolet radiation of the hot stars as well as the large-scale spiral density disturbances in the Milky Way (Sect. 11.2.5).

According to the model of the "galactic fountain", exploding supernovae produce a hot gas in the disk (at around 10^6 K), which expands and spreads through "tubes" out through both sides of the disk and into the galactic halo. There, it cools off by radiation losses and condenses into "clouds", which again fall back onto the disk. Observations in the 21 cm line, using UV lines, and in the X-ray region are compatible with these ideas. However, the heating and dynamics of the interstellar medium, which is permeated by magnetic fields and continuously subjected to shock waves, is only poorly understood in detail.

The Local Interstellar Medium. To conclude this section, we ask how the interstellar medium in the immediate neighborhood of our Solar System is constituted. Given the average density distribution in the galactic disk, we would expect roughly the same amounts of atomic and of molecular gas with a particle density of about $5 \cdot 10^5$ m^{-3}, mixed with dust (mass fraction about 1%). In fact, the Sun is found to be within an irregularly-shaped "bubble" of mainly ionized, very tenuous ($5 \cdot 10^4$ m^{-3}) and hot ($T \simeq 10^6$ K) gas. This "coronal" gas (Sect. 10.3.4), that can be detected directly through its X-ray emissions which reach us from all directions, fills a region of around 200 pc diameter, with the Sun at roughly 50 pc from its edge, and is probably left over from a supernova explosion which took place some 10^5 yr ago. Embedded in the hot gas are some small concentrations of "warm" gas at $T \simeq 800$ K and around 10^5 m^{-3} (cf. Sect. 10.2.1), of which one with a diameter of 5 pc lies near the Sun in the direction towards Sagittarius. Cooler and denser H I clouds and molecular clouds are to be found only beyond 50 to 150 pc from the Sun.

The existence of a local hot "bubble" has by the way proven to be a decisive advantage for observations in the *extreme ultraviolet*, especially with the EUVE satellite (Sect. 5.4.3). Only owing to this coincidentally "transparent" environment near the Sun, with its extremely low density of *neutral* hydrogen ($\leq 5 \cdot 10^3$ m^{-3}), has it been possible to observe the EUV radiation of several nearby stars, mainly hot white dwarfs.

11.2.3 The Galactic Orbits of the Stars. Local Mass Density

How well does our previous assumption of circular galactic orbits hold up? If we allow noticeable eccentricities e of the stellar orbits, we find that relative velocities of 100 km s^{-1} and more would occur in the immediate neighborhood of the Sun. We can thus understand the phenomenon of high-velocity stars, as J. H. Oort remarked in 1928.

Stellar Orbits in the Galactic Disk. We initially limit our considerations to stellar orbits within the galactic plane; these orbits are then determined by their galactic velocity components U and V or by the analogous velocity components relative to the Sun's vicinity,

$$U' = U \quad \text{and} \quad V' = V - 220 \, \text{km s}^{-1} . \tag{11.20}$$

The position coordinates of stars which are accessible to precise observation can namely be set equal to those of the Sun with sufficient accuracy. In a diagram with the coordinates U' and V', and for a given galactic force or potential field, we can therefore draw for example curves of constant eccentricity e, curves of constant apogalactic distance R_1, etc. Calculations of this type were first carried out by F. Bottlinger in 1932 for a $(1/R^2)$-force field; Fig. 11.12 shows a corresponding *Bottlinger diagram* for a force field which is adjusted to better fit the actual Milky Way. In a field which deviates from the $1/R^2$ law, the stellar orbits are in general not closed curves and e is not an orbital element in the strict sense, as it is in the case of planetary orbits.

The "normal" stars in our vicinity have small values of U' and V', i.e. they all move (like the Sun) in the normal sense on nearly circular orbits. The velocity vectors of the high-velocity stars, in contrast, indicate that these stars move on orbits of large eccentricity e around the galactic center, in some cases in the normal sense, in some cases with a retrograde motion (see below).

Motion Perpendicular to the Disk. The motions perpendicular to the plane of the Milky Way of the stars in our neighborhood can be understood to a great extent, according to J. H. Oort (1932, 1960), by considering the distribution of mass density ϱ in our region of the galactic disk to be planar. We then take into account only the dependence on the distance z of the mass density from the galactic plane; we can also consider the W components (Sect. 11.1.3) of the stellar velocity vectors independently of their motions (U and V) parallel to the galactic plane. These stars undergo oscillations through the galactic plane and perpendicular to it with periods of about 10^8 yr.

Analysis of the W-velocity distribution again exhibits clearly the two types of stars mentioned above: the disk stars with $|\overline{W}| \simeq 12$ km s^{-1}, and the high-velocity stars with considerably larger values of W. While the disk stars indeed describe nearly circular, planar orbits, the high-velocity stars mostly move on strongly eccentric orbits which are also inclined with respect to the galactic plane.

The stellar dynamics also lead us again to the fundamental concept of star populations introduced in 1944 by W. Baade. We prefer, however, to postpone their discussion until Sect. 11.2.6.

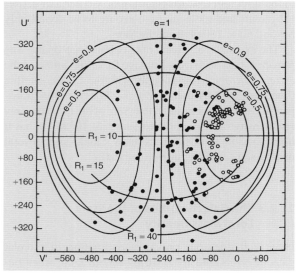

Fig. 11.12. A Bottlinger diagram for stars with spatial velocities > 100 km s^{-1}. The galactic velocity components in [km s^{-1}] U' (to the anticenter) and V' (in the direction of the rotation) are plotted relative to the neighborhood of the Sun; the axes correspond to the absolute velocity components U and V. The eccentricity e of the orbit and its apogalactic distance R_1 in [kpc] can be read off from the two families of curves. (This diagram is based on older, somewhat higher values of R_0 and V_0 instead of (11.17) and (11.18).) • Stars with an ultraviolet excess, $\delta(U-B) > +0.15$ mag (Sect. 9.2.2), i.e. metal-poor stars of halo Population II; these are all high–velocity stars with large spatial velocities. ○ Stars with $\delta(U-B) < 0.15$ mag; these stars make up the transition region from halo Population II to the disk population, i.e. to stars with more nearly circular orbits. (After O. J. Eggen)

Local Mass Density. The distribution of the stellar density perpendicular to the plane is connected with the gravitational field of the galactic disk on the one hand, and with the velocity distribution of the W-components on the other, in an analogous manner to the relation between the density distribution of molecules in an atmosphere and the gravitational field as well as the Maxwell–Boltzmann velocity distribution or the temperature. In the case of the galaxy, however, the gravitational field is itself directly related to the matter density ϱ by Newton's law of gravitation (or the Poisson equation). Therefore, J. H. Oort in 1960 was able to estimate the overall matter density in the galactic plane near the Sun; newer determinations yield:

$$\varrho = 0.7 \cdot 10^{-20} \, \text{kg m}^{-3} = 0.1 \, \mathcal{M}_\odot \, \text{pc}^{-3} \quad (11.21)$$

with an uncertainty of about 20%. We compare this matter density with the overall density of the stars which have been observed in our immediate neighborhood (within 20 pc), which is about $0.04 \, \mathcal{M}_\odot \, \text{pc}^{-3}$, and the contribution of the interstellar matter which is around $0.04 \, \mathcal{M}_\odot \, \text{pc}^{-3}$. Here, the M dwarfs and the white dwarfs make up the major portion of the stellar mass density, while the contribution of the brown dwarfs, whose number is comparable, is estimated to be only about $0.003 \, \mathcal{M}_\odot \, \text{pc}^{-3}$. Owing to the uncertainties both in the number of stellar objects and in the local density determined from gravitation, we cannot at present say just how much dark matter contributes; its effect could be estimated only by Oort's method. Recent analyses of the observational data obtained by the astrometric satellite Hipparcos indicate that no noticeable missing mass is "left over" in the mass balance, which would have to be attributed to dark matter in the disk. The contribution of unknown dark matter in the galactic halo (Sect. 11.2.4) to the disk is estimated to be about $0.01 \, \mathcal{M}_\odot \, \text{pc}^{-3}$ and is thus unimportant for the mass balance in the vicinity of the Sun.

Stellar Orbits in the Halo. The halo contains the globular clusters and numerous field stars, which make themselves apparent initially as high-velocity stars through their high spatial velocities of up to 300 km s^{-1} relative to the Sun. If their velocity components U and V in the galactic plane are plotted in a Bottlinger diagram (Fig. 11.12), one can see that these stars move around the center of the galaxy on extended elliptical orbits, some of them even with a retrograde motion. Their large velocity components W perpendicular to the plane of the galaxy show that – in contrast to the nearly coplanar orbits of the stars in the disk – the orbital inclinations of the high-velocity stars are nearly randomly distributed.

The shape of the nearly spherical halo shows that its angular momentum per unit mass, or more precisely, its component $h = V \cdot R$ in the direction of the axis of rotation of the Milky Way, is small. We compare it with the Sun and the galactic disk:

$$\begin{array}{cccc} & \text{Halo} & \text{Disk} & \text{Sun} \\ h \, [\text{km s}^{-1} \, \text{kpc}] & 150 & 1540 & 1870 \end{array} \quad (11.22)$$

The contributions of the bulge and the thick disk (Sect. 11.2.4) to the angular momentum are negligible. The halo rotates (in the same direction as the disk) with a typical overall velocity of about 30 km s^{-1}. These results are clearly of importance in connection with the formation and evolution of the Milky Way galaxy (Sect. 12.5.4).

11.2.4 The Mass Distribution in the Milky Way Galaxy

If the entire mass $\mathcal{M}$ which determines the circular orbit of the Sun were concentrated in the center of the galaxy, then the following relation would hold, as for the motions of the planets, according to (2.61):

$$V_0^2 = \frac{G\mathcal{M}}{R_0} \, . \quad (11.23)$$

From this relation, as a first approximation, one obtains for the mass of the Milky Way $\mathcal{M} \simeq 2 \cdot 10^{41}$ kg $\simeq 10^{11} \, \mathcal{M}_\odot$. From Kepler's 3rd law (2.57), for a "point mass" $\mathcal{M}$, the rotational velocity $V(R)$ would obey the proportionality $V(R) \propto R^{-1/2}$, or the angular velocity would be $\omega(R) \propto R^{-3/2}$. A glance at the rotational curve for our galaxy (Fig. 11.9) shows that the assumption of a gravitational potential $\propto 1/R$ gives rather poor agreement with the observed curve.

The Mass Distribution. The mass distribution or the density distribution can be obtained from models which are constructed in such a way that their gravitational potentials yield the observed rotation curve. Here, simplifying assumptions such as rotational symmetry and neglect of the spiral structure are made. For a spherically-symmetric mass distribution, $V(R)$ depends only on the mass $\mathcal{M}(R)$ which is located *within R*,

$$V^2(R) = \frac{G\mathcal{M}(R)}{R} \, . \quad (11.24)$$

Our estimate above would thus correspond to the mass within a sphere having a radius equal to that of the Sun's orbit, $R_0 = 8.5$ kpc. In the case of a homogeneous ellipsoid, $V(R)$ is determined only by the mass within its surface (which is an equipotential surface).

The often-used *mass model* of the Milky Way Galaxy due to M. Schmidt (1965), which in spite of its simplicity reproduces the kinematic observations sufficiently well, is based on an inhomogeneous, strongly flattened ellipsoid of rotation (eccentricity $e = 0.999$, or flattening ratio $c/a = \sqrt{1-e^2} = 0.05$) having a mass of $1.7 \cdot 10^{11} \, \mathcal{M}_\odot$, and a small central point mass of $7 \cdot 10^9 \, \mathcal{M}_\odot$. The overall mass of the Milky Way is thus $1.8 \cdot 10^{11} \, \mathcal{M}_\odot$ in this model; roughly half of this mass lies within the distance R_0 from the Sun to the galactic center.

Components. More precise models take into account *several* components in the mass distribution of the Milky Way, which can be distinguished from each other on the basis of their dynamics and star populations (see also Fig. 11.4):

1. The *disk* with a mass density which decreases exponentially on going outwards radially and is $\propto \exp(-R/R_d)$, where $R_d \simeq (3.5 \pm 0.5)$ kpc, on a scale height of about 0.3 kpc;
2. The flattened central *bulge* ($c/a \simeq 0.6$) with a radius of 2.5 kpc in the galactic plane – it can hardly be observed in the optical region but it can readily be seen in the infrared ($\lambda \geq 2.2 \, \mu$m);
3. The barely flattened halo of radius around 20 kpc, with the globular clusters and high-velocity stars; and
4. The *core* of the Milky Way ($R \leq 0.1$ kpc).

The bulge and the halo are often combined as the *spheroidal component*, although it is not yet clear just how the halo merges inwards to the bulge.

Recently, it has been realized that besides the (young) thin disk, also a *thick disk* of similar radial extension but with a scale height of about 1 kpc exists. The latter rotates by about 40 km s^{-1} more slowly, and contains an intermediate stellar population, whose density (within the midplane of the galaxy) is only about 1/10 that of the young disk.

We shall return to the components of our Milky Way galaxy in Sects. 12.1.6 and 12.1.7, where they will be compared with the corresponding components of other normal spiral galaxies, which we can observe from *outside* the galaxy. The central region of the Milky Way, i.e. that part within $R \leq 3$ kpc, will be described in detail in Sect. 11.3.

The Dark Halo. The rotation curve in the Milky Way is more difficult to observe for $R > R_0$ than within the Sun's orbit. Recent measurements in the optical spectral range of H II regions and open star clusters and in the radiofrequency range (using H I- and CO lines, recombination lines in H II regions) show that $V(R)$ is essentially *flat* out to $R \simeq 30$ kpc, perhaps even increasing slightly to about 300 km s^{-1}. The Keplerian situation has thus not been attained out to this distance, and therefore the total mass of the system does not lie within this radius. Based on these observations and on the analysis of the motion of the Sun within the local group of galaxies (Sect. 12.4.1), as well as on theoretical considerations concerning the stability of rotating flat disks, the following picture of our galaxy has crystallized out since about 1975:

The long-known "visible" subsystems, the central region and the (inner) halo of globular clusters, are probably embedded in an enormous, spheroidal *outer halo* (or galactic corona) of 60 to 100 kpc radius, that consists of "dark matter" of still unknown composition, and which makes up the main portion of the overall mass of the system, about 3 to 10 times as much as the visible portions.

We shall return to the question of this halo, with its dark populations, in connection with the discussion of rotation curves of other galaxies in Sect. 12.1.5.

11.2.5 The Dynamics of the Spiral Arms

Most galaxies which contain a rotating disk also have spiral arms embedded in the disk; the S0 galaxies are an exception (Sect. 12.1.2). The spiral arms are highlighted by short-lived, bright OB stars, while the older, redder stars are more uniformly distributed throughout the whole disk.

The naive idea that a spiral arm always contains the *same* stars, gas clouds, etc., is refuted by considering that such a formation would be destroyed in the course of a few rotations of the disk, i.e. in several 10^8 yr, by the differential rotation.

B. Lindblad attempted many years ago to explain the persistence of the spiral structure over longer periods of time, in terms of a sectored system of *density waves*. In his model, the spiral arms of a galaxy are assumed to contain different stars, gas, etc. at different times, in a similar manner to an ocean wave whose crest is formed by constantly changing water particles. In order to justify the model, Lindblad and his coworkers carried out extensive calculations of the orbits of individual stars in the average gravitational field (potential field) of a galactic disk. These investigations were not well received, probably in part because Lindblad maintained for a long time that the spiral arms curve "forward", i.e. with the concave edge ahead. However, in many spiral galaxies, the "forward" and "backward" directions can be clearly distinguished, e.g. by observing the dark clouds, and it is found that the spiral arms in fact always "drag *behind*".

Density-Wave Theory. In the years following 1964, the density wave theory of the spiral structure was revived by C. C. Lin et al., now however in the form of a continuum theory. If there is a location in the disk of a galaxy with, for example, a somewhat higher gas density, it will cause a perturbation (potential well) in the potential field describing the differential rotation of the disk. This in turn influences the velocities of the stars and thus the mass density of the "stellar gas". In order that the potential field, the gas density, and the stellar density all be mutually consistent, a (rather complicated) system of differential equations must be obeyed; it describes the structure and propagation of density waves in the galactic disk. From the manifold of possible solutions, Lin and his coworkers selected *one*, corresponding to a quasistatic spiral structure. This structure moves, rigidly so to speak, with a constant angular velocity Ω_p. If this constant is fixed, then the shape of the potential well can be calculated, and it then forms the basis for the spiral arms. For example, in our Milky Way system, observations give $\Omega_p \simeq 13.5 \text{ km s}^{-1} \text{ kpc}^{-1}$; i.e. in our neighborhood, where the angular velocity of the stars is $\omega_0 = 26 \text{ km s}^{-1} \text{ kpc}^{-1}$ (11.18), the density wave moves around the galaxy with about *half* the velocity of the stars.

The observable spiral arms (Fig. 11.13) are then produced, according to W. W. Roberts (1969), by the fact that the interstellar *gas* flows along the density wave from the concave edge (in our case at 115 km s^{-1}). A compression occurs in going towards the potential minimum, and it can be recognized initially by means of interstellar dust. The compression of the gas (and of the magnetic field) furthermore leads to the appearance of a spiral-shaped *shock wave*, which in turn promotes Jeans instabilities (Sect. 10.5.3) and thereby the forma-

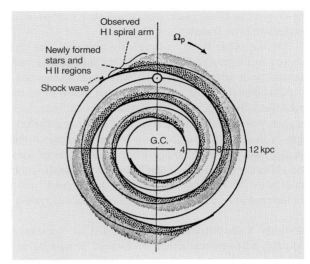

Fig. 11.13. The spiral structure of our galaxy ($\odot$ = Sun). The quasistationary spiral density wave according to C. C. Lin et al. (two arms) moves around at an angular velocity of $\Omega_p \simeq 13.5 \text{ km s}^{-1} \text{kpc}^{-1}$. It is thus continually overtaken by the galactic matter, which moves about twice as fast (gas, stars, ...). This matter is compressed in the region of the potential well of the density wave. In the gas, dark clouds are produced by the compression, as well as an increased intensity of synchrotron radiation and a shock wave (*solid curve*). In the course of 10^7 yr, young bright stars and H II regions are formed in this "shocked" gas. At a larger distance from the shock wave, older stars follow. The major portion of the potential well is filled with neutral hydrogen (H I; 21 cm radiation). From W. W. Roberts (1969)

tion of *young stars* and *H II regions*. A narrow strip is indeed observed at the convex edge of the shock wave, containing bright blue stars and H II regions. It is followed by a broad, diffuse band with older stars and star clusters; finally, the old disk population is distributed nearly homogeneously. Quantitatively, Lin estimates the amplitudes of the gravitational fluctuations, and those of the gas and stellar densities, to be about 5% of their average values.

In particular, the irregular spiral structures of the late Hubble type of galaxies (Sect. 12.1.2) suggests that in addition to a global density wave, some *stochastic* process is acting on a smaller distance scale in stellar genesis.

The density wave theory of the spiral structure has no doubt provided many new impulses, in particular for the evaluation of 21 cm observational data. Essential discrepancies between the theory and observations have so far not been found.

On the other hand, it is readily apparent that the theory in its present form is still very incomplete, and leaves open several questions: why is only *one* quasistatic density wave observed? What determines its rotational frequency Ω_p? Under what conditions does a galaxy *not* develop a spiral structure? What is the origin of the initial *excitation* and of the *damping* of the density waves? For the excitation, tidal interactions with neighboring galaxies (e.g. the Magellanic Clouds in the case of the Milky Way) or a rotating, "bar-shaped" mass in the central region of the galaxy have been suggested.

***N*-Body Calculations.** Given this list of questions, a quite different mathematical method is gaining interest: using large computers, several research groups, beginning with R. H. Miller, K. H. Prendergast and W. J. Quirk in 1968 and F. Hohl in 1970, among others, have simulated the motions of several 10^5 stars under the influence of their mutual gravitation as an N-body problem, and have generated images of a series of the numerous successive states of such a system. The significance of this technique is to be found not least in the fact that it is now possible to experiment with various types of galaxies. In the cases thus far investigated, a spiral structure forms out of an initially homogeneous disk in the course of less than one rotation. It is not quasistatic, but instead changes and renews itself continually. It can be assumed that while the hypothesis of the quasistatic spiral structure is not strictly correct, it is still a reasonable first approximation.

11.2.6 Stellar Populations and Element Abundances

We shall now attempt to bring some order into the confusing multiplicity of observations from our Milky Way (and from other galaxies) by making use of the concept of stellar *populations*.

Taking advantage of the observationally favorable conditions of the wartime blackout, W. Baade, using the 2.5 m telescope on Mount Wilson, succeeded in 1944 in resolving the central region of the Andromeda galaxy, M 31, as well as those of its elliptical companions, M 32 and NGC 205 (Fig. 12.1) into stars from the edge inwards. The brightest stars in this region were found not to be the blue OB stars, as in the spiral arms, but instead red giants, similar to those in the globular clusters of the Milky Way. Baade recognized that various galaxies or parts of galaxies are occupied by different star populations with quite different color–magnitude diagrams. He initially distinguished:

1. *Stellar Population I*, with a color–magnitude diagram similar to that of stars in our vicinity (Fig. 6.9). The brightest stars are *blue* OB stars with $M_V \simeq -7$ mag.
2. *Stellar Population II*, with a color–magnitude diagram similar to that of the globular clusters (Figs. 9.7, 8). The brightest stars here are the *red giants*.

It furthermore became clear that the stellar populations – or parts of them – differ also in terms of stellar dynamics and statistics, that Population I consists of young stars, while Population II consists of old stars, and that a portion of the latter, the halo population, contains *metal-poor* stars.

Classification. In the course of time, a clarification and refinement of the classification scheme of the stellar populations became necessary. This was worked out by J. H. Oort and others in 1957 at a conference on stellar populations in Rome. Table 11.1 gives a summary of the classification of stellar populations, with some more recent additions; they are characterized primarily by the distribution of the orbits of their stars within the galaxy. The age and metal abundance are, to be sure, additional parameters which are correlated with this one-dimensional sequence, but the correlation is not uniquely determined. On the contrary, in detailed investigations, they must be obtained separately. Following some methodological remarks about the empirical determination of these parameters, we will study the three main populations in more detail in the following paragraphs.

Age. The age of a star cluster or any sort of generically unified group of stars can be found (as we have already seen in Sects. 9.1.3 and 2.3) from its color–magnitude diagram by comparison with theoretically computed isochrones. However, such calculations depend on knowledge of the chemical compositions of the stars. Furthermore, the treatment of mass loss, the mixing of original matter with the products of nuclear

11.2 The Dynamics and Distribution of Matter

Table 11.1. Stellar Populations in the Milky Way galaxy

	(Old) Halo Population II	Middle Population II	Disk Population	Older Population I	Extreme Population I
	Spheroidal Component			Disk Component	
Typical representatives	Subdwarfs, globular clusters, RR Lyr variables with $P > 0.4$ d, bright red giants		"Normal" stars in the central region and the disk, Planetary Nebulae, Novae, RR Lyr variables with $P < 0.4$ d		Bright blue OB stars, open star clusters and associations, interstellar matter
Mean distance from the galactic plane $\bar{z}$ [pc]	2000	700	400	160	120
Average velocity perpendicular to the plane $\overline{W}$ [km s^{-1}]	75	25	18	10	8
Concentration towards the Galactic Center	strong		considerable		weak
Metal abundance ε relative to the Sun	$\simeq 10^{-3}$ to 1		mainly $\simeq 1$ ($\frac{1}{3}$ to 3)		$\simeq 1$
Age of the stars [10^9 yr]	$\simeq 12$ to 15		$\simeq 5$ to 10		$\lesssim 5$

fusion reactions, and also convection during stellar evolution, particularly in the region of the giant branches, are still fairly uncertain and limit the accuracy of age determinations.

Metal Abundance. The quantitative analysis of spectra with a high dispersion gives us knowledge of the chemical composition of the stars. This method is tedious and mainly limited to brighter stars; however, it gives the most detailed information. The results of such spectral analyses were already described in Sect. 7.2.7: since the relative abundance ratios of the heavy elements (nuclear charge $Z > 6$) for stars of the spiral arm, disk, and halo populations of our galaxy is in the main everywhere the same as in the Sun (Table 7.5), except for some individual elements, it seems justified for most purposes to refer to *the* abundance of the heavier elements or simply to the metal abundance ε (9.1).

The metal abundance of fainter stars can be obtained from broadband UBV photometry, to be sure without the possibility of taking the anomalies of individual elements into account. As we saw from the example of the globular clusters (Sect. 9.2.2), the UV excess $\delta(U-B)$ in the two-color diagram is closely related to the metal abundance ε. Before using the two-color diagram, we must correct the color indices for interstellar reddening.

Metal indices which are nearly independent of interstellar reddening are often used, such as the Q index for UBV photometry (10.5). The primary significance of the metal indices, aside from their use for faint stars, lies in the fact that they make it possible to extract at least a rough measure of the metal abundance from the superposed spectra of distant globular clusters, galaxies, and portions of them.

After these preparatory remarks, let us return to the stellar populations of our Milky Way galaxy (Table 11.1). It is convenient to begin with the two extreme main populations, halo Population II and the spiral arm Population I, and then to discuss the disk population. The transition populations will then need no further special treatment.

Halo Population II. The halo is, as we have seen, a nearly spherical subsystem of the spheriodal component of our Milky Way, with a radius of about 20 kpc. (We consider here only the inner, stellar halo; the hypothetical dark population of the outer halo will not be discussed.)

The halo population, globular clusters and field stars, is characterized by color–magnitude diagrams corresponding to those of the globular clusters (Fig. 9.8). Here, there are no bright OB stars; instead, the brightest stars are red giants of $M_V \simeq -3$ mag. Furthermore, the cluster variables or RR Lyrae with periods of > 0.4 d. are also typical. The classical Cepheids (such as δ Cep) of Population I correspond here to the rarer W Virginis stars, with periods of 14 to 20 d.

All the stars and star clusters of the halo population are more or less metal-poor. Among the most metal-poor *stars* (of less than 1/100 of the solar metal abundance) and the highest spatial velocities are the relatively rare K giants HD 122563, with $\varepsilon \simeq 1/500$, and CD $-38°$ 245 with $\varepsilon \simeq 5 \cdot 10^{-5}$, as well as the stars in the neighborhod of the main sequence, HD 140283 with $\varepsilon \simeq 1/200$, and G 64-12 ($=$ Wolf 1492) with $\varepsilon \simeq 1/3000$ (Sect. 7.2.7). Metal abundances of about 1/10 (relative to the Sun) are found rather frequently.

It would now seem obvious to ask if there is a correlation between these metal abundances and the galactic orbital elements or equivalent quantities. It is found that there is in fact a strong correlation between very small metal abundances and a large eccentricity e, as well as a large inclination i of the galactic orbit, or a high velocity component W perpendicular to the galactic plane. At larger metal abundances, above about 1/10 of the solar value, or for orbital eccentricities $e < 0.5$ or velocity components $|W| < 50$ km s^{-1}, the correlation is much weaker, or perhaps disappears entirely. There is thus, both in terms of metal abundance and in terms of galactic orbits, a rather large intermediate population, which forms the transition between the extreme halo population and the disk population (see below). Whether it is reasonable to define a special classification "intermediate Population II" remains an open question.

In the region of the globular clusters, the two-color diagrams and likewise the metal indices show that M 92 (Fig. 9.7) is just as metal-poor as the extreme subdwarfs (metal abundance $\varepsilon \leq 1/100$), while other globular clusters make a continuous transition to lower "metal defects". The color–magnitude diagrams (Fig. 9.8) become more similar on going from M 92, with increasing metal abundance, to that of the old open star cluster NGC 188, which has normal metal abundance. In this process, the giant branch moves downwards and ends earlier, and the horizontal branch gradually disappears.

Regarding the positions of the globular clusters in the sky, it can be recognized that in general, their metal abundances increase with decreasing distance from the center of the galaxy. Some globular clusters near the galactic center appear to have only small metal deficiencies, even on the basis of their color indices, or even a small overabundance; presumably they belong more properly to the rapidly rotating bulge rather than to the barely rotating halo system.

The age of the halo population can be determined from the color–magnitude diagrams of the globular clusters (Figs. 9.7, 8). From them (see Sect. 9.2.3), by comparison with theoretical evolution diagrams, it is found that that the main part of the globular clusters have an age of about 12 to $15 \cdot 10^9$ yr with an uncertainty of several 10^9 yr. Some few clusters are younger, by 3 to $4 \cdot 10^9$ yr.

Spiral Arm Population I. The extreme or spiral arm Population I is characterized by its bright ($M_V \simeq -6$ mag) blue O and B stars in young open clusters and associations, whose chemical compositions are very similar to that of the Sun (Fig. 9.3). All these structures were formed rather recently from dense regions of interstellar matter. Our earliest ancestors could have observed the formation of the bright blue stars in Orion from their treetop-observatories! The galactic orbits of the young stars of Population I and of the interstellar matter are, as we have seen, nearly circular (Sect. 11.2.3).

In the extreme Population I, radial abundance *gradients* are observed. Thus e.g. in the H II regions, the abundances of N/H and O/H decrease from within towards the outside of the galaxy at a rate of the order of $\Delta \log \varepsilon / \Delta R \simeq -0.1$ kpc^{-1}.

The Disk Population. The disk population of the Milky Way galaxy consists (on superficial examination) of stars having roughly normal metal abundances with the well-known color–magnitude diagrams of our vicinity (Fig. 6.9) and roughly circular galactic orbits. Only on more detailed investigation do the relationships to the Halo Population II and the spiral arm Population I become apparent, along with other important indications of the evolution of the galaxy.

An initial estimate of the age of the disk population can be obtained from the color–magnitude diagrams of the open star clusters (Figs. 9.2, 3). For one of the oldest, NGC 188, an age of 6 to at most $10 \cdot 10^9$ yr is found; M 67 is only slightly younger. There are also many field stars of the disk population which fit into the color–magnitude diagrams of the oldest open clusters. Even though the maximum age of the disk population is not very precisely known, it must in any case be $\leq 10 \cdot 10^9$ yr; it is thus clearly younger than most of the globular clusters. There are numerous younger star clusters (Fig. 9.3); therefore, the history of the disk population includes the whole time period up to the formation of spiral arm Population I. In the spiral arms, associations and clusters of younger stars are continually being formed from interstellar matter; a portion of these groups "flows apart" and thus continuously replaces the evolutionary losses of the disk population.

A more exact investigation of the metal abundances ε of the disk stars shows that they differ only slightly among themselves, but that they do scatter within a factor of 3 to (at most) 5. In particular, the narrow-band photometric observations of B. Strömgren, B. Gustafsson, P. E. Nissen and others, along with spectroscopic analyses, have shown that the metal abundance of the Hyades is about 1.5 times larger than that of the Sun, while the Pleiades have practically the solar mixture, with only a factor 1.1 overabundance.

The *scatter* of the metal abundances ε of the disk stars is on the one hand due to their dependence on the *age* of the stars, and on the other to radial abundance *gradients* within the disk. The age effect is difficult to see; it is in the range of $\Delta \log \varepsilon \simeq 0.5$ in 10^{10} yr. Since the formation of the disk, ε has increased by only about a factor of 3. Numerous investigations from about 1970 of our Milky Way and of other galaxies have shown that the metal abundance of the disk components – like that of the spiral arm population, see above – decreases continuously from the center outwards; in the Milky Way, by about $\Delta \log \varepsilon / \Delta R \simeq -0.05$ kpc^{-1}, i.e. by a factor of 3 from the galactic center to the vicinity of the Sun. In the galactic central region, the metal abundance is about 2 to 3 times higher than the solar value, and we find "super *metal-rich*" stars there.

11.3 The Central Region of the Milky Way

The central regions of our Milky Way galaxy are not accessible to optical observations due to the dense dark clouds in the Scorpio–Sagittarius area. Only with radio telescopes of high resolution in the centimeter wavelength range, and with the development of infrared astronomy, has it been possible to investigate this region since the late 1960's.

In the infrared, at shorter wavelengths ($1.25 \leq \lambda \leq 5$ μm), emissions from the stars dominate the overall radiation, while in the far infrared, we measure mainly the radiation from warm dust. The radiofrequency range and the infrared spectral lines give us information about the gaseous components.

The *visual* extinction in the direction of the galactic center ($b = 0°$) is $A_V \simeq 31$ mag, corresponding to an attenuation factor of $2.5 \cdot 10^{12}$ (!); on the average, it decreases to about 10 mag if we move away from the galactic center by 1° of galactic latitude. In the *infrared* at a wavelength of $\lambda = 2.2$ μm, we find an extinction of $A_{2.2 \, \mu m} \simeq 3$ mag, which however still allows observations of the central region.

As an aid to orientation, we first show the relation between some angles on the sphere and the corresponding length scales at a distance from the galactic center of $R_0 = 8.5$ kpc (11.17). The correspondence is:

1°	1′	1″	10^{-3}″
150 pc	2.5 pc	0.04 pc	
		8500 AU	8.5 AU.

We will start with a discussion of the region more distant from the galactic center and then approach the center itself; it coincides with the "point-like", unique radio source Sgr A*.

11.3.1 The Galactic Bulge ($R \leq 3$ kpc)

Within a distance of about 3 kpc from the galactic center, the structure and kinematics of the Milky Way change noticeably compared to the outer parts: we can no longer observe a spiral structure, and the density of stars increases sharply on going inwards, while the density of atomic and molecular gas decreases rapidly. A first indication of "unusual" processes in the central region of the Milky Way was provided by the 21 cm observations of neutral hydrogen in the galactic disk, which showed strong deviations from the circular orbits typical of the outer disk, with the *expanding* 3 kpc arm and its counterpart behind the galactic center (Sect. 11.2.2).

We begin with a description of the stellar component in the galactic bulge, and then discuss the interstellar components.

Stars. Following the pioneering investigations of E. E. Becklin, G. Neugebauer, F. J. Low, D. E. Kleinmann and others in 1968/69, broadband infrared observations at 2.2 and 2.4 µm have been carried out with steadily improving sensitivity and angular resolution. The galactic bulge has also been observed in the infrared with the COBE satellite (Sect. 5.2.1). From the surface luminosity, we can recognize a strong increase in the stellar density in the central region of the galaxy (within $R \leq 1.5$ kpc or within a galactic longitude of $|l| \leq 10°$ from the center). The distribution of stars in the bulge can be approximated as a flattened ellipsoid of rotation with a radius of about 2.5 kpc and an axial ratio of 0.6, or a semiaxis perpendicular to the plane of the galaxy of 1.5 kpc. According to the model of S. Kent (1992), the density of matter within $R \leq 0.9$ kpc decreases as $\varrho(R) \propto R^{-1.8}$. This corresponds to a radial mass increase of $\mathcal{M}(R) \propto R^{1.2}$. Further out, the density decreases more steeply ($\propto R^{-3.7}$).

Characteristic, easily recognizable objects in the bulge are the bright K and M giants, the OH/IR stars, and the planetary nebulae. The stars in the bulge rotate in the same direction as the galactic disk, but more slowly than the matter which lies further out. They have metal abundances in the range of $1/10 < \varepsilon < (3\ldots10)$, and thus contain more heavy elements than the region near the Sun. With ages between about 1 and $15 \cdot 10^9$ yr, the stars in the bulge are on the average younger than the halo population. Their total *mass* is about $10^{10}\,\mathcal{M}_\odot$.

Interstellar Matter. Between roughly 1.5 and 3 kpc, there is no measurable concentration of either atomic or of molecular hydrogen; as we go further in, at first the density of H I increases sharply, then somewhat closer to the center ($R \leq 0.6$ kpc) also that of H_2.

The 21 cm cm observations show complicated motions of the interstellar atomic gas for which we still have no unique model; the explanation of the observed radial velocities is rendered difficult by the fact that there is no way to determine the distance to the H I regions. G. W. Rougoor and J. H. Oort originally (1960) interpreted the observations in terms of a thin nuclear disk (or a ring), rotating at 250 km s^{-1}, with a radius of ≤ 0.7 kpc. Newer observations of the H I radiation have been interpreted by W. B. Burton and H. S. Liszt (1978/80) in terms of either (a) a gaseous disk which is rotating and expanding at the same time, inclined by about 20° to the galactic plane, and having a radius of 1.5 kpc, or else (b) by a (likewise inclined) rotating *bar* having a length of around 2 kpc, in which the gas moves on strongly elliptical orbits.

Independently of the precise model used, about $3 \cdot 10^7\,\mathcal{M}_\odot$ of H I with a density of the order of $\leq 1 \cdot 10^6$ m^{-3} must be present within a distance of 2 kpc from the galactic center.

11.3.2 The Nuclear Region of the Galactic Bulge ($R \leq 300$ pc)

Even the first isophotic map of the central region, made at $\lambda = 3.75$ cm (8.0 GHz), by D. Downes et al. in 1966 using a 36 m paraboloid antenna with an angular resolution of 4.2′, shows near the center the radio source *Sagittarius A*, whose position relative to many other sources leaves no doubt that it represents the *nucleus* of our galaxy and includes the galactic "center point".

Radiofrequency Emissions. In Fig. 11.14, we show observations with the 100 m telescope in Effelsberg made by W. J. Altenhoff et al. in 1978 at a wavelength of $\lambda = 6$ cm and a resolution of 2.6′. At this wavelength, nonthermal and thermal sources, i.e. synchrotron and free–free radiation, contribute with comparable intensities to the radio emissions, although the thermal radiation is more sharply concentrated towards the galactic equator. The emissions in the radio continuum are not symmetric with respect to the galactic plane; the extended structure of about 200 pc dimensions above Sgr A is quite noticeable.

The 45 pc long "arc" perpendicular to the equator consists of many finer filaments and emits synchrotron radiation; it is connected with Sgr A by a "bridge". At higher resolution, e.g. with the Very Large Array, one can distinguish three components within the extended *Sgr A complex* ($R \leq 25$ pc) (Fig. 11.15): (a) the nonthermal, dish-shaped source *Sgr A East*, and (b) the H II region *Sgr A West*, as well as, within it (c) the *ultracompact source Sgr A**, a strong nonthermal, "point-like" source.

11.3 The Central Region of the Milky Way

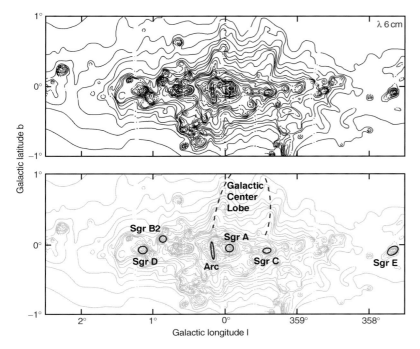

Fig. 11.14. The central region of the Milky Way at $\lambda = 6$ cm (5.0 GHz) (*upper map*) and an overview (*lower map*). Observations by W. J. Altenhoff et al. (1978) with the 100 m radio telescope in Effelsberg at an angular resolution of 2.6′. Sagittarius A contains the nonthermal central source of our galaxy; the thermal source (giant H II region) Sagittarius B2 is associated with a large molecular cloud complex. (P. G. Mezger, W. J. Duschl, R. Zylka, 1996)

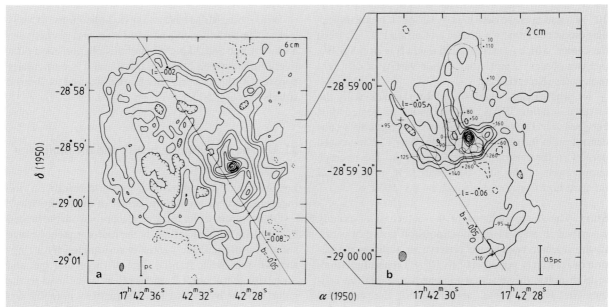

Fig. 11.15a,b. Sagittarius A: isophotes of the continuum radiation (**a**) of the whole source at $\lambda = 6$ cm (5 GHz) with a resolution of $5'' \cdot 8''$; and (**b**) of the central region of the component Sgr A West at $\lambda = 2$ cm (15 GHz), with a resolution of $2'' \cdot 3''$. The images are based on observations with the Very Large Array by R. D. Ekers et al. (1983). The *circles* and *crosses* in (**b**) mark the "[Ne II] clouds" observed by J. H. Lacy et al. in 1980, with their radial velocities indicated in [km s^{-1}]

Sgr A East is probably the remains of a (supernova-like?) explosion which occurred some 50 000 yr ago within a dense molecular cloud; this hypothesis is supported by the X-radiation from this region. The source Sgr A West, with Sgr A*, will be discussed in detail in the next section.

The giant H II regions Sgr B, C and D in the central area, which are noticeable as strong thermal sources in the cm range, are large molecular clouds and are associated with regions of star formation. These sources are embedded in a thinner ionized medium (with an electron density around 10^7 m^{-3}).

Gas and Dust. The mass of *atomic* hydrogen in the nuclear region is only about $10^7\,\mathcal{M}_\odot$. In contrast, observations of the spectral lines from various molecules, in particular from CO, which occurs together with H$_2$ and indicates the distribution of the latter, and of OH, CS, H$_2$CO, as well as HCN, which are emitted by the most dense regions, show that *molecular* gas predominates in these areas. It forms a thin (≤ 50 pc) layer having a mass of $\simeq 10^8\,\mathcal{M}_\odot$, of which about half is concentrated in giant molecular clouds with masses of 10^4 to $10^6\,\mathcal{M}_\odot$ and average densities $\geq 10^6$ m^{-3}. In the molecular cloud complex at a distance of around 200 pc from the center, which coincides with the H II region Sgr B2, one can observe a particularly wide variety of different molecules (Sect. 10.2.3).

As was already seen from observations with the infrared satellite IRAS, the emissions of heated dust between $\lambda = 12$ and 100 µm show – along with a concentration in the plane of the Milky Way – an increased density in the central region, $R \leq 400$ pc. The average temperature of the dust lies near 28 K, and its total infrared luminosity corresponds to $1.3 \cdot 10^9\,L_\odot$.

Stars. The star population of the outer bulge, already described in Sect. 11.3.1, continues on going inwards. In the nuclear region, we observe in addition a fraction of relatively young (10^7 to 10^8 yr) stars of medium to high mass; this fraction increases towards the center. The average metal abundance of the stars and of the interstellar gas in the nucleus of the bulge is about twice as high as that in the neighborhood of the Sun.

All together in the central region of the bulge, we find a mass of about $4 \cdot 10^9\,\mathcal{M}_\odot$.

Magnetic Fields. In the nuclear region of the bulge, using the polarization of the emitted radiation and the Zeeman splitting of the molecular lines, we can detect magnetic fields of the order of $2 \cdot 10^{-7}$ T (2 mG), which are considerably stronger than those in the galactic disk (Sect. 10.4.1). Within the large molecular clouds, they are oriented for the most part parallel to the plane of the galaxy, but in the intermediate medium, they are perpendicular to it.

The Bar Structure of the Milky Way. As can be seen both from the 21 cm observations of H I in the outer bulge and also from the molecular lines, there is an asymmetry in the radial velocities which can be interpreted as a bar structure. Furthermore, newer analyses of the surface luminosity at 2.2 µm give indications for a bar structure in the distribution of stars, as well.

According to dynamics calculations, a rotating bar could on the one hand maintain the region outside 1.5 kpc (out to 3 kpc) free of interstellar matter, and on the other it could be responsible for the observed matter inflow, of the order of $1/100\,\mathcal{M}_\odot$ yr^{-1} within 150 pc.

While our Milky Way shows indications of only a weak bar structure consisting of stars and gas between a few 0.1 kpc and 2 kpc, we find this phenomenon relatively frequently in other galaxies, and it is particularly apparent in the systems which are classified as "barred spirals" (Sect. 12.1.2).

11.3.3 The Circumnuclear Disk and the Minispiral ($R \leq 10$ pc)

Let us now look somewhat more carefully at the innermost few pc around the center of the Milky Way!

Observations of far infrared radiation from the heated dust and the molecular lines in the radiofrequency range show that the galactic center is surrounded by a ring-shaped, somewhat warped disk of molecular gas and dust between about 1.7 and 7 pc; it is inclined to the galactic plane by around 25°. The matter within this *circumnuclear disk* is noticeably clumped; the concentrations have masses of typically 30 $\mathcal{M}_\odot$ and are extremely opaque to optical radiation ($A_V \simeq 30$ mag). This disk rotates at 100 km s^{-1} in the same sense as the Milky Way and contains roughly $10^4\,\mathcal{M}_\odot$ of H$_2$ and dust. Its gas temperature lies at ≤ 400 K, and that of the dust is about 60 K. Through this circumnuclear disk, and also through the nuclear region itself ($R \leq 150$ pc), matter flows *inwards* with a radial velocity of about 20 km s^{-1} and a mass transport rate of the order of $10^{-2} \, \mathcal{M}_\odot$ yr^{-1}. The luminosity in the infrared is roughly $10^7 \, L_\odot$.

From considerations of the dynamics, one can conclude that within a radius of 1.7 pc from the center there is a mass of $4 \cdot 10^6 \, \mathcal{M}_\odot$.

At the inner edge of the disk, at a radius of $R \simeq 1.7$ pc, the density drops off rapidly towards the H II region Sgr A West. In high-resolution images made with the VLA – e.g. at $\lambda = 2$ cm (Fig. 11.15b) – one can see that about 40% of the thermal emissions of ionized hydrogen originate from a small, spiral structure, the "minispiral". Its outer arcs coincide with the inner edge of the circumnuclear disk; the source Sgr A* lies approximately in (seen from the Earth, somewhat in front of) the center of symmetry of this structure.

The total mass of ionized hydrogen is about 260 $\mathcal{M}_\odot$, of which roughly 10 $\mathcal{M}_\odot$ is in the minispiral. The average electron density in the H II region is 10^9 m^{-3}, while in the minispiral, it is 10^{10} m^{-3}. In order to ionize this gas (and to heat the dust), a total luminosity of about $8 \cdot 10^8 \, L_\odot$ is required, an amount which is consistent with the luminosity of the bright stars in the central star cluster of the Milky Way (Sect. 11.3.4).

Dynamics. The radial velocities of the H I recombination lines such as H 109α and especially the forbidden infrared fine structure transitions, such as that of [Ne II] at $\lambda = 12.8 \, \mu$m, [Ar III] at $\lambda = 9.0 \, \mu$m, [S III] at $\lambda = 18.7 \, \mu$m, or [O III] at $\lambda = 51.8 \, \mu$m, give us a first insight into the dynamics of the near neighborhood of the galactic center. For example, in the [Ne II] line, we can observe cloudlike structures in the emission regions which are located along the radio spirals (Fig. 11.15b). The surprisingly high radial velocities of these "[Ne II] clouds", up to 250 km s^{-1}, indicate a central mass of 10^6 to $10^7 \, \mathcal{M}_\odot$ (assuming circular orbits around the center). As the existence of the spiral structure shows, the dynamics near the central region are however probably more complex; furthermore, the motion of the gas can be influenced not only by gravitational forces, but also by magnetic fields, stellar winds, or frictional forces, so that this mass determination cannot be considered to be very conclusive.

In contrast to the gas, the *stars* are affected only by gravitational forces. The observation of their motions, as close as possible to the galactic center, is thus of decisive importance for the clarification of the nature of the central source.

11.3.4 The Innermost Region ($R \leq 1$ pc) and Sgr A*

In the innermost 1 pc, the radiation of the *stars* in the medium and near infrared ($\lambda < 30 \, \mu$m) exceeds that of the heated dust. By means of high-resolution observations, E. E. Becklin and G. Neugebauer showed in 1975 that at a wavelength of $2.2 \, \mu$m, a considerable fraction of the observed radiation is due to only a few *individual* bright supergiants of late spectral types ($T_{\text{eff}} \simeq 4000$ K), each having a luminosity of about $10^4 \, L_\odot$; their contribution to the *overall* stellar luminosity of around $8 \cdot 10^7 \, L_\odot$ however is only about 1% of the total, as was later recognized. The stars are concentrated in a dense, richly populated cluster (Fig. 11.16) at the galactic center, in which the ultracompact source Sgr A* dominates the radiofrequency emissions.

The Central Star Cluster. Among the brightest stars in $2.2 \, \mu$m radiation in the nuclear region ($R \leq 1$ pc), we find 24 hot, massive supergiants with effective temperatures of 25 000 K and luminosities $\geq 10^6 \, L_\odot$. These stars produce almost the whole overall luminosity of $10^8 \, L_\odot$; they form a subgroup with a strong concentration towards the galactic center, whose density distribution falls by one-half at 0.2 pc ("core radius"). The remaining ≥ 200 stars which are bright in the $2.2 \, \mu$m range are K and M supergiants. The main contribution to the mass is made by some 10^6 cool giant and dwarf stars. The core radius of the whole central star cluster lies around 0.38 pc; the mass density at its center is $\varrho_c = 4 \cdot 10^6 \, \mathcal{M}_\odot$ pc^{-3}, and is thus comparable with that of the most dense globular clusters in the Milky Way.

Stellar Motions and the Central Mass. Since the 1990's, it has become possible (especially thanks to R. Genzel, A. Eckart and coworkers) to measure both the radial velocities as well as the proper motions of numerous stars near the galactic center. The *radial velocity* can be determined from spectra in the $2.2 \, \mu$m

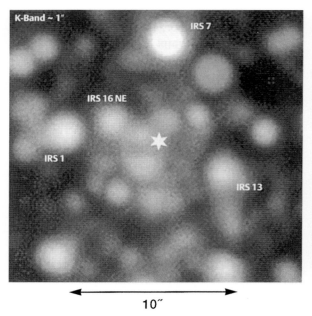 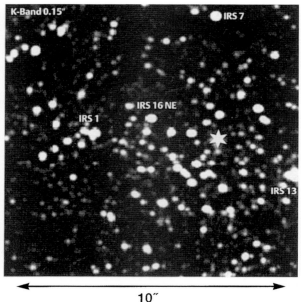

Fig. 11.16. An infrared image in the K band ($\lambda \simeq 2.2\,\mu$m) of the central $10''$ or 0.4 pc of the dense star cluster at the center of our Milky Way galaxy. *Left:* a "conventional" image, with an angular resolution limited by air turbulence to about $1''$. *Right:* An image obtained with the speckle camera "Sharp" of the Max Planck Institute for Extraterrestrial Physics, Garching, at the 3.5 m New Technology Telescope of the European Southern Observatory (ESO) by A. Eckart and R. Genzel in 1996. It has a diffraction-limited resolution of around $0.15''$. Some bright infrared sources (IRS = *Infrared Source*) are indicated; the position of the ultracompact radio source Sgr A*, at the dynamic center of the Milky Way, is marked by an *asterisk*. (With the kind permission of *Sterne und Weltraum*)

range using the emission lines of He I and H I for the hot supergiants and the absorption bands of CO for the cool stars, in to a distance of about $1''$ from the center. Using a speckle camera for the near infrared, the positions of stars can be determined with a precision of around $\pm 0.1''$ or ± 850 AU (!) (Fig. 11.16), and thus their *proper motions* very close to the central object in the Milky Way can be measured. Since the observed velocities at a distance of about $0.15''$ from the center are higher than $1000\,\mathrm{km\,s^{-1}}$, a measurable proper motion at a distance of 8.5 kpc can be registered within a few years. Recently, A. Ghez et al. (2000) succeeded in measuring for the first time the *changes* with time in the proper motions, i.e. orbital *accelerations*, for three stars with a projected distance of about 0.1 to $0.3''$ from the galactic center. The values, of a few $10^{-3}{''}$ per year, correspond to some $10^{-3}\,\mathrm{m\,s^{-2}}$ and are, incidentally, not very different from the acceleration of the Earth in its orbit around the Sun. This can be understood qualitatively if we consider the larger orbital dimensions of about 10^3 AU to be compensated by the greater central mass $\mathcal{M}_0$, which compared to the mass of the Sun is a factor of $(10^3)^2 = 10^6$ times larger, since the centrifugal acceleration – with the simplifying assumption of a circular orbit of radius r – is equal to the gravitational acceleration, $G\mathcal{M}_0/r^2$ (Sect. 2.3.6).

The observed star motions can be well reproduced by the gravitational field of a mass distribution which consists of a central star cluster with a mass of about $0.5 \cdot 10^6\,\mathcal{M}_\odot$ and a core radius (see above) of 0.38 pc, together with a

central point mass of $\quad \mathcal{M}_0 = 2.6 \cdot 10^6\,\mathcal{M}_\odot$. (11.25)

The orbital period P of a star with an orbital semimajor axis of $a = 0.004$ pc $= 850$ AU around this central mass is found from Kepler's third law, $a^3/P^2 \propto \mathcal{M}_0$ (2.57), to be about 15 yr. Therefore, we should be able to observe the complete elliptical orbits of some stars around the galactic center in the not-too-distant future.

Sagittarius A*. Within the positional accuracy of about $\pm 0.1''$, the ultracompact source Sgr A* at

$l = -0°3'20.7''$ $\alpha(2000) = 17\,\text{h}\,45\,\text{min}\,39.97\,\text{s}$
$b = -0°0'\ 2.9''$ $\delta(2000) = -29°\,0'\,34.9''$

is without doubt *the* dynamic center of our Milky Way galaxy.

The radiation flux from Sgr A* in the frequency range of about 1 to 3000 GHz is proportional to $\nu^{1/3}$ and falls off rapidly above and below this range. The spectrum can be explained in terms of optically thin synchrotron radiation (Sect. 12.3.1) from relativistic electrons with energies of the order of 100 MeV and a density of $10^{10}\,\text{m}^{-3}$, in a magnetic field of about 10^{-3} Tesla.

The variable luminosity of Sgr A* in the radiofrequency and infrared regions is about $300\,L_\odot$; for the optical and ultraviolet regions, one estimates $\leq 500\,L_\odot$, and in the X-ray region, $\leq 200\,L_\odot$, so that all together, at most a value of $10^3\,L_\odot$ results.

Observations using very long-baseline interferometry at longer radio wavelengths show that the apparent *size* of the ultracompact source is proportional to the square of the wavelength. It is therefore determined by scattering of the waves in the interstellar medium, i.e. the "point source" is not resolved. Only at a wavelength of $\lambda = 3.5$ mm were T. P. Krichbaum et al. as well as A. E. E. Rogers et al. in 1994 able to approach the resolution limit very closely or even to reach it: Sgr A* is an elliptical source with an (average) apparent angular radius of $\leq 1.2 \cdot 10^{-4''}$ or a radius of $R^* \leq 1$ AU. If it were located at the Sun's position, the central galactic source would extend out only to the Earth's orbit! Its average mass density is $\varrho_0 \geq 6 \cdot 10^{21}\,M_\odot\,\text{pc}^{-3}$, i.e. a factor of 10^{15} higher than the density in the center of the central star cluster.

If we express the measured radius of Sgr A* in terms of the Schwarzschild radius (8.83) of a mass of $M_0 = 2.6 \cdot 10^6\,M_\odot$, we find that $R^* \leq 20\,R_S$, with $R_S = 7.5 \cdot 10^9$ m $= 0.05$ AU. The average mass density ϱ_0 of the ultracompact central source is thus only a factor of $20^3 = 8000$ smaller than the (formal) average density if we were to distribute the mass M_0 within the Schwarzschild radius. Since ϱ_0 is 10^{15} times larger than the densities of the most dense known star clusters, there is hardly another explanation than to assume that a *black hole* of mass M_0 exists at the center of our Milky Way galaxy.

The size of Sgr A* of only $\leq 20\,R_S$ allows no room for an extended ($\simeq 100$–$1000\,R_S$) accretion disk, as is found in the nuclear regions of many active galaxies (Sect. 12.3.6). The matter flow into the black hole is correspondingly modest; it is estimated to be only $10^{-5}\,M_\odot\,\text{yr}^{-1}$. On the other hand, some $10^{-2}\,M_\odot\,\text{yr}^{-1}$ flow inwards through the circumnuclear disk (Sect. 11.3.3), and this mass can not simply "come to a halt". The velocity field in the central region can thus not be stationary, but must instead be variable. The characteristic (dynamic) time scale is of the order of 10^5 yr; it corresponds to about the time which the matter takes to reach the center from a distance of a few pc.

The Activity of the Galactic Center. As we shall see in Sect. 12.3.6, many galaxies, in particular the radio galaxies and the quasars, exhibit in their compact central regions a strong "activity" in the form of high-energy phenomena, which lead e.g. to variable, nonthermal radiation emissions. The energy source for this activity is most probably the accretion of matter from the galaxy into a massive black hole of up to $10^{10}\,M_\odot$, which takes place via an accretion disk.

The processes in the center of the Milky Way galaxy are qualitatively similar to those in the nuclei of active galaxies, but however the degree of activity is considerably less. The Milky Way at present contains only a *"microquasar"* in its center, whose activity is comparable at most to the weak activity in the nuclei of the Seyfert galaxies (which make up about 10% of all spiral galaxies). The current phase of extremely weak nuclear activity could however come to an end, if a stronger mass flow into the central black hole were to develop.

12. Galaxies and Clusters of Galaxies

The leap into deep space outside our Milky Way Galaxy, into the realm of the distant galaxies (or the extragalactic nebulae, as they were formerly called), and the beginnings of a cosmology based on observations, will be considered throughout history to be one of the most important achievements of the 20th century.

We have already mentioned (in Chap. 9) the classical catalogue of Ch. Messier (M) dating from 1784, as well as the catalogues of J. L. E. Dreyer, which grew out of the work of W. and J. Herschel, the New General Catalogue (NGC) of 1890, and the Index Catalogue (IC) of 1895 and 1910, whose notation is still used today. The brightest galaxy, apart from the Magellanic Clouds in the southern sky, is the *Andromeda Galaxy* (Fig. 12.1); it was already observed in 1612 by S. Marius and has, for example, the catalogue numbers M 31 or NGC 224.

In more recent times, we can list the Shapley–Ames Catalogue of 1932, which covers the entire sky and contains 1249 galaxies brighter than about 13.5 mag, and the Revised Shapley–Ames Catalogue by A. R. Sandage and G. A. Tammann (1981). The Revised New General Catalogue (RNGC) of J. W. Sulentic and W. G. Tifft (1973) is based on the Palomar Observatory Sky Survey. The Reference Catalogue of Bright Galaxies by G. and A. de Vaucouleurs (1964, and 1976 with A. Corwin) contains data for 4364 galaxies brighter than 16 mag. We also should mention the Catalogue of Galaxies and Clusters of Galaxies by B. A. Vorontsov–Velyaminov et al. (1962/74). The Hubble Atlas of Galaxies by A. R. Sandage (1961) gives an excellent overview of the subject, as does the Atlas of Peculiar Galaxies by H. Arp (1966).

In Sect. 12.1, we begin with an overview of methods for distance determination and of the various types of galaxies, and then discuss the normal galaxies, i.e. mainly the spiral galaxies like our Milky Way, and the elliptical galaxies. Then we turn to more exotic galaxies, treating the infrared galaxies and the starburst galaxies (where a phase of intensive star formation has occurred relatively recently) in Sect. 12.2, and the radio galaxies and quasars in Sect. 12.3, with their strong activities in the central regions. Often, galaxies occur in groups or clusters, which in turn may belong to even larger structures, the superclusters; we shall consider these various systems of galaxies in Sect. 12.4. Finally, in Sect. 12.5, we treat the problems of the formation and the evolution of the galaxies and clusters of galaxies or galaxy clusters, which are closely connected to the interactions of the galaxies with each other.

12.1 Normal Galaxies

Before we consider the types of normal galaxies in detail, we first deal in Sect. 12.1.1 with the methods for determining the distances to the galaxies, and give an overview in Sect. 12.1.2 of the variety of forms which galaxies can take, along with their classification and absolute magnitudes. In Sect. 12.1.3, we consider the distribution of luminosities among the galaxies, and in Sect. 12.1.4, their diameters in relation to their radial brightness distributions. In Sect. 12.1.5, we discuss the dynamic behavior of stars and gas within a galaxy, and the mass distribution which can be derived from it, including the "dark matter", whose nature remains unknown. The various stellar populations of the galaxies, and their chemical compositions, are treated in Sect. 12.1.6; and finally, in Sect. 12.1.7, the galactic gas and dust components.

12.1.1 Distance Determination

The position in the cosmos of the "spiral nebulae" was the subject of heated debate among astronomers in the 1920's. We can refer here only briefly to the major contributions of H. Shapley, H. Curtis, K. Lundmark, and many others. Then, in 1924, E. Hubble succeeded in resolving stars in the outer portions of the Andromeda Galaxy, M 31, and in some other galaxies; and, as a basis for photometric distance determinations, in identifying various objects with known absolute magnitudes.

In the following paragraphs, we give a selection of the most important objects which can be used for the determination of distances to the galaxies.

The Cepheids. Their light curves (Fig. 12.2) with periods of 2 to 80 d were at first naturally interpreted in terms of the "general" period–luminosity relation. Only

12.1 Normal Galaxies

Fig. 12.1. The Andromeda Galaxy, M 31 = NCG 224, and its companions, the elliptical galaxies M 32 = NGC 221, and NGC 205 *(lower left)*. The inclination of the Andromeda Galaxy is 78° and its distance from the Sun is 690 kpc. The image was made with the Mt. Palomar 48″ Schmidt telescope

in 1952 did W. Baade and others recognize, in part due to discrepancies concerning the absolute magnitudes of red giant stars, that the "classical" Cepheids of Population I (δ Cep stars), thus in particular the Cepheids with long periods in the Andromeda Galaxy, are about 1.5 mag *brighter* than the corresponding Cepheids of Population II (W Vir stars). This led to an "expansion" of all extragalactic distances by roughly a factor of 2. At present, the relation between the period Π and the mean brightness M_V is well established, apart from small corrections which depend on the color $(B-V)$ or on the amplitude and the metal abundance of the star (Sect. 7.4.1). Observing values of M_V of -2 to -7 mag with the Hubble Space Telescope, we can penetrate into deep space out to distances of around 20 Mpc ($65 \cdot 10^6$ light years), i.e. out to the Virgo cluster (Sect. 12.4.2); cf. Fig. 12.3.

Novae. Along with the Cepheids, Hubble was able to identify novae, whose light curves are exactly similar to those of galactic novae, and use them for distance determinations. The much brighter "nova" S And, with $V_{max} \simeq 6$ mag, observed in 1885 by E. Hartwig, was later recognized to have been a supernova (SN 1885A). Novae are, indeed, much brighter than the Cepheids at their maxima, with M_V up to -10 mag, but their brightness *decay* must be measured, so that they permit distance determinations out to only about 20 Mpc.

The Brightest Stars and Globular Clusters. The absolutely brightest stars, such as the luminous blue variables with $M_V \simeq -10$ mag (Sect. 7.4.4), can also be used for distance determinations; however, some of the "brightest stars" later turned out to be groups of stars or H II regions.

The brightest globular clusters have magnitudes of about $M_V \simeq -10$ mag in the Milky Way (Sect. 9.2.1), up to -12 mag (in other large galaxies) and thus also permit distance determinations out to several 10 Mpc.

H II Regions. The angular diameters of the bright H II regions determined from Hα images were found after precise calibration to be an excellent aid to the measurement of great distances of up to ≤ 100 Mpc. This method is however hardly used now because of its uncertainty (dependence on the type of galaxy).

Supernovae. Because of their bright absolute magnitudes, these variables offer the possibility of reaching out to much greater distances, up to several Gpc (Gigaparsec). They were recognized as a separate phenomenon from the novae in 1934 by W. Baade and F. Zwicky. In particular, they allow distance determinations for galaxy *clusters*, in which the scarcity of supernova explosions in each individual galaxy is balanced by the large number of galaxies in the cluster.

The most suitable as "standard candles" are the supernovae of type Ia in elliptical galaxies, since their light suffers only a slight extinction in the parent galaxy and their light curves are all very similar (Sect. 7.4.7). Their absolute magnitude at maximum is, to be sure, not independent of the calibration of the extragalactic distance scale (7.110).

A distance determination using supernovae which is independent of the Hubble constant can be carried

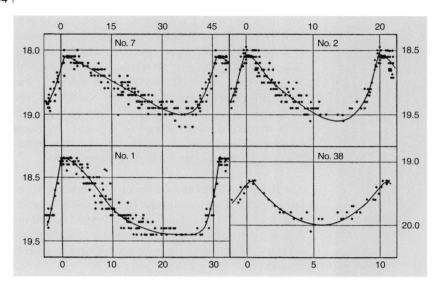

Fig. 12.2. Light curves from four Cepheids in the Andromeda Galaxy, after E. Hubble (1929). The *abscissa* is in days and the *ordinate* is the apparent photographic magnitude m_{pg} in units of [mag]

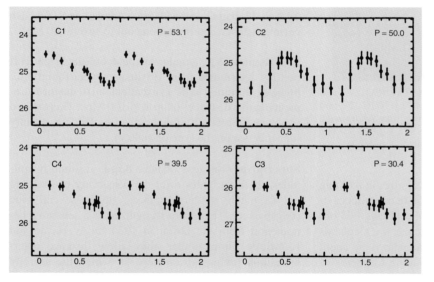

Fig. 12.3. Light curves of four Cepheids within the spiral galaxy M 100 in the Virgo Galaxy Cluster. Observations with the Hubble Space Telescope by W. L. Freedman et al. (1994). *Abscissa:* relative phases; *ordinate:* apparent visual magnitude m_V in units of [mag]; the pulsation period P in [d] is given at the *upper right* in each case. Compare the light curves of these Cepheids, at a distance of around 17 Mpc, with the historic observations by E. Hubble (Fig. 12.2) of the Cepheids in the Andromeda galaxy at a distance of 0.7 Mpc. (With the kind permission of Nature **371**, 757, © 1994 Macmillan Magazines Ltd., and of the authors)

out by a method originally suggested by W. Baade and A. J. Wesselink for the determination of the absolute radii of pulsating variables (Sect. 7.4.1). It requires a measurement of the magnitude at two times t_1 and t_2, and in addition, a determination of the radial velocity V_r on the basis of spectral data: the radiation F_λ emitted from the surface of a star of radius R is, from (4.40) and (4.54), diluted at a distance r by the factor $(R/r)^2$ and attenuated along the line of sight depending on the optical thickness τ_λ, so that a radiation flux

$$f_\lambda = F_\lambda \left(\frac{R}{r}\right)^2 e^{-\tau_\lambda} \qquad (12.1)$$

is observed on the Earth. The ratio of radiation fluxes observed at the times t_1 and t_2 is then:

$$\frac{f_{\lambda,1}}{f_{\lambda,2}} = \frac{F_{\lambda,1}}{F_{\lambda,2}} \left(\frac{R_1}{R_2}\right)^2 . \qquad (12.2)$$

Near the maximum of the light curve, we can estimate F_λ to a sufficiently good approximation from the theory of stellar atmospheres using the spectral energy distribution or the color of the supernova, and thus obtain the ratio R_1/R_2. At the same time, we can calculate the difference of the radii by integration over the radial velocity (7.102):

$$R_2 - R_1 = \int_{t_1}^{t_2} V_r(t)\, dt \qquad (12.3)$$

$$\simeq \frac{V_{r,1} + V_{r,2}}{2}(t_2 - t_1) \;.$$

We thus know both R_1 and R_2, and the distance r is then obtained from (12.1), providing that we can also estimate τ_λ.

Galaxies. In order to probe out further, past the range of the distance criteria calibrated for individual objects in the galaxies, we can, by relaxing our accuracy requirements somewhat, make use of the properties of *entire* galaxies such as their surface luminosities, their morphological luminosity classes (Sect. 12.1.2), or the width of the 21 cm emission line from neutral hydrogen in a galaxy, which is determined by its mass and thus its absolute luminosity (Sect. 12.1.5). Furthermore, the luminosity of the brightest galaxy in a cluster (Sect. 12.4.2) and the action of mass concentrations as *gravitational lenses* (Sects. 8.4.3 and 12.3.4) can also be employed for distance determinations.

Red Shift. Finally, we are left with only the cosmological red shift z (13.2) for estimating the greatest extragalactic distances.

A Hubble constant of $H_0 = 65$ km s^{-1} Mpc^{-1} (13.4) yields

$$r \simeq 4600 \cdot z \text{ Mpc} \qquad (12.4)$$

for small values of z. At larger red shifts, r also depends on the cosmological model (13.46) via the deceleration parameter q_0.

Andromeda Galaxy. For M 31 we obtain, after correction of the (small) interstellar absorption, a true distance modulus $(m - M)_0 = 24.5$ mag, or a distance of 690 kpc. (In his pioneering publication in 1926, E. Hubble had given the distance as 263 kpc.)

The resolution of the central portion of the Andromeda Galaxy, and of the neighboring smaller galaxies M 32 and NGC 205 (Fig. 12.1), was first achieved by W. Baade in 1944 using extremely refined photographic techniques. The brightest stars here are red giants of $M_V \simeq -3$ mag, while the brightest blue stars of the spiral arm Population I are lacking.

Hubble's investigations showed definitively that galaxies such as M 31 are greatly similar to our Milky Way galaxy.

12.1.2 Classification and Absolute Magnitudes

Before we discuss the properties of galaxies in detail, let us first ask how their apparent *distribution* in the sky comes about.

The zone around the galactic equator up to about $b = \pm 20°$, which appears to be nearly free of galaxies and is called the "*zone of avoidance*", is, as we have seen, due to a thin layer of absorbing matter in the equatorial plane of the Milky Way. Thus, for example, the nearby large galaxy Maffei 1, at $b = -0.6°$ and a distance of only about 2 Mpc, remained hidden behind interstellar matter with an extinction of $A_V \simeq 5$ mag and was discovered only in 1970. In 1994, the large barred spiral galaxy Dwingeloo 1 at $b \simeq -0.1°$ and a distance of 3 Mpc was discovered by means of 21 cm observations and then was also found in the infrared and the red spectral regions.

Relatively often, *groups* of two, three,... galaxies are observed, which clearly belong together physically. Our Milky Way forms the *Local Group* together with the Andromeda Galaxy, M 31, its companions, and several additional galaxies (Sect. 12.4.1).

Then, at higher galactic latitudes, where galactic dark clouds are absent, the distribution of distant galaxies in the sky becomes quite inhomogeneous. They form in some cases *clusters of galaxies* (galaxy clusters), first noticed by M. Wolf, such as the Coma Cluster (in Coma Berenices), which contains several thousand galaxies (Sect. 12.4).

Morphological Classification. First, let us take a closer look at the great variety of forms in which galaxies occur! They were classified by E. Hubble, and his scheme, with some improvements, forms the basis of the Hubble

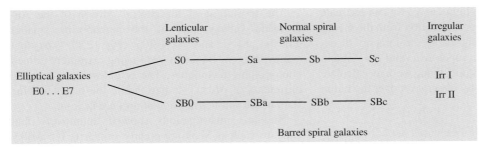

Table 12.1. Classification of galaxies

Atlas of Galaxies (A. Sandage, 1961); it is also precisely defined there. It must be clear from the outset, as is also the case with the Harvard sequence of stellar spectral types, that such a purely descriptive scheme by no means necessarily represents an evolutionary sequence.

In the Hubble–Sandage scheme, the galaxies are classified exclusively on the basis of their *shapes* as seen in *blue*-sensitive photographic plates (Table 12.1):

The *elliptical galaxies*, E0 to E7, have rotationally symmetrical shapes without noticeable structure (Fig. 12.4). The observed ellipticity is naturally in part a result of the projection of the true spheroid onto the celestial sphere, as seen by the observer. The sub-classification of the E galaxies depends on their *apparent* ellipticities, ranging from E0 (circular) to E7 (greatest ellipticity, $(a-b)/a \simeq 0.7$, where a and b are the apparent semimajor and semiminor axes). The true ellipticity is in general greater than the apparent ellipticity. The statistics of apparent ellipticities indicate that the true ellipticities of the E-galaxies are fairly evenly distributed. The surface luminosity of these galaxies decreases uniformly on going outwards from the galactic center.

From the elliptical galaxies, there is a continuous transition through the *lenticular galaxies* S0, which show indications of a disk and a circular absorption structure, to the *spiral galaxies* S. These are without exception more flattened.

The sequence branches at the S0 and S types: the *normal* spiral galaxies S have a central region or core from which the spiral arms grow out more or less symmetrically. In the *barred spirals* SB, first a straight "bar" emerges from the core, and the spiral arms are attached to it almost at right angles (Fig. 12.5). We also distinguish among the lenticular galaxies type SB0, in which a bar is recognizable, but usually does not extend out past the bright core region.

Between the spiral arms in both the normal and the barred galaxies, there are still large numbers of stars, so that the arms do not stand out strongly against this background on photometric recording curves.

The sequence of types S or SB0, a, b, c is characterized by a relative decrease in the size of the *central region*, accompanied by more open curves of the *spiral arms*. For example, the Andromeda Galaxy, M 31 (Fig. 12.1), is a typical Sb spiral; our Milky Way is on the border between Sb and Sc.

Fig. 12.4. The elliptical galaxy NGC 4697, of type E5. Photograph made with the 5 m Mt. Palomar telescope, from the Hubble Atlas of Galaxies

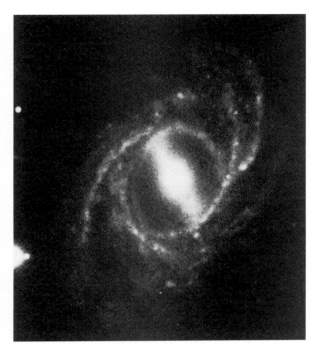

Fig. 12.5. The barred spiral galaxy NGC 2523, of type SBb(r), photographed with the 5 m Mt. Palomar telescope, from the Hubble Atlas of Galaxies

The Hubble Atlas divides the S and SB spirals into two further subclasses, depending on whether the spiral arms are attached directly to the central region or the ends of the bar (suffix s), or are tangential to an inner ring (suffix r).

At the centers of many galaxies, there is a clearly bounded, bright condensation region which is starlike in appearance, the *galactic nucleus*. Its angular diameter is e.g. for M 31 only about 1″, corresponding to about 3 pc. The nuclei of galaxies and the "activity" which is associated with them will be discussed further in Sect. 12.3.6.

The Sc galaxies (such as M 33 in the Local Group) merge continuously into the *irregular galaxies*, Irr I. These relatively rare systems exhibit no rotational symmetry and no readily-visible spiral arms, etc. The best-known examples are our neighbors, the Large Magellanic Cloud (LMC; Fig. 7.40) and the Small Magellanic Cloud (SMC), in the southern sky at a distance of about 50 kpc. Precise observations have shown that the irregular distribution of luminosity in the Irr I systems is caused only by the bright blue stars which are particularly noticeable on blue-sensitive photographs, and the gas nebulae which surround them. The substrate of fainter red stars shows, in agreement with radioastronomical observations of the arrangement and velocity distribution of hydrogen based on the 21 cm line, that the major portion of the mass of e.g. the Magellanic Clouds has a much more regular, flattened shape and undergoes considerable rotation.

The *extension of the Hubble sequence* introduced by G. de Vaucouleurs, continuing past Sc (or SBc) to the types Sd, Sm, and Im (or SBd, SBm, and IBm), takes into account the continuous transition from Sc to Irr I. In this scheme, for example, the Magellanic Clouds are classified as SBm (LMC) and IBm or Im (SMC).

A. Sandage and R. Brucato (1979) redefined the group of the *irregular* galaxies Irr II, introducing a type called *amorphous* galaxies, which is characterized (in blue-sensitive photographs) by a smooth, unresolved shape without spiral arms, sometimes interrupted by dust absorption structures. Occasionally, weak filaments can also be seen. For example, M 82 (Fig. 12.6) belongs to this type. In recent times, it has become clear that the amorphous and other particular galaxies emit most of their radiation in the far infrared. We shall return to these *infrared galaxies* in Sect. 12.2.1.

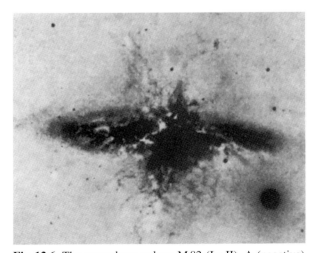

Fig. 12.6. The amorphous galaxy M 82 (Irr II). A (negative) Hα photograph taken by A. R. Sandage (1964) with the 5 m telescope on Mount Palomar

The morphological classification of galaxies according to W. W. Morgan (1958) is based on their brightness concentrations. In this book, we use only the following types from this *Yerkes System*: *D galaxies* have a bright, elliptical core with an unusually extended outer shell, whose brightness decreases more slowly in going outwards than for the E-galaxies. The supergiants among them (see below), the *cD galaxies*, are found as the brightest objects at the centers of many galaxy clusters (Sect. 12.4.2). The *N galaxies* are notable for their "star-shaped", very bright cores against the much fainter background of the rest of the galaxy (Sect. 12.3.5).

We discuss further possibilities for classifying the galaxies, in particular spectroscopic classification schemes, in Sects. 12.1.6 and 12.3.5. We emphasize again that in the Hubble sequence, the galaxies are classified only according to their shapes.

galaxies (whose absolute blue magnitudes are fainter than -16 mag) with a significant spiral structure, i.e. of type Sa–c, since only Sd and Sm dwarf systems were known, and they are similar to the irregular galaxies. In the mid-1990's, however, several dwarf galaxies with well-developed spiral arms were finally discovered; they were hardly noticeable on the usual images due to their small disks and low surface luminosities.

In the case of the elliptical galaxies, the *surface brightness* at the center of the image (proportional to the number of stars along the line of sight) is a measure of the luminosity. In particular, it is found that the brightest E-galaxy of a cluster (the socalled "first-ranked E-galaxy") always has the same absolute magnitude, $M_V = -23.3$ mag, within a very narrow range of variation.

Absolute Magnitudes. The study of the distances to individual galaxies and their grouping into galaxy clusters has shown that the absolute magnitudes and diameters of the galaxies are spread over a wide range within a given Hubble type.

In the class of the spirals, barred spirals, and irregular galaxies, S. van den Bergh found in 1960 that a strong development of the spiral arms and analogous structures is correlated with increasing luminosity. He based a morphological *luminosity classification* on this observation (with notation analogous to that used for stars); it allows a unified calibration according to absolute magnitudes for the given region of Hubble types (see upper right).

Thus, for example, the Andromeda Galaxy, M 31, is classified as Sb I-II, and M 33 as Sc II-III.

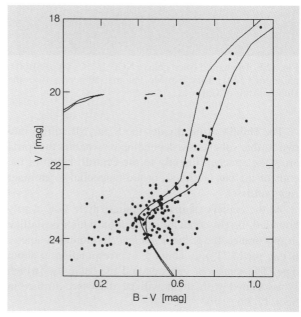

Fig. 12.7. The color–magnitude diagram of the Draco dwarf galaxy. Photometry was carried out using a CCD camera with the Canada-France-Hawaii 3.6 m telescope. The Draco system (distance 60 kpc, diameter 0.3 kpc) has many similarities to a metal-poor globular cluster, as is shown by the comparison to the color–magnitude diagrams of M 92 and M 3 (*solid curves*; the blue giant branch is shown for M 92, cf. Fig. 9.8). However, it is about 10 times larger. From P. B. Stetson et al. (1985); courtesy of the Publications of the Astronomical Society of the Pacific

Luminosity class		Average absolute magnitude M_{pg} [mag]
I	Supergiants	-21.2
II	Bright Giants	-20.3
III	Normal Giants	-19.2
IV	Subgiants	-17.8
V	Dwarf Galaxies	-14.5

While absolutely bright systems are more common among the spiral galaxies, in the class of irregular galaxies, in contrast, the dwarfs predominate. For a long time, it appeared that there was no system among the dwarf

Fig. 12.8. The irregular dwarf galaxy IC 1613 (Im IV). A red-sensitive image made by W. Baade (1953) with the Hale Telescope. At the lower right are several H II regions. In comparison to the Large Magellanic Cloud, IC 1613 has a visual luminosity about 30 times fainter

The overall range of absolute magnitudes is considerably greater for the elliptical galaxies than for the spiral galaxies. Starting with the giant E galaxies, there is a decrease of luminosities amounting to nearly six orders of magnitude in passing through the normal E galaxies and going to the faintest dwarf E galaxies and the spheroidal dwarf galaxies, which are not sharply distinguished from the dwarf E galaxies. An example of a spheroidal dwarf galaxy is the Draco system, with $M_V = -8.6$ mag, which can best be compared to a very large globular cluster; this also accords with its color–magnitude diagram (Fig. 12.7).

In an analogous manner, we compare the *Large Magellanic Cloud*, of $M_V = -18.1$ mag (SBm III; Fig. 7.40) with the irregular dwarf galaxy IC 1613 (Im IV), which also belongs to the Local System and has $M_V = -14.8$ mag (Fig. 12.8). It likewise contains a series of bright OB stars and H II regions.

Small dwarf galaxies often appear to be physically associated with neighboring giant galaxies. Recognizing such mini-galaxies is very difficult even in the Local System (Sect. 12.4.1); outside it, they are practically undetectable.

12.1.3 The Luminosity Function

The luminosity function Φ (Fig. 12.9) contains information about the abundances of galaxies of different absolute magnitudes or luminosities. The distributions determined by E. Hubble (1936) and E. Holmberg (1950) are today of historical interest only. The luminosity function derived by F. Zwicky in 1957 from galaxy cluster data made it clear for the first time that the *dwarf galaxies* (with magnitudes fainter than $M_B \simeq -17$ mag) represent in numbers a considerable fraction of the population of the cosmos, although they have very little weight on the basis of their luminosities and masses.

The number Φ of galaxies (of all types) in a given volume of space with luminosities between L and $L + dL$ can be expressed empirically, following for example P. Schechter (1976), in the form

$$\Phi(L)\, dL = \mathrm{const}_L \cdot \left(\frac{L}{L^*}\right)^\alpha e^{-L/L^*}\, d\left(\frac{L}{L^*}\right)\,. \quad (12.5)$$

For the case that $L \gg L^*$, Φ decreases exponentially, while for faint luminosities, $L \ll L^*$, it follows a power law with the exponent α.

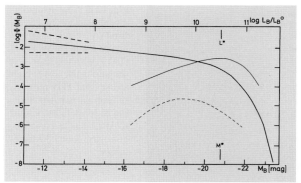

Fig. 12.9. The luminosity function of the galaxies, $\Phi(M_B)$, after P. Schechter (12.3, 5), normalized according to J. E. Felten (1978) for a Hubble constant of $H_0 = 50 \text{ km s}^{-1} \text{ Mpc}^{-1}$. The upper abcissa scale gives the monochromatic luminosity L_B of a galaxy in the blue, at $\lambda = 440$ nm, in solar units, $L_B^\odot = 5 \cdot 10^{23}$ W nm^{-1}. (———) Distribution of field galaxies and galaxies in non-compact clusters ($\alpha = -1.25$); dashed curves at the low-luminosity end: $\alpha = -1.5$ and $\alpha = -1.0$. The upper curve, with $\alpha = -1.5$, corresponds to the luminosity function derived by F. Zwicky in 1957, $\Phi(M) \propto 10^{+0.2M}$. (———) Galaxies in compact clusters, ($\alpha = 0$). (– – –) The first, historical distribution function of E. Hubble (1936), corresponding roughly to a Gaussian distribution, with an arbitrary normalization, reduced approximately to the current value of H_0

If we relate the distribution function not to L, but instead to the absolute magnitude M, we find owing to $\Phi(L)\,dL = \Phi(M)\,dM$ and $dM = -2.5\log(e)\,dL/L$ (6.19) the following relation for the number of galaxies in the interval M to $M + dM$:

$$\Phi(M)\,dM = \text{const}_M \cdot \Xi^{\alpha+1} e^{-\Xi}\,dM \qquad (12.6)$$

with $\Xi = 10^{-0.4(M-M^*)}$.

The brightness distribution in the blue of both the field galaxies and of the galaxies in most of the large (not compact) clusters is well reproduced by a luminosity function of the form of (12.5) or (12.6), with $\alpha \simeq -1.25(\pm 0.25)$ and a characteristic absolute magnitude $M_B^* \simeq -20.8$ mag (corresponding to $M_V^* \simeq -21.7$ mag) or to a characteristic monochromatic luminosity at $\lambda = 440$ nm of $L^* \simeq 3 \cdot 10^{10} L_B^\odot$ with $L_B^\odot = 5 \cdot 10^{23}$ W nm^{-1}. Only the number of the very brightest cD-galaxies in the centers of galaxy clusters is not very well represented by this function.

The *compact* clusters (Sect. 12.4.2), appear, in contrast to the field galaxies, to follow a Schechter distribution with $\alpha \simeq -0.8\ldots 0$. The abundance distribution of the various Hubble types in different galaxy clusters depends, as we shall see in Sect. 12.4.4, on the compactness of the cluster.

The main contribution to the luminosity of all galaxies together, in an arbitrary color system, comes from galaxies with $L \simeq L^*$, i.e. in the blue, from systems which roughly correspond to our own Milky Way or to M 31. The luminosity density $\mathcal{L}_B$ averaged over field and cluster galaxies probably lies within a factor of 3 of

$$\mathcal{L}_B = \int_0^\infty L_B\,\Phi(L_B)\,dL_B \simeq 10^8\,L_B^\odot\,\text{Mpc}^{-3}\ . \qquad (12.7)$$

The uncertainty is due for the most part to the contributions from those galaxies whose surface luminosities are lower than that of the night sky (Low Surface Brightness Galaxies) and which are therefore underrepresented in observations due to selection effects.

12.1.4 Brightness Profiles and Diameters

While nearby galaxies can generally be resolved into their components and individual objects within them can to some extent be distinguished, the only "internal" observational quantity which is accessible for the majority of distant galaxies is the distribution of their surface brightness as projected onto the celestial sphere, in various color systems. As long as the mass–luminosity relation does not vary too strongly, the brightness function also gives us some initial insight into the mass distribution within a galaxy.

Diameter. Since a galaxy does not have a sharp edge, its diameter or radius is not well defined. In practice, its extent is determined in a rather arbitrary way from an *isophote*, i.e. a curve of constant surface brightness, which can readily be measured. As a rule, a *blue* luminosity of 25.0 mag per square arcsec is chosen; this corresponds to about one tenth of the brightness of the night sky, and is also roughly the limiting brightness of the blue plates used for the Palomar sky survey. In contrast, the *Holmberg radius* of a galaxy is defined by an isophote of blue or photographic brightness of 26.5 mag per square arcsec.

The diameters of the galaxies cover a wide range: while dwarf galaxies lie in the region of 0.1 to 10 kpc, the large (normal) galaxies have diameters of up to 30 kpc (spirals) or 50 kpc (E galaxies).

Brightness Profiles. Another method to define the *effective* radius of a galaxy is based on empirically determined laws for the radial distribution of the surface brightness $I(R)$.

In an *elliptical* galaxy as well as in a *spheroidal* system such as the bulge in the Milky Way and other spiral galaxies or also in a globular star cluster, the brightness profile according to E. Hubble, G. de Vaucouleurs and I. R. King can be approximately represented by formulas of the type

$$\frac{I(R)}{I_0} = \frac{1}{(1+R/R_c)^2} \qquad (12.8)$$

or

$$\frac{I(R)}{I_e} = \exp\left\{-7.67\left[(R/R_e)^{1/4} - 1\right]\right\}, \qquad (12.9)$$

if R is not too near to the center or too far out. I_0 is the central intensity; the "core radius" R_c gives the distance from the center at which the intensity has decreased to 1/4 of its central value. I_e is the surface brightness at R_e, where half of the luminosity of the system lies within R_e; from (12.9), $I_0 \simeq 2150\, I_e$.

For the *disks* of the spiral galaxies, an *exponential* distribution of surface brightness is observed at large distances from the center:

$$I(R) = I_0\, e^{-R/R_d} . \qquad (12.10)$$

Elliptical Galaxies. Their effective radii, defined by (12.9), lie in the range of 1 to 10 kpc, and the associated surface brightnesses correspond to around 18 to 21 mag per square arcsec.

The smooth radial dependence of their brightness and their elliptical isophotes with the same axis ratio at first led to the assumption that the elliptical galaxies represent relatively simple systems which are flattened by their rotation. Only at the end of the 1970's was it shown by precise brightness measurements, as well as determinations of their rotational velocities (Sect. 12.1.5) that a considerably more complex description was necessary for the elliptical galaxies: the isophotes within a galaxy can have different shapes and orientations of their semimajor axes; often, the galaxies are not even axially symmetric. There is in fact evidence that perhaps all the elliptical galaxies are genuine *triaxial* systems. Furthermore, their rotation is too slow to have been the cause of their flattened shapes.

Spiral Galaxies. All the S galaxies exhibit the characteristic brightness profile which is known from the Milky Way and the well-studied Andromeda Galaxy, consisting of two main components, the flat disk with an exponential decrease in brightness on going outwards, giving the main contribution to the luminosity, and a spheroidal system consisting of a bulge and a halo. Aside from their typical bar structure in the central region, the SB galaxies also have a similar brightness profile to that of the normal spiral galaxies.

The length scale R_d of (12.10) varies among the early types (S0 to about Sbc) between 0.5 and 5 kpc; for the Milky Way, it is $R_d \simeq 3.5$ kpc. For the later types of galaxies, the maximum values of R_d are around 1 kpc.

According to K. C. Freeman (1970), all spiral galaxies have a notably small scatter of their central surface brightnesses in the blue, with $I_0(B) = (21.65 \pm 0.3)$ mag per square arcsec, after correction for the bulge by extrapolation inwards and subtraction. As a result, the exponential disk would be characterized by only *one* scale length R_d, and its luminosity $\propto I_0 r_0^2$ would likewise be determined uniquely by R_d.

12.1.5 Dynamics and Masses

In the spectra of *entire* galaxies, the existence of *absorption* lines verifies that they, like the Milky Way and neighboring galaxies which can be resolved, consist in the main of *stars*. *Emission lines* from H II regions are observed in the arms of spiral galaxies and, still more noticeably, in the irregular systems. The very broad emission lines from the nuclei of Seyfert galaxies and the amorphous systems like M 82, as well as some giant elliptical galaxies (Sects. 12.2.1 and 12.3.3), clearly have quite a different origin.

The Doppler shifts of the absorption and emission lines (the latter can be more readily measured) give information about the *radial velocity* and the *rotation* of a galaxy. H. W. Babcock (1939) and N. U. Mayall (1950) carried out early radial velocity measurements on H II regions in the nearby Andromeda Galaxy.

Considerable progress in extragalactic radioastronomy was made possible by the construction of the Dutch Synthesis-Radiotelescope in Westerbork, established on the initiative of J. H. Oort. Since 1971, observations of the 21 cm line of H I (and in the continuum) have been carried out with it; thus, the rotational velocities of a large number of galaxies could be determined radioastronomically, and their spiral arms could be resolved in some cases.

Rotation of the Spiral Galaxies. We consider first the rotation of the thoroughly-investigated *Andromeda Galaxy, M 31*. The uncorrected measurement of its radial velocity yields -300 km s^{-1} for the center of the galaxy, i.e. it is approaching us. Since M 31 has the galactic coordinates $l = 121°$ and $b = -21°$, most of this measured velocity is the reflection of our own galactic rotational velocity. After subtracting -300 km s^{-1} from the components of the radial velocity which are symmetric to the galactic center, taking into account the inclination of the rotation axis (78° with respect to the line of sight), and assuming circular orbits, we arrive at the *rotational* velocities. The optical measurements from H II regions and the radioastronomical data using the 21 cm line (Fig. 12.10) show in the main good agreement. The latter are sensitive to parts of the galaxy which are farther out from the center than are the optical observations.

The rotational velocity $V(R)$ as a function of the distance R from the galaxy's center exhibits a steep increase on going outwards, with a maximum of 225 km s^{-1} at $R = 0.4 \text{ kpc}$, then decreases to a minimum at $R \simeq 2 \text{ kpc}$, and finally increases again to a flat maximum of $\simeq 280 \text{ km s}^{-1}$ at $R \simeq 10 \text{ kpc}$. Further out, $V(R)$ decreases only slightly, and remains essentially flat to the outermost point which can be measured with the 21 cm line, at $R_0 \simeq 30 \text{ kpc}$.

In the neighborhood of the "starlike" nucleus of M 31, with a radius of around 3.5 pc (1″), A. Lallemand,

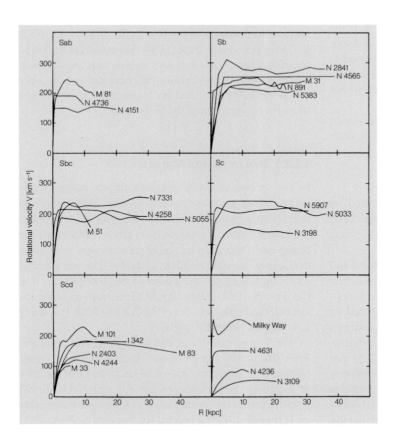

Fig. 12.10. Rotational velocities V in spiral galaxies of different Hubble types as a function of the distance R from the galaxy's center, taken from radioastronomical observations of the 21 cm neutral hydrogen line (N = NGC). From A. Bosma, Astron. J. **86**, 1825 (1981); reproduced with the kind permission of the author and the Astronomical Journal

M. Duchesne and M. F. Walker in 1960 were the first to carry out measurements using an electronic image converter. The maximum velocity in the nucleus is 150 km s^{-1}. We will return to a discussion of the central region of M 31 in connection with the nuclei of "active" galaxies in Sect. 12.3.6.

Sufficiently far *outside* all attractive masses, we expect the Kepler orbit around a galaxy of total mass $\mathcal{M}$ to be given by (2.60):

$$V^2 = \frac{G\mathcal{M}}{R}, \qquad (12.11)$$

i.e. the rotational velocity decreases on going outwards as $V \propto R^{-1/2}$. Observations since the 1970's, especially those using the 21 cm line (Fig. 12.10), however indicate that for many galaxies, out to the furthest observable radii of 30 to 50 kpc, the rotation curves are essentially flat, that is, the Keplerian case is *not* attained. The deviations of the rotation curves of some galaxies such as M 51 and M 81 at smaller radii are probably due to perturbations by their companions.

The Masses of the Spiral Galaxies. Although we cannot directly determine the total masses of galaxies with the present state of the observational art, we can at least find their masses within a well-defined radial distance from their centers. For this purpose, we can use e.g. the radius R_0 out to which radial velocity measurements can be carried out, or a corresponding distance from normal photographic images in the visible region, e.g. the Holmberg radius.

For a *spherically* symmetric mass distribution, the rotational velocity $V(R)$ at a radius R depends only on the mass $\mathcal{M}(R)$ within R:

$$V^2(R) = \frac{G\mathcal{M}(R)}{R}. \qquad (12.12)$$

Using this expression, a flat rotation curve, $V(R) = $ const, corresponds to a linear increase of the mass distribution with radius, $\mathcal{M}(R) \propto R$, or a decrease of the mass density proportional to R^{-2}. A rigid rotation $V(R) \propto R$, as is observed to a good approximation in the inner regions of spiral galaxies, corresponds to an increase of $\mathcal{M}(R) \propto R^3$.

From (12.12), for M 31 within the Holmberg radius of 16 kpc, with a velocity of 230 km s^{-1}, we obtain a mass of $2 \cdot 10^{11}\,\mathcal{M}_\odot$; and inside the furthest radioastronomical observation point (Fig. 12.10) at $R_0 \simeq 30$ kpc, we find a mass which is roughly twice as large, $\mathcal{M}(R_0) \simeq 4 \cdot 10^{11}\,\mathcal{M}_\odot$. According to the 21 cm observations, only about 3% of the mass consists of neutral hydrogen. The "lens" at the center of M 31, inside about 0.4 kpc, has a mass of about $6 \cdot 10^9\,\mathcal{M}_\odot$. For all of that portion of the Andromeda Galaxy which lies inside the Holmberg radius, the mass–luminosity ratio is roughly 8 solar units.

For an arbitrary (non-spherically symmetrical) mass distribution, in contrast to (12.12), the mass *outside* R also influences the rotational velocity $V(R)$. In order to determine the mass distribution, models are constructed which neglect the spiral structure, e.g. using concentric ellipsoids or disks. For these, the potential field and thus the distribution of rotational velocities, $V(R)$, can be calculated. By fitting these model calculations to the observational data, essentially the *surface density* $\mu(R)$ of the mass distribution can be obtained, i.e. the mass contained in a cylinder of cross-sectional area 1 pc^2 and perpendicular to the central plane.

The mass distribution of the spiral galaxies can be well reproduced, neglecting the nucleus itself, by superposing a thin *disk* and a *spheroidal component* which is only slightly flattened (Fig. 11.4). The inner portions of the spheroidal component correspond to the bright central region which can be seen in optical images (the central bulge); outside $R \gtrsim 3 \ldots 5$ kpc, the disk dominates the mass distribution. The *surface density* decreases exponentially, like the surface brightness [Eq. (12.9)], $\propto \exp(-R/R_\mathrm{d})$, since the mass-luminosity relation does not vary strongly within the disk. It falls from 10^3 to $10^4\,\mathcal{M}_\odot$ pc^{-2} in the central region to $\lesssim 10\,\mathcal{M}_\odot$ pc^{-2} in the outermost zones which can still be observed by radioastronomical methods.

As was already indicated by the brightness function, this basic dynamic structure, consisting of a spheroid and a disk, is the same for all spiral galaxies, including our own Milky Way. Going along the Hubble sequence from Sa to Sd and Sm, the mass of the spheroidal component relative to that of the disk decreases, and the mass–luminosity ratio $\mathcal{M}/L_\mathrm{B}$ (within the Holmberg radius) also decreases from about 10 to 2 solar units. Within a galaxy, the mass–luminosity ratio increases with increasing radius.

Dark Matter. Finally, let us turn to the parts of the spiral galaxies which are furthest out from their centers.

The flatness of their rotation curves out to the greatest distances which can still be observed ($R_0 \simeq 30\text{--}50$ kpc) forces us to the conclusion, practically independently of the details of the mass distributions, that a considerable portion if not most of the galactic masses lies *outside* R_0. An acceptable model that would explain the observed $V(R)$ curves would be a roughly spherical mass distribution with a radius of the order of 100 kpc and an overall mass of $\gtrsim 10^{12}\,\mathcal{M}_\odot$. This *dark halo*, also known as the galactic corona, is the continuation outwards of the spheroidal central region and of the (inner) halo of globular clusters. With a mass–luminosity ratio of 100–1000 solar units, it must have a completely different composition from e.g. the known mixture of stars, gas, and dust within the Holmberg radius. What *sorts* of objects may account for the major part of the mass of this halo, whose existence has been inferred only from dynamical considerations, is at present still unknown; we can only speculate on their nature. Current observational techniques would not detect e.g. very cool dwarf stars, planet-like objects, or solid particles with diameters of more than a few cm. In addition, low-energy neutrinos (Sect. 13.3.2) or "exotic" particles postulated by modern elementary-particle theories would be possible carriers of the mass of a dark halo, assuming that they have nonzero rest masses ($\gtrsim 10\,\text{eV}/c^2$).

A massive galactic halo would also explain the relatively large masses of the galaxy clusters, as derived from the scatter in the velocities of their members (Sect. 12.4.2).

The Tully–Fisher Relation. For more distant spiral galaxies whose rotation curves cannot be spatially resolved, the observed Doppler width ΔV_0 of the profile of the *21 cm line*, integrated over the whole galaxy and corrected for its inclination, can be used to estimate their masses; this method was suggested by R. B. Tully and J. R. Fisher in 1977. ΔV_0 is roughly equal to twice the maximum rotational velocity, and therefore from (12.12),

$$(\Delta V_0)^2 \propto \frac{\mathcal{M}}{R}\,. \tag{12.13}$$

When the distance r to the galaxy is known, its observable apparent angular diameter is $\alpha = R/r$, so that the overall mass is given by:

$$\mathcal{M} \propto (\Delta V_0)^2 \alpha r\,. \tag{12.14}$$

Conversely, the Tully–Fisher relation is used to determine the *distances* to galaxies when the mass can be replaced by the luminosity, e.g. from known mass–luminosity ratios. In practice, correlations between ΔV_0 and the absolute magnitudes are established and calibrated empirically using data from nearby galaxies; for this purpose, *infrared* magnitudes are preferable to blue magnitudes owing to absorption by internal dust in the galaxies.

Dynamics of the Elliptical Galaxies. In the case of the brighter, *large* elliptical galaxies ($M_B \leq -20.5$ mag), it is found from the widths of the spectral lines, which are determined by the radial velocities along the line of sight, that in contrast to the spiral galaxies, the large-scale rotation is eclipsed by the *statistical* motions of the stars in the common gravitational field. A measure of the statistical motions is the mean squared velocity $\langle v^2 \rangle$ or the velocity *scatter* $\sigma = \sqrt{\langle v^2 \rangle}$, where v is measured relative to the average velocity. σ depends strongly on the *anisotropy* of the stellar velocity field.

The velocity dispersion in the center of an elliptical galaxy is strongly correlated with its luminosity L (S. M. Faber and R. E. Jackson, 1976):

$$L \propto \sigma_c^n \quad \text{with} \quad 3 \leq n \leq 5\,. \tag{12.15}$$

In the brighter E galaxies, with a velocity dispersion of several $100\,\text{km s}^{-1}$, the rotational velocities are noticeably smaller; on the average, one observes $v_{\text{rot}}/\sigma \simeq 0.2$. In order to explain the flattening of these galaxies by rotation, a factor of 2 to 3 higher rotational velocity would be required. Therefore, the shape of the elliptical galaxies must be determined by the anisotropy of the stellar velocities or by the pressure of the "stellar gas", which opposes the forces of gravitation. This leads to a *"triaxial"* structure of the galaxy, an ellipsoid with three different semimajor axes.

Owing to the long range of the gravitational forces, direct "collisions" of the stars with each other are extremely rare; instead, the behavior of the "stellar gas" is determined by the *collective* gravitational interactions in the overall system ("collision-free" gas). The relaxation time, i.e. the characteristic time between two stellar collisions, is of the order of 10^{15} yr and is thus considerably longer than the age of the Universe, around 10^{10} yr.

The absolutely fainter ($M_B \geq -20.5$ mag), *smaller* elliptical galaxies rotate more rapidly than the larger systems; their maximum rotational velocities and their velocity dispersions are however of the same order of magnitude. The dynamics of these smaller E galaxies, which are dominated by rotation, thus correspond more closely to those of the central regions (bulges or spheroidal components) of the spiral galaxies, if we leave out of consideration the possible barred structures of the latter. The similarity of the two systems extends to their stellar populations (Sect. 12.1.6), so that to a first approximation, we can regard the bulge of a spiral galaxy as a small elliptical galaxy.

In the case of the elliptical *dwarf* galaxies ($M_B \geq -18$ mag), we observe on the average a ratio $v_{\rm rot}/\sigma$ which lies between those of the large and the smaller E galaxies.

The Masses of the Elliptical Galaxies. For galaxies in which the stellar velocity dispersion exceeds the rotational velocity, the total mass $\mathcal{M}_G$ can be estimated by applying the virial theorem, which states that on the average over time, the kinetic energy of the galaxy equals $(-1/2)\times$ its potential energy:

$$\mathcal{M}_G \langle v^2 \rangle = G \int_0^{R_G} \frac{\mathcal{M}(R)\, d\mathcal{M}(R)}{R} = a \frac{G\mathcal{M}_G^2}{R_G} \,. \quad (12.16)$$

Here, R_G is the radius of the galaxy and $\mathcal{M}(R)$ is its mass within a sphere of radius R. The constant a for a homogeneous sphere is equal to 3/5 (2.51). The order of magnitude of the total mass of a galaxy is thus given by

$$\sigma^2 = \langle v^2 \rangle \simeq \frac{G\mathcal{M}_G}{R_G} \,. \quad (12.17)$$

The motions in galaxy clusters can also be used to estimate the masses of elliptical galaxies, again by applying the virial theorem (Sect. 12.4.2).

The (visible) matter in the elliptical galaxies lies in a broad range of masses from about $10^6\, \mathcal{M}_\odot$ in the dwarf systems up to around $10^{12}\, \mathcal{M}_\odot$ in the giant galaxies. One of the most massive galaxies yet observed is the giant E galaxy M 87 ($=$ Vir A $=$ NGC 4486) at a distance of about 15 Mpc. Measurements of its velocity dispersion in the optical region have been made out to an angular radius of $1'$, corresponding to a distance of 4.4 kpc from the center; from them, the mass within 4.4 kpc is found to be roughly $2 \cdot 10^{11}\, \mathcal{M}_\odot$.

With the Einstein satellite, the first X-ray emissions from M 87 were observed; they indicated hot gas ($\leq 3 \cdot 10^7$ K) out to a distance of $100'$ (0.44 Mpc; cf. Sect. 12.4.3). In order to bind this X-ray halo gravitationally to the galaxy, according to model calculations a mass of at least $3 \cdot 10^{13}$ to $10^{14}\, \mathcal{M}_\odot$ would be required within a radius of 0.44 Mpc. Out to this radius, no flattening in the mass–radius function $\mathcal{M}(R)$ can be observed. The hot gas itself, which is readily observable in the X-ray region, contributes only about 5% of this mass. Therefore, analogously to the spiral galaxies, M 87 must have a massive, extended *dark* halo of unknown composition. Many other elliptical galaxies also show evidence for dark matter, which is however more difficult to detect than in the spiral galaxies, owing to the more complex stellar motions in the elliptical galaxies.

12.1.6 Stellar Populations and Element Abundances

We shall now attempt to construct color–magnitude diagrams for other galaxies, as we did for the star clusters in our own galaxy, and then to study their different stars and stellar populations, in order finally to gain an insight into the formation and evolution of the galaxies.

In practice, we are faced with the problem of "light quantum economy". In the past two decades, the construction of larger telescopes and especially the development of better and better detectors (CCDs), as well as the extension of the accessible spectral regions, have all allowed considerable progress to be made in the study of galaxies. A further improvement in observational capabilities in the 1990's resulted from of the availability of the 2.4 m Hubble Space Telescope.

Apart from our nearest cosmic neighbors, we must in general content ourselves with *integrated* spectra for the investigation of stellar populations and element abundances in other galaxies, particularly in their central regions. Since the spectrum of a galaxy results from the superposition of many stellar spectra, in the shorter wavelength range the hot, blue stars will dominate the spectrum, while in the longer wavelength range, the cooler red stars will predominate.

Spectral Classification. The classification of spectra from the galaxies is based mostly on images of their central regions or of the whole galaxy, and represents an attempt to gain information about the stellar populations of the galactic system and to correlate these populations with the galaxy's classification in the Hubble sequence.

The early classification system of W. W. Morgan and N. U. Mayall (1957) is limited for the most part to the range $\lambda = 385$ to $410\,\text{nm}$ and distinguishes the following types:

A Systems have broad Balmer lines and spectra corresponding to the spectral type A of the stars. A typical example is the galaxy NGC 4449, similar to the Magellanic Clouds; altogether, the Hubble types Irr I, Sc, and SBc occur in this group.

F systems correspond in the violet to spectral type F. The Sc galaxy M 33 is a typical representative; Sb galaxies also occur in this group.

K systems exhibit spectra which can be interpreted as a superposition of those of normal, not metal-poor G8 to early M giant stars (CN criterion), with fainter F8 to G5 stars. The prototype is M 31. In addition to the large Sb and Sa spirals, the corresponding barred spirals and elliptical giant galaxies such as M 87 are in this group.

Color Indices and Energy Distributions. Closely related to the spectral type is the color index $B - V$ of the galaxies. Spiral galaxies which we see from their edges exhibit absorption and reddening due to their *own* interstellar matter, and this must be corrected for, along with the absorption and reddening in our own galaxy. The *true* color index $(B - V)_0$ shows a close correlation with the Hubble type and likewise with the spectral type of the galaxies. $(B - V)_0$ is on the average 1.0 mag for S0 and giant E galaxies, 0.8 for Sb, and 0.4 mag for Irr galaxies. The corresponding color index $(U - B)_0$ is about $+1.0, 0$, and -0.2 mag, respectively. In the elliptical galaxies, $(B - V)_0$ is about 0.1 mag redder for the absolutely bright systems than for the dwarf systems. Within the spiral galaxies, we observe an increasingly blue color index on going outwards from the center.

Measurements of the energy distribution in an extended spectral range, e.g. using the extended Johnson color system or narrow-band spectrophotometry (Sect. 6.3.2), show that an important contribution to the integrated light from many galaxies falls in the range $\lambda \gtrsim 800\,\text{nm}$. Thus, the *red* and *infrared* spectral regions play an important role in the investigation of stellar populations in galaxies.

At the other end of the spectrum, the intensity below $\lambda \lesssim 400\,\text{nm}$ decreases strongly due to the large number of absorption lines. With decreasing wavelength, stars of earlier types become more and more predominant in the integrated spectra; finally, in the far ultraviolet ($\lambda \lesssim 160\,\text{nm}$), the integrated radiation comes almost exclusively from a relatively small number of OB stars. For nearby galaxies, the IUE satellite was able to register the energy distribution down to about $120\,\text{nm}$ with a spectral resolution of $0.6\,\text{nm}$.

At this point we should mention that all measurements of the energy distribution, colors, etc., are referred to the (fixed) wavelength scale in the observer's system. However, due to the *red shift z* of the galaxies, the wavelength of their light is increased by the factor $(1 + z)$, so that their radiation is, for one thing, "diluted" by this effect, and for another, a shorter-wavelength range (in the system of the emitting galaxy) enters the band chosen for the detector (Sect. 13.2.1). In the case of very distant galaxies, the energy distribution must be corrected for this effect (which was historically referred to as the "K term").

Population Analysis. A further refinement of the analysis of the superposed spectra of galaxies has been achieved since the 1960's using the photoelectric scan technique. For example, H. Spinrad and B. J. Taylor (1969/71) have measured important lines and bands in selected regions of a few nm width between 330 and 1070 nm, on the one hand for galactic spectra, and on the other for stars of known spectral type, luminosity class, metal abundance, etc. They attempted by computer data analysis to characterize a mixture of stars or of stellar populations in such a way that all the narrow-band magnitudes would be correctly reproduced. The obvious question of how far such a decomposition can be carried in detail requires further work.

More recently, due to the development of highly sensitive detector systems in the optical region, stellar populations in galaxies can be investigated on a large scale by *multichannel* spectrophotometry, at bandwidths of only 0.1 to 0.5 nm, i.e. with a spectral resolution corresponding to photographic spectra with about $100\,\text{Å}\,\text{mm}^{-1}$ dispersion.

Important absorption lines or bands for the population analysis of galaxies are e.g. those of CN at $\lambda = 388\,\text{nm}$, Mg I at $\lambda = 520\,\text{nm}$, Na I at $\lambda = 589\,\text{nm}$, and the weaker groups of lines from Fe I = 527 and 533 nm as well as in the infrared from CO at $\lambda = 2.3$ μm and H_2O at $\lambda = 1.9$ μm.

Spiral Galaxies. The spiral galaxies such as M 31 are basically similar to the Milky Way, as we might have expected. Along the Hubble sequence from Sa to Sc and Sd, the proportion of gas, dust, and young stars increases, and the disk component becomes larger in comparison to the central region. The central regions, which are easier to observe in other galaxies than in our own, resemble small elliptical galaxies, judging by the stars they contain. As in the elliptical galaxies, the absolutely brighter (more massive) systems have, on the average, a redder color and a higher metal abundance; the similarities extend also to their dynamic behavior (Sect. 12.1.5).

In M 31, as in the Milky Way system, we find a *metal-rich* population (with stronger CN absorption) towards the center. Concerning the *globular clusters* in M 31, S. van den Bergh was able to verify that they are metal-poor by using their color indices; more precise investigations show, however, that they are somewhat less metal-poor than our own globular clusters.

Irregular (Irr I) Galaxies. Among the irregular galaxies (of types Sm, Im), our immediate neighbors, the *Magellanic Clouds* (LMC and SMC), have been most thoroughly investigated. Their most prominent feature is an extensive Population I with gaseous nebulae, blue OB stars, etc. There is, however, also an old population with red giants, etc. Furthermore, numerous globular clusters have been observed, and their color–magnitude diagrams, at least the upper parts, are known.

The *Globular Cluster* system in the Magellanic Clouds is quite different from that in our Milky Way Galaxy, in terms of the types of its members and its kinematics. This shows that it is by no means a good assumption that galaxies of different Hubble types will contain only those populations which we already know from our own galaxy. There are only a very few "genuine" metal-poor globular clusters with an age similar to those in our galaxy. On the contrary, clusters of a medium age of several 10^9 yr predominate; all the globular clusters are more metal-poor than $\varepsilon \lesssim 1/3$. There are also *young*, massive globular clusters, only 10^7 to 10^8 yr old, with bright *blue* stars, which have no counterparts in our Milky Way system. In contrast to our system, in the Large Magellanic Cloud even the older globular clusters do not form a halo kinematically, but rather move within a disk. Remarkably, the rotational axis of the disk of the older clusters seems to be tilted by about 50° relative to that of the younger clusters and the interstellar gas (observations of the 21 cm line); this is probably a result of the gravitational interaction of the Cloud with our galaxy.

The chemical compositions have been determined for a few *gaseous nebulae* and for several *stars* in the Magellanic Clouds, especially supergiants and blue OB stars. In the LMC, the metal abundance of the stars is only about a factor of 3 smaller than in our galactic Population I, while the relative abundances of the heavier elements are essentially the same as in Population I. On the other hand, the stars of the SMC exhibit a somewhat smaller metal abundance of around 1/5 of the solar values. The formation of stars and their enrichment in heavier elements must have therefore followed rather different paths in the Magellanic Clouds and in the Milky Way.

Elliptical Galaxies. These consist in the main of *old* stars, of which the brightest are red giants. It was already noticed by W. Baade in his early observations that a few systems contain small filaments of dark matter, in which bright blue stars are also found. Newer investigations show that probably all the E galaxies contain a faint *disk* of stars. Our main interest is naturally directed towards the *metal abundances* of the E galaxies, as functions of the luminosities or the (roughly proportional) masses of the galaxies. Many spectrophotometric investigations since about 1970 have shown that the giant E galaxies have normal to somewhat elevated metal abundances, but that with decreasing luminosity, the metal abundance also decreases down to values which approach those of the metal-poor globular clusters. In the dwarf E galaxies, we see apparently (almost?) only metal-poor halo populations.

Although the dwarf elliptical galaxies (dE) and the spheroidal dwarf galaxies (dSph) which continue the series to smaller luminosities ($M_B > -14$ mag) represent "large globular clusters" on the basis of their stellar populations, there are still noticeable differences between them. For one thing, the dwarf galaxies have a more uniform distribution of stars and their central densities are a factor of 100 below those of the more compact globular clusters; for another, their color–magnitude diagrams and the distribution of metal abundances differ in detail. The Fornax system (dSph) in the Local Group (Sect. 12.4.1) shows itself clearly to be a "true" galaxy, since it contains globular star clusters.

The populations of the dE and dSph galaxies consist of intermediate and old stars with a mean metal abundance of $\varepsilon \simeq 1/30$ of the solar value.

Blue Compact Dwarf Galaxies. These systems are, judging by their sizes and spectra, giant *H II regions*; but morphologically, they do not form a homogeneous group. We find irregular and elliptical galaxies, tiny barred spirals, and galaxies interacting with each other, with two or more central regions. Their absolute magnitudes are fainter than $M_V \simeq -17$ mag, and their masses are at most several $10^9 \mathcal{M}_\odot$. From observations of the 21 cm line of H I, a high proportion of gas ($\leq 20\%$) can be derived. The spectroscopic studies of L. Searle and W. L. W. Sargent in 1972, among others, show that some of these galaxies, for example II Zw 40 (notation from F. Zwicky's catalogue of compact galaxies), exhibit on the one hand underabundances of N, O and Ne in their gaseous components, but on the other hand, they have a large population of blue stars. These galaxies are probably "at present" in a stage of intensive star formation, after having remained stable, without stellar genesis, for a long period of time, and can be counted among the starburst galaxies (Sect. 12.2.2).

Abundance Ratio He/H. To conclude this section, we consider the question (which, as we shall see, is very important for cosmology) of the abundance ratio of the two lightest elements, helium and hydrogen, in the different stellar populations. For the spiral arm population and the disk population, He/H is known from analyses of the spectra of O and B stars. The analysis of interstellar gas gives results quite comparable to those for these young stars; here, the He/H ratio can be found from the recombination radiation of gaseous nebulae or H II regions, e.g. in the Orion Nebula, and from radiofrequency transitions between terms of very high principal quantum numbers (Sect. 10.3.1). All of these methods yield roughly He/H $\simeq 1/10$ (atom ratio). The same value is obtained for the Sun and for many planetary nebulas, although the latter represent an advanced stage of stellar evolution. During the formation of their gas shells, unburned matter from the outermost layers of the central star is often cast off. From data for the H II regions of our own galaxy, a weak radial gradient can be derived: He/H decreases from 0.11 in the inner parts of the disk to about 0.07 in the outer zones.

For the halo population, the determination of the He/H ratio meets with the serious difficulty that no helium lines can be expected at the low temperatures of its main sequence and giant stars. We are thus forced to rely on data from the planetary nebulae and some other late evolutionary stages, where there is always a danger that products of nuclear fusion reactions could have gotten into the outer layers. Planetary nebula K 648, which is a member of the globular cluster M 15 and certainly belongs to Population II, shows that even in an object with a low metal abundance, the He/H ratio remains $\simeq 0.1$, like that of the stars of Population I and the disk. This and some other similar observations, along with the theory of the color–magnitude diagrams of the globular clusters, indicates that the two lightest elements H and He were formed in quite a different manner from the heavier elements, of $Z \geq 6$.

As we shall see in Sect. 13.2.4, helium was produced for the most part in the early phases of the cosmos, before the formation of the galaxies and stars; only a small amount was formed, like the heavier elements, within the stars. From observations of H II regions which are far removed from the center of our galaxy, of the gaseous nebulae in irregular galaxies, and of the (metal-poor) blue compact dwarf galaxies, we conclude that the *primordial* helium had an abundance in the range

$$(\text{He/H})_0 = 0.07\text{–}0.08 \quad \text{or}$$
$$Y_0 = 0.22\text{–}0.25 \,, \tag{12.18}$$

where Y_0 is the mass fraction of helium. Therefore, only a fraction $\Delta(\text{He/H}) \lesssim 0.03$ or $\Delta Y \lesssim 0.06$ of the current helium abundance in the extreme Population I is of *stellar* origin.

12.1.7 The Distribution of Gas and Dust

Now that we have come to know the interstellar medium in our own Milky Way Galaxy (Sects. 10.1–4), we turn to the properties and the distribution of gas and dust in *other* galaxies. The variety of observational methods which detect radiations in all parts of the electromagnetic spectrum are in principle the same as those used for the investigation of galactic interstellar matter. As a result of the greater distances to the other galaxies, however, the angular resolution of the instruments and the sensitivity of the detectors must fulfill even more stringent requirements.

We have already discussed the observation of neutral gas using the 21 cm line of H I in Sect. 12.1.5 in connection with the determination of the rotation curves and the mass determination of galaxies. The ionized gas of the H II regions can be observed not only through its emission lines in the optical and infrared regions, in particular the Hα emissions, but also through its thermal (free–free) radiation in the radiofrequency range. The element abundances derived from the H II regions of the spiral galaxies were already discussed in Sect. 12.1.6.

The nonthermal radiofrequency radiation of the normal galaxies – mainly synchrotron radiation from energetic electrons moving in the interstellar magnetic fields – will be treated later in Sect. 12.3.2, together with the observations of radio galaxies.

Molecular hydrogen H_2, which is not accessible to radio astronomy, can – as in the Milky Way – be observed indirectly via the 2.6 mm line of carbon monoxide. New possibilities for extragalactic astronomy were opened up in the 1980's by the availability of large telescopes in the millimeter wavelength range for the observation of the CO lines, as well as through the IRAS satellite (1983) and the ISO (1995/98), which give us a view into the molecular clouds and the regions of star formation in the medium and far infrared.

Gas in Spiral Galaxies. The gas and dust content of the spiral galaxies in general increases on going from Hubble type Sa to Sm/Im; the mass fraction of neutral hydrogen increases from about 0.02 to 0.10.

The *large-scale* distribution of *atomic* gas and dust within the S galaxies is characterized by a strong concentration into a very flat disk, as in our Milky Way

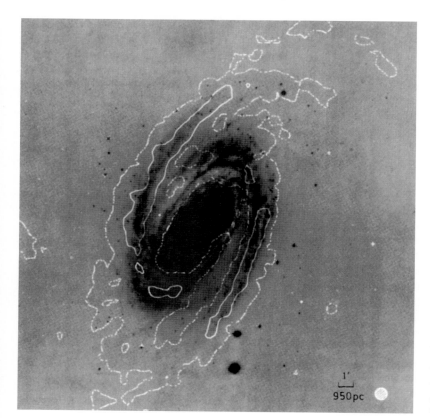

Fig. 12.11. Neutral hydrogen in the spiral galaxy M 81 = NGC 3031 (Sb I–II). From observations of the 21 cm line by A. H. Rots and W. W. Shane (1974) with the Westerbork Synthesis Radiotelescope. The isophotes for column densities of 3 and $10 \cdot 10^{24}$ H atoms m^{-2} are overlaid onto the optical photograph. The galaxy is inclined at an angle of 35° to the line of sight; its distance from the Sun is 3.25 Mpc. The angular resolution of 50″ corresponds to 800 pc

Galaxy. In the Sa and Sb galaxies, as in the Milky Way, we find a minimum in the hydrogen density in the inner part ($R \leq 3$ kpc) of the disk; for later Hubble types, this "gap" is filled in. The layer of H I extends out past one to two Holmberg radii from the center in the early types, and considerably further in the later types.

In galaxies with a suitable inclination, for example M 51 (Sbc I–II), M 101 (Sc I), or M 81 (Sbc I–II), we can investigate the large-scale *spiral structures* by looking at them "from above" (Fig. 12.11); this is of course not possible for our own Milky Way system (Sect. 11.2.2). In those galaxies which we see by chance from the edge on, we can frequently observe a *deflection* and *fanning out* of the H I layer, similar to that known in our own galaxy (Fig. 12.12).

The distribution of *molecular* hydrogen – obtained indirectly from radioastronomical observations of the CO molecule – shows a heterogeneous picture for the spiral galaxies. In many galaxies, the H_2 is strongly concentrated in the central region and decreases continuously within the disk on going outwards. On the other hand, molecular gas is found for M 31 and M 81 only in the disk, and the Milky Way shows a minimum in the H_2 density at 1 to 4 kpc. However, in considering these results it must be remembered that with the current state of observational techniques in the millimeter range, the number of galaxies which can be observed with sufficient resolution is still relatively small and selection effects probably favor those systems with central concentrations.

Within the spiral galaxies, there is no large-scale correlation between the distribution of H_2 and that of atomic hydrogen, H I. However, molecular hydrogen is closely connected with the regions of nonthermal radio emissions and emissions in the far infrared. This can be explained by the fact that hot OB stars are being formed in the molecular clouds, and at the end of their short evolution, they produce high-energy particles and nonthermal radiation in their supernova remnants following supernova explosions (Sect. 10.3.3). The temperature of the interstellar dust, and thus its infrared emission, is determined by the radiation from the stars (see below).

Dust in Spiral Galaxies. The dust in spiral galaxies is warmed to about 10 to 50 K by the background radiation field, as we have seen from the example of our own system (Sect. 10.1.4). Within and in the neighborhood

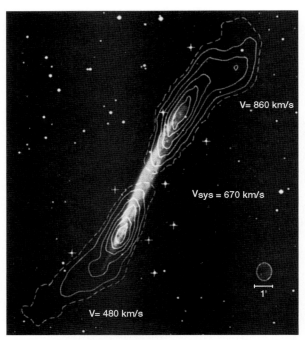

Fig. 12.12. Deflection of the layer of neutral hydrogen in the Sc galaxy NGC 5907, seen nearly on edge. Observations by R. Sancisi (1976) with the Westerbork Systhesis Radiotelescope at an angular resolution of $51'' \cdot 61''$. The isophotes from the 21 cm line of H I are overlaid onto the optical image (a IIIa-J photograph by P. G. van der Kruit and A. Bosma). In order to emphasize the outer parts of the disk, with a radius of 50 kpc, only two velocity channels are shown, centered approximately on the maximum rotational velocity (± 190 km s^{-1}). The radial velocity of the galaxy's center is 670 km s^{-1}

of H II regions, higher temperatures, up to the order of 100 K, are attained due to absorption of the ultraviolet radiation from young OB stars.

Since the absorbed energy is re-emitted in the far *infrared*, we can make use of infrared astronomy to find *regions of stellar genesis*, even when thick dust layers in the molecular clouds allow no optical radiation to pass through.

Infrared observations in nearby spiral galaxies have confirmed the close connection between the radiation of the blue OB stars, the dust zones which can be seen in the optical region, and the emissions of the warm dust in the far infrared. In the Andromeda Galaxy, the infrared radiation – as was first shown by the IRAS satellite in the 60 μm band – comes mainly from a ring which is

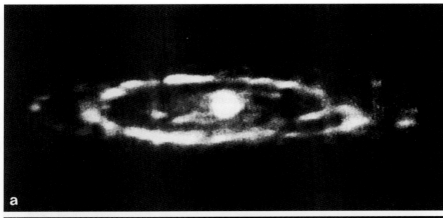
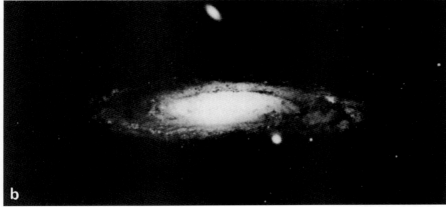

Fig. 12.13a,b. The Andromeda Galaxy, M 31. (**a**) Emission of the warm dust in the far infrared. Observations by the infrared satellite IRAS in the 60 μm wavelength band. (**b**) A blue-sensitive photograph with the Mt. Palomar 1.2 m Schmidt telescope for comparison; from H. J. Habing et al., Astrophys. J. **278**, L59 (1984). (Reprinted courtesy of The University of Chicago Press, The American Astronomical Society, and the authors)

identifiable with gas/dust structures that can be seen in the visible, and from the central region (Fig. 12.13).

Observations in the far infrared from satellites (IRAS, ISO), combined with measurements in the millimeter wavelength range, give us information about the *spectral distribution* of the dust emissions from 12 μm to 1.3 mm and make it possible to determine the dust temperatures T_D and the relative mass fraction of the dust. In the normal spiral galaxies, the spectrum is dominated by the dust components of the disk, with T_D between about 10 and 30 K. The largest contribution to the mass of the dust is that of the *coldest* components.

The luminosity L_{IR} of the S galaxies in the far infrared is correlated with the blue luminosity L_B and increases on going from early to later Hubble types. The strong variations of L_{IR}/L_B among galaxies of the same type are surprising; for example, in M 31, L_{IR}/L_B is only about 0.03, while this ratio is about 0.5 in our own galaxy. In the later Hubble types and especially among the irregular galaxies, we observe values of $L_{IR}/L_B \geq 1$. We will discuss the infrared galaxies, with their extremely high values of L_{IR}, in Sect. 12.2.

The infrared luminosity is however *strictly* correlated with the mass of molecular gas, so that we can characterize the efficiency of star formation in a galaxy globally by the ratio $L_{IR}/\mathcal{M}(H_2)$. In the galaxies with "normal" star formation – such as our Milky Way and other spiral galaxies – this ratio lies at around 5 solar units. In galaxies with heightened rates of star formation (Sect. 12.2), or those with strong activity in their nuclear regions such as the quasars (Sect. 12.3.4), $L_{IR}/\mathcal{M}(H_2)$ is considerably higher.

The Magellanic Clouds. These prototypical irregular galaxies have been thoroughly investigated; they have a high proportion of gas, around 10% of the mass in the

LMC and possibly up to 50% in the SMC. Its distribution and dynamics are complex, especially in the SMC, and reflect the influence of the gravitational interaction with our Milky Way Galaxy on these satellite galaxies. Our knowledge of the molecular gas in the Magellanic Clouds is still incomplete, in particular since the conversion factor relating the observed CO intensity to the H_2 abundance is different from that in the Milky Way. In the Large Cloud, around 2% of the interstellar matter is present in the form of H_2; in the SMC it is $\leq 10\%$.

As we have seen in Sect. 12.1.6, the global chemical composition of the Magellanic Clouds and thus also of their interstellar matter differs from that in the Milky Way. The Large Cloud exhibits an underabundance of metals compared to the solar mixture of around 1/3; the Small Cloud has an underabundance of 1/10. This gives rise to a corresponding underabundance of dust in these galaxies. As shown by the differences in the ultraviolet and the infrared regions of the extinction curve A_λ compared to the Milky Way (Fig. 10.4), the dust also has different properties; in particular, the graphite feature at $\lambda = 220\,\mu m$ is weaker in the Magellanic Clouds.

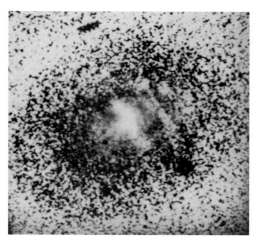

Fig. 12.14. Dust layers in the elliptical galaxy NGC 1052 (type E3, $M_B = -21.1$ mag). The *dust* appears *bright* in this (B–I) image due to the image processing. The picture was made from CCD images taken by W. B. Sparks et al. in 1985 with the Danish 1.5 m telescope on La Silla (Chile). NGC 1052 contains about $10^8\,\mathcal{M}_\odot$ of neutral hydrogen gas. Along with dust and H I, optical emission lines have also been observed in this "active" E-galaxy, for example from [O II], as well as continuum radiation in the radiofrequency range

Elliptical Galaxies. In the E and S0 galaxies – apart from a few cases – we observe no noticeable dust absorption in the optical region, nor in the radio emissions of neutral hydrogen in the 21 cm line. This, along with the lack of young, blue OB stars, indicates that in these systems, star formation *no longer* takes place at a rate worthy of mention.

To be sure, these galaxies are not completely free of interstellar gas, as previously assumed; this has been shown by more sensitive observational techniques as well as by observations in other spectral ranges.

The X-ray emissions of large elliptical galaxies show that they contain very *hot, ionized* gas with temperatures of 10^7 K, representing a mass fraction which is comparable to that of the interstellar matter in the spiral galaxies. Stars, however, can hardly be formed from such a hot, tenuous gas.

Layers of *dust* have also been detected in some E galaxies, using modern image processing techniques with CCD images in different colors; however, their optical thicknesses are only ≤ 0.1 in the visual range (Fig. 12.14).

In the spheroidal dwarf galaxies, so far no interstellar matter has been found. In these systems of very low masses, the escape velocity is too small, so that the gas does not remain gravitationally bound.

12.2 Infrared and Starburst Galaxies

In the normal spiral galaxies, the infrared luminosity, due predominantly to thermal emission from dust heated by stellar radiation, is at most 50% of the luminosity in the optical spectral region (Sect. 12.1.7). In contrast, in the irregular galaxies (Irr II systems) and in the amorphous galaxies (Sect. 12.1.2), among others – as for example in the case of the prototypical M 82 – we observe much stronger emission of infrared radiation.

In the sky survey with the infrared satellite IRAS in 1983, about 50 000 distant galaxies were discovered in the 60 µm band whose ratios L_{IR}/L_B attained values up to the order of 100, and thus exceed that of e.g. M 82 by a large factor. Many of these *infrared galaxies* could be found in the optical region only after their discovery by IRAS, on highly sensitive CCD images. The observations with the ISO satellite in 1995/98, which extend out further into the far infrared than those of IRAS, showed many more infrared galaxies. In these objects also, as in the case of M 82, the energy distribution in the infrared corresponds to the radiation from warm dust. This intense infrared radiation – and the higher dust temperatures than in the normal spiral galaxies which it implies – indicates a particularly high rate of star formation.

12.2.1 Infrared Galaxies

We denote a galaxy as an infrared galaxy when its radiation occurs mainly in the spectral region between about 5 and 500 µm, or when its ratio of infrared brightness to that in the blue, L_{IR}/L_B, is greater than one.

The luminosities of the infrared galaxies span a broad range from about 10^6 up to several $10^{13}\,L_\odot$. Galaxies of more than $10^{11}\,L_\odot$ are termed *luminous*, and those of over $10^{12}\,L_\odot$ *ultra*luminous infrared galaxies. Among all the galaxies having $> 10^{11}\,L_\odot$, the infrared galaxies are the predominant population group. The ultraluminous infrared galaxies are, along with the quasars (Sect. 12.3.4), the absolutely brightest stellar systems in the cosmos; their abundance is about twice as great as that of the quasars.

For the brightest IRAS galaxy, with a red shift (13.1) of $z = \Delta\lambda/\lambda = 2.3$, a luminosity of $3 \cdot 10^{14}\,L_\odot$ was at first derived. Later, it became clear that in this particular case, a galaxy which lies closer to us and by chance on the same line of sight acts as a gravitational lens and amplifies the brightness of the distant galaxy by a factor of 10 or more above its "real" value (cf. Sects. 8.4.3 and 12.3.4).

We will now describe several relatively near infrared galaxies in which details can be observed: these are M 82, at a distance of 3.3 Mpc, NGC 6240, and Arp 220 (this notation refers to H. Arp's Atlas of Peculiar Galaxies, 1966), which are at distances of about 150 and 100 Mpc, respectively, and are both among the most luminous of the nearby infrared galaxies.

Messier 82 = NGC 3034 (Fig. 12.6), which together with the large spiral M 81 (Fig. 12.11) belongs to a nearby group of galaxies, shows distinct filaments on Hα images out to about 4 kpc from its disk, whose velocities decrease on going from the center outwards. M 82 was originally interpreted as an exploding galaxy by C. R. Lynds and A. R. Sandage (1963), with large gas masses being flung out from its center at velocities of several $1000\,\mathrm{km\,s^{-1}}$. Although velocities of only around $100\,\mathrm{km\,s^{-1}}$ are observed, the higher values were assumed to result from the small inclination of the galaxy. The explosion model however was found *not* to be tenable, after N. Wickramasinghe and A. R. Sandage found in 1972 that not only the optical continuum, but also the Hα emissions themselves are partially polarized. Today, we interpret the Hα emission as radiation from the disk which is scattered by dust particles that lie in part far out from the galaxy and move over an extended region; this gives rise to Doppler shifts of the order of $100\,\mathrm{km\,s^{-1}}$. Observations from around 1980 on made it clear that M 82 radiates energy mainly in the far infrared, around $\lambda = 100\,\mu\mathrm{m}$, corresponding to a dust temperature of about 45 K.

The luminosity of about $3 \cdot 10^{10}\,L_\odot$ originates in the main with the large number of OB stars in the star formation regions near the galactic nucleus, which are hidden by dense dust clouds from optical observations. The luminosity in the infrared exceeds that in the blue ($L_{IR}/L_B \simeq 4$), so that M 82 – as already mentioned – can be classified as an infrared galaxy; its light is dominated by strong dust absorption.

M 82 moves together with its dust halo through a still more extended, thinner dust cloud within the M 81 group. This dust distribution, along with kinematic considerations and the bridge of neutral hydrogen between M 82 and M 81, discovered through observations of the 21 cm line, indicate that about $2 \cdot 10^8$ yr ago, there was a near collision between the two galaxies, and the associated tidal interactions must have initiated the high rate of star formation in M 82.

NGC 6240. In this ultraluminous infrared galaxy (Fig. 12.15), in the optical range we observe two *separate* nuclei at a distance of $1.8''$ (0.9 kpc) apart, with differing radial velocities. With the Hubble Space Telescope, these nuclei could be resolved into further condensation regions of 0.1 to $0.2''$ (around 100 pc) diameter, which suggest high rates of star formation. The principal portion of the infrared and radiofrequency radiations as well as the unusually strong emissions in the H_2 line at $2.12\,\mu\mathrm{m}$ come from the central region around the two nuclei. In the case of NGC 6240, we are dealing

Fig. 12.15. The infrared galaxy NGC 6240. A photograph by J. Fried (1982) using a CCD camera with the 2.2 m telescope at Calar Alto (red filter, 5 min exposure time). This galaxy, at a distance of 150 Mpc, shows two distinct nuclei with a separation of 1.8″, as well as noticeable "tidal tails". The radial velocities of the nuclei differ by 150 km s^{-1}

with a collision between two galaxies whose nuclei have not yet coalesced. Here, again, their interactions during this close encounter have apparently caused a burst of star formation.

Arp 220 = IC 4553. In this galaxy, the nuclear region is hidden from view in the optical range by strong dust absorption. In the near infrared, one can observe two nuclei at a diatance of 0.8″ from one another, and with the Hubble Space Telescope, a very complex structure can be seen in the center. Arp 220 exhibits great similarities to NGC 6240, so that we can in this case likewise attribute the strong infrared luminosity of the galaxy and the accompanying high rate of star formation to the fusion of two galaxies.

12.2.2 Starburst Activity

A strongly elevated rate of star formation is probably characteristic of all infrared galaxies. In general, we define a *starburst galaxy* as a system whose luminosity is dominated by a phase of intensive star formation, which is too high to be maintained over the whole lifetime of the galaxy.

For the total luminosity of a starburst galaxy, we thus have $L \simeq L_{\mathrm{IR}} \simeq L(\text{starburst}) \simeq L(\text{OB stars})$. Their global rate of star formation (10.43) is $> 10 \, \mathcal{M}_\odot \, \mathrm{yr}^{-1}$, which in contrast in normal galaxies such as the Milky Way is in the range of a few solar masses per year (Sect. 10.5.6). The ultraluminous infrared galaxies attain rates of some 100 to 1000 $\mathcal{M}_\odot \, \mathrm{yr}^{-1}$. The *duration* of a starburst lies between 10^7 and 10^8 yr.

As we saw in Sect. 10.5.2, star formation regions – and thus also the phase of starburst activity in galaxies – are characterized by young, bright OB stars, which, along with their associated H II regions, are often not observable in the optical spectral range due to the dense dust absorption of the molecular clouds, and can be detected only by means of the thermal radiation of the surrounding dust which is heated by their stars. The occurrence of OB stars (blue color), of H II regions (recombination lines, e.g. from hydrogen), and of strong infrared radiation alone are naturally not sufficient for the identification of a starburst galaxy. Instead, this identification must be supported by additional considerations, in order to distinguish it from the weaker "normal" galactic star formation.

We thus find starburst galaxies at the one end of the scale as *blue compact dwarf galaxies*, i.e. as "giant H II regions" (Sect. 12.1.6), and as *Markarian galaxies*, which show – in spectrograms of low dispersion – a strong ultraviolet excess with respect to normal galaxies. At the other end of the scale, we find the *infrared galaxies* ranging out to the ultraluminous infrared galaxies with luminosities of up to $L_{\mathrm{IR}} \simeq 10^{14} \, L_\odot$.

The infrared galaxies introduced in the previous section, M 82, NGC 6240, and Arp 220, are likewise starburst galaxies. While probably every infrared galaxy is also a starburst galaxy, conversely it is not true that every starburst galaxy is also an infrared galaxy.

Galactic Activity. We can distinguish *two* kinds of galactic activity: on the one hand, the compact *nu-*

clei of galaxies (AGN = *a*ctive *g*alactic *n*uclei) exhibit a variety of high-energy phenomena and nonthermal radiation processes, which we will learn about in Sects. 12.3.4–6. Here, the driving force is the gravitation in the neighborhood of a massive central black hole. On the other hand, there is also, – as we have just seen – starburst activity, with an increased rate of stellar genesis, which is caused by tidal interactions and is likewise often concentrated in the central regions of the galaxies.

Especially the weaker forms of AGN, such as that of the Seyfert galaxies of type Sy2 (Sect. 12.3.5), have similar properties to the starburst regions of galaxies and are not always clearly distinguishable from them. To what extent a certain amount of starburst-like activity always accompanies AGN is not yet clear.

High-resolution ISO spectra in the medium infrared range would seem to offer a possibility of distinguishing starburst galaxies from active galactic nuclei: forbidden emission lines from highly ionized atoms such as [Ne V], [Ne VI] and [Si IX] are typical of AGN, but appear to be lacking in the spectra of star formation regions. According to this criterion, for example, Arp 220, with its strong [Ne II] lines, is a starburst galaxy.

The Cause of Starburst Activity. All observations support the hypothesis that starburst activity is caused by *tidal forces* between colliding galaxies. This interaction is, from (2.69), proportional to $1/r^3$ (r = mutual distance) and is thus particularly strong in *near* "collisions".

The large range of observed infrared luminosities is due to the different types of galaxies involved as well as to the strengths of their interactions, which range from small perturbations to fusion of the two systems. The extreme radiation of the ultraluminous infrared galaxies ($L \geq 10^{12} L_\odot$) is probably always caused by a fusion process between two large galaxies.

We will return to the interactions between galaxies in Sect. 12.4.4, where we treat the galaxy clusters, and in Sects. 12.5.3, 5 dealing with galactic evolution; here, we mention only that such interactions are not at all rare events. They play an important role in the formation of galaxies and in the early phases of their evolution. Perhaps *every* galaxy has experienced in the course of its further evolution a strong perturbation due to the tidal forces of another galaxy (at least) once, and has entered an intensive phase of star formation as a result.

12.3 Radio Galaxies, Quasars, and Activity in Galactic Nuclei

A great number of galaxies, especially large elliptical galaxies, are strong radio sources. In these radio galaxies, emissions in the radiofrequency range predominate over those in the optical region. The *nonthermal* spectra and polarization of these emissions indicate that they are synchrotron radiation, produced by the motion of highly energetic electrons in a magnetic field. We shall begin in Sect. 12.3.1 with the fundamentals of the theory of synchrotron radiation, and give in Sect. 12.3.2 a brief summary of the sources of nonthermal radiation in the Milky Way and in other galaxies. Then, in Sect. 12.3.2, we treat the radio galaxies, discussing Cyg A and Cen A in detail. In general, an "activity" of varying strength can be observed in the *nuclei* of many galaxies; it expresses itself mainly through the production of nonthermal radiation in all wavelength ranges and of energetic particles. In Sect. 12.3.4, we begin with the quasars, the most luminous galaxies with the strongest activities. Then, in Sect. 12.3.5, we turn to the Seyfert galaxies, which have active nuclei. The core of our own Milky Way, where we can observe many well-resolved details, although its activity is rather weak, was already introduced in Sect. 11.3. Finally, in Sect. 12.3.6, we show how all galactic activity phenomena are based on common underlying physical processes, and discuss the origins of the enormous quantities of energy released from quasars and other galactic sources.

12.3.1 Synchrotron Radiation

For the description of extended radiation sources, in addition to the surface brightness (radiation intensity) I_ν, the radiation temperature or *brightness temperature* T_B (cf. also Sect. 6.1.3) is also often used. This is the temperature which a *black body* would be required to have in order to emit precisely the observed radiation intensity I_ν in the radiofrequency range, according to the Rayleigh–Jeans law (4.63), (which is sufficiently accurate in this spectral region):

$$T_B = \frac{c^2}{2k\nu^2} I_\nu \qquad (12.19)$$

or

$$I_\nu \, [\text{W m}^{-2} \, \text{Hz}^{-1} \, \text{sr}^{-1}]$$
$$= 3.08 \cdot 10^{-28} \left(\nu \, [\text{MHz}]\right)^2 T_B \, [\text{K}] \,.$$

For optically thin thermal radiation, we thus find $T_B \propto \nu^{-2}$.

The spectral *energy distribution* of radio sources, i.e. the frequency dependence of the intensity I_ν or of the radiation flux per unit frequency F_ν, is often approximated by a power law:

$$I_\nu \text{ oder } F_\nu \propto \nu^{-\alpha} \,. \qquad (12.20)$$

The exponent α is then called the spectral index (some authors prefer the opposite choice of sign for α). For thermal radiation – e.g. that of H II regions – from an optically thin layer, one finds $\alpha \simeq 0$; thermal radio continua (from optically thin sources) can thus be recognized by the fact that their spectra are nearly independent of ν (cf. Fig. 10.9). From an optically thick layer, one would find $\alpha = -2$.

The *thermal* part of the radiofrequency radiation of the Milky Way was already described in earlier sections. It consists of line emissions from neutral hydrogen (H I regions) at $\lambda = 21$ cm (Sect. 10.2.2) and from the CO molecule at $\lambda = 2.6$ mm, as well as from many other molecules (Sect. 10.2.3). The 21 cm observations of other galaxies were discussed in Sect. 12.1.7. We can observe also the continuum due to free–free radiation from ionized hydrogen, especially in the H II regions and gaseous nebulae, as well as in planetary nebulae.

In contrast, the radiofrequency radiation of the Milky Way, which is spread over the whole sky, and likewise that of the strong radio sources, has a spectrum described approximately by:

$$I_\nu \propto \nu^{-0.8} \quad \text{or} \quad T_B \propto \nu^{-2.8} \,, \qquad (12.21)$$

which, even after absorption and self-absorption are taken into account, *cannot* be attributed to thermal emission. The measured brightness temperatures of $> 10^5$ K for long wavelengths are also hardly permit an interpretation in terms of thermal radiation. Therefore, in 1950, H. Alfvén and N. Herlofson proposed the mechanism of *synchrotron radiation* or *magneto-bremsstrahlung* to explain the nonthermal radio continua. This idea was then taken up and developed by I. S. Shklovsky, V. L. Ginzburg, J. H. Oort and others. It was known to physicists that relativistic electrons (i.e. electrons with velocities $v \simeq c$, whose energy E is considerably greater than their rest energy $m_e c^2 = 0.511$ MeV) moving on circular orbits in the magnetic field of a synchrotron emit intense, continuous radiation in the direction of their motion; its spectrum extends into the far UV. This continuum can be distinguished from free–free radiation or bremsstrahlung by the fact that the acceleration of the electrons is due not to atomic electric fields, but instead to a macroscopic magnetic field $\boldsymbol{B}$.

Theory of Synchrotron Radiation. The theory of synchrotron radiation was developed in 1948/49 by V. V. Vladimirsky and J. Schwinger. It is based on the following considerations:

An electron which is orbiting around the field lines at the cyclotron frequency ω_C (10.31) emits radiation (according to the laws of electrodynamics) in a cone of opening angle

$$\theta \simeq \frac{m_e c^2}{E} = \frac{1}{\gamma} \,, \qquad (12.22)$$

where γ is the Lorentz factor (4.14) (cf. Fig. 12.16). This radiation cone passes the observer rapidly, like the light beam from a lighthouse, so that it produces a series of light flashes, each of duration Δt. The spectral decomposition or, mathematically speaking, the Fourier analysis of this spectrum, taking the relativistic Doppler effect into account (4.18), yields a continuous distribution, with its maximum at the circular frequency

$$\omega_m \simeq \frac{1}{\Delta t} \propto \gamma^3 \omega_C \,. \qquad (12.23)$$

Exact calculation gives for the frequency at the maximum:

$$\nu_m = 0.29 \, \frac{3}{4\pi} \frac{eB_\perp}{m_e} \gamma^2$$
$$= 6.9 \cdot 10^{-2} \frac{eB_\perp}{m_e} \left(\frac{E}{m_e c^2}\right)^2 \,, \qquad (12.24)$$

where $B_\perp$ is the component of the magnetic flux density perpendicular to the direction of motion of the electron with charge $-e$ and energy E. The total radiation at all frequencies is proportional to $B_\perp \nu_m \propto B_\perp^2 E^2$.

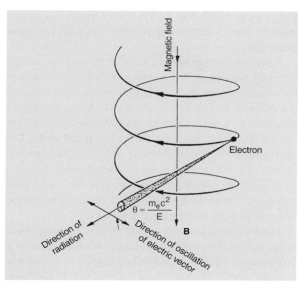

Fig. 12.16. The synchrotron radiation from a relativistic electron in a magnetic field B

Calculating in units which are appropriate for application to cosmic radio sources, e.g. ν_m in [GHz], $B_\perp$ in $[10^{-10}\,\text{T} = 10^{-6}\,\text{G}]$ and E in [GeV], we thus find

$$\nu_m\,[\text{GHz}] = 4.7 \cdot 10^{-3}\, B_\perp\,[10^{-10}\,\text{T}]\,(E\,[\text{GeV}])^2\,. \tag{12.25}$$

Thus, a characteristic galactic magnetic field of 10^{-10} T yields the following electron energies E at various observed characteristic frequencies ν_m or wavelengths λ_m:

ν_m [GHz]	10^{-2}	1	10^2	10^4	10^6
λ_m	30 m	30 cm	3 mm	30 μm	300 nm
E [GeV]	1.5	15	$1.5 \cdot 10^2$	$1.5 \cdot 10^3$	$1.5 \cdot 10^4$

(12.26)

Even for the production of radiation in the radiofrequency region, and quite certainly for visible light, electrons with energies in the range of those of *cosmic rays* (Sect. 10.4.2) are required.

If the energy distribution of the electrons can be represented by a power law

$$n(E)\,dE = \text{const} \cdot E^{-p}\,dE\,, \tag{12.27}$$

then for the emission coefficient of the synchrotron radiation, we find

$$j_\nu \propto B_\perp^{(p+1)/2}\,\nu^{-(p-1)/2}\,, \tag{12.28}$$

and also the same dependence on $B_\perp$ and ν for the intensity I_ν (in the optically-thin case, which usually applies).

For some some compact radio sources, we must also take into account the absorption of the synchrotron radiation (synchrotron self-absorption and free–free absorption); this leads to a flattening out of the spectrum at its low-frequency end.

Changes in the energy distribution $n(E)$ of the electrons as a result of the *energy losses* on ionization (at low energies) or through synchrotron emission and inverse Compton scattering (at high energies) also lead to deviations from the simple power law (12.28).

The Lifetimes of Relativistic Electrons. It is important for the interpretation of radio sources to know the lifetimes of the relativistic electrons. The time $t_{1/2}$ in which an electron loses half of its energy E to synchrotron radiation, $\dot{E} \propto B_\perp \nu_m$, is proportional to $E/\dot{E}$. If we use (12.24) to eliminate the energy $E \propto \nu_m^{1/2} B_\perp^{-1/2}$, we find $t_{1/2} \propto B_\perp^{-3/2} \nu_m^{-1/2}$. The precise value, again in the "cosmic units" used above, is:

$$t_{1/2}\,[\text{a}] = 5.7 \cdot 10^8\,\left(B\,[10^{-10}\,\text{T}]\right)^{-3/2}$$
$$\times \left(\nu_m\,[\text{GHz}]\right)^{-1/2}\,. \tag{12.29}$$

If, for example, we observe synchrotron radiation at 1 GHz ($\lambda = 30$ cm), then in a magnetic field of 10^{-10} T on the one hand, electrons of energy 15 GeV would have been required (12.26); on the other hand, these electrons could emit synchrotron radiation of this frequency only during $6 \cdot 10^8$ yr without being "refreshed". For emissions in the optical range, e.g. at 10^6 GHz (300 nm), the lifetime of the 10^4 GeV electrons is only $6 \cdot 10^5$ yr.

Magnetic Fields and Energy Density. Radio astronomy and likewise X-ray and gamma-ray astronomy hold the possibility of yielding information about the presence of *high-energy* electrons in cosmic objects. This is of great fundamental importance, since the original directional distribution of charged particles arriving at the Earth (cosmic-ray particles) can no longer be determined, due to the complex deflections caused by terrestrial and interplanetary magnetic fields. It is, to be sure, a serious drawback that we still know very little about the magnetic fields which also enter into equation (12.28). We cannot decide without further information how to attribute the observed galactic yield of radiofrequency radiation to the factor $B_\perp^{(p+1)/2}$ and to the density of cosmic electrons $n(E)$ (in the appropriate energy range).

However, with the plausible assumption that the energy densities of the synchrotron electrons plus the "associated" protons and of the magnetic field, $B^2/(2\mu_0)$, are roughly equal, we can obtain the *total energy density*, or more precisely, a lower limit for it.

After these initial theoretical considerations, we turn to the *observation* of the nonthermal radiofrequency radiation of our Milky Way system and other normal galaxies, initially leaving their central regions and nuclei out of our discussion.

12.3.2 The Non-thermal Radiofrequency Emissions of Normal Galaxies

When in 1939 G. Reber investigated the distribution of radiofrequency radiation on the celestial sphere at 167 MHz or $\lambda = 1.8$ m with his radio telescope, which had only a moderate angular resolution, he immediately noticed the concentration of the radiation intensity towards the galactic plane and the center of the galaxy. Then, in 1950, M. Ryle, F. G. Smith and B. Elsmore were able to show that several of the known brighter galaxies emit radiofrequency radiation with about the strength that would be expected from their similarity to our own galaxy. Shortly thereafter, R. Hanbury Brown and C. Hazard succeeded in making a rough determination of the brightness distribution of the Andromeda Galaxy (M 31) in the radiofrequency range; it was found to be very similar to the Milky Way galaxy in this spectral range as well as in the optical region. Finally, the development of more powerful radiotelescopes with high angular resolution (interferometry, aperture synthesis; see Sect. 5.2.1) allowed rapid progress in the investigation of the radio emissions from galaxies in the 1960's.

The Milky Way. In our Milky Way system, we can observe a number of *discrete* nonthermal sources, whose more important types we have already met in previous chapters: the remnants of supernovae, on the one hand the pulsars and their surrounding nebulae (such as the Crab Nebula or the Vela remnant), and on the other the extended ring or circular nebulae (such as the remnant of Tycho's supernova, Fig. 10.15, or the Cygnus Arc); also, the *flare stars*, which are red dwarf stars with eruptions at irregular time intervals.

After subtracting out these discrete sources, an essentially smooth, *diffuse* component remains, which in the range of m and dm wavelengths forms a *thick disk* of the order of several kpc in thickness. Along the galactic equator, several steps in the intensity distribution can be discerned; they are due to the spiral arms. The spectrum of these components, $I_\nu \propto \nu^{-\alpha}$, has spectral indices of $\alpha = 0.4$ to 0.9 in the frequency range from 0.2 to 3 GHz. We interpret this component as synchrotron radiation, produced by the motion of electrons in the large-scale galactic magnetic field.

In addition to the disk component of nonthermal radiation, in the meter wavelength range we observe another component, which is distributed relatively uniformly over the whole celestial sphere; following J. E. Baldwin, it has been ascribed to the *galactic halo*. Measurements with increasingly better angular resolution have however shown that more and more of this "halo component" can be ascribed to discrete galactic and particularly extragalactic radio sources, which are closely spaced in the sky, so that it is at present questionable whether "anything will be left over" for the galactic halo once the many individual discrete sources have been resolved.

The determination of the spatial distribution of the radio sources in our Milky Way system is very difficult because we ourselves are *within* the system.

Spiral Galaxies. We can gain more immediate information about the distribution of radio emissions from *neighboring* galaxies. In those spiral galaxies which we observe from edge-on, there is frequently an ellipsoidal *disk*, which in the continuum, e.g. at 21.2 cm, is noticeably thicker than the intensity distribution of the 21 cm line from H I. Only rarely has a *halo* of nonthermal radiation been observed (Fig. 12.17).

In some favorable cases, the *spiral arms* can also be investigated in the radio continuum: the ridge line, i.e. the intensity maximum of the spiral arms, does not coincide exactly with that seen in blue-sensitive photographs, which corresponds to the maximum density of stars; this can be most clearly seen in the case of M 51 (Sbc I-II). Instead, the radio arms lie along the trailing edges of the "optical" arms, in the area of the

12.3 Radio Galaxies, Quasars, and Activity in Galactic Nuclei

The central regions of the galaxies are probably not the source of many of the synchrotron electrons, since no correlation between the intensities of the nonthermal disk components and those in the central regions is observed. Furthermore, it seems that the amplification of synchrotron radiation in the normal spiral arms (e.g. M 51) is due to compression of the plasma already present, including its magnetic field and synchrotron electrons.

12.3.3 Radio Galaxies

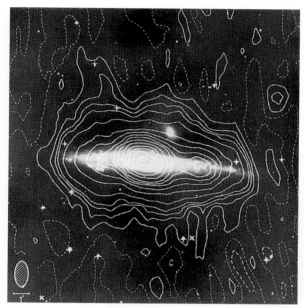

Fig. 12.17. The nonthermal radio halo of the spiral galaxy NGC 4631 (Sc/SBm); from observations by R. D. Ekers and R. Sancisi (1977) with the Westerbork Synthesis Radiotelescope. The isophotes at $\lambda = 0.49$ m (0.61 GHz) are superimposed onto the visible image

In 1946, J. S. Hey and his coworkers detected the first radio source in the Swan, *Cygnus A*, initially on the basis of its intensity fluctuations. Later, simultaneous measurements made at widely separated points showed that these fluctuations result from scintillation in the ionosphere. Like optical scintillation, which is observed for stars but not for planets (with their larger angular diameters), radio scintillation is seen only in the case of sources with a sufficiently small angular diameter. In the quasistellar radio sources or quasars (Sect. 12.3.4), whose angular diameters are less than 1″, another kind of scintillation has been discovered; it originates in the interplanetary plasma.

By 1954, it had become possible to determine the positions of several of the stronger radio sources with sufficient precision that W. Baade and R. Minkowski were able to associate them with visible objects.

In the following paragraphs, we describe the radio structures of the two strong sources Cyg A and Cen A in detail.

Cygnus A. The second-strongest radio source in the northern hemisphere, Cyg A, has been attributed to a surprisingly faint object of photographic magnitude 17.9 mag. Its spectrum exhibits Hα and several forbidden emission lines with a *red shift* of $z = \Delta\lambda/\lambda \simeq 0.056$, corresponding to 16 800 km s^{-1}, along with a weak continuum. It is therefore an *extragalactic* object at a distance of about 350 Mpc.

The optical image of Cyg A shows that it is a giant E-galaxy (cD galaxy) with a dark absorption band caused by dust. W. Baade and R. Minkowski at first interpreted this optical image as two colliding galaxies. The stars would be only slightly disturbed by the collision, but

dust and gas concentrations which are marked by dark clouds and H II regions. D. D. Mathewson et al. have interpreted this, following a suggestion of W. W. Roberts, in terms of the density-wave theory of the spiral arms (Sect. 11.2.5), according to which a compression of the gas and the magnetic field lines is to be expected in these regions. The resulting higher density of the synchrotron electrons and the intensification of the magnetic field would cause an increase in the strength of the radio emissions. The intensity of the radio arms decreases on going outwards from the galaxy's center.

The Origin of the Synchrotron Electrons. The distribution of synchrotron radiation from the spiral arms gives us information about the origin of the relativistic electrons. The emission from a thick disk and the observed proportionality of the mean intensity to the average optical luminosity lead us to suspect that a major portion of the synchrotron radiation originates from the old disk population (Sect. 11.2.6). To what extent supernovae and their remnants can serve as sources and can explain the observed intensity is still not clear.

interstellar gas would be swept out of both galaxies and excited to cause emission of radiofrequency radiation. Later investigations however failed to confirm this hypothesis.

The radiofrequency radiation from Cyg A = 3 C 405 (3 C refers to the Cambridge Catalogue of Discrete Radio Sources) originates for the most part not within the galaxy itself, but in two extended components which are located in nearly symmetric positions far out from its center. This was shown in 1953 by R. Hanbury Brown, R. C. Jennison and M. K. Das Gupta using the correlation interferometer at the Jodrell Bank Observatory. A weaker radio source coincides with the position of the optical galaxy. In Fig. 12.18, we show a more re-

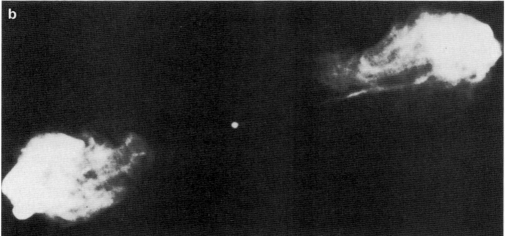

Fig. 12.18a,b. The radio galaxy Cyg A = 3C 405. Radio observations at 5 GHz ($\lambda = 6$ cm) with the VLA in a photographic representation, corresponding to two different "exposure times". The central source coincides with a cD-galaxy observed in the optical region, with a red shift of $z \simeq 0.06$. In (**a**) the structures in the outer emission regions, in particular the "*hot spots*", are easily distinguished; in (**b**) the *jet* which reaches from the center towards the NW (*right*) out to the outer emission region can be seen. The overall extent of the radio double structure is about 200 kpc; the angular resolution of $0.4''$ corresponds to about 0.6 kpc. From R. A. Perley et al., Astrophys. J. **285**, L 35 (1984). (Reprinted courtesy of The University of Chicago Press, The American Astronomical Society, and the authors)

cent radio image of Cyg A, taken with the Very Large Array. The double structure has an overall length of about 0.2 Mpc. In the outer radio sources, we find smaller regions with very strong radio emission, which have relatively sharply defined outer boundaries. These "hot spots" were discovered in 1974 by P. J. Hargrave and M. Ryle with the Cambridge 5 km Aperture Synthesis Radiotelescope (at 5 GHz). Furthermore, a *beam* or *jet* was found, ejected out from the center of one of the outer sources. The overall *radio luminosity* of Cyg A is about 10^{38} W, corresponding to $2 \cdot 10^{11} L_\odot$.

Centaurus A. The nearest radio galaxy to us, at a distance of 5 Mpc, is Centaurus A = NGC 5128 in the southern hemisphere. Here, we can recognize a number of interesting details. The *optical* image initially shows NGC 5128 to be a giant elliptical galaxy (type E0 or S0), surrounded by a band of dark matter perpendicular to its axis (similar to that which originally made Cyg A appear to be two galaxies). Hα observations of H II regions in the band of dust show a rotation curve corresponding to a normal spiral galaxy with a mass of several $10^{11} \mathcal{M}_\odot$. NGC 5128 thus represents an unusual "mixture" of two types of galaxies: a nearly spherical E-galaxy with a diameter of about 10 kpc, which is penetrated (or surrounded) by a rotating disk about 1 kpc thick and filled with dust.

The first *radio* map of Cen A in the continuum at 1.4 GHz made by B. F. C. Cooper, R. M. Price and D. J. Cole in 1965 already permitted the identification of two pairs of plasma clouds, whose separations from the optical galaxy are about 1 Mpc and 10 kpc, respectively (Fig. 12.19). Observations at high resolution with the Very Large Array at 4.9 and 1.5 GHz made in 1983 by J. O. Burns, E. D. Feigelson and E. J. Schreier showed additional structures in Cen A (Fig. 12.20). The isophote chart of the inner pair of radio sources has an S-shaped structure, roughly symmetrical around a central point. This twisting continues in the outer pair. Near the center, we find a *jet* about 1.5 kpc long, consisting of several "knots", which joins the center to the northern inner radio source across a gap of 0.4 kpc. It points in the direction of the axis of the optical galaxy, perpendicular to the dust band. A southern counterpart to this jet is not present. The jet was, incidentally, first discovered in the *X-ray spectrum* by the Einstein Satellite. Since the knots observed in the radiofrequency range

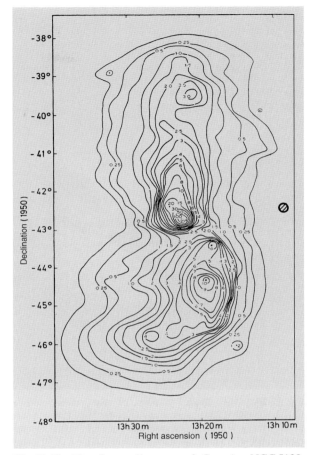

Fig. 12.19. The first radio map of Cen A = NGC 5128: isophotes of the radio continuum at 1.42 GHz (λ = 21 cm), after B. F. Cooper et al. (1965). The brightness temperatures in [K] are shown. The shaded circle of 14′ diameter indicates the angular resolution of the antenna

coincide with those in the X-ray image, it is tempting to interpret the X-ray emission also as synchrotron radiation. To generate it, synchrotron electrons with more than 10^5 GeV (12.25) and certainly also heavier particles of comparable energies must be present. Some of the knots are recognizable in the *optical* region as a nonthermal continuum with emission lines. Within the central radio component, observations using long baseline interferometry at 2.3 GHz were finally able to resolve an additional *inner* jet about 0.05″ or 1 pc long, extending in the same direction as the outer jet.

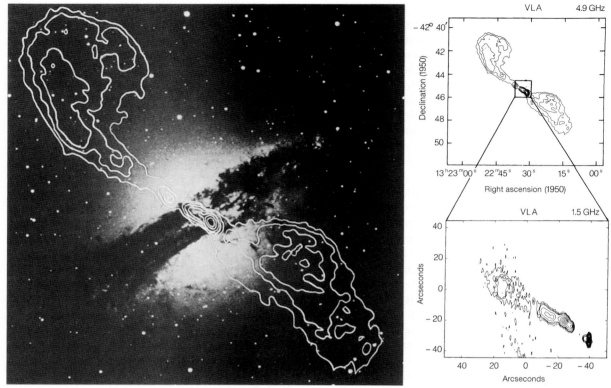

Fig. 12.20. The inner portion of the radio source Cen A = NGC 5128; observations with the Very Large Array at 4.9 GHz ($\lambda = 6$ cm) and a resolution of $15.6'' \cdot 10.1''$ (*upper right*), and at 1.5 GHz ($\lambda = 20$ cm) and $3.65'' \cdot 1.05''$ (*lower right*). The 4.9 GHz isophotes are superimposed at the *left* onto a blue photograph of the central giant elliptical galaxy (E0 or S0 + Spec) taken by J. Graham with the 4 m telescope of the Cerro Tololo Inter-American Observatory. Radio images from J. O. Burns et al., Astrophys. J. **273**, 128 (1983). (Reprinted courtesy of The University of Chicago Press, The American Astronomical Society, and the authors)

Observations of the giant elliptical galaxy M 87 = NGC 4486, the radio source *Virgo A* (cf. also Sect. 12.1.5) first showed that the *optical* continuum with its UV excess (compared to normal galaxies) is also generated by the synchrotron mechanism. A jet was observed to emerge from the center of this galaxy, emitting polarized blue light which could be synchrotron radiation. Later, this jet was detected also in the radio and X-ray ranges.

Structure. Cyg A, with its characteristic double structure of radio emission lying symmetric to the central cD galaxy, is the prototype of a radio galaxy. In general, we define radio galaxies by the fact that their luminosities in the radiofrequency range are greater than those in the optical region.

The *double structure* in the radio emissions is found in more than two-thirds of all strong radio galaxies. Its size lies mainly in the range 0.1 to 0.5 Mpc. The record is held by the giant radio galaxy 3C 236, at a distance of 600 Mpc from the Sun, with a length of 5.6 Mpc. As in the case of Cyg A, we frequently find a central, *compact* ($< 1''$) radio component, particularly in observations at high frequencies; it can usually be identified with the active nucleus of the galaxy (Sect. 12.3.6), and its radiation flux exhibits strong time variations, unlike that of the extended emission regions.

Often, the morphological *classification* of the radio galaxies (after B. L. Fanaroff and J. M. Riley, 1974) is used: in type FR I, the radio emission of cm waves is weaker than in type II and is more strongly concentrated towards the center of the galaxy. Type FR II includes the galaxies of higher radio luminosity with marked double structures, roughly symmetrical and showing "hot spots" at their edges. Frequently, collimated jets are observed. The prototype of a radio galaxy of type FR II is Cyg A, while Cen A is classified as type I.

Radio Spectra and Energy. The spectra of the radio galaxies are nonthermal, i.e. they obey a power law $I_\nu \propto \nu^{-\alpha}$ (12.20) with an average spectral index (in the spectral range between 6 and 11 cm wavelength) of $\langle\alpha\rangle \simeq 0.8$ for the extended double sources and $\langle\alpha\rangle \simeq 0$ for the compact sources. Here, again, we can identify the nonthermal radio emissions as synchrotron radiation from relativistic electrons, whereby the spectrum of the compact sources is flattened due to absorption.

The energy contained in the synchrotron electrons and the magnetic field of a galaxy can be estimated using the theory of synchrotron radiation (Sect. 12.3.1). For the radio galaxies, energies in the range of 10^{49} to 10^{54} J are typical; the magnetic fields lie between 10^{-10} and $2 \cdot 10^{-8}$ Tesla.

The Central Galaxies. Surveys of all the optically identified radio galaxies show that about half, such as Cyg A and Cen A, are giant E/S0 galaxies; the other half are quasars (Sect. 12.3.4). The distinction between the two is however not as sharp as it appears at first glance. Careful investigations of the core regions of Cyg A and some other objects have shown that within them, behind a massive dust ring, a quasar is "hidden". In the case of Cyg A, the dust extinction in the visible is estimated to be 30 to 50 mag! The relations of the different kinds of activity in galactic nuclei to one another will be taken up in more detail in Sect. 12.3.6.

Many radio galaxies emit strong optical lines from their galactic *nuclei*. We distinguish the *B*road *L*ine *R*adio *G*alaxies (BLRG), which show broad allowed emission lines around $8000\,\mathrm{km\,s^{-1}}$ (and narrower forbidden lines) from the *N*arrow *L*ine *R*adio *G*alaxies (NLRG), whose allowed and forbidden lines have widths around $500\,\mathrm{km\,s^{-1}}$. With respect to the emission lines from their nuclei, the quasars and the Seyfert galaxies also show similarities (Sect. 12.3.6).

Radio Jets. While jets have been found in the optical and X-ray regions in only a few radio galaxies and quasars, radio observations, in particular with the Very Large Array, have shown that radio jets are a widespread phenomenon and that relativistic particles are probably transported from the centers to the outer zones of radio galaxies by these jets. The fact that jets are often observed in only *one* direction, as e.g. in the case of Cen A, can probably be understood in terms of the relativistic Doppler effect, which intensifies the radiation from particles which are moving towards us (Sect. 12.3.4).

When the supersonic plasma rays collide with a denser medium (gas in galaxy clusters or intergalactic gas), a shock wave is formed, and it is presumably seen as a "hot spot" in the radiofrequency region. The effective collimation of the jets around their axes over great distances is surprising, as is their stability over periods of 10^6 to 10^7 yr; this is not yet well understood. In some cases, e.g. in the radio galaxy NGC 6251, we see a linear alignment of the jets over dimensions of 1 pc up to 1 Mpc. Crooked or "bent" jets as in Cen A can be explained by perturbations from accompanying galaxies or by precessional motions of the central galactic axis.

An important limitation of any model of a radio galaxy is the *lifetime* of the synchrotron electrons. We estimate for example in Cyg A a magnetic field of $1.5 \cdot 10^{-8}$ T (see above). Then, according to (12.29), the electrons whose synchrotron emissions are observed at 5 GHz can emit this radiation only during a period of about 10^5 yr; however, their time of flight from the center to the edge of the galaxy, a distance of 0.1 Mpc, is in the most favorable case (propagation along straight lines at the velocity of light) equal to $3 \cdot 10^5$ yr. In many cases, the previously suggested explanation of the radio galaxies in terms of a single explosion in the center 10^4 to 10^7 yr ago, causing ejection of clouds of plasma along the axis, meets with difficulties. Instead, there is probably a continuous "energy replacement" along the length of the jets, and an acceleration of the electrons to relativistic energies *in situ* within the shock waves of the hot spots.

In some of the jets from radio galaxies, velocities are observed which are apparently greater than the velocity of light. We will explain this phenomenon, which also occurs in connection with quasars, together with the treatment of those objects in Sect. 12.3.4.

We shall postpone further discussion of the question of the energy source, and first consider the observations of the *nuclei* of the various types of galaxies. These are the locations of "activity", which varies strongly from galaxy to galaxy and also with time; the significance of this activity was pointed out by V. A. Ambarzumian as early as 1954, in a series of articles which were hardly noticed then.

12.3.4 Quasars (Quasistellar Objects)

In the 1960's, it was discovered that a number of by no means faint radio sources from the 3rd Cambridge Catalogue of Discrete Radio Sources (3 C) could be attributed to optical objects that were not distinguishable from stars on photographs taken with the Mt. Palomar 5 m telescope. Their *diameters* of $< 1''$ as determined optically and radio-astronomically made it clear from the start that these new objects have unusually high surface brightnesses in both spectral regions. In 1962/63, M. Schmidt investigated their spectra; they show a *continuum* and strong *emission lines*, similar to those of the radio galaxies already known. The new factor was the enormous *red shifts*, $z = \Delta\lambda/\lambda$, which, when interpreted in terms of the Hubble relation (13.2), indicated that these objects are very distant blue *galaxies*. Their absolute visible magnitudes surpass those of the normal giant galaxies by factors of up to 100; their radio luminosities correspond roughly to that of Cyg A. At this stage of the investigations, the name *quasar* (for quasistellar radio source) was coined.

In 1965, A. Sandage discovered that there are many more quasistellar galaxies, which are optically indistinguishable from the quasistellar radio sources in terms of their compact structures, their high surface brightnesses, and their blue color, but which emit no (or only very weak) radiofrequency radiation. A distinction was thus often made between QSR's = Quasistellar Radio Sources and QSO's = Quasistellar Objects (without radiofrequency emissions). Today, it is clear that both represent the same type of galaxy, of which a small fraction suffers from the "radio disease"; thus, *only* the name "quasars" or "quasistellar objects" is now used, and, as necessary, mention is made of whether or not they emit strong radiofrequency radiation ("radio-loud" or "radio-quiet", respectively).

Discovery and Red Shift. The quasar which is nearest to us (not counting the BL Lac objects, see below), discovered as the first quasar by Maarten Schmidt, is 3 C 273, with $m_V = 12.8$ mag and a red shift of $z = 0.158$ or a distance of about 730 Mpc. All of the other quasars are fainter than 16 mag. Initially, several other quasars were discovered on the basis of their radio emissions. As a result of their *blue* color, quasars with z values which are not too large can also be located in the optical range. In two-color diagrams (U − B, B − V), they can be readily distinguished from stars, since they fall together with the Seyfert nuclei and the N galaxies (Sect. 12.3.5) in a narrow band roughly along the black-body line (Fig. 6.12). To locate quasars with higher z values, optical techniques in the *red* spectral region and furthermore observations in the *infrared* range are employed.

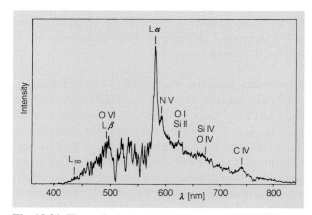

Fig. 12.21. The optical spectrum of the quasar PKS 2000−330 ($m_V \simeq 19$ mag), with a red shift of $z = 3.78$: relative intensity *vs.* observed wavelength. This image was made with the Anglo-Australian 3.9 m telescope, at a resolution of 1 nm. The identification of the strongest lines and the Lyman edge is indicated. In the rest system, Lα is at 121.6 and C IV at 155 nm. The numerous absorption lines on the short-wavelength side of Lα originate from individual hydrogen clouds along the line of sight ("Lα forest"). From B. A. Peterson et al., Astrophys. J. **260**, L 27 (1982). (With the kind permission of The University of Chicago Press, The American Astronomical Society and of the authors)

For several years, the quasar with the largest known red shift was PKS 2000-330 (magnitude $m_V \simeq 19$ mag), with $z = 3.78$. Then, after the mid-1980's, more and more quasars with larger red shifts were discovered in short order, including, up to the year 2000, some 200 quasars with $z > 4$ and several with $z > 5$.

The Sloan Digital Sky Survey (SDSS), begun in 1998 and making use of photometric surveys in the optical region and the near infrared with the 2.5 m telescope of the Apache Point Observatory in New Mexico, has proven to be particularly successful in finding quasars with the highest red shifts. In 2000, the quasar with the (abbreviated) notation SDSS J1044 − 0125 was discovered at a red shift of $z = 5.8$; its apparent brightness in the infrared is about 22 mag, and its absolute brightness (in its rest system at $\lambda = 145$ nm) is -27.2 mag.

In the spectrum of, for example, the quasar PKS 2000–330 ($z = 3.78$), the Lα resonance line of hydrogen is shifted from its rest wavelength at $\lambda_0 = 121.6$ nm to $\lambda = (1+z)\lambda_0 = 581.2$ nm (Fig. 12.21); for SDSS J1044 − 0125 ($z = 5.8$), it is shifted to 826.9 nm. We can thus readily observe the strong ultraviolet emission lines from the distant quasars with $z \geq 4$, including the Lyman line and C IV at $\lambda_0 = 155$ nm, without difficulty in the in the optical and near infrared ranges.

In optical and radio-astronomical sky surveys, several thousand quasars have now been discovered. While the first quasar catalogue (in 1971) included only about 200 objects, for example the (9th) catalogue of M.-P. Véron-Cetty and P. Véron (2000) lists over 13 000 quasars with their measured positions, red shifts, colors, radiation fluxes in the radiofrequency range, etc.

The quasars are also strong X-ray sources, so that the sky survey with the ROSAT X-ray satellite identified among the roughly 60 000 sources around half as candidates for quasars (and active galactic nuclei), whose precise nature must still be verified through spectroscopic observations.

In the case of red shifts which are no longer small compared to one, we cannot use the simple Hubble relation, z proportional to distance, to determine the distance; instead, we need a generalized relation which is found from the theory of General Relativity, given a model for the Universe (13.46).

From the statistics of quasars, we can conclude that their *spatial density* has a marked maximum for red shifts between 2 and 4, and then drops off steeply at larger values of z.

Luminosities. The absolute magnitudes M_V of the quasars are in the range from -24 to -32 mag (for a Hubble constant of $H_0 = 65$ km s^{-1} Mpc^{-1}, Sect. 13.1.1). This corresponds to *visual* luminosities of 10^{12} up to several 10^{14} $L_\odot$ ($4 \cdot 10^{38}$ to around 10^{41} W). Since a substantial portion of their radiation is emitted in other spectral regions as well, the *total* luminosities are considerably larger. Thus, for example, for an "average" quasar such as 3 C 273 the total luminosity is $L \simeq 10^{14}$ $L_\odot$, while in the visual range, only about $5 \cdot 10^{12}$ $L_\odot$ is emitted. The highest observed luminosities range up to several 10^{15} $L_\odot$. The quasars thus represent the *absolutely brightest* stellar systems in the Universe. We usually observe only the extremely bright *nuclei* of these distant galaxies.

The Mother Galaxies. The associated galaxies are recognizable in normal optical images only in the cases of the nearest quasars, such as 3 C 273, as faint, extended nebulous spots; in general, they are washed out by the much brighter nucleus. Only since the mid-1980's, using extremely sensitive CCD detectors (with a large dynamic range!) and modern image-enhancement techniques, has it become possible to detect the associated galaxies of *all* the quasars with $z \leq 0.5$ (Fig. 12.22).

From the luminosity distributions, after subtracting out the quasar itself, absolute magnitudes of about -21 to -23 mag and diameters of around 40 to 150 kpc are derived for the galaxies; these are typical of *large* spiral or elliptical galaxies. Often, deformed structures, double galactic nuclei and other anomalies are observed, indicating strong gravitational interactions. The observations suggest that the quasars with strong radiofrequency emissions possibly belong to E galaxies, and those with no radiofrequency emissions to S or also E galaxies.

Spectral Energy Distributions. In the case of 3 C 273, owing to its relative nearness, almost the entire spectrum can be observed, in particular also the mm and infrared regions. On the whole, the radiation flux over 14 orders of magnitude from the radio up to the gamma-ray range can be represented by a power law, $F_\nu \propto \nu^{-\alpha}$ (12.20), with a spectral index of $\alpha \simeq 1$. It is therefore tempting to

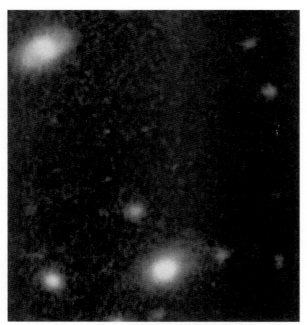

Fig. 12.22. A direct photograph of the quasar QSO 0054 + 144 ($z = 0.171$) taken with the CCD camera of the 2.2 m telescope on Calar Alto by J. Fried in 1983 (red filter, 1 h exposure). The size of the section is about $90' \cdot 90'$; the quasar is the brightest object in the lower half of the picture. The bright quasar is surrounded by a nebulous spot, its mother galaxy, whose luminosity distribution is typical of an elliptical galaxy. (Picture courtesy of J. Fried)

interpret the whole electromagnetic spectrum of these objects as being due to nonthermal synchrotron radiation. However, more precise observations have shown that a power law with a fixed value of α is a good approximation only in a limited range of wavelengths, and that clear deviations from the power-law dependence occur. Furthermore, the spectral energy distribution of the *majority* of the quasars is *not* well reproduced by a global power law. On the contrary, it is seen that the continuous spectrum of the quasars consists of four main components, with relative intensities which vary from object to object:

1. In the *radiofrequency* range, the quasars exhibit a *nonthermal* spectrum, which without doubt can be attributed to *synchrotron* radiation. Its spectral index lies – like that of the radio galaxies (Sect. 12.3.3) – in the range between 0 and 1, whereby the flat spectrum ($\alpha \simeq 0$) probably results from the superposition of radiation from several compact components with different synchrotron self-absorption.

2. The *"infrared bump"* between about 2 and 300 μm is *thermal* radiation from *dust*, which is heated by the (shorter-wavelength) radiation field of the quasar to $T_D \geq 50$ K.

3. The broad *"blue bump"* between around 10 and 300 nm is likewise due to *thermal* radiation. This region is ascribed in the main to the hot portions (10^4 to 10^5 K) of an accretion disk around the central black hole (Sect. 12.3.6).
   Many quasars also show a still smaller UV bump at around 300 nm.

4. In the *X-ray* and the *gamma-ray* regions, the continuum emission follows essentially a power law, and is thus of *nonthermal* origin. To be sure, the energetic X-ray and gamma-ray photons are not generated directly by the synchrotron process, but instead by inverse Compton scattering (4.143) of less energetic photons on relativistic electrons.

A *stellar* component has so far *not* been observed in the nuclei of quasars.

The main portion ($\geq 90\%$) of the luminosity of all quasars comes from the region between about 3 nm (0.4 keV) and 300 μm. In this broad range, from the soft X-ray region out to the far infrared, thermal radiation is predominant; to what extent synchrotron radiation also contributes to the continuum remains an open question. On the other hand, to explain the emissions in the radiofrequency and the hard X-ray and gamma-ray regions, relativistic electrons and nonthermal processes must be invoked.

In detail, *simultaneous* observations in as wide a wavelength range as possible, carried out with considerable effort for selected, nearby quasars – among others, for 3 C 273 – have shown a complex time variability of the intensity and the spectral index, which we cannot treat further here. Furthermore, owing to the different instrumental angular resolutions in the different spectral ranges, the observed radiation flux is due to different emission regions, which leads to uncertainties in the determination of the luminosity, especially in the case of very distant quasars.

It is surprising that the *spectral dependence* of the energy distribution from the infrared to the X-ray region is essentially the same for *all* quasars (except for the BL Lac objects, see below), whereby of course the observed spectrum must be corrected to the rest system of the quasar depending on its red shift. Only in the radio and in the gamma-ray regions are there noticeable differences between individual quasars.

In the radio-quiet quasars, which make up around 90% of all the quasars, the emission decreases sharply for $\lambda \geq 100\,\mu$m towards the radiofrequency range. Their radio luminosity is on the average about a factor of 10^3 weaker than that of the radio-loud quasars (such as e.g. 3 C 273).

In the gamma-ray region, in particular the special group of the BL Lac objects are notable for their strong emissions.

Absorption Lines. In the spectra of many quasars, in addition to the broad emission lines, numerous faint, sharp absorption lines are observed (Fig. 12.21), which for more distant quasars can often be grouped according to their relative intensities into several systems with different absorption red shifts z_{abs}; as a rule, $z_{\mathrm{abs}} \leq z$, where z is the red shift of the quasar as determined from its emission lines. If z_{abs} differs only slightly from z, the absorption must be due to matter in the immediate neighborhood of the quasar, while differences $|z - z_{\mathrm{abs}}|/z \geq 0.01$ can be attributed to absorbing matter along the line of sight between the observer and the quasar. For the more distant quasars, there are numerous sharp absorption lines on the short-wavelength side of the Lα emission line (Fig. 12.21). These are for the most part identifiable as Lα absorption at different red shifts ("Lyman α forest"), which originates in a large number of hydrogen concentrations in the intergalactic medium and in protogalaxies, and in the halos of galaxies along the line of sight (Sect. 12.5.2).

Structure. Observations in the radiofrequency range at high angular resolution show that many quasars consist of two or more components, usually a compact central source ($\ll 1''$) and a more extended, long, thin source (jet), similar to the structure of the central region of Cyg A (Figs. 12.18 and 12.23).

In the case of the nearest quasar, 3 C 273, optical and radio images both show that a thin *jet* (component 3 C 273 A) stretches out to a distance of about $20''$ or 80 kpc. The Hubble Space Telescope was able to re-

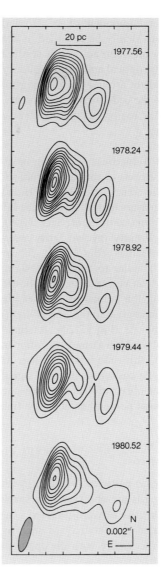

Fig. 12.23. The quasar 3 C 273: observations of the compact component B by very long-baseline interferometry at 10.65 GHz ($\lambda = 2.8$ cm) in the years 1977 to 1980. The expansion of the source of $0.76 \cdot 10^{-3}{''}\,\mathrm{yr}^{-1}$ corresponds to an apparent velocity of about 9 times the velocity of light (for $H_0 = 65\,\mathrm{km\,s^{-1}\,Mpc^{-1}}$). The shaded ellipse indicates the resolution of $4.2 \cdot 1.2$ milliseconds of arc. From T. J. Pearson et al., Nature **290**, 365 (1981). (Reprinted by permission of Macmillan Magazines Ltd., London, and of the authors)

solve individual knots and a twisted spiral structure. Since the jet cannot have propagated outwards at more than the velocity of light, its age must be at least of the order of 10^6 yr.

The actual radio quasar consists of a radio halo 80 pc across (3 C 273 B); within it, there are several smaller, variable components ($\lesssim 10$ pc) and a compact, unresolved component of $\lesssim 0.0004''$ or $\lesssim 1.5$ pc. In the frequency region around $\gtrsim 10$ GHz, the compact component emits the greatest portion of the radiation flux;

the *small size* of the central component of 3 C 273 and other quasars, which can evidently be regarded as the real sources of their energy and their activity, is verified by the *time variability* of the optical and radiofrequency radiation (with an amplitude of around 0.2 to 1 mag); it has characteristic times of the order of weeks to years. The diameter of the radiating region can thus be at most a few tenths of a light year.

Apparent Superluminal Velocities. Very long-baseline interferometric observations of the time variations in the structure of the radio sources, for 3 C 273 B and some other more distant quasars and radio galaxies (as well as galactic stellar gamma sources, cf. Sect. 7.4.8), reveal motions which apparently involve *faster-than-light* or superluminal velocities (Fig. 12.23). These apparent velocities perpendicular to the line of sight, $v_\perp > c$ or $\beta_\perp = v_\perp/c > 1$, can be explained as follows:

Let a source move with the velocity v or $\beta = v/c$ towards the observer (at an angle δ to the line of sight) from P_1 to P_2 in the time Δt (Fig. 12.24). A signal emitted at P_2 reaches the observer at a time Δt_{obs} after an earlier signal emitted at P_1; this time is shorter than Δt by the propagation time $v\Delta t \cos\delta/c$ of the first signal over the distance $P_1 P_1'$. The observer then measures the apparent velocity component perpendicular to the line of sight to be

$$\beta_\perp = \frac{\beta \sin\delta \Delta t}{\Delta t_{\text{obs}}} = \frac{\beta \sin\delta}{1 - \beta \cos\delta} \,. \quad (12.30)$$

Now if a beam of *relativistic* particles is moving with $\beta \simeq 1$ almost directly towards the observer (at a *small* angle δ to the line of sight), its forward-directed synchrotron radiation is emitted by a source (as seen by the observer) moving on the sphere with the apparent superluminal velocity:

$$\beta_\perp \simeq \frac{2}{\delta}, \quad v_\perp \simeq \frac{2c}{\delta} \quad ((1-\beta) \ll \delta^2/2) \quad (12.31)$$

($\sin\delta \simeq \delta$, $\cos\delta \simeq 1 - \delta^2/2$). For e.g. 3 C 273 B, the increase of the distance between two components with $v_\perp/c \simeq 9$ observed in 1977/80 corresponds to a beam angle of $\delta \simeq 2/9$ or around $12°$.

Relativistic Intensity Enhancement. If matter is moving in a jet with the relativistic velocity $\beta = v/c \simeq 1$ towards us, its radiation intensity will be enhanced relative to that of a source at rest in the frame of the observer ("boosting").

We can give a simple expression for the effect of the relativistic motion on the *intensity* or the surface brightness I_ν (4.27); this is not possible for the radiation flux (4.34). Since – as we simply state here without proof – it is not the intensity itself, but rather I_ν/ν^3 which is relativistically invariant (with respect to the Lorentz transformation, (4.13)), an observer measures an intensity at the frequency ν_{obs} of

$$I_{\text{obs}} = \left(\frac{\nu_{\text{obs}}}{\nu}\right)^3 I_\nu \quad (12.32)$$

when the intensity I_ν was emitted at the frequency ν in the rest system of the source. If the source is moving at an angle δ relative to the line of sight, the frequency shift is given by the relativistic Doppler effect (4.18):

$$\frac{\nu_{\text{obs}}}{\nu} = \frac{1}{\gamma(1 - \beta \cos\delta)}, \quad (12.33)$$

where $\gamma = 1/\sqrt{1-\beta^2} = E/m_e c^2$ is the Lorentz factor (4.14) (E = energy of the relativistic electrons); $\delta = 0°$ denotes motion straight towards the observer. The observed intensity relative to that emitted in the rest system of the source is then

$$\frac{I_{\text{obs}}}{I_\nu} = \frac{1}{[\gamma(1 - \beta \cos\delta)]^3} \quad (12.34)$$

$$\simeq \begin{cases} 8\gamma^3 & \text{for } \delta = 0° \\ \gamma^{-3} & 90° \\ \frac{1}{8}\gamma^{-3} & 180° \,. \end{cases}$$

In evaluating this expression, we have set $1+\beta \simeq 2$ and used the identity $1/\gamma^2 = 1 - \beta^2 = (1-\beta)(1+\beta)$.

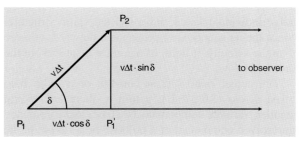

Fig. 12.24. Explanation of the apparent faster-than-light velocities

Since the emission of synchrotron radiation is limited to a small solid angle of the order of $1/\gamma$ (12.13), it is thus strongly anisotropic.

If we now choose a Lorentz factor of e.g. $\gamma = 5$ or $\beta = 0.98$, then the enhancement factor for a jet which is directed straight at the observer is equal to 1000, and for an angle of $\delta = 10°$, it is still 110. A jet moving perpendicular to our line of sight appears to be attenuated by a factor of 125, and one which is directed away from us, by a factor of 1000. This relativistic effect explains fundamentally the frequent observation of only *one* jet from the radio sources.

In a more precise calculation, we would have to take into account the finite bandwidth of the measurement and the spectral dependence of the intensity. For a nonthermal spectrum, $I_\nu \propto \nu^{-\alpha}$ for example the exponent 3 in (12.34) would have to be replaced by $(3+\alpha)$. It is also often expedient to recalculate the observations in terms of a common frequency ν in the rest system of the source.

Our considerations furthermore show that for radiation sources moving at *relativistic* velocities, the "usual" estimate of the luminosity from the apparent magnitude (6.43), which is based on the assumption of isotropic emission, can lead to greatly exaggerated values. Also, because of the time dilation (4.17), the observed time scales of brightness variations are shorter by the factor γ than the true variations in the rest system of the source.

Energy Requirements. The total energy which must be available to a quasar can be estimated from the thermal radiation of the accretion disk and the heated dust as well as the synchrotron radiation emitted. The energy of the relativistic electrons, the associated ions and the magnetic field can be obtained as in the case of the radio galaxies (Sect. 12.3.3) under the assumption of approximate equipartition.

All together, we find a characteristic energy of $\geq 10^{55}$ J, which corresponds to a relativistic rest energy $\mathcal{M}c^2$ of several $10^8 \mathcal{M}_\odot$! The total *emission* of a typical quasar of around $10^{14} L_\odot = 4 \cdot 10^{40}$ W could thus be supplied from this energy reserve for at most 10^7 to 10^8 yr.

The process of energy production itself will be discussed in Sect. 12.3.6, together with the global topic of activity phenomena.

Cosmological Nature. As a result of the enormous energy requirement of the quasars, some astronomers have questioned the determination of their distances on the basis of Hubble's red shift relation. The large red shifts are interpreted by them in some other manner, and the quasars are taken to be "local" phenomena, which for example might have been formed by ejection of matter from other galaxies. In particular, these astronomers point out the frequent occurrence of quasar and galaxy pairs and groups with very small angular spacings, which thus apparently belong together, but which have very different red shifts. A satisfactory statistical investigation of such coincidences has yet to be performed.

On the other hand, a series of other observations speaks convincingly in favor of the *cosmological* interpretation of the red shifts of the quasars:

For one thing, many quasars also belong to groups and clusters in which the normal galaxies have the *same* red shifts as the quasars.

Then, especially, the *direct* observation of the mother galaxies of nearby quasars clearly supports the cosmological interpretation of the red shift. Their luminosities agree with those of nearer large E- and S-galaxies precisely when the Hubble relation is applied to the distance determination of the quasars from their z values.

The "Lα forest" in the spectra of more distant quasars (Fig. 12.21), which is interpreted as due to concentrations of neutral hydrogen in the intergalactic medium along the line of sight *in front of* the quasars (Sect. 12.5.2) also supports their cosmological nature.

Finally, in every respect (energy flux and content, radio structure, activity in the nucleus, belonging to galaxy clusters), there is a continuous transition from the Seyfert galaxies (Sect. 12.3.5) and the radio galaxies (Sect. 12.3.3) to the quasars. This is also apparent in the Hubble diagram (Fig. 13.3).

Gravitational Lenses. A further indication of the great distances to the quasars is given by the "*twin quasars*" QSO 0957 + 561 A and B discovered in 1979 by D. Walsh, R. F. Carswell and R. J. Weymann, as well as similar objects observed more recently. The two quasars A and B, both with magnitudes of $m_V \simeq 17$ mag, are only 6″ apart and have *identical* red shifts ($z = 1.41$) and spectra. It is tempting to interpret them as images of a *single* quasar, whose light rays have passed through a "gravitational lens" on the way to us, i.e. they have been bent in such a way by the gravitational field of a massive galaxy that two images resulted. We have already discussed the fundamentals of gravitational lenses in Sect. 8.4.3. If we take the *extended* mass distribution of the foreground galaxy into account, we find that a gravitational lens can produce three (or more) images, whose brightnesses are in general not the same.

For $0957 + 561$, the B component has in the meantime been resolved into two images with a spacing of $\simeq 0.1''$, and furthermore the galaxy responsible for the deflection, at $z_L = 0.36$ (i.e. at a distance of $\simeq 1800$ Mpc) has been identified on the line connecting A and B, and its mass has been determined to be about $10^{12} \mathcal{M}_\odot$.

At present, we know of over 20 clearly identified gravitational lenses for quasars with a wide variety of arrangements of the multiple images; some Einstein rings have also been observed, mostly in the radiofrequency range. We mention the "cloverleaf" $H 1413 + 117$, with four images at spacings of $1''$ and a red shift of $z = 2.55$, as well as the unusual "Einstein cross" $Q 2237 + 0305$ (Fig. 12.25), with a diameter of $1.8''$ and $z = 1.69$. In its case, the lens is in the center at only $z_L = 0.04$ and is surrounded by four symmetrically arranged images of the quasar.

Fig. 12.25. The Einstein cross, $Q 2237 + 0305$, with a diameter of $1.8''$. A massive spiral galaxy at a red shift of $z = 0.04$ acts as a gravitational lens for a distant quasar which lies behind it (at $z = 1.69$) and produces four images which are arranged roughly symmetrically around its center. These two pictures, made in August 1991 (*left*) and August 1994 (*right*) with the William Herschel 4.20 m telescope on La Palma, show the changes in the brightness of the quasar images within three years' time. These were caused by foreground stars acting as micro-gravitational lenses (Sect. 8.4.3). (Picture: G. Lewis and M. Irwin, William Herschel Telescope, © Royal Astronomical Society)

Deflection by a gravitational lens leads to *time-of-flight differences* in the different wavefronts which reach the observer from a quasar. A noticeable variation in the brightness of the quasar thus appears to the observer at different times in the different images. Now, if it is possible to identify the intensity variations in the different images as due to *one and the same* event in the source (and this is, in fact, the difficult part), then from the time difference and the angular spacing of the images, the *distance* to the quasar can be determined independently of the red shift z. This method is therefore independent of the Hubble relation (13.2) and can be used to determine H_0.

In practice, the clear-cut identification of "associated" brightness variations in the different images is difficult. For the first time, a time-of-flight difference of 1.1 yr was detected in the case of QSO $0957 + 561$.

Another determination of H_0 was carried out in 1996 using the "fourfold quasar" $PG 1115 + 08$, with a red shift of $z = 1.72$ (gravitational lens: $z_L = 0.29$). Here, the time-of-flight differences between the various images lie in the range of 10 to 25 d. Both quasar systems yield a value for the Hubble constant of 50 to 60 km s^{-1} Mpc^{-1} with an uncertainty of about 25 to 30%.

The *amplification* of the intensity by gravitational lenses (8.90) raises the question as to what extent the high luminosities of the quasars may be simply "artificial". Are only a few objects affected, or have the statistics of the quasars been markedly influenced by lens effects? A reliable answer to this question is at present not available.

BL Lacertae Objects. Closely related to the quasars are the BL Lacertae objects, named for BL Lac $= 2200 + 420$ ($m_V = 14.4$ mag, $z = 0.069$), an object which was for a long time "unrecognized" and listed as a variable star in the catalogues. The characteristic features of BL Lac objects are their extremely faint emission lines superimposed onto a structureless optical continuum in their spectra, the strong ($\leq 30\%$) time-variable linear polarization of their emissions in the optical range, and their striking irregular variability of ≥ 1 mag over all wavelength ranges on a time scale of months down to days and hours, at the short-wavelength end even minutes. During strong outbursts (flares), their brightness increases up to 4 mag.

From the mm range out to the X-ray region, the continuum obeys approximately a power law; the radio spectrum is very flat ($\alpha \simeq 0$). The BL Lac objects are similar in many of their properties to the radio-loud quasars with their flat radio spectra. For some objects, in the radiofrequency range and – with the Hubble Space Telescope – in the optical region, jets up to several 100 pc long with knot-like concentrations have been observed. The BL Lac objects are found mainly at relatively small red shifts ($z < 0.2$); their mother galaxies are probably elliptical galaxies.

At higher red shifts of $z = 0.1 \ldots \geq 2$, we find the *OVV quasars* (OVV = Optically Violently Variable), a related group of radio-loud, strongly variable quasars with flat radio spectra, which however in contrast to the "classic" BL Lac objects exhibit very broad, strong emission lines in the optical spectral region.

The phenomena which occur in connection with the BL Lac objects and OVV quasars are often summarized under the term *blasar* (a neologism derived from *BL Lac object* and qu*asar*). At present, we know of about 350 such objects. The origin of the blasar phenomena is a *relativistic jet*, which is directed almost exactly towards us and thus produces both the high luminosities – owing to relativistic amplification of the apparent radiation intensity – and the short time scales of the variability, due to the time dilation (see above). For a large fraction of the BL Lac objects, apparent superluminal velocities are also observed. The object of this type which is nearest to us is Mrk 421 (Mrk: Markarian galaxy, Sect. 12.3.5), at $z = 0.031$.

With the Compton Gamma-Ray Observatory a further related group, that of the *gamma-blasars*, was discovered in 1991; in the meantime, it numbers more than 100 members. These objects are characterized by an extremely high luminosity in the gamma-ray region between 20 MeV and 30 GeV, which exceeds that of their radiation in all the other spectral regions. Furthermore, one observes rapid variations, strong polarization in the optical region, and compact radio emissions with a flat spectrum. Under the assumption of isotropic emission, their gamma-ray luminosities would be several $10^{15} L_\odot$; since, however, in the case of these objects the jet is pointing almost precisely towards us, the true luminosities are substantially lower.

The first energetic outbursts in the TeV range (1 TeV = 10^{12} eV) were detected from the two bright, relatively near blasars Mrk 421 and Mrk 501 ($z = 0.034$) in the 1990's by means of the Cherenkov radiation (Sect. 5.3.3) which their high-energy gamma quanta excite on entering the Earth's atmosphere.

12.3.5 Seyfert Galaxies

Nuclear activity in the optical and radio spectral regions can be clearly detected in the class of otherwise normal (mostly spiral) galaxies that were discovered in 1943 by C. K. Seyfert; they were first observed optically.

Line Spectra and Classification. The "Seyfert nuclei" exhibit a strong *emission* spectrum, in which the Balmer lines of hydrogen as well as He I and II and other allowed transitions are all characterized by broad linewidths. The *forbidden* lines, mostly from higher excitation states of N II, O II–III, Ne III–IV, S II and others, have the *same* widths in some Seyfert galaxies (type Sy 2); in others (type Sy 1), they are considerably narrower than the allowed lines (Fig. 12.26). The widths at half maximum for the Sy 1 galaxies correspond to a Doppler effect of about 5000 km s^{-1} in the allowed and 500 km s^{-1} in the forbidden lines; in the Sy 2 galaxies, all the linewidths lie in the range of 300 to 1000 km s^{-1}.

The classification criterion described by Seyfert in 1943 is not the only possible one. Objects which have an (optically) bright, starlike nucleus surrounded by a fainter, nebulous shell are termed *N galaxies* (W. W. Morgan, 1958). Probably all Seyfert galaxies are at the same time N-galaxies, but certainly not *vice versa*. Galaxies which show a strong ultraviolet excess (relative to normal objects), are termed *Markarian galaxies* (Mrk). The relation of the criterion chosen by B. E. Markarian in 1967 to the Seyfert criterion is evident.

We will consider the interpretation of the spectra in Sect. 12.3.6.

The apparently brightest Seyfert galaxies are NGC 4151 (Sy 1), with a nucleus of $m_V \simeq 12$ mag, and NGC 1068 = M 77 (Sy 2), with $m_V = 10.5$ mag. The Sy 1 galaxies are absolutely brighter objects than those of type Sy 2; their nuclei attain absolute visual luminosities of 10^{35} to $\gtrsim 10^{38}$ W or $2 \cdot 10^{8}$ to $\gtrsim 2 \cdot 10^{11} L_\odot$.

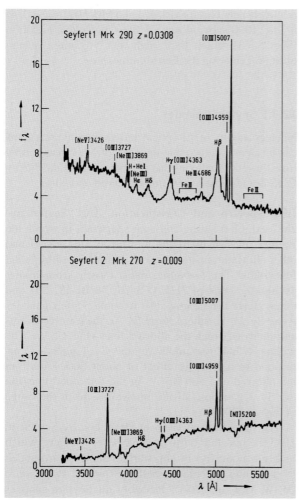

Fig. 12.26. The spectra of Seyfert galaxies (type Sy 1 and Sy 2), after H. Netzer (1982). The radiation flux (outside the Earth's atmosphere) f_λ, in $[10^{-17}\,\mathrm{W\,m^{-2}\,nm^{-1}}]$, is plotted against the observed wavelength $\lambda = (1+z)\lambda_0$ in [Å] (z = red shift)

X-ray luminosities of the Sy 1 galaxies exceed their optical luminosities. NGC 4151 and the quasar 3 C 273 are the only discrete extragalactic *gamma-ray sources* which were identified before the Compton Gamma-Ray Observatory was available.

In the *infrared*, radiation from heated dust dominates the emissions of the Seyfert galaxies; their luminosities in the infrared are comparable to or somewhat greater than those in the optical region. In the *radiofrequency range*, the Seyfert galaxies represent compact sources, which emit nonthermal synchrotron radiation. Observations with the Very Large Array and very long-baseline interferometry at angular resolutions of $0.001''$ to $\lesssim 1''$ indicate that for some galaxies, the radio emissions show a double or triple structure, with some sources of only a few 0.1 pc diameter; in other galaxies, the radio source which coincides with the optical nucleus is lacking in noticeable structure. In comparison to the radio galaxies, the radio sources in the Seyfert galaxies are weaker and much less extended ($\lesssim 0.1$ to 1 kpc).

In the Seyfert-2 galaxy NGC 1068 (= 3 C 71), we find a jet-like radio structure of several 10 pc length. In NGC 1275 = 3 C 84, which was identified with the strong radio source Per A in 1954 by W. Baade and R. Minkowski, three collinear compact components (and some fainter structures) are observed in the cm wavelength region at a resolution of 10^{-3} seconds of arc; the emissions in the mm wavelength range seem to originate from a different source. The radiofrequency radiation varies strongly with time. In NGC 1275, in addition to the activity in the nucleus, numerous long, extended filaments are seen in Hα images (R. Lynds, 1970) out to about 75 kpc from the center, with velocities of several 1000 km s^{-1}; they exhibit a certain superficial similarity to those of the Crab Nebula.

Continuum Emissions. The *optical* continuum includes also a nonthermal contribution, which is superposed onto the radiation from the stars and increases towards the ultraviolet. The Sy 1 galaxies are strong *X-ray sources* and also exhibit a nonthermal spectrum in the hard X-ray region (up to $\lesssim 50$ keV), with $I_\nu \propto \nu^{-0.7}$, so that in their nuclei, extremely energetic synchrotron electrons must be present. Extrapolation of the power law to higher energies suggests that the

Variability. A variability of the emissions in all spectral regions from the radio to the X-ray range on time scales of months to years is characteristic of the nuclei of all Seyfert galaxies; the amplitudes of the variation are in the range from 0.2 to 1 mag. In the Sy 1 galaxies, an increase in the (nonthermal) optical continuum is correlated with an increase in the intensity of the allowed emission lines. Observations of NGC 4151 also show variations in the line*widths*, so that possibly the type of a Seyfert galaxy can also change with time.

Liners. As a result of the development of high-sensitivity detectors with good spectral and spatial resolutions, beginning in the 1980's it became possible to detect weak *emission lines* in the nuclei of most spiral galaxies, even in "normal" galaxies such as M 81 or M 51, whose intensity ratios differ from those observed in H II regions, planetary nebulae or supernova remnants. A characteristic feature of these spectra is the occurrence of Hα lines with broad wings and strengths comparable to that of the neighboring [N II] line at $\lambda = 658$ nm. In some of these "liners" (*l*ow *i*onization *n*uclear *e*mission-line *r*egions), as the emission sources are called, a weak nonthermal continuum can be discerned, so that here, we presumably are dealing with a weaker form of the activity seen in the Seyfert galaxies.

An *activity* at the modest level of that in the nucleus of our Milky Way galaxy (Sect. 11.3.4) is not observable at present in other normal galaxies.

The activity phenomena in the "normal" galaxies give the impression of being somewhat fragmentary or poorly-developed counterparts of those seen in the Seyfert galaxies. One can thus conclude that all or at least most of the galaxies are subject to the "Seyfert disease" at times. Since about 1% of all galaxies exhibit the Seyfert phenomena, their total duration must be of the order of 10^8 yr (perhaps with interruptions).

12.3.6 Activity in Galactic Nuclei

The observations of the nuclear regions of "normal" galaxies, Seyfert galaxies, radio galaxies, quasars, etc. described in the preceding sections allow us to conclude that although their activities differ quantitatively, they exhibit so many basic similarities that we are clearly dealing here with a single, *unified* phenomenon.

The following are characteristics of the *activity* in galactic nuclei or in an active galactic nucleus (AGN): (a) the presence of a *compact* nucleus, which is brighter than in other galaxies of the same Hubble type; (b) a radiation *continuum* from the nucleus, reaching from the infrared to the X-ray region, which on the whole can be described by a power law $F_\nu \propto \nu^{-\alpha}$ with α of the order of 1; (c) *emission lines* from the nucleus, which show signs of *nonstellar* excitation, having for example intensity ratios different from those in H II regions; (d) *nonthermal radio* emissions; and (e) *variability* of the continuum and/or of the emission lines.

While only some of these characteristics occur in the weaker forms of activity, we observe in the quasars, the radio galaxies, and many of the Seyfert galaxies the whole palette of activity phenomena.

Unified Model. Although the understanding of the precise physical nature of galactic activity and its causes is still fraught with unsolved problems, we can nevertheless attempt to put together a picture of the essential processes of activity in galactic nuclei.

The various phenomena of activity in galactic nuclei can initially be ordered *empirically* according to two properties, namely according to the strength of their (nonthermal) *radio emissions*, and according to the width and intensity of the optical and ultraviolet *emission lines* from their central regions (Table 12.2). In this scheme, the quasars represent the continuation of the Sy 1 galaxies or the galaxies with broad emis-

Table 12.2. The main groups of Active Galactic Nuclei (AGN), arranged according to their radio emissions and the properties of their optical emission lines resulting from the different viewing angles relative to their axes. NLRG and BLRG are radio galaxies with narrow or broad emission lines, resp.; $\rightarrow$: the direction of increasing luminosity

		Emission lines in the optical region and UV			Mother galaxy
		narrow	broad	weak	Type
Radio emissions	weak	Sy 2	Sy 1 (N) $\rightarrow$ radio-quiet QSO		S, E
	strong	NLRG	BLRG $\rightarrow$ radio-loud QSO	BL Lac Objects	E
		Viewing angle δ to the (jet)axis			
		$> 20°$	$< 20°$	$\simeq 0°$	

sion lines (BLRG) towards higher luminosities, while the Sy 2 galaxies or the galaxies with narrow emission lines (NLRG) apparently have no successors among the quasars. (It is however under discussion that the "type 2 quasars" might be well "hidden" within the ultraluminous infrared galaxies (Sect. 12.2.1)).

The radio luminosity is probably essentially an intrinsic physical property. It can, however, be amplified to some extent through relativistic effects – as we discussed in the case of the quasars (Sect. 12.3.4) – when we observe a jet from nearly head-on. This intensity enhancement is particularly strong in all wavelength ranges in the case of the BL Lac objects (blasars).

In contrast, the differences in the emission line spectra of the galactic nuclei can be almost exclusively attributed to different *viewing angles* δ between our line of sight and the axis of rotation of the mother galaxy or the axis of a jet pair, since, depending on the viewing angle, parts of the central region can be hidden from our view by dust absorption.

We will now introduce the most important "building blocks" of the unified model and then discuss some selected aspects in more detail. The unified model of an active galactic nucleus includes the following components:

1. At the center of the galaxy, there is a compact mass of 10^6 to $10^{10}\,\mathcal{M}_\odot$, for which the most plausible assumption is that it consists of a *black hole* (Sect. 8.4.4).
2. A relatively small (≤ 1 pc), hot *accretion disk* around the black hole is the *central source* of the strong *continuum* radiation in the optical and ultraviolet regions.
3. "Gas clouds" orbit around the black hole; they are ionized and excited by the central source and can be observed by means of their *emission lines*. Since the central regions of the galactic nucleus cannot be spatially resolved, the clouds which lie closer to the center exhibit *broad* lines due to their higher orbital velocities, while *narrow* lines are characteristic of the more distant clouds.
4. The central region is surrounded by a dense ring of molecular gas and dust within the symmetry plane of the mother galaxy; it is extremely opaque to optical radiation. This *dust torus*, whose axis coincides with the galactic axis of rotation and the direction of the jets, prevents direct observation of the nuclear region if we are observing the galaxy essentially on edge ($\delta < 20°$). The dust in the torus is heated by the central source and supplies the principal part of the observed infrared luminosity of the nucleus.
5. Two tightly focused jets are ejected from the central region in opposite directions along the axis. These jets are moving at nearly the velocity of light in the case of the radio-loud sources and emit nonthermal synchrotron radiation.

Radio Emissions. The large-scale structure of a galaxy with an active nucleus is typified by its nonthermal radio emissions; the different forms taken by their isophote charts can be understood within the unified model: as we have already discussed for the case of the radio galaxies in Sect. 12.3.3, the energy of the central source, which coincides with the compact radio nucleus, is transported outwards by two oppositely-directed *jets* of relativistic particles or plasma to the outermost regions. In some cases, interaction of the jets with the surrounding medium produces intense radio emissions in "hot spots". The spectra from the radio to the X-ray regions are mainly ascribable to synchrotron radiation from relativistic electrons. The variety of apparent forms of the radio emissions is essentially determined by the *angle* δ of the jets relative to the line of sight, and by their intrinsic *strengths*. At small δ values, only *one* jet is observable (12.34). In the case of the BL Lac objects, we are looking directly into the emerging jet ($\delta \simeq 0°$).

Emission Lines. The gas in the neighborhood of the central sources is ionized by their *continuum* radiation in the optical, ultraviolet, and X-ray regions; it can then be observed by means of its emission lines. As in the H II regions or in planetary nebulae, the degree of ionization and excitation depends on the strength of the ionizing UV flux ($\lambda \leq 91.1$ nm) and on the electron density n_e of the gas; however, here the ionization is due not to stars, but rather to an intense, broadband radiation field of the form $\nu F_\nu \simeq$ const. We further assume that the "clouds" which orbit the central source cover only about one tenth of it. A photoionization model of this kind, whose details we cannot treat here, can explain the essential properties of the emission lines from the various types of active galactic nuclei.

In this model, *narrow lines* are emitted in those clouds which have low electron densities

($n_e \simeq 10^{10}$ m^{-3}) and are relatively distant from the center (several 100 pc to 1 kpc). In contrast, *broad* lines are emitted by clouds with electron densities of $n_e \simeq 10^{13}$ to 10^{16} m^{-3} and electron temperatures of around 10^4 K. In the Sy 1 galaxies and broad-line radio galaxies, these clouds are about 0.1 pc from the center; in the quasars, they are at ≥ 1 pc.

In the Sy 2 galaxies and the radio galaxies with narrow emission lines, our view of the clouds with high orbital velocities, i.e. the component with broad lines, is blocked by the dust-filled torus; we view these galaxies from an angle essentially perpendicular to their axes. Only in those cases in which the light emitted parallel to the axis is scattered or reflected in our direction – and polarized in the process – do we observe the broad lines. In the Sy 1 galaxies and the BLRG, we view at an angle to the axis which is not too large ($\delta < 20°$), so that we can observe both the narrow and the broad line components.

In the case of the BL Lac objects, the extremely weak emission lines can be explained by the fact that only very little gas is present near the central source.

The question of the *abundances* of the chemical elements in the gas clouds is of considerable interest. It has been found that the Seyfert nuclei and the quasars contain fairly *normal* cosmic matter. The derived elemental abundances differ by no more than a factor of 3 from those of the solar mixture. Stars must have therefore formed in the relatively early phases of galactic evolution, and nucleosynthesis (Sect. 8.2.5) took place in them on a large scale.

The Energy Sources. The primary question is of course that of the origin of the quantity of energy, of $\geq 10^{55}$ J, which must be contained in active galactic nuclei, and of their correspondingly high luminosities. We have already seen in connection with the quasars that nuclear energy, with a mass conversion efficiency of about 1% (Sect. 8.1.3), is by no means sufficient to explain it. Even if, as a limiting possibility, the whole rest energy $\mathcal{M}c^2$ of a mass $\mathcal{M}$ could be made "available", it would require that about $10^8 \mathcal{M}_\odot$, i.e. a considerable portion of the mass of the galactic nucleus, must be completely converted into radiation.

The only process currently known to physics which can free a considerable fraction of the rest energy of a cosmic mass $\mathcal{M}$ is the release of (potential) *gravitational energy* by contraction or collapse. For example (this is to be understood in the sense of only a rough estimate!), if the originally thinly-spread matter contracts to a homogeneous sphere of radius R, then the gravitational energy released is

$$E_G \simeq \frac{G\mathcal{M}^2}{R} . \quad (12.35)$$

If we now require that $E_G \simeq \mathcal{M}c^2$, then R must be of the order of the Schwarzschild radius R_S (8.82), i.e. the collapsed matter must assume an extremely compact structure. For $10^8 \mathcal{M}_\odot$, the radius, from (8.83), would have to be $R_S \simeq 10^{-5}$ pc $\simeq 2$ AU. In realistic estimates, an efficiency of less than 100% must be assumed for the release of the rest energy, so that the required masses would be more nearly in the range of 10^9 to $10^{10} \mathcal{M}_\odot$.

Possibilities under discussion for the massive central objects are, among others, extremely dense star clusters or black holes (Sect. 8.4.4). The observational evidence for massive black holes in the active nuclei of galaxies will be summarized below; arguments for the existence of a black hole with a mass of a few $10^6 \mathcal{M}_\odot$ at the center of the Milky Way galaxy were already given in Sect. 11.3.4.

From observations (with very long-baseline interferometry as well as of the time variability of the radiation), the upper limit for the size of active galactic nuclei is found to be ≤ 10 AU, in general agreement with the above estimate.

Accretion Luminosity. The radiated energy or luminosity of several $10^{14} L_\odot$ ($4 \cdot 10^{41}$ W) must finally be maintained by accretion of matter onto the compact central object at a sufficiently high rate $\dot{m}$. Where this matter comes from is still unexplained. Near collisions of galaxies may play an essential role; the tidal interaction between the galaxies (Sect. 12.5.3) could allow large amounts of gas (and stars) from the inner parts of the galaxy to flow into the central region. We can estimate the required accretion rate $\dot{m}$, whereby it is not important for our purposes to distinguish whether the material is falling onto a sphere or onto an accretion disk. According to (8.62), the accretion luminosity from matter falling to within a distance of R from the central mass is given by:

$$L_a = \frac{G\mathcal{M}\dot{m}}{R} = \eta \cdot \dot{m}c^2 , \quad (12.36)$$

where $\eta = G\mathcal{M}/(Rc^2) = R_S/(2R) \leq 0.5$ is the efficiency for the conversion of mass into radiation and R_S is once again the Schwarzschild radius. The matter in this process cannot, however, fall to within R_S of the central mass, since then it would take its energy "along" into the black hole and not radiate it outwards. More precise considerations thus give a maximum efficiency for a spherically symmetrical gravitational field (Schwarzschild metric) of $\eta = 0.08$, and for a rotating black hole (Kerr metric, Sect. 8.4.2), depending on its angular momentum, of $\eta = 0.06\ldots 0.42$. If we assume for the purposes of this estimate – rather arbitrarily – a value of simply

$$\eta = 0.1 , \qquad (12.37)$$

then matter must fall onto the central source at a rate of

$10\ldots 500 \; \mathcal{M}_\odot \; \mathrm{yr}^{-1}$ for quasars,
$\leq 0.1 \; \mathcal{M}_\odot \; \mathrm{yr}^{-1}$ for Seyfert galaxies,
$\leq 10^{-3} \; \mathcal{M}_\odot \; \mathrm{yr}^{-1}$ for our Milky Way galaxy

in order to supply the observed luminosity. In a time period of 10^8 yr, in the extreme case of the quasars, for example, $10^{10} \; \mathcal{M}_\odot$ would have to be accreted by the central source.

A Central Black Hole? In concluding this section, we ask to what extent the observations of other galaxies (not limiting ourselves to the active galaxies) give evidence for the existence of a black hole at their centers.

Since a black hole can be detected only by means of its strong gravitational effects, there are only the following more or less direct criteria for its existence:

1. an extreme density concentration of stars due to the deep potential well of the black hole, observed as a sharp maximum in the brightness distribution, that cannot be explained by known mass concentrations such as massive globular star clusters; and
2. high velocities of the stars or the gas, of the order of $\geq 1000 \; \mathrm{km\,s}^{-1}$, observed along a line of sight which passes as close as possible, i.e. at only a few Schwarzschild radii, to the center.

For the application of these two criteria, it is of decisive importance that the angular resolution of the observations be sufficiently high to identify the contribution to the light or to the velocities originating in the immediate neighborhood of the black hole. For example, with a galaxy at a distance of 10 Mpc, an angle of $0.1''$ already corresponds to 5 pc; this makes it clear that even very large earthbound telescopes or the Hubble Space Telescope can be expected to gather convincing evidence for a central black hole from only the very nearest galaxies.

We cannot describe the many direct and indirect observations of galactic nuclei in detail here; instead, we must content ourselves with the result that at present, there are strong indications – if not yet a conclusive "proof" – that a central massive black hole is present in over a dozen relatively nearby galaxies: thus, within the Local Group (Table 12.3), we can observe for example in the Andromeda galaxy M 31, a normal large spiral galaxy, and also in the smaller elliptical galaxy M 32 (Fig. 12.1) in both cases a compact mass concentration at the center, of about $3 \cdot 10^7$ and $3 \cdot 10^6 \, \mathcal{M}_\odot$, respectively. Candidates for massive black holes having masses of several $10^9 \, \mathcal{M}_\odot$ are found in the large elliptical galaxies NGC 4261 and M 87.

The occurrence of double nuclei in a few galaxies, along with the observation of gas disks with unusual orientations, of regions with opposite rotation directions within some galaxies, etc. are indications of the fusion of galaxies ("cannibalism") or tidal perturbations in near collisions of two systems. Presumably, such interactions can "feed" the black hole with a sufficiently large flow of matter and thus produce the activity of the galactic nucleus, at least for a limited time.

12.4 Clusters and Superclusters of Galaxies

Galaxies frequently belong to larger systems which are held together by the mutual gravitational attraction of their members. We can identify large or rich *clusters* containing up to several thousand galaxies, and smaller *groups* of decreasing size down to the *binary galaxies*. Among the galaxies brighter than 21 mag, some 10 000 clusters and/or groups are known. A few noticeable concentrations of galaxies were studied as early as 1902/06 by M. Wolf, and later (around 1924) by C. Wirtz; systematic surveys of the distribution of galaxies on the celestial sphere were carried out from about

1930 on by H. Shapley, E. Hubble, and others. The catalogue of G. O. Abell (1958), with 2712 rich galaxy clusters, and the Catalogue of Galaxies and Clusters by F. Zwicky et al. (1960/68) are both based on the Palomar Sky Survey.

In addition to the galaxies, many clusters also contain a tenuous, very hot gas, whose thermal radiation in the X-ray region can be observed. The ROSAT X-ray satellite discovered about 5000 galaxy clusters in its sky survey, on the basis of their X-ray emissions.

In Sect. 12.4.1, we first deal with our immediate neighborhood, the Local Group. Then, in Sect. 12.4.2, we take up the classification of the various galaxy clusters, and their masses; in Sect. 12.4.3, we deal with the intracluster gas. Especially in the central regions of the denser clusters, "collisions" and near-misses of galaxies are frequent, and the primary result of these events is a sweeping-out of the gas from the galaxies involved. Occasionally, two galaxies also merge or fuse to form a larger system. In Sect. 12.4.4, we give an overview of the interactions between galaxies and the evolution of the galaxy clusters to which they give rise.

As we shall see in the concluding Sect. 12.4.5, the the galaxy clusters themselves are not uniformly distributed in space, but instead form connected structures, mostly flat or long and filamentary: the *superclusters* of galaxies. These superclusters and clusters define the large-scale structure of the Universe, making up a network of connected filaments and disks which surrounds vast, roughly spherical, nearly empty regions.

12.4.1 The Local Group

The Local Group is a loose collection of about 30 galaxies within a region of around 2 Mpc in diameter. By far its brightest and most massive members are our Milky Way galaxy and the Andromeda galaxy, M 31. Several "satellite galaxies" are bound to each of these large spiral galaxies. In Table 12.3, we give a summary of the members of the Local Group, without making a claim to completeness. In some cases, the physical membership of a galaxy in the group or its status as a satellite is uncertain. It is notable that on the one hand, the three brightest systems of the Local Group are spiral galaxies, and on the other, the dwarf galaxies are the most numerous subgroup among its members.

Table 12.3. The Local Group: the Milky Way and the Andromeda galaxy (M 31) with their satellite galaxies, along with some other "isolated" members. The type and the absolute visual magnitude of the galaxies are given (LMC, SMC: Large and Small Magellanic Cloud, WLM: Wolf-Lundmark-Melotte System)

	Type	M_V [mag]
Milky Way	Sb/Sc I–II	−20.9
LMC	Im III–IV (SBm III)	−18.5
SMC	Im IV–V	−17.1
Sagittarius	dSph	−13.8
Fornax	dSph (dE3)	−13.1
Leo II	dSph (dE0)	−10.1
Sculptor	dSph (dE3)	−9.8
Carina	dSph (dE4)	−9.4
Ursa Minor	dSph (dE5)	−8.9
Draco	dSph (dE3)	−8.6
M 31 = NGC 224	Sb I–II	−21.2
M 32 = NGC 221	E2	−16.5
NGC 205 = M 110	Sph (dE5)	−16.4
NGC 185	dSph (dE3)	−15.6
NGC 147	dSph (dE5)	−15.1
Andromeda I	dSph (dE3)	−11.8
Andromeda II	dSph (dE2)	−11.8
Andromeda III	dSph (dE5)	−10.2
Additional Members:		
M 33 = NGC 598	Sc II–III	−18.9
IC 10	dIm	−16.3
NGC 6822	Im IV–V	−16.0
IC 1613	Im V	−15.3
WLM = DDO 221	Im IV–V	−14.4
Leo I	dSph (dE3)	−11.9
Aquarius = DDO 210	Im V	−10.9
Pisces = LGS 3	dIm/dSph	−10.4
Cetus	dSph	−10.1
Phoenix	dIm/dSph	−9.8
Tucana	dSph (dE5)	−9.6
Sextans	dSph	−9.5

Satellite galaxies of the Milky Way are, in addition to the Magellanic Clouds (Fig. 7.40), among others the Systems in Sculptor, in Ursa Minor and in Draco, which are all closer than 100 kpc to us. As companions to M 31, the E galaxies NGC 205 and M 32 have long been known (Fig. 12.1). In the neighborhood of the Andromeda galaxy, S. van den Bergh found three spheroidal galaxies in 1972 (Andromeda I–III). They have absolute magnitudes of $M_V \simeq -11$ mag and diameters of 0.5 to 0.9 kpc, and are basically similar to "our" Sculptor system. Even in the 1990's, four more members of the Local Group were discovered: Sextans, Tucana, Sagittarius, and Cetus. We give below some details about the interesting Sagittarius system.

The discoveries of dwarf galaxies only in recent years raises the suspicion that even within our Local Group, the luminosity function (12.5) is not yet completely known at its fainter end.

The Local Group belongs – together with a series of other such groups – to the *Virgo cluster*, whose center lies at a distance of about 20 Mpc from us.

Sagittarius System. This system was discovered only in 1994 by R. Ibata et al. during an investigation of the kinematics of the galactic bulge. It is visible against the foreground of the star fields of the bulge only as a faint concentration, but its "collective motions" are immediately noticeable due to their clearly different velocity, as is its intermediate star population (only about 10^{10} yr old, with about 1/25 of the solar metallicity). It contains several globular star clusters, and its mass is comparable to that of the Fornax system.

Sagittarius is only about 25 kpc from us and lies (from our viewing angle) on the other side of the galactic center, at a distance of roughly 16 kpc from it. This system is thus only about half as far from us as the Magellanic Clouds, and it is our nearest neighboring galaxy.

Its elongated shape, about 3 kpc in length with irregular clumps, indicates that Sagittarius is being torn apart by the tidal action of the Milky Way; we are therefore observing a satellite galaxy in a state of dissolution. After a few 10^8 yr, then, the stars of the Sagittarius system will have been "taken over" by our galaxy. The significance of such galactic fusions resulting from tidal forces for the evolution of the Milky Way and other galaxies will be treated further in Sect. 12.5.3.

12.4.2 The Classification and the Masses of the Galaxy Clusters

It is not easy to distinguish galaxy clusters, especially the more open ones, from chance density fluctuations in the general "field" of galaxies. G. O. Abell defined a measure for the population of a cluster, its *richness*, by counting all the galaxies within an arbitrarily chosen interval of magnitudes (2 mag from the third-brightest galaxy) and within a fixed distance from the center of the cluster. For this distance, he chose $1.7/z$ minutes of arc (z = red shift), corresponding to 3 Mpc for a Hubble constant of $H_0 = 50$ km s^{-1} Mpc^{-1} (Sect. 13.1.1).

Classification. Following Abell, we divide the galaxy clusters into regular and irregular clusters. The *regular clusters* (corresponding approximately to the *compact* clusters in the scheme of F. Zwicky), contain large numbers of galaxies, have a roughly spherical shape, and exhibit a strong concentration of the galaxies towards the center of the cluster. They contain predominantly elliptical and lenticular galaxies (ca. 70%–80%), of which the brightest (cD and giant E galaxies) are found mainly near the center. The prototype is the Coma cluster (= Abell 1656) at a distance of about 130 Mpc, with several thousand galaxies in a region of about 8 Mpc diameter.

The *irregular clusters* show no noticeable concentration towards their centers and do not have symmetric shapes; sometimes, they exhibit several local density concentrations. These irregular clusters, which also include the *galaxy groups*, contain galaxies of all the Hubble types; in particular, in contrast to the regular clusters, they have a high percentage (ca. 50%) of S galaxies. In the inner region of the Virgo cluster, with a diameter of about 3 Mpc and whose center is at a distance of around 20 Mpc from the Sun, we can find more than a thousand galaxies. The Local Group falls within the outer zone of this cluster.

There are three other classification schemes in use for galaxy clusters, which on the whole are parallel. A. Oemler (1974) defines the *cD clusters*, with predominantly bright elliptical galaxies, corresponding roughly to Abell's regular clusters, and the *spiral-rich clusters* (corresponding to the irregular clusters); in addition, as an intermediate type, he introduced the *spiral-poor clusters*, dominated by S0 galaxies. The cluster types I to III according to L. P. Bautz and W. W. Morgan (1970) correspond to a series of clusters ranging from those dominated by a bright central galaxy (BM I) to the clusters without a predominant galaxy (BM III). H. J. Rood und G. N. Sastry (1971) give a finer classification scheme, which is based on the type and arrangement of the ten brightest galaxies of the cluster.

Masses. The diameters of the larger galaxy clusters are in the range of 3 to 10 Mpc and are thus comparable with the sizes of the largest known radio galaxies such as 3 C 236 (Sect. 12.3.3). The mass $\mathcal{M}$ of a galaxy cluster can be estimated with the help of the virial theorem, in a manner analogous to the mass determination of elliptical galaxies (12.17):

$$\sigma^2 = \langle v^2 \rangle \simeq \frac{G\mathcal{M}}{R} \ . \quad (12.38)$$

Here, σ is now refers to the dispersion of radial velocities of the galaxies within the cluster, and R is its radius. With typical values, $\sigma = 1500 \text{ km s}^{-1}$ and $R \simeq 5 \text{ Mpc}$, we obtain for the total mass:

$$\mathcal{M}(\text{galaxy cluster}) \simeq 10^{15} \, \mathcal{M}_\odot \ . \quad (12.39)$$

This corresponds to an average mass-luminosity ratio for a galaxy cluster of $\langle \mathcal{M}/L \rangle \simeq 200 \, \mathcal{M}_\odot/L_\odot$. On the other hand, as was noted as early as 1933 by F. Zwicky, the summation of the "visible" masses of all the individual galaxies gives a much smaller total mass, only about one-tenth as large. The hot gas observed in galaxy clusters by means of its X-ray emissions (see the following section) also contributes only about a tenth of the virial mass (12.38). This discrepancy can, however, be understood if the galaxies are assumed to have massive, dark halos, as is suggested e.g. by the observed rotation curves of many spiral galaxies (Sect. 12.1.5).

Gravitational Lenses. Due to their large masses, the galaxy clusters act as gravitational lenses (Sect. 8.4.3) for the background galaxies lying along the line of sight behind them, at around twice their distance from us. These background galaxies appear to be stretched out into luminous *arcs* which are grouped on a circle around

Fig. 12.27. The gravitational lens effect. A color image made with the Wide Field Planetary Camera 2 of the Hubble Space Telescope. The massive galaxy cluster 0024 + 1654 (yellowish color) in Pisces, with a red shift of $z = 0.39$, is seen at the center of the image (which covers an area of about $1.3' \cdot 1.3'$). It acts as a gravitational lens for the light from a galaxy behind it, at $z = 1.39$. This extended source appears as a *multiple image* in the form of extended *blue arcs*, one near the center of the cluster and four others on a circle around the outside of the cluster. The image amplification of the lens enables us to discern individual structures within the distant galaxy, indicating that it is an irregular blue galaxy in a very early stage of its evolution (Sect. 12.5.5). (Image: W. N. Colley, E. Turner, J. A. Tyson and NASA, 1996, © Association of Universities for Research in Astronomy AURA/Space Telescope Science Institute STScI)

the cluster and reflect its mass distribution. Since the mid-1980's, several galaxy clusters with this lens effect have been discovered; they have red shifts of $z = 0.2$ to 0.4 (Fig. 12.27). The order of magnitude of the degree of distortion is given by (8.92). The blue light from the arcs and their spectra confirm the great distances ($z \geq 0.8$) of their sources and thereby the interpretation in terms of a lens effect.

The masses derived from the gravitational lenses agree rather well with those from the galactic motions and from the X-ray-emitting gas, and likewise indicate a predominant fraction of dark matter in the mass of the galaxy clusters.

12.4.3 The Gas in Galaxy Clusters

The galaxy clusters form a class of strong extragalactic, extended X-ray sources, which were first observed by the Einstein satellite and later, with higher sensitivity and resolution, by ROSAT. Their spectra tell us that these X-ray emissions represent the *thermal radiation* of an extremely hot gas in the temperature range of 10^7 to 10^8 K (1 to 10 keV). The optical thicknesses in the X-ray region are very small ($\lesssim 10^{-3}$), and the gas densities lie in the range of 10^2 to 10^3 m^{-3}.

Emission Structure. The X-ray emission exhibits complex behavior which varies from cluster to cluster. At the one extreme, we find widely scattered, irregular emission regions, often containing individual galaxies; at the other end of the scale, we observe a relatively smooth emission curve, frequently with a stronger concentration around a bright galaxy near the center of the cluster (Fig. 12.28). The structure of the X-ray emission is clearly correlated to the cluster type: thus, Abell's regular clusters such as the Coma cluster and Abell 85, with a low content of spiral galaxies, have X-ray emissions with a smooth contour and a relatively high X-ray luminosity, $L_X \gtrsim 10^{37}$ W. In contrast, the emissions from the irregular clusters which are rich in spiral galaxies, such as the Virgo cluster or

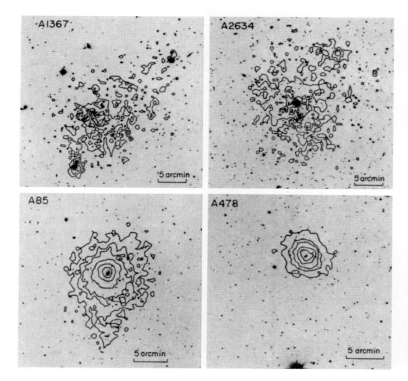

Fig. 12.28. X-ray emission from galaxy clusters: observations with the Einstein Satellite by C. Jones et al. (1979). Lines of equal X-ray intensity are superposed onto the optical photographs from the Palomar Sky Survey. The clusters Abell 1367 and 2634 have irregular contours which often include individual galaxies, while Abell 85 and 478 show relatively smooth X-ray emissions, strongly concentrated around a bright galaxy in the center of the cluster

Abell 1367, are less regular and have lower luminosities ($L_X \lesssim 10^{37}$ W). The gas in the irregular clusters is somewhat cooler than the hot gas in the regular clusters, at 1 to $4 \cdot 10^7$ K as opposed to $\gtrsim 6 \cdot 10^7$ K. Furthermore, the X-ray emissions from the clusters containing a bright cD galaxy (such as Abell 85) are more strongly concentrated towards the center of the cluster than is the case in the clusters which lack a dominant galaxy.

More precise investigations using ROSAT have shown some cases in which smaller galaxy clusters are falling into larger ones.

Temperature. In a galaxy cluster which is in dynamic equilibrium, both the individual galaxies and the atoms of hot gas move in the same *common* gravitational potential, and the spatial extent of each of the components is determined by its mean kinetic energy or temperature. From the observations of the velocity dispersion σ of the galaxies and the gas temperature T in many clusters, it is known that the ratio β of the kinetic energy of the galaxies per unit mass, $\sigma^2/2$, to that of the gas, $(3/2)kT/(\mu m_H)$,

$$\beta = \mu m_H \sigma^2 / (3kT) , \qquad (12.40)$$

is of the order of one ($\mu \simeq 0.6$ is the average atomic weight of the completely ionized gas, and m_H is the mass of a hydrogen atom). This means that the hot gas and the "galaxy gas", in which we can look upon the individual galaxies as "atoms", have about the same temperature and the same relative density distribution. For example, $\sigma = 1000$ km s^{-1} corresponds to a temperature of $T \simeq 10^8$ K.

To be sure, probably only the regular clusters are in dynamic equilibrium, but for all clusters down to the galaxy groups, a correlation has been observed between the velocity dispersion σ of their galaxies and both their gas temperatures T and their X-ray luminosities L_X.

Observations of continuum and line emissions (e.g. Fe XXIV+XXV at 6.7 keV, and O VIII at 0.65 keV) in some nearby galaxy clusters have shown that each of the individual galaxies is surrounded by its own X-ray halo. In the Virgo cluster, particularly the giant elliptical galaxies M 87 and M 86 are surrounded by gas at temperatures from several 10^6 to $3 \cdot 10^7$ K; it is thus somewhat cooler than the general cluster gas.

Mass of the Gas. The mass contained in the hot gas makes up around 10 to 40% of the overall mass (including that of the dark matter). The fraction of mass due to the gas is thus in many cases larger than that from the galaxies.

The Sunyaev–Zeldovich Effect. An indirect method of detecting hot gas in galaxy clusters was suggested by R. A. Sunyaev and Ya. B. Zeldovich in 1970/72. A photon of the 3 K cosmic background radiation (Sect. 13.2.2) obtains a slightly increased energy, with a relative contribution on the order of $kT/(m_e c^2)$ ($m_e c^2 = 511$ keV, the rest energy of the electron; T is the temperature of the hot gas), through scattering from an electron in the cluster gas. For example, at $T = 10^8$ K or $kT = 10$ keV, this relative energy increase is 0.02. The scattering cross-section of this *inverse Compton scattering* is the Thomson scattering coefficient $\sigma_T = 6.65 \cdot 10^{-29}$ m^2 (4.132), so that a photon moving through a distance L in a hot gas of electron density n_e experiences an energy or frequency increase given by:

$$\frac{\Delta \nu}{\nu_0} \simeq \left(\frac{kT}{m_e c^2} \right) \sigma_T n_e L . \qquad (12.41)$$

This energy shift corresponds to a decrease of the radiation temperature of $T_0 \simeq 3$ K on the low-frequency side of the maximum of the 3 K radiation, according to the Rayleigh-Jeans law (4.63), of $\Delta T_0/T_0 = -2\Delta \nu/\nu_0$. Therefore, the intensity of the background radiation in the center of a galaxy cluster (with $n_e = 3 \cdot 10^3$ m^{-3} and $L = 0.5$ Mpc) is decreased by an amount corresponding to $\Delta T_0 \simeq 3 \cdot 10^{-4}$ K. This extremely small Sunyaev-Zeldovich effect could actually be observed for several clusters, among them Abell 478 (Fig. 12.28).

Head–Tail Galaxies. The existence of a hot gas in galaxy clusters can also be concluded from the structure of a particular class of radio galaxies, the head-tail galaxies (Fig. 12.29), which occur in dense clusters. The supersonic motion of the galaxy through the cluster gas pushes the ejected relativistic electrons and magnetic fields backwards. From the observed geometry of the Mach cone, the velocities and pressures in the flow field can be estimated, and thus also the temperature and density of the cluster gas. The results agree with those obtained from X-ray emission.

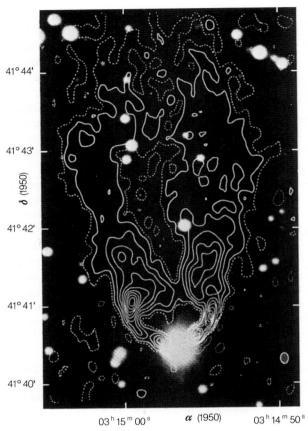

Fig. 12.29. The head-tail galaxy NGC 1265: observations at $\lambda = 6$ cm with the Westerbork Synthesis Telescope. This overlay of the radio isophotes onto the optical image allows us to see that the radiofrequency emission regions were evidently ejected from the elliptical galaxy at the head. From K. J. Wellington et al., Nature **244**, 502 (1973). (Reprinted by kind permission from Macmillan Magazines Limited, London, and of the authors)

12.4.4 Interacting Galaxies. The Evolution of Galaxy Clusters

Especially in the denser central regions of the regular clusters, many close interactions ("collisions") of galaxies take place within the Hubble time H_0^{-1} (13.5), and they must have substantially affected the structure and galactic content of these clusters since their formation. The various types of galaxy clusters probably represent different phases of a dynamic evolutionary process.

In the first instance, a collision results in the loss of interstellar gas from the affected galaxies, and therefore finally also in the loss of their spiral arms. The S galaxies thus become lenticular galaxies, possibly via intermediate types such as the *anemic* (A) galaxies, which have very weak spiral arms and were observed in the Virgo cluster by S. van den Bergh in 1976.

The interstellar gas, heated by the collision, finally collects for the most part at the minimum of the gravitational potential in the center of the cluster; only particularly massive galaxies can retain the gas through their own gravitation. If a galaxy passes through this cluster gas at a supersonic velocity, its interstellar gas is likewise swept out, so that additional spiral galaxies in the cluster can be "annihilated" in this way. The denser regular clusters, with their stronger X-ray emissions and their small fraction of spiral galaxies, thus represent a dynamically later stage of development than the irregular galaxy clusters.

Analysis of the 6.7 keV X-ray emission line of Fe supports this picture of the origin of the hot cluster gas. In nearly all galaxy clusters, practically independently of the details of the model assumptions, an *iron abundance* is found from this emission line which is about half the solar value. This relatively high abundance can be understood within the framework of our ideas about the formation of the chemical elements (Sect. 12.5.4) only if we assume that the gas came mostly from the stars within the galaxies, i.e. it is not of primordial origin.

Compared to the interstellar gas, the majority of the *stars* are considerably less influenced by a meeting of galaxies. However, in the case of an approximately central collision between two galaxies, the result can be that they merge into a single larger system. This "*galactic cannibalism*" was probably decisive for the formation of the cD galaxies at the centers of galaxy clusters; the occasional occurrence of two or more "nuclei" in these galaxies provides observational support for this hypothesis.

12.4.5 Superclusters of Galaxies

The very noticeable concentration of the apparently bright galaxies within the northern galactic hemisphere, especially within the constellation Virgo, led early to speculations that here, the galaxy clusters belong to a still larger system, called the *Local Supercluster* or the *Virgo Supercluster*; in particular, G. de Vaucouleurs studied this supercluster in detail in the 1950's. G. O. Abell noted that many of the rich galaxy clusters in his catalogue seem to be members of superclusters. A quantitative indication of the occurrence of superclusters was given by the investigations carried out by P. J. E. Peebles et al. of the two-point correlations of the distribution of galaxies projected onto the celestial sphere. A breakthrough in the observation of superclusters and a convincing demonstration of their existence was, however, made only in 1976, when the development of sensitive detectors made it possible to measure radial velocities from the spectra of many distant, faint galaxies and thus to investigate also the *radial* structure of the cluster distribution.

A typical supercluster contains, along with several galaxy groups, also a few rich clusters (about 2 to 6), and forms a strongly flattened system with a diameter of around 100 Mpc, lacking in axial symmetry or a central concentration. The frequently-occurring filamentary or chainlike arrangement of the galaxies is quite noticeable.

Thus, the *Virgo Supercluster* consists of the Virgo cluster itself, about 3 Mpc across and at a distance of 20 Mpc, of an irregularly-shaped disk of around 50 Mpc diameter with an axis ratio of 6 : 3 : 1, and of a "halo" of several galaxy groups, including the Local Group at its edge. The *Coma Supercluster* (Fig. 12.30), at a distance of about 140 Mpc, containing among others the Coma cluster and Abell 1367, forms a flattened system with a diameter of about 100 Mpc.

The characteristic time which a galaxy moving at a velocity of $1000\,\mathrm{km\,s^{-1}}$ requires to cross a supercluster of 100 Mpc diameter is 10^{11} yr; it is thus longer than the age of the Universe. The superclusters are therefore not relaxed systems, and the virial theorem cannot be used to estimate their masses. The direct summation of the masses of the individual members yields a typical mass for a supercluster of the order of magnitude of $10^{16}\,\mathcal{M}_\odot$; the overall luminosity is around $10^{14}\,L_\odot$.

A precise investigation of the structure of the superclusters is made difficult by the fact that their radial extension cannot be directly read off from the red shifts of their member galaxies. Instead, the nonuniform dis-

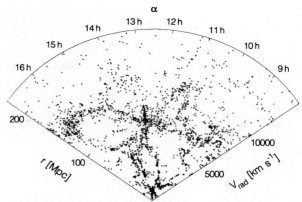

Fig. 12.30. The spatial distribution of 2500 galaxies out to a limiting blue magnitude of $m_B = 15.5$ mag, from the CfA Sky Survey (Harvard–Smithsonian Center for Astrophysics). The data are represented in a wedge diagram: the radial velocity $V_{\mathrm{rad}} = cz$, the distance r, and right ascension α are plotted for a declination range between $26.5°$ and $44.5°$. The distances are referred to a Hubble constant of $H_0 = 65\,\mathrm{km\,s^{-1}\,Mpc^{-1}}$. The group of galaxies at 13 h is the Coma cluster. The "Great Wall" is notable: it is a concentration of galaxies in a thin disk ($\lesssim 10$ Mpc thickness) at a distance of 100 to 150 Mpc, stretching over the whole range of right ascension. (With the kind permission of M. J. Geller and J. P. Huchra, Center for Astrophysics, Cambridge, MA)

tribution of the galaxy clusters causes an acceleration of the galaxies towards the mass concentrations and thus invalidates the Hubble relation. For example, the Local Group is apparently falling towards the Virgo cluster at several $100\,\mathrm{km\,s^{-1}}$.

Large-Scale Structures. The superclusters of galaxies are themselves not isolated systems in space, but instead touch each other at their edges and thus make up the large-scale structure of the Universe, in the form of a network of connected flat disks and filaments (Fig. 12.31). These enclose *voids*, roughly spherical regions which are nearly free of galaxies and have a characteristic diameter of about 50 Mpc.

In the sky survey carried out in 1989 by M. J. Geller and J. P. Huchra of the Harvard–Smithsonian Center for Astrophysics, an extremely extended, contiguous structure was discovered, called the "Great Wall": it is a relatively thin ($\lesssim 10$ Mpc) sheet which extends over at least 100 Mpc by 270 Mpc (Fig. 12.30) and contains a density of galaxies well above the average value,

Fig. 12.31. The Hubble Deep Field. A section of area $1.3' \cdot 0.9'$ from the deep-field image of an area in the sky (of roughly $2.6' \cdot 2.6'$) that is free of bright foreground stars and nearby galaxies; taken with the Wide Field Planetary Camera 2 of the Hubble Space Telescope. The full Hubble image is shown on p. 311 as the *frontispiece* to Part IV. This compound color picture was made from three images taken with B, R, and I filters. (Picture: R. Williams and the Hubble Deep Field Team, and NASA, 1996, © Association of Universities for Research in Astronomy AURA/Space Telescope Science Institute STScI)

at $\Delta\varrho/\varrho \simeq 5$. Its mass is estimated to be $2 \cdot 10^{16} \mathcal{M}_\odot$. Since this wall stretches out past the limits of surveys carried out up to the present, both in right ascension and in declination, its true extension and mass are probably still larger.

In the intervening years, other similar large structures have been discovered, including a "superfilament" in Aquarius, which consists of more than 20 galaxy clusters strung together in a chain and stretching over nearly 400 Mpc (24 000 to 48 000 km s^{-1}), in a direction which is essentially radially away from us.

One of the most massive large-scale concentrations of galaxies is the "*Great Attractor*", with a mass of around $5 \cdot 10^{16} \mathcal{M}_\odot$ and a density which is twice that of the average density in the Universe. This mass concentration is responsible for a major part of the gravitational force acting on the Milky Way, the Local Group, and the Virgo cluster, among others; it produces systematic velocities of the order of 500 km s^{-1} relative to the Hubble expansion. The Great Attractor is located at a distance of around 100 Mpc behind the clusters in Hydra and Centaurus, with its center roughly at the galactic coordinates $l \simeq 320°$ and $b \simeq 0°$. As a result of the strong dust extinction in the plane of the Milky Way, it is not very noticeable in the optical spectral region. The massive galaxy cluster Abell 3627 (with a mass of $\simeq 5 \cdot 10^{15} \mathcal{M}_\odot$) is near to its center of gravity and could perhaps even form the core of the Great Attractor.

The goal of the British–Australian–American "2° field of view" survey (2dFGRS = Two degree Field Galaxy Redshift Survey), which makes use of an efficient spectrograph at the Anglo–Australian 3.9 m Telescope, is to determine the spatial distribution of 250 000 galaxies. This survey includes red shifts up to $z \simeq 0.2$, i.e. it extends around 4 times deeper into space than previous surveys. The initial results, based on somewhat more than 100 000 galaxies, show that there is *no* hierarchy of spatial structures in the Universe on a scale of more than a few 100 Mpc.

12.5 The Formation and Evolution of the Galaxies

The subject of the formation and evolution of galaxies is still fraught with riddles and unsolved problems. Even a superficial examination of the Hubble Atlas of Galaxies (A. Sandage, 1961), or of the Atlas of Peculiar Galaxies (H. Arp, 1966), and most especially of the Hubble Deep Field (Fig. 12.31), the in-depth image of the most distant galaxies obtained with the Hubble Space Telescope under the leadership of R. E. Williams in 1995, should serve as a "modest deterrent" to overeager theoreticians.

We shall therefore limit ourselves here to an initial presentation of the most important observations and the processes which must play an important role in the development of the galaxies: we begin in Sect. 12.5.1 with some remarks about the formation of the clusters of galaxies and the galaxies themselves, insofar as this problem is not treated under the topic of cosmology (Sect. 13.3.3). We then discuss in Sect. 12.5.2 the observational evidence for the intergalactic medium; here, the interpretation of the red-shifted Lyman α resonance line of neutral hydrogen in the absorption spectra of the quasars plays a key role. In Sect. 12.5.3, we describe the most important effects of gravitational interactions during near collisions of galaxies. These range from small perturbations up to the merging of two galactic systems, and can be of decisive importance for galactic evolution, especially in its earliest phases. Sect. 12.5.4 is devoted to the dynamic and chemical evolution of the Milky Way and other galaxies. Finally, in Sect. 12.5.5, we consider the young galaxies with high red shifts which we can observe in the Hubble Deep Field.

12.5.1 The Formation of the Galaxies and the Clusters of Galaxies

Galaxies contain mass concentrations of up to $10^{12}\, \mathcal{M}_\odot$; the characteristic masses of galaxy clusters are of the order of $10^{15}\, \mathcal{M}_\odot$, and of the superclusters of galaxies, $10^{16}\, \mathcal{M}_\odot$ (Sects. 12.1.5 and 12.4.2, 5). As we shall see, the average density of the matter in the Universe is about $9 \cdot 10^{-28}$ kg m^{-3} (Table 13.2); the uncertainty in this table is at least a factor of 3. The question of the extent to which *intergalactic* matter, i.e. matter outside the galaxy clusters, is distributed within the superclusters or in general in the Universe will be considered in the next section.

The formation of galaxies and clusters of galaxies can be fundamentally understood in terms of a gravitational instability according to the Jeans criterion (Sect. 10.5.3), which we must now apply to the expanding Universe. Compared to the static case, the growth of density perturbations is slower here. The heart of the problem, however, lies in the fact that we know neither the time at which the structures formed, nor the state of the cosmic matter then – in particular its average density – with sufficient accuracy.

First Structures. The question of the formation of the galaxies and galaxy clusters is closely connected with that of the origin of the large-scale structure of the Universe: the lattice-like arrangement of the superclusters and clusters of galaxies in filaments and flat disks, enclosing vast empty spaces (Sect. 12.4.5). According to current ideas about the evolution of the cosmos, (Sects. 13.3.3, 4), the *seeds* of this structure were planted at a very early evolutionary stage of the Universe, about 10^{-35} s after the Big Bang, when the strong fundamental interaction separated from the electroweak interactions as a result of symmetry breaking (Table 13.5), thus causing an extremely rapid expansion (inflation) of the cosmos. The origin of the structure may even date to quantum fluctuations at a still earlier time.

However, galaxies and galaxy clusters could be formed only much later, after the beginning of the expansion of the Universe; at around 10^6 yr (corresponding to a red shift of $z \simeq 1000$), owing to recombination of the free electrons with protons (at $T \simeq 3000$ K), the close coupling of the radiation field to matter was released and density fluctuations could grow rapidly. The galaxies began to form at a still later time.

Fluctuations of the 3 K Radiation. The fact that no sharp inhomogeneities were present in the cosmos before this time is demonstrated by the observed high degree of isotropy of the microwave background radiation (Sect. 13.2.2), which shows that before $z \simeq 1000$, no notable inhomogeneities were present. However, the extremely difficult and costly measurements of this radiation in the 1990's with the COBE satellite (Sect. 5.2.2) revealed small fluctuations of the radiation temperature up to a few $10\,\mu$K with angular extensions of the order of $10°$ (Fig. 12.32). These variations of $\Delta T/T_0 \simeq 10^{-5}$ around $T_0 = 2.73$ K (13.48) reflect relative *density* fluctuations $\Delta \varrho / \varrho$ of the *same* order of magnitude. The fact that the temperature fluctuations of the background radiation were observed to be independent of frequency supports their interpretation as "nucleation points" for the formation of galaxies.

The size of these fluctuations on the celestial sphere is to be sure much too large to connect them directly with the structures which are visible today. Later investigations with various kinds of earthbound instruments (carried by balloons) have shown that the fluctuations continue down to significantly smaller angular size ranges with about the same amplitudes. Their mass spectrum and origins are still very uncertain.

Among the many observations of the fluctuations on small angular scales, the American–Italian BOOMERANG experiment (*B*alloon *O*bservations *O*f *M*illimetric *E*xtragalactic *R*adiation *AN*d *G*eomagnetics) is notable both for its high angular resolution, of the order of $10'$ at a good signal/noise ratio, as well as for covering a significant portion of the celestial sphere. The measurements, with 16 low-temperature bolometers at the four frequencies 90, 150, 240, and

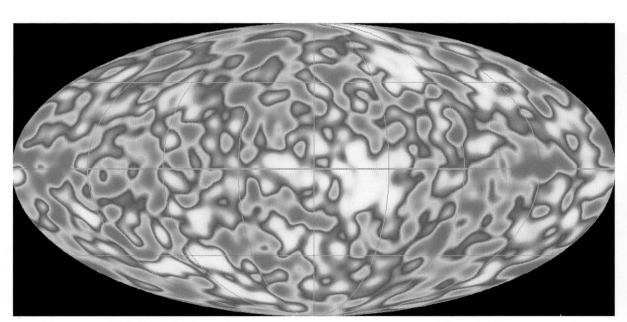

Fig. 12.32. A celestial map of the fluctuations in the microwave background radiation (Sect. 13.3.2), smoothed on a scale of $7°$ over the celestial sphere, in galactic coordinates (the plane of the Milky Way is indicated by the horizontal line in the middle, with Sagittarius at the center). This image represents the average over more than 2 years of measurements made with the Differential Microwave Radiometer (DMR) of the Cosmic Background Explorer (COBE) at 53 and 90 GHz ($\lambda = 5.7$ and 3.3 mm, resp.). The contribution from the Milky Way and the dipole component (13.51) resulting from the motion of the Earth relative to the 3 K radiation have been subtracted. (© COBE Science Working Group, NASA Goddard Space Flight Center and NASA Space Science Data Center)

400 GHz (corresponding to wavelengths of 3.3, 2.0, 1.25, and 0.75 mm) were carried out from a balloon which remained for more than 10 days at an altitude of 38 km above the Antartic in the (Southern Hemisphere) Summer of 1998/99. The observations at four frequencies made it possible to eliminate foreground emissions and effects of the interstellar dust. The *power spectrum* obtained, i.e. the dependence of the temperature fluctuations ΔT on their angular size $\Delta \phi$, is independent of the frequency band chosen and is thus clearly attributable to the fluctuations of the microwave background radiation.

Instead of $\Delta \phi$, we could also give the multipole order l of the associated spherical harmonic functions (compare the analogous multipole expansion of the sound waves in helioseismology, Sect. 7.3.8, Fig. 7.29). On the celestial sphere, the order l corresponds to an angle of $\Delta \phi \simeq 180°/l$.

The results of Boomerang cover the range of l values between about 50 and 600, while measurements with the COBE satellite included only structures with $l < 20$. Even the smallest fluctuations detected by Boomerang, with $l \simeq 600$ or $\Delta \phi \simeq 20'$, correspond to only the largest structures which have been observed in the cosmos to date, the "honeycomb" of the superclusters of galaxies (Sect. 12.4.5).

The dependence of the power spectrum of the fluctuations of the background radiation measured by Boomerang and other experiments is not uniform at greater angular resolutions, but rather it shows noticeable structures, in particular a strong maximum at $\Delta \phi \simeq 1°$ or $l \simeq 200$, with an amplitude of about 70 μK. Structures of this order of magnitude reflect the density or temperature fluctuations in the early Universe shortly before the decoupling of radiation from matter (Sect. 13.3.3). The spectrum and the propagation of these sound waves depends on the physical state of the Universe and its evolution, so that the analysis of the fluctuations observed today can yield important conclusions about the parameters of the cosmological model (Sect. 13.2.2).

Growth of the Density Fluctuations. Numerical model calculations, whose details we cannot consider here, show that the density perturbations in the Universe, which later gave rise to the galaxies and clusters of galaxies, increased in strength at the time of recombination. Their rate of growth would have been much too slow, however, if we take into consideration only "ordinary" (baryonic: cf. Sect. 13.3.2) matter. In order to obtain agreement with observations, we have to assume that a considerable fraction of *dark* matter was present. This is also required by a variety of other observations (rotation curves and X-ray emissions of galaxies, virial mass of clusters of galaxies). Furthermore, the model calculations show that for a rapid growth of the structures on the observed scale, "cold" dark matter, i.e. particles of a still unknown type with *non*relativistic velocities, was essential. In contrast, "hot" dark matter such as e.g. massive neutrinos would not serve, since such particles would move at nearly the speed of light.

The density fluctuations of the dark matter could have undergone a strong increase even before the recombination phase, and they were considerably stronger than those of the ordinary-matter mixture of H and He, so that the latter could collect in the gravitational potential of the dark matter. In this picture, the existence of dark halos around the clusters of galaxies and the galaxies would be readily understandable.

After some 10^7 yr (corresponding to $z \simeq 100$) at the earliest, the density fluctuations would have become so strong that the actual formation of the galaxies or the protogalaxies could begin to occur. At this time, the Jeans mass (10.38, 39) was about $2 \cdot 10^5 \mathcal{M}_\odot$, and all greater masses were gravitationally unstable.

We thus must date the formation of the galaxies at an early stage of the expansion of the Universe, at a time when all the dimensions were a factor of $(1+z) \lesssim 100$ smaller (13.42) and the mean density was therefore $(1+z)^3 \lesssim 10^6$ times higher ($\lesssim 5 \cdot 10^{-22}$ kg m^{-3}) than today's values.

A *lower limit* for the average matter density in the cosmos at the time of the formation of the galaxies can be obtained from the largest known red shifts of $z \simeq 6$ for the quasars and the most distant galaxies. At a time when galaxies obviously already existed, the average density was a factor of $(1+z)^3 \simeq 350$ times higher than it is now.

The Shape of the Condensations. Ya. B. Zeldovich recognized as early as 1970 that it would be improbable for the original density fluctuations to have had spherical symmetry. He showed that density fluctuations of

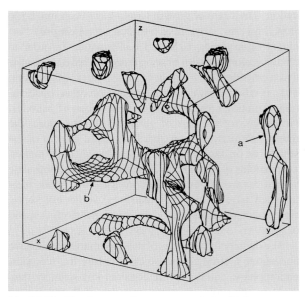

Fig. 12.33. The large-scale structure of the Universe: surfaces of equal matter density in the early universe, based on numerical simulations of the growth of density fluctuations. Filaments (a) and flat disks ("pancakes", b) have formed and enclose large empty regions. The edge of the cube shown corresponds to 50 Mpc. From J. Centrella and A. L. Melott, Nature **305**, 196 (1983). (Reprinted by permission of Macmillan Magazines Limited, London, and of the authors)

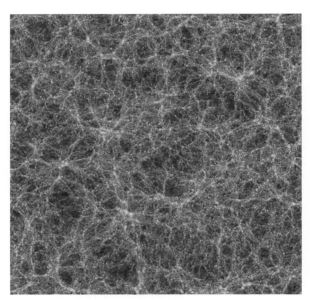

Fig. 12.34. A numerical simulation of the large-scale honeycomb structure of the present-day cosmos in a volume of area 480 Mpc · 480 Mpc perpendicular to the line of sight and of 40 Mpc depth. The N-body calculations were done within the international Virgo project on two parallel supercomputers (Cray T3D) in Garching and Edinburgh, and took into account $256^3 \simeq 1.7 \cdot 10^7$ "particles", each with a mass of $4.5 \cdot 10^{11} \, \mathcal{M}_\odot$ in a closed standard model, which was dominated by *cold*, i.e. nonrelativistic *dark* matter, with a density parameter of $\Omega_0 = 1$ (13.27), a cosmological constant of $\Lambda = 0$ (13.33), and a Hubble constant of $H_0 = 50 \, \text{km s}^{-1} \, \text{Mpc}^{-1}$ (13.4). (© Max-Planck-Gesellschaft, Munich 1998)

a general shape will collapse first in the direction of their shortest diameter, and therefore form flat disks ("pancakes"). Further compression can lead to cigar-shaped formations. Early numerical simulations made it seem at least plausible that the characteristic structural framework of the superclusters and clusters of galaxies would have formed (Fig. 12.33).

The growth of density fluctuations in the expanding Universe depends decisively on the *assumptions concerning the nature of the dark matter* (as well as on the parameters of the cosmological model, cf. Sect. 13.2.5). While "hot" (relativistic) dark matter would lead to diffuse structures, contradicting the observations, models with a predominance of "*cold*" dark matter yield realistic density fluctuations (Fig. 12.34). However, within the standard model (Sect. 13.1.3), the largest structures are still somewhat too small; models with a cosmological constant of $\Lambda \neq 0$ give better agreement.

We must content ourselves here with these estimates and cannot delve deeper into the various approaches to the explanation of the formation of the galaxies, the galaxy clusters, and the superclusters. A comprehensive theory has not yet been formulated. This is perhaps not astonishing, since we do not even understand the structure of *present-day* galaxies in detail, as our discussion of their rotation curves and dark halos has shown.

12.5.2 The Intergalactic Medium and Lyman α Systems

We first ask to what extent the observations indicate the existence of a general, *diffuse* intergalactic medium, and then consider the numerous Lyman α absorption lines in the quasar spectra and their interpretation in terms of "*discrete*" concentrations of matter.

A Diffuse Medium? If the intergalactic matter contains *neutral* hydrogen, we can expect a strong absorption in its Lyman-α resonance line. Supposing the intergalactic gas to be located at a distance corresponding to a red shift of z, then we will observe the Lyman-α line, whose rest wavelength is $\lambda_0 = 121.6$ nm, at the red-shifted wavelength $\lambda = (1+z)\lambda_0$.

J. E. Gunn and B. A. Peterson noticed in 1965 that a diffuse distribution of intergalactic neutral hydrogen would give rise to an extended absorption region stretching from λ_0 to $\lambda = (1+z)\lambda_0$. A continuous absorption of this type has up to now not been observed, and an extremely low concentration as an upper limit for the density of H I follows from this. If there is in fact any diffuse intergalactic matter containing hydrogen, by far the most abundant element, then it must be almost completely ionized. The observations would also be compatible with a very hot intergalactic gas having an average density of around $4 \cdot 10^{-27}$ kg m^{-3}.

On the other hand, this gas would be discernible through an extragalactic background radiation in the X-ray and perhaps the gamma-ray range. Recent observations with the X-ray satellite Chandra however show that at least 75% of the background in the hard X-ray range can be resolved into a large number of discrete sources such as quasars, Seyfert galaxies, and galaxies with active nuclei. The extragalactic background in the gamma-ray range which is observed between 30 MeV and 100 GeV can be attributed almost entirely to numerous unresolved gamma blasars (Sect. 12.3.4).

Current observations thus hardly leave room for a general intergalactic medium.

Lyman α Systems. In the spectra of distant quasars, a number of sharp absorption lines of Lα are observed on the short-wavelength side of the red-shifted Lα emission line (Fig. 12.21); their red shifts z are smaller than those of the quasar itself. These lines can be explained by the absorption due to "cloudy structures" along the line of sight between the quasar and the observer.

From the strengths of the lines which are shifted to $\lambda = (1+z)\lambda_0$, we can determine the *column density* N [H atoms m^{-2} = 10^4 H atoms cm^{-2}] of neutral hydrogen,

$$N(\text{H I}) = \int n(\text{H I}) \, d\ell \, , \qquad (12.42)$$

which is defined as the integral over the particle density n along the line of sight. Our Milky Way system would have a column density of interstellar H I – e.g. for observations made in a direction perpendicular to the disk – of the order of 10^{24} m^{-2}, as is found from the average particle density of $4 \cdot 10^5$ m^{-3} and the disk thickness of about 200 pc (Sect. 11.2.2).

A column density of 10^{16} m^{-2} is already sufficient to detect a matter concentration through its Lα absorption in the quasar spectra. Observed column densities cover the large range from 10^{16} up to about 10^{26} m^{-2}. Above roughly 10^{22} m^{-2}, the Lα line becomes so strong that it exhibits broad *damping wings*, i.e. the lines fall on the damping or square-root portion of the growth curve (Fig. 7.12), where the column density is proportional to the square of its equivalent width. These cases are termed "damped Lyman α systems".

Although the Lα line is readily observable, the relative fraction of neutral hydrogen in the "Lyman α clouds" is vanishingly small, only about 10^{-4} to 10^{-7}. In fact, the clouds are highly ionized, mainly by the strong ultraviolet radiation of the quasars, as indicated by model calculations. The cosmic matter must have again been ionized in recent times, at red shifts of $z \leq 6$, following the recombination at $z \simeq 1000$ (Sect. 13.3.3).

Concerning the physical properties of the "Lyman α clouds" and their sizes, we still know very little. In systems with *high* H I column densities, i.e. with Lα lines on the damping portion of the growth curve, we are probably observing the earliest phases of the evolution of a galaxy, the *protogalaxy*. In these, in addition to the strong Lα line, we often observe the Lβ as well as the red-shifted absorption lines of heavier elements such as Mg II at $\lambda = 279.8$ nm (with relatively small z) and C IV at $\lambda = 155.0$ nm (at higher z). The average metal abundance in the Lyman α forest is estimated to be about 1/500 of the solar value, so that we are not dealing here with primordial matter; instead, nucleosynthesis in *stars* (of a Population III?) must have already occurred.

In the H I concentrations at *low* column densities, we are possibly no longer seeing isolated "clouds", but instead parts of the more or less continuous honeycomb structure of the Universe, which is "penetrated" by our line of sight to the background quasar and thus becomes observable.

The Lyman α systems contribute at most a few percent of the critical density $\varrho_{c,0}$ (Table 13.2) to the average density of the cosmos.

12.5.3 Interacting Galaxies

In near collisions of two galaxies at a distance r, their mutual tidal interactions, whose strength is proportional to $1/r^3$ (2.69) strongly influence both the stars and especially the interstellar gas within the galaxies. The almanacs of peculiar or interacting galaxies by H. C. Arp (1966) and B. A. Vorontsov-Velyaminov (1959 and 1977) contain numerous examples of double and multiple systems of interacting galaxies with common shells, connecting bridges, or long, drawn-out tails of stars and glowing gas (Fig. 12.35 and the picture opposite the title page). In radio-astronomical observations, also, we find indications of interactions, for example from the 21 cm line of neutral hydrogen, among other things in the form of connecting bridges or common gas shells. In recent times, in particular observations made with the Hubble Space Telescope, with its high angular resolution, have yielded important contributions to the study of the phenomenon of galactic interactions.

Surveys of the galaxies which searched for the various morphological indications of gravitational disturbances, i.e. for bridges, filaments, tails, deformations, ring-shaped galaxies, shell structures and counter-rotating stars in elliptical systems, double and multiple cores, etc., have shown that interactions are not rare occurrences, but instead represent a widespread phenomenon. Probably, in fact, strong interactions right up to and including the merging of galaxies played *the* decisive role in the formation of the current shapes of the galaxies and in the early phases of their evolution. No doubt every galaxy was also – at some point in its later evolution – subjected to strong tidal perturbations from another galaxy passing nearby, so that the conventional idea of the evolution of galaxies as essentially isolated systems requires considerable revision.

The great variety of typical indications for mutual gravitational interactions have been reproduced very well by numerical N-body simulations which were begun in the 1970's in work by A. and J. Toomre.

An indication of interactions in the neighborhood of the Milky Way is the *Magellanic stream*, six H I clouds

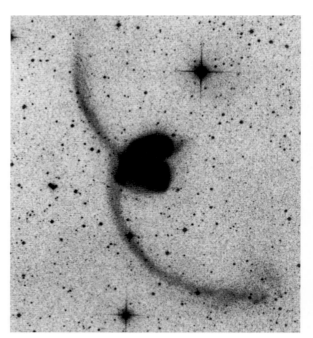

Fig. 12.35. The interacting galaxies NGC 4038/9 (= Arp 244 = Vorontsov–Velyaminov 245, the "antennae") in a photograph taken by O. Pizarro and H. E. Schuster with the 1.62 m Schmidt camera of the European Southern Observatory (ESO). The projected distance between the ends of the "antennae" is about 160 kpc, and that between the centers of the two galaxies is about 9 kpc. (From S. Laustsen et al., 1987)

which follow the Large Magellanic Cloud on its orbit around the Milky Way. These clouds were presumably "drawn out" of the satellite galaxy by tidal forces from our galaxy and are now falling into the Milky Way on a spiral orbit, whose last perigalactic point occurred about $1.4 \cdot 10^9$ yr ago.

The two *Magellanic Clouds* themselves, which are surrounded by a common shell of H I, will also plunge into the Milky Way on spiral orbits in some 10^9 yr. The elongated shape of the Small Cloud and its relatively high rate of star formation are likewise indications of the tidal forces from the Milky Way.

Finally, in Sect. 12.4.1 we described the elongated, irregular spheroidal dwarf galaxy *Sagittarius*, which is only 16 kpc from the center of our galaxy and is in the process of dissolution as a result of the tidal forces from the Milky Way.

The tidal interactions of galaxies can apparently set off a very high rate of *star formation* over a short period of time, especially in the dense galactic central regions. As examples, we have met up with the "dusty" infrared or starburst galaxies (Sect. 12.2.2) as well as the blue compact dwarf galaxies (Sect. 12.1.6).

The interactions between galaxies are particularly important, as we saw in Sect. 12.4.4, for the evolution of the galaxy *clusters*. In dense clusters, the morphological type of the galaxies can be modified by these interactions. Spiral galaxies lose their gas and presumably become S0 galaxies; in the centers of the clusters, galactic fusion of several galaxies leads to the formation of the giant cD galaxies.

12.5.4 Evolution of the Galaxies

The conventional scheme of the evolution of the galaxies assumes that (at some time between $z \simeq 100$ and 5) individual gas clouds were formed with masses in the range $\leq 10^{11}$ to $10^{12}\,\mathcal{M}_\odot$, which then continued to evolve into their presently observed state as essentially isolated "cosmic islands". As we shall see in Sect. 12.5.5, observations of the earliest evolutionary stages of the galaxies (at red shifts of $z \geq 2\ldots3$) show however that today's normal, large S and E galaxies probably formed by fusion of many smaller blue galaxies, i.e. they did not begin as gas clouds (protogalaxies) which contained their present large masses. Furthermore, the mutual gravitational interactions of the galaxies (Sect. 12.5.3) decisively influence their evolution, in particular through near collisions by which, for a limited time, a very intensive phase of star formation is initiated.

Here, we can treat only the most important aspects of a theory of galactic evolution; in fact, no complete theory exists at present. In this section, we leave the earliest evolutionary stages out of consideration and concentrate on those phases after the time when the galaxy had already attained its current mass. We leave open the questions of how the "protogalaxy" might have looked, what elemental mixture it might have contained, and how it was formed.

Dynamic Evolution. The evolution of the galaxies is closely connected to the distribution of *angular momentum* and to the *formation of stars*.

At a steady low rate of star formation, a gaseous protogalaxy with a large amount of angular momentum forms a disk on contracting. We then have a (spiral) galaxy; the collapse to form a galactic disk will be considered in detail below. In the other extreme case, when all the gas is consumed right at the beginning by formation of stars, the protogalaxy changes its dimensions only slightly, since the stars, in contrast to gas, cannot radiate away their kinetic energy, and, due to energy conservation, the sum of the kinetic and potential energies must remain constant. In this case, an *elliptical* galaxy is formed.

The structure of the *disks* is, at least in their outer portions, surprisingly similar for all types of galaxies from S0 to Sm. The physical distinguishing feature of the Hubble types S0, Sa, Sb, etc. seems to be mainly (or entirely?) the fraction of their overall mass which is still present in *gaseous* form. This fraction varies from less than 1% in the regularly-shaped S0 disks to about 10% in irregular galaxies. The formation of spiral arms is evidently connected directly to the gas in the disk.

The *spheroidal* components of the spiral galaxies were clearly formed from matter with *a small amount* of angular momentum. As the dynamic calculations of R. B. Larson and B. M. Tinsley (1974/78) demonstrate, the structure of our Milky Way and other spiral galaxies must have been produced in a *two-step* process: in a first phase with very rapid star formation, the spheroidal component is formed within a few free-fall times, while in the following phase, with considerably slower star formation, the remaining gas can collapse to a disk before any more stars are formed.

Observations of the halo population of the Milky Way however show different groups of globular clusters with different properties (age, metal abundance) and suggest that at least a part of the spheroidal component must have originated through "cannibalism" by our galaxy, which in the course of time "consumed" a series of smaller satellite galaxies, such as Sagittarius, (Sect. 12.4.1), that then within $\leq 10^9$ yr became mixed into the previously existing halo population. Some of the brighter globular clusters observed today could thus be the "remains" of the nuclear regions of dwarf galaxies.

The *elliptical* and S0 galaxies obviously evolved from protogalaxies with low angular momenta. Among them, the elliptical and spheroidal *dwarf* systems (Sect. 12.1.6) probably developed directly from the original protogalaxies of smaller mass. The beginning of star formation, which leads to supernova explosions, removes the interstellar gas from the galaxy at an early evolutionary stage and practically prevents further star formation. The *giant* E galaxies, on the other hand, would seemed to have formed through merging of several smaller protogalaxies (Sect. 12.5.5).

Collapse to a Galactic Disk. In the disks of the S galaxies, in which all the stars rotate in the same sense around the center, there is a *large* angular momentum, as is to be expected from their degree of flattening. It is thus tempting to ask whether, and how, an (exponential) galactic disk can have formed from a more or less spherical cloud.

The calculation of the dynamics of the collapse of a spherical protogalaxy to a galactic disk was first performed by O. J. Eggen, D. Lynden-Bell and A. R. Sandage in 1962. As the simplest model of such a *protogalaxy*, we consider a sphere of gas which is rotating rigidly at the constant angular velocity ω. Let $\mathcal{M}(r) = (4\pi/3)\varrho r^3$ be the mass within a sphere of radius r, ϱ the density, R the overall radius, and $\mathcal{M} = \mathcal{M}(R)$ the overall mass. Then the ratio η of the components of centrifugal force and gravitational force perpendicular to the axis of rotation at the point P (which has the coordinates r and θ; cf. Fig. 12.36) is:

$$\eta = \frac{\omega^2 r \sin\theta}{G\mathcal{M}(r) \sin\theta/r^2} = \frac{1}{r}\left[\frac{(\omega r^2)^2}{G\mathcal{M}(r)}\right], \quad (12.43)$$

where G is the gravitational constant. Furthermore, the *specific angular momentum* (per unit of mass) at the distance r from the axis of rotation is given by

$$h(r) = \omega r^2 . \quad (12.44)$$

Now, in a thought experiment, we allow our sphere to contract *radially*, in such a way that each mass element maintains its angular momentum. In particular, for a spherical shell of radius r, the specific angular momentum ωr^2 and the enclosed mass $\mathcal{M}(r)$ inside the shell remain constant and the factor in square brackets in (12.43) does not change. But owing to the factor $1/r$, η increases until a new equilibrium is established between the force components perpendicular to the axis of rotation. The component of gravitational force parallel to the axis, however, remains unchanged. The protogalaxy must therefore become flattened, i.e. it must

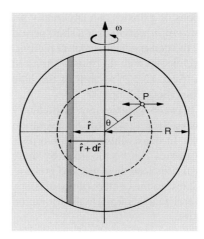

Fig. 12.36. The collapse of a spherical protogalaxy to a disk

collapse in the direction of the rotational axis. The energy released in this process is mostly dissipated; in this way, a thin disk is finally formed. In order that a gravitational equilibrium $\eta = 1$ (gravitation = centrifugal force) be maintained everywhere within the disk, a radial redistribution of the mass must naturally take place. The rotation will then in general no longer be uniform.

We cannot follow these complex processes in detail, but we can expect that every mass element maintains its angular momentum. Thus, the fraction $d\mathcal{M}(h)/\mathcal{M}$ of the mass whose angular momentum per unit mass lies in the interval h to $h + dh$ must be the same size in the finished galaxy as in the protogalaxy. Instead of $d\mathcal{M}(h)/\mathcal{M}$, we could of course just as well consider the fraction of the mass $\mathcal{M}(\leq h)/\mathcal{M}$, after integration over h, in which the angular momentum per unit mass is less than a given value h.

In order to calculate these ratios, we take for lack of more precise knowledge a homogeneous spherical protogalaxy of radius R with the constant density ϱ. The angular velocity ω can, from (12.43), be adjusted in such a way that in any case, within the whole equatorial plane, a gravitational equilibrium is established with

$$\omega^2 = G\mathcal{M}/R^3 . \quad (12.45)$$

Then, from (2.67), the total angular momentum $J = I\omega = (2/5)\mathcal{M}R^2\omega$ (I = moment of inertia) and the specific total angular momentum is

$$H = \frac{2}{5}\omega R^2 = \frac{2}{5}(G\mathcal{M}R)^{1/2} . \quad (12.46)$$

12.5 The Formation and Evolution of the Galaxies

In order to calculate the *distribution* of the angular momentum, we cut a cylinder out of the sphere, parallel to the axis of rotation, with inner and outer radii $\hat{r}$ and $\hat{r}+d\hat{r}$ and height $2\sqrt{R^2-\hat{r}^2}$. Its mass is then (see Fig. 12.36)

$$d\mathcal{M} = \varrho \cdot 2R\sqrt{1-(\hat{r}/R)^2} \cdot 2\pi\hat{r}\,d\hat{r} \qquad (12.47)$$

and its angular momentum is

$$h\,d\mathcal{M} = \omega\hat{r}^2\,d\mathcal{M} \qquad (12.48)$$
$$= \omega\varrho \cdot 2R\sqrt{1-(\hat{r}/R)^2} \cdot 2\pi\hat{r}^3\,d\hat{r}\;.$$

According to (12.44), we now have $dh = 2\omega\hat{r}\,d\hat{r}$. Thus, with $\mathcal{M} = (4\pi/3)\varrho R^3$ and (12.46), we obtain the fraction of the mass whose specific angular momentum lies in the range h to $h + dh$:

$$\frac{d\mathcal{M}(h)}{\mathcal{M}} = \frac{3}{5}\left(1 - \frac{2}{5}\frac{h}{H}\right)^{1/2}\frac{dh}{H}\;. \qquad (12.49)$$

The fraction of the mass $\mathcal{M}(\leq h)$ for which the specific angular momentum is $\leq h$ can be obtained from this by integration:

$$\frac{\mathcal{M}(\leq h)}{\mathcal{M}} = 1 - \left(1 - \frac{2}{5}\frac{h}{H}\right)^{3/2}\;. \qquad (12.50)$$

In Fig. 12.37, the two distribution functions $d\mathcal{M}(h)/\mathcal{M}$ and $\mathcal{M}(\leq h)/\mathcal{M}$ are plotted as functions of the normalized specific angular momentum.

It was shown in 1964 by D. J. Crampin and F. Hoyle, and then in 1970 by J. H. Oort, that some Sb and Sc galaxies, and also our Milky Way system, in fact do exhibit angular momentum distributions corresponding to (12.49) and (12.50); this indicates their possible formation from a uniformly rotating, homogeneous spherical protogalaxy.

The *time* τ required for a galactic collapse can be estimated in an order-of-magnitude calculation by taking the rotational period on a circular orbit of radius R:

$$\tau \simeq \frac{2\pi}{\omega} = \alpha \cdot 2\pi\sqrt{\frac{R^3}{G\mathcal{M}}} \qquad (12.51)$$

with $\alpha \simeq 0.2\ldots 0.5$. For $\mathcal{M} = 1.5 \cdot 10^{11}\mathcal{M}_\odot$ and $R = 25$ kpc, one thus obtains several 10^8 yr. This time is, of course, roughly equal to the free fall time (8.41) or the period of the fundamental oscillation of a gaseous sphere (7.100), i.e. $\tau \simeq 1/\sqrt{G\varrho}$. Arguments in connection with the formation of the elements (see below),

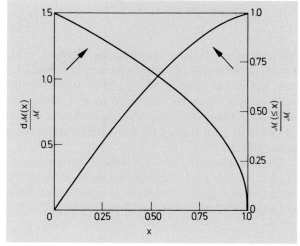

Fig. 12.37. The distribution of angular momentum. The following quantities are plotted as functions of $x = h/(\omega R^2) = h/(5H/2)$ (h is the specific angular momentum per unit mass, referred here to its maximum value $\omega R^2 = 5H/2$): (1) the fraction of the mass $d\mathcal{M}(x)/\mathcal{M}$ for which x lies between x and $x + dx$ (*left-hand scale*), and (2) the fraction of the mass $\mathcal{M}(\leq x)/\mathcal{M}$ for which x is smaller than the value on the abscissa (*right-hand scale*). In the formation of a disk galaxy from a spherical protogalaxy with conservation of angular momentum, $d\mathcal{M}(x)/\mathcal{M}$ and $\mathcal{M}(\leq x)/\mathcal{M}$ are conserved for each element of mass as functions of x

however, support the idea that the formation time of the disk was probably closer to 10^9 yr.

We can make further statements for *exponential* disks according to K. C. Freeman's investigations (1970): their surface luminosities folllow (12.10), and thus their surface densities μ obey the distribution law

$$\mu(\hat{r}) = \mu_0\,e^{-\hat{r}/\hat{r}_d}\;. \qquad (12.52)$$

From $\mu(\hat{r})$, one obtains the total mass:

$$\mathcal{M}_d = \int_0^\infty \mu_0\,e^{-\hat{r}/\hat{r}_d} \cdot 2\pi\hat{r}\,d\hat{r} = 2\pi\hat{r}_d^2\,\mu_0\;. \qquad (12.53)$$

By the way, e.g. 80% of the total mass is within $\hat{r} \leq 3\hat{r}_d$.

We are now in a position to calculate the angular velocity of the rotation, $\omega(\hat{r})$, and then the distribution of angular momentum per unit mass, $h_d(\hat{r})$, as well as the fraction $\mathcal{M}_d(\leq h_d)/\mathcal{M}_d$ of mass for which the specific angular momentum is $\leq h_d$. By numerical integration, the total specific angular momentum is

$$H_d = 1.11\,(G\mathcal{M}_d \hat{r}_d)^{1/2}\,. \qquad (12.54)$$

We can now compare the function $\mathcal{M}_d(\le h_d)/\mathcal{M}_d$ for exponential disk galaxies with that calculated for our "theoretical" protogalaxy, (12.50); taking the results of K. C. Freeman, the agreement is excellent. This is a strong indication that such galaxies could in fact have been formed by the collapse of spherical protogalaxies.

It is also interesting to compare the length scale factor $\hat{r}_d$ of the disk from (12.52) with the radius R of our hypothetical protogalaxy for $\mathcal{M}_d = \mathcal{M}$ and $H_d = H$. From (12.46) and (12.54) we obtain immediately:

$$\hat{r}_d/R = 0.13\,. \qquad (12.55)$$

The *radial* contraction of the protogalaxy is thus considerable. The velocities in protogalaxies must therefore have been correspondingly smaller than the currently observed rotational velocities in the disk galaxies, owing to conservation of angular momentum.

The total specific angular momentum of a galactic disk, from (12.53) and (12.54), is:

$$H_d \propto \mu_0^{-1/4}\mathcal{M}_d^{3/4}\,. \qquad (12.56)$$

It thus depends practically *only* on the mass $\mathcal{M}_d$ for disk galaxies, like the total angular momentum, since μ_0 hardly varies.

Following this brief introduction to the dynamic aspects of galactic evolution, we turn to the problem of the development of the abundance distribution of the elements.

Chemical Evolution. Initially, we assume that a protogalaxy containing nearly pure hydrogen (plus 10% helium) was present in the early Universe, $15 \cdot 10^9$ yr ago, and that the first stars immediately began to form in it. In their interiors, according to the ideas of E. M. and G. R. Burbidge, W. A. Fowler and F. Hoyle (Sect. 8.2.5), *nuclear processes* took place, which produced the heavy elements. Some branches of stellar evolution lead to the continuous or explosive (supernovae) release of stellar matter to the interstellar medium. Thus, a second generation of more metal-rich stars can be formed, and this cycle continues through further stellar generations.

In the dwarf E galaxies, this process must have been arrested sooner than in the giant galaxies, and in the S galaxies, the formation of heavy elements must have been essentially concluded by the time of the collapse of the disk. In all the sufficiently large galaxies, we find that, independently of the formation of a disk and independently of the mass fraction of interstellar gas which is still present today, the order of magnitude of the extent to which the protomatter has been converted into heavy elements, and the mixing ratios of those elements, are much the same. Within many S galaxies, we can also observe an increase in the abundance of heavy elements in going from the edge of the disk towards the center.

In our *Milky Way system*, we can obtain a more precise picture of the chronology of formation of the heavy elements:

At *present*, stars are formed from the interstellar medium which is concentrated in a flat disk, with a metal content $Z = Z_\odot \simeq 0.02$ (8.46), at an overall rate of a few $\mathcal{M}_\odot$ per year and with a mass spectrum $\psi(\mathcal{M})$ (Sect. 10.5.6). In particular, the more massive stars enrich the interstellar gas with heavy elements at the end of their evolution, e.g. through supernova explosions. In this process, a part of their mass which has been modified by nuclear reactions remains in the remnant stars (neutron stars), and is thus removed from the cycle of interaction between the interstellar medium and many generations of stars.

The observed element abundances in the spheroidal component (halo population) show a maximum metal content of about $Z_\odot/6$, so that an increase of the metal abundance from practically zero up to this value must have occurred in at most several 10^9 yr. This time estimate is based on age determinations of the globular clusters and is longer than the collapse time (12.51) of the disk estimated from the dynamic considerations of O. J. Eggen, D. Lynden-Bell and A. R. Sandage.

With the consolidation of the disk, of mass around $7 \cdot 10^{10}\,\mathcal{M}_\odot$, the synthesis of elements essentially came to its conclusion, since the oldest open star clusters already have roughly the normal composition $Z_\odot$.

According to Table 8.2, as a result of their lifetimes only relatively massive stars, with masses $\ge 10\,\mathcal{M}_\odot$, contribute to the formation of the heavy elements. Stars with masses $\le 1\,\mathcal{M}_\odot$, on the other hand, have not evolved away from the main sequence since the formation of the Milky Way, so that for example among the G dwarfs, metal abundances in the *entire* range $0 \leq Z \leq Z_\odot$ should be found. If the formation of the Population II stars had corresponded even roughly to the luminosity function of open star clusters (Sect. 10.5.6), then there should still be a large number of long-lived *metal–poor* dwarf stars present today; this is in contradiction to the observations. In fact, extremely metal-poor stars are very rare.

This "*G-dwarf problem*", and also the time evolution of the average metal abundance (with a rapid increase at the beginning and a flattening-out during the last $5 \cdot 10^9$ yr), as well as the radial abundance gradient within the disk, cannot be understood in terms of simple models based on a constant rate of star formation over space and time within a closed system. Attempts to solve these problems therefore begin with hypotheses (in part *ad hoc* assumptions) such as (a) a radically different mass function $\psi(\mathcal{M})$ at the beginning, as a result of which in particular many more massive stars were formed; (b) a very rapid initial enrichment of the protogalaxy in heavy elements, through a "Population III" which no longer exists today, before the formation of the oldest halo stars; or (c) the transfer of primordial or nuclear-processed matter from the halo to the disk over a long period of time following the collapse which formed the disk.

Nuclear Energy Release. In this connection, we emphasize how small is the nuclear energy release rate in the Milky Way system and other galaxies in their present states. In (8.45), we related the hydrogen consumption of a star to its mass–luminosity ratio. If we take the latter to be about 10 solar units, we find that our galaxy, in its *present* state, would have consumed only a few percent of its hydrogen after 1 to $2 \cdot 10^{10}$ yr. The production of a major portion of the heavy elements within the first 10^9 yr by stellar evolution therefore implies that the Milky Way previously had a much greater luminosity than it does today.

From this estimate, we could expect to find numerous *protogalaxies* – in the infrared at large red shifts – with about 100 times the luminosity of the present-day galaxies. Such systems are however *not* observed in the deep sky surveys of the youngest galaxies. This result is in agreement with the hypothesis (Sect. 12.5.5) that the larger S galaxies were formed only by the merging of several smaller systems with lower luminosities.

Cosmochronology. These ideas about the formation of the heavy elements, especially in the Milky Way, make it possible from the nuclear physics point of view to determine a *cosmic time scale* (see also Sect. 13.2.5). In this cosmochronology, long-lived radioactive nuclides of different half-lives are used as "chronometers", similar to the radioactive dating in the Solar System (Sect. 3.2.2). Examples are ^{187}Re (half-life $5 \cdot 10^{10}$ yr), ^{232}Th ($1.4 \cdot 10^{10}$ yr), ^{238}U ($4.5 \cdot 10^9$ yr); in addition, also ^{244}Pu ($8.0 \cdot 10^7$ yr) and ^{129}I ($1.6 \cdot 10^7$ yr) are used.

A first guess is obtained by assuming that most of the heavy elements were produced within a short time during the formation of the Milky Way system. We then take two radionuclides which are not replaced by the decay of any parent element and which decay with greatly differing half-lives, e.g. the uranium isotopes ^{235}U and ^{238}U, whose present abundance ratio is 1 : 138. In previous times, this ratio must have been much greater. If we extrapolate back to an initial ratio somewhat larger than one, we arrive at an age of formation of between 7 and $8 \cdot 10^8$ yr ago. For a more precise age determination, the *time dependence* of the production of elements during the evolution of the galaxy is required, in particular a model for the production of isotopes by the r-process, since all the nuclides used as chronometers were generated by it. The uncertainties in the model, in some of the nuclear-physics quantities, and in the observation and interpretation of the isotopic ratios of meteorites leave open a relatively wide range of values between about 11 and $20 \cdot 10^9$ yr for the *age* of the Milky Way galaxy, and therefore of all of the galaxies. These times are, in any case, compatible both with the age of the Milky Way as obtained from the color–luminosity diagrams of the globular clusters, some 12 to $14 \cdot 10^9$ yr (Sect. 9.2.3), and with the age of the Universe as estimated from the expansion of the galaxies (Friedmann time, Table 13.1), in the range of 11 to $20 \cdot 10^9$ yr.

A nearly direct determination of the age of the Milky Way Galaxy was recently made by R. Cayrel and coworkers (2001) by analysis of the lines of the radioactive elements U and Th in the spectrum of an old, extremely metal-poor star. This first determination of the uranium abundance for a Population II star (using the U II line at $\lambda = 385.9$ nm) yields an age of $(12.5 \pm 3) \cdot 10^9$ yr and thus decreases the uncertainty in the determination of the age of the Milky Way.

To conclude this section we now summarize briefly the fundamental open questions concerning the evolution of the galaxies. Although we understand the essential features of the galactic cycle connecting the different generations of stars and interstellar matter, and although we can reconstruct to a considerable extent their "history", in particular that of our Milky Way galaxy, from present-day observations, we are still far from having a complete quantitative picture of the dynamic and chemical evolution of the galaxies of different Hubble types.

For one thing, we know nothing about the nature of the "invisible" major portion of the matter in many galaxies whose presence is evident from the dynamic calculations. The process of stellar genesis and its dependence on the physico-chemical state of the interstellar medium is also still unclear to a great extent. It has become apparent, especially from infrared observations, that the rate of star formation is strongly increased by tidal interactions between the galaxies. Furthermore, our knowledge of the later phases of stellar evolution, and along with it the enrichment of the interstellar gas in heavy elements, still has serious gaps. The role played by the galactic nuclei and their activity in galactic evolution is also still unclear. Finally, our knowledge of the early evolutionary stages of galactic evolution (which we shall discuss in more detail in the next section) remains very rudimentary.

12.5.5 Galaxies in the Early Universe

In contrast to the stars, the galaxies were apparently all formed at about the same time, around $2 \cdot 10^{10}$ yr ago, although not necessarily with their present-day masses. Young galaxies or protogalaxies can therefore be observed only as very distant, faint objects (with large red shifts), whose light – emitted in the early stages of galactic evolution – is only now reaching us. The observation of galaxies or quasars e.g. with $z \simeq 5$ currently allows us to look back into the past up to about 95% of the age of the Universe (Table 13.1), and there are indications of evolutionary effects in the most distant systems (bluer color, other types and sizes, galactic merging processes, ...). According to theoretical estimates, we are not too far from being able to observe the birth of the galaxies.

In the 1980's it became clear that among the more distant galaxies, there were numerous faint *blue* objects; in the 1990's, systematic sky surveys were begun, using larger telescopes and modern spectrographs and searching for galaxies with red shifts in the range of $z = 0.1 \ldots 1.3$, which exhibit irregular shapes and blue, compact regions.

A milestone in the investigation of young galaxies was the deep-space, four-color image of the *Hubble Deep Field* (Fig. 12.31) acquired in 1995 by the Hubble Space Telescope: in order to avoid the strong extinction within the Milky Way, a region of the sky at a high galactic latitude was chosen, in which furthermore neither bright stars nor bright galaxies or strong radio sources are present. This field in Ursa Major, of dimensions $2.7' \cdot 2.7'$, at the coordinates $\alpha = 12\,\text{h}\,36\,\text{min}\,49.4\,\text{s}$ and $\delta = +62°\,12'\,58''$ (Epoch Y 2000), was photographed in four colors with an overall exposure time of 10 days(!). The complete image is a composite made up of sections with shorter exposure times. The four broad-band filters have their maxima at the wavelengths λ of 300 (U), 450 (B), 606 (R), and 814 nm (I). The limiting brightnesses detected were about 29 mag in the blue and nearly 30 mag in the infrared. In 1998, the Space Telescope was used to obtain corresponding deep images from fields in the southern hemisphere, in the opposite direction on the celestial sphere. At first view, they show very similar distributions of galaxies and galactic types to those in the field in the northern hemisphere.

Complementary observations of the objects in the Hubble Field were then carried out, using the Hubble Space Telescope itself, and also larger earthbound telescopes such as the 10 m Keck telescope, in particular to determine the red shifts of as many as possible of the distant galaxies.

Determination of the Red Shifts. Up to now, z has been measured for only about 100 galaxies in the Hubble Field; only ca. 10% of the nearly 3000 galaxies are bright enough for a direct z determination using a spectrograph.

Larger red shifts of $z \geq 2.3$ can be estimated from the colors of the galaxies in the Hubble Deep Field, by making use of the strong decrease of brightness (roughly a factor of 10) on the short-wavelength side of the *Lyman edge*; it is due to the stellar energy distribution and in particular to absorption from H I within the galaxy and in the intergalactic medium. Since the

Lyman edge is observed at its red-shifted wavelength $\lambda = 91.2\,(1+z)$ nm, a galaxy at e.g. $z = 3$ can no longer be seen through the U filter, one at $z = 5$ also no longer with the B filter, and finally, a galaxy at $z \geq 6$, can be observed only with the I filter.

Galaxies in the Hubble Deep Field. From the first analysis of the data from the Hubble Deep Field, the majority of the nearly 3000 galaxies is found to have red shifts of $z \leq 3$; based on their brightness decreases at the Lyman edge, a few tens of objects with $3 \leq z \leq 4$ and a similar number with $4 \leq z \leq 5$ were found. Finally, there are a few candidates for extremely high red shifts of $z \geq 5$. With the VLA radiotelescope, 6 sources have been observed which could all be attributed to relatively bright elliptical galaxies and early spirals with $z \simeq 0.2\ldots 1.2$. The infrared satellite ISO found 15 infrared or starburst galaxies in the Hubble Deep Field.

The *morphological type* of the distant galaxies can be determined only by the Space Telescope with its angular resolution of around $0.1''$; large earthbound telescopes can be used only down to about 24 mag owing to air motion. Even at $z > 1$ we can already observe some larger elliptical galaxies and spiral galaxies, whereby the latter are presumably somewhat rarer than today. We observe no S galaxies with regular, symmetrical spiral structures. Instead, they exhibit an irregular, "flaked" appearance with blue condensations which indicate intensive star formation. Their irregular appearance is intensified by the fact that we are actually observing these systems in the far ultraviolet due to their red shifts.

A notable feature of the galaxies of the Hubble Deep Field is the strong increase in the number of *irregular* systems (Irr I) with increasing z; of these, many have much more blue color than the Magellanic Clouds. While today ($z = 0$), irregular types and perturbed systems make up only about 10% of all galaxies, their fraction for $z \simeq 0.5$ is 30%, and for $z \simeq 2$, it is 40%. With this high percentage of peculiar systems, the normal Hubble classification (Table 12.1) becomes essentially meaningless for the young galaxies.

The ultraviolet luminosities of the blue irregular galaxies (in their rest systems), at high red shifts of ($z \geq 2.5$), range typically from 10^9 to $10^{10}\,L_\odot$. Their diameters are only of the order of a few kpc (at $z \simeq 3$); thus, even after taking the factor $(1+z)$ into account, they are relatively small. Either they represent only the nuclear regions of later normal-sized galaxies, or else several such systems later merged to form larger galaxies. The latter suggestion is supported by the large fraction of such objects observed, much greater than today, so that probably not all of them "survived". In any case, we are dealing here with "youthful" galaxies which must still evolve into the present-day S and E galaxies.

The Rate of Star Formation. From their brightness in various wavelength regions, which is determined essentially by the contributions from the *stars*, the time development of the rate of star formation can be estimated (Sect. 10.5.6) – quoted for example as $[\mathcal{M}_\odot\,\mathrm{Mpc}^{-3}\,\mathrm{yr}^{-1}]$. This rate increases from its present ($z = 0$) value with increasing z, and at $z \simeq 1.5$ it reaches a maximum of about 10 times the current rate; it then falls off strongly at still higher z values. Between $z = 4$ and 5, it is about the same as today. This time dependence reflects simultaneously the production of *heavy elements* by the stars in the galaxies; we may therefore expect that times earlier than about $z \simeq 5$ make no contribution worthy of mention to the synthesis of the elements.

The Earliest Evolutionary Stages. From the observations at high red shifts, we can make out the "new" picture of the evolution of the galaxies, which however is as yet by no means certain in all its details:

The galaxies originally formed as relatively *small* irregular protogalaxies. A portion of them merged together to form larger systems; some of them, however, finished their evolution as metal-poor elliptical or spheroidal dwarf galaxies, after losing their gas in early stages (following an intensive phase of star formation) due to the "galactic wind" from the supernova explosions.

The large elliptical and spiral galaxies formed (all of them, or the majority?) through merging processes in the course of their further evolution. They are fundamentally different systems, especially in terms of their interior distribution of angular momentum; the reasons for these differences are not yet known in detail. The formation of the large elliptical galaxies and perhaps also of the spheroidal systems of the spiral galaxies most likely took place at red shifts of $z \geq 3$. In contrast, the disks of the S galaxies were probably formed at only $z \simeq 3\ldots 1$, and they are therefore younger.

13. Cosmology: the Cosmos as a Whole

Having studied our Milky Way Galaxy, various other galaxies of the most diverse kinds, and the arrangement of the galaxies into clusters and superclusters, we now turn to the Universe as a *whole*, its spatial structure and its evolution in time. We begin, in Sect. 13.1, with E. Hubble's discovery of the expansion of the Universe, and then discuss the possible cosmological models on the basis of the theory of General Relativity. Now, in order to determine *the* correct model of our Universe and its parameters, we must make use of a wide variety of different observations. To this end, we investigate somewhat more carefully the propagation of electromagnetic radiation over cosmic distances and describe the isotropic background radiation at 3 K, discovered by A. A. Penzias and R. W. Wilson, and the hot, radiation-dominated initial state of the Universe where the lightest elements, in particular helium, were produced (Sect. 13.2). Finally, in Sect. 13.3, we give an overview of the most important relevant facts in order to arrive at a picture of the evolution of the Universe from its earliest stages, which are closely connected with our knowledge of the fundamental physical interactions, through the production of the 3 K radiation, on to the formation of the galaxies and galaxy clusters, and finally to the present-day state of the cosmos.

13.1 Models of the Universe

Five years after his determination of the distances to far-away galaxies, E. Hubble in 1929 made a second discovery of enormous importance: the proportionality of the red shifts of the galaxies to their distances. On the one hand, this opened up the possibility of determining distances in the cosmos; and on the other, the multiplicity of possible cosmological models was now limited to those with *expansion*. We begin in Sect. 13.1.1 with a discussion of the expanding motion of the galaxies and of the fundamental Hubble constant, which determines the distance scale in the cosmos. In Sect. 13.1.2, we first treat cosmological models in the framework of Newtonian cosmology, which already yields some essential properties, and then in Sect. 13.1.3 we discuss the various relativistic cosmological models which are permissible within the theory of General Relativity. Since matter strongly predominates over radiation in our present-day Universe, we collect in Sect. 13.1.4 the equations which are especially appropriate to the so-called matter cosmos.

13.1.1 The Expanding Universe

The *red shift* of the lines in the spectra of distant galaxies

$$z = \Delta\lambda/\lambda_0 = (\lambda - \lambda_0)/\lambda_0 \qquad (13.1)$$

(λ_0: laboratory wavelength or wavelength in the rest system of the galaxy, λ: measured wavelength) increases *proportionally* to their *distances* r. If this red shift is interpreted as a Doppler effect, then for the velocity at which the galaxy is receding, $v = dr/dt$, we find the Hubble relation:

$$v = c(\Delta\lambda/\lambda_0) = cz = H_0 r \,. \qquad (13.2)$$

(We limit ourselves initially to the case $z \ll 1$; otherwise, we would have to make a relativistic calculation of the Doppler effect).

Apparent exceptions for nearby galaxies (for example, the Andromeda Galaxy is *approaching* us at 300 km s^{-1}) could be readily explained as the reflection of the rotation of our own Milky Way Galaxy. Measurements of the 21 cm line of hydrogen have yielded an excellent confirmation of the wavelength dependence of the effect assumed in Hubble's relation.

The Hubble Constant. The value of the Hubble constant H_0 must be determined from observations. This means essentially that *distance* determinations must be carried out for suitable distant objects; the velocities or red shifts can readily be obtained spectroscopically with sufficient precision. We mention only a few methods of cosmic distance determination: the most important technique uses the period–luminosity relation of various kinds of pulsating stars (Sects. 7.4.1 and 12.1.1). Furthermore, "standard candles" are used, i.e. objects of known, constant and high luminosities such as supernovae of types Ia or also II (Sect. 7.4.7) or planetary nebulae (Sect. 10.3.2). Another technique is based on the differences in the transit time of light from various images, e.g. of a quasar, which are projected by a gravitational lens (Sects. 8.4.3 and 12.3.4).

For H_0, Hubble himself in 1929 obtained the value 530 km s^{-1} Mpc^{-1}. Following W. Baade's revision of the cosmic distance scale (distinguishing between the Cepheids of Populations I and II), A. R. Sandage in 1958 calculated the most probable value to be 75 km s^{-1} Mpc^{-1}. Extensive investigations in the following years using a variety of methods yielded values on the one hand around 50 km s^{-1} Mpc^{-1} (A. R. Sandage, G. A. Tammann et al.), and on the other of about 100 km s^{-1} Mpc^{-1} (G. de Vaucouleurs et al.), and finally also values around 75 km s^{-1} Mpc^{-1}. The differences in the individual determinations of H_0 lie outside the respective error limits (of about 10 to 15%), which have been estimated for the different methods. The use of the Hubble Space Telescope, with its enormous range, has reduced the discrepancies in H_0, but not removed them, so that the resolution of this uncertainty in the determination of one of the most fundamental parameters remains a high-priority goal of observational cosmology.

According to the most recent measurements, H_0 must lie in the range between 55 and 75 km s^{-1} Mpc^{-1}. We can write the Hubble constant as

$$H_0 = 50\,\tilde{h} \quad \text{km s}^{-1}\,\text{Mpc}^{-1} \tag{13.3}$$

with

$$1.1 \leq \tilde{h} = \frac{H_0}{50\,\text{km s}^{-1}\,\text{Mpc}^{-1}} \leq 1.5 \;.$$

In this book, we shall use the value:

$$H_0 = 65\,\text{km s}^{-1}\,\text{Mpc}^{-1} \quad \text{or} \quad \tilde{h} = 1.3\;. \tag{13.4}$$

Distance Determination. Conversely, (13.2) is often employed – in the absence of a better method – in order to estimate the distance r to a galaxy which cannot be telescopically resolved, by using the measured red shift of its spectrum. At the present state of the art, the value used for H_0 should always be indicated.

The Hubble Time. Relation (13.2) can be interpreted initially in a naive way as implying that an *expansion* of the Universe from a relatively small volume began at some earlier time τ_0.

If a particular galaxy, now at a distance r, was ejected with a velocity v in the course of the expansion, it would have required a time τ_0 to move through the distance r; this time is the same for all galaxies:

$$\tau_0 = r/v = 1/H_0\;. \tag{13.5}$$

The Hubble time τ_0 is thus equal to the reciprocal of the Hubble constant, $1/H_0$. If, as is usual, we quote H_0 in units of [km s^{-1} Mpc^{-1}] and τ_0 in [yr], then we find

$$\tau_0\,[\text{yr}] = \frac{978 \cdot 10^9}{H_0\,[\text{km s}^{-1}\,\text{Mpc}^{-1}]} = 19.6 \cdot 10^9\,\tilde{h}^{-1}\;. \tag{13.6}$$

While Hubble's original value of H_0 led to an age of the Universe which was much too short in comparison to the age of the globular clusters, etc., the newer numerical value (13.3) yields:

$$13 \cdot 10^9\,\text{yr} \leq \tau_0 \leq 18 \cdot 10^9\,\text{yr}\;. \tag{13.7}$$

Corresponding to 65 km s^{-1} Mpc^{-1} (13.4), we use here

$$\tau_0 = 15.0 \cdot 10^9\,\text{yr} = 4.7 \cdot 10^{17}\,\text{s}\;. \tag{13.8}$$

13.1.2 Newtonian Cosmology

The kinematics of the expanding universe contained in (13.2) would seem at first glance to represent a throwback to heliocentric concepts. That is however not the case! If, namely, (13.2) is written as a vector equation:

$$\boldsymbol{v} = H_0 \boldsymbol{r}\;, \tag{13.9}$$

where the origin of the coordinate system is taken to lie in our galaxy, then an observer in another galaxy at the distance $\boldsymbol{r}_1$ and with a velocity $\boldsymbol{v}_1$ relative to us would find

$$\boldsymbol{v} - \boldsymbol{v}_1 = H_0\,(\boldsymbol{r} - \boldsymbol{r}_1)\;, \tag{13.10}$$

where $\boldsymbol{v}_1 = H_0 \boldsymbol{r}_1$. A universe expanding according to (13.9) thus offers the *same* aspect to all observers in different galaxies; i.e. our kinematic model of the universe is *homogeneous* and *isotropic*. It may be shown that (13.9) is the *only* velocity flow field which fulfills these conditions under the additional assumption of irrotational flow (curl $\boldsymbol{v} = 0$).

This initially purely kinematic model was extended by E. Milne and W. H. McCrea in 1934 into a Newtonian cosmology; they investigated the kinds of flow which can occur in a medium (the "galaxy gas") within the framework of Newtonian mechanics when homogeneity, isotropy, and irrotational flow are assumed. (The requirement of isotropy throughout the medium implies homogeneity, but not *vice versa*.)

The Equation of Motion. If we consider a finite, expanding "cosmic sphere" with a radius $R(t)$ at a time t, then a galaxy which is e.g. on its surface will be attracted by the mass $\mathcal{M}$ within the sphere according to Newton's law of gravitation. The equation of motion is then

$$\frac{d^2 R}{dt^2} = -\frac{G\mathcal{M}}{R^2} \qquad (13.11)$$

with

$$\mathcal{M} = \frac{4\pi}{3} R(t)^3 \varrho(t) = \text{const} , \qquad (13.12)$$

where $\varrho(t)$ is the mass density at the time considered. The use of Newtonian mechanics is justified as long as $G\mathcal{M}/Rc^2 \ll 1$ is fulfilled (8.76). The problems which arise for *infinitely* extended systems cannot be dealt with here.

Conservation of Energy. If we multiply the above equation (13.11) by $\dot{R} = dR/dt$, it can be integrated without difficulty, leading to the equation of conservation of energy:

$$\frac{1}{2}\left(\frac{dR}{dt}\right)^2 - \frac{G\mathcal{M}}{R} = h \qquad (13.13)$$

with $h = \text{const}$, or

$$\left(\frac{\dot{R}}{R}\right)^2 = -\tilde{k}\frac{c^2}{R^2} + \frac{8\pi}{3} G\varrho(t) . \qquad (13.14)$$

The pressure forces, which we have ignored here for simplicity, would not change the result obtained. In anticipation of a later comparison with relativistic calculations. we have written the constant as $h = -\tilde{k}c^2/2$ ($c =$ velocity of light).

The Hubble Function. It is easy to convince oneself that at a particular time t, the same red shift law would be observed from every galaxy within our sphere. Since the distance r between two galaxies is proportional to the *scale factor* $R(t)$, the relative velocity is given by

$$v = \frac{\dot{R}}{R} r = H(t) r , \qquad (13.15)$$

where we call $H(t) = \dot{R}/R$ the *Hubble function*. Denoting all quantities which refer to the *present* time $t = t_0$ by a subscript 0, we find for the Hubble constant in (13.2):

$$H_0 = \frac{\dot{R}_0}{R_0} . \qquad (13.16)$$

Deceleration Parameter. In order to fully characterize a model of the Universe, we require in addition to $H(t)$ or H_0 a second quantity describing the gravitational acceleration due to the mass $\mathcal{M}$ which acts inwards and opposes the expansion of the Universe; this is the so-called deceleration or delay parameter

$$q(t) = -\left(\frac{\ddot{R}}{R}\right) \Big/ \left(\frac{\dot{R}}{R}\right)^2 = -\frac{1}{H(t)^2}\frac{\ddot{R}}{R} \qquad (13.17)$$

or its present-day value

$$q_0 = -\frac{1}{H_0^2}\frac{\ddot{R}_0}{R_0} = \frac{4\pi G \varrho_0}{3 H_0^2} \qquad (13.18)$$

from (13.11, 12). The delay parameter q_0 relates the acceleration $-\ddot{R}_0$ to a "unit acceleration", which would lead in the Hubble time $\tau_0 = 1/H_0$ starting from zero velocity to the expansion velocity $R_0 H_0$ currently observed at a distance R_0.

Cosmological Models. The solution of our equations leads to models of the Universe which, starting from a point (singularity) of infinitely high density, lead either to a *monotonic expansion* (total energy $\mathcal{M}h \geq 0$) or to a periodic *oscillation* between $R = 0$ and $R_{\max}$ (for $h < 0$). Static models are not possible within the framework of (13.11).

Generally speaking, Newtonian mechanics extends the purely kinematic cosmological model which we treated first, in the sense that it considers the Hubble constant H_0 to be a function of time t. In a periodic universe, for example, an age of red shifts would be followed by an age with violet shifts and conversely. We shall not discuss further here the question of the choice among the numerous different models possible within a Newtonian cosmology, nor the fundamental difficulties of such a cosmology; instead, we turn immediately to a *relativistic* cosmology.

13.1.3 Relativistic Cosmology

After the formulation of the modern field theory of gravitation by A. Einstein (1916), in the form of his *theory of General Relativity* (Sect. 8.4.1), Einstein himself and W. de Sitter in 1917 constructed some special models of the Universe, in which the curvature of space was assumed to be independent of time. Einstein showed that his field equations (extended to include the Λ-term, see below) have a static cosmological solution; de Sitter found a matter-free expanding Universe as their solution.

Referring to the older radial velocity measurements of V. M. Slipher (ca. 1912), C. Wirtz had already in 1924 noticed their increase with distance and related it to de Sitter's relativistic cosmological model, five years before Hubble published his proportionality between the distance and the radial velocity.

In 1922–24, A. Friedmann succeeded in formulating an important generalization to spaces with time-dependent curvatures. His publications went unheeded for a time, until about 1927–30 when G. Lemaître, A. S. Eddington, and others took up the investigation of models of an *expanding* Universe on the basis of the theory of General Relativity.

The Cosmological Postulate. We assume from the beginning that the cosmological postulate holds, i.e. that the Universe must be *homogeneous* and *isotropic* throughout. This requirement is in agreement with all of those portions of the Universe which are accessible to observation, if we consider sufficiently large regions, so that we can average over the largest structures such as the superclusters of galaxies and the "cosmic honeycomb", i.e. over dimensions of more than a few 100 Mpc (Sect. 12.4.5).

The Metric. Assuming the cosmological postulate to be valid, then – following H. P. Robertson and A. G. Walker – we can cast the four-dimensional line element ds (8.77) in the form:

$$ds^2 = c^2 dt^2 \qquad (13.19)$$
$$- R(t)^2 \left[\frac{dr^2}{1-kr^2} + r^2(d\theta^2 + \sin^2\theta \, d\varphi^2) \right].$$

Here, r, θ, and φ are dimensionless Lagrange coordinates, which remain constant for a galaxy that is following the expansion of the Universe. The time dependence of the geometry thus depends exclusively on the scale factor $R(t)$. $R(t)$ determines the *radius of curvature* of the three-dimensional space, defined analogously to the radius of curvature of a two-dimensional surface. The constant k, which can take on the values 0 or ± 1, fixes the sign of the spatial curvature that is the same everywhere at a given time.

The constant k corresponds to

$k=0$	*Euclidean* space,
$k=+1$	closed *spherical* or elliptical space of finite volume,
$k=-1$	open *hyperbolic* space

(For the Euclidean, flat space $k=0$, the radius of curvature becomes infinite.)

These three kinds of spaces or geometries are best imagined by considering their two-dimensional analogs (Fig. 13.1); in particular, a closed, expanding universe can be represented by the surface of a balloon which is being inflated.

Friedmann–Lemaître Equations. In the Robertson–Walker metric (13.19), the number of Einsteinian field equations (8.79) which relate the metric g_{ik} of the Universe to its material contents, extended to include the term Λg_{ik} with the cosmological constant Λ, is reduced (in a way which we cannot treat in detail here) to two differential equations for $R(t)$:

$$\left(\frac{\dot{R}}{R}\right)^2 = -k\frac{c^2}{R^2} + \frac{8\pi G}{3}\varrho + \frac{c^2}{3}\Lambda, \qquad (13.20)$$

$$2\frac{\ddot{R}}{R} + \left(\frac{\dot{R}}{R}\right)^2 = -k\frac{c^2}{R^2} - \frac{8\pi G}{c^2}P + c^2\Lambda. \qquad (13.21)$$

Here, we have assumed the Universe to be filled with a "cosmological gas" whose energy–momentum tensor is determined by the local *mass density* $\varrho = \varrho(t)$ and the pressure $P = P(t)$. In contrast to the Newtonian case, ϱ now includes the equivalents of all energy

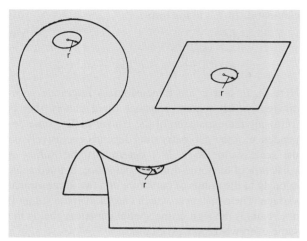

Fig. 13.1. Surfaces (i.e. two-dimensional spaces) with curvatures of $k > 0$, $k = 0$, and $k < 0$ as models of curved spaces.

Curvature k of the surface:	$k > 0$	$k = 0$	$k < 0$
Geometry:	Spherical or elliptical	Euclidean	Hyperbolic
Circumference of the circle:	$< 2\pi r$	$= 2\pi r$	$> 2\pi r$
Area of the circle:	$< \pi r^2$	$= \pi r^2$	$> \pi r^2$

forms, corresponding to the theory of General Relativity (Sect. 8.4.1); furthermore, it also contains the unidentified dark matter whose existence is indicated by the dynamics of galaxies and clusters of galaxies (Sects. 12.1.5 and 12.4.2). An alternative form of equation (13.21) can be obtained by subtraction of (13.20):

$$\frac{\ddot{R}}{R} = -\frac{4\pi G}{3}\left(\varrho + 3\frac{P}{c^2}\right) + \frac{c^2}{3}\Lambda . \qquad (13.22)$$

For a *vanishing* cosmological constant $\Lambda = 0$, Equation (13.20) is seen to be identical with the differential equation (13.14) from Newtonian cosmology. For systems with negligible pressure P, (13.22) also becomes identical to (13.11), so that we have exactly the same choice of cosmological models as in Newtonian cosmology. However, it is now *a priori* forbidden for velocities greater than the speed of light to occur. Relativistic cosmology is the first theory to make possible a self-consistent and noncontradictory description of the Universe as a whole.

The Cosmological Constant. A comparison of (13.20) with the Newtonian equations (13.9) and (13.14) further shows us that a *positive* constant Λ corresponds to a *repulsive* acceleration, which thus acts in opposition to gravitation.

The Λ term was introduced by Einstein in 1917 into his gravitational equations; the purpose of this additional term was to make it possible to construct a *static* cosmological model. We can readily see by referring to the Friedmann–Lemaître equation (13.20) that $\dot{R} = 0$ results from $k = \Lambda R^2$ and $\Lambda = 4\pi G\varrho/c^2$. Einstein originally held this model to be the only reasonable assumption. This point of view, however, became obsolete after Hubble's discovery of the red shift law.

The cosmological constant, which has the dimensions [m^{-2}], can be expressed formally as an *energy density* $\varrho_\Lambda c^2$ or as its equivalent mass density ϱ_Λ, where

$$\Lambda = \frac{8\pi G}{c^2}\varrho_\Lambda \qquad (13.23)$$

holds. Denoting the matter density ϱ for clarity by ϱ_M, we can write the Friedmann–Lemaître equations (13.20, 22) in the form

$$\left(\frac{\dot{R}}{R}\right)^2 = -k\frac{c^2}{R^2} + \frac{8\pi G}{3}(\varrho_M + \varrho_\Lambda), \qquad (13.24)$$

$$\frac{\ddot{R}}{R} = -\frac{4\pi G}{3}\left[(\varrho_M + \varrho_\Lambda) + 3\left(\frac{P}{c^2} - \varrho_\Lambda\right)\right] \qquad (13.25)$$

Following W. H. McCrea (1951) and Ya. B. Zeldovich (1968), the cosmological constant Λ can be interpreted within the framework of quantum field theory in terms of the energy density of the vacuum, $\varrho_{\text{vac}}c^2$, i.e. $\varrho_\Lambda = \varrho_{\text{vac}}$. The associated (negative!) pressure of the quantum vacuum is $P_{\text{vac}} = -\varrho_{\text{vac}}c^2$. The vacuum, to be sure, contains by definition no real particles, but according to quantum field theory it can be seen as an energetic ground state filled with fluctuating matter fields or virtual particles. This state can have a nonvanishing energy density and possibly can contribute to gravitation. It is at present however not possible to calculate even an approximate value for ϱ_{vac} or Λ from quantum field theory.

It is also currently very difficult to estimate the cosmological constant or the "dark energy" ϱ_Λ, whose nature is still a complete riddle, from *observations*, since the Λ term is apparent only over very great distances. From (13.20) and using (13.16), we can estimate that this term becomes important when

$c\sqrt{|\Lambda|} \sim \dot{R}/R \sim H_0$, i.e. when the distances become greater than about $1/\sqrt{|\Lambda|}$. Cosmological observations over many years have yielded only upper limits of around $|\Lambda| \le 10^{-52}$ m^{-2}. Owing to this uncertainty, frequently a value of simply $\Lambda = 0$ was assumed (see the standard model, below), particularly since the other cosmological parameters are also not too precisely known.

Beginning about 1995, observations of supernovae (of type Ia) gave clear indications of a (positive) cosmological constant $\Lambda \ne 0$ (Sect. 13.2.5). Together with recent observations of the fluctuations of the 3 K background radiation (Sect. 13.2.2), these data give a density parameter (13.29) of $\Omega_{\Lambda,0} \simeq 0.7$ and thus of $\Lambda \simeq +10^{-52}$ m^{-2}; this corresponds to a considerable contribution of the Λ term over distances of more than $1/\sqrt{\Lambda} \simeq 10^{26}$ m $\simeq$ 3000 Mpc.

A gravitationally acting, nonvanishing energy density of the vacuum also contributes – independently of any assumptions about the nature of the cosmological constant – to the energy–momentum tensor T_{ik} (8.78, 79). Thus, the quantum vacuum, and formally the cosmological constant, might have played an important role in the very early evolutionary stages of the Universe (inflationary expansion; see Sect. 13.3.4). We note finally that in Einstein's field equations (8.79), Λ was introduced as a *constant*. In the framework of newer cosmological models, it has been speculated that $\Lambda = \Lambda(t)$ could instead *vary with time*.

The Critical Density. In the following, we define the critical density

$$\varrho_c(t) = \frac{3H(t)^2}{8\pi G} \qquad (13.26)$$

and the dimensionless *density parameter*

$$\Omega(t) = \Omega_M(t) + \Omega_\Lambda(t) \qquad (13.27)$$

with the corresponding parameters for the matter

$$\Omega_M(t) = \frac{\varrho(t)}{\varrho_c(t)}, \qquad (13.28)$$

and for the cosmological constant (13.23)

$$\Omega_\Lambda(t) = \frac{\varrho_\Lambda(t)}{\varrho_c(t)} = \frac{c^2 \Lambda}{3H(t)^2}. \qquad (13.29)$$

If we express $\dot{R}$ and $\ddot{R}$ using (13.15, 17) in terms of $H(t)$ and $q(t)$ in the Friedmann–Lemaître equations, and introduce the density parameters, we then obtain for the "curvature term"

$$k\frac{c^2}{H^2 R^2} = \Omega - 1 = \Omega_M + \Omega_\Lambda - 1 \qquad (13.30)$$

and for the deceleration parameter

$$q = \frac{1}{2}\Omega_M - \Omega_\Lambda \qquad (13.31)$$

or with (13.30), also

$$2q - 1 = \frac{c^2}{H^2 R^2} - 3\Omega_\Lambda, \qquad (13.32)$$

where for the last expression we have assumed cosmological models with vanishing pressure, $P = 0$ (see below). From (13.31), there is a *deceleration* of the expansion of the Universe (i.e. a deceleration parameter $q > 0$) only if $\Omega_\Lambda < \Omega_M/2$. We will however retain the historical name "deceleration parameter" even for the case that $\Omega_\Lambda > \Omega_M/2$ (and thus $q < 0$).

From the first of the Friedmann–Lemaître equations (13.20), it follows for continuously expanding models that there is a remarkable connection, $c^2 \Lambda = 3H(\infty)$, between the cosmological constant and the Hubble parameter $H(\infty)$ in the limit $t \to \infty$.

The great variety of cosmological models for $\Lambda \ne 0$ (Friedmann–Lemaître models), which are compatible with the *current* values H_0, $\Omega_{M,0}$ and $\Omega_{\Lambda,0}$ cannot be discussed here. Instead, we restrict ourselves in the following to the so-called standard models, with

$$\Lambda = 0. \qquad (13.33)$$

Standard Models. In relativistic cosmology within the standard model ($\Lambda = 0$), as in Newtonian cosmology, $R(t)$ is initially determined by the Hubble constant H_0 and the deceleration parameter q_0. The latter also determines the type of spatial curvature, i.e. the value of k or the basic structure of the spatial universe. Setting $\Lambda = 0$, we obtain from (13.30) and (13.31) the relation

$$2q - 1 = \frac{\varrho_M}{\varrho_c} - 1 = \Omega_M - 1 = k\frac{c^2}{H^2 R^2}. \qquad (13.34)$$

From it (noting that $c^2/(H^2 R^2)$ is always positive), we can read out the following three possibilities:

Deceleration parameter	Curvature	Density		Time dependence of $R(t)$
$0 \leq q \leq \frac{1}{2}$	$k = -1$	$\varrho_M < \varrho_c$	$\Omega_M < 1$	monotonic
$q = \frac{1}{2}$	$k = 0$	$\varrho_M = \varrho_c$	$\Omega_M = 1$	increasing
$q > \frac{1}{2}$	$k = +1$	$\varrho_M > \varrho_c$	$\Omega_M > 1$	finite (cycloidal)

The present-day matter density $\varrho_{M,0}$ thus, together with H_0, determines the deceleration parameter q_0 and therefore also the type of solution of the cosmological model.

In the *present-day* state of the Universe, the mass density of the matter concentrated in the galaxies exceeds that of the radiation (3 K background radiation, Sect. 13.2.2) by far, so that here, the neglect of the pressure in the Friedmann–Lemaître equations is justified.

We speak generally of a *matter cosmos* when the density of *non*relativistic particles, i.e. practically the rest-mass density, predominates and its pressure is negligible. On the other hand, we speak of a *radiation cosmos*, when the mass or energy density of *relativistic* particles, such as photons, predominates.

13.1.4 The Matter Cosmos

If we disregard the early stages of the Universe (up to about $2 \cdot 10^6$ yr after the singularity, cf. (13.64)), the matter cosmos represents a good approximation for a model of the Universe. This is also true in particular for those epochs in which the most distant quasars, with $z \simeq 6$, emitted the radiation that we are just detecting today; thus we can employ models corresponding to (13.34) for comparison to the observations.

For the matter cosmos, we expect conservation of mass, (13.12),

$$\varrho_M(t) R(t)^3 = \varrho_{M,0} R_0^3 = \text{const}, \quad (13.35)$$

so that we can write (13.20) for $\Lambda = 0$ in the form:

$$\dot{R}^2 = -kc^2 + \frac{A}{R} \quad (13.36)$$

where $A = 8\pi G \varrho_{M,0} R_0^3 / 3$ is a constant.

In the case $k = 0$ ($q = 1/2$), we obtain by integration with $8\pi G \varrho_{M,0}/3 = H_0^2$ (13.26) the *Einstein–de Sitter cosmos* as solution:

$$\frac{R(t)}{R_0} = \left(\frac{3}{2} H_0 t\right)^{2/3}; \quad (13.37)$$

this is a cosmological model of a monotonically expanding universe.

For sufficiently small R in (13.36), the term kc^2 is negligible compared to A/R, so that (13.37) represents a good approximation even for models with $k \neq 0$.

In the cosmological models with negative curvature, $k = -1$ ($q < 1/2$), $R(t)$ increases continually and monotonically; for large R or t, $R(t)$ becomes proportional to t. However, the time dependence of the scale factor in models with $k = +1$ ($q > 1/2$) is represented by a cycloid; in these models, the expansion is eventually followed by a contraction.

In Fig. 13.2, we have plotted $R(t)$ for several cosmological models with $\Lambda = 0$ for different deceleration parameters q_0; they are all characterized at the present time t_0 by the *same* Hubble constant H_0 and the same scale factor R_0. According to (13.16), H_0 determines the (common) tangent to the curves at $t = t_0$. Between t_0 and the intercept of this "Hubble line" with the abscissa is the time interval called the *Hubble time*, $\tau_0 = 1/H_0$ as in (13.5). The *beginning* of our cosmological epoch is determined by the intercept of the curve of $R(t)$ with the abscissa. For $\Lambda = 0$, it precedes t_0 by a time interval which is termed the *Friedmann time*, τ_F (Table 13.1). There can be no galaxy, star cluster, etc. whose age is greater than τ_F. For the Einstein–de Sitter cosmos, from (13.37) we obtain $\tau_F = (2/3)\tau_0$.

Table 13.1. The Friedmann time τ_F and the "look-back time" $t(z = 5)$ for various values of the Hubble constant H_0 and the deceleration parameter q_0. The Hubble time $\tau_0 = 1/H_0$ is equal to τ_F when $q_0 = 0$

q_0	τ_F [10^9 yr]			$t(z = 5)/\tau_F$
	$H_0 = 50$	$H_0 = 75$ [km s^{-1} Mpc^{-1}]	$H_0 = 100$	
0	19.6	13.0	9.8	0.83
0.1	16.7	11.1	8.3	0.90
0.5	13.1	8.7	6.5	0.93
1.0	11.2	7.5	5.6	0.94
2.0	9.3	6.2	4.6	0.95

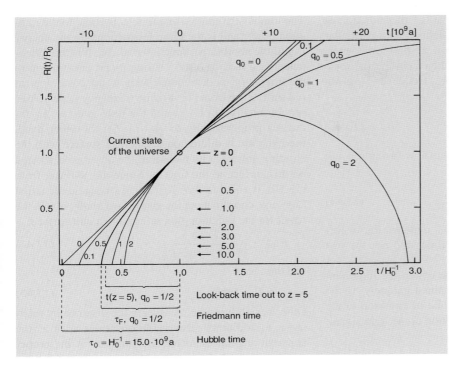

Fig. 13.2. Matter-dominated cosmological models ($q_0 \leq 1/2$ are open, $q_0 > 1/2$ are closed universes for $\Lambda = 0$). The distance between two galaxies is proportional to the scale factor $R(t)$. All models have the *same* Hubble constant $H_0 = 65$ km s^{-1} Mpc^{-1}, or the same tangent to the $R(t)$ curve at the present time. The relationship between $R(t)$ and the red shift z is indicated by the arrows. The Hubble time τ_0 and, for the Einstein–de Sitter cosmos ($q_0 = 1/2$), the Friedmann time τ_F and the look-back time $t(z = 5)$ are shown below. The time scale t is shifted here by $(\tau_0 - \tau_F)$ compared to the scale used in (13.36) and (13.37)

13.2 Radiation and Observations. Element Synthesis in the Universe

In Sect. 13.2.1, we will first investigate the propagation of radiation in the Universe and discuss its red shift as well as the measurement of distances based on the luminosities of galaxies. The isotropic 3 K radiation discovered by A. A. Penzias und R. W. Wilson, which is interpreted as the remnant of the Big Bang, points to an early, very hot stage of the Universe which was dominated by radiation. We describe the observations of this background radiation in Sect. 13.2.2, and then, in Sect. 13.2.3, we consider the models for such a radiation cosmos. In Sect. 13.2.4, we treat the synthesis of the elements in the Universe, a topic which is closely connected to the observations of the 3 K radiation. The most important observations which give us some information as to *which* model corresponds to the real (present-day) Universe will then be presented in Sect. 13.2.5. Concluding in Sect. 13.2.6, we discuss the significance for the cosmological models of the stellar radiation in the cosmos.

13.2.1 The Propagation of Radiation

According to the theory of General Relativity, electromagnetic radiation propagates along *zero geodesic* world lines, $ds^2 = 0$ (8.80). In the line element of the Robertson–Walker metric (13.19), the Lagrange coordinates of galaxies which are moving apart, neglecting proper motions and simply as a result of the expansion of the Universe, remain constant in time. If we consider a light signal which was emitted at a time t_e from a galaxy with the dimensionless radial coordinate r, and which reaches us (at $r = 0$) today (at t_0), then its propagation along $ds^2 = 0$ can be described by

$$c \frac{dt}{R(t)} = \text{const} \tag{13.38}$$

or

$$c \int_{t_e}^{t_0} \frac{dt}{R(t)} = \int_0^r \frac{dr'}{\sqrt{1-kr'^2}} \tag{13.39}$$

$$\simeq r + \frac{k}{6} r^3 + \ldots$$

where we have for simplicity assumed that only the radial coordinates of the two galaxies are different.

The Red Shift. The *frequency* ν_e of the radiation at the time of its emission can be defined by a particular number N of oscillations within a short time interval Δt_e. A present-day observer then measures the *same* number N in an interval Δt_0, or a frequency ν_0 given according to (13.38) by

$$\frac{\nu_0}{\nu_e} = \frac{\Delta t_e}{\Delta t_0} = \frac{R(t_e)}{R_0} \,. \tag{13.40}$$

The *red shift* z defined by (13.1) is now (using $c = \nu \lambda$) given by

$$1 + z \equiv 1 + z_e = \frac{\lambda_0}{\lambda_e} = \frac{\nu_e}{\nu_0} \,. \tag{13.41}$$

(In contrast to (13.1), here λ_0 denotes the wavelength observed at the present time, and not the rest wavelength.)

From (13.40) and (13.41), we obtain the important result that the red shift is determined uniquely by the ratio of the scale factor at the time of emission to the scale factor at the time of absorption of the light signal:

$$1 + z = \frac{R_0}{R(t)} \,. \tag{13.42}$$

We can now, for a given q_0, find the corresponding R for every z, and with the appropriate cosmological model, the "look-back time" $t(z)$, i.e. the time that a light or radio signal would require to travel from the galaxy in question to the Earth (Fig. 13.2). In this way, we obtain among other things a lower limit $t(z \simeq 6)$ for the age of the most distant galaxies or quasars (Sect. 12.3.4).

The Luminosity Distance. Now, in order to obtain a form of the Hubble relation (13.2) which is valid for *large* distances or red shifts, we must relate not only z but also the *distance* to the parameters of the cosmological model. The "proper distance" $r_E \simeq rR(t)$ is unsuitable for this purpose, since it is not directly observable. On the other hand, we can introduce for example a "*luminosity distance*" r_L, which can be derived from observations of the apparent magnitude of a galaxy, by defining r_L for large distances using the usual relation (6.43) between the observed radiation flux f and the absolute luminosity L of the galaxy, taken to be known:

$$f = \frac{L}{4\pi r_L^2} \,. \tag{13.43}$$

The luminosity emitted at a time t_e is equal to the radiation power and is therefore proportional to the energy $h\nu_e$ of the photons per time interval Δt_e. It is thus attenuated by a factor $(1+z)^2$ during its propagation up to the time t_0 of its detection, owing to the cosmological red shift, since from (13.40), the frequency ν_e decreases proportionally to $(1+z)$ and the time interval Δt_e increases proportionally to $(1+z)$. On the other hand, the radiation is diluted on a spherical surface with the emitting galaxy at its midpoint; the observer is located on this surface at the time t_0. Since this surface, from (13.19), is equal to $4\pi r^2 R_0^2$, where r is again the radial Lagrange coordinate of the galaxy, we finally obtain the result for the radiation flux referred to a unit surface:

$$f(t_0) = \frac{L(t_e)}{(1+z)^2 \, 4\pi r^2 R_0^2} \tag{13.44}$$

and therefore

$$r_L = (1+z) r R_0 \,. \tag{13.45}$$

Other possibilities for measuring the distance by relating it to additional observable quantities such as the apparent angular diameter, the parallax, or the proper motions can only be mentioned here.

The Generalized Hubble Relation. The relationship between the red shift and the distance is now determined by equations (13.39), (13.42), and (13.45), together with the Friedmann–Lemaître equations. We give the result here for the generalized Hubble relation (in a universe dominated by matter, for $\Lambda = 0$):

$$cz \cdot \psi(q_0, z) = H_0 r_L \tag{13.46}$$

with

$$\psi(q_0, z) = 1 + \frac{1 - q_0}{q_0} \left[1 - \frac{1}{q_0 z} \left(\sqrt{1 + 2q_0 z} - 1 \right) \right]$$
$$= 1 + \tfrac{1}{2}(1 - q_0) z + \ldots \qquad (q_0 z \ll 1) \,.$$

Corresponding to (13.44), the luminosity distance r_L is obtained from observations of galaxies of known absolute luminosities L. For $z \ll 1$, we obtain again from (13.46) the original Hubble relation, (13.2).

If we express the brightness as usual in magnitudes (6.35), we find the following relation between the apparent bolometric magnitude m_{bol} and the red shift

$$m_{\text{bol}} = 5 \log cz + 5 \log \psi(q_0 z) + M_{\text{bol}} \tag{13.47}$$
$$\quad - 5 \log H_0 + 25 \qquad [\text{mag}] \,,$$

where c is measured in [km s^{-1}] and H_0 in [km s^{-1} Mpc^{-1}]. The absolute magnitude M_{bol} depends on L as given in (6.39). For small $q_0 z$, the term $5 \log \psi(q_0, z) \simeq 1.086 (1 - q_0)z$.

In practice, instead of the bolometric magnitudes, the brightness integrated over a limited range of wavelengths is measured. We cannot deal here with all the required corrections to (13.47), which depend on z and q_0.

13.2.2 The Microwave Background Radiation

Our interest now naturally turns to the earliest stage of cosmic evolution, the *Big Bang*. Fundamental research on this topic (beginning about 1939) is inseparably connected with the names of G. Lemaître and G. Gamow.

Here, we first skip over the very earliest phases, and consider the Universe from about 200 s after the singularity, when it had cooled off to around 10^9 K. At this temperature, the synthesis of the chemical elements from protons and neutrons could begin, as we shall see in detail in Sect. 13.2.4, since then deuterium, once formed, would no longer be destroyed by reactions with energetic photons. Deuterium represents the starting nucleus for the gradual construction of all the heavier elements. G. Gamow in 1948 originally attempted to place the synthesis of *all* the heavier elements in this time period. This idea however proved to be untenable, since the formation of nuclides would come to a halt already at mass number $A = 5$, for which no stable nucleus exists. At this stage, practically only isotopes of H and He were formed; the synthesis of heavier nuclei occurred at considerably later times in the interiors of stars (Sect. 8.2.5).

Predictions. Gamow already noticed that we could also obtain *direct* information, so to speak, about this phase of the "Primeval Fireball", in addition to the elemental abundances. Shortly after this stage, namely, the interaction between radiation and matter had become so weak that the *radiation field* of the cosmos could only expand *adiabatically* along with the rest of the Universe. L. Boltzmann had shown long before that the radiation field of a black body radiation cavity at a temperature T remains black during adiabatic expansion, and that furthermore, the product T^3 times the volume V of the cavity remains constant. However, after the completion of formation of the atoms of H and He, their particle number nV was also conserved, i.e. $T^3 \propto n$ would have to decrease. From nuclear-physics considerations, Gamow began with $T = 10^9$ K and $n \simeq 10^{24}$ m^{-3} at the time of the formation of the elements, and took an average value for the present-day Universe of $n_0 \simeq 1$ m^{-3}. After expansion to a 10^{24}-fold volume, he concluded, the present-day Universe must be filled with *blackbody radiation* at a temperature of the order of 10 K. The detection of this radiation at that time was, however, out of the question, given the state of the art of radio-astronomical observational techniques.

The Discovery. Making use of the enormously improved technology in radio astronomy in the meantime, R. H. Dicke, P. J. E. Peebles, P. G. Roll and D. K. Wilkinson, guided by newer calculations, began in 1964 to search for the cosmic background radiation. However, before they had completed their measurements, this microwave radiation was discovered coincidentally in 1965 by A. A. Penzias and R. W. Wilson; it was immediately recognized as such by Dicke et al.

Penzias and Wilson detected weak radiofrequency radiation at $\lambda = 7.35$ cm (4.08 GHz) using a large, low-noise horn antenna, which was originally constructed for purposes of communications via the Echo satellite. After subtraction of the contribution from the Earth's atmosphere and the receiver noise, an isotropic, unpolarized component remained, which was independent of the time of day or the season and had an unexpectedly high excess-antenna temperature of (3 ± 1) K.

It was later confirmed that this radiation was in fact the remnant of the Big Bang, through the observation that it follows *Planck's law* (4.61) in the entire accessible wavelength region, and is isotropic and unpolarized. The high degree of isotropy, with only very small deviations (see below), confirms the cosmological postulate (Sect. 13.1.3) which lies at the root of the cosmological models.

"Optical Observations". In fact, the background radiation had already been observed indirectly in 1941 by A. McKellar: from the intensity ratios of the *interstellar* absorption lines of the CN radical at $\lambda = 387.46$

and 387.39 nm in the spectrum of the O star ζ Oph, he derived an excitation temperature of 2.3 K for the first rotational level, lying $4.7 \cdot 10^{-4}$ eV (corresponding to $\lambda = 2.64$ mm) above the ground state; this was a completely incomprehensible result at the time.

The modern value of "optical observations" of the 3 K radiation from lines of CN, CH and CH$^+$ agrees very well with (13.48).

Spectrum and Temperature. The intensity $I_\nu = B_\nu(2.73 \text{ K})$ of the background radiation has its maximum at $\nu = 180$ GHz or $\lambda = 1.7$ mm; on the short-wavelength side, it falls off rapidly. At long wavelengths, above $\lambda \geq 30$ cm, the galactic nonthermal radio emissions predominate over the 3 K radiation (Sect. 12.3.2). Below $\lambda \leq 3$ mm, the observations must be conducted outside the Earth's atmosphere. After initial measurements from balloons and rockets, from 1989 on for several years, the COBE satellite (Sect. 5.2.1) was available to carry out extended, very precise observations.

The temperature of the background radiation determined from all the measurements is

$$T_0 = (2.73 \pm 0.01) \text{ K} . \qquad (13.48)$$

(We again denote the present-day state of the Universe by an index 0.) A still higher accuracy is attained by the absolute measurements in the far infrared alone; they give $T_0 = (2.728 \pm 0.002)$ K. The spectrum agrees with a Planck distribution (for $\lambda \leq 10$ cm) within about 10^{-4} of its relative intensity; the polarization lies at a level below 10^{-5}. So far, no "dents" in the spectral shape have been detected; they would be an indication of deviations from the adiabatic expansion of the radiation field, i.e. of energy inputs or dissipation.

The Energy and Photon Densities. The energy density of a black-body radiation field at a temperature T is given by the Stefan–Boltzmann law (4.66, 71), $u = aT^4$ with $a = 7.56 \cdot 10^{-16}$ J m^{-3} K^{-4}. For the microwave background radiation, at $T_0 = 2.73$ K, we find for the corresponding mass density:

$$\varrho_{\gamma,0} = \frac{u_{\gamma,0}}{c^2} = 4.7 \cdot 10^{-31} \text{ kg m}^{-3} . \qquad (13.49)$$

This is negligible compared to the present matter density (Table 13.2) i.e. the assumption of a *matter cosmos* (Sect. 13.1.4) for the present-day Universe is clearly justified.

We also consider the photon density n_γ in the background radiation, which is found from the energy density u through division by the mean energy of the photons, $\langle h\nu \rangle = 2.70 \, kT$ (4.76), to be $n = bT^3$ with $b = 2.02 \cdot 10^7$ m^{-3} K^{-3} (4.74). At present, its value is

$$n_{\gamma,0} = 4.1 \cdot 10^8 \text{ photons m}^{-3} , \qquad (13.50)$$

so that at the present particle density of 0.6 m^{-3} in the Universe (Table 13.2), there are around 10^9 photons of the 3 K radiation for every hydrogen atom.

Deviations from Isotropy. The only anisotropy in the intensity or the temperature of the 3 K radiation observed up to now has an angular dependence proportional to $\cos \theta$ ("dipole characteristic") with a relative amplitude of $\leq 10^{-3}$. It is a result of the *motion of the Earth* relative to the coordinate system fixed with respect to the background radiation. Due to the Lorentz transformations, an observer moving at a velocity v relative to this coordinate system measures a Planck distribution in every direction θ, but its temperature depends on θ, as a result of the Doppler effect (4.18):

$$\frac{T(\theta)}{T_0} = \frac{\sqrt{1-(v/c)^2}}{1-(v/c)\cos\theta}$$
$$\simeq 1 + \frac{v}{c}\cos\theta + \ldots . \qquad (13.51)$$

The maximum temperature rise is seen in the direction of the Earth's motion, or that of the Sun ($\theta = 0$). The observed value is

$$\frac{\Delta T_{\max}}{T_0} = 1.23 \cdot 10^{-3} \quad \text{or}$$
$$\Delta T_{\max} = (3.37 \pm 0.03) \cdot 10^{-3} \text{ K} \qquad (13.52)$$

corresponding to a velocity of $v = (371 \pm 0.5)$ km s^{-1} towards a point with the coordinates

$$\alpha = 11.20 \text{ h} , \quad \delta = -7.0° \quad \text{or}$$
$$l = 264.1° , \quad b = 48.3°$$

(with an uncertainty of about $\pm 0.2°$). The main effect comes from the rotation of the Sun around the center of the Milky Way galaxy. If we relate the velocities to the center of mass of the *Local Group* (Sect. 12.4.1), we find that it is moving at about 630 km s^{-1} relative to the cosmological coordinate system. The direction of the motion, which cannot be determined very precisely (only to about $\pm 3°$), is towards the galactic coordinates $l = 276°$, $b = 30°$, around 50° away from the center of the Virgo cluster.

The intensity variation of the 3 K radiation due to Compton scattering during transmission through the hot gas in the galaxy clusters was already mentioned in Sect. 12.4.3.

Deviations of the background radiation from the Planck distribution and from isotropy on much smaller angular scales than that of the dipole component could give us important information about "perturbations" in the density distribution during the early phases of the Universe, about the formation of the galaxies, and about the cosmological contribution from dark matter. Measurements with COBE over several years starting in 1992 showed fluctuations in the temperature of around $\Delta T/T_0 \simeq 10^{-5}$ over areas of $\geq 7°$ on the celestial sphere. Later observations of selected regions of the sky using earthbound instruments showed variations of the same order of magnitude over considerably smaller scales.

The significance of these data was already discussed in Sect. 12.5.1 in connection with the formation of the galaxies and clusters of galaxies. In recent times, the precision of the measurements has been improved to such an extent that they permit far-reaching conclusions about the parameters and the evolution of the Universe to be drawn in the field of *cosmology* as well.

The time development of the density and temperature disturbances from the time of the "formation" of the microwave background radiation up to the present can be calculated on the basis of cosmological models. The details of the emission of the 3 K radiation thus give us a "snapshot" of the acoustic waves in the past. The model calculations, in particular those for an inflationary Universe (Sect. 13.3.4), show a series of maxima in the power spectrum $\Delta T(l)$ of the fluctuations of the 3 K radiation. The measurements from the Boomerang experiment and other balloon observations (Sect. 12.5.1) indeed show the first two maxima with

Multipole order	$\simeq 200$	$\simeq 500$
Angular size	$\simeq 0.9°$	$\simeq 0.4°$
Amplitude ΔT [μK]	70	40 .

The position and amplitude of these maxima give important information about the physical state of the cosmos and thereby contribute to the detemination of the fundamental cosmological parameter (in ways that we cannot describe in detail here).

The position of the marked *first* maximum is determined essentially by the curvature or the energy density in the Universe. Roughly, $l_{\max,1} \simeq 200/\sqrt{\Omega_0}$, where Ω_0 is the total density parameter (13.27) at present. The observations are thus – within a precision of about 10% – compatible with a flat Universe,

$$\Omega_0 = \Omega_{M,0} + \Omega_{\Lambda,0} \simeq 1, \qquad (13.53)$$

as would be required of an inflationary cosmos (see (13.77)).

The *second*, surprisingly weak and not so precisely determined maximum is mainly dependent on the fraction of baryonic matter in the overall matter in the cosmos: the greater the ratio of the two density parameters, $\Omega_{B,0}/\Omega_{M,0}$, the higher its amplitude in the power spectrum. Current analyses show that the baryonic density $\Omega_{B,0}$ derived in connection with the synthesis of the elements in the cosmos (Sect. 13.2.4) must be considerably increased in order to be in agreement with the observations of the 3 K fluctuations.

The investigation of the anisotropies of the background radiation has already made important contributions to our knowledge of the Universe, and we can expect further decisive impulses for cosmology in the coming years from even more precise measurements of the fluctuations of the 3 K radiation.

A systematic overview of the parameters of the present-day Universe will be given in Sect. 13.2.5.

13.2.3 The Radiation Cosmos

Although in the present-day Universe, the matter density predominates, the radiation field density was dominant during the early phases of the Universe, since on expansion, the density of the radiation field decreases more rapidly than that of (nonrelativistic) matter (see below).

Conservation Laws. In order to obtain a model for this radiation cosmos, we first derive a conservation law by rearranging the two Friedmann–Lemaître equations (for $\Lambda = 0$): multiplication of (13.20) by $R(t)^3$

and differentiation with respect to time, followed by subtraction of (13.21) after multiplying it by $\dot{R}R^2 = (1/3)\mathrm{d}(R^3)/\mathrm{d}t$ yields the condition for energy conservation

$$\frac{\mathrm{d}}{\mathrm{d}t}(\varrho c^2 R^3) + P\frac{\mathrm{d}}{\mathrm{d}t}(R^3) = 0, \quad (13.54)$$

which corresponds formally to the 1st Law of Thermodynamics, $\mathrm{d}U + P\mathrm{d}V = 0$. In order to evaluate it, we require also the equation of state, $P = P(\varrho)$.

For a *matter* cosmos, with a dominant *baryon* density ϱ_B and negligible pressure, $P_\mathrm{B} = 0$, we again obtain (13.20) for the conservation of mass,

$$\varrho_\mathrm{B}(t) R(t)^3 = \mathrm{const}. \quad (13.55)$$

The term "baryons" includes neutrons, protons, and hyperons. The ("ordinary") matter in the *present-day* Universe consists predominantly of hydrogen, so that the baryon density is roughly equal to that of the H atoms. Baryonic matter however makes up only a small fraction of the total present-day matter density $\varrho_0 = \varrho_{\mathrm{M},0}$, as follows from the theory of element synthesis in the cosmos (Sect. 13.2.4). The predominant *dark* matter (Sects. 12.1.5 and 12.4.2), which does not interact with the photon gas, likewise obeys the law of conservation of mass (13.55); to what extent it contributes to the pressure is an open question.

In a *radiation* cosmos (index r), $\varrho_\mathrm{r} \gg \varrho_\mathrm{B}$, the density and the pressure are due not only to the photons (index γ), but rather to *all* the relativistic particles. The equation of state of a relativistic gas (4.46) applies, i.e. the pressure is equal to 1/3 of the energy density,

$$P_\mathrm{r} = \tfrac{1}{3}u_\mathrm{r} = \tfrac{1}{3}\varrho_\mathrm{r} c^2, \quad (13.56)$$

and therefore cannot be neglected. From integration of (13.44), we obtain

$$\varrho_\mathrm{r}(t) R(t)^4 = \mathrm{const}. \quad (13.57)$$

The fact that ϱ_r decreases on expansion of the Universe by an *additional* factor of $R(t)$ relative to ϱ_B is a consequence of the energy decrease $\propto R^{-1}$ due to the red shift, which must be considered in addition to the decrease from the increasing volume, $\propto R^3$, which affects both densities.

In contrast to the mass or energy density of the radiation, the *photon density* n_γ is, from (4.74), proportional to $R(t)^{-3}$. The ratio η of photon density to baryon density thus remains *constant* during the expansion of the Universe, and is of the order of magnitude of

$$\eta = \frac{n_\gamma}{n_\mathrm{B}} \simeq \frac{n_{\gamma,0}}{n_{\mathrm{B},0}} \simeq 10^9. \quad (13.58)$$

Here, we have assumed that $n_{\gamma,0} = 4.1 \cdot 10^8\,\mathrm{m}^{-3}$ (13.50), and, corresponding to the conditions for cosmological element synthesis, that $n_{\mathrm{B},0} \simeq \varrho_{\mathrm{B},0}/m_\mathrm{H} \simeq 0.05\,\varrho_{\mathrm{c},0}/m_\mathrm{H}$ (13.69 and Table 13.2).

With the aid of (13.42), we can express (13.55) and (13.57) in terms of the red shift:

$$\varrho_\mathrm{B} = \varrho_{\mathrm{B},0} \cdot (1+z)^3, \quad (13.59)$$
$$\varrho_\mathrm{r} = \varrho_{\mathrm{r},0} \cdot (1+z)^4. \quad (13.60)$$

The Scale Factor $R(t \le t^*)$. Now, starting from the present-day densities of matter and of the 3 K radiation, we can estimate the time t^* at which $\varrho_\mathrm{r} = \varrho_\mathrm{B}$, i.e. the time before which the condition for a radiation cosmos was fulfilled. From (13.59, 60), t^* is given by:

$$\frac{R_0}{R(t^*)} = 1 + z^* = \frac{\varrho_{\mathrm{B},0}}{\varrho_{\mathrm{r},0}} = \alpha \frac{\varrho_{\mathrm{B},0}}{\varrho_{\gamma,0}}. \quad (13.61)$$

For the evaluation of this equation, we require the value of $\alpha \equiv \varrho_{\gamma,0}/\varrho_{\mathrm{r},0}$. It is determined by the present-day energy density of the *neutrino* background radiation, which in turn depends on assumptions about the various types of neutrinos (Sect. 13.3.2). For the "standard cosmos", with three (massless) neutrino types (the electron neutrino, the muon neutrino or neutretto, the tau neutrino, and their respective antiparticles), we find $\alpha \simeq 0.6$. With $\varrho_{\mathrm{B},0}/\varrho_{\gamma,0} \simeq 10^3$, we obtain as a first estimate of the associated red shift of $z^* \simeq 600$.

Our estimate, which is based on (13.57) and not on the more general (13.54), does not hold for the very early phases of the Universe ($t \le 1\,\mathrm{s}$). Above $T \ge 10^{10}\,\mathrm{K}$, there are numerous kinds of relativistic particles and the ratio α is not constant in time (13.72).

The relationship between R and t in the radiation cosmos is then obtained from the Friedmann–Lemaître equation (13.20) (for vanishing Λ), where in the radiation cosmos ($t \le t^*$) the curvature term is negligible due to $R(t) \le 10^{-3} R_0$:

$$\dot{R}^2 = \frac{8\pi G}{3}\varrho_\mathrm{r} R^2 = \frac{8\pi G}{3}\frac{K}{R^2}. \quad (13.62)$$

Since, according to (13.57), $K = \varrho_r R^4 \simeq \varrho_{\gamma,0} R_0^4$ is constant, the integration yields:

$$\frac{R(t)}{R_0} = \left(\frac{t}{t_r}\right)^{1/2}$$

with

$$t_r = \left(\frac{32\pi G}{3}\frac{\varrho_{\gamma,0}}{\alpha}\right)^{-1/2} \simeq 2 \cdot 10^{19}\text{ s}. \quad (13.63)$$

From this result, radiation or relativistic particles dominated the Universe up to the time

$$t^* = t_r \left(\frac{R(t^*)}{R_0}\right)^2 \simeq 2 \cdot 10^6 \text{ yr} \quad (13.64)$$

after the singularity.

Temperature Dependence. Equation (13.57) also holds separately for each single type of relativistic particles, insofar as they have already "decoupled" from the remaining matter, i.e. so long as they expand adiabatically along with the latter without significant interactions.

The *photons* interact with matter mainly through electrons. If the temperature in the universe sinks in the course of its evolution to about 3000 K, then the number of free electrons decreases drastically owing to recombination of hydrogen, and the radiation field decouples from the (nonrelativistic) matter. We can thus estimate the time t_γ after which the radiation field has expanded adiabatically, and finally cooled off to the present-day temperature of $T_0 = 2.7$ K:

Due to $\varrho_\gamma \propto T^4$, the time dependence of the temperature follows from (13.60) and (13.42):

$$\frac{T(t)}{T_0} = \frac{R_0}{R(t)} = 1 + z. \quad (13.65)$$

This result indicates that our Universe became practically transparent to radiation at a red shift of $z \simeq 1000$, or, from (13.61), at a time of about $t_\gamma \simeq 8 \cdot 10^5$ yr. The observation of the microwave background radiation thus allows us a view into the past, back to the state of the Universe about $8 \cdot 10^5$ yr after the Big Bang.

The temperature dependence $T \propto (1+z)$ was recently confirmed by observation of an isolated gas cloud at $z = 2.34$ on the basis of its absorption lines in the spectrum of a quasar with an emission red shift of 2.57: from the fine-structure lines of C I and II and rotational transitions in H_2, excitational temperatures between 6 and 14 K could be derived (this can be considered to be a "modern version" of the first "optical" background observation by A. McKellar; see above). From (13.65), the theoretical value is $(1 + 2.34) \cdot 2.73$ K $= 9.1$ K.

13.2.4 Element Synthesis in the Universe

The model chosen for the early, radiation-dominated Universe, which is based essentially on the present-day temperature of the microwave background radiation, naturally also determines the course of the nuclear processes in the synthesis of the chemical elements; the background radiation was, in fact, predicted by G. Gamow et al. from theoretical considerations of just this process of element synthesis. Right after the discovery of the 3 K radiation, P. J. E. Peebles (1966), as well as R. V. Wagoner, W. A. Fowler and F. Hoyle (1967) carried out detailed calculations of element synthesis. According to these and more recent calculations, for a ratio $n_\gamma/n_B \simeq 10^9$ (13.58), the formation of the elements from protons and neutrons began about 220 s after the singularity, when the temperature had fallen to $0.9 \cdot 10^9$ K or $kT = 0.1$ MeV, since then the deuterium formed, with its low binding energy of 2.2 MeV, would not be destroyed again by photons and energetic particles. In the main, ^{4}He nuclei and to a lesser extent ^{2}D, ^{3}He and ^{7}Li were formed. The "bottleneck" at the mass numbers $A = 5$ and 8, where no stable nuclides exist, could not be overcome under the conditions which dominated the cosmos; heavier nuclei were formed mainly in the course of stellar evolution (Sect. 8.2.5).

Helium. The free neutrons were almost entirely used up in the synthesis of ^{4}He. If we denote the densities of neutrons and protons immediately before the beginning of element synthesis by n_n and n_p, then $n_n/2$ helium nuclei could have been formed, since a ^{4}He nucleus contains two neutrons, while $(n_p - n_n)$ protons were left over. The helium abundance in numbers of atoms

$$\frac{n(^4\text{He})}{n(\text{H})} \simeq \frac{n_n}{2(n_p - n_n)} \quad (13.66\text{a})$$

or in mass fractions

$$\frac{Y}{X} \simeq \frac{2n_n}{n_p - n_n} \quad (13.66\text{b})$$

with $X + Y = 1$ thus depends, over a wide range of values of the density parameter, practically only on the ratio n_n/n_p.

Above $T \geq 10^{10}$ K ($kT \geq 1$ MeV), i.e. before the decoupling of the neutrinos (Sect. 13.3.3), the neutrons and protons were in thermodynamic equilibrium, and then according to the Boltzmann formula (4.78),

$$\frac{n_n}{n_p} = \exp\left(-\frac{\Delta mc^2}{kT}\right) \quad (13.67)$$

holds. $\Delta mc^2 = 1.293$ MeV corresponds to the mass difference between the neutron and the proton.

The ratio n_n/n_p decreased monotonically during the expansion of the Universe, starting from an initial value of 1. Below 10^{10} K, the reactions which convert protons into neutrons and *vice versa* became more and more rare, until finally only the β decay of the neutrons $n \rightarrow p + e^- + \bar{\nu}_e$, with a half-life of 617 s, remained. The neutron to proton ratio then decreased further until, at the beginning of element synthesis, it had the value $n_n/n_p \simeq 0.14$. According to (13.66), this leads to a helium abundance of

$$\frac{n(^4\text{He})}{n(\text{H})} \simeq 0.08 \quad \text{or} \quad Y \simeq 0.25 . \quad (13.68)$$

This value agrees rather well with observations (12.18).

Deuterium. In contrast to ^{4}He, the formation of *deuterium* depends sensitively on the baryon density. All of the ^{2}D is of cosmological origin, however a portion is converted to ^{3}He in the stars through hydrogen fusion. The abundance of deuterium derived from observations of H I clouds at high red shifts ($z \simeq 3$), an atomic ratio of $n(^2\text{D})/n(^1\text{H}) \simeq 2 \cdot 10^{-5}$, must lie close to the primordial value. From this, we find a *baryonic* density $\varrho_{B,0}$, or a contribution of baryons to the density parameter Ω_0 of the present-day Universe (13.27), of only

$$\Omega_{B,0} = \frac{\varrho_{B,0}}{\varrho_{c,0}} \simeq 0.05 , \quad (13.69)$$

so that in our Universe, *non*baryonic matter should predominate (Sect. 13.2.5).

^{3}He and ^{7}Li. From meteorites, one finds the abundance of ^{3}He at the time of the formation of the Solar System to be $n(^3\text{He})/n(\text{H}) \simeq 10^{-5}$; from the spectra of the old Population II stars, a lithium abundance of $n(^7\text{Li})/n(\text{H}) \simeq 1.2 \cdot 10^{-10}$ can be derived. Both the abundance determinations of these light nuclides as well as their interpretation are currently too uncertain to allow us to make clear statements about the parameters of the Universe.

13.2.5 Observed Parameters of the Present-Day Universe

Now, which observations are available (some in practice, some only in principle) for determining *the* actual parameters of our Universe? While during the 1990's the standard model (13.33), in part with an initial inflationary phase, provided the framework for these considerations, more recently the observation of supernovae at high red shifts (see below) and the measurements of fluctuations in the microwave background radiation (Sect. 13.2.3) have set a new discussion in motion, which has by no means reached a conclusion at present.

The Hubble Constant. We have already mentioned (in Sect. 13.1.1) the measurement of the Hubble constant H_0 from the red shift z of the spectra of those galaxies whose distance r can be obtained independently (using Cepheids up to $z = 0.05$, supernovae of type Ia at $z \gtrsim 1.0$). The large-scale expansion of the entire observable Universe can be represented in a *Hubble diagram* (Fig. 13.3), in which, in accord with (13.46, 47), the red shift z or cz is plotted against the apparent magnitude, e.g. m_V or m_I. In order to determine from this diagram (luminosity-) distances and thereby H_0, as well as the deceleration parameter q_0 for objects with the highest values of z, we require "standard candles" such as supernovae of type Ia or galaxies with known, fixed, and the largest possible absolute magnitudes; otherwise, the scatter in the Hubble diagram would be too great. Every type of standard candle must in the end be calibrated with reference to the Cepheids, which can still be observed in galaxies of the Virgo and Fornax clusters using the Hubble Space Telescope (Fig. 12.3).

The *radio galaxies*, whose prototype is Cyg A (Sect. 12.3.3), extend the relation between z and the apparent magnitudes of the normal giant galaxies. With $M_V \simeq -23$ mag, they are well suited to serve as standard candles, since on the one hand, owing to their

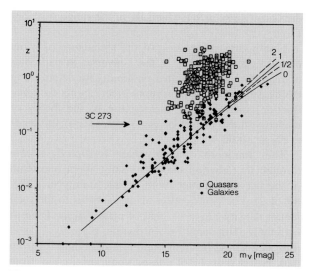

Fig. 13.3. A Hubble diagram for radio galaxies and quasars, from H. Kühr (1987). The red shift z is plotted against the (uncorrected) apparent visual magnitude m_V. The Hubble relation (13.47) is adjusted to the observations for small z (line of slope 0.2) and is drawn for delay parameters of $q_0 = 0, 1/2,$ 1 and 2. In contrast to the radio galaxies, the quasars are not suitable as "standard candles"

strong radio emissions, they can be recognized even at great distances ($z \geq 5$), and on the other hand, they can be investigated spectroscopically even when they are faint, due to their spectral emission lines.

For a long time, the quasars could be observed at larger z values than could galaxies. Currently, thanks to the progress in observational techniques, the galaxies no longer trail behind the quasars in this respect.

The quasars – with or without radio emissions – can be observed clearly even at great distances owing to their high luminosities. The largest known red shift of a quasar (as of 2001) is $z \simeq 5.8$ (Sect. 12.3.4). The quasars and galaxies with active nuclei however exhibit considerable *scatter* on the Hubble diagram, owing to their widely differing and variable core-region activities; they lie between the line representing ordinary giant galaxies and a region of up to 6 mag greater brightness (Fig. 13.3). They are therefore *not* suitable as standard candles. No quasar, however, has a smaller magnitude than would be indicated by the line for ordinary galaxies at its z value. The cosmological interpretation of the red shift of quasars is therefore supported by their positions in the Hubble diagram.

In recent times, H_0 has also been determined from the propagation time differences for light from the various images of a quasar projected by gravitational lenses (Sect. 12.3.4).

The Deceleration Parameter and the Average Density. In the *standard model* (13.33), with a vanishing cosmological constant, the average matter density ϱ_M or the density parameter $\Omega = \Omega_M$ is connected with the spatial curvature k or the deceleration parameter q via (13.34), so that here, the determination of *one* of these two parameters is sufficient. On the other hand, in the cosmological models with $\Lambda \neq 0$, the deceleration parameter q and the density parameter $\Omega = \Omega_M + \Omega_\Lambda$ (or k) must be determined *independently* of each other from the observations. According to (13.30) and (13.31), q and Ω can also be replaced by Ω_M and Ω_Λ.

The *determination of q_0* from the spread in the curves in the Hubble diagram (Fig. 13.3) is very difficult and uncertain. We must at the outset consider that for *very* distant galaxies of every type, in the long time interval which their light has required to reach the Earth, their brightnesses, colors, etc. will have changed markedly as a result of their *evolution* (Sect. 12.5.4), so that a straightforward comparison with objects in our neighborhood, which we see in their *present-day* states, is not possible. Taking the large uncertainties in the empirical determination of q_0 into account, we can conclude only that $-1.3 < q_0 < +2$.

More promising is the Hubble diagram for *supernovae Ia* with *large* red shifts z as "standard candles"; they became accessible to observation in the mid-1990's. Their light curves (brightness, color and their time variations) of course must be referred to the rest system (e) of each supernova; among other things, an observer (o) experiences (according to (13.40) and (13.41)) a time interval Δt_e as if it were "stretched" to $\Delta t_o = (1+z)\Delta t_e$. To be sure, in this method, a possible evolution of the luminosity due to differences in the chemical compositon of the original star is neglected. The observations of SN Ia with $z \gtrsim 1$ can best be fitted with cosmological models having a *negative* (!) deceleration parameter q_0 in the range of ca. -0.3 to -0.8, which implies a nonzero cosmological constant Λ (see below).

We now turn to the determination of the present-day average *matter density* $\varrho_{M,0}$ or of the density parameter $\Omega_{M,0}$. Initially, from the observed mean blue (B) luminosity densities in the Universe, $\mathcal{L}_B^\odot \simeq 10^8 \, L_B^\odot \, \text{Mpc}^{-3}$ (12.7), together with the mass-luminosity ratios $\mathcal{M}/L_B \simeq 200$ of *galaxies* and *galaxy clusters* derived from their dynamics (Sects. 12.1.5 and 12.4.2), we find a mean mass density of $\varrho_{M,0} \simeq 2 \cdot 10^{10} \, \mathcal{M}_\odot \, \text{Mpc}^{-3} \simeq 10^{-27} \, \text{kg m}^{-3}$. With the critical density $\varrho_{c,0} = 7.9 \cdot 10^{-27} \, \text{kg m}^{-3}$ (13.26), this corresponds to a density parameter of $\Omega_{M,0} \simeq 0.1 \ldots 0.2$. This value includes also the "dark matter", which is known only through the dynamics; it is at best accurate to within only a factor of 2. The uncertainty in the Hubble constant H_0 does not enter into this estimate of the density parameter $\Omega_{M,0}$.

We must furthermore take into account a possible contribution to $\varrho_{M,0}$ from *intergalactic* matter (the hot gas *within* the clusters of galaxies was already included in the above value). Even a mean particle density of $5 \, \text{m}^{-3}$ would give $\Omega_0 = 1$ and thus a closed universe. A hot intergalactic gas ($T \simeq 5 \cdot 10^8$ K), which would contribute to the very diffuse extragalactic X-ray emission and would lead to a density parameter of $\Omega_0 \leq 0.5 \ldots 0.9$, cannot at present be excluded. In contrast, an intergalactic medium consisting of neutral or ionized, relatively cool hydrogen is not consistent with observations.

Although the contribution of the 3 K background radiation to the mean baryonic density, of about $10^{-3} \varrho_{B,0}$ (Sect. 13.2.2), is negligible, the background due to low-energy cosmic *neutrinos* (Sect. 13.3.3) may well play an essential role, *if* the neutrinos turn out to have a nonzero rest mass. This possibility, which has been suggested by elementary-particle theories (going beyond the standard model; cf. Sect. 8.2.2), has yet to be confirmed experimentally. Since the particle density of the background neutrinos, $10^8 \, \text{m}^{-3}$, is similar to that of the photons of the 3 K radiation, even a very small neutrino mass of e.g. $5 \, \text{eV} c^{-2} \simeq 10^{-35}$ kg (cf. Table 13.3) would already amount to 0.2 of the critical density $\varrho_{c,0}$. Under discussion as possible candidates for dark matter are in addition several "exotic" particles, which are predicted by extended elementary-particle theories. We cannot go into the details here.

Also gravitational lens effects, model calculations of the evolution of the clusters of galaxies and especially the precise determinations of the fluctuations in the 3 K background radiation (Sect. 13.2.2), permit estimates of the density parameter which lie in the range of $0.2 \leq \Omega_{M,0} \leq 1$.

Finally, the observation of the abundance of some of the lighter elements in the Universe, in particular of *deuterium*, offers another possibility for determining the density parameter (or, more precisely, its *baryonic* fraction) indirectly, by making use of the theory of nucleosynthesis in the Universe (Sect. 13.2.4); this leads to $\Omega_{B,0} \simeq 0.05$.

The "luminous" matter which is *directly visible* through its radiation makes up only about one tenth of the baryonic matter (with an uncertainty of a factor of 2). The contribution of the *baryons* to the total matter density $\Omega_{M,0}$, i.e. that of matter in the form of extremely faint stars, brown dwarfs, planets,..., is at most 0.1. Thus, the contribution of more than 0.1 from dark matter must be of *non*-baryonic origin, e.g. from neutrinos or "exotic" elementary particles.

From the combination of all the observations, we find a value range of roughly $0.2 \leq \Omega_{M,0} \leq 1.0$ for the present-day density parameter, with a large uncertainty.

Age. The Friedmann time τ_F (Table 13.1), which gives the *maximum age* of cosmic objects, can be obtained from age determinations of the globular star clusters (Sect. 9.2.3) and from the decay of long-lived radioactive nuclides (Sect. 12.5.4). These methods lead to an age in the range of 11 to $20 \cdot 10^9$ yr.

Parameters. Let us now attempt, in spite of all the uncertainties, to specify a cosmological model. We must take the large error limits into account: those of the Hubble constant H_0, the deceleration parameter q_0, the density parameter Ω_0 with its two contributing terms $\Omega_{M,0}$ and $\Omega_{\Lambda,0}$, and the Friedmann time τ_F.

In Table 13.2, we have collected the most important parameters, derived from observations, which determine the present-day state of the Universe. Here, we have chosen a value for the *Hubble constant* (13.3) of $H_0 = 65 \, \text{km s}^{-1} \, \text{Mpc}^{-1}$, i.e. we have set $\tilde{h} = 1.3$. With a different choice of H_0, but the same values of q_0 and Ω_0, the Hubble and Friedmann times τ_0 and τ_F would scale proportionally to $\tilde{h}^{-1}$; the critical density $\varrho_{c,0}$ or $n_{c,0}$ would scale as $\tilde{h}^2$; and the baryonic density $\Omega_{B,0}$ as $\tilde{h}^{-2}$. The parameters q_0 and $\Omega_{M,0}$ do not depend upon $\tilde{h}$, while $\Omega_{\Lambda,0}$ – via $\varrho_{c,0}$ – is proportional to $\tilde{h}^{-2}$.

Table 13.2. Parameters of the present-day Universe corresponding to the Standard Model ($\Lambda = 0$) and to a Friedmann–Lemaître Model ($\Lambda \neq 0$)

		Standard models		Friedmann–Lemaître model
Hubble constant	H_0 [km s^{-1} Mpc^{-1}]		65	
Hubble time	τ_0 [a]		$15.0 \cdot 10^9$	
Critical density	$\varrho_{c,0}$ [kg m^{-3}]		$7.9 \cdot 10^{-27}$	
	$n_{c,0}$ [m^{-3}]		5	
Cosmological constant	Λ [m^{-2}]	0	0	$6 \cdot 10^{-53}$
Deceleration parameter	q_0	0.1	0.5	-0.6
Density parameter				
Total	$\Omega_0 = \Omega_{M,0} + \Omega_{\Lambda,0}$	0.2	1.0	1.0
Matter	$\Omega_{M,0}$	0.2	1.0	0.3
Quantum vacuum	$\Omega_{\Lambda,0}$	0	0	0.7
Baryons	$\Omega_{B,0}$	0.05	0.05	> 0.05 (?)
Matter density	$\varrho_{M,0}$ [kg m^{-3}]	$1.6 \cdot 10^{-27}$	$7.9 \cdot 10^{-27}$	$2.4 \cdot 10^{-27}$
	$n_{M,0}$ [m^{-3}]	0.6	5	0.9
Photon density	$\varrho_{\gamma,0}$ [kg m^{-3}]	$4.7 \cdot 10^{-31}$	$4.7 \cdot 10^{-31}$	$4.7 \cdot 10^{-31}$
	$n_{\gamma,0}$ [m^{-3}]	$4.1 \cdot 10^8$	$4.1 \cdot 10^8$	$4.1 \cdot 10^8$
Friedmann time (Age of the Universe)	τ_F [a]	$12.8 \cdot 10^9$	$10.1 \cdot 10^9$	$18.9 \cdot 10^9$

Before it became possible to carry out observations of distant type Ia supernovae, as well as precise measurements of the fluctuations of the microwave background radiation, the Universe could be described with sufficient accuracy by a *standard model* (13.33), in which the density parameter $\Omega_0 = \Omega_{M,0}$ could be determined only within a relatively broad range from about 0.2 to 1.0.

Theoretical considerations relating to an *inflationary* Universe (Sect. 13.3.4) suggest that the limiting case of a *flat*, Euclidean space ($k = 0$) is most likely, corresponding to $\Omega_0 = 1$ or $q_0 = 1/2$, in which at the present time dark, nonbaryonic matter predominates. The recent observations of the 3 K fluctuations (13.53) also indicate a flat Universe, independently of the assumption of a standard model.

According to the standard model, we are living in an open hyperbolic or a Euclidean Universe which is continually expanding.

At the present state of the observations, however, the power spectrum of the fluctuations of the microwave background radiation (Sect. 13.2.2) is not consistent with the baryonic density (13.69) determined from the theory of cosmic element synthesis.

The observations of *type Ia supernovae* at large red shifts have great significance for the determination of the parameters of the present-day Universe. These yield clear indications of a *negative* deceleration parameter of around $q_0 = -0.6$, as already mentioned. This value can be explained within a *Friedmann–Lemaître* model with a cosmological constant $\Lambda \neq 0$ and a corresponding contribution $\Omega_{\Lambda,0}$ to the density parameter.

If we now combine this value $q_0 \simeq -0.6$ with the result (13.53) from the measurements of the 3 K fluctuations, then we find using (13.31)

$$\Omega_{M,0} \simeq 0.3 \,, \qquad \Omega_{\Lambda,0} \simeq 0.7 \,, \qquad (13.70)$$

i.e. the contribution of the quantum vacuum predominates by far over that of (baryonic and dark) matter and is thus the determining factor for the dynamics of the Universe. The nature of this "dark energy" of the quantum vacuum is still completely unknown. According to this model, the Universe is expanding continually, and is at present in a phase of accelerated expansion ($q_0 < 0$).

The discussions of the state of the Universe are at present evidently still far from conclusive, so that we must be content with the rather imprecisely determined values of the parameters as given in the summary Table 13.2.

13.2.6 Olbers' Paradox

The simple observation that the night sky is *dark* allows far-reaching conclusions to be drawn about the large-scale structure of the Universe. This was already realized by J. Kepler (1610), E. Halley (1720), J.-P. Loy de Cheseaux (1744), and H. W. M. Olbers (1826). The de Cheseaux–Olbers paradox states the following:

If the Universe were spatially and temporally infinite and (more or less) uniformly filled with stars, then, in the absence of absorption, the entire sky would be illuminated with an intensity corresponding to the average surface brightness of the stars, i.e. about that of the Sun. The fact that this is not the case cannot be explained in terms of interstellar absorption alone, since the absorbed energy would not be lost.

In order to eliminate the paradox, however, it is sufficient to take into account the *limited* period of time ($\leq 10^{12}$ yr) during which the stars can maintain their luminosities (E. R. Harrison, 1964). The stars would need about 10^{23} yr just to "fill up" the Universe with their light. The red shift of the radiation due to the expansion of the Universe plays, in contrast, only a secondary role.

A simple energy estimate also argues against a cosmic radiation field with a temperature of the order of 10^4 K: if *all* the baryonic matter in the present-day Universe, at a density of $\varrho_{B,0}$, were to be converted *completely* into radiation, i.e. if $\varrho_{\gamma,0}$ (13.49) were increased by a factor of around 10^3 according to (13.69) with $\varrho_{c,0} = 7.9 \cdot 10^{-27}$ kg m^{-3} (Table 13.2), then, owing to $\varrho_\gamma \propto T^4$ and $T_0 = 2.7$ K, a temperature of only about $2.7 \cdot 10^{3/4} \simeq 15$ K would be reached; this temperature is obviously much lower than the 10^4 K observed at a star's surface.

13.3 The Evolution of the Universe

After having reached some conclusions about the state of the Universe in the distant past (excepting the very earliest epochs) by making use of the observations of galaxies and of the 3 K radiation to construct cosmological models, we shall now describe the evolution of the cosmos as a continuous process from the initial extremely hot and dense phases up to the present time. We do this within the framework of the *standard model*, which is based on General Relativity and the cosmological postulate of large-scale homogeneity and isotropy. We also describe briefly extensions of the standard model such as the *inflationary* cosmos and some other cosmological approaches.

Before we discuss the evolution of the standard model in detail in Sect. 13.3.3, we make some remarks in Sect. 13.3.1 about the first 10^{-43} s of the cosmos after the singularity (Big Bang), during which time the theory of General Relativity, which forms the basis of the cosmological models, cannot yet be applied. It is also appropriate to give a preliminary survey of the important results from elementary-particle physics which determine the evolution of the model in the earliest phases of the cosmos, and we do this in Sect. 13.3.2. Finally, we consider models which go beyond the standard model: in Sect. 13.3.4, we describe inflationary cosmology, which offers fundamental solutions to some problems which remain unsolved within the standard model. Then, in Sect. 13.3.5, we make some remarks about several other cosmologies.

13.3.1 The Planck Time

At arbitrarily short times after the singularity, we can*not* apply the standard model, since at very high mass concentrations and short distances the theory of General Relativity, which treats space-time as a continuum, would have to be replaced by a quantum theory of gravity. The latter has at present not yet been completely formulated.

Following M. Planck, we can obtain the time τ_P or the associated length l_P and mass M_P below which Einstein's theory of gravitation loses its validity by composing natural units out of the three constants G, c, and h or $\hbar$ which characterize the three fundamental theories of physics, General and Special Relativity and quantum mechanics:

$$\tau_P = \left(\frac{G\hbar}{c^5}\right)^{1/2} \simeq 5.4 \cdot 10^{-44} \text{ s},$$

$$l_P = c\tau_P \simeq 1.6 \cdot 10^{-35} \text{ m}, \qquad (13.71)$$

$$M_P = \left(\frac{c\hbar}{G}\right)^{1/2} \simeq 2.2 \cdot 10^{-8} \text{ kg}$$

$$\simeq 1.2 \cdot 10^{19} \text{ GeV}c^{-2}.$$

The corresponding Planck density is given by $\varrho_P = M_P/l_P^3 = 5.2 \cdot 10^{96}$ kg m^{-3}. Within the relativistic cosmological models, we can make no statements about times which are shorter than the *Planck time* τ_P.

The Planck mass and length can be estimated to an order of magnitude by setting the size (i.e. the Schwarzschild radius GM_P/c^2) of a black hole of mass M_P equal to the length which is given by the Heisenberg uncertainty relation $M_P c\, l_P \simeq \hbar$ by fixing the conjugate momentum to the value $M_P c$ (this corresponds to the Compton wavelength $\hbar/(M_P c)$).

13.3.2 Elementary Particles and Fundamental Interactions

The great progress which has been made in the area of elementary particle physics since the 1960's has had an important influence on cosmology. In the early phases of the Universe, we are dealing with energies of at least 10^{19} GeV, which are far beyond our experimental capabilities; thus, we must depend to a large extent on the *theory* of elementary particles and their interactions. The currently existing and planned particle accelerators reach energies of only several 10^4 GeV!

According to the *standard* theory of particle physics, the particles of matter are made up of *fermions*, i.e. of particles with a spin of $(1/2) \cdot \hbar$, while the forces or interactions between them are transmitted by the exchange of *bosons* with a spin of $1 \cdot \hbar$.

Elementary Particles. The basic constituents of matter or elementary particles, which show no indications of an internal structure, are considered in the standard theory to comprise the leptons and the quarks with their corresponding antiparticles.

The relatively light *leptons* are known to occur in three families (or generations), each of which is made up of a pair of particles, one electrically charged and one neutral fermion, with their respective antiparticles (Table 13.3). Experiments have shown that there are exactly *three* such lepton families. The first includes the *electron* e^- and its antiparticle, the positron e^+, as well as the electron neutrino ν_e with its antiparticle $\bar{\nu}_e$. The second family contains the muons μ^-, μ^+ with their associated muon neutrinos ν_μ and $\bar{\nu}_\mu$. The third family consists of the tau mesons τ^-, τ^+ and the tau neutrinos ν_τ and $\bar{\nu}_\tau$. The neutrinos are massless or nearly massless particles.

In Sect. 8.2.2, we have described recent (1998) experiments which have given indications of neutrino oscillations and thus of a nonzero neutrino mass, in connection with the discussion of the neutrinos emitted by the Sun.

The *quarks*, whose electric charges are found to be multiples of one-third of the proton's charge, are distinguished in part by a color quantum number (color charge) which has three possible values. Like the leptons, they form pairwise (with their antiparticles) three families, which are associated to those of the leptons. The first family includes the up and the down quark (u and d) and their antiquarks. The nucleons, the protons and neutrons, are composed of these particles, and thus they are the basic constituents of "ordinary" matter. The second family is comprised of the charmed and the strange quarks (c and s), each with its corresponding antiparticle. The third family contains the top quark t, which was discovered in 1995 as the last of the six quarks, and the bottom quark b. In Table 13.3, we have collected the masses and electric charges of the quarks, without distinguishing between the current and the constituent masses.

The quarks do not occur as free particles, but rather only in pairs or triplets within the strongly-interacting *hadrons*:

A *baryon* (or antibaryon) contains three quarks (or antiquarks). The structure of e.g. the proton is thus p = {uud}, that of the neutron is n = {ddu}, and that of the hyperon is Λ_0 = {uds}.

A *meson* is composed of a quark–antiquark pair; e.g. the pion corresponds to π^+ = {u$\bar{\mathrm{d}}$}, the kaon to K^+ = {u$\bar{\mathrm{s}}$}, and the J/ψ = {c$\bar{\mathrm{c}}$}. The B_c^+ meson = {c$\bar{\mathrm{b}}$} was discovered in 1998 as the last of the 15 mesons which can be constructed from the 5 quarks (The top quark is too short-lived to serve as a partner in pair formation.)

Table 13.3. The elementary particles of matter. Masses are in [GeVc^{-2}], and charges Q are in units of the proton's charge

1st Family	2nd Family	3rd Family	Q
Leptons:			
ν_e Electron neutrino $< 5 \cdot 10^{-9}$	ν_μ Muon neutrino $< 2 \cdot 10^{-4}$	ν_τ Tau neutrino < 0.024	0
e Electron $5.11 \cdot 10^{-4}$	μ Muon 0.106	τ Tau meson 1.78	-1
Quarks:			
u up quark 0.005	c charmed quark 1.5	t top quark 175	2/3
d down quark 0.01	s strange quark 0.15	b bottom quark 4.7	$-1/3$

Fundamental Interactions. The fundamental interactions or forces between the various particles are summarized in Table 13.4. The *electromagnetic* interaction is mediated according to quantum electrodynamics by the exchange of massless *photons*. For the *weak* interaction, which is responsible for e.g. the β decay of atomic nuclei, the exchange of the massive vector *bosons* $W^\pm$ and Z^0 with masses of 80 and 92 GeV c^{-2}, respectively, mediates the interaction. Through the theoretical work of S. Glashow, A. Salam, and S. Weinberg in the 1960's, it became clear that the electromagnetic and the weak interactions can be regarded as different aspects of a single, "unified" *electroweak* interaction whose field quanta $W^\pm$ and Z^0 were observed from 1983 on. Therefore, above $m(W^\pm, Z^0)c^2 \simeq 100$ GeV, the electromagnetic and the weak interactions are of comparable strength; below this energy, a "symmetry breaking" takes place.

The *strong* interaction, which according to quantum chromodynamics is mediated by the exchange of eight massless *gluons*, each carrying a color charge, acts on only the quarks, but not on the leptons. The gluons also couple to the W and Z bosons, so that the weak interaction also involves the quarks.

The nuclear binding force is, by the way, not an elementary force, but rather is the indirect result of the interactions between the quarks.

The standard theory of elementary particle physics includes the strong interaction (quantum chromodynamics) and the unified electroweak forces or their exchange bosons together with the elementary particles listed in Table 13.3. Gravitation, which acts on all of the elementary particles, is added on in the form of a classical field theory (General Relativity) independently of the other forces.

Although the validity of the standard theory has been brilliantly confirmed by experiments in high-energy physics, several fundamental questions remain open. In particular, the masses and coupling of the particles and their symmetries are unexplained. Within the standard theory, the masses are determined by the interactions of the particles with a Higgs field, associated with the *Higgs bosons*. The mass of the Higgs boson that mediates the masses of the W and Z bosons must be greater than 65 MeVc^{-2} (this follows from the experiments). The search for this hypothetical particle plays a key role in the verification of the standard theory.

The Grand Unification. The goal of a complete elementary particle theory is to attribute *all* forces to a common origin, whereby the standard theory would be absorbed into a higher-order theory. The first step towards this goal is the combination of the strong and the electroweak forces in a Grand Unified Theory (GUT). This requires the introduction of additional field quanta, the *X bosons*, with masses of the order of 10^{14} GeVc^{-2}. Above $m(X)c^2 \simeq 10^{14}$ GeV, the strong and the electroweak forces would thus also be similar. Within the framework of the GUT it is predicted that the proton should be *unstable* and decays (to be sure very slowly) according to $p^+ \rightarrow e^+ + \pi^0 \rightarrow e^+ + 2\gamma$ or related reactions. An experimental verification of this decay has not yet been forthcoming; the limiting value for the half-life of the *proton decay* lies above a few 10^{32} yr.

Finally, it is speculated that above about 10^{19} GeV, *complete* unification of the gravitational force with

Table 13.4. An overview of the fundamental interactions and their exchange quanta. The strengths of the interactions or their effective coupling constants are energy-dependent; the values quoted are for low energies, ($\ll 100$ GeV). $\alpha = e^2/(2\varepsilon_0 hc) \simeq 1/137$ is the fine-structure constant

Interaction	strong	electroweak		gravitational
		electromagnetic	weak	
Strength	$\simeq 1$	$\alpha \simeq 10^{-2}$	$\simeq 10^{-5}$	$\simeq 10^{-39}$
Acts on	quarks, indirectly on hadrons	electrically charged particles	leptons and quarks (hadrons)	all particles
Exchange quanta	gluons, X-particles	photons	vector bosons	(gravitons)
Mass [GeVc^{-2}]	0, 10^{14}?	0	80 ($W^\pm$), 92 (Z^0)	0

13.3.3 Cosmic Evolution According to the Standard Model

In the very early Universe, the temperatures and densities were so high that the photons and the great variety of relativistic particles were in *thermodynamic equilibrium*. As long as the mean thermal energy $kT \gg mc^2$, it has to be assumed from conservation of energy that every elementary particle of rest mass m can be converted into every other particle. Creation and annihilation of particle–antiparticle pairs and the interactions with other particles thus keep any particular type of particle of mass m in equilibrium with a noticeable abundance above the energy mc^2.

As the average energy in the Universe decreased due to its expansion to a value less than the equivalent mass mc^2, particles of mass m which had decayed or been annihilated could no longer be replaced. Cosmic evolution is thus characterized by a "dying off" of the various types of particles, beginning with the most massive ones, until around 3 s after the singularity, almost only photons and neutrinos are still present and the fraction of baryons is only about 10^{-9} (13.58).

The temperature T as a function of *time* or the average thermal energy kT in the radiation cosmos is, from (13.63) and (13.65), given approximately by

$$t \, [\text{s}] \simeq \frac{1.5}{\left(T \, [10^{10} \, \text{K}]\right)^2} \qquad (13.72)$$

$$\simeq \frac{1}{\left(kT \, [\text{MeV}]\right)^2} \quad \text{for} \quad t \leq 2 \cdot 10^6 \, \text{yr}$$

if we leave out of consideration details (order α) which are different for the various types of relativistic particles.

The fundamental parameters of the Universe, which agree with all the available empirical data, are listed in Table 13.2 and (13.48, 58). For the cosmological constant, we have assumed $\Lambda = 0$ (13.33).

Now in order to describe cosmic evolution, we make use of Table 13.5, where the *masses* of the most important particles are shown along with the time dependence of the temperature and the thermal energy, so that we can read off the sequence of annihilations of the various types of particles. The early evolution of the Universe is usually divided into different *eras* which are characterized by the type of particle predominant during that time.

The Era of the Great Unification. Following the singularity, after a time equal to the Planck time (Sect. 13.3.1) had elapsed, we can expect that a completely symmetrical state existed from 10^{-43} up to 10^{-35} s, as in the Grand Unified Theory: all the interactions except for gravitation had the same strength. In the extremely hot and dense plasma, quarks, gluons, X-particles, leptons, photons, and their antiparticles are all present with about equal abundances and are continually being interconverted.

If now the average energy in the Universe (as a result of its expansion) decreases to below the equivalent mass of the hypothetical X bosons, which mediate the conversion of quarks into leptons and *vice versa*, i.e. 10^{14} GeVc^{-2}, then the decaying X-particles are no longer replaced and the strong and electromagnetic interactions separate; their symmetry is broken.

The decay of the X-particles could have produced an extremely small relative *excess* of quarks over antiquarks of the order of 10^{-10} to 10^{-9}; in later phases of the cosmos, this excess then leads to a corresponding ratio of baryons to photons, which finally can be observed today by comparing the present density of matter and of the 3 K radiation (13.58). The formation of the galaxies, and thus our own existence, is therefore a result of a small asymmetry in the decay of the X-particles only 10^{-35} s s after the singularity. In this connection, the experimental confirmation of the theoretically predicted proton decay (Sect. 13.3.2) would be of decisive importance.

The Quark Era. Within this era, which followed the decay of the X-particles, a further symmetry breaking occurs at about 10^{-10} s or 100 GeV (corresponding to the masses of the $W^\pm$ and Z^0 bosons); at this time, the electromagnetic and the weak interactions separate. Only after this era do the elementary particles have the properties that we observe today.

The Hadron Era. At energies around 1 GeV, the quarks, antiquarks and gluons no longer exist as free

Table 13.5. The series of different eras and the temperature as a function of time during cosmic evolution, according to the standard model. Particles and antiparticles of rest mass m annihilate each other when the mean energy becomes $kT \leq mc^2$. Temperatures below $T \leq 5 \cdot 10^9$ K refer to the photon gas

Era	t [s]	T [K]	kT [GeV]	m [GeVc^{-2}]						
				Field quanta, Quarks		Hadrons		Leptons		
						Baryons	Mesons			
"Grand Unification"	10^{-43}	10^{32}	10^{19}							Planck time
	10^{-35}	10^{27}	10^{14}	10^{14}?	X?					Symmetry breaking: strong – electroweak
	10^{-10}	10^{15}	10^{2}	175	t					Symmetry breaking: electromagnetic – weak
				92	Z					
				80	W			⋮		
Quark Era							5.3 B			
							3.1 J/ψ			
				4.7	b					
				1.5	c	1.67 Ω		1.78	τ	
						⋮				
	10^{-6}	10^{13}	1			1.12 Λ				Transition: quarks → hadrons
						0.940 n				
Hadron Era						0.938 p				
							0.50 K			
	10^{-4}	10^{12}	10^{-1}	0.15	s		0.14 π	0.11	μ	
				0.01	d					
Lepton Era				0.005	u					
	1	10^{10}	10^{-3}							Decoupling of neutrinos
	3	$5 \cdot 10^{9}$	$5 \cdot 10^{-4}$					$5 \cdot 10^{-4}$	e	Element synthesis (^{4}He)
	$2 \cdot 10^{2}$	$9 \cdot 10^{8}$								
Photon Era										
	$2 \cdot 10^{13}$	$3 \cdot 10^{3}$	$z \simeq 1000$							Decoupling of photons
	$6 \cdot 10^{13}$	$2 \cdot 10^{3}$	600							Formation of the galaxies
	$1 \cdot 10^{15}$	$3 \cdot 10^{2}$	100							
Matter Era										
	$4 \cdot 10^{17}$	2.7	0							today 3 K 2 K

particles. The hadron era begins with the annihilation of the quarks and antiquarks, at about 10^{-6} s. The excess of quarks from the decay of the X-particles is transformed into an excess of baryons relative to antibaryons. When the energy density in the cosmos decreases further, the hyperons and heavy mesons first mutualy annihilate with their corresponding antiparticles, followed by the neutrons and finally the protons and the lighter mesons.

The Lepton Era. After the lightest hadrons, the π mesons, had decayed at about 10^{-4} s, the lepton era began. It was dominated by electrons, positrons, neutrinos, and photons. The muons already annihilate at the beginning of this era, and the few (10^{-9}) baryons play no role in its energy density.

Around 1 s, the interactions of the (electron) *neutrinos* with the other particles become so weak that they are no longer able to maintain thermal equilibrium. This decoupling of the neutrinos is the essential factor in determining the proton–neutron ratio, which in turn determines the abundance of the helium which later formed.

The neutrino temperature T_ν is proportional to the temperature of the radiation field,

$$T_\nu = \sqrt[3]{\frac{4}{11}} T = 0.71\, T\,, \qquad (13.73)$$

and has continued to decrease up to the present time; it is now $T_{\nu,0} = 1.9$ K. At about 3 s, the lepton era ends with the annihilation of the electrons and positrons, except for a very small fraction of the electrons; these, together with the protons, guarantee charge neutrality of the Universe.

The Photon Era. In the following photon or radiation era, after about 200 s, the *synthesis of the light* *elements* begins (Sect. 13.2.4); in this process, the free neutrons are almost completely used up by the formation of ^{4}He. Protons, helium nuclei, and electrons are about 10^{-9} times less abundant than photons (and the decoupled neutrinos). Near the end of the photon era, around $8 \cdot 10^5$ yr, the photons decouple from matter, after the free electrons are removed by recombination with the protons to form neutral hydrogen atoms. In contrast to the neutrinos, the photons can gain energy from the $e^\pm$ annihilation process, so that their temperature remains higher than that of the neutrinos and reaches the current value of 2.7 K (Sect. 13.2.2).

The Matter Era. Although there are about 10^9 times more photons than baryons in the Universe, after around $2 \cdot 10^6$ yr, the mass or energy density of the photons becomes smaller than that of the baryons (13.64), and the matter era begins. Only after the decoupling of radiation and matter could large-scale *density fluctuations* in the cosmos occur, beginning at "nucleation centers" which dated from a much earlier time (Sect. 12.5.1). These fluctuations continued to grow, so that from about $3 \cdot 10^7$ yr (at red shifts of $z \simeq 100\ldots 10$), they had grown so strong that the formation of the galaxies and galaxy clusters could take place, at a stage when the average density of the cosmos was still about 10^3 to 10^6 times higher than today.

Finally, after $(10\ldots 13) \cdot 10^9$ yr, the Friedmann time (for $\Lambda = 0$), we have reached the *present* state of the Universe.

Future Evolution. How will cosmic evolution continue in the future? This depends decisively on the average density ϱ_0 or the density parameter $\Omega_0 = \Omega_{M,0} + \Omega_{\Lambda,0}$ which, as we have seen, can be determined observationally only within a large uncertainty; we recall the problems of the "dark matter" in the galaxies, a possible intergalactic medium, the neutrino mass, and the "dark energy" $\Omega_{\Lambda,0}$. As the best estimate of the density parameter, compatible with all the observations and within the standard model, we have found a value within the range $0.2 \lesssim \Omega_0 \lesssim 1$. Accordingly, we are in an *open* universe which is continually expanding and whose density is decreasing.

The temperatures of the radiation field and the neutrino background will continue to decrease. New stars will no longer be formed after the interstellar gas has been consumed, and the stars now present will have finished their evolution within $\leq 10^{12}$ yr. After still much longer times, of the order of $\geq 10^{32}$ yr, the nucleons will have also mostly decayed, according to the GUT models; even black holes, which may have been formed in the early stages of the Universe, will finally radiate away. We end up with a cold, dark, eternally expanding universe, containing as "normal" matter only photons and neutrinos, out of which all the stars and galaxies have disappeared.

Problems of the Cosmological Models. After this brief overview of cosmic evolution, we turn to some remaining questions which in particular the standard model of relativistic cosmology leaves open.

Lacking a quantum theory of gravitation, we can only speculate about the first 10^{-43} s after the singularity; and the age-old question, "What came *before* the creation of the world?" cannot be answered within the cosmological models that we have discussed.

The high degree of *isotropy* of the 3 K radiation also remains unexplained within the standard model. It is, to be sure, taken into account in the cosmological postulate, but we have no physical explanation for it, since, for example, parts of the Universe which lie on opposite sides of the celestial sphere as seen from the Earth would have had no time to interact with each other (with at most the velocity of light) up to the time of the decoupling of the radiation at around $8 \cdot 10^5$ yr.

Also, the fact that our Universe does not deviate *very much* from the Euclidean case $k = 0$ or $\Omega_0 = 1$ is by no means self-evident. On the contrary, this fact requires an extraordinarily precise "fine tuning" of the parameter Ω in the early stages of the cosmos, so that it could remain in the neighborhood of one during about 10^{60} Planck times, unless we choose to regard its value as purely coincidental.

13.3.4 The Inflationary Universe

Finally, we discuss briefly the promising ideas of A. Guth (1981) and A. D. Linde (1982), who solved or at least mitigated several open problems within the standard model by means of their inflationary Universe:

Within the first 10^{-36} s, during the era of the grand unification, a "*phase transition*" occurred in the Universe, accompanying the symmetry breaking of the fundamental interactions (Table 13.5). The vacuum, originally rich in energy through the Higgs field, takes on its present low-energy state and in principle can supply the kinetic energy for an extremely rapid, exponential expansion of the cosmos, the *inflation*. Following this process, around $t \simeq 10^{-33}$ s, its evolution then continued in a manner corresponding to the standard model (Sect. 13.3.3), in which the Higgs field transfers its energy to the elementary particles (X bosons, quarks, leptons, and photons) in the early Universe.

More precise calculations of the fluctuations which appeared in the cosmos make it seem improbable, however, that the Higgs field itself was sufficiently strongly coupled to matter to bring about the inflation. Rather, this is attributed to a new hypothetical scalar field, the *inflaton field*.

We can convince ourselves that for a vacuum energy density $\varrho_{\text{vac}} c^2 \gg \varrho c^2$ which is assumed to be constant, and for $k = 0$, the Friedmann–Lemaître equation (13.24) with $\varrho_\Lambda = \varrho_{\text{vac}}$ has the form

$$\left(\frac{\dot{R}}{R}\right)^2 = \frac{8\pi G}{3} \varrho_{\text{vac}} = \text{const}, \qquad (13.74)$$

with an *exponential* expansion as its solution

$$R(t) = R_1 \, e^{H \cdot t}, \qquad (13.75)$$

where $R_1 = R(t = 10^{-33}$ s$)$. For the "short-time Hubble constant"

$$H = \sqrt{\frac{8\pi G}{3} \varrho_{\text{vac}}} = \sqrt{\frac{c^2 \Lambda}{3}} \qquad (13.76)$$

we must require a value of $H \simeq 10^{35}$ s^{-1}. During the inflationary phase, which lasts only about 10^{-33} s, the dimensions of the cosmos then enlarge by the enormous factor of $10^{30\ldots45}$.

In (13.76), we have expressed ϱ_{vac} formally using (13.23) in terms of the cosmological constant Λ, which thus is connected to the physics of the fundamental interactions via inflationary cosmology: for a very short time the Higgs or the inflaton field acts as a positive Λ.

This inflationary model can in principle explain the isotropy of the 3 K background radiation, since the current observable Universe occupied such a small space before the inflation that all of its parts could interact with each other. In addition, the enormous expansion insured that any original curvature or deviation from the Euclidean case was quickly "leveled out" and thus according to (13.30),

$$\Omega_M + \Omega_\Lambda = 1 \qquad (13.77)$$

was obtained. However, problems are still encountered in the quantitative description of the end of the inflationary phase and its transition to the standard model.

Since, during the initial phase transition, very small regions quickly attain enormous dimensions and can no longer interact with one another, it would be possible within the framework of the inflationary model that the cosmological postulate no longer holds and that isotropy and homogeneity are fulfilled for only a portion of the cosmos. Our observable Universe might then be merely a "bubble" within a larger manifold, i.e. it might represent simply *one* universe among many.

13.3.5 Other Cosmologies

We cannot give here the details of other possible cosmologies; instead, we must content ourselves with a few additional remarks.

The Complete Cosmological Postulate. The cosmological models which we have discussed so far start with the cosmological postulate of spatial homogeneity and isotropy. Should we not also require *homogeneity of the time scale* $-\infty < t < +\infty$? Should not the Universe in principle be the same "from eternity to eternity"? This complete cosmological postulate, which has been suggested by numerous philosophers and theologians, was the basis of the *Steady State* Universe theory developed in the years following 1948 by H. Bondi and T. Gold, then by F. Hoyle et al. Formally, this theory can also be considered to be an open relativistic model with $k = -1$ and $q_0 = -1$; in contrast to "conventional physics", however, it must assume the existence of a mechanism which allows the continual production of hydrogen within the Universe, since this element is required to replace the matter which has "expanded away" and to "fuel" the stars. The whole problem of the creation of the Universe is avoided in the steady-state model, but it also provides no explanation for the 3 K radiation. The observed indications of an evolution of the quasars and the galaxies are also not compatible with the complete cosmological postulate.

"Cosmological Physics". A. S. Eddington, P. Jordan, P. A. M. Dirac and others have attempted to approach the problem of of the combination of physics and cosmology from quite a different viewpoint. Taking the elementary constants of *physics* on the one hand:

> e, h, c, the electron mass m_e, the proton mass m_p and the gravitational constant G,

and those of *cosmology* on the other:

> the Hubble time $\tau_0 \simeq 6 \cdot 10^{17}$ s and the average matter density $\varrho_0 \simeq 10^{-27}$ kg m^{-3},

a number of *dimensionless* constants can be derived. A *first* group of numbers having values near one is obtained, including the fine structure constant $\alpha = e^2/(2\varepsilon_0 hc) \simeq 1/137$ and $m_p/m_e = 1836$. A *second* group of the order of 10^{40} is found: e.g. the ratio of the electrostatic to the gravitational attraction of a proton for an electron, $e^2/(4\pi\varepsilon_0 G m_p m_e) \simeq 0.2 \cdot 10^{40}$, or the ratio of the length $c\tau_0$ (roughly equal to the radius of the Universe in a closed model) to the Compton length of the proton, $c\tau_0/h/(m_p c) \simeq 14 \cdot 10^{40}$; and the number of nucleons in a volume of $(c\tau_0)^3$, of the order of magnitude of $\varrho_0 c^3 \tau_0^3/m_p \simeq (0.1 \cdot 10^{40})^2$.

These facts can be summarized by the statement that the deceleration parameter $q_0 \simeq G\varrho_0\tau_0^2$ of relativistic cosmology (13.18) is of the order of magnitude of 1. This means that the real world cannot be far removed from a Euclidean universe, which, as we have already remarked in Sect. 13.3.3, is by no means self-evident.

If we consider the order-of-magnitude equality of the numbers around 10^{40} to be an essential fact, then, since the age of the Universe τ_0 occurs in them, we can speculate about a possible *time dependence* of the gravitational "constant" G and other physical constants. Recent experiments, however, leave almost no room for a variation in the value of G; the upper limit of the possible relative time variation is found to be $|(\Delta G/G)/\Delta t| < 4 \cdot 10^{-12}$ yr^{-1}.

This "cosmological physics" appears in view of the progress in recent times in both particle physics and in cosmology to be of only historical interest today. It is perhaps more promising to search for an explanation of the relationship of the physical constants to the cosmological parameters within the framework of inflationary cosmology (Sect. 13.3.4), or in the theories which unify the three fundamental interactions between the elementary particles and a "super unification" theory which includes gravitation (Sect. 13.3.2).

The Anthropic Principle. Relativistic cosmology offers us an entire catalogue of *possible* models of the cosmos. Why, then, did precisely *this* Universe, with its particular parameters, come into being? We can still not give an answer to this question. The anthropic principle (R. H. Dicke, 1961) represents an interesting attempt in this direction: our Universe is as it is because we live in it. Simply because of our existence, because there are astronomers who can observe the Universe, there was necessarily a limitation (or even determination?) of the imaginable parameters: it must have been possible for the stars and the galaxies to have formed, for elements such as carbon to have been synthesized, and for planets to exist on which the conditions for the genesis and evolution of life were fulfilled for a sufficiently long period of time. Conversely, if the cosmos had been very different in its earlier periods, there would be no one today to observe it.

With these remarks we close our discussion of the large-scale structure of the cosmos, its physical state and its evolution, which has led us out to the boundaries of present-day science. In the final chapter of this book, we turn again to our Solar System, in order to learn about its origins and its development.

14. The Cosmogony of the Solar System

We turn back now from the depths of interstellar space to our own Solar System, and the old question of how it came into existence. The daring thought that this question cannot be answered by the handing-down of ancient myths, but only through our own probing, was proposed in France as early as 1644 by René Descartes in his whirlpool theory. In Germany, still by 1755, Immanuel Kant had to publish the first edition of his "Allgemeine Naturgeschichte und Theorie des Himmels" anonymously, for fear of the (protestant) theologians; in it, he treated the origin of the Solar System for the first time "according to Newtonian principles". Kant assumed a rotating, flattened *primordial nebula*, from which the planets and later their satellites were formed. The description offered independently somewhat later by S. Laplace in 1796 in his popular "Exposition du Système du Monde" was based on a similar hypothesis.

We shall not deal with the details and differences among these historical approaches, but instead in Sect. 14.1 we turn immediately to a brief summary of the most important facts relevant to the modern ideas about the formation of the Sun and the Solar System. In Sect. 14.2, we then consider — somewhat self-interestedly — the development of the Earth and in particular the origins and evolution of life on the Earth. The internal planetary structure, the sequence of geological eras, and the contintental drift on the Earth were already treated in Sect. 3.2.1.

14.1 The Formation of the Sun and of the Solar System

In the following sections, we begin first in Sect. 14.1.1 with a summary of the most important properties of the Solar System which are relevant to its cosmogony; we have already introduced them in detail in Chap. 3. Then, in Sect. 14.1.2, we turn to the common origin of the Sun and our planetary system from a rotating disk of gas and dust, and deal with further details of the origin of the meteorites (Sect. 14.1.3) and of the Earth–Moon system (Sect. 14.1.4). In Sect. 14.1.5, we take up the particularly interesting question of the existence of planetary systems around *other* stars.

14.1.1 A Survey of the Solar System

We collect here briefly the most important facts which are to be explained by a cosmogony of the Solar System (cf. also Table 2.2 and Chap. 3):

1. The *orbits* of the planets are nearly circular and coplanar. The direction of revolution is the same for all the planets (direct); it is also the same as the direction of rotation of the Sun. The orbital radii represent *approximately* a geometric series

$$a_n = a_0 k^n \qquad (14.1)$$

with $n = -2, -1, 0$ (for the Earth), $1\ldots 6$; $a_0 = 1$ AU and $k \simeq 1.8$ (the asteroids are considered together as one object); see Fig. 14.1. This relation was discovered by J. D. Titius (1766) and J. E. Bode (1772); its significance for the cosmogony of the Solar System is a subject of controversy.

2. The *rotation* of the planets is for the most part in the direct sense (exceptions: Venus and Uranus), with the axis, i.e. the rotational angular momentum, essentially parallel to the orbital angular momentum and to the axis of rotation of the Sun (exceptions: Uranus and Pluto).

3. The Sun contains 99.9% of the mass, but only 0.5% of the *angular momentum* of the overall system, while, conversely, the planets (mainly Jupiter and Saturn) have only 0.1% of the mass, but 99.5% of the angular momentum in the form of orbital angular momentum $\sum a\mathcal{M}v$ (2.53).

   As can readily be calculated, the angular momentum of the planetary orbits (Jupiter, Saturn) is $3.2 \cdot 10^{43}$ kg m^2 s^{-1}, and that of the Sun (with some uncertainty due to lack of knowledge of the angular velocity in the solar interior) is $1.7 \cdot 10^{41}$ kg m^2 s^{-1}.

4. The *earthlike* planets (Mercury, Venus, Earth, Mars and the asteroids) have relatively high densities (3900 to 5500 kg m^{-3}), while the *major* planets have low densities (700 to 1600 kg m^{-3}). The former consist (like the Earth) in the main of metals and stone,

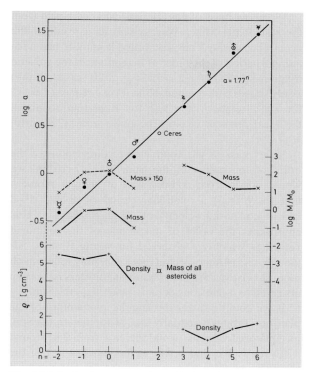

Fig. 14.1. In the *lower part*, the orbital semimajor axis a and the mass $\mathcal{M}$ of the planets (excepting Pluto) are shown relative to those of the *Earth* (values = 1). The average densities ϱ are in [g cm^{-3}] (*left-hand scale*). For the asteroids, the orbital semimajor axis of 1 Ceres and the estimated total mass are plotted. The line drawn in the *upper part* corresponds to the formula $\log a = 0.247\, n$ or $a = 1.77^n$

while the latter contain barely modified solar matter (hydrogen, helium, hydrides) with a solid, earthlike core (like Jupiter and Saturn); or they consist of a stone-ice mixture (like Uranus and Neptune).

5. *Pluto*, with a density similar to that of the major planets, is distinguished from them by its small mass and its sharply inclined, eccentric orbit. We can regard it as the largest member of the transneptunian planetoids in the Kuiper belt (Sect. 3.5) and shall therefore leave it out of consideration in the following.

6. In the case of the *satellites*, we must distinguish between the inner satellites, which (and this must be explained by our cosmogony) have orbits with small eccentricities, slight inclinations to the equatorial plane, and revolution in the direct sense; and the outer satellites, with their considerably larger eccentricities and orbital inclinations. The latter were probably retroactively "captured" in the early period of the Solar System, while the former may be seen as the "original property" of their respective planets. The earthlike planets have longer rotational periods and few satellites, while the major planets exhibit relatively rapid rotation and, even neglecting the captured satellites, have a large number of moons and in some cases striking ring systems.

In its composition and structure, our Moon follows the scheme of the earthlike planets, while the large satellites of Jupiter and Saturn are in part like the Moon, but in part consist of a mixture of ice and silicates.

7. The numerous *impact craters* which are observed on the surfaces of planets and satellites from Mercury out to the Neptune system indicate a considerably larger number of planetesimals in the early period of the Solar System than are present today (these are small, solid objects up to a few km in size).

8. From precise radioactive *age determinations*, we know that the Earth, the Moon, and the meteorites, and thus without doubt the Sun and the whole Solar System, were formed $4.5 \cdot 10^9$ yr ago during a relatively brief time interval.

The remarkable *regularities* in the structure of the Solar System support an evolution of the whole system in the sense of the theories of I. Kant, S. Laplace and later of C. F. von Weizsäcker, D. ter Haar, G. Kuiper and others. In contrast, "catastrophe theories", which attribute the origin of the planets to the action of a star which passed close to the Sun (J. Jeans, R. A. Lyttleton, and others) appear to be excluded from the beginning. In addition to the improbablility of such a near collision of two stars, these theories do not allow us to understand how a filament torn from the Sun by tidal forces could have formed the planets.

We shall therefore discuss the cosmogony of the Solar System directly in connection with the formation of stars from the interstellar medium as the result of a gravitational instability (Sect. 10.5.3). Here, the unusual distribution of angular momentum in the system makes it appealing to search for a common origin of the Sun and the planets out of a rotating disk of gas and dust.

14.1.2 The Protoplanetary Disk and the Formation of the Planets

All stars are formed, as we have seen in Sect. 10.5.2, from an interstellar gas and dust cloud of $\geq 10^3$ solar masses which contracts and then divides up into individual stars. Even single stars like our Sun originally belonged to such multiple systems, associations, or clusters.

In the following paragraphs, we first consider the spherical collapse of an interstellar cloud to a star and then take into account the angular momentum or rotation of the cloud.

Spherical Collapse. The process of the formation of a star of $1 \mathcal{M}_\odot$ was calculated by R. B. Larson in 1969, under the assumption of spherical symmetry, i.e. without angular momentum.

The initiation of the collapse of a cloud of radius R is determined by *Jeans' criterion* for gravitational instability (10.37):

$$R \leq R_J = 0.4 \, G \mathcal{M} \frac{\mu m_u}{kT} \, . \tag{14.2}$$

(Larson chose a constant which was a factor of two larger).

With a mass of $\mathcal{M} = \mathcal{M}_\odot = 2 \cdot 10^{30}$ kg, a molecular weight of $\mu = 2.5$, and an initial temperature estimated to be of the order of $T \simeq 10$ K, one finds that a cloud of $1 \mathcal{M}_\odot$ becomes unstable only when its radius has reached a value of

$$R \simeq 1.6 \cdot 10^{15} \text{ m} \simeq 11\,000 \text{ AU} \simeq 0.05 \text{ pc} \, . \tag{14.3}$$

Its average density is then

$$\varrho \simeq 1.1 \cdot 10^{-16} \text{ kg m}^{-3} \quad \text{or}$$
$$n \simeq 3 \cdot 10^{10} \text{ molecules m}^{-3} \, . \tag{14.4}$$

This corresponds to the densities found in molecular clouds (Sect. 10.2.3). The collapse of such a cloud, up to the time of the formation of the Sun and the planetary system, requires several 10^5 yr, as can be estimated using (12.51).

The Angular Momentum Problem. Before we describe the results of numerical collapse calculations including angular momentum, we first want to give some estimates which provide insight into the problem of the angular momentum of cosmic masses and especially of the Solar System.

When our mass of gas (14.3, 4) was decoupled from the rest of the Milky Way system as a result of the occurrence of the Jeans instability, it initially had a rotational period of the order of magnitude of the galactic rotational period of $\tau_{\text{gal}} = 2\pi/\omega_{\text{gal}} \simeq 10^9$ yr. More precisely stated: with rotation of the form $v \propto R^{-n}$ ($n = -1$ for rigid rotation, $n = 1/2$ for Keplerian rotation), the local angular velocity becomes $\omega = (1/2)|\text{curl}\,v|$ $= (1-n)\omega_{\text{gal}}/2$.

If the mass of gas contracted to dimensions of the order of the radius of the Sun, $R_\odot = 7 \cdot 10^8$ m, we would obtain from conservation of angular momentum ($\propto R^2/\tau_{\text{gal}}$) a rotational period of about 0.07 d and an equatorial velocity of about 700 km s^{-1}. This approaches the order of magnitude of the rotational velocities which are observed for young B stars. Although stars of later spectral types generally rotate much more slowly, such stars in young star clusters have been observed to rotate considerably faster than the Sun.

If we could transfer the angular momentum of the planetary orbits (essentially that of Jupiter) to the Sun, its equatorial velocity would increase from 2 to around 370 km s^{-1}. We thus arrive at the conclusion that the enormous angular momentum of the orbital motion of the planets is nothing other than a major fraction of the angular momentum which *any* object of around $1 \mathcal{M}_\odot$ would have originally acquired from the rotation of the galaxy.

The Sun, however, must have either transferred its own angular momentum to the interstellar medium later through magnetic coupling via the solar wind (R. Lüst and A. Schlüter, 1955), or, as numerical calculations suggest, may never have acquired it; instead, it may have been transferred outwards within an accretion disk (cf. Sects. 8.2.7 and 10.5.4) by turbulent or magnetic friction already during the collapse.

The Accretion Disk. The flat accretion disk which surrounded the proto-Sun, the "solar nebula", consisting of a mixture of gas and dust similar to that known from the interstellar medium, must clearly have had the same chemical composition as the Sun itself (Table 7.5). On the other hand, its matter can never have been contained *within* the Sun, since, for example, the Earth and the meteorites contain roughly a hundred times the concentration of *lithium* as does the Sun, where about 99% of this element has been eliminated in the course of time by nuclear processes. The solar nebula must thus have been formed about the same time as the Sun itself.

The fact that the formation of such a system is not an unusual coincidence is demonstrated by the observations of less massive companions and of dust rings or disks around some neighboring stars, which we will discuss in more detail in Sect. 14.1.5. The question of *what conditions* are necessary for a planetary system or a double or multiple star system to come into being cannot be uniquely answered. In any case, the formation of a planetary system cannot presuppose any very special initial conditions for the gravitational collapse of an interstellar cloud.

Complex (two-dimensional) hydrodynamic model calculations of the gravitational collapse of a cloud taking into account its *angular momentum*, performed in the 1980's by P. Bodenheimer, G. E. Morfill, W. Tscharnuter, H. J. Völk and others with the goal of explaining consistently the origin of the Sun and the planetary system "at one stroke", initially indicated that even a very small initial overall angular momentum (ratio of rotational to gravitational energy $\simeq 10^{-5}$) would prevent the formation of a single star. Instead, a rotating *ring* is formed, which presumably later becomes unstable and breaks up into a double or multiple star system. If, however, a sufficiently strong *turbulent friction* is assumed within the collapsing cloud, then a single central star is formed with gradually increasing mass, surrounded by a flat, rotating *accretion disk* (Fig. 14.2). Within the disk, matter flows inwards, while angular momentum is transported outwards as a result of viscosity.

The *temperature* of the disk is relatively low, and its radial dependence varies only slowly at first: from about 1000 K in the neighborhood of Mercury's orbit, it decreases on going outwards to several 10 K at the edge of the disk (at around 40 AU). At the radius of the Earth's orbit, for example, the temperature is about 300 K, and it is near 150 K at the orbit of Jupiter. These temperatures agree well with the values obtained from infrared observations of the circumstellar (protoplanetary?) disks around young stars (Sect. 14.1.5).

Formation of the Planets. A central question now remains: how were the earthlike and the major planets, the meteors, etc. formed out of the accretion disk? A first possibility is formation as a result of *gravitational instabilities* within the disk, i.e. formation of planets in a manner similar to that of stars. The Jeans condition (10.39) for the beginning of the condensation of, for example, one *Jupiter mass* at $T \simeq 150$ K requires a density of about 10^6 kg m^{-3}, which appears plausible for an evolutionary phase somewhat later than that depicted in Fig. 14.2. While the formation of the large outer planets, which consist of nearly unchanged solar matter, could have been caused or at least favored by such an instability, for the inner planets it remains to be explained how, following the production of a solid core by sedimentation of the condensed matter, the light elements such as hydrogen and helium, which make up about 99% of the overall mass, could have escaped. The asteroids and other small objects in the Solar Sys-

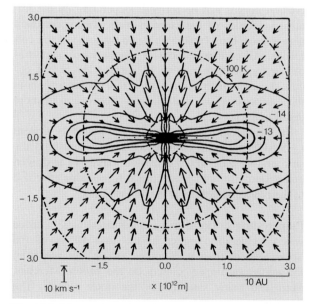

Fig. 14.2. The gravitational collapse of a rotating protostellar cloud (ratio of rotational to gravitational energy $1.2 \cdot 10^{-4}$) of mass $3\,\mathcal{M}_\odot$ with turbulent viscosity, after G. E. Morfill et al. (1985). The time depicted is about $3.3 \cdot 10^4$ yr after the beginning of the collapse, when the central mass has increased to $0.5\,\mathcal{M}_\odot$. A meridional section is shown, with the axis of rotation in the z-direction: the contours (——) represent lines of constant density (log ϱ and ϱ are in [g cm^{-3}]), the dot-dashed lines (– · –) are lines of constant temperature, and the arrows ($\rightarrow$) indicate the velocity field. (Reproduced with the kind permission of the University of Arizona Press, Tucson AZ)

tem, finally, have masses which are much too small for them to have been formed by a gravitational instability; however, they could have resulted from the breakup of a larger body.

The observation of the *solid* cores of 10 to 20 Earth masses in the major planets, independently of their overall masses, finally led to the currently accepted theory of the formation of the Solar System. It begins with the *dust* component in the accretion disk, which, as we observed in the interstellar medium, makes up about 1% of the overall mass. Mutual inelastic collisions cause the dust particles within the turbulent gaseous disk to accrete within 10^7 yr, the "lifetime" of a gas-dust disk (Sect. 14.1.5), into larger and larger clumps and to form the *planetesimals*, irregularly shaped solid bodies with dimensions of the order of 1 to 10 km, which are concentrated in the midplane of the disk. Later, over periods of the order of 10^8 yr, *solid protoplanets* or planets which contain no gas can be formed. Depending on the relative velocities of the planetesimals, we expect them also to have been broken up again by some of the collisions. All of these processes release large quantities of energy, so that melting of surface layers can occur. The *meteorites* contain important documents about these processes; we shall discuss their origin in more detail in Sect. 14.1.3.

In the outer, cooler region of the planetary system (≥ 5 AU), the solid cores could collect and bind the surrounding *gas*, which consisted mainly of hydrogen and helium, through gravitational forces; this gave rise to the typical structure of the major planets.

Planetesimals. The density of planetesimals in space was considerably higher in the first 10^9 yr of the Solar System than it is today, as we can see from the numerous impact craters on the Moon, Mercury, Mars and many satellites of the outer planets. The large objects in the planetary system thus moved for a long period of time after their formation through a medium with strong *frictional forces* caused by numerous clumps of stone of varying sizes. This was probably the source of the small eccentricities and inclinations to the plane of the ecliptic of the planetary orbits, as well as the corresponding orbital inclinations of the (regular) satellites of the major planets. Older, purely celestial-mechanical calculations describing the origin and evolution of the Solar System are not accurate, due to their neglect of this friction. With the end of the intense bombardment by planetesimals, our Solar System also reached the end of its accretion phase.

We can today regard the majority of the asteroids and meteors as the "remains" of the planetesimals, together with the two moons of Mars, the many small satellites of Jupiter, Saturn, and Uranus, and the comets which come from the outermost parts of the Solar System (the Kuiper belt).

14.1.3 The Origin of the Meteorites

The point of view held previously, that the meteorites represent *the* cosmic material with the cosmic element abundance distribution, has, as we have already seen, long since been supplanted by a thorough study of their mineralogical, chemical, and isotopic structures.

We gave an overview of the main types of meteorites and their finer classification in Fig. 3.31. The iron meteorites are the most strongly differentiated; however, we can expect the most far-reaching conclusions from the *chondrites*. As we saw in Chap. 3, the chondrules, small spherules of around a millimeter in diameter, contain various silicates, including iron-containing silicates (in particular the enstatites) embedded in a more fine-grained matrix of similar composition.

Condensation of Matter. A starting point for the interpretation of the extraordinary variety of meteoric structures can be taken to be the condensation and separation of solids from the gas phase of solar composition, a very effective chemical fractionation process. It was the great achievement of H. C. Urey (1952) to apply the experience and methods of physical chemistry for this process to the cosmogony of the Solar System. Calculations were carried out by J. W. Larimer and E. Anders (1967/68), who discussed two limiting cases for the time dependence: (1) Thermodynamic equilibrium; the cooling process takes place so slowly that formation of solid solutions in diffusion equilibrium is possible. (2) Rapid cooling, so that the condensed elements and compounds cannot interdiffuse.

At a pressure of about 10 Pa (10^{-4} bar), the condensation of solar matter takes place (under both assumptions) according to the following steps:

$T < 2000$ K:	nonvolatile compounds of Ca, Al, Mg, Ti, ...
1350 to 1200 K:	magnesium silicates, nickel-iron
1100 to 1000 K:	alkali silicates
680 to 620 K:	iron sulfide (FeS, troilite) and other sulfides, then Pb, Bi, Tl, In, ...
400 K:	from iron and water vapor, Fe_3O_4 is formed
400 to 250 K:	hydrated silicates

At $T \simeq 1000$ K, around 90% of the chondritic material has condensed. The sequence of condensation as a function of temperature can be considered only as a first orientation, since the details of the mineralogical and chemical processes which took place in the meteorites must have been much more complex.

Fractionation and Heating. Analysis of the "ordinary" chondrites reveals several processes involving different components which took place at differing temperatures between about 1800 and 400 K:

Material with a high content of Ca, Al, Ti, ... was separated out partially at $T \geq 1300$ K, at least in the region of formation of the meteors and the earthlike planets. Its abundance is relatively enriched in the Earth–Moon system, and in contrast is reduced in ordinary and enstatitic chondrites. A metal-silicate fractionation influenced many types of meteorites and certainly also the earthlike planets at temperatures from about 1000 to 700 K. Remelting and degassing at around 600 to 450 K led to a reduced abundance of volatile elements in some meteorites and in the Earth–Moon system.

Laboratory experiments have shown that the *chondrules* were probably formed by a brief period of heating to 1800 K within a few minutes, followed by cooling to a few 100 K. The causes of the required heating processes are not yet clear.

The accretion of meteoric matter took place in the main within a relatively narrow range of temperatures around 450 K and at low pressures of about 1 Pa.

We cannot treat here the further details of these processes, especially since it has not yet proved possible to fit together those processes which have been recognized to be important into a self-consistent astronomical picture.

Carbonaceous Chondrites. The rare carbonaceous chondrites are of special interest with regard to the origin of the Solar System and also the beginning of life on the Earth. They are classified according to their content of volatile elements (H, C, S, O, ...) into the types C1, C2, and C3. These differences correspond on the one hand to the relative amounts of the high-temperature fractions of the chondrules and of similar iron particles which were formed at around 1200 K, and on the other hand to the low-temperature fractions of the fine-grained matrix in which they are embedded, which was formed around only 450 to 300 K.

The *abundance distribution* of the chemical elements in this matrix and in the C1 meteorites agrees, apart from the highly volatile elements, with that of the Sun (Table 7.5). The cosmic abundance of extremely rare elements can thus be determined from them, even when these elements can hardly be detected spectroscopically in the Sun. Many of these elements are of great importance for the understanding of nucleosynthesis. The relative abundances of the isotopes are essentially the same in the meteorites, on the Earth, and in the Sun (where however only a few are precisely known). The C1 chondrites, consisting entirely of this dark matrix, contain up to 4% carbon, mainly in the form of organic compounds. They can be most readily compared with terrestrial coals or humic acids.

Some researchers let themselves be tempted at first by these carbonaceous chondrites into making the assumption of some sort of extraterrestrial life. This problem, along with the related question of terrestrial contamination, was finally resolved with the aid of the C2 chondrite which fell to Earth near Murchison in Australia in 1969; it was possible to collect its fragments within a few months. The *Murchison* meteorite contains among other compounds the amino acids, which play an essential role in terrestrial life forms. But although in terrestrial organisms practically only the optically active (levorotating or dextrorotating) L-configuration occurs, both the mirror-image isomers (L- and D-forms) are roughly equally represented in the meteorite: it contains an optically *in*active *racemic* mixture. The meteorite also contains numerous additional amino acids which do not occur in living organisms. The investigation of these amino acids, as well as of the (mostly straight-chain) hydrocarbons from carbonaceous chondrites, has shown that these compounds must have been formed in the early phases of the Solar System, without the intervention of mysterious life forms.

How can we understand this process? One possibility is offered by the Fischer–Tropsch synthesis (1923), in which carbon monoxide, CO, and hydrogen, H_2, can be combined in the presence of suitable catalysts to yield *hydrocarbons*, mainly of the straight-chain type C_nH_{2n+2} (alkanes). It can be assumed that in the solar nebula, not all of the CO would have been converted into CH_4 (as would be expected in thermal equilibrium below 650 K), since this reaction proceeds extremely slowly. The remaining CO could have combined with H_2 at 380 to 400 K, with catalysis by the Fe_3O_4 and hydrated silicates which were then present, in a kind of Fischer–Tropsch synthesis. In the presence of NH_3, according to E. Anders and others, the biologically important amino acids and the many other organic compounds found in carbonaceous chondrites would also have been synthesized.

We shall return to the exciting question of the origin of life in Sect. 14.2.4. First, however, we consider further the formation of the meteors.

Time Scales. A precise investigation of time dependences yielded the discovery that many parts of meteorites contain Xe isotopes, which (as was confirmed by laboratory experiments) can have been formed only by decay of the "extinct" radioisotopes ^{129}I (with a half-life of $1.6 \cdot 10^7$ yr) and ^{244}Pu (half-life $8.3 \cdot 10^7$ yr; spontaneous fission). This supports the idea that the formation of all meteors, including the carbonaceous chondrites, took place within a time of 10^6 to at most 10^7 yr.

It is thus possible to determine the irradiation age of numerous meteorites, i.e. the time during which the meteorite was subjected to irradiation by cosmic rays (Sect. 3.7.3). In many cases, the irradiation age gives us the time at which the piece was broken out of a larger object (protometeor) and thus exposed to cosmic radiation. The irradiation ages of the tough iron meteorites lie mostly between a few 10^8 and 10^9 yr, while the much more brittle stone meteorites attained their present size in the main only 10^6 to $4 \cdot 10^7$ yr ago. According to mineralogical indications, which permit conclusions to be drawn about the gravitational forces in the protometeors, their sizes can be estimated to have been between 50 to 250 km, with iron cores of 10 km diameter.

These results, along with their "formation temperatures" and their surface reflectivities (Sect. 3.7.3),

indicate that the major portion of the meteors was formed in the *asteroid belt*. The asteroids, on the other hand, are "hindered" planets, so that the meteorites give us insight into the evolutionary phases of our Solar System during which the protoplanets were being formed (Sect. 14.1.2).

14.1.4 The Formation of the Earth–Moon System

Following our discussion of the origin of the planets and the meteors, we now turn to the especially interesting and enlightening question of the origin and evolution of the Earth–Moon system. The results of older astronomical and geophysical research are complemented here by the fascinating findings from the landings of manned and robotic spacecraft on the Moon (1966–1973).

Our Earth–Moon system must have been formed in a quite different way from the satellite systems of the other planets: in all the other planet/satellite systems, with the exception of Pluto/Charon, the ratios of mass and angular momentum of the satellite to those of the planet are namely $\ll 1$, while the mass ratio Moon : Earth is 1 : 81.3, and the orbital motion of the Moon takes up 83% of the angular momentum of the whole system.

Tidal Friction. In order to learn something about the earlier states of the Earth–Moon system, we begin with terrestrial observations and, following the direction indicated by the classic works of G. H. Darwin (1897), return to the phenomenon of tidal friction which we have already mentioned briefly (Sect. 2.4.6). The two tidal maxima which the Moon continually drags around the Earth, both in the oceans and in the solid material of the planet, cause a braking action on the Earth's rotation. In particular, after subtracting all the other effects, an investigation of past eclipses leads us to an estimate of the resulting increase of the length of the day, amounting to 1.64 ms per century. The angular momentum released by the Earth can be taken up only by the orbital motion of the Moon, i.e. the period of revolution and the orbital radius of the Moon are increasing.

This effect, which from an astronomical viewpoint is vanishingly small, however becomes considerable over the course of geological times. In 1963, J. W. Wells and C. T. Scrutton noted that the calcium carbonate shells of *corals* (and other lifeforms) living in waters with strong tides show fine bands, which reflect the periods of the year, the synodic month, and the days. While recent corals confirm the well-known astronomical data, the investigation of petrified corals has shown that for example in the middle Devonian, i.e. about $370 \cdot 10^6$ yr ago, 1 year had about 400 (then current) days, 1 synodic month was about 30.6 days, and one day was about 22 hours, in sufficiently good agreement with an extrapolation of the present values.

Recently, C. P. Sonnett et al. (1996) were able to estimate from the layer structure of *sediments* that the length of the day in the late Proterozoian, about $900 \cdot 10^6$ yr ago (Table 3.3), was only about 18 h, so that a year consisted of about 487 d.

But evidently even the geological data offer us only the "recent history" of the Moon; for the distant past, we must rely on *theoretical* extrapolations. These have been carried out several times following G. H. Darwin's first efforts, with the result that the the Moon must have come very near to the Earth (within a few Earth radii) about 1.5 to $2.5 \cdot 10^9$ yr ago, which would have caused a tidal wave of genuinely apocalyptic proportions. The geological evidence strongly suggests that this event never occurred, at least not in the past $4 \cdot 10^9$ yr. This is, however, not an argument against the theory of tidal friction as such, but only against its extrapolation using the present-day value of the friction. In fact, we know hardly anything about the structure of the oceans and thus about the strength of the tidal friction over the course of geologic time. Along with the actual tidal friction, a braking action (or also an acceleration) due to the impacts of asteroids or meteors, which were much more numerous in the early period of the Solar System, could have been significant.

The Lunar Surface. We now approach our problem of the evolution of the Earth–Moon system from the opposite direction and attempt to evaluate the results of the lunar landings in terms of this subject.

The entire lunar surface is covered with a layer of fine dust and fragments of varying sizes, in part baked into breccia; it is called the regolith. This material was clearly formed by the impacts of numerous small and large meteorites. The most outstanding features of the current lunar surface (apart from a few volcanic structures), the maria and the craters, were created by the impacts of planetesimals or meteorites ranging up to the size of asteroids onto the originally solid crust. Only later were they covered for the most part by enormous basalt flows. In the Mare Imbrium, which has a diameter of 1150 km and originally had a depth of about 50 km, we can distinguish three stages of this lava flooding process, which occurred about $3.9 \cdot 10^9$ yr ago. Both the points of origin of the lava flows, and also the "drowned craters" which emerge here and there from the otherwise smooth surface of the hardened lava, indicate that the lava flooding generally had nothing to do with the meteoric impacts themselves. Radioactive age determinations confirm that the lava flows are several 10^8 yr younger than the meteoric impacts. Local melting on impact played a minor role. Instead, the mare basalt (with a density of ca. 3300 kg m^{-3}) later flowed up through cracks in the stone crust, from which we can conclude that at that time, the Moon itself was still partially molten at a depth of 200 to 400 km. In comparison to the rocks of the highlands and several intermediate levels, the mare basalt shows clear indications of an additional chemical differentiation; in particular, it has a lower content of Al_2O_3.

If we compare the number of craters of differing diameters in the highlands (Fig. 3.5), the surfaces of maria of differing ages, and the floors of the largest craters themselves, we find that the intensity of the cosmic bombardment decreased by several orders of magnitude during the first 10^9 yr following the formation of the Moon (we could just as well say following the origin of the Solar System), and that from then on (from $3.8 \cdot 10^9$ yr ago), the decrease continued much more slowly. The fact that there are so few meteoritic craters on the Earth is now readily understandable: the present crust of the Earth was formed only after the supply of meteors had been nearly exhausted. The observations of the Moon, as well as those of Mercury, Mars, Venus, and many satellites of the outer planets, which show craters quite comparable to those on the Moon, indicate that in the early period of the Solar System, at least out to the orbit of Neptune, the plane of the planetary orbits was filled rather densely with objects of the order of ≤ 100 km in size, the planetesimals.

When the first pictures of the *dark side* of the Moon became available, it caused some surprise that large impact craters are apparently more common on the hemisphere facing the Earth. From this we can conclude that the rotation and revolution of the Moon were already synchronous about $3.8 \cdot 10^9$ yr ago, just as they are now.

The Formation of the Moon. We cannot take up the details of lunar petrology and of selenology here. Instead, we turn to the cardinal question of the origin of the Moon: was the Moon formed by *splitting off* from a rapidly rotating primeval Earth, or was it *captured*?

It is clear from the outset that the capture of a "finished" Moon is relatively unlikely on the basis of celestial-mechanical limitations as well as of age determinations of the lunar rocks. We need also not take seriously the popular version of the splitting-off theory, according to which the Pacific Ocean is the scar left by the departing Moon; this is contradicted by the results of plate tectonics. A common origin of the Earth and the Moon as a "double planet", whereby the Moon formed from a ring which circled the primeval Earth, is also not probable in view of the different chemical compositions of the two bodies. In this case, furthermore, the rapid rotation of the Earth would be difficult to explain.

The *splitting-off theory* (G. H. Darwin) is supported by the agreement of the average density of the Moon (3340 kg m^{-3}) with that of the upper layers of the Earth. On the other hand, the objection was raised early that the present value of the angular momentum of the Earth–Moon system would not have sufficed to make the primeval Earth rotate rapidly enough at any point to cause the centrifugal force to be greater than the gravitational force (required length of the day $\simeq 2.7$ h). We can reply that the whole process of formation of the Earth–Moon system took place during a period of intense bombardment by meteorites or asteroids, and that a considerable exchange of angular momentum must have occurred during this time. A collision with one or more large planetesimals could have initiated the separation of the Moon.

On the other hand, the *capture* of the Moon is supported by its underabundance of volatile elements relative to that in the Earth's mantle, which indicates that the Moon was formed at a different place in the Solar System, somewhat nearer to the Sun. Another point favoring this theory is the mass ratio of the satellite to the planet, which is unusual within the Solar System. However, we still have the difficulty of explaining the relatively low global abundance of iron on the Moon. It would seem very improbable that the lunar matter was *previously* fractionated in a gravitational field. A further problem of the capture theory, that the Moon would have to have been slowed down drastically in the course of a near collision with the Earth in order to have been captured at all, does not seem to be too serious: this braking could have occurred as the result of a collision with a large planetesimal while the Moon was in the vicinity of the Earth.

The two theories, which were originally formulated only rather schematically, appear today in consideration of the high density of asteroids in the early period of the Solar System to be not so very divergent.

The Impact Theory. Since the mid-1980's, the impact theory of the origin of the Moon has begun to be generally accepted; it proposes that the Moon was formed somewhat more than $4.4 \cdot 10^9$ yr ago as the result of a collision with a *single* large celestial body (which had a mass of about 15% of that of the Earth, comparable to Mars).

Towards the end of the accretion phase of many smaller planetesimals onto the (inner) planets, there must have been many more large objects than exist today, including a few the size of Mars; they were then accreted or destroyed in the course of time by the (present) planets. The collision theory assumes that both in the case of the primeval Earth as well as in the colliding object, the metallic core and the stony mantle had already formed. In a relatively slow, grazing collision, the material from the colliding object would be mixed with that of the Earth's crust and ejected. While the core material of the impacting object would finally remain on the Earth, the initially mostly molten Moon would have coalesced out of the mixture of silicates from the Earth's mantle and those from the impacting object, on an orbit outside the Roche limit of around 2.9 Earth radii (3.11). In the lunar matter, the material from the impacting object would predominate, while the Earth's mantle would have been only more or less "contaminated" by the collision; thus on the one hand the similarities, and on the other the clear differences in the chemical compositions of Moon rocks and the Earth's crust would have come about. Since a large body could have transferred sufficient angular momentum to the Earth, this theory also solves the problem of its relatively rapid rotation. In contrast to the older theories, the impact theory thus renders plausible the unique role of the Earth–Moon system within the Solar System and explains both the similarities and the differences in the minerals on the Moon from those on the Earth.

Finally, we remark that indications of collisions with large celestial bodies can also be found elsewhere in the Solar System. For example, the large inclination of the rotational axis of Uranus relative to its orbital plane (Table 3.1) is probably due to such an event.

14.1.5 Planetary Systems Around Other Stars

The formation of our Solar System was very probably not a unique, accidental occurrence. Rather, we can expect that numerous stars in the cosmos are orbited by planets. While we have a surfeit of detailed observations of our own system, the present-day observational techniques are hardly sufficient to observe planets directly even around the stars in our immediate neighborhood. The great strides made recently, particularly in spectroscopic techniques, have made it possible in the past few years to detect exoplanets around many stars; these are mostly about the size of Jupiter (Sect. 6.5.7).

Fundamentally, the observation of *other* star-planet systems of various kinds in different phases of their evolution would give important information regarding the formation and evolution of *our own* Solar System as well.

Exoplanets. Since the discovery of 51 Peg B in 1995, the number of planets found around nearby, normal stars has increased rapidly; at present (June 2001), we already know of about 60 exoplanets, leaving planet-like companions of the quite different pulsars (Sect. 7.4.7) out of consideration. In view of the rapid development of this area of research, we refer the reader to "The Extrasolar Planets Encyclopedia" on the Internet at http://www.obspm.fr/planets, where the most recent information about extrasolar planets and their properties can be found.

The central stars of most of the currently known exoplanets are not very different from our Sun; their spectral classes lie in the main between F6 and G8. Based on spectroscopic observations of several hundred nearby stars, it is estimated that about 5% of all stars which are similar to the Sun are orbited by a Jupiter-like companion planet.

The fraction of planetary systems in which a *large* planet is circling the central star at a *small* distance is noticeable; these systems have a quite different structure from that of our Solar System. We know of several exoplanets with orbital periods of ≤ 3 d and orbital radii of ≤ 0.04 AU, which thus orbit their stars at only about a tenth of the distance of Mercury from our Sun. Such systems, to be sure, will be preferentially discovered, since their orbital periods are short and the amplitudes of their radial velocities are high owing to the relatively large masses of the planets.

In the spectroscopic determination of the mass, as we showed in Sect. 6.5.7, only $\mathcal{M} \sin i$, i.e. only a *lower* limit for the mass, can be derived as a result of the unknown orbital inclination i.

We cannot discuss systematically the properties of the exoplanets here; instead, we limit ourselves to the description of a few particularly interesting systems:

In the cases of some stars, e.g. HD 83443 (spectral class K0V) and HD 16141 (G5IV), it was possible to detect companions with a minimum mass of $0.2\,\mathcal{M}_J$, and thus with masses of the same order of magnitude as that of Saturn ($\mathcal{M}_S \simeq 0.3\,\mathcal{M}_J$).

The companion of 16 Cyg B is remarkable for its extreme orbital eccentricity ($e = 0.67$), which gives rise to a "sawtooth" radial velocity curve, i.e. strong deviations from a sine curve. 16 Cyg B itself, with its companion, is a member of a triplet star system; neither the principal component A (likewise similar to the Sun), at around 1100 AU, nor the more distant component C have shown indications of having planet-like companions.

The Sun-like star HD 209458, at a distance of 47 pc from the Solar System, is notable for the fact that its companion (with $\mathcal{M} \geq \mathcal{M}_J$ and an orbital period of 3.5 d) as seen from the Earth passes in front of the disk of the star and decreases its brightness by 1.8% in passing. This *occultation* by the planet makes it possible to determine from analysis of the light curve both the inclination of the orbit and the radius of the planet. The results are $i = 85.2°$ and a radius of 1.54 Jupiter radii.

The first planetary *system* was discovered in 1999 by G. Marcy and R. P. Butler around the F8 V star υ And (HR 458) at a distance of 13.5 pc. After a first planet around this star had been discovered a few years before, they were able to identify two additional planets with orbits further out in the same orbital plane. Their radii, orbital periods, and eccentricities as well as their masses are:

a [AU]	P [d]	e	$\mathcal{M}$ [$\mathcal{M}_J$]
0.06	4.6	0.03	≥ 0.7
0.8	241	0.2	≥ 2.1
2.5	1267	0.4	≥ 4.6

All three of the companions have larger masses than about $1\,\mathcal{M}_J$, and orbits which are considerably closer to υ And than Jupiter is to our Sun. Astrometric observations from the Hipparcos satellite together with determinations of the radial velocities give an upper limit for the mass of the outermost planet of $\mathcal{M} \leq 10\mathcal{M}_J$.

Gliese 876, an M-dwarf with a mass of $0.3\,\mathcal{M}_\odot$, belongs to the most common type of stars in the Milky Way. It is only 4.7 pc from us and is the first star of a different type from our Sun which has been discovered to have a planet-like companion. It was later found that Gliese 876 is circled by *two* large planets with masses of ≥ 0.6 and $\geq 2.0\,\mathcal{M}_J$ at distances of 0.13 and 0.21 AU with their orbital periods in nearly a (2:1) resonance (30.1 and 61.0 d).

The G5 star HD 168443, at a distance of 37.9 pc, is orbited by two companions with orbital periods of 58 d and 5.85 yr, orbital radii of 0.29 and 2.9 AU, and eccentricities of 0.5 and 0.2, respectively. While the inner companion has a mass of $\geq 7.2\,\mathcal{M}_J$, the outer one has a notably larger mass of at least $17\,\mathcal{M}_J$. This mass lies above the lower limit for deuterium burning, so that we are probably dealing here with a brown dwarf.

Planet or Brown Dwarf? Distinguishing between a large, Jupiter-like gas planet and a brown dwarf star is difficult and not always conclusive, especially when only one companion object can be observed.

We can expect that all planets are formed from a protoplanetary gas-dust disk, while brown dwarfs condense from a concentration of the interstellar medium, together with their stellar companions, like the stars in binary and multiple star systems. However, the history of the formation of a system cannot necessarily be read out from its observational data.

The most important criterion which is applied to distinguish planets from brown dwarfs is the *mass* of the substellar companion. Often, only objects of less than about $13\,\mathcal{M}_J$, in which deuterium fusion can no longer take place (Sect. 8.3.1), are considered to be candidates for planets. Since the masses determined spectroscopically from the fluctuations of the radial velocity of the star are only lower limits, it must also be attempted in each case – e.g. from the rotation of the star (Sect. 6.5.7) or from astrometric measurements – to obtain an estimate for the upper limit of the mass. Distinguishing a planet from a brown dwarf on the basis of mass alone is however problematic.

Originally, another criterion for distinguishing planets and brown dwarfs was their *orbital eccentricities*. Since the planets of our Solar System follow *circular* orbits owing to the effects of viscous forces in the protoplanetary disk, orbits with small eccentricities were expected for exoplanets also. However, the discovery of more and more exoplanets with considerable orbital eccentricities led to the dropping of this criterion.

A theoretical explanation for the origins of the observed great variety of extrasolar planetary systems and for the differences from our own Solar System has yet to be given.

Protoplanetary Disks. The infrared satellite IRAS discovered infrared excesses at $\lambda \geq 25\ldots 60\,\mu\text{m}$ for several A stars such as α Lyr (Sect. 6.3.2), α PsA, and β Pic; these result from circumstellar dust and can be regarded as indications of proto-planetary systems. In the case of β Pic, a flat, dust-filled gas disk with a diameter of ≥ 400 AU was then directly observed on CCD images.

More recently, a number of circumstellar disks with diameters of the order of that of our Solar System have been identified on the basis of their strong infrared emissions, around young stars such as the T Tau stars (Sect. 10.5.1) and around protostars in regions of star formation (Sect. 10.5.5). It was possible with the Hubble Space Telescope to directly resolve circumstellar disks in the Orion Nebula (Fig. 10.28).

The radii of the circumstellar disks observed thus far lie between roughly 10 and 300 AU; their temperatures fall in the range from about 50 to 400 K, with a decrease in temperature on going towards the perimeter of the disk. Their masses range from 0.001 to $\leq 1\,\mathcal{M}_\odot$. Since these disks occur only around younger stars, we can estimate their maximum "lifetimes": a disk breaks up after $\leq 10^7$ yr, whereby part of its matter could go into the formation of planets and the rest must fall into the central star or escape from the system.

The central question however remains open: whether planets indeed form from these disks, i.e. whether we are in fact observing *protoplanetary* disks.

14.2 The Evolution of the Earth and of Life

The history of the Earth, with its atmosphere and its oceans, is very closely connected with the origin and evolution of life. Like cosmogony, this problem was also the domain of mythological fantasies for a long period of time. After the successful synthesis of urea in 1828 by Friedrich Wöhler removed the imaginary boundary between inorganic and organic chemical compounds, contributions from the fields of astrophysics, geology, and biochemistry have admittedly not yet solved the problem of the origin of life, but have at least placed it firmly within the realm of scientific investigation.

We turn first in Sect. 14.2.1 to the interesting questions of the formation of the oceans and of the Earth's atmosphere. In Sect. 14.2.2, we then give a brief introduction to the molecular-biological fundamentals of life, followed in Sect. 14.2.3 by a discussion of the prebiotic molecules as its material prerequisite. In Sect. 14.2.4, we take up the problems of the origins and the evolution of life on the Earth.

14.2.1 The Development of the Atmosphere and of the Oceans

Both the oceans (hydrosphere) and the Earth's atmosphere can*not* have belonged to the original structure of the planet; we have seen, indeed, that in the early stages of the Earth–Moon system (as also in the meteorites), the light elements were expelled from the planets and other objects, at least in the inner regions of the Solar System. The well-known fact that the noble gases are very rare in the Earth's atmosphere, although helium and neon are among the most common elements on the Sun, shows that our atmosphere cannot be a direct remnant of the solar nebula. Instead, the atmosphere and the oceans were formed in a secondary process, probably from *volcanic* exhalations, which yield in particular H_2O and CO_2 as well as traces of SO_2, CO, H_2, N_2, ...

The most abundant argon isotope, ^{40}Ar, was produced on the other hand by the decay of ^{40}K in the Earth's crust (potassium feldspars), and helium was formed as a result of the α-decay of the known radioactive elements.

The atmospheres of Venus and Mars, consisting mainly of CO_2, are also of secondary origin. In the atmospheres of the major planets, in contrast, ammonia (NH_3) and methane (CH_4) were formed directly in the course of chemical fractionation from the original solar matter.

The Hydrosphere. The Earth is the only planet whose surface temperature allowed the formation of a hydrosphere, i.e. of large quantities of liquid H_2O. Through reactions with silicates in the oceans, this in turn permitted most of the carbon dioxide to be removed from the atmosphere in the course of time. Today, we find the major portion of the original CO_2 in the form of potassium and magnesium carbonates in sedimentary rocks.

Atmospheric Oxygen. The *protoatmosphere* of the Earth contained as yet *no* free oxygen, since that element was completely bound up in compounds as oxides, silicates, etc. The formation of O_2, and thereby also of ozone (O_3) in the optically thin atmosphere began when water vapor, H_2O, was decomposed by solar ultraviolet radiation (photodissociation) into $2H + O$. But this process could have produced at most about 10^{-3} of the present oxygen concentration: a thicker layer of oxygen (along with the ozone which accompanies it) namely absorbs the short-wavelength solar radiation, so that the amount of gas from solar irradiation is self-limiting.

The additional oxygen in our atmosphere can only have been produced through *photosynthesis* by *lifeforms*, i.e. in connection with their development. The production capacity of the *present-day* flora can be made clear by considering the estimate that all the oxygen in the Earth's atmosphere passes through the process of photosynthesis or carbon dioxide assimilation in a period of only 2000 yr! In this process, with the aid of the energy $h\nu$ from solar radiation, energy-rich glucose is produced and oxygen is released according to the overall reaction

$$6\,CO_2 + 6\,H_2O + h\nu = C_6H_{12}O_6 + 6\,O_2 \,. \quad (14.5)$$

In the complex intermediate reactions, a decisive role is played by *chlorophyll*. A further idea of the carbon metabolism of the Earth's plants is given by the enormous mass of the fossil fuel (hydrocarbon) deposits.

The history of the Earth's atmosphere and of the Earth itself is thus closely connected with the origin and evolution of life.

14.2.2 Fundamentals of Molecular Biology

Living organisms are extremely complex physicochemical structures. Even the simplest lifeform makes up a *system*, in which two essential functions interact:

1. The ability to *reproduce*, from complex molecules to entire organisms, is anchored in the genetic material, which contains the information and the control mechanisms that allow a molecule or organism of a given kind to be "copied". "Errors" in this replication mechanism, which occur as a result of chemical influences or ionizing radiation, or even through the thermal motions of the molecules, can produce *mutations*. The origin of a mutation is thus simply a problem in probability theory. Once it *has* occurred, then the modified molecule or organism will continue to be *exactly* reproduced. The mutations, and furthermore the many possibilities of genetic recombination, form the basis for the *evolution* of lifeforms, according to the mechanism of *natural selection* discovered by Charles Darwin in 1859. Many mutations lead, as might be expected, to nonfunctional structures which again quickly "disappear". The structures which are at least capable of functioning within themselves are immediately exposed to the effects of their environment and (later) of other lifeforms. In this struggle for existence, to use the somewhat unfortunate catchphrase coined by H. Spencer, that molecular complex or organism will win, which has the highest rate of reproduction. The latter thus provides a clearcut measure of the selection value of a particular mutation.
2. A mechanism for the delivery of the necessary *energy* and of course the necessary *materials* must be associated with the control system. The construction of a complex form with particular structures, which is thus "improbable" or not likely to occur through coincidence alone, presupposes a certain amount of *information* or "know how". It is thus connected according to the fundamentals of thermodynamics with a decrease in the *entropy* (entropy is proportional to the logarithm of the probability of a state), or an increase in the negative entropy or negentropy of the system. The latter can however only occur through an input of energy, whereby in the environment (from which the energy is taken), a corresponding increase in entropy must occur, according to the 2nd Law of Thermodynamics.

Modern biochemistry has succeeded in the past decades in obtaining a clear overview of the most important groups of substances and processes which make possible the wonder of life. Here, again, we meet up with the interaction of information and function, analogous to the legislative and the executive branches of a government, in a heirarchy of reaction cycles, which altogether make possible the formation and reproduction of functional macromolecular or living systems.

In all organisms, the nucleic acids occur as carriers of information or of the genetics of the molecules, while the structure and function of the organisms are determined by proteins, which, with varying degrees of specialization, carry out their functions according to the "instructions" of the nucleic acids. The fundamental structures and functions are basically the same throughout the whole organism.

We first consider the proteins in somewhat more detail, then we will take up the nucleic acids.

Proteins and Amino Acids. The proteins, the most important carriers of function, e.g. the recognition and processing of particular substances, catalysis, regulation of reaction rates, etc., are macromolecules in which up to several thousand amino acids are bound together in a certain sequence by rather rigid peptide bonds into a *peptide chain* (Fig. 14.3). It is very remarkable that in the proteins of *all* lifeforms on Earth, only about 20 particular amino acids occur, although chemically, many more are possible.

The polypeptides can be folded in a variety of ways and can, in particular, be rolled up into balls (globulins). The shape of the ball is determined uniquely through the sequence of amino acids. The resulting shape of the

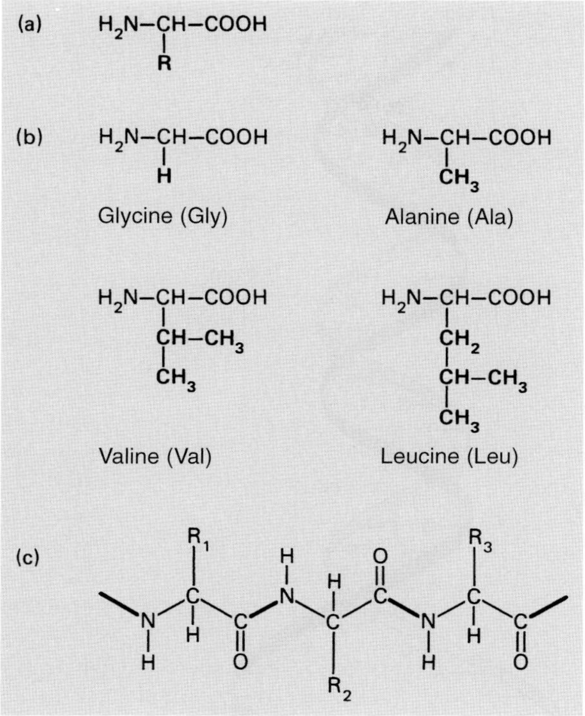

(a) H₂N–CH–COOH
       |
       R

(b) H₂N–CH–COOH   H₂N–CH–COOH
       | |
       H   CH₃
    Glycine (Gly)   Alanine (Ala)

    H₂N–CH–COOH   H₂N–CH–COOH
       | |
      CH–CH₃   CH₂
       | |
      CH₃   CH–CH₃
                                |
                               CH₃
    Valine (Val)   Leucine (Leu)

(c) [peptide chain structure with R₁, R₂, R₃ side groups]

Fig. 14.3a–c. Amino acids and proteins. (**a**) The basic structure of the majority of natural amino acids, which differ only in the nature of the side group R. (**b**) Examples of the protein-component amino acids with their symbols. (**c**) A protein molecule (polypeptide) is formed by joining from around 50 up to several 1000 amino acids in a linear chain, by rather rigid peptide bonds (indicated as heavy lines). $R_1, R_2, \ldots$ are the amino acid side groups, e.g. H in glycyl, CH_3 in alanyl etc.

molecule and its distribution of interactive forces allow its activity to be *specialized* with respect to certain substances or chemical reactions. An especially important group of proteins in this sense are the *enzymes*.

As can be readily calculated, various arrangements of altogether 20 amino acids in chains of about 1000 links can *theoretically* yield so many different proteins that even one set of samples would fill the entire known Universe! We can thus readily understand the enormous *variety* of organic lifeforms; it must, in fact, be limited by some ordering principle.

Nucleic Acids. The carriers of the information, "how and what is to be produced", are the nucleic acids.

The discovery of their role in the replication of organisms was the beginning of modern molecular biology. O. T. Avery and coworkers recognized in 1944 that the *desoxyribosenucleic acids* (abbreviated DNA) are the material carriers of genetic information. J. D. Watson and F. H. Crick then succeeded in 1952 in constructing a spatial molecular model of DNA, the famous *double helix*, and thus attained some insight into the close relationship between the information localized in the DNA and the mechanism of replication.

The nucleic acids (polynucleotides) are macromolecules formed by joining *nucleotides* in a particular sequence. These in turn consist of a sugar molecule (a pentose with five carbon atoms), a phosphate group, and a heterocyclic nitrogen base.

In the desoxyribosenucleic acids, the sugar is 2-desoxyribose; evidently with regard to the above-mentioned limitation of the variety in protein formation, only *four* different nucleotides occur, containing the purine bases *adenine* and *guanine* and the pyrimidine bases *thymine* and *cytosine*, which are denoted for short by their initials A, G, T, and C (Fig. 14.4).

The *DNA molecule* consists of two strands of nucleotides, which together form a double helix. The bases are joined pairwise as in Fig. 14.5 by hydrogen bonds. There are thus four *base pairs* A–T, G–C, T–A, and C–G. The information about the production of a protein is now coded in the sequence of the bases in a DNA strand, so to speak in a "genetic language", using the four symbols A, G, T, C. According to the principle of base pairing, the structure of *one* strand in the double helix completely determines that of the *other* strand and thus forms the basis for the replication of a DNA molecule, the most important "elementary process" of biology: the double helix is opened up like a zipper by breaking of the relatively weak hydrogen bonds in the base pairs, then the corresponding free bases attach themselves to each base in the strand, resulting in two identical double helices (Fig. 14.5).

The genetic information is in fact not read off directly from a DNA strand itself, but instead is first transferred to a corresponding molecule of a *ribonucleic* acid (RNA) and transported by it to the place where protein synthesis occurs. RNA has a similar structure to DNA, except that in the nucleotides, desoxyribose is replaced by ribose and the base uracil (U) takes the place of thymine (T). In uracil, the CH_3 group on the aromatic

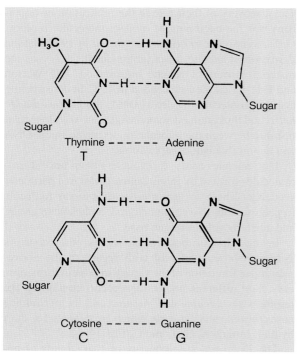

Fig. 14.4. Base pairing in desoxyribosenucleic acid. A pyrimidine base (*left*) and a purine base (*right*) are joined by hydrogen bonding (*dashed lines*) and thus fit into the DNA helix (Fig. 14.5). (Reproduced with the kind permission of the G. Fischer Verlag, Stuttgart)

ring of thymine (Fig. 14.4) is replaced by H. Thus, the "DNA symbols" AGTC are transcribed as UCAG in the "RNA code". Fascinating as they are, we cannot go into the details of the transcription process here; instead, we turn directly to the deciphering of the genetic code.

The Genetic Code. In order to specify *one* of the 20 amino acids uniquely using the 4-symbol code, at least 3 symbols must be combined into a "word" (triplet). In fact, each amino acid is specified by a sequence of three nucleotides in the DNA or in the corresponding RNA after transcription. For example, the sequence G–C–U specifies alanine. The sequence of triplets along the DNA chain thus serves so to speak as a "blueprint" for the production of a corresponding protein with a particular series of amino acids. Since the genetic code could spell altogether $4^3 = 64$ words, while only 20 amino acids need to be specified, nature

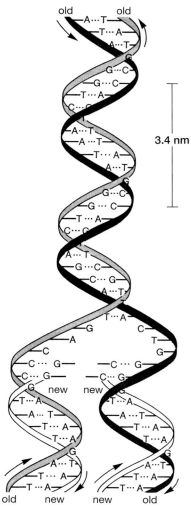

Fig. 14.5. The DNA molecule is a double helix, i.e. it has the form of a spiral staircase. The "balustrades" of the staircase are formed by the phosphate and desoxyribose (saccharide) groups (not shown). The "steps" of the staircase correspond to the base pairs A–T and G–C (Fig. 14.4). In the replication process, the two complementary strands of the double helix are separated like an opened zipper (*upper part*). Then each base attaches itself to a new base according to the pairing scheme described, and the result (*below*) is two identical DNA double helices

"allows" some amino acids to be specified by several different words; furthermore, some of the words are used as "punctuation marks".

A section of DNA strand consisting of about 1000 nucleotide pairs, containing the information required to produce a particular protein, is called a *gene*, the carrier of inheritance. The genetic code is *universal*, i.e. it is the same for all living organisms on Earth. Since chemical processes such as DNA replication and protein synthesis must be accomplished in the organism rapidly and with the least possible energy consumption, they are catalyzed by special proteins, the *enzymes*, which themselves are again produced using the "know how" of the DNA. It is, to be sure, a still-unsolved problem to identify the *beginnings* of these closed cycles, but the mechanism of the genetic code, which we have indicated only schematically here, makes the replication of particular systems of molecules and thus of particular species of organisms fundamentally understandable. It also explains the origin of mutations through disturbances in the word sequence along the DNA chain.

14.2.3 Prebiotic Molecules

Having described the molecular-biological fundamentals of the structure, the functions, and the mechanisms of self-reproduction in present-day living organisms, we turn to the question of the origin of life on Earth and its evolution, first investigating the material preconditions for this development.

The complex molecules which are typical of the structure of living matter, especially the nucleic acids and the proteins, are not stable in the presence of oxygen. Their formation from inorganic substances is therefore not possible on the Earth under present conditions without the intervention of living organisms. In the "beginning", they could only have been formed in an oxygen-*free* atmosphere. On the other hand, such molecules could not have survived for long on the surface of the primeval Earth without the protective oxygen and ozone layers in the atmosphere, owing to the very strong irradiation in the far ultraviolet ($\lambda < 290$ nm). However, they could have collected for example in the pores of minerals or on the bottoms of shallow bodies of water of ≤ 10 m depth.

The similarity of the genetic code and its application in essentially the same organic molecules for the DNA and for the proteins in lifeforms from the most primitive bacteria to human beings indicate that the origin of life occurred at a single time and place, where the most important macromolecules were already present.

The formation of such high-molecular weight prebiotic molecules "all by themselves" is not so extremely improbable as one might at first think. Beginning in 1953, S. L. Miller, led by the considerations of H. C. Urey, was able to prepare complex organic molecules, in particular amino acids, by subjecting a mixture of methane (CH_4), ammonia (NH_3), and water vapor (H_2O), a kind of protoatmosphere, to electrical discharges or irradiation with UV or gamma radiation. According to current ideas (Sect. 14.2.1), the early atmosphere was, to be sure, not a strongly reducing methane atmosphere, but rather consisted mainly of CO_2 and H_2O with at most traces of CH_4 and NH_3; but experiments have shown that even under these conditions, small amounts of the relevant organic molecules can be formed. However, it is questionable whether concentrations would result which were sufficiently high for life to have come about.

A quite different, interesting possibility for the formation of complex molecules is suggested by the investigation of the interstellar medium and of meteorites. As we have seen in Sect. 10.2.3, radio astronomers have observed surprisingly complex organic molecules in the dense interstellar clouds from which stars are later formed. On the other hand, a rich variety of amino acids and other substances which are necessary for the origin of living organisms are found in the carbonaceous chondrites. These meteorites, with their organic matter, were formed about $4.5 \cdot 10^9$ yr ago, along with the Sun and all the other objects in the Solar System, from the "solar nebula", a rotating disk of gas and dust (Sect. 14.1.3). Today, we also know that the Earth, along with the Moon and the other planets, was subjected to an intense bombardment by meteorites (or planetesimals) for about 10^9 yr after its formation; among them, as among present-day meteorites, there must have been a considerable fraction of carbonaceous chondrites. Since many of these were slowed down by the Earth's atmosphere and also fell into water, sometimes after fragmentation, it is quite possible that the prebiotic molecules came for the most part from these meteorites.

Organic molecules, however, are still far from being living organisms! We cannot say how the first lifeforms were structured and how they came into being. Even the *simplest* living organisms or *"biomolecules"* must, in contrast to lifeless matter, be able to maintain a metabolism and especially to reproduce themselves; in this process, occasional "errors", or mutations, occur. From the palette of different mutations, the best-adapted species were *selected* in the process of evolution.

Presumably, the *ribonucleic acids* assumed a key role at the threshold of the origin of life. Recent investigations make it seem likely that RNA can both reproduce itself *without* the presence of proteins, and also can act as a catalyst for all of the reactions which were necessary for early lifeforms, in particular for the synthesis of proteins from amino acids. If sufficient amounts of proteins are already present, then the alternating interactions of the nucleic acids and the proteins which is typical of present-day life can begin to take place.

Behind the question of the origin of life and of its evolution stands the fundamental problem of whether our current physics and natural sciences comprise a sufficient basis for understanding the origin of biomolecules capable of self organization, or whether we need to assume a mysterious "vis vitalis" or something similar. In 1971, M. Eigen succeeded in explaining the Darwinian principle of "survival of the fittest" in terms of purely physico-chemical fundamentals. The actual origin of life, as well as all the mutations, must be regarded as "coincidences"; all the rest is physics. Eigen first determined that no solution can be hoped for on the basis of classical (equilibrium or quasi-equilibrium) thermodynamics. The interactions of proteins and nucleic acids necessary for every life-process are so complex, even in the most primitive forms imaginable, that their occurrence under such conditions must be regarded as out of the question. "Life" is indeed possible only with the aid of an energy input, metabolism, etc.; i.e. as a (quasi-) stationary irreversible thermodynamic process. In a series of fundamental publications during the 1970's, I. Prigogine, M. Eigen, H. Kuhn, H. Haken, and others were able to describe the essential features of the origin and evolution of self-organizing molecules as *necessary* properties of systems of nonlinear reaction kinetics equations, which are *far* from the state of thermodynamic equilibrium and are subjected to strong energy flows. The origin of biomolecules can be attributed to characteristic *instabilities* of these systems.

The further development of the living organisms takes place necessarily in discrete steps, which each begin with a new "invention": the production of a membrane around the nucleus of a cell, the principle of sexuality with its greater security and variability of reproduction, the combination of different types of cells with particular cooperative functions, etc. We can unfortunately not go into these very interesting considerations further here or justify them in more detail.

14.2.4 The Development of Lifeforms

Making use of the current status of research in the fields of palaeontology, geology, geochronology, etc., we shall now attempt to draw a consistent picture of the early history of the Earth and its atmosphere in connection with the development of life, based on the works of E. S. Barghoorn, J. W. Schopf, P. E. Cloud and others. For most of the details, we must of course refer the reader to the specialized literature on this subject. We shall first give a rough outline of the geological sequence during the Precambrian era, and of the basic forms of living organisms.

The Precambrium. This period includes the longest portion of the Earth's history, from its formation $4.5 \cdot 10^9$ yr ago up to the beginning of the Cambrian era $0.59 \cdot 10^9$ yr ago (Table 3.3). The Precambrian is subdivided into the *Archaic* (the Archaizoic era, which ended $2.5 \cdot 10^9$ yr ago) and the *Proterozoic* era which followed it.

The transition into the Cambrian marked the beginning of the Phanerozoic age – the time of "appearing life" – with the sequence of geological eras which we have already shown in Table 3.3.

The Principal Groups of Lifeforms. We divide all lifeforms into the prokaryotic and the eukaryotic organisms. The *prokaryotes* are single-celled organisms *lacking* a well-defined nuclear membrane and chromosomes. They include on the one hand the eubacteria, to which the true bacteria as well as the cyanobacteria (previously incorrectly termed "blue algae") belong. On the other hand, they include also the archaebacteria, whose representatives such as the sulfur-, salt-, or methane-bacteria are still living today under extreme conditions (in hot sulfurous springs, concentrated brine or sewage), frequently as autotrophes.

All the other lifeforms are *eukaryotic*: single- or many-celled organisms *with* nuclear membranes and chromosomes. This group contains all of the plants (including the single- and many-celled algae), the protozoa (i.e. single-celled animals), the metazoa, and the higher animals.

Recent research on RNA and protein sequences has shown how closely related the various lines are. Thus, the eubacteria probably first branched off from a still unknown primeval form, and later the archaebacteria separated from the eukaryotic organisms:

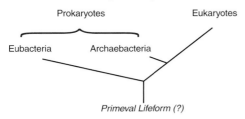

The absolute dating of these branching events is still controversial.

The Oldest Lifeforms. We first ask the question: what do we know about the earliest lifeforms? In sediments as old as 3.5 to $3.6 \cdot 10^9$ yr from South Africa and Australia, we find definite traces of living organisms: *microfossils* such as the Ramsay spheres (yeast-like organisms) and *stromatolites*, layers built up from the calcium carbonate deposits of cyanobacteria (cf. Table 3.3). The currently oldest known, radioactively dated rock deposits are the Isua formations in Southwest Greenland, with an age of $3.8 \cdot 10^9$ yr. Their sediments contain carbonaceous inclusions, which were interpreted by H. Pflug and M. Schidlowski as fossil structures of cyanobacteria and other microorganisms which have been modified by later metamorphosis of the rocks containing them. According to this interpretation, which is to be sure not uncontroversial, there was already a widespread, differentiated world of bacterial flora on the Earth as long as $3.8 \cdot 10^9$ yr ago, and certainly at least $3.5 \cdot 10^9$ yr ago.

As we have seen, the Earth–Moon system originated about $4.5 \cdot 10^9$ yr ago. However, while the rocks of the lunar highlands on the Moon date back to this time, the oldest rocks on the Earth are considerably younger. The Isua sediments verify the existence of a widespread hydrosphere as early as $3.8 \cdot 10^9$ yr ago, and they themselves are probably derived from ancient rock strata dating back about 4.0 to $4.1 \cdot 10^9$ yr. Since, on the one hand, we can hardly assume that the Earth's crust was solid earlier than roughly $4.2 \cdot 10^9$ yr ago, and on the other, the first traces of life date from 3.8 to $3.5 \cdot 10^9$ yr ago, only the amazingly short period of around 400 to at most 800 million years remains for the decisive steps in the origin of the first lifeforms and their evolution to the stage of bacteria to have taken place! The structural materials for this process, amino acids and other prebiotic molecules, were probably "delivered" for the most part by the impacts of carbonaceous chondrites and comets together with "cosmic" dust from space. It may also have been important for the origin of life that around 10^9 yr after the formation of the Solar System, the intense bombardment by planetesimals was drastically reduced, and that the oceans and the protoatmosphere, consisting mainly of CO_2 and H_2O, had already begun to develop.

An oxygen-*free* atmosphere was a prerequisite for the origin of organic life; many substances essential for life are attacked by oxygen. However, there was originally also no ozone layer which would have protected the early lifeforms from lethal solar ultraviolet radiation. The most primitive organisms must therefore have developed either under a protective layer of water, or else in pores in rocks which let in some light but not too much of the deadly UV radiation.

These first organisms were certainly *heterotrophic*, i.e. they lived on the organic substances already present, as do present-day animals and those plants which have no chlorophyll. However, *autotrophic* organisms of different kinds, which are self-sufficient and independent of other living forms, making use of inorganic starting materials, developed very early from bacteria and expanded rapidly owing to *photosynthesis* in green plants (14.5). As early as $3.5 \cdot 10^9$ yr ago, a primitive form of photosynthesis was "invented" by the cyanobacteria.

Evolution and O_2 Production. Parallel to the development of the plants, a general enrichment in oxygen of the hydrosphere and the atmosphere took place.

In the period about $2.1 \cdot 10^9$ yr ago, there were, according to the remains that we can study today, only prokaryotes. The small amount of oxygen produced by these early prokaryotes remained in the hydrosphere and was mostly used up in converting ferrous to ferric oxides. The *banded* iron ore deposits, which occur together with marine sediments, showed a noticeable increase around $2.5 \cdot 10^9$ yr ago, but were also laid down over the whole time period from 3.8 to $1.8 \cdot 10^9$ yr ago. Their origin can be interpreted (P. E. Cloud, 1973) as a result of the formation of Fe^{2+} solutions under an oxygen-free atmosphere; these were then transported into the seas and there, by the action of oxygen-producing bacteria, were periodically precipitated as nearly insoluble Fe^{3+} compounds. In this manner, enormous amounts of O_2 were bound up in iron oxides.

We find the first *eukaryotes* (algae) about $2.1 \cdot 10^9$ yr ago; they were aerobic, i.e. they reguired O_2 in order to live. Their origins and their connection to the prokaryotes are still largely unexplained.

Only between about 2.2 and $1.9 \cdot 10^9$ yr ago did the oxygen content of the atmosphere increase from less than 1% to about 15% of the current value. This *excess* free oxygen made it possible for O_2-processing enzymes to develop and quickly provided the *animals* with a high-yield energy source (respiration!). The first *multicelled* (if still very small) animals appeared around $1.0 \cdot 10^9$ yr ago.

There was a fundamental change when the oxygen content of our atmosphere reached the threshold of a few percent of the present value. This oxygen sufficed to promote the formation of an ozone layer, which absorbed the lethal UV radiation so strongly that now the surface of the water and later (at the beginning of the Cambrian era) the land became habitable.

The last 300 million years of the Proterozoic era, from about $0.8 \cdot 10^9$ yr ago onward, were a very disturbed period for life on the Earth. Together with strong tectonic activity, substantial ice ages alternated with very warm periods. The last of these ice ages, the Varanger ice age, probably brought with it the most difficult conditions for life in the Earth's history. It was followed by the *Ediacara* period (from ca. 630 to $590 \cdot 10^6$ yr ago), with an enormous blossoming of life, especially in the development of the animal world. In this period, the first larger many-celled animals (metazoa) with a complex internal organization appeared, at first only in soft-bodied forms.

Phanerozoic. From the Cambrian era onwards, i.e. $590 \cdot 10^6$ yr ago, in the phanerozoic age, the great variety of metazoa with shells arose. At this point, the series of *geological periods* described in Table 3.3 began; their petrified lifeforms (fossils) have provided us with a detailed record of the further evolution of the plant and animal kingdoms. About $410 \cdot 10^6$ yr ago, from the Silurian to the Devonian eras, the first widespread forests developed, and in the Carbonaceous era, whose fossilized plants are still serving us as fuel today, the current level of oxygen was no doubt reached, if not even surpassed for periods of time.

Periods of Mass Extinction. There have been repeated critical periods of change in the evolution of the plant and animal kingdoms, each lasting several million years, in which at first the variety of different species was drastically reduced, and then numerous new species appeared; for example, the transition from the Permian to the Triassic eras about $250 \cdot 10^6$ yr ago, between the Tirassic and the Jurassic ca. $210 \cdot 10^6$ yr ago, or near the end of the Cretaceous ($65 \cdot 10^6$ yr ago; see Table 3.3). The causes of such periods of massive extinction, in which the rate of mutations was evidently much higher than before and after, are not clear. Owing to their rarity, supernova explosions are hardly a possibility; other speculations include increased UV irradiation, a stronger intensity of cosmic rays as a result of the polarity reversals of the Earth's magnetic field, the impacts of large planetesimals or other processes.

With some certainty, we can state the cause – or at least one of the decisive influences – of at least one transition period between the Cretaceous and the Tertiary, $65 \cdot 10^6$ yr ago, during which the extinction of the dinosaurs, the ammonites, and of at least 50% of the other species occurred: the impact of a giant meteorite or asteroid of about 10 km diameter, and possibly the increased volcanic activity which it triggered, caused catastrophic, long-lasting changes in the conditions for life on the Earth. The observation of a high concentration of iridium and osmium in a thin layer of clay which is found between the Cretaceous and the Tertiary deposits and is spread over nearly the whole Earth first led L. and A. Alvarez and their coworkers in 1980 to the hypothesis that the enrichment in these metals, which are normally quite rare in the Earth's crust but are common in certain meteorites, could in fact be an indication of the impact of an enormous meteorite. The associated crater structure was then found in 1992, buried under thick sediments of limestone, near Chicxulub in Yucatán, Mexico (Sect. 3.7.1). Its radiologically-determined age of $65 \cdot 10^6$ yr, from the ratio of ^{40}Ar to ^{39}Ar, agrees very well with that of the geological period. Tektites found in Mexico and Haiti can also be attributed to this impact on the basis of radioactive dating.

In contrast to this transition period at the end of the Cretaceous era, we can find no direct evidence for the impact of a celestial object as the cause of the greatest known catastrophe in the course of evolution, between the Permian and the Triassic. At that time, more than 85% of all species in the oceans and about 70% of those on land died out over a period of 10^4 to 10^5 yr. The surface of the Earth at that time was essentially covered by the giant Panthalassa Ocean, much larger than today's Pacific Ocean, and a single "supercontinent", *Pangaia*, which combined all of the present-day continents in one (cf. Fig. 3.3). At its edge, for the most part enclosed by the contintent, was the warm Tethys Sea, rich in species. The cause of the nearly complete extinction of all the higher animals about $250 \cdot 10^6$ yr ago was probably not a single event; rather, the complex fossil record indicates the combined effects of several processes which are not understood in detail. For one thing, near the end of the Permian, a slow but substantial drop in the sea level gave rise to a drastic decrease in the available maritime habitat. Extreme fluctuations of the climate and in particular intense, long-lasting volcanic eruptions led among other things to an increase in the carbon dioxide content of the atmosphere and thereby to global warming and oxygen deficiency in the air and in the water. A very rapid rise of the sea level at the beginning of the Triassic ended this transition period.

The Evolution of Humans. Finally, in the most recent past, geologically speaking, the evolution of humans occurred; as with other organisms, in a series of steps and under the influence of natural selection. Beginning with common ancestors, the evolution of human beings (hominids) (Table 14.1) branched off from that of the hominoid apes (of the genus Pan: the chimpanzees), probably in the late Tertiary, 5 to $6 \cdot 10^6$ yr ago.

Following the discovery of Australopithecus afarensis ("Lucy", about $3.6 \cdot 10^6$ yr old) by D. C. Johanson und T. Gray in 1974 in the Afar Region of Ethiopia, the fossilized remains of still older individuals were found: the ca. $4.4 \cdot 10^6$ yr old Ardipithecus ramidus, by T. D. White and B. Asfaw (1992/93), less than 100 km from the site where A. afarensis was found; and *Aus-*

Table 14.1. The appearance of the hominids. The times are given in units of 10^6 yr ago

Ardipithecus ramidus	4.4
Australopithecus anamensis	4.2–3.9
afarensis	3.9–3.0
Homo habilis	2.5–1.6
erectus	1.8–0.1
sapiens	≤ 0.8

tralopithecus anamensis, found by M. G. Leakey et al. in 1995 on Lake Turkana in northern Kenia, with an age of about $4.1 \cdot 10^6$ yr.

The Ardipithecus ("ground-ape") and the Australopithecines ("tree-apes"), of which in the meantime seven species are known, can be regarded as the oldest verified traces of the hominids. Ancient man of the species Australopithecus already had an upright gait and could make and use tools, but had a brain volume only about as great as that of the hominoid apes, and had no speech as we understand it.

The further evolution of the hominids was characterized by an enormous development of the brain, in particular of the forebrain, leading from *early* humans, Homo habilis and Homo erectus, finally to the *modern* humans, Homo sapiens.

The Principle of Evolution. The whole evolution of life, if we may be allowed to reduce it to its simplest elements, is evidently based on the principle of producing more and more complex systems by the storage and passing-on of more and more *information*. These systems are then increasingly successful in applying their ordering principle (production of negentropy) as opposed to the "natural" tendency of the 2nd Law of Thermodynamics towards the establishment of a statistical-thermodynamical equilibrium, i.e. to maximum disorder (entropy production). This evolution leads even in the prokaryotes away from the fixation of genetic information towards the development of instinctive, i.e. program-controlled behavior schemes, then further to memory and the beginnings of intelligent behavior. The development of humans, just a few million years ago, was driven essentially by their enormously improved techniques for storing and using information, in comparison to simpler organisms, at first through language, then through writing (of all kinds), and most recently through the invention of the electronic computer and information storage devices.

This ever-increasing "battle against the law of entropy" is *necessarily* coupled to an increasing *energy consumption*. A species is, as one says, in a flow-equilibrium. That is, it continually takes in energy from its food supplies, in the case of plants from sunlight; this is in part necessary for the (re)production of the system and in part leads to waste heat (as in any thermal energy source), which must be discarded. Humankind can only continue its development along the lines described above if we succeed in making more and more energy "usable" from natural sources. Along with the muscle power of humans and animals, the forces of wind and water were used. The next great step was the application of the fossil fuels, coal and petroleum, and finally also the application of nuclear power.

The Future of the Sun. Concerning the question of the future development of life, we can make no detailed statements. But to the question of the future of the Sun, and thus of life on the Earth, the theory of stellar evolution (Sect. 8.2) permits us to give a quite specific answer. The Sun is moving (over a period of a few billion years) towards the upper right on the Hertzsprung–Russell diagram; it will become a red giant star (Fig. 8.8). Its radius will grow enormously in this process, and its luminosity will correspondingly increase (Table 14.2). After about $1.1 \cdot 10^9$ yr, the greenhouse effect will be so strong that life in its present forms on Earth would be greatly disturbed. After $3.5 \cdot 10^9$ yr, the solar luminosity will have increased by about 40%, and the temperature on the Earth will rise to well above the boiling point of water, so that the oceans will evaporate. This will doubtless mean the *end* of all organic life on the Earth. Finally, after about $7.8 \cdot 10^9$ yr, the radius of the Sun will be about equal to that of the Earth's present-day orbit; by this time, however, the strong solar wind from the red giant will have reduced its mass and thus the Sun's gravitational attraction by so much that the Earth will be moving on a larger orbit, at a radius of around 1.7 AU.

Life on Other Celestial Bodies? This exciting question is at present only reasonable if we understand life to mean the occurrence of organisms whose structures have a certain similarity to those of terrestrial lifeforms. We need not belabor the point that the environmental conditions permitting such life are fixed within rather narrow limits.

In our Solar System, only Mars can be considered as a possibility for harboring life. Its thin atmosphere in fact contains small amounts of the required gases, but it offers no protection from the UV radiation of the Sun; its temperature lies somewhat lower than that of the Earth. It therefore appears possible that at most extremely primitive organisms might be found there, but even this is not very probable according to the present state of our knowledge.

In our Milky Way and in other galaxies, there are numberless stars which are very similar to our Sun. There is nothing to contradict the assumption that some of these stars also have planetary systems, and it seems quite plausible that here and there in such a system, a planet offers similar conditions on its surface to those found on the Earth. Why should not life have evolved there also?

In our near neighborhood within the Milky Way, in recent years several planets have been detected (Sect. 14.1.5); of these, however, so far none is similar to the Earth in size or physical conditions.

A first survey of the radiofrequency emissions from nearby star systems to search for signals from "extraterrestrial civilizations" (Project SETI: Search for Extraterrestrial Intelligence) was begun in 1960 by F. D. Drake; other projects with the same goal have followed.

Up to now, this *bioastronomy* has been equally unsuccessful in finding evidence for "earthlike" planets around other stars and in detecting "artificial" radio signals from outer space.

* * *

Table 14.2. The future evolution of the Sun according to I.-J. Sackmann et al. (1993). (a) Today; (d) End of the Main Sequence stage; (e) Tip of the first giant branch; (f) the first helium-shell flash

	Time from today $[10^9$ yr$]$	Luminosity $L/L_\odot$	Radius $R/R_\odot$	Mass $\mathcal{M}/\mathcal{M}_\odot$
a	0.0	1.0	1.0	1.0
b	1.1	1.1	1.0	1.0
c	3.5	1.4	1.2	1.0
d	6.4	2.2	1.6	1.0
e	7.7	2350	166	0.7
f	7.8	3000	$\simeq 200$	0.6

Outlook. From the study of cosmic structures and evolutionary processes, we have returned to the problems of the Earth, of life and of our existence. The circle of our considerations is thus completed. Equally fascinating as the insight into cosmic events is the process of the acquisition of human knowledge itself. The boundary of our knowledge is constantly moving out into regions which were previously unknown or at least not understood. This process seems to be connected with a kind of adjustment of the human intellect and its thought processes to the areas of existence which have only become accessible through just this intellect. We owe in fact to just a few great figures of history the finding of new forms of thinking and of dealing with problems which themselves only became clear through their genius. All great discoveries and deeds contain a certain illogical element.

Our categories of knowledge are arranged, so to speak, in layers, which evidently are the result of the basic areas of interaction of humanity with its environment. Even the most primitive human uses different concepts and ways of thinking when he deals with non-living natural objects, with his animals and other living creatures (which are beneath him in the evolutionary chain), with his own emotional life and that of his fellow humans, or finally with overreaching intellectual or especially religious subjects. These different areas of experience, which at first were almost heedlessly allowed to have their parallel but separate existences, have come into closer and closer contact in the course of time, and mankind has long since suffered inner and outer struggles in an effort to unite them into a whole, out of the unprovable conviction that this must indeed be possible.

No one can finally refute the view held by many primitive religions that on a certain level, the Universe is simply a result of the arbitrary decisions of some sort of gods and demons. In fact, however, the "other" hypothesis has proved repeatedly to be more fruitful.

The problems of human life and living with each other are intertwined with those of the understanding of natural phenomena. We see them on the one hand from our *subjective* point of view, and speak of desire and duty, of love and hate, of justice and injustice. On the other hand, they are a part of the *objective framework* of the Possible, governed by the known (or still unknown) laws of Nature. The great conflicts of guilt and fate, of human goals and our ability to attain them, or however they may be termed, arise out of this remarkable duality of the subjective and the objective worlds.

It has been the conviction of truly religious men and women in all periods of history that these tensions could gradually, bit by bit, be resolved, presuming that humanity could lay aside its old prejudices and "renew itself", a process which is perhaps not so different from the renewal and reformation of natural science in its great epochs.

Whether humankind turns its gaze *outwards* and probes ever deeper into the further reaches of Nature, or looks *inwards* and finds there new realms for humanity, over and over again we view with amazement and joy a

"New Cosmos".

Appendix

A.1 Units: the International and the Gaussian Unit Systems

We use the *International System of Units* (Système *I*nternational d'Unités) throughout this book; at the same time, the more important quantities are also given in the Gaussian system. It is expedient to introduce "astronomical units" as well (see inside back cover).

SI Base Units

Meter	m	(length, distance)
Kilogram	kg	(mass)
Second	s	(time)
Ampère	A	(electric current)
Kelvin	K	(thermodynamic temperature)
Mole	mol	(amount of substance)
Candela[1]	cd	(luminous intensity)

Prefixes for Orders of Magnitude

10^1	Deca	D	10^{-1}	deci	d
10^2	hecto	h	10^{-2}	centi	c
10^3	kilo	k	10^{-3}	milli	m
10^6	Mega	M	10^{-6}	micro	μ
10^9	Giga	G	10^{-9}	nano	n
10^{12}	Tera	T	10^{-12}	pico	p
10^{15}	Peta	P	10^{-15}	femto	f
10^{18}	Exa	E	10^{-18}	atto	a
10^{21}	Zetta	Z	10^{-21}	zepto	z
10^{24}	Yotta	Y	10^{-24}	yocto	y

---

[1] Not used in this book.

Some Relations Between SI and Gaussian Electromagnetic Units

The Gaussian system of "mixed" cgs units uses both electrostatic units (esu) and electromagnetic units (emu) in their respective areas. In this system, the permittivity and the permeability of vacuum are dimensionless and have the value 1.

X: Quantity in the international unit system (SI)

$\tilde{X}$: Quantity in the Gaussian system.

Here, we consider only the *vacuum*, so that the magnetic flux density $\boldsymbol{B}$ and the magnetic field strength $\boldsymbol{H}$ are related by

$$\boldsymbol{B} = \mu_0 \boldsymbol{H} \quad \text{or} \quad \tilde{\boldsymbol{B}} = \tilde{\boldsymbol{H}}.$$

Electric charge: $\quad \tilde{e} = \dfrac{e}{\sqrt{4\pi\varepsilon_0}}$

Electric field strength: $\quad \tilde{\boldsymbol{E}} = \sqrt{4\pi\varepsilon_0}\, \boldsymbol{E}$

Magnetic flux density: $\quad \tilde{\boldsymbol{B}} = \sqrt{\dfrac{4\pi}{\mu_0}}\, \boldsymbol{B} = \sqrt{4\pi\varepsilon_0}\, c \boldsymbol{B}$

ε_0 = electric, μ_0 = magnetic field constant (see table inside back cover), with $\varepsilon_0 \mu_0 = 1/c^2$, c = speed of light.

Force on a moving charge e: $\boldsymbol{F} = e\boldsymbol{E} + e\boldsymbol{v} \times \boldsymbol{B} = \tilde{e}\tilde{\boldsymbol{E}} + \dfrac{\tilde{e}}{c}\boldsymbol{v} \times \tilde{\boldsymbol{B}}$

Energy density in vacuum: $w = \dfrac{1}{2}\left(\varepsilon_0 E^2 + \dfrac{B^2}{\mu_0}\right) = \dfrac{1}{8\pi}(\tilde{E}^2 + \tilde{B}^2)$

Poynting vector: $\boldsymbol{S} = \boldsymbol{E} \times \boldsymbol{H} = \dfrac{1}{\mu_0}\boldsymbol{E} \times \boldsymbol{B} = \dfrac{c}{4\pi}\tilde{\boldsymbol{E}} \times \tilde{\boldsymbol{B}}$

Cyclotron frequency: $\omega_C = \dfrac{eB}{m} = \dfrac{\tilde{e}\tilde{B}}{mc}$

Bohr radius: $a_0 = 4\pi\varepsilon_0 \dfrac{\hbar^2}{m_e e^2} = \dfrac{\hbar^2}{m_e \tilde{e}^2}$

Classical electron radius: $r_e = \dfrac{1}{4\pi\varepsilon_0} \dfrac{e^2}{m_e c^2} = \dfrac{\tilde{e}^2}{m_e c^2}$

Various Units

Length:

Ångström $\quad\quad\quad\quad\quad\quad$ 1 Å $= 10^{-10}$ m $= 10^{-8}$ cm

Mass:

Ton $\quad\quad\quad\quad\quad\quad\quad\quad$ 1 t $= 10^3$ kg

Atomic mass unit m_u $\quad\quad$ 1 u $= 1.6605 \cdot 10^{-27}$ kg

A.1 Units: the International and the Gaussian Unit Systems

Time:
Minute	1 min	$= 60$ s
Hour	1 h	$= 60$ min $= 3600$ s
Day	1 d	$= 24$ h $\quad = 86\,400$ s
Year	1 yr	$\simeq 3.156 \cdot 10^7$ s
		independently of the exact definition

Frequency:
Hertz 1 Hz $= 1\,\text{s}^{-1}$

Angle:
Radian 1 rad $= 1\,\text{m}\,\text{m}^{-1}$ (dimensionless)
$= 57.2958° = 3437.74' = 206\,264.81''$
Degree $1°$ $= \pi/180\,\text{rad} = 1.7453 \cdot 10^{-2}$ rad
Minute of arc $1'$ $= (1/60)° = 2.9089 \cdot 10^{-4}$ rad
Second of arc $1''$ $= (1/60)' = 4.8481 \cdot 10^{-6}$ rad

Solid angle:
Steradian 1 sr $= 1\,\text{m}^2\,\text{m}^{-2}$ (dimensionless)
$= (180/\pi)^2$ square degrees
$= 3282.8$ square degrees

Force:
Newton 1 N $= 1\,\text{m}\,\text{kg}\,\text{s}^{-2} = 10^5\,\text{dyn} = 10^5\,\text{cm}\,\text{g}\,\text{s}^{-2}$

Pressure:
Pascal 1 Pa $= 1\,\text{m}^{-1}\,\text{kg}\,\text{s}^{-2} = 1\,\text{N}\,\text{m}^{-2}$
$= 10\,\text{dyn}\,\text{cm}^{-2}$
Bar (atm.) 1 bar $= 10^5$ Pa

Energy:
Joule 1 J $= 1\,\text{m}^2\,\text{kg}\,\text{s}^{-2} = 1\,\text{N}\,\text{m} = 1\,\text{W}\,\text{s}$
$= 10^7\,\text{erg} = 10^7\,\text{cm}^2\,\text{g}\,\text{s}^{-2}$
Electron volt 1 eV $= 1.6022 \cdot 10^{-19}\,\text{J} = 1.6022 \cdot 10^{-12}$ erg
Energy equivalents:
  kT for $T = 1$ K $1\,\text{K} \cdot k = 1.3807 \cdot 10^{-23}\,\text{J} = 8.6174 \cdot 10^{-5}$ eV
  kilogram $1\,\text{kg} \cdot c^2 = 8.9876 \cdot 10^{16}$ J
  atomic mass unit $m_u c^2$ $= 1.4924 \cdot 10^{-10}\,\text{J} = 931.49$ MeV
  proton mass $m_p c^2$ $= 1.5033 \cdot 10^{-10}\,\text{J} = 938.27$ MeV
  electron mass $m_e c^2$ $= 8.1872 \cdot 10^{-14}\,\text{J} = 0.5110$ MeV

Power:
Watt 1 W $= 1\,\text{m}^2\,\text{kg}\,\text{s}^{-3} = 1\,\text{J}\,\text{s}^{-1} = 1\,\text{V}\,\text{A}$
$= 10^7\,\text{erg}\,\text{s}^{-1}$

Appendix

Temperature:

Celsius temperature $\quad t[°C] \quad$ = absolute temperature $T[K] - 273.15$ K

Temperature equivalent $\quad 1\,\text{eV}k^{-1} = 11\,605$ K
of the electron volt

Electric charge:

Coulomb $\quad\quad\quad\quad\quad\quad$ 1 C $\quad$ = 1 A s = $2.9979 \cdot 10^9$ esu

Electric potential:

Volt $\quad\quad\quad\quad\quad\quad\quad\quad$ 1 V $\quad$ = 1 m² kg s⁻³ A⁻¹ = $3.3356 \cdot 10^{-3}$ esu

$\quad\quad\quad\quad\quad\quad\quad\quad\quad\quad$ i.e. $\quad$ = 1 m² kg s⁻³ A⁻¹ = $3.3356 \cdot 10^{-3}$ esu

Magnetic flux density (in vacuum):

Tesla $\quad\quad\quad\quad\quad\quad\quad$ 1 T $\quad$ = 1 kg s⁻² A⁻¹ = 1 V s m⁻²
$\quad\quad\quad\quad\quad\quad\quad\quad\quad\quad\quad\quad$ = 10^4 G (Gauss)

A.2 Names of the Constellations

Standard abbreviations, Latin names (with their genitive) and English names of the constellations

And	Andromeda	Andromedae	Andromeda
Ant	Antlia	Antliae	air pump
Aps	Apus	Apodis	bird of paradise
Aql	Aquila	Aquilae	eagle
Aqr	Aquarius	Aquarii	water bearer
Ara	Ara	Arae	altar
Ari	Aries	Arietis	ram
Aur	Auriga	Aurigae	cart driver
Boo	Bootes	Bootis	Bootes
Cae	Caelum	Caeli	grave pick
Cam	Camelopardalis	Camelopardalis	giraffe
Cap	Capricornus	Capricorni	mountain goat
Car	Carina	Carinae	keel
Cas	Cassiopeia	Cassiopeiae	Cassiopeia
Cen	Centaurus	Centauri	centaur
Cep	Cepheus	Cephei	Cepheus
Cet	Cetus	Ceti	whale
Cha	Chamaeleon	Chamaeleontis	chameleon
Cir	Circinus	Circini	drawing compass
CMa	Canis Maior	Canis Maioris	large dog
CMi	Canis Minor	Canis Minoris	small dog
Cnc	Cancer	Cancri	crab
Col	Columba	Columbae	dove
Com	Coma Berenices	Comae Berenices	Berenice's hair
CrA	Corona Austrina	Coronae Austrinae	southern crown
CrB	Corona Borealis	Coronae Borealis	(northern) crown
Crt	Crater	Crateris	beaker
Cru	Crux	Crucis	southern cross
Crv	Corvus	Corvi	crow
CVn	Canes Venatici	Canum Venaticorum	hunting dogs
Cyg	Cygnus	Cygni	swan
Del	Delphinus	Delphini	dolphin
Dor	Dorado	Doradus	swordfish
Dra	Draco	Draconis	dragon
Equ	Equuleus	Equulei	foal
Eri	Eridanus	Eridani	Eridanus
For	Fornax	Fornacis	furnace
Gem	Gemini	Geminorum	twins
Gru	Grus	Gruis	crane
Her	Hercules	Herculis	Hercules
Hor	Horologium	Horologii	clock
Hya	Hydra	Hydrae	hydra
Hyi	Hydrus	Hydri	southern hydra
Ind	Indus	Indi	Indian
Lac	Lacerta	Lacertae	lizard
Leo	Leo	Leonis	lion
Lep	Lepus	Leporis	hare
Lib	Libra	Librae	scales
LMi	Leo Minor	Leonis Minoris	small lion

Lup	Lupus	Lupi	wolf
Lyn	Lynx	Lyncis	lynx
Lyr	Lyra	Lyrae	lyre
Men	Mensa	Mensae	mesa (table)
Mic	Microscopium	Microscopii	microscope
Mon	Monoceros	Monocerotis	unicorn
Mus	Musca	Muscae	fly
Nor	Norma	Normae	square
Oct	Octans	Octantis	octant
Oph	Ophiuchus	Ophiuchi	snake bearer
Ori	Orion	Orionis	Orion
Pav	Pavo	Pavonis	peacock
Peg	Pegasus	Pegasi	Pegasus
Per	Perseus	Persei	Perseus
Phe	Phoenix	Phoenicis	phoenix
Pic	Pictor	Pictoris	artist
PsA	Piscis Austrinus	Piscis Austrini	southern fish
Psc	Pisces	Piscium	fishes
Pup	Puppis	Puppis	poop deck
Pyx	Pyxis	Pyxidis	compass
Ret	Reticulum	Reticuli	net
Scl	Sculptor	Sculptoris	sculptor
Sco	Scorpius	Scorpii	scorpion
Sct	Scutum	Scuti	shield
Ser	Serpens	Serpentis	serpent
Sex	Sextans	Sextantis	sextant
Sge	Sagitta	Sagittae	arrow
Sgr	Sagittarius	Sagittarii	archer
Tau	Taurus	Tauri	bull
Tel	Telescopium	Telescopii	telescope
TrA	Triangulum Australe	Trianguli Australis	southern triangle
Tri	Triangulum	Trianguli	triangle
Tuc	Tucana	Tucanae	toucan
UMa	Ursa Maior	Ursae Maioris	big bear (big dipper)
UMi	Ursa Minor	Ursae Minoris	small bear (little dipper)
Vel	Vela	Velorum	sails
Vir	Virgo	Virginis	virign
Vol	Volans	Volantis	flying fish
Vul	Vulpecula	Vulpeculae	little fox

Notation for the use of names of the constellations: non-variable *stars* are denoted by Greek (in some cases also by Roman) letters or numbers together with the genitive of the Latin name of their constellation, usually in the form of the standard abbreviation using three letters, e.g. β UMa = Beta Ursae Maioris or 48 UMa = 48 Ursae Maioris, ι Her = Iota Herculis, 1 Car = 1 Carinae, a Cen = a Centauri.

Variable stars are denoted by capital letters R, S, ... , Z; RR, RS, ... ZZ; AA, ... , AZ; BB, ... , QZ and the genitive of the constellation (334 possibilities; J is not used); additional variables in a constellation are denoted by V 335 etc., e.g. RR Lyr, W Vir, SS Cyg, V 1057 Cyg.

For *strong radio* and *X-ray sources*, the Latin name of the constellation is used in the nominative case, together with capital letters and numbers, e.g. Tau A = Taurus A, Her X-1 = Hercules X-1, Sco X-3 = Scorpius X-3.

Selected Exercises

1. Coordinate Systems (Sects. 2.1 and 11.1)

1.1 The relationship between the horizontal axis system (azimuth A and altitude h or zenith distance $z = 90° - h$)) and the equatorial system (declination δ and hour angle t or right ascension α = sidereal time − t) is contained in the *polar triangle* (or nautical triangle): pole P – zenith Z – celestial object O, with the sides z, $90° - \varphi$ and $90° - \delta$ (φ: altitude of the pole).

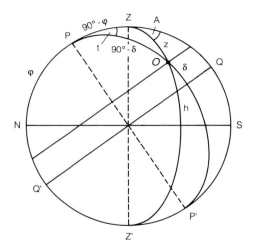

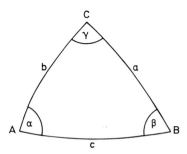

Give formulas for the calculation of z given φ, δ, and t; and of δ, given φ, z, and A.

What are the corresponding equations for going from the equatorial system (α, δ) to (a) ecliptic coordinates and to (b) galactic coordinates?

1.2 Where and when does the Sun rise in Heidelberg (geographic latitude $\varphi = 49.41°$) on the longest day of the year? What are the azimuth and the sidereal time of rising and setting of Arcturus (α Boo, $\alpha = 14$ h 15 min 39.6 s, $\delta = +19°10'57''$) in New York ($\varphi = 40.7°$)? How long does the star remain above the horizon?

Solve the exercise (a) without considering refraction in the Earth's atmosphere, and (b) with a refraction of $34'\,50''$ at the horizon (Sect. 2.1.1). We define the instant of sunrise by the time at which its upper rim becomes visible.

Hint: from *spherical trigonometry*, we know that in a triangle ABC with sides a, b, and c (expressed as angles) and the opposite angles α, β, and γ, the following relations hold:

$$\sin a : \sin b : \sin c = \sin \alpha : \sin \beta : \sin \gamma$$
$$\cos a = \cos b + \cos c + \sin b \sin c \cdot \cos \alpha$$
$$\cos \alpha = -\cos \beta \cos \gamma + \sin \beta \sin \gamma \cdot \cos a .$$

2. Resolving Power and Brightness (Sects. 3.4 and 5.1)

The apparent visual magnitude of Jupiter's moon J1 Io is (at a middle opposition of the planet) $V = +4.8$ mag. With the unaided eye, under favorable conditions stars down to about 6 mag can still be seen. Why was Io discovered only after the invention of the telescope by G. Galilei? How much brighter than Io is Jupiter?

Hint: Calculate the maximum angular distance of Io from Jupiter and the apparent radius of the planet (Table 3.4), and compare these values with the resolving poser of the human eye (5.2). The difference in brightness between Jupiter and Io can be estimated to an order of magnitude by the ratio of the disk size of the planet and its moon.

3. Radioactive Heat Production (Sect. 2.2, Chap. 3, and Sect. 7.2)

The production of heat in the Earth's crust by radioactivity is primarily due to the decay of ^{238}U, ^{232}Th, and ^{40}K. The specific energy release q for natural uranium (99.3% ^{238}U) is $9.4 \cdot 10^{-5}$, for thorium ^{232}Th, it is $2.6 \cdot 10^{-5}$, and for natural potassium (0.02% ^{40}K), $3.6 \cdot 10^{-9}$ J kg^{-1} s^{-1}.

3.1 Calculate q from the cosmic element abundances (Table 7.5) for (a) cosmic matter containing all the elements, and (b) chondritic material, for which it can be assumed for simplicity that it consists of practically only olivine ((Mg, Fe)$_2$SiO$_4$) and traces of K, Th, and U in the cosmic element ratios. Compare the result of (b) with the known value for chondrites, $q = 5.2 \cdot 10^{-12}$ J kg^{-1} s^{-1}. Which radioactive decay chain dominates the production of heat?

3.2 Calculate from q the heat flux Q at the surface of a homogeneous sphere of radius R and density ϱ. How large would Q be for the Earth if it was composed completely of chondritic material? How large would Q be for a layer of granite ($q = 1.1 \cdot 10^{-9}$ J kg^{-1} s^{-1}) 10 km thick covering the entire surface of the Earth?

3.3 The energy balance of an object in the Solar System at a distance r from the Sun is given by (3.2). What is the ratio of the radioactive heat flux Q to the radiative input from the Sun, $S(r)$ for an albedo of $A = 0.5$ as a function of r and R?

How large would the radius R have to be in order for Q to predominate over the radiative input $S(r)$? Are there objects in the Solar System for which $Q \geq S(r)$ holds? Discuss numerical examples for objects in the asteroid belt, near the orbit of Saturn, and at the limit of the Oortian Cloud of comets (Sect. 2.2).

3.4 Using the half–lives given in Sect. 3.2.2, calculate the abundances of ^{238}U, ^{232}Th, and ^{40}K at the time of formation of the Solar System $4.5 \cdot 10^9$ yr ago. What was the relative contribution of these radioactive nuclides at that time to the heat production in stone?

4. The Atmospheres of Planets and Moons (Sect. 2.4 and Chap. 3)

A celestial object can hold its own atmosphere when the thermal velocity $\bar{v}$ of the molecules (3.23) is very small compared to their escape velocity $v_e = \sqrt{2GM/R}$ (2.61). Show that, assuming the acceleration of gravity $g = GM/R^2$ to be independent of the altitude in the atmosphere, the following relation holds:

$$\left(\frac{\bar{v}}{v_e}\right)^2 = \frac{H}{R},$$

where H is the scale height or equivalent height (3.17) of the atmosphere. Make up a list of the planets and their moons ranked according to v_e and according to $v_e/\bar{v}$, using the data given in Chap. 3, and compare it with the observed existence of atmospheres. What is the resulting minimum value of the ratio $v_e/\bar{v}$ which allows the object to retain its own atmosphere?

5. "Cosmic Collisions" (Sects. 3.7 and 14.2)

5.1 Employ the present flux of particles at 1 AU distance from the Sun from Fig. 3.32 to estimate how large is the probability of collision of the Earth with a meteorite or asteroid of radius 1, 10, and 100 km.

5.2 An asteroid with a radius of 10 km and a density of 3000 kg m^{-3} is assumed to collide with the Earth at a relative velocity of 50 km s^{-1}. Assuming that its mass remains "together" during the collision, estimate:

a) How large is the momentum of the asteroid in comparison to the orbital momentum of the Earth?

b) How large is the maximum angular momentum which could be transferred to the Earth in comparison to the its rotational angular momentum? How much of a change in the rotational period would be produced by such a collision? How long would it take for tidal forces (Sect. 2.4.6) to cause the same change?

c) Assume that all the kinetic energy of the asteroid could be converted into thermal energy of the molecules of the Earth's atmosphere. How high would the temperature of the atmosphere rise? Compare this kinetic energy with the total thermal energy of the atmosphere, with the total dissociation energy of its molecules (N$_2$: 9.8 eV, O$_2$: 5.1 eV), with the rotational energy of the Earth, and with the energy radiated to the Earth from the Sun in one year.

5.3 Assume that the mass of the asteroid were distributed evenly over the entire surface of the Earth. How thick would the resulting layer be?

Assume that the mass of the asteroid were distributed homogeneously throughout the Earth's atmosphere up to an altitude of 10 km in the form of dust particles of 1 μm radius. Estimate the resulting extinction coefficient (10.10, 11) using $Q_{\text{ext}} \simeq Q_{\text{sca}} \simeq 1$, as well as the change in the Earth's albedo and its atmospheric temperature. Compare the mass of the asteroid with that of the Earth's atmosphere.

6. Asteroids (Sect. 2.4 and Chap. 3)

6.1 A small asteroid at a distance 3 AU from the Sun is supposed to have a radius of 10 km and an albedo of $A = 0.03$ (similar to that of the carbonaceous chondrites). How high is its effective temperature for very rapid and for very slow rotation? Where is the maximum in its emitted radiation, assuming it to be black body radiation?

Compare this effective temperature with the values given in Table 3.1 for the planets.

6.2 Is a *binary system* of two asteroids bound by gravitation possible? Assume both asteroids to have the same mass $\mathcal{M}$ and calculate numerically using Kepler's 3rd law the relationship between their distance and their orbital period (a) for $\mathcal{M} = 10^{21}$ kg (corresponding to the largest asteroids); (b) for $\mathcal{M}$ of an asteroid with a radius of 10 km and a density of 3000 kg m^{-3}; and (c) for $\mathcal{M} = 1$ kg(!).

How large are the forces on such "binary asteroids" at 3 AU distance from the Sun due to other bodies in the Solar System?

6.3 In the story "The Little Prince" by A. de Saint-Exupéry (1943), the prince visits a tiny planet which has just enough room for a street lamp and a lamplighter. The planet has been rotating faster and faster from year to year, so that the lamp now has to be lit and extinguished once every minute.

Assume the planetoid to be a sphere of radius 1.25 m with a density of 3000 kg m^{-3}. How large are the gravitational and the centrifugal accelerations at its equator? How long would it take for a stone to fall from an altitude of 1 m onto its surface?

Is the centrifugal force sufficiently strong to overcome the frictional force between the standing lamplighter (mass 70 kg) and the surface of the planetoid?

What is the situation concerning the stability of the planetoid at a rotational period of 1 min, if the tensile strength of its material is 10^8 Pa?

7. Distances and Spatial Velocities of the Stars (Sects. 2.2, 6.2, 6.3 and 11.1)

7.1 How large is the angular distance on the sphere between the two brightest stars in Gemini, Castor (α Gem) and Pollux (β Gem)? How large is their true spatial distance in [pc]? The coordinates (α, δ) for the Epoch 2000 and the parallaxes (p) of the two stars are

α Gem : $\alpha = 7$ h 34 min 35.9 s ,
$\delta = +31° 53' 18''$, $p = 0.067''$;

β Gem : $\alpha = 7$ h 45 min 18.9 s ,
$\delta = +28° \ 1' 34''$, $p = 0.094''$.

7.2 In a star catalogue, the coordinates of Sirius (α CMa) for the Epoch 1900 are given as $\alpha = 6$ h 40 min 44.6 s and $\delta = -16° 34' 44''$. Calculate its position in the year 2000, taking into consideration the precession from Table 2.1, and compare the result with the entry in the Bright Star Catalogue. How much of a contribution does the proper motion of Sirius make (it is $-0.545''$ yr^{-1} in right ascension and $-1.211''$ yr^{-1} in declination)? The radial velocity of Sirius is -8 km s^{-1} and its parallax is 0.378''. By how much does its distance from the Sun change in 100 yr due to its radial velocity? Calculate the galactic coordinates (l, b) of Sirius and verify the result using Fig. 11.1. How large are the components of its spatial velocity in the Milky Way?

8. Sirius A and B – the Fundamental Properties of the Stars (Sects. 6.3, 6.5 and 8.3)

The visual binary system α CMa = Sirius has a parallax of $p = 0.378''$. Relative to the brighter Sirius A ($m_V = -1.46$ mag), its fainter companion, the white dwarf Sirius B ($m_V = +8.44$ mag) moves on an elliptical orbit with a semimajor axis of $a = 7.5''$ and a period of $P = 50.1$ yr. The ratio of the semimajor axes of the orbits of the two components around their common center of gravity is $a_B/a_A = 2.1$.

8.1 How large are the masses $\mathcal{M}$ of the two stars? What are the greatest and the least distances of approach of the two components in [AU]? How large are the average orbital velocities in the center of mass system?

8.2 Analysis of the spectra of the two stars yields the following effective temperatures and gravitational accelerations at their surfaces:

$$T_{\text{eff,A}} = 9980 \text{ K}, \quad T_{\text{eff,B}} = 22\,500 \text{ K},$$
$$g_A = 204 \text{ m s}^{-2}, \quad g_B = 7.1 \cdot 10^6 \text{ m s}^{-2}$$

and thus, as bolometric corrections (6.37), B.C.(A) $\simeq$ 0.2 mag and B.C.(B) $\simeq$ 2.4 mag. Calculate the luminosities L, the radii R, and the average densities $\bar{\varrho}$ of the two components. To what extent is the mass–luminosity relation (6.51) fulfilled by the two components?

8.3 Interferometry measurements yield an apparent angular diameter for Sirius A of 0.0059″. Compare this value with the spectroscopically-determined diameter. For white dwarfs, the mass–radius relation (8.73) is given by $R/R_\odot = 0.0128(\mathcal{M}/\mathcal{M}_\odot)^{-1/3}$. What radius does this relation yield for Sirius B?

9. Radiative Power and Maximum Luminosity of the Stars (Sects. 2.3 and 4.3)

A star reaches its maximum luminosity, the *Eddington luminosity* L_E, when the outwardly-directed radiation acceleration g_{rad} at its surface is equal to its gravitational acceleration $g = G\mathcal{M}/R^2$ (6.53); for $g_{\text{rad}} > g$, there are thus no stable stars (A.S. Eddington, 1921).

9.1 The radiation acceleration is equal to the momentum which is transferred to a unit mass per second through the absorption of radiation; cf. (2.14), where for radiation the momentum is equal to energy/c and c is the velocity of light. Show that the following relation holds:

$$g_{\text{rad}} = \frac{1}{c} \int_0^\infty k_{\nu,M} F_\nu \, d\nu,$$

where F_ν denotes the monochromatic radiation flux (4.35, 36) and $k_{\nu,M}$ is the mass extinction coefficient (4.109). In a completely ionized hydrogen plasma, the radiation force is transferred to the free electrons mainly by Thomson scattering with the frequency-independent cross section σ_T (4.132). Show that in this case, the Eddington luminosity is given by

$$L_E = \frac{4\pi c G \mathcal{M}}{\sigma_{T,M}}$$

(where $\sigma_{T,M} = \sigma_T \cdot n_e/\varrho$). How large is L_E in units of the solar luminosity for $\mathcal{M} = 1\,\mathcal{M}_\odot$?

9.2 In the brightest supergiants, absolute bolometric magnitudes of $M_{\text{bol}} \simeq -10$ mag are observed. Calculate their maximum masses under the assumption that M_{bol} corresponds to the Eddington luminosity, and discuss them in connection with the empirical mass–luminosity relation (6.51).

10. Excitation, Ionization and Continuous Absorption in the Sun's Atmosphere (Sects. 4.3, 7.1, and 7.2)

In the photosphere of the Sun, in the layer $\bar{\tau} = 0.1$ the temperature is $T = 5070$ K, the electron pressure is $P_e = 0.38$ Pa, and the gas pressure is $P_g = 4360$ Pa (Table 7.4). What is the electron density n_e and the total particle density n_{tot} in this layer?

10.1 Calculate, using the Boltzmann formula (4.88), the occupation numbers of the first two excited energy states (principal quantum numbers $n = 2$ and 3) of neutral hydrogen relative to the ground state ($n = 1$). The energies are given by

$$\chi_n \text{ [eV]} = 13.60 \left(1 - \frac{1}{n^2} \right)$$

(7.2), and the associated statistical weights are $g_n = 2n^2$. How great is the degree of ionization $n(\text{H II})/n(\text{H I})$ of hydrogen and how large is the relative fraction of negative hydrogen ions, $n(\text{H}^-)/n(\text{H I})$? Use the Saha formula (4.92, 7.12) and take only the states $n = 1$ to 3 into account in the partition function Q_0 of H I. For H II, Q is equal to 1. The H$^-$ ion has an "ionization energy" (electron affinity) of 0.75 eV and only one bound state (ground state), $Q(\text{H}^-) \equiv g_0(\text{H}^-) = 1$.

10.2 Calculate the contributions of H I and H$^-$ to the continuous absorption κ_λ of the solar photosphere in the layer $\bar{\tau} = 0.1$ at the wavelength $\lambda = 364$ nm, i.e. on the short wavelength side of the Balmer series limit (compare Fig. 7.7). The atomic absorption or photoionization cross sections of H I are given by (4.130), where $\kappa_\lambda = \alpha_{n\kappa}(\lambda) \cdot n_n = \kappa_{\lambda,\mathrm{at}} \cdot n_n$ (4.109). Which energy states n of H I need to be taken into account for calculating the photoionization cross section at the wavelength 364 nm? For the H$^-$ ion, $\alpha_{0\kappa}(364\,\mathrm{nm}) \simeq 2.4 \cdot 10^{-21}$ m^2.

How strong is Thomson scattering from free electrons (4.132) in comparison to κ_λ?

11. Radiation Transport: Thermal Layering of a Stellar Atmosphere (Sects. 4.3 and 7.2)

The radiation transport equation for an atmosphere of parallel plane layers (7.39) in the "grey limit", i.e. for an absorption coefficient which is independent of the frequency or the wavelength, has the form:

$$\cos\theta \frac{\mathrm{d}I}{\mathrm{d}\tau} = I - B,$$

where τ is the frequency-independent optical depth, $I = \int_0^\infty I_\nu \,\mathrm{d}\nu$ and $B = \int_0^\infty B_\nu \,\mathrm{d}\nu = (\sigma/\pi)T^4$.

Consider only two directions of radiation propagation, one outwards, $\cos\theta = +1$, with the intensity I^+, and one inwards, $\cos\theta = -1$, with I^-. In this greatly simplified *two-ray approximation*, the mean intensity (4.31) is given by $J = \frac{1}{2}(I^+ + I^-)$ and the radiation flux (4.34) by $F = I^+ - I^-$.

11.1 Show that on adding the two radiation-transport equations for the two directions, $\cos\theta = \pm 1$, one obtains energy conservation, i.e. that in radiation equilibrium $J = B$, the radiation current F is independent of the depth τ ($F = \sigma T_{\mathrm{eff}}^4$).

11.2 Subtracting the two equations gives a differential equation for J. Solve this equation with the boundary condition that at the surface of the star ($\tau = 0$), no radiation goes inwards (i.e. $I^-(\tau = 0) = 0$), and compare the resulting temperature layering $T(\tau)$ with the grey layering (7.42).

12. Supernovae and Pulsars (Sects. 2.3, 2.4, 4.3, 6.3, 7.4 and 8.2)

12.1 The *bolometric* magnitude $L(t)$ of the bright supernova SN 1987A of Large Magellanic Cloud has decreased exponentially with time t during the period from about 120 d to 800 d after the outburst.

For this time interval, determine $L(t)$ approximately from the figure. How large is the decrease in magnitudes per day? Find the half-life $\tau_{1/2}(\mathrm{SN})$ of the light curve and compare it with the half-life of the radioactive decay of ^{56}Co into ^{56}Fe, $\tau_{1/2}(^{56}\mathrm{Co}) = 77.28$ d.

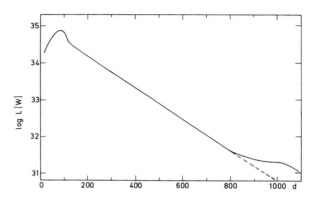

12.2 How many solar masses of ^{56}Co would be required to explain the observed luminosity $L(t)$ in the exponential part of the curve as due to the decay of ^{56}Co? In the decay of a ^{56}Co nucleus, on the average 3.6 MeV of energy is released.

12.3 After about 800 d, the bolometric light curve flattens out and, after e.g. 1030 d following the explosion, it has a value of $2 \cdot 10^{31}$ W, well above the extrapolated exponential decrease. A possible explanation for this would be an additional energy input from the pulsar or rotating neutron star (not yet directly observable) at the cost of its rotational energy $E_{\mathrm{rot}} = \frac{1}{2}J\omega^2$. Calculate the moment of inertia J of the neutron star assuming that it is a homogeneous sphere (2.67) with a radius $R = 10$ km and a mass of $\mathcal{M} = 1.4\,\mathcal{M}_\odot$. The pulsar in the Crab Nebula (Sect. 7.4.7) changes its period $P = 2\pi/\omega$ according to the relation $P/\dot{P} = 2500$ yr ($\dot{P} = \mathrm{d}P/\mathrm{d}t$); about 1/100 of the decrease in its rotational energy E_{rot} emerges as pulsar radiation. Calculate $E_{\mathrm{rot}}/100$ and compare the result with the observed luminosity of SN 1987A, 1030 d after its explosion.

12.4 The neutron star formed by a supernova explosion has a surface temperature T of over $2 \cdot 10^6$ K for up to 100 yr after the outburst.

At what wavelength is the maximum of the radiation from a black body at $T = 2 \cdot 10^6$ K? How large is the corresponding luminosity if the radius of the neutron star is $R = 10$ km?

The X-ray satellite ROSAT could just detect a "point source" at a flux of $2 \cdot 10^{-7}$ Wm^{-2} (in the energy range from 0.1 to 0.3 keV), or $6 \cdot 10^{-7}$ Wm^{-2} (for 0.5–2 keV). Could the thermal X–radiation from a neutron star formed by the explosion of SN 1987A with $R = 10$ km and $T = 2 \cdot 10^6$ K be observed by ROSAT (distance to the Large Magellanic Cloud: 50 kpc)?

13. Thermonuclear Reactions (Sects. 7.1 and 8.1)

13.1 Two positively-charged nuclei (with charges $Z_1 e$ and $Z_2 e$) separated by a distance r repel each other by a force corresponding to the *Coulomb potential*:

$$V(r) = \frac{1}{4\pi\varepsilon_0} \frac{Z_1 Z_2 e^2}{r}$$

(8.15). A nuclear reaction is possible only if the two nuclei approach each other to within $r_0 = 10^{-15}$ m (Fig. 8.1). How high is the Coulomb barrier $B = V(r_0)$? How great is its thickness $\Delta = \delta - r_0 \simeq \delta$ for a particle with the kinetic energy $E = \frac{1}{2} mv^2 = \frac{3}{2} kT$, where δ is given by $V(\delta) = E$? Express δ in units of the de Broglie wavelength $\Lambda = \hbar/(mv)$. (For simplicity, we assume that the nucleus is at rest.)

According to G. Gamow, the probability for *tunnelling* through a Coulomb barrier is $P(v) = \exp(-2\pi\eta)$ (8.16), where η is given by:

$$\eta = \frac{1}{4\pi\varepsilon_0} \frac{Z_1 Z_2 e^2}{\hbar v} .$$

How does η depend on δ? Calculate B, δ, and $P(v)$ for the reaction ^{12}C(p, γ) ^{13}N from the CNO triple cycle for $T = 10^7$ and $2 \cdot 10^7$ K.

13.2 For more precise calculations, the *screening* of the nuclear charges by the electrons in the plasma must be taken into account. In the Debye theorie, the electrostatic potential then becomes:

$$\tilde{V}(r) = V(r) \mathrm{e}^{-r/r_\mathrm{D}}$$

(8.17), with the Debye length r_D for a completely ionized hydrogen plasma of particle density n given by:

$$r_\mathrm{D} = \sqrt{\frac{\varepsilon_0 kT}{e^2 n}} .$$

How do B, δ, and $P(v)$ change when the screened potential $\tilde{V}(r)$ is used? Take $n = 10^{32}$ and 10^{35} m^{-3}.

13.3 The *rate* of a thermonuclear reaction is essentially determined by the velocity distribution $\phi(v)$ of the nuclei, as well as by $P(v)$. We can take a Maxwell–Boltzmann distribution (4.86) for $\phi(v)$. At what value of v are the two opposed exponential functions $\exp(-b/v)$ and $\exp[-(v/v_0)^2]$ equal (where b and v_0 are constants)? For what v does the maximum of the integrand $\sigma(v) v \phi(v)$ in (4.127), the socalled "Gamow peak", occur? Assume the interaction cross-section to be given by $\sigma(v) = \mathrm{const} \cdot P(v)/v^2$.

14. Energy Release by Accretion
(Sects. 2.3, 7.4, 8.2, 8.4 and 12.3)

If a mass m is brought from a great distance onto the surface of a sphere having a mass $\mathcal{M}$ and radius R, then the gravitational energy

$$\Delta E = -\frac{G\mathcal{M}m}{R}$$

will be released (8.27). The accretion of matter at a rate $\dot{\mathcal{M}} = \mathrm{d}m/\mathrm{d}t$ can, if ΔE is completely converted to radiation, therefore give rise to an *accretion luminosity*:

$$L_\mathrm{a} = \left|\frac{\mathrm{d}E}{\mathrm{d}t}\right| = \frac{G\mathcal{M}\dot{\mathcal{M}}}{R}$$

(7.109).

14.1 Calculate L_a in units of the solar luminosity for a mass flux of $\dot{\mathcal{M}} = 10^{-10}\, \mathcal{M}_\odot\, \mathrm{yr}^{-1}$, typical of close binary systems (Sects. 7.4.5 and 7.4.6), onto (a) a white dwarf, and (b) a neutron star, each having $\mathcal{M} = 1\, \mathcal{M}_\odot$.

14.2 In cataclysmic variables, the mass current to the white dwarf is not directed radially onto the dwarf, but instead into an accretion disk, whereby each mass particle moves inwards on a spiral orbit. Assume that its orbit at each moment can be approximated by a circular orbit with a Kepler velocity according to (2.60). Show using (2.59) that when the "innermost" orbit, close to the surface of the white dwarf is reached, only the luminosity $\frac{1}{2}L_a$ is obtained. How can the full accretion luminosity L_a be produced?

14.3 How large is L_a when a mass $\mathcal{M}$ of matter is accreted up to its Schwarzschild radius R_S (8.82)? (Here, make the greatly simplified assumption that Newtonian gravitational theory can be applied down to R_S.) How does L_a depend on $\mathcal{M}$? In Sect. 12.3.6, the radiation from the nuclei of active galaxies is estimated to be $\eta \dot{\mathcal{M}} c^2$ where η is the efficiency (12.36). How does this expression relate to L_a?

14.4 The Eddington luminosity L_E (see Exercise 9) gives an upper limit for the rate of continuous matter fall, $\dot{\mathcal{M}}$, since when $L_a \geq L_E$, radiation pressure prevents further accretion. How large is the maximum rate for accretion onto a white dwarf of mass $1\,\mathcal{M}_\odot$, onto a neutron star of $1\,\mathcal{M}_\odot$, and onto a galactic nucleus (Schwarzschild radius R_S) of $10^9\,\mathcal{M}_\odot$?

15. The Microwave Background Radiation and Interstellar Absorption Lines
(Sects. 7.1, 7.2, 10.2, and 13.2)

In the spectrum of the bright O star ζ Oph, the following weak absorption lines of the CN radical have been observed, which originate from the two lowest levels of CN with the rotational quantum numbers $J = 0$ and 1 (see figure):

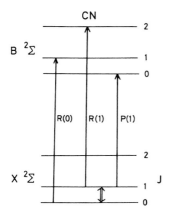

Line	Wavelength λ [nm]	Equivalent Width W_λ [pm]	Relative f-Values
$R(0)$	387.46	0.73	1
$R(1)$	387.39	0.21	2/3
$P(1)$	387.57	0.11	1/3

15.1 Calculate using (7.35) the energy of the level $J = 1$ and the wavelength λ_0 and frequency ν_0 which correspond to the transition $(J = 0) \leftrightarrow (J = 1)$. The rotational constant is $B = 1.8910\,\text{cm}^{-1}$.

15.2 Determine the excitation temperature ("rotational temperature") from the ratio of the occupation numbers $n_1/n_0 = n(J = 1)/n(J = 0)$ according to the Boltzmann formula (4.88). The lines can be assumed to be unsaturated, so that W_λ is proportional to the absorption coefficient κ_λ and thus to the product $nf\lambda_0^2$ (7.32).

15.3 Is the excitation into the level $J = 1$ caused by the microwave background radiation at $T = 2.7\,\text{K}$? Verify the answer by solving the kinetic equilibrium (4.113) for a 2-level atom with $J = 0$ and $J = 1$ (cf. Fig. 7.4), taking the following processes into account:

a) Spontaneous emission with an Einstein coefficient of $A_{10} = 1.2 \cdot 10^{-5}\,\text{s}^{-1}$;
b) Stimulated emission, $B_{10} J_\nu$, and absorption, $B_{01} J_\nu$ (4.122), where J_ν is given by the intensity of the 2.7 K radiation at the frequency ν_0 (Kirchhoff–Planck function); and
c) Excitation and deexcitation by electron impact (4.128), with $C_{01} = \alpha n_e$ and $\alpha = 2.1 \cdot 10^{-12}\,\text{m}^3\,\text{s}^{-1}$.

Assume $n_e = 10^5\,\text{m}^{-3}$ for the electron density in the interstellar medium (H I clouds), corresponding to a degree of ionization of $n_e/n_H = 5 \cdot 10^{-4}$. What would the equilibrium be like for an electron density 10 times higher?

15.4 Compare the radiation field of the microwave background with the general stellar radiation in the Milky Way, when the latter is assumed to be diluted black-body radiation at $T = 10^4$ K and a dilution factor (10.14) of 10^{-15}. How large would the average distance between stars have to be (at a mean stellar radius of $2\,R_\odot$) to make the intensity of the stellar radiation equal to that of the 3 K radiation, (a) at λ_0 and (b) at the maximum of the Planck distribution?

16. Identification of Spectral Lines and the Red Shift (Sects. 7.1, 12.3, and 13.1)

The optical spectrum of a radio source exhibits six broad emission lines, ordered in decreasing intensity, at wavelengths of $\lambda = 563.2$, 323.9, 579.2, 503.2, 475.3, and 459.5 nm. Try to identify these lines as Balmer lines from neutral hydrogen. What red shift z or radial velocity is found from the Hubble relation (13.2), and to what distance does this correspond? What would one predict for the wavelengths of $H\alpha$ and $L\alpha$ with this z? Are there plausible identifications for the remaining lines? Is the identification definite, or could the observed lines also be attributed to e.g. the Lyman series?

Hint: The laboratory wavelengths of the hydrogen lines can be found in spectroscopic tables, or (7.2) can be used. The principal quantum number of the lower level is $n = 1$ for the Lyman series and $n = 2$ for the Balmer series. For further identifications, consider the emission lines of other objects, e.g. the Orion Nebula, Fig. 10.7. (Here, we are dealing with 3C 273, the first quasar, discovered by Maarten Schmidt.)

17. The Distribution of Galaxies (Sects. 12.4 and 13.2)

In the galaxy survey carried out by the Center for Astrophysics in Cambridge, Mass., a large structure, spread out over (at least) 7.5 h in right ascension and 36° in declination (the "Great Wall") was found; in it, the density of galaxies is about 5 times higher than their average density in the cosmos (Fig. 12.30). This structure has an average surface velocity of 8700 km s^{-1} with a variation of only about 500 km s^{-1} in the radial direction.

17.1 Calculate the volume and the mass of the "Great Wall" as functions of the Hubble constant H_0 and the density parameter $\Omega = \varrho/\varrho_c$ (13.27). How many galaxies like our Milky Way would correspond to this mass?

17.2 Estimate the volume of the region which is free of galaxies in Fig. 12.30, assuming it to be spherical with its center at $\alpha \simeq 15$ h and $v_{\rm rad} \simeq 7400$ km s^{-1}.

Literature and Sources of Data

General Astronomical Data

An exhaustive collection of data is available from the *Centre de Donées astronomiques de Strasbourg (CDS)* at the URL http://cdsweb.u-strasbg.fr/ on the World Wide Web.

General Remarks Concerning Literature

Information about individual journal articles etc. can be found electronically in the *NASA Astrophysics Data System*, ADS Abstract Service, at the URL http://adswww.harvard.edu/.

This bibliography is limited to the most important handbooks and data collections, journals, and books, which are intended to help the reader to gain a deeper understanding of particular problems and areas of research.

Conference Proceedings

Most topical conferences publish reports of the work presented in the form of conference proceedings. These volumes are in general *not* listed in the bibliography of each chapter in this book; we simply mention the following series:

Rewiews in Modern Astronomy (Astronomische Gesellschaft, Hamburg)

Publications of the International Astronomical Union. Proceedings of Symposia Series (Kluwer Academic Publ., Dordrecht; after 1998: Astronomical Society of the Pacific, San Francisco)

Proceedings of IAU Colloquia (various publishers)

Astronomical Society of the Pacific Conference Series (ASP, San Francisco)

The Astronomy and Astrophysics Review (Springer, Berlin, Heidelberg)

Handbooks, Review Articles, Data and Formula Tables, and Sky Charts

Annual Review of Astronomy and Astrophysics (Annual Reviews, Palo Alto, Calif.)

Mitton, J.: *A Concise Dictionary of Astronomy* (Oxford University Press, Oxford 1991)

Murdin, P. (ed.): *Encyclopedia of Astronomy and Astrophysics* (Institute of Physics, Bristol 2000)

Schaifers, K., Voigt, H. H. (ed.): *Landolt-Börnstein, Zahlenwerte und Funktionen aus Naturwissenschaften und Technik.* Neue Serie, Gruppe VI, Band 2, *Astronomie und Astrophysik* (Springer, Berlin, Heidelberg 1981 [Subvolume a], 1982 [Subvolumes b and c])

Voigt, H. H. (ed.): *Landolt-Börnstein, Numerical Data and Functional Relationships in Science and Technology.* New Series, Group VI, Vol. 3, *Astronomy and Astrophysics* (Springer, Berlin, Heidelberg 1996 [Subvolume b], 1999 [Subvolume c])

Cox, A. N.: *Allen's Astrophysical Quantities* (AIP, Springer, New York, Berlin 2000)

Lang, K. R.: *Astrophysical Formulae*, Vols. 1 and 2, 3rd edn. Astronomy and Astrophysics Library (Springer, Berlin, Heidelberg 1999)

Norton, A. P., Ridpath, I. (ed.): *Norton's Star Atlas and Reference Handbook* (Longman, Harlow 1998)

Tirion, W., Sinnott, R. W.: *Sky Atlas 2000.0* (Cambridge University Press, Cambridge 1999)

Important Journals

a) Popular Journals

Astronomy (Kalmbach, Waukesha)

Sky and Telescope (Sky Publishing Corp., Cambridge, Mass.)

Scientific American (Scientific American Inc., New York)

Sterne und Weltraum (Verlag Sterne und Weltraum, Hüthig, Heidelberg)

b) Professional Journals

The Astronomical Journal (American Institute of Physics, New York)

Astronomy and Astrophysics (Editions de Physique, Les Ulis)

The Astrophysical Journal (University of Chicago Press, Chicago)

Astrophysics and Space Science (Kluwer Academic Publ., Dordrecht)

Astronomy & Geophysics (Institute of Physics Publ., Bristol)

Icarus. International Journal of Solar System Studies (Academic Press Inc., New York)

Monthly Notices of the Royal Astronomical Society (Blackwell Scientific Publications, Oxford)

Nature (Macmillan Journals Ltd., London)

New Astronomy (Elsevier Science, Amsterdam)

Publications of the Astronomical Society of Japan (Universal Academic Press, Tokyo)

Publications of the Astronomical Society of the Pacific (San Francisco, Calif.)

Science (American Association for the Advancement of Science, Washington, D.C.)

Solar Physics (Kluwer Academic Publ., Dordrecht)

Space Science Reviews (Kluwer Academic Publ., Dordrecht)

Introductions to the Whole of Astronomy, Handbooks and General Treatments

Audouze, J., Israel, G. (eds.): *The Cambridge Atlas of Astronomy* (Cambridge University Press, Cambridge 1988)

Harwit, M.: *Astrophysical Concepts.* Astronomy and Astrophysics Library (Springer, Berlin, Heidelberg 1998)

Karttunen, H., Kröger, P., Oja, H., Poutanen, M., Donner, K. J. (eds.): *Fundamental Astronomy* (Springer, Berlin, Heidelberg 1996)

Kippenhahn, R., Steinberg, J. (translator): *100 Billion Suns* (Princeton University Press 1993)

Longair, M. S.: *Theoretical Concepts in Physics* (Cambridge University Press, Cambridge 1984)

Pasachoff, J. M.: *Astronomy: From the Earth to the Universe* (Saunders, Philadelphia 1983)

Roy, A. E., Clarke, D.: *Astronomy (Principles and Practice); Astronomy (Structure of the Universe)* (Hilger, Bristol 1988, 1989)

Sagan, C.: *Cosmos* (Ballantine Books, New York 1993)

Shu, F. H.: *The Physical Universe* (University Science Books, Mill Valley 1982)

Bibliographies of the Individual Chapters

References are listed for groups of thematically related chapters. As a rule, each book appears only once, so that the reader should look also under related topics.

The History of Astronomy
(Historical Introductions to Parts I, II, III and IV)

Ashbrook, J.: *The Astronomical Scrapbook: Skywatchers, Pioneers, and Seekers in Astronomy* (Sky Publishing, Cambridge 1984)

Becker, F.: *Die Geschichte der Astronomie* (Bibliographisches Institut, Mannheim 1980)

Gingerich, O. (ed.): *Astrophysics and Twentieth-Century Astronomy to 1950* (Cambridge University Press, New York 1984)

Hearnshaw, J. B.: *The Analysis of Starlight* (Cambridge University Press, Cambridge 1987)

Hoskin, M. (ed.): *Cambridge Concise History of Astronomy* (Cambridge University Press, Cambridge 1999)

King, H. C.: *The History of the Telescope* (Dover, New York 1979)

Neugebauer, O.: *Astronomy and History* (Springer, Berlin, Heidelberg 1983)

Pannekoek, I.: *A History of Astronomy* (Dover Publications, Sky Publ. Corp. 1989)

Pedersen, O.: *Early Physics and Astronomy* (Cambridge University Press, Cambridge 1993)

Smith, R.: *The Expanding Universe: Astronomy's Great Debate 1900–1931* (Cambridge University Press, Cambridge 1982)

Swerdlow, N. M.: *The Babylonian Theory of the Planets* (Princeton University Press, Princeton 1998)

Verschuur, G. L.: *Interstellar Matters* (Springer, Berlin, Heidelberg 1989)

Thurston, H.: *Early Astronomy* (Springer, Berlin, Heidelberg 1994)

Classical Astronomy (Chap. 2)

Brown, E. W.: *An Introductory Treatise on the Lunar Theory* (Dover, New York 1960)

Green, R.: *Spherical Astronomy* (Cambridge University Press, Cambridge 1985)

Roy, A. E.: *Orbital Motion* (Hilger, Bristol 1982)

Taff, L.: *Celestial Mechanics* (Wiley, New York 1985)

The Physical Structures of the Objects in the Solar System (Chap. 3)

Beatty, J. K., Peterson, C. C., Chaikin, A. L. (eds.): *The New Solar System* (Cambridge University Press, Cambridge 1999)

Black, D. C., Matthews, M. S. (eds.): *Protostars and Planets II* (University of Arizona Press, Tucson 1985)

Bergstrath, J. T., Miner, E. D., Matthews, M. S. (eds.): *Uranus* (University of Arizona Press, Tucson 1991)

Binzel, R. P., Gehrels, T., Matthews, M. S. (eds.): *Asteroids II* (University of Arizona Press, Tucson 1989)

Bougher, S. W., Hunten, D. M., Phillips, R. J. (eds.): *Venus II* (University of Arizona Press, Tucson 1998)

Brandt, J. C., Chapman, R. C.: *Introduction to Comets* (Cambridge University Press, Cambridge 1983)

Bronshten, V.: *Physics of Meteoritic Phenomena* (Reidel, Dordrecht 1983)

Chamberlain, J. W., Hunten, D. M.: *Theory of Planetary Atmospheres: An Introduction to Their Physics and Chemistry*, 2nd edn. (Academic, San Diego, Calif. 1987)

Cole, G. H. A.: *Physics of Planetary Interiors* (Hilger, Bristol 1984)

Crovisier, J., Encrenaz, T.: *Comet Science* (Cambridge University Press, Cambridge 2000)

Cruikshank, D. P. (ed.): *Neptune and Triton* (University of Arizona Press, Tucson 1996)

Dodd, R. T.: *Meteorites* (Cambridge University Press, Cambridge 1982)

Dressler, A. J.: *Physics of the Jovian Magnetosphere* (Cambridge University Press, Cambridge 1983)

Gehrels, T. (ed.): *Asteroids* (University of Arizona Press, Tucson 1979)

Gehrels, T. (ed.): *Protostars and Planets* (University of Arizona Press, Tucson 1979)

Gehrels, T. (ed.): *Jupiter* (University of Arizona Press, Tucson 1976)

Gehrels, T., Matthews, M. (eds.): *Saturn* (University of Arizona Press, Tucson 1984)

Greenberg, R., Brahic, A. (eds.): *Planetary Rings* (University of Arizona Press, Tucson 1984)

Houghton, J. T.: *The Physics of Atmospheres* (Cambridge University Press, Cambridge 1986)

Hubbard, W.: *Planetary Interiors* (Van Nostrand Reinhold, New York 1984)

Hunten, D. M., Colin, L., Dunahue, T. M., Moroz, V. I. (eds.): *Venus* (University of Arizona Press, Tucson 1983)

Kieffer, H. H., Jakosky, B. M., Snyder, C., Matthews, M. S. (eds.): *Mars* (University of Arizona Press, Tucson 1993)

Levy, E., Lunine, J. I. (eds.): *Protostars and Planets III* (University of Arizona Press, Tucson 1993)

Lewis, J., Prinn, R.: *Planets and Their Atmospheres: Origin and Evolution* (Academic, New York 1983)

Mannings, V., Boss, A. P., Russell, S. S. (ed.): *Protostars and Planets IV* (University of Arizona Press, Tucson 2000)

McSween, H. Y., Jr.: *Meteorites and their Parent Planets* (Cambridge University Press, Cambridge 1999)

Melchior, P.: *The Physics of the Earth's Core* (Pergamon, Oxford 1986)

Morrison, D. (ed.): *Satellites of Jupiter* (University of Arizona Press, Tucson 1982)

Stern, S. A, Tholen, D. J. (eds.): *Pluto and Charon* (University of Arizona Press, Tucson 1997)

Weissman, P., McFadden, L., Johnson, T. (eds.): *Encyclopedia of the Solar System* (Academic Press, London 1998)

Whipple, F. L.: *The Mystery of Comets* (Cambridge University Press, Cambridge 1986)

Wilkening, L. L. (ed.): *Comets* (University of Arizona Press, Tucson 1982)

Radiation, Instruments and Observational Techniques (Chaps. 4 and 5)

Burke, B. F., Graham-Smith, F.: *An Introduction to Radio Astronomy* (Cambridge University Press, Cambridge 1996)

Christiansen, W. N. Högbom, J. A.: *Radiotelescopes* (Cambridge University Press, Cambridge 1985)

Hirsh, R.: *Glimpsing an Invisible Universe. The Emergence of X-Ray Astronomy* (Cambridge University Press, Cambridge 1983)

Jaschek, C.: *Data in Astronomy* (Cambridge University Press, Cambridge 1989)

Jones, R., Wykes, C.: *Holographic and Speckle Interferometry* (Cambridge University Press, Cambridge 1983)

Kitchin, C. R.: *Astrophysical Techniques* (Institute of Physics, Bristol 1998)

Kitchin, C. R.: *Telescopes and Techniques* (Springer, Berlin, Heidelberg 1996)

Kleinknecht, K.: *Detectors of Particle Radiation* (Cambridge University Press, Cambridge 1986)

Kraus, J. D.: *Radio Astronomy* (Cygnus-Quasar Books, Powell, Ohio 1986)

Lena, P., Lebrun, F., Mignard, F.: *Observational Astrophysics* (Springer, Berlin, Heidelberg 1998)

Longair, M. S.: *High Energy Astrophysics*, Vol. 1, *Particles, Photons and their Detection*, Vol. 2, *Stars, the Galaxy and the Interstellar Medium* (Cambridge University Press, Cambridge 1992, 1994)

Malin, D., Murdin, P.: *Colours of the Stars* (Cambridge University Press, Cambridge 1984)

Roddier, F. (ed.): *Adaptive Optics in Astronomy* (Cambridge University Press, Cambridge 1999)

Rohlfs, K., Wilson, T. L.: *Tools of Radio Astronomy*. Astronomy and Astrophysics Library (Springer, Berlin, Heidelberg 2000)

Shu, F. H.: *The Physics of Astrophysics*, Vol. I, *Radiation* (Univ. Science Books, Mill Valley 1991)

Sullivan, W. T.: *The Early Years of Radio Astronomy* (Cambridge University Press, Cambridge 1984)

Verschuur, G. L.: *The Invisible Universe Revealed* (Springer, Berlin, Heidelberg 1987)

The Sun and the Stars: Fundamental Properties, Spectra, Atmospheres, and Activity; and Variable Stars (Chaps. 6 and 7)

Arnett, D.: *Supernovae and Nucleosynthesis* (Princeton University Press, Princeton 1996)

Bahcall, J. N.: *Neutrino Astrophysics* (Cambridge University Press, Cambridge 1989)

Bode, M. F., Evans, A. (eds.): *Classical Novae* (Wiley, Chichester 1989)

Böhm-Vitense, E.: *Introduction to Stellar Astrophysics,* Vol. 1, *Basic Stellar Observations and Data,* Vol. 2 *Stellar Atmospheres* (Cambridge University Press, Cambridge 1989)

Bray, R. J., Loughhead, R. E., Durrant, C. J.: *The Solar Granulation* (Cambridge University Press, Cambridge 1984)

Cannon, C. J.: *The Transfer of Spectral Line Radiation* (Cambridge University Press, Cambridge 1985)

Chandrasekhar, S.: *Radiative Transfer* (Dover, New York 1950, 1960)

Cowan, R. D.: *The Theory of Atomic Structure and Spectra* (University of California Press, Berkeley 1981)

Cravens, T. E.: *Physics of Solar System Plasmas* (Cambridge University Press, Cambridge 1997)

Dalgarno, A., Layzer, D.: *Spectroscopy of Astrophysical Plasmas* (Cambridge University Press, Cambridge 1987)

Durrant, C. J.: *The Atmosphere of the Sun* (A. Hilger, Bristol, Philadelphia 1988)

Griem, H. R.: *Principles of Plasma Spectroscopy* (Cambridge University Press, Cambridge 1997)

Gray, D. F.: *The Observation and Analysis of Stellar Photospheres* (Cambridge University Press, Cambridge 1992)

Hoffmeister, C., Richter, G., Wenzel, W.: *Variable Stars* (Springer, Berlin, Heidelberg 1985)

Jaschek, C., Jaschek, M.: *The Classification of Stars* (Cambridge University Press, Cambridge 1987)

Jaschek, C., Jaschek, M.: *The Behavior of Chemical Elements in Stars* (Cambridge University Press, Cambridge 1995)

Jokipii, J. R., Sonett, C. P., Giampapa, M. S. (eds.): *Cosmic Winds and the Heliosphere* (University of Arizona Press, Tucson 1997)

Kaler, J. B.: *Stars and Their Spectra. An Introduction to the Spectral Sequence* (Cambridge University Press, Cambridge 1989)

Kenyon, S. J.: *The Symbiotic Stars* (Cambridge University Press, Cambridge 1986)

Kitchin, C. R.: *Stars, Nebulae and the Interstellar Medium* (Hilger, Bristol 1987)

Lang, K. R.: *Sun, Earth and Sky* (Springer, Berlin, Heidelberg 1995)

Lang, K. R.: *The Sun from Space* (Springer, Berlin, Heidelberg 2000)

Lamers, H. J. G. L. M., Cassinelli, J. P.: *Introduction to Stellar Winds* (Cambridge University Press, Cambridge 1999)

Lyne, A. G., Graham-Smith, F.: *Pulsar Astronomy* (Cambridge University Press, Cambridge 1998)

Mihalas, D.: *Stellar Atmospheres* (Freeman, San Francisco 1978)

Mihalas, D., Mihalas, B. W.: *Foundations of Radiation Hydrodynamics* (Oxford University Press, New York 1984)

Morgan, W. W., Keenan, P. C., Kellman, E.: *An Atlas of Stellar Spectra* (University of Chicago Press, Chicago 1943)

Morgan, W. W., Abt, H. H., Tapscott, J. W.: *Revised MK Atlas for Stars Earlier than the Sun* (Kitt Peak National Observatory, Tucson, Ariz. 1978)

Oxenius, J.: *Kinetic Theory of Particles and Photons.* Springer Ser. Electrophys., Vol. 20 (Springer, Berlin, Heidelberg 1986)

Petit, M.: *Variable Stars* (Wiley, Chichester 1987)

Priest, E., Forbes, T.: *Magnetic Reconnection – MHD Theory and Application* (Cambridge University Press, Cambridge 2000)

Pringle, J. E., Wade, R. A.: *Interacting Binary Stars* (Cambridge University Press, Cambridge 1985)

Reid, I. N., Hawley, S. L.: *New Light on Dark Stars* (Springer, Berlin, Heidelberg & Praxis Publ. Ltd., Chichester 2000)

Schatzman, E. L., Praderie, F.: *The Stars* (Springer, Berlin, Heidelberg 1993)

Schrijver, C. J., Zwaan, C.: *Solar and Stellar Magnetic Activity* (Cambridge University Press, Cambridge 2000)

Seitter, W. C.: *Atlas für Objektiv-Prismenspektren. Bonner Spektralatlas I, II* (Dümmlers, Bonn 1970, 1975)

Shu, F. H.: *The Physics of Astrophysics,* Vol. II, *Gas Dynamics* (Univ. Science Books, Mill Valley 1992)

Stix, M.: *The Sun – An Introduction* (Springer, Berlin, Heidelberg 1991)

Sturrock, P. A. (ed.): *Physics of the Sun;* Vol. 1, *The Solar Interior;* Vol. 2, *The Solar Atmosphere;* Vol. 3, *Astrophysics and Solar-Terrestrial Relations* (Reidel, Dordrecht 1985)

Tandberg-Hanssen, E., Emslie, A. G.: *The Physics of Solar Flares* (Cambridge University Press, Cambridge 1988)

Thorne, A., Litzén, U., Johansson, S.: *Spectrophysics – Principles and Applications* (Springer, Berlin, Heidelberg 1999)

Warner, B.: *Cataclysmic Variable Stars* (Cambridge University Press, Cambridge 1995)

Zirin, H.: *Astrophysics of the Sun* (Cambridge University Press, Cambridge 1988)

Structure and Evolution of the Stars. Theory of General Relativity (Chap. 8)

Audouze, J., Vauclair, S.: *An Introduction to Nuclear Astrophysics* (Reidel, Dordrecht 1980)

Chandrasekhar, S.: *An Introduction to the Study of Stellar Structure* (1939) (Dover, New York 1957)

Clayton, D. D.: *Principles of Stellar Evolution and Nucleosynthesis* (University of Chicago Press, Chicago 1983)

Collins II, G. W.: *The Fundamentals of Stellar Astrophysics* (Freeman 1989)

Cowley, Ch. R.: *An Introduction to Cosmochemistry* (Cambridge University Press, Cambridge 1995)

Cox, J. P.: *Theory of Stellar Pulsation* (Princeton University Press, Princeton 1980)

Eddington, A. S.: *The Internal Constitution of the Stars* (1926) (Dover, New York 1959)

Glendenning, N. K.: *Compact Stars* (Springer, Berlin, Heidelberg 2000)

Hansen, C. J., Kawaler, S. D.: *Stellar Interiors. Physical Principles, Structure, and Evolution* (Springer, Berlin, Heidelberg 1994)

Kippenhahn, R., Weigert, A.: *Stellar Structure and Evolution* (Springer, Berlin, Heidelberg 1994)

Rolfs, C. E., Rodney, W. S.: *Cauldrons in the Cosmos. Nuclear Astrophysics* (The University of Chicago Press, Chicago, London 1988)

Rose, W. K.: *Advanced Stellar Astrophysics* (Cambridge University Press, Cambridge 1998)

Schneider, P., Ehlers, J., Falco, E. E.: *Gravitational Lenses* (Springer, Berlin, Heidelberg 1999)

Schwarzschild, M.: *Structure and Evolution of the Stars* (Princeton University Press, Princeton 1958 und Dover, New York 1965)

Shapiro, S. L., Teukolsky, S. A.: *Black Holes, Dwarfs, and Neutron Stars: The Physics of Compact Objects* (Wiley, New York 1983)

Will, C. M.: *Theory and Experiment in Gravitational Physics* (Cambridge University Press, Cambridge 1985)

Star Clusters. The Structure and Dynamics of the Milky Way (Chaps. 9 and 11)

Binney, J., Merrifield, M.: *Galactic Astronomy* (Princeton University Press, Princeton 1998)

Bok, B. J., Bok, P. F.: *The Milky Way* (Harvard University Press, Cambridge, Mass. 1981)

Chandrasekhar, S.: *Principles of Stellar Dynamics* (Dover, New York 1960)

Mihalas, D., Binney, J.: *Galactic Astronomy – Structure and Kinematics of Galaxies* (Freeman, San Francisco 1981)

Scheffler, H., Elsässer, H.: *Physics of the Galaxy and Interstellar Matter* (Springer, Berlin, Heidelberg 1987)

Spitzer, L., Jr.: *Dynamical Evolution of Globular Clusters* (Princeton University Press, Princeton 1987)

Interstellar Matter and Star Formation (Chap. 10)

Aller, L. H.: *Physics of Thermal Gaseous Nebulae* (Reidel, Dordrecht 1984)

Bohren, C. F., Huffman, D. R.: *Absorption and Scattering of Light by Small Particles* (Wiley, New York 1983)

Dyson, J. E., Williams, D. A.: *The Physics of the Interstellar Medium* (Institute of Physics Publ., Bristol 1997)

Gaisser, Th. K.: *Cosmic Rays and Particle Physics* (Cambridge University Press, Cambridge 1990)

Gurzadyan, G. A.: *The Physics and Dynamics of Planetary Nebulae* (Springer, Berlin, Heidelberg 1997)

Hartmann, L.: *Accretion Processes in Star Formation* (Cambridge University Press, Cambridge 1998)

Kwok, S.: *The Origin and Evolution of Planetary Nebulae* (Cambridge University Press, Cambridge 2000)

Osterbrock, D. E.: *Astrophysics of Gaseous Nebulae and Active Galactic Nuclei* (Univ. Science Books, Mill Valley, Calif. 1989)

Ramana Murthy, P. V., Wolfendale, A.: *Gamma-Ray Astronomy* (Cambridge University Press, Cambridge 1993)

Spitzer, L., Jr.: *Physical Processes in the Interstellar Medium* (Wiley, New York 1998)

Whiltet, D. C. B.: *Dust in the Galactic Environment* (Institute of Physics Publ., Bristol 1992)

Galaxies and Clusters of Galaxies (Chap. 12)

Arp, H.: *Atlas of Peculiar Galaxies* (California Institute of Technology, Pasadena 1966)

Bergh, S. van den: *Galaxies of the Local Group* (Cambridge University Press, Cambridge 2000)

Bertin, G., Lin, C. C.: *Spiral Structure in Galaxies – A Density Wave Theory* (The MIT Press, Cambridge, Mass. 1996)

Bertin, G.: *Dynamics of Galaxies* (Cambridge University Press, Cambridge 2000)

Binney, J., Tremaine, S.: *Galactic Dynamics* (Princeton University Press, Princeton 1987)

Combes, F., Boissé, P., Mazure, A,: *Galaxies and Cosmology* (Springer, Berlin, Heidelberg 1995)

Frank, J.; King, A. P., Raine, D. J.: *Accretion Power in Astrophysics* (Cambridge University Press, Cambridge 1992)

Krolik, J. H.: *Active Galactic Nuclei* (Princeton University Press, Princeton 1999)

Longair, M. S.: *Galaxy Formation* (Springer, Berlin, Heidelberg 1998)

Pagel, B. E. J.: *Nucleosynthesis and Chemical Evolution of Galaxies* (Cambridge University Press, Cambridge 1997)

Robson, I.: *Active Galactic Nuclei* (Wiley, New York 1996)

Sandage, A.: *The Carnegie Atlas of Galaxies* (Carnegie Institution of Washington 1996)

Verschuur, G. L., Kellermann, K. I. (eds.): *Galactic and Extragalactic Radio Astronomy*. Astronomy and Astrophysics Library (Springer, Berlin, Heidelberg 1988)

Cosmology (Chap. 13)

Barrow, J. D., Tipler, F. J.: *The Anthropic Cosmological Principle* (Oxford University Press 1986)

Berry, M.: *Principle of Cosmology and Gravitation* (Cambridge University Press, Cambridge 1989)

Coles, P., Lucchin, F.: *The Origin and Evolution of Cosmic Structure* (Wiley, New York 1995)

Harrison, E. R.: *Cosmology* (Cambridge University Press, Cambridge 2000)

Hawking, S. W.: *A Brief History of Time* (Bantam Doubleday Dell, New York 1998)

Hoyle, F., Burbidge, G., Narlikar, J.V.: *A Different Approach to Cosmology* (Cambridge University Press, Cambridge 2000)

Klapdor-Kleingrothaus, H. V., Zuber, K.: *Particle Astrophysics* (Institute of Physics Publ., Bristol 1999)

Liddle, A., Lyth, D.: *Cosmological Inflation and Large Scale Structure* (Cambridge University Press, Cambridge 2000)

Misner, C. W., Thorne, K. S., Wheeler, J. A.: *Gravitation* (Freeman, San Francisco 1973)

Partridge, R. B.: *3 K: The Cosmic Microwave Background Radiation* (Cambridge University Press, Cambridge 1995)

Peacock, J. A.: *Cosmological Physics* (Cambridge University Press, Cambridge 1999)

Peebles, P. J. E.: *Principles of Physical Cosmology* (Princeton University Press, Princeton 1993)

Rees, M.: *New Perspectives in Astrophysical Cosmology* (Cambridge University Press, Cambridge 2000)

Weinberg, S.: *Gravitation and Cosmology* (Wiley, New York 1972)

Cosmogony of the Solar System (Chap. 14)

Atreya, S. K., Pollack, J. B., Matthews, M. S. (eds.): *Origin and Evolution of Planetary and Satellite Atmospheres* (University of Arizona Press, Tucson 1989)

Dirk, S. J.: *The Biological Universe* (Cambridge University Press, Cambridge 1996)

Dyson F. J.: *Origins of Life* (Cambridge University Press, Cambridge 1999)

Encrenaz, T., Bibring, J.-P.: *The Solar System* (Springer, Berlin, Heidelberg 1995)

Jakosky, B.: *Search for Life on Other Planets* (Cambridge University Press, Cambridge 1998)

Kerridge, J. F. (ed.): *Meteorites and the Early Solar System* (University of Arizona Press, Tucson 1988)

Monod, J.: *Chance and Necessity* (Random House, New York 1972)

Suess, H. E.: *Chemistry of the Solar System. An Elementary Introduction to Cosmochemistry* (Wiley, New York 1987)

Taube, M.: *Evolution of Matter and Energy on a Cosmic and Planetary Scale* (Springer, Berlin, Heidelberg 1985)

Wasson, J.: *Meteorites: Their Record of Early Solar-System History* (Freeman, Oxford 1985)

Woolfson, M. M.: *The Origin and Evolution of the Solar System* (Institute of Physics, Bristol 2000)

Acknowledgements

Figures

Frontispiece S. Laustsen, C. Madsen, R. M. West: *Entdeckungen am Südhimmel. Ein Bildatlas der Europäischen Südsternwarte (ESO)* (Springer-Verlag, Berlin, Heidelberg und Birkhäuser Verlag, Basel, Boston 1987) Fig. 78

Frontispiece to Part I, 3.28 D. Lynch & T. Puckett, The Puckett Observatory

2.3–11, 15 Seydlitz: *Allgemeine Erdkunde*, 5. Teil, 7. Aufl. (Schroedel, Hannover 1961)

2.24 After A. Unsöld: Phys. Bl. **5**, 205 (1964)

2.25 After K.-P. Wenzel, R. G. Marsden, D. E. Page, E. J. Smith: Astron. Astrophys. Suppl. **92**, 207 (1992) Fig. 1

2.26 NASA bzw. D. Mondey (ed.): *The International Encyclopedia of Aviation* (Crown, New York 1977) p. 414

2.27, 3.5 Photo. NASA

2.28 G. Hunt, P. Moore: *Saturn* (Herder, Freiburg 1983) p. 16 (Fig. 2)

3.2 Drawnn after W.-H. Ip, W. I. Axford: In: Landolt-Börnstein, Neue Serie VI/2a (Springer, Berlin, Heidelberg 1981)

3.3 A. Unsöld: *Evolution kosmischer, biologischer und geistiger Strukturen* (Wissenschaftliche Verlagsanstalt, Stuttgart 1983) and J. D. Phillips: Oceanus **17**, 24 (1973/74)

3.4 Photo. Lick Observatory, from Sky Telesc. **26**, 342 (1963)

3.6 E. Brüche, E. Dick: Phys. Bl. **26**, 351 (1970) Fig. 7

3.7 H. Wänke, F. Wlotzka: Universitas **26**, 850 (1971)

3.8 G. E. McGill: Nature **296**, 14 (1982) Fig. 2

3.9, 13, 16, 21 NASA/JPL/Caltech

3.10 NASA: Naturwissenschaften **59**, 395 (1972) Fig. 5

3.11, 14, 15, 18, 20, 24 Photo. NASA/Raumfahrt-Bildarchiv H. W. Köhler, Augsburg

3.12 Sojourner (TM), Mars Rover (TM) and spacecraft design and images © 1996/97 Caltech

3.14 Drawn after V. I. Moroz: Space Sci. Rev. **29**, 3 (1981); A. Seif, D. B. Kirk: J. Geophys. Res. **82**, 4364 (1977), W. B. Hanson, S. Sanatani, D. R. Zuccaro: J. Geophys. Res. **82**, 4351 (1977); J. T. Houghton: *The Physics of Atmospheres* (Cambridge University Press, Cambridge 1977)

3.17 Photo. B. Lyot and H. Camichel, Observatoire Pic du Midi

3.19 Photo. T. Herbst et al.: Sterne Weltraum **33**, 673 (1994)

3.22 Photo. H. Camichel, Observatoire Pic du Midi

3.23 NASA: G. Briggs, F. Taylor: *The Cambridge Photographic Atlas of the Planets* (Cambridge University Press, Cambridge 1982) p. 228

3.25, 26 Photo. Jet Propulsion Laboratory, California Institute of Technology

3.27, 7.41, 12.8 Photo. Hale Observatories

3.29 P. Swings, L. Haser: *Atlas of Representative Cometary Spectra* (Universitaire de Liège, Liège 1956) Plate IV

3.30 W. Gentner: Naturwissenschaften **50**, 192 (1963) Fig. 1

3.31 E. Anders: Acc. Chem. Res. (1968)

3.32 H. Fechtig, C. Leinert, E. Grün: In: Landolt-Börnstein. Neue Serie VI/2a (Springer, Berlin, Heidelberg 1981)

Frontispiece to Part II, 5.14 Photo. European Southern Observatory ESO

II.1 After B. Rossi: In: *Electromagnetic Radiation in Space*, ed. by J. G. Emming (Reidel, Dordrecht 1967) p. 164

4.1–3, 7.6, 15 A. Unsöld: *Physik der Sternatmosphären*, 2nd ed. (Springer, Berlin, Göttingen, Heidelberg 1955) Figs. 1, 2, 4, 51, 180

4.8 M. S. Longair: *High Energy Astrophysics* (Cambridge University Press, Cambridge 1981) Fig. 4.2

5.4 S. W. Burnham: Publ. Yerkes Observatory of the University of Chicago **1** (1900)

5.5 Photo. Mt. Wilson and Palomar Observatories

5.6 *Das Weltall* (Time-Life, München 1964) p. 37

5.9, 6.8 H. N. Russell, R. S. Dugan, J. W. Stewart: *Astronomy II* (Ginn, New York 1927)

5.10 Photo. G. Weigelt and K. Hofmann (1996)

5.11 N. P. Carleton, W. F. Hoffmann: Phys. Today **31**, No. 9, 30 (1978)

5.12, 13, 8.9, 10.1, 8, 12 Photo. European Southern Observatory ESO

5.15, 16 H. Tüg: Sterne Weltraum **16**, 366 (1977)

5.18 E. H. Schröter, E. Wiehr: Sterne Weltraum **24**, 319 (1985)

5.19 J. C. Brandt and the HRS Investigation Definition and Experiment Development Teams: In: *The Space Telescope Observatory*, ed. by D. N. B. Hall, NASA CP-2244 (1982) p. 76

5.20 Photograph from the IUE satellite, image no. SWP 5176, observer: R. Wehrse (1979)

Acknowledgements

5.21 Photo. G. Hutschenreiter; Max-Planck-Institut für Radioastronomie, Bonn

5.22 P. Thaddeus: Phys. Today **35**, No. 11, 36 (1982)

5.23 Photo. D. Fiebig

5.24 M. Garcia-Munoz, G. M. Mason, J. A. Simpson: Astrophys. J. **217**, 859 (1977)

5.25 M. Simon: Private communication; see also J. A. Esposito et. al.: 19th International Cosmic Ray Conference, Vol. 3, ed. by F. C. Jones, J. Adams, G. M. Mason (NASA Conf. Publ. 2376, 1985) p. 278

5.26 M. Gottwald: Sterne Weltraum **22**, 466 (1983)

Frontispiece to Part III, 7.40 Photo. European Southern Observatory ESO

6.1 M. Minnaert, G. F. W. Mulders, J. Houtgast: *Photometric Atlas of the Solar Spectrum* (Schnabel, Kampfert, Helm, Amsterdam 1940) Detail

6.3 H. Neckel: Space Sci. Rev. **38**, 187 (1984) and Sterne Weltraum **23**, 297 (1984)

6.6 H. L. Johnson, W. W. Morgan: Astrophys. J. **114**, 523 (1951)

6.7 Drawn after H. Tüg, N. M. White, G. W. Lockwood: Astron. Astrophys. **61**, 676 (1977); C. Jamar, D. Macau-Hercot, A. Monfils, G. I. Thompson, L. Houziaux, R. Wilson: ESA SR-27 (1976); A. D. Code, M. Meade: Astrophys. J. Suppl. **39**, 195 (1979); R. L. Kurucz: Astrophys. J. Suppl. **40**, 1 (1979)

6.9 H. Jahreiss: In: Landolt-Börnstein, Neue Serie VI/3C (Springer, Berlin, Heidelberg 1999) p. 131

6.10 W. W. Morgan, P. C. Keenan, E. Kellman: *An Atlas of Stellar Spectra* (University of Chicago Press, Chicago 1943) Detail

6.11 Drawn after Th. Schmidt-Kaler: In: Landolt-Börnstein, Neue Serie VI/2b (Springer, Berlin, Heidelberg 1982) p. 1

6.12 W. Becker: In: *Stars and Stellar Systems III* (University of Chicago Press, Chicago 1963), p. 254; Effective temperatures after I. Bues

6.13 R. H. Baker: *Astronomy*, 6th edn. (Van Nostrand, New York 1955)

6.14 Drawn after D. M. Popper: Annu. Rev. Astron. Astrophys. **18**, 115 (1980)

6.15a J. Adam: Private communication (1987)

6.15b C. Hoffmeister, G. Richter, W. Wenzel: *Veränderliche Sterne*, 2nd ed. (Springer, Berlin, Heidelberg 1984) Fig. 124

6.16 R. A. Hulse, F. H. Taylor: Astrophys. J. **195**, L51 (1975)

6.17 M. Mayor, D. Queloz: Nature **378**, 355 (1995) Fig. 4

7.2 After P. W. Merrill: Pap. Mt. Wilson Observ. **IX**, 118 (1965)

7.5, 10, 11 A. Unsöld: Angew. Chem. **76**, 281 (1964)

7.13 R. E. Danielson: Astrophys. J. **134**, 280 (1961)

7.14 R. Howard: In: *Illustrated Glossary for Solar and Solar-Terrestrial Physics*, ed. by A. Bruzek, C. I. Durrant (Reidel, Dordrecht 1977) p. 7

7.16 J. Houtgast: Rech. Astron. Utrecht **13**, 3 (Utrecht 1957)

7.17 C. de Jager: *Handbuch der Physik*, Vol. 52, *Astrophysik III* (Springer, Berlin, Göttingen, Heidelberg 1959) p. 136

7.18 G. van Biesbroeck: In: *The Sun*, Vol. 1, ed. by G. P. Kuiper (University of Chicago Press, Chicago 1953) p. 604

7.19, 22 J. A. Eddy: *A New Sun. The Solar Results from Skylab* (NASA, Washington, D. C. 1979) pp. 97, 162

7.20 D. H. Menzel, J. G. Wolbach: Sky Telesc. **20**, 252 (1960) Fig. 5

7.21 T. Royds: Mon. Not. R. Astron. Soc. **89**, 255 (1929)

7.23 Cape Observatory: Proc. R. Inst. G. B. **38**, No. 175, Plate I (1961)

7.24 L. Golub, M. Herant, K. Kalata, I. Lovas, G. Nystrom, F. Pardo, E. Spiller, J. Wilczynski: Nature **344**, 842 (1990)

7.25 H. P. Palmer, R. D. Davies, M. I. Large: *Radio Astronomy Today* (Manchester University Press, Manchester 1963) p. 19

7.26 J. M. Wilcox: Space Science Lab., Univ. of Calif., Berkeley **12**, 53 (1971) Fig. 2

7.27 Drawn after J. R. Jokipii, B. Thomas: Astrophys. J. **243**, 1115 (1981) Eq. 1

7.28 K. G. Libbrecht, M. F. Woodard: Nature **345**, 779 (1990) Fig. 1

7.29 After R. Kippenhahn, A. Weigert: *Stellar Structure and Evolution* (Springer, Berlin, Heidelberg 1990) Fig. 40.1

7.30 J. Schou & SOI International Rotation Team: In: *New Eyes to See Inside the Sun and Stars*, ed. by F.-L. Deubner et al. (Kluwer, Dordrecht 1998) p. 141

7.31 After H. W. Duerbeck, W. C. Seitter: In: Landolt-Börnstein, Neue Serie VI/2b (Springer, Berlin, Heidelberg 1982) p. 197

7.32 C. Hoffmeister, G. Richter, W. Wenzel: *Veränderliche Sterne*, 2nd ed. (Springer, Berlin, Heidelberg 1984) Fig. 10

7.33 After J. L. Linsky, B. M. Haisch: Astrophys. J. (Lett.) **229**, L27 (1979) Fig. 1

7.34 T. P. Snow: In: *Mass Loss and Evolution of O-Type Stars*, IAU Symp. No. 83, ed. by P. S. Conti, C. W. H. de Loore (Reidel, Dordrecht 1979) p. 65

7.35 E. L. Robinson: Annu. Rev. Astron. Astrophys. **14**, 119 (1976)

7.36 J. Trümper, W. Pietsch, C. Reppin, W. Voges, R. Staubert, E. Kendziorra: Astrophys. J. (Lett.) **219**, L 109 (1978) Fig. 2

7.37 W. H. G. Lewin, P. C. Joss: Nature **270**, 211 (1977) Fig. 2

7.38 G. W. Clark: Sci. Am. **237**, 42 (Oct. 1977)

7.39 J. C. Wheeler: In: *Supernovae*, eds. J. C. Wheeler, T. Piran, S. Weinberg (World Scientific, Singapore 1990) p. 1

7.42 F. R. Harnden Jr., F. D. Seward: Astrophys. J. **283**, 279 (1984)

7.43 I. F. Mirabel, L. F. Rodríguez, B. Cordier, J. Paul, F. Lebrun: Nature **358**, 215 (1992) Fig. 2

7.44 C. Wolf: Sterne Weltraum **37**, 842 (1998) Fig. 3

8.3 E. Meyer-Hofmeister: In: Landolt-Börnstein, Neue Serie VI/2b (Springer, Berlin, Heidelberg 1982) p. 152, Fig. 3

8.5 T. Kirsten: Phys. Bl. **39**, 313 (1983)

8.6 R. Kippenhahn, H. C. Thomas, A. Weigert: Z. Astrophys. **61**, 246 (1965)

8.7a,b E. Meyer-Hofmeister: In: Landolt-Börnstein, Neue Serie VI/2b (Springer, Berlin, Heidelberg 1982) p. 152, and M. Patenaude: Astron. Astrophys. **66**, 225 (1978) Figs. 1, 3

8.8 Drawn after A. V. Sweigart, P. G. Gross: Astrophys. J., Suppl. **36**, 405 (1978); D. Schönberner: Astron. Astrophys. **79**, 108 (1979)

8.10 Drawn after J. R. Wilson, R. Mayle, S. E. Woosley, T. A. Weaver: Proc. 12th Texas Symp. on Relativistic Astrophysics, ed. by M. Livio, G. Shaviv, Ann. N.Y. Acad. Sci. **470**, 267 (1985)

8.12 R. Kippenhahn, K. Kohl, A. Weigert: Z. Astrophys. **66**, 58 (1967)

Frontispiece to Part IV, 12.31 R. E. Williams, B. Blacker, M. Dickinson et al.: Astron. J. **112**, 1335 (1996) Fig. 8

9.1 J. C. Duncan: *Astronomy* (Harper, New York 1950) p. 408

9.2a H. L. Johnson: Astrophys. J. **116**, 646 (1952)

9.2b O. J. Eggen, A. Sandage: Astrophys. J. **158** 672 (1969)

9.3 After A. Sandage, O. J. Eggen: Astrophys. J. **158**, 697 (1969)

9.4 Drawn after P. M. Hejlesen: Astron. Astrophys., Suppl. **39**, 347 (1980) Table 1

9.5 Photo. Mt. Wilson and Palomar Observatories. In: O. Struve: *Astronomie* (de Gruyter, Berlin 1962) S. 326, Fig. 26.3

9.6 P. Guhathakurtha, B. Yanny, D. P. Schneider, J. N. Bahcall: Astron. J. **111**, 267 (1996) Fig. 2

9.7 A. Sandage: Astrophys. J. **162**, 852 (1970) Figs. 13, 14

9.8 A. Sandage: Astrophys. J. **252**, 574 (1982) Fig. 5

10.4 Drawn after B. D. Savage, J. S. Mathis: Annu. Rev. Astron. Astrophys. **17**, 73 (1979) Table 2

10.5 D. S. Mathewson, V. L. Ford: Mem. R. Astron. Soc. **74**, 143 (1970)

10.6 After D. C. Morton: Astrophys. J. **197**, 85 (1975)

10.7 Photo. Mt. Wilson and Palomar Observatories: In: P. W. Merrill: *Space Chemistry* (University of Michigan Press, Ann Arbor 1963) p. 122

10.9 C. G. Wynn-Williams, E. E. Becklin: Publ. Astron. Soc. Pac. **86**, 5 (1974) Fig. 4

10.10 L. Goldberg, L. H. Aller: *Atoms, Stars and Nebulae* (Blackiston, Philadelphia 1946) p. 182

10.11 C. R. O'Dell, K. D. Handron: Astron. J. **111**, 1630 (1996) Fig. 2

10.13 See also S. S. Murray, G. Fabbiano, A. C. Fabian, A. Epstein, R. Giacconi: Astrophys. J. (Lett.) **234**, L 69 (1979)

10.14 B. Aschenbach, R. Egger, J. Trümper: Nature **373**, 587 (1995)

10.15 R. M. Duin, R. G. Strom: Astron. Astrophys. **39**, 33 (1975)

10.16a R. H. Becker, S. S. Holt, B. W. Smith, N. E. White, E. A. Boldt, R. F. Mushotzky, P. J. Serlemitsos: Astrophys. J. (Lett.) **234**, L 73 (1979) Fig. 2

10.16b S. H. Pravdo, R. H. Becker, E. A. Boldt, S. S. Holt, R. E. Rothschild, P. J. Serlemitsos, J. H. Swank: Astrophys. J. (Lett.) **206**, L 41 (1976) Fig. 1

10.17 C. J. Burrows, J. Krist, J. Hester et al.: Astrophys. J. **452**, 680 (1995) Fig. 4

10.18 R. E. Lingenfelter: Astrophys. Space Sci. **24**, 89 (1973)

10.19 M. Grewing: In: Landolt-Börnstein, Neue Serie VI/2c (Springer, Berlin, Heidelberg 1982) p. 134, Fig. 4 and P. Meyer: *Origin of Cosmic Rays*, IAU Symp. No. 94, ed. by G. Setti, G. Spada, A. W. Wolfendale (Reidel, Dordrecht 1981) p. 7

10.20 H. A. Mayer-Hasselwander, K. Bennett, G. F. Bignami, R. Buccheri, P. A. Caraveo, W. Hermsen, G. Kanbach et al.: Astron. Astrophys. **105**, 164 (1982)

10.21 G. Kanbach: Annual Report 1993 Max-Planck-Institut für extraterrestrische Physik, p. 23 (1994) Fig. 2.18

10.22 I. Iben Jr.: Astrophys. J. **141**, 1010 (1965)

10.23 M. Walker: Astrophys. J., Suppl. **2**, 376 (1956)

10.24a M. L. Kutner, K. D. Tucker, G. Chin, P. Thaddeus: Astrophys. J. **215**, 521 (1977)

10.24b P. Thaddeus: In: *Symposium on the Orion Nebula to Honor Henry Draper*, ed. by A. E. Glassgold, P. J. Huggins, E. L. Schucking, Ann. N.Y. Acad. Sci. **395**, 9 (1982)

10.25 B. Zuckerman: Nature **309**, 403 (1984)

10.26 I. Appenzeller: In: Landolt-Börnstein, Neue Serie VI/2b (Springer, Berlin, Heidelberg 1982) p. 357 [after

Acknowledgements

I. Appenzeller, W. Tscharnuter: Astron. Astrophys. **40**, 397 (1975)]

10.27 R. Mundt: In: *Nearby Molecular Clouds*, ed. by G. Serra, Lecture Notes in Physics, Vol. 217 (Springer, Berlin, Heidelberg 1985) p. 160

10.28 C. R. O'Dell, Zheng Weng, Xihai Hu: Astrophys. J. **410**, 696 (1993) Fig. 3

10.30 R. Wielen, H. Jahrreis, R. Krüger: In: *The Nearby Stars and the Stellar Luminosity Function.* IAU Coll. No. 76, ed. by A. G. D. Philip, A. R. Upgren (L. Davis, Schenectady, N.Y. 1983) p. 163

11.1 G. Westerhout: University of Maryland, USA

11.4 J. H. Oort: In: *Stars and Stellar Systems*, Vol. 5, ed. by A. Blaauw, M. Schmidt (University of Chicago Press, Chicago 1965) p. 484

11.5 H. Scheffler: In: Landolt-Börnstein, Neue Serie VI/2c (Springer, Berlin, Heidelberg 1982) p. 191

11.9 R. Wielen: In: Landolt-Börnstein, Neue Serie VI/2c (Springer, Berlin, Heidelberg 1982) p. 208, Fig. 4

11.10 J. H. Oort: In: *Interstellar Matter in Galaxies*, ed. by L. Woltjer (Benjamin, New York 1962)

11.11 M. A. Gordon, W. B. Burton: Astrophys. J. **208**, 346 (1976) Fig. 4

11.12 O. J. Eggen: R. Obs. Bull. **84**, 114 (1964)

11.13 W. W. Roberts: Astrophys. J. **158**, 132 (1969) Fig. 7

11.14 P. G. Mezger, W. J. Duschl, R. Zylka: The Astron. Astrophys. Rev. **7**, 289 (1996) Fig. 8

11.15 R. D. Ekers, J. T. van Gorkum, U J. Schwarz, W. M. Goss: Astron. Astrophys. **122**, 143 (1983) Figs. 1, 5

11.16 A. Eckart, R. Genzel: Sterne Weltraum **37**, 224 (1998) Fig. 2

12.1, 4, 5 Photo. Mt. Wilson and Palomar Observatories, from A. Sandage: *The Hubble Atlas of Galaxies* (Carnegie Institution of Washington, D.C. 1961)

12.2 E. Hubble: Astrophys. J. **69**, 120 (1929)

12.3 W. L. Freedman, B. F. Madore, J. R. Mould et al.: Nature **371**, 757 (1994) Fig. 1

12.6 A. R. Sandage: Sci. Am. **211**, 39 (Nov. 1964)

12.7 P. B. Stetson, D. A. Vandenberg, R. D. McClure: Publ. Astron. Soc. Pac. **97**, 908 (1985) Fig. 7b

12.9 Drawn after P. Schechter: Astrophys. J. **203**, 297 (1976); J. E. Felten: Astron. J. **82**, 861 (1978)

12.10 A. Bosma: Astron. J. **86**, 1825 (1981) Fig. 3

12.11 A. H. Rots, W. W. Shane: Astron. Astrophys. **31**, 245 (1974) Fig. 1b

12.12 R. Scancisi: Astron. Astrophys. **53**, 159 (1976) Fig. 1

12.13 H. J. Habing, G. Miley, E. Young, B. Baud, N. Bogess, P. E. Clegg, T. de Jong, S. Harris, E. Raimond, M. Rowan-Robinson, B. T. Soifer: Astrophys. J. (Lett.) **278**, L 59 (1984)

12.14 W. B. Sparks, J. V. Wall, D. J. Thorne, P. R. Jorden, I. G. van Breda, P. J. Rudd, H. E. Jorgensen: Mon. Not. R. Astron. Soc. **217**, 87 (1985)

12.15 J. W. Fried, H. Schulz: Astron. Astrophys. **118**, 166 (1983)

12.17 R. D. Ekers, R. Sancisi: Astron. Astrophys. **54**, 973 (1977) Fig. 1

12.18 R. A. Perley, J. W. Dreher, J. J. Cowan: Astrophys. J. (Lett.) **285**, L 35 (1984)

12.19 B. F. C. Cooper, R. M. Price, D. J. Cole: Aust. J. Phys. **18**, 602 (1965)

12.20 J. O. Burns, E. D. Feigelson, E. J. Schreier: Astrophys. J. **273**, 128 (1983) Figs. 2, 11

12.21 B. A. Peterson, A. Savage, D. L. Jauncey, A. E. Wright: Astrophys. J. (Lett.) **260**, L 27 (1982)

12.22 T. Gehren, J. Fried, P. A. Wehniger, S. Wyckoff: Astrophys. J. **278**, 11 (1984)

12.23 T. J. Pearson, S. C. Unwin, M. H. Cohen, R. P. Linfield, A. C. S. Readhead, G. A. Seielstad, R. S. Simon, R. C. Walker: Nature **290**, 365 (1981) Fig. 1

12.25 G. F. Lewis, M. J. Irwin, P. C. Hewett, C. B. Foltz: Mon. Not. R. Astron. Soc. **295**, 573 (1998) Fig. 1

12.26 H. Netzer: In: Landolt-Börnstein: New Series. VI/2c (Springer, Berlin, Heidelberg 1982) p. 300, Fig. 1

12.27 W. N. Colley, J. A. Tyson, E. L. Turner: Astrophys. J. **461**, L 83 (1996) Fig. 1

12.28 C. Jones, E. Mandel, J. Schwarz, W. Forman, S. S. Murray, F. R. Harnden Jr., Astrophys. J. (Lett.) **234**, L 21 (1979) Fig. 2

12.29 K. J. Wellington, G. K. Miley, H. van der Laan: Nature **244**, 502 (1973)

12.30 M. J. Geller, J. P. Huchra: Science **246**, 897 (1989)

12.32 C. L. Bennett, A. Kogut, G. Hinshaw et al.: Astrophys. J. **436**, 423 (1994)

12.33 J. Centrella, A. L. Mellot: Nature **305**, 196 (1983)

12.34 Max-Planck-Gesellschaft, Press Information PRI C 2/SP 2/98(13)

12.35 S. Laustsen, C. Madsen, R. M. West: *Entdeckungen am Südhimmel. Ein Bildatlas der Europäischen Südsternwarte (ESO)* (Springer-Verlag, Berlin, Heidelberg and Birkhäuser Verlag, Basel, Boston 1987) Fig. 92

13.3 H. Kühr: Dissertation, Universität Bonn (1980); H. Kühr, A. Witzel, I. I. K. Pauliny-Toth, U. Nauber: Astron. Astrophys., Suppl. **45**, 367 (1981)

Acknowledgements

14.1 Partly after W. Gentner: Naturwissenschaften **56**, 174 (1969) Fig. 3

14.2 G. E. Morfill, W. Tscharnuter, H. J. Völk: In: *Protostars and Planets II*, ed. by D. C. Black, M. S. Matthews, (University of Arizona Press, Tucson 1985) p. 493

14.4 S. Fittkau: *Kompendium der organischen Chemie* (G. Fischer, Stuttgart 1980) p. 231

14.5 Th. Wieland, G. Pfleiderer (Eds.): *Molekularbiologie* (Umschau, Frankfurt/M. 1969) p. 46, Fig. 3

Tables

3.3 K. Krömmelbein: Private Communication (1967), R. Kraatz: Private communication (1984), W. B. Harland, A. V. Cox, P. G. Llewellyn, C. A. G. Pickton, A. G. Smith, R. Walters: *A Geologic Time Scale* (Cambridge University Press, Cambridge 1982)

7.5, 6 Sun: H. Holweger: Private Compilation (1985); Meteorites: E. Anders, M. Ebihara: Geochim. Cosmochim. Acta **46**, 2263 (1982)

8.1 Z. Abraham, I. Iben Jr.: Astrophys. J. **170**, 157 (1971)

12.3 After S. van den Bergh, Publ. Astron. Soc. Pac. **112**, 529 (2000)

14.2 I.-J. Sackmann, A. I. Boothroyd, K. E. Kraemer: Astrophys. J. **418**, 457 (1993)

Subject Index

A

aberration
– of light   26, 27
– spherical   124
absorption   117
absorption coefficient   107
– atomic   201
– continuous   208, 209, 346
– line-   119, 204, 208
– 21 cm line   337
absorption cross sections
– atomic   117
absorption lines
– identification of   183
– in the solar spectrum   167, 169
– interstellar   335, 336, 354
– of galaxies   416
– theory of   211
abundance distribution of the elements
   see chemical composition
acceleration of gravity   32, 167
– in stellar atmospheres   205
accretion disk   365, 372
– emission of radiation by   298
– in AGN   444
– magnetic fields   373
– physics of   297
– protoplanetary   504
– protostellar   372, 373
– Sgr A*   401
– thin   297
– viscosity of   298
accretion luminosity   298, 445
accretion, of matter   255, 256, 297, 372, 496
achondrites   91
active galactic nuclei (AGN)
   see galactic nucleus, active
active optics   133, 316
adaptive optics   316
adenine   507
age
– of globular clusters   326

– of meteorites   495
– of the Earth   495
– of the Moon   55, 495
– of the Solar System   55
age dating, radioactive
– of Earth's crust   53
– of meteorites   92
air shower   154, 157
Alfvén velocity   222
Allende   see meteorites
Almagest   7, 173
α process   292
altitude   11, 12
altitude formula, barometric   50
AM Herculis stars   255
Am stars   see metallic-line stars
amino acids   506, 507
amplifiers   149
angular momentum
– conservation of   29, 30, 34
– in stellar evolution   369, 370, 496
– of Earth   36
– of Earth-Moon system   37
– of galaxies   see galaxies, angular momentum
– of the galactic halo   389
– of the Solar System   496
– of the Sun   494, 496
angular velocity   28
annihilation radiation   267
antenna temperature   144, 145
anthropic principle   493
anticoincidence circuit   156
aperture ratio, of a telescope   124, 128
aperture synthesis   99, 146
apex, of solar motion   379
aphelion   13, 14
Ap stars   218
Ariel   see Uranus, satellites of
Aries point   12, 14
association   see stellar association
asteroid belt   500
asteroids   43, 44, 69
– near Earth   70
– orbits of   70

– properties of   71
asteroids, individual
– Braille   71
– Ceres   8, 23, 71
– Chiron   70, 87
– "companion" of the Earth   70
– Eros   70, 71
– Gaspra   71
– Hidalgo   87
– Ida   71, 72
– Mathilde   71
– Pallas   71
– Pholus   87
– Trojans   70
– Vesta   71
asteroseismology   166, 242
astronomical unit   22, 26, 27
astronomy, computers in   160
asymptotic giant branch   287, 288, 294, 324, 326, 363
atmosphere of Earth   see Earth's atmosphere
atomic elementary processes   199, 201, 344
atomic spectroscopy   195
azimuth   11

B

baryons   480, 482, 484, 487
beam width   144
Becklin–Neugebauer object   366–368, 370
Bernoulli's equation   237
B^2FH theory   see nucleosynthesis
Big Bang   477
binary star pulsar   see pulsar
binary star systems (binary stars)   365
– accretion   255
– cataclysmic   see variables
– close   189, 190, 251, 266, 267, 291, 294, 295, 308, 322
– contact system   190, 295
– equipotential surface   190
– mass (function)   186, 187

Subject Index

– periods of   257
– primary component   295
– pulsars   190
– semidetached system   190
– spectroscopic   187, 192
– SS 433   259
– velocity curve   187, 191, 193
– visual   186
– X-ray   257, 360
bioastronomy   514
bipolar nebulae, matter flows in   367, 368, 371, 372
BL Lac objects   440, 441, 445
black body   109, 185
– photon density   111
– total radiation density   111
– total radiation of   110
black-body radiation   109, 163
– Universe   *see* microwave background radiation
black hole   165, 258, 267, 270, 290, 302, 308, 309, 316, 401, 444, 446
blasar   441
blue straggler   326
bolometer   152
bolometric correction   178
Boltzmann formula   111, 112, 198
Boltzmann statistics   111
boson   487, 488
Bottlinger diagram   388, 389
Bp stars   *see* Ap stars
bremsstrahlung   121, 362
brightness   *see* magnitude
brightness temperature   144, 425
brown dwarf   192, 194, 299, 364, 504
BY Draconis variables   249

C

calendar   16
Callisto   *see* Jupiter, satellites of
carbon fusion (burning)   277, 289, 291
carbon stars   180, 293
carbonaceous chondrites   *see* chondrites
cascade
– electromagnetic   121, 157
– nucleon   153, 157
cavity radiation   *see* black-body radiation

CCD   *see* detector
celestial equator   11
celestial pole   10, 11, 14
celestial sphere   10–12
center of gravity   34
centrifugal force   28, 29
Cepheids   244–246, 312, 381, 394, 402, 404
Ceres   *see* asteroids, individual
Chandler period   14
Chandrasekhar limiting mass   *see* white dwarfs
Charon   *see* Pluto
chemical composition
– abundance distribution   218
– normal (solar)   199
– of carbonaceous chondrites   91, 217, 499
– of cosmic radiation   356, 358
– of galaxies   415
– of galaxy clusters   452
– of interstellar dust   334
– of interstellar matter   337, 341, 344
– of quasars   445
– of Seyfert nuclei   445
– of stellar atmospheres   199, 205, 216, 218
– of stellar interiors   280
– of the Sun   199, 216, 217, 282
Cherenkov radiation   157
chondrites   91, 499, 511
– carbonaceous   91, 92, 499
chromosphere   167, 225, 227
clock
– atomic   16
cluster variables   *see* RR Lyrae stars
clusters of galaxies   *see* galaxy clusters
CNO triple cycle   275, 276, 282, 285, 292, 293, 296, 363
coal sack, dark cloud   330
collapse, of stars   309, 369
collision processes   116
color excess   331, 332
color index   175, 176, 185
– reddening-independent   331
color–magnitude diagrams (CMD)   321, 322, 365, 380, 392, 408, 409, 418
– evolutionary paths in   286, 321, 364
– explanation of   313, 319, 325

– isochrones   321
color temperature   176
column density   215, 459
comets   23, 87, 511
– breakup   25
– compositions of   89
– evolution of   89
– families   25
– head of   87
– nomenclature   24
– nucleus of   88
– orbits of   24
– periods of   24
– spectrum of   87, 88
– tail of   87, 89
comets, individual
– Cunningham   88
– Hale-Bopp   24, 88
– Halley   24, 44, 88, 89
– Mrkos   87
– Shoemaker-Levy 9   25, 75
Compton scattering   120
– inverse   363, 436, 451
conductivity, electrical   221, 225
conjunction   20, 21
conservation of energy   30
– in planetary orbits   35
constellation   10
convection   220
corona   164, 225, 228
– Fraunhofer (F-)   228
– galactic   *see* Milky Way, halo
– solar   *see* solar corona
corona lines   *see* solar corona
coronograph   226
cosmic background radiation
   see microwave background radiation
cosmic radiation   328, 340, 350, 355, 363, 427
– abundance distribution   358
– dwell time in the Milky Way   359
– electron component of   356, 359
– energy spectrum of   356
– galactic   356
– nucleon component of   356
– origins of   359
– primary   355
– solar   355
cosmochronology   465
cosmological constant   304, 472, 473, 483, 485, 492

Subject Index

cosmological models 468, 470, 474, 484
cosmological postulate 471
– complete 492
cosmology 313, 317
– Newtonian 469
– relativistic 471
cosmos *see* Universe
Coulomb barrier 274
counterglow *see* gegenschein
Crab Nebula 261, 264, 265, 349
– central star 264
– pulsar PSR 0531+21 264, 350
– X-ray emissions from 350
cross section *see* interaction cross section
culmination 11
curve of growth 213, 214
cyclotron emission 255–257
cyclotron frequency 255, 256, 357, 426
cyclotron radius 360
cytosine 507

D

damping profile 202, 214, 459
damping, of a spectral line 202
dark clouds 328, 330
– coal sack 329
– Tau MC1 340
Darwinian principle 506, 510
deceleration parameter, in cosmology 473, 483, 485
declination 11, 12
deflection of light, by the Sun 304
Deimos *see* Mars, satellites of
δ Cephei stars *see* Cepheids
δ Scuti stars 244, 246
density parameter 484
detector
– CCD (charge-coupled device) 136–138
– Cherenkov 153
– digicon 137
– for gravity waves 309
– for radiofrequency range 149
– gamma-ray 155
– infrared 151
– neutrino 283, 285

– optical 135
– particle 152
– photoconductive 151
– photon-counting 137
– plastic 153
– semiconductor 137
– sensitivity of 149, 175
– solid-state 153, 158
– vidicon 137
– X-ray 158
deuterium
– Big Bang 482, 484
– burning 299, 504
diffraction grating 138
diffusion processes *see* Ap stars
dilution factor 336, 344
dipole radiation
– electric 197
– magnetic 197, 337, 345
dipole, magnetic 55
disk, in galaxies *see* galactic disk
dispersion measure 264
distance determination 162, 245, 337, 469
– of galaxies 260, 402, 414
– of stars *see* star parallaxes
– spectrophotometric 380
distance modulus 331, 380, 381
DNA *see* nucleic acid
Doppler effect 26, 27, 103, 185, 203, 220, 229, 240, 338, 438
Doppler profile 203, 214
Doppler width 203, 215, 221
DQ Herculis stars 255
dust *see* interstellar *or* interplanetary dust
dust cloud, circumstellar 177
dwarf (star) 181, 282
dwarf cepheids 244, 246

E

Earth 48, 52
– age of 53
– angular momentum of 36
– density, mean 33
– evolution of 505
– flattening of 52
– geological periods of 512
– interior of 49, 52

– magnetic field of 55, 357
– mass of 33
– meteoritic craters on 90
– orbital velocity of 27
– plate tectonics 56, 57
– radius, equatorial 17, 27
– rotation of 27, 37
– surface temperature 51, 66
Earth's atmosphere 67, 505
– exosphere 68
– extinction of 175
– magnetosphere 68
– optical window 99
– oxygen, enrichment in 511
– ozone layer 67
– protoatmosphere 509, 511
– radiofrequency window 143
– refraction 12
– transmission 99
Earth–Moon system 36, 500
eclipse
– lunar 19
– solar 19
eclipsing variables 187, 296
ecliptic 13, 14
Eddington–Barbier approximation 209, 213
effective temperature
– of planets 48
– of stars 178
– of stellar atmospheres 205
– of the Sun 170
Einstein coefficients 117, 118, 198, 201
Einstein's field equations *see* General relativity theory
electromagnetic radiation 101
electron 487
electron degeneracy 182, 272, 287, 290, 299–301
electron–positron pairs
– and instability, of massive stars 290
– annihilation radiation 122, 267
– creation 120
electron pressure 113, 200, 300
electron temperature 115, 346, 350
electron volt 102, 195
elementary particle physics 317, 487, 488
elements, abundance distribution *see* chemical composition

Subject Index

elements, formation of
  see nucleosynthesis
elongation 21
emission 107
– spontaneous 117
– stimulated 117, 119
emission coefficient 107, 119
emission measure 344
energy
– kinetic 30
– potential 31, 32
energy levels, of atoms 195, 221
– quantum numbers of 116
energy, conservation of
  see conservation of energy
epicycle theory 7, 38
e-process 292
equation of continuity 237
equation of state
– of degenerate electron gas 300
– of ideal gas 50, 237, 272, 287
– of planetary interiors 49
– of stellar interiors 272
equation of time 15
equinox 14
equivalent height 50, 51
equivalent width 212–214
era of the Great Unification 489
escape velocity 35
eukaryotes 511, 512
Euler equation 237, 364
Europa see Jupiter, satellites of
evolution time, of main-sequence stars
  282, 286, 374
excitation 198
– by electron impact 118
– thermal 112, 198, 199
exoplanets 192, 193, 266, 317, 503
expansion age, of OB associations
  319, 366
extinction 107, 333
– Earth's atmosphere 108
– interstellar 178, 328, 330–332, 337,
  378

F

falling stars see meteors
Fermi–Dirac degeneracy see electron
  degeneracy
Fermi mechanism 356, 360

Fischer–Tropsch synthesis, in
  meteorites 500
flare see solar eruption
flare stars 249, 251
flash spectrum see chromosphere
fluorescence
– Bowen 345
flux density
– magnetic 220
– radiation see radiation flux
focal ratio see aperture ratio
formation of galaxies 455
fragmentation, and star formation
  369
Fraunhofer lines see absorption lines
free fall time 279, 364
free–free radiation
  see bremsstrahlung
Friedmann–Lemaître equations
  471–473, 492
Friedmann time 474, 475, 484, 485,
  491
f value see oscillator strength

G

galactic see also Milky Way
galactic center see Milky Way
  (Galaxy), center of
galactic coordinates 377, 378
galactic disk 411, 413, 461, 464, 467
– exponential 463
– formation by collapse 462
galactic nucleus 360, 407, 412,
  433–435
– active (AGN) 425, 442, 443
– unified model for 443
galactic star clusters see open star
  clusters
galaxies see also Milky Way, 313,
  316, 453
– absorption lines 411
– activity of 316, 424, 434, 443
– amorphous 407
– anemic 452
– angular momentum of 461–463
– barred spiral (SB) 406, 407
– BLRG 433, 444
– brightness profiles of 411
– cD 408, 410, 452

– central region 406
– chemical compositions of 415
– classification of 314, 405, 406, 433,
  467
– D 408
– diameter of 410, 467
– disk see galactic disk
– distribution on the celestial sphere
  405
– dwarf 408, 409, 415, 417, 447
– dwarf, blue compact 418, 424, 461
– dwarf, spheroidal 417, 422, 447,
  461, 467
– dynamics of 411, 414
– elliptical (E) 406, 408, 411, 414,
  415, 417, 422, 461, 467
– emission lines 411, 433, 441, 443,
  444
– energy distribution 416
– giant E 409, 414, 415, 433, 462
– halo see galactic halo
– head–tail 451, 452
– in the early Universe 466
– infrared 407, 422–424, 461
– interacting 316, 327, 417, 452, 460,
  461
– irregular 407, 409, 417, 467
– jets in 315, 431–433, 439, 441
– lenticular 406
– luminosity of 408, 410, 421, 467
– Markarian 424, 441
– masses of 413–415
– N 408, 441
– NLRG 433, 444
– nonthermal radio emissions 428,
  443, 444
– normal 402, 428
– nuclei of see galactic nucleus
– radio 315, 429–433, 445, 482
– rotation of 411–414
– Seyfert 441, 442, 445, 446
– spectra of 411, 415, 416
– spectrophotometry of 416
– spheroidal component 413, 461
– spiral (S) 406, 408, 411–413, 417,
  419, 420, 428, 461, 462, 467
– starburst 418, 422, 424, 425, 461
– surface density of 413
– variability of 442
galaxies, evolution of 455, 461, 466,
  467

– chemical 464
– dynamic 461
galaxies, formation of *see* formation of galaxies
galaxies, halo of 415
– dark 414, 449
– radio emissions of 428
galaxies, individual
– Andromeda I–III 447
– Arp 220 (IC 4553) 423–425
– Cen A 157, 431–433
– Cetus 448
– Cyg A 429, 430, 432, 433
– Draco 408, 409, 447
– Dwingeloo 1 405
– Fornax 418
– IC 1613 409
– LMC *see* Magellanic Cloud
– M 31 392, 402–405, 407, 408, 412, 413, 416, 420, 446, 447
– M 32 392, 403, 405, 446, 447
– M 33 407, 416
– M 51 413, 420, 428, 443
– M 81 413, 419, 420, 423, 443
– M 82 407, 423, 424
– M 86 451
– M 87 327, 415, 432, 446, 451
– M 100 404
– M 101 420
– Maffei 1 405
– Mrk 270 442
– Mrk 290 442
– NGC 205 392, 403, 405, 447
– NGC 1052 422
– NGC 1068 441, 442
– NGC 1265 452
– NGC 1275 442
– NGC 2523 407
– NGC 4038/9 460
– NGC 4151 441, 442
– NGC 4261 446
– NGC 4449 416
– NGC 4631 429
– NGC 4697 406
– NGC 5907 420
– NGC 6240 423, 424
– NGC 6251 433
– NGC 6769/71 frontispiece
– Sagittarius 448, 460, 461
– Sculptor 447
– Sextans 448
– SMC *see* Magellanic Cloud
– Tucana 448
– Ursa Minor 447
– II Zw 40 418
– 3C 236 449
galaxies, nuclei of *see* galactic nucleus
galaxy clusters 316, 405, 410, 446, 449
– cD 448
– classification of 448
– Coma 448, 450, 453
– compact 410, 448
– evolution of 452, 461
– formation of 455
– gas in 450–452
– Great Attractor 454
– groups 405, 448
– irregular 448
– Local Group 405, 447, 448, 478
– masses of 449, 451
– regular 448
– Virgo 448, 450–452
– X-ray emissions from 450
galaxy groups *see* galaxy clusters, groups
gamma bursters 165, 267
– distribution of 269
– energy source of 270
– identification of 269
– properties of 268
gamma radiation 262, 436, 437
– blasars 362, 441
– diffuse 362
– extragalactic 362, 442
– from galactic center 267
– from solar eruptions 234
– from X-ray binaries 258, 267
– galactic 360
– origin of 121
– production of 362
– spectral lines 122, 351, 363
gamma-ray astronomy 100, 315, 328
gamma sources, individual
– Cyg X-3 360
– Geminga 266
– GRB 000301C 270
– GRB 970228 270
– GRB 970508 270
– GRB 980425 270
– GRB 990123 270
– GRO J0422+32 267
– GRO J1655-40 267
– GRS 1915+105 267
– SGR 0526-66 268
– SGR 1900+14 268
– SN 1987A 262
– 1E 1740.7-2942 268
Ganymede *see* Jupiter, satellites of
gaseous nebulae, luminous *see* H II regions
G-dwarf problem 465
gegenschein (counterglow) 94
General relativity theory 303, 309, 313, 468
– field equations 303, 471
– gravitational waves 191
– Λ term *see* cosmological constant
– precession of the periastron, of binary star pulsars 191, 305
– precession of the perihelion, of Mercury 191, 302, 304, 305
– tests of 304
genetic code 508
geomagnetic field 55
gf value *see* oscillator strength
giant (star) 181, 199, 218, 287, 364
– red 180, 286, 293, 324, 392
globular (star) clusters 218, 312, 322, 380, 383, 394, 403, 411
– age of 313, 326
– blue 327
– color–magnitude diagrams 324–326
– in other galaxies 326
– luminosity function of 376
– X-ray burster 323
globular clusters, individual
– M 15 266
– 47 Tuc 266
gluon 488
Grand Unified Theory *see* elementary particle physics
granulation, Sun 219, 220
grating spectrograph 138–141, 168
gravitation, law of 31
gravitational constant 32
– time dependence of 493
gravitational energy 278
– in stellar interiors 363, 364
– of galactic nuclei 445
– of the Sun 277

Subject Index

– rate of release 279
– stellar interiors 277
gravitational lens 306, 317, 423, 439, 440, 449, 483, 484
– micro- 270, 307, 440
gravitational radius
  see Schwarzschild radius
gravitational red shift 305
gravitational waves 166, 291, 308
greenhouse effect 51
Grotrian diagram 116, 117, 195, 197
groups of galaxies see galaxy clusters, groups
guanine 507

H

hadron 487
hadron era 489
halo, of galaxies see galaxies, halo of
halo, of the Milky Way see Milky Way, halo
Hayashi line 364
heliopause 239
helioseismology 166, 240, 242
heliosphere 45, 239, 240
helium 344
– in disk population 418
– in halo population 418
– in the Sun 227
– primordial (Big Bang) 418, 477, 481
helium flash 287, 288, 326
helium fusion (burning) 276, 285–288, 296, 326
helium stars 293, 296
Helmholtz–Kelvin time 278, 363, 364, 370
Herbig Ae and Be stars 366
Herbig–Haro objects 372
– HH 46/47 371, 372
Hertzsprung gap 322
Hertzsprung–Russell diagram (HRD) 163, 181, 243, 281, 286–288, 319, 370
– evolutionary paths on 286, 287
high-energy astronomy 315
– instruments for 152
– physical processes in 119
high-velocity stars 218, 381, 387, 388

Holmberg radius 410
horizon 11, 12
horizontal branch 324, 326
hour angle 11
Hubble constant 440, 468, 470, 473, 474, 482, 484, 485
Hubble Deep Field 454, 466, 467
Hubble diagram 482, 483
Hubble relation 468
– generalized 476
Hubble–Sandage scheme
  see galaxies, classification of
Hubble Space Telescope (HST)
  see telescope, Hubble Space
Hubble time 469, 474, 475, 485
hydrodynamics 237
hydrogen fusion (burning) 165, 254, 273, 275, 281, 296, 299, 326, 364, 370
hydrogen, 21 cm line 315, 328, 337, 383, 414, 419, 420
hydrostatic equilibrium 287
– in planetary atmospheres 50
– in stellar atmospheres 207
– in stellar interiors 271, 364
– of planets 48
– stability of 279
H I clouds 328, 337–339
H II regions 328, 334, 342, 346, 366, 381, 403, 418
– compact 344
– giant 344, 398
– infrared lines from 345
– metastable states in 344, 345
– radio emissions from 345
– recombination in 344
– recombination lines from 344
– Sgr A 396, 397, 399
– Sgr B 398
– Sgr B2 398
– spectra of 344, 346
– 30 Dor 262

I

image formation, in a telescope 124
image intensifier 136
impact craters see planets, impact craters
inflation see Universe, inflationary

infrared astronomy 312, 315
instability strip (pulsation) 246, 324
instrumental profile 212
intensity see radiation intensity
interaction cross section 114, 115
– for photoionization 118
interactions, fundamental 488
interferometer
– correlation 130, 179
– Michelson 129, 130, 179
– radio 146
– radio, very long baseline- (VLBI, VLBA) 146, 148
– speckle 130, 179
intergalactic medium 458
interplanetary dust 94
interplanetary magnetic field 239, 356, 357
interstellar absorption bands 334
interstellar dust 330, 337, 341
– formation of 334
– in galaxies 420–422
– polarization by 332–334
– properties of 332
– radiation emitted by 334
– temperature of 334, 421
interstellar matter 312, 328
– chemical composition of 337
– dust see interstellar dust
– dynamics of 387
– energy density of 358
– graphite particles in 334
– hot gas 328, 354, 387, 422
– in galaxies 418, 419
– ionization of 336
– magnetic field of 332, 355
– molecule formation in 340
– molecules in 339, 340
– near the Sun 387
– neutral hydrogen in 337, 338
– polycyclic aromatic hydrocarbons in 335
– silicates in 335
interstellar medium see interstellar matter
Io see Jupiter, satellites of
ionization 198
– by electron impact 112, 118
– photo- see photoionization
– thermal 112, 198, 199
ionization energy 113, 120, 196

Subject Index

ionosphere 67
iron meteorites 90, 91

J

Jansky (Jy, flux unit) 144
Jeans mass 457
Jeans gravitational instability 368, 455, 496
jet
– from galaxies *see* galaxies, jets in
– from protostar 372
– gamma-ray sources 267
Julian century 16
Julian day 16
Jupiter 25, 48, 72, 73, 75
– atmosphere of 72
– crash of comet onto 74
– inner structure of 74
– magnetic field of 74
– rings of 75
– satellites of 44, 75, 77

K

Kepler's laws 7, 14, 33, 34
kinetic (statistical) equilibrium 116
– occupation probabilities 199
Kirchhoff's Law 110
Kirchhoff–Planck function 109, 206
Kleinmann–Low object 366, 367
Kuiper ring (or belt) 25, 86

L

Lagrange point 70, 80, 190, 226, 295
Larmor frequency 221, 357
law of mass action 113
lepton 487
lepton era 490
life 513, 514
– development of 510
– evolution of humans 513
– mass extinction of 512
– material preconditions for 509
– molecular-biological fundamentals of 506
– oldest lifeforms 511
– on other celestial bodies 514
– origin of 509, 510

light deflection, by the Sun 305
light gathering power, of a telescope 129
line profile 214
liner *see* galaxies, emission lines
lithium criterion 299
LMC *see* Magellanic Cloud
Local Standard of Rest (LSR) 379, 382
look-back time, in cosmology 474–476
Lorentz factor 103, 357, 426, 439
Lorentz profile *see* damping profile
Lorentz transformation 103
low-mass stars 370
luminosity
– of galactic nuclei 445
– of galaxies 408
– of stars 179
– of the Sun 170
luminosity class 181, 408
luminosity criteria, for stars 182, 184
luminosity distance, of galaxies 476
luminosity function 373, 375
– initial 374, 375
– of galaxies 409, 410, 448
– of globular clusters 376
– of open star clusters 374
– of stars 373, 378
Luminous Blue Variables (LBV) *see* variables (stars)
lunar *see* Moon
Lyman-α systems 458, 459

M

Magellanic Cloud 407, 417, 421, 460
– chemical composition of 417
– globular clusters in 417
– Large (LMC) 262, 326, 407, 409, 422
– Small (SMC) 326, 407, 422
Magellanic stream 460
magnetohydrodynamics 164
magnitude 162, 174
– absolute 178, 181, 182
– apparent 174
– bolometric 178
main sequence 163, 181, 185, 188, 324

– zero age 280, 281, 286, 320, 321, 325
main-sequence star *see* dwarf (star)
Mars 21, 22, 43, 52, 65
– atmosphere of 69
– interior of 53
– life on 65
– satellites of 66
– surface of 63, 64
– surface temperature 51
maser 119, 366
– OH- 247, 341
mass function, initial (IMF) 375
mass–luminosity relation 188, 189, 280, 326, 375
matter density 485
matter era 491
Maxwell–Boltzmann velocity distribution 112, 199
Maxwell's equations 221
mean free path 114, 115
Mercury 43, 47, 52
– exosphere 66
– interior of 53
– magnetic field of 55, 57
– rotation of 60
– surface of 60
meridian 11, 15
meson 487
metal abundance 323, 324, 393–396, 417, 418, 459, 464, 465
metallic-line stars 218, 248
meteorites 25, 89, 498, 509
– abundance distribution of 91
– Allende 92, 93
– classification of 91, 92
– impact craters from 90
– irradiation age of 92, 500
– isotopic anomalies in 92
– maximum age of 92
– Murchison 92, 93, 499
– origin of 93, 498
meteoroid 89
meteors 25, 90
micrometeorites 94
microturbulence 203, 215, 220
microwave background radiation 317, 360, 451, 477, 479
– anisotropy of 478
– discovery of 477
– fluctuations of 456, 479

Subject Index

– mass density of 478
– optical observation of 477
– photon density of 478
– temperature of 478, 481
Mie theory 333
Milky Way (Galaxy) 316, 406, 446, 447, 464
– abundance gradients in 394, 395
– age of 394, 465
– atomic and molecular hydrogen in 386
– bar structure of 396, 398
– bulge of 380, 381, 390, 395, 411
– central region of 395, 397
– disk of 380, 381, 387, 390
– dust in 384
– dynamics of 312, 381, 388
– element synthesis in 465
– flat disk of 386
– gamma radiation from 361–363
– halo of, gaseous 354
– halo, dark (outer) 314, 380, 390
– halo, globular clusters 380, 381, 389, 390
– halo, radio 428
– interstellar matter in 383, 396, 464
– ionized gas in 386
– magnetic field in 355, 357
– mass distribution in 389
– matter density in the Sun's vicinity 388
– molecular hydrogen in 386
– neutral hydrogen in 385, 396
– nonthermal radio emissions of 428
– nucleus of *see* Milky Way (Galaxy), center of
– populations in 392, 393
– radiofrequency radiation from 426
– rotation curve of 383, 384, 390
– rotation of, differential 338, 382–384
– spheroidal component of 390, 413
– structure of 312, 377, 380, 381, 390
– thick disk 390, 428
– vicinity of the Sun 374, 388
Milky Way (Galaxy), center of 316, 390
– activity 401
– central star cluster 399, 400
– circumnuclear disk 398
– distance from the Sun 383, 395

– dust in 398
– dynamics 399
– gas in 398
– magnetic field 398, 401
– mass 399
– minispiral 399
– radiofrequency radiation 396, 401
– Sgr A* (ultracompact component) 399, 401
– stars in 398, 399
mira stars *see* variables (stars)
MK classification *see* stellar spectra, classification of
MMT (Multiple/Monolithic Mirror Telescope) 132
model atmosphere 209, 215
moldavites 92
molecular clouds 315, 328, 339, 362, 366, 369, 370, 373
– giant 339
– IRC +10°216 340
– Orion (Ori MC1) 340, 363, 366
– Sgr B2 339, 340
– Taurus (Tau MC1) 340
molecular weight, mean 50, 300
molecules 339
– carbon monoxide 328, 339, 366, 367
– hydrogen 339
– isotopes 204, 341
– prebiotic 509
– rotational energy of 204
– spectra of 204
moment of inertia 35
moment, magnetic 55
momentum (linear), conservation of 28, 29
month
– draconitic 18
– sidereal 17, 18
– synodic 17, 18
Moon 52
– abundance of elements on 58, 60
– age of 55, 59
– craters on 57, 58, 501
– dust on 58, 59
– exosphere 66
– formation of 502
– horizontal parallax 17, 19
– ice on 59
– librations 19

– line of nodes 18, 20
– maria 8, 59, 501
– nutation 18
– phases of 17, 18
– radius 17
– rills on 58
– rotation 19
– seismic waves on 53
– stellar occultation by 20
– Terra regions 58
moving clusters 172
multiplet 197
– f values, relative 202
muon 487
Murchison *see* meteorites

N

nadir 11
nebulium 328, 342, 345
neon fusion (burning) 277, 289
Neptune 9, 23, 48, 83
– satellites and ring system of 84, 85
Nereid *see* Neptune, satellites and ring system of
neutrino 277, 291, 484, 490
– detection of 283
– electron 487
– from SN 1987A 262, 291
– mass of 262, 284, 484, 487
– muon 487
– tau 487
neutrino astronomy 282
neutron 293, 481, 487
– Big Bang 482
– in stellar interiors 277
– p-process 293
– r-process 293, 465
– s-process 277, 292–294
neutron star 165, 256, 290, 296, 301, 309, 348
– gravitational energy of 302
– limiting mass of 258, 290, 302
– magnetic field of 256, 257, 264
– mass of 257
– matter in 301
– pulsar 262
– remnant star, from supernova explosion 261, 290
– rotating 264

– structure of   262, 265, 302
– thermal radiation from   302
neutronization   290, 302
Newton's law of gravitation
    see gravitation, law of
Newton's Laws, mechanics   28
noise power   145, 149
non-LTE   see thermodynamic
    equilibrium, deviations from
nova   253, 291, 296, 403
– accretion disk   253–255
– dwarf   255
– energy release   254
– pre-   253
– recurrent   254
– X-ray   258
novae, individual
– Cyg 1992   255
– DQ Her (Nova Her1934)   255
– T Pyx   254
– U Sco   254
nuclear emulsion   153
nuclear reactions, in stellar interiors   218, 271
– energy production   273–275, 289
nuclei of galaxies   see galactic
    nucleus
nucleic acid   507, 508, 510
nucleosynthesis   459, 464, 465, 467
– Big Bang   292, 314, 481, 491
– explosive   293, 294, 296
– in stars   291, 292, 314, 363
nucleotide   507
null geodesic   304
nutation   14

O

OB association   see stellar association
OH/IR stars   247
Ohm's law   221, 222
Olbers' paradox   485
onion-layer structure   289, 294
Oort Cloud   24
Oort constants   382, 383
opacity, Rosseland   206, 272
opposition   21
optical thickness (depth)   108, 212, 213, 346
– for 21 cm line   338

Orion   342, 344, 363, 366–368, 372, 373
Orion Nebula   366, 374
oscillator strength   118, 201

P

parallax   see stellar parallax
parsec   171
particles, energetic   121
partition function   112
P Cyg profile   251, 252, 347
penumbra, of sunspot   222
peptide chain   506
perihelion   13, 14, 23
period–luminosity relation   245, 312, 380, 383
Phanerozoic   512
phase angle, of planets   21
Phobos   see Mars, satellites of
photodisintegration   277, 290
photographic plate   135
photoionization
– X-ray region   119, 120
photoionization model, for galactic nuclei   444
photometry   136
photomultiplier   136
photon   102, 481, 488, 491
– high-energy   119
photon density   106, 485
photon era   490
photosphere   163, 167, 168, 207
photosynthesis   505, 511
pion   153, 362, 487
plages, chromospheric   227
Planck function
    see Kirchhoff–Planck function
Planck time (mass, length)   486
planetary atmospheres   50, 51, 66
– chemical composition   51, 67
– convection   50
– escape velocity   51
– exosphere   51
– hydrostatic equilibrium   50
– magnetosphere   51
– temperature gradient   50
planetary nebulae   142, 288, 346–348
– central star of   288, 294, 346
– condensations in   347

– expansion of   346
planetary orbits   20, 22, 34, 38, 304, 494
– eccentricity   23
– motion, direct and retrograde   20, 38
– nodes   23
– orbital elements   22, 23
– period   22, 23
planetary system   see Solar System
planetary systems, around other stars
    see exoplanets
planetesimals   495, 498, 501, 509
planetoids   see asteroids
planets   46, 66, 504
– albedo of   47, 48
– earthlike   494
– effective temperature   48
– energy balance of   48
– hydrostatic equilibrium   48
– impact craters   66, 495
– major   72, 494
– mass distribution, internal   47, 53
– moments of inertia of   52
– orbits of   see planetary orbits
– physical structure of   47
– plate tectonics   62, 66
– rotation of   46, 494
– small   see asteroids
– structure, inner   48
– surfaces of   60
– transneptunian   9, 86
plasma frequency   234
Pluto   9, 23, 84, 85, 495
polar altitude   11, 12
position, of stars   14
positron   487
p-process   see neutron
prebiotic molecules   see molecules
Precambrium   510
precession   6, 14, 15, 36
pre-main-sequence stars   251, 364, 365
profile function   117, 201, 202
prokaryotes   510, 511, 513
prominences   231, 232
proper motion   172, 173, 379
proportional counter   153, 158
protein   507
protogalaxy   317, 457, 459, 461, 462, 464, 465, 467

Subject Index

proton 481, 487
– decay of 488
proton–proton chain 275, 281, 292
protostar 339, 364–366, 369–373
pulsar 262, 301, 360, 362
– age of 264, 266
– binary star 166, 191, 296, 304, 309
– Crab 157, 264, 265, 350
– Geminga 264, 266
– millisecond 191, 266
– period of 262, 264
– spatial velocities of 266
– Vela 264
pulsation period, of stars 241, 245

Q

quadrature 21
quadrupole radiation, electric 197, 345
quantum vacuum 485
quantum yield 136, 137
quark 487
quark era 489
quasar 315, 317, 433, 434, 443, 446, 483
– absorption lines of 437, 459
– cosmological explanation 439
– emission lines of 434
– luminosity of 435
– mother galaxies of 435
– OVV- 441
– radio emissions of 436
– relativistic intensity enhancement of 438
– spatial density of 435
– spectral energy distribution of 435
– structures of 437
– total energy of 439
– variability of 436, 438, 439
– X-ray emissions of 436
quasars, individual
– BL Lac (2200 + 420) 440
– H 1413 + 117 (cloverleaf) 440
– Mrk 421 441
– Mrk 501 441
– PG 1115 + 08 440
– PKS 2000–330 434, 435
– Q 2237 + 0305 (Einstein cross) 440

– QSO 0957 + 561 (twin quasar) 307, 439, 440
– QSO 0054 + 144 436
– SDSS J1044 − 0125 435
– 3C 273 434, 435, 437, 438, 442

R

radial velocity 26, 163, 172, 245, 338, 379, 383
radiation
– black-body 109
– Cherenkov 153
– flux density 105
– nonthermal 110, 121, 426
– thermal 109, 302, 426, 436
radiation density 106
radiation equilibrium 163, 206, 364
– in stellar atmospheres 206
– in stellar interiors 272
radiation era see photon era
radiation flux (density) 105, 206, 210, 426
radiation intensity 104, 209, 213
– from a layer of matter 108, 109
– mean 105, 116
– relativistic enhancement of 438
radiation power 144
radiation pressure 107, 253
radiation processes 116, 118
radiation temperature 144
radiation theory
– fundamentals of 104, 106
– units 106
radiation transport equation 107
– for stellar interiors 272
– in stellar atmospheres 206
radio astronomy 312, 314, 315
radio pulsar see pulsar
radiotelescope see telescope
rate equations 116, 200
rate of star formation 467
Rayleigh–Jeans law 110, 425
R Coronae Borealis stars 247
reaction rate 114–116
recombination 118, 336
recombination lines 399
red shift 476
– in cosmology 317, 468, 476, 480
– of galaxies 405, 416, 466–468, 483

– of quasars 434, 483
reddening, interstellar 330
reflection nebulae 330
reflector (mirror telescope) 124, 126
refractor (lens telescope) 124, 125
relativistic astrophysics 303
resolving power
– of grating spectrograph 139
– of telescope 128, 144
rest energy 104
rest mass 103
right ascension 11, 12
Robertson–Walker metric 471, 475
Roche surface 190, 295
Roche's stability limit see stability, of satellites
rocket equation 39
rotation measure 355
rotation, Earth 17
r-process see neutron
RR Lyrae stars 244, 246, 324, 380, 394
RS Canum Venaticorum stars 251, 296
Russell–Vogt theorem 280
RV Tauri stars 245
RW Aurigae stars see T Tauri stars

S

Sagittarius
– Sgr A see H II regions
– Sgr A* (ultracompact component) see Milky Way (Galaxy), Sgr A*
– Sgr B2 see H II regions or molecular clouds
Saha equation 113, 198
Saros cycle 20
satellites, artificial 39
– space stations 41
– Sputnik 1 40
– Vostok 1 41
satellites, astronomical
– ASCA 40, 159
– BeppoSAX 268
– Chandra 40, 159, 264, 459
– Compton Gamma Ray Observatory, (CGRO) 40, 155, 156, 234, 266, 267, 269, 361–363

Subject Index

– Copernicus (OAO-3) 40, 141, 177, 335
– COS-B 40, 155, 156, 266, 361
– Cosmic Background Explorer (COBE) 40, 148, 456
– Einstein Observatory (HEAO-2) 40, 158
– European X-Ray Observatory Satellite (EXOSAT) 40, 158
– Extreme Ultraviolet Explorer (EUVE) 40, 157, 160, 387
– Far Ultraviolet Spectroscopic Explorer (FUSE) 40, 142
– Ginga 40
– HALCA 40, 147
– HIPPARCOS 40, 171
– Hubble Space Telescope *see* telescope, Hubble Space (HST)
– Infrared Astronomical Satellite (IRAS) 24, 40, 151, 177, 420–422
– Infrared Space Observatory (ISO) 40, 151, 421, 422, 425
– International Ultraviolet Explorer (IUE) 40, 141, 142, 184, 255, 354
– Orbiting Solar Observatory (OSO) 41, 155
– ROSAT 40, 159, 250, 266, 350, 351, 435, 450
– Solar Maximum Mission 226
– Submillimeter Wave Astronomy Satellite (SWAS) 148
– Transition Region And Coronal Explorer (TRACE) 226
– UHURU 157
– X-Ray Multi-Mirror Mission (XMM) Newton 40, 159
Saturn 43, 48, 77
– atmosphere of 78
– interior of 78
– magnetosphere of 78
– rings of 78–80
– satellites of 78–80
scattering 107, 119
– in stellar atmospheres 208
– Thomson 119, 120
Schmidt camera 138
Schwarzschild metric 304
Schwarzschild radius 304
scintillation detectors 153, 158
S Doradus stars 251
seasons 13

seismic waves 52
selection rules, for dipole radiation 197, 204
sidereal time 11
silicon diode 137
six color system 177
Skylab mission 41, 158, 226
Sloan Digital Sky Survey (SDSS) 435
SMC *see* Magellanic Cloud
SOHO (Solar and Heliospheric Observatory) 226, 240
solar activity 164
– activity centers 223
– cycle of 223
– dynamo theory of 225
solar atmosphere 209
– mechanical energy current 231
– shock waves in 231
solar constant 48, 51, 170
solar corona 167, 228
solar eruption (flare) 233
solar motion 379, 382
solar nebula 494, 496, 509
solar parallax 26, 27, 167
solar spectrum 167, 212
– center–limb variation 163, 168, 169
– continuum 169
– energy distribution of 169, 170
– radiation intensity of 168
– total radiation flux 169
Solar System 371, 494
– accretion disk 496
– age of 55, 495
– cosmogony of 495, 497
– Jupiter–Sun system 192
– proto 496
solar wind 96, 236, 238, 240, 357
source function 108, 200, 210
space probes
– Apollo 42
– Cassini 44
– Clementine 42
– Deep Space 1 71
– Galileo 43, 44, 71, 72, 74, 77
– Giotto 44
– Helios 45
– Luna 42
– Lunar Prospector 43, 58
– Magellan 43, 60, 62
– Mariner 43, 60, 63, 66

– Mars 43
– Mars Climate Orbiter 43
– Mars Global Surveyor 63, 64
– Mars Pathfinder 63–65
– Mars Polar Lander 43
– NEAR Shoemaker 44, 71
– Phobos 43, 66
– Pioneer 43, 72, 78, 80, 239
– Pioneer Venus 43, 61, 69
– Ulysses 40, 45, 94
– Vega 44, 68
– Venera 43, 60, 62
– Viking 43, 63, 64, 66, 69
– Voyager 43, 72, 73, 76–78, 80, 83, 85, 239
space research
– astronomical satellites 40
– Moon 42
– Solar System 43
Space Shuttle 41
spallation 358
spark chamber 153
spatial velocity, of stars 172, 378, 379
special relativity theory 102
speckle interferometry *see* interferometer
spectral analysis, Kirchhoff and Bunsen 162, 163
spectral classification *see* stellar spectra, classification of
spectral type (or class) 180, 183
spectrograph
– échelle 141
– of Hubble Space Telescope 141, 143
– optical 138
spectroheliogram 226
spectrometer 150, 152, 156, 158, 159
– X-ray 158
spectrum synthesis 212
speed of light *see* velocity of light
spheroidal component *see* Milky Way *or* galaxies
spicules, chromospheric 227
spiral arms 381, 406, 452
– 3 kpc arm 386
– density waves in 391
– dynamics of 390
– of the Milky Way 385
s-process *see* neutron
stability, of satellites 80

Subject Index

star clusters, galactic *see* star clusters, open
star clusters, individual
– h and χ Persei 318, 319, 321, 322
– Hyades 173, 318, 319, 324
– Liller 1 259
– M 3 325, 376
– M 11 319
– M 13 = NGC 6205 322, 323
– M 15 323, 418
– M 67 319
– M 92 324, 325, 394
– NGC 188 319–322, 325, 394
– NGC 1866 (LMC) 327
– NGC 2264 365
– NGC 2362 319, 321
– NGC 3680 319
– NGC 6624 258
– NGC 7789 319
– ω Cen 322, 323
– Pleiades 318, 319, 322, 330
– Praesepe 319, 320, 322
– 47 Tuc 322, 325, 326
star clusters, open (galactic) 318, 365, 380
– age of 322
– color–magnitude diagrams of 281, 288, 319–321
– luminosity function of 374
– Praesepe 320
star formation (stellar genesis) 315, 328, 363, 366, 369, 417, 424, 461
– burst of 418, 424
star formation rate 376
star formation regions 313, 316, 319, 344, 366, 367, 420, 424
star parallaxes 162, 171, 172
– stream 172
star populations *see* stellar populations
Stark effect 203
stars
– activity of 248
– apparent magnitudes of 174
– brightness of 248
– center–limb darkening 185
– chromospheres of 164, 248
– collapse of 289
– companion, with substellar mass 192
– coronas of 164, 249–251, 253
– effective temperature of 178, 189

– galactic orbits of 387, 389
– high-velocity 312
– luminosity class of 183
– luminosity of 189, 273
– magnitudes of 178, 179, 189
– mass losses of 249, 252
– masses of 162, 188
– oldest 218
– radius of 179, 189
– rotation of 185, 188, 249, 250
– structure of *see* stellar interior
– surface gravity of 189
– variable *see* variables (stars)
stars, individual
– α Boo (Arcturus) 217
– α Cen 171, 179, 329
– α CMa (Sirius) 187
– α CMa B (Sirius' companion) 182
– α Cru 329
– α^2 CVn 247
– α Cyg (Deneb) 174, 182, 217, 251
– α Gem (Castor) 188
– α Lyr (Vega) 174–177, 209, 210, 217, 504
– α Ori (Betelgeuse) 179, 180, 250
– α PsA 504
– α Sco (Antares) 180
– α UMa 250
– Barnard's Star (BD + 4°3561) 192
– β Cen 329
– β Cet 250
– β Dra 250
– β Gem 217
– β Lyr 296
– β Per (Algol) 162, 187, 296
– β Pic 504
– β Vir 217, 218
– CD –38°245 394
– δ Cep 394
– δ Ori 335, 367
– η Boo 242
– η Car 131, 251, 342
– ε Ori 367
– ε Vir 217, 218
– FG Sge 294
– G 64-12 (Wolf 1492) 394
– Gliese 229 194
– Gliese 876 504
– HD 16141 503
– HD 83443 503
– HD 114762 193

– HD 122563 217, 394
– HD 140283 217, 218, 394
– HD 161817 217, 218
– HD 168443 504
– HD 209458 503
– ι Her 217
– o Cet (Mira) 180, 243, 245
– P Cyg 251, 253
– R Cas 180
– R Dor 179, 180
– R Leo 180
– S And 403
– S Dor 251
– S CrA 370
– θ^1 Ori 366, 373
– θ^1 Ori C 374
– τ Sco 209, 210, 217, 218, 253
– V 1500 Cyg (Nova Cyg 1975) 254
– V 1974 Cyg (Nova Cyg 1992) 254
– van Maanen 2 217
– W Hya 180
– υ And (HR 458) 503
– ζ Oph 336
– ζ Ori 367
– ζ Per 217
– ζ Pup 253
– ζ UMa (Mizar) 162, 187
– 10 Lac 217
– 16 Cyg B 503
– 51 Peg 193, 503
– 61 Cyg 162, 171
starspots 247, 249
statistical weight 111, 113, 198
Stefan–Boltzmann radiation law 110
stellar association 319, 366, 381
stellar atmosphere 164
– absorption coefficients of 207
– chemical composition of 205, 217
– effective temperature of 205, 206
– energy transport in 205
– grey 206
– hydrostatic equilibrium 207
– non-grey 207
– number of absorbing atoms, effective 214
– radiation equilibrium in 206
– radiation pressure in 207
– structure of 205
– suface gravity of 205
– temperature distribution of 206
stellar collapse 279

Subject Index

stellar disk, mean intensity of   106
stellar evoluion
– initial phases   363
stellar evolution   164, 165, 281, 365, 391
– angular momentum in   370
– close binary stars   295
– final stages of   288, 298
– initial phases of   369
– late phases   287, 346
– mass loss   286–288, 347
– onion-layer structure   289
– towards helium burning   285
stellar interior   164
– absorption coefficient in   273
– energy production   276
– energy transport   272, 273
– equation of state   271
– fundamental equations   280
– gravitational energy   277
– hydrostatic equilibrium   271
– ionization   273
– molecular weight in   272
– neutrinos   274
– radiation field   273
– stability   279, 290
– structure of   271, 298
– temperature of   272, 278
stellar parallax   379
stellar photometry   162, 174, 179
stellar populations   312, 392, 393, 415, 416
– disk population   314, 394
– halo population (II)   314, 323, 392, 393, 464
– Population I   246, 321, 392
– Population II   246, 326, 376, 392, 465
– Population III   465
– spiral-arm Population (I)   314, 394
stellar position   173
stellar spectra   162, 163, 168, 210, 249
– analysis of   215
– center–limb variation of   179
– classification of   163, 180, 183, 198
– color temperature from   176
– energy distribution   175, 211
– true continuum in   212
stellar statistics   312, 378
stellar wind   164, 250–252, 347

stone meteorites   91
Strömgren photometry   177, 331
Strömgren sphere   342
stratosphere   67
subdwarf   218, 325
subgiant   184
Sun   158, 209, 250
– acoustic waves   231, 240, 241
– bursts   234
– chemical composition   216, 217, 282
– density, mean   167
– distance modulus of   178
– effective temperature of   170
– energy transport, mechanical   231
– faculae   222
– flux tubes, magnetic   224, 225, 230
– future evolution of   514
– gravitational energy   277
– helium in   282
– hydrogen burning   281
– hydrogen convection zone   220
– isotope ratios in   217
– luminosity of   170, 281
– magnetic fields on   219
– magnitude of   178
– mass of   35, 167
– mean density of   35
– neutrino radiation from   166, 283, 284
– radiation temperature of   170
– radio emissions, non-thermal   234
– radio emissions, thermal   230
– radius of   35, 167
– rotation of   167, 223, 242, 243
– structure of, internal   282
– temperature of interior   272
– X-ray emission of   230
sunspots   222
– cycle   223
– evolution of   223
– groups of   222
– relative number of   223
Sunyaev–Zeldovich effect   451
superclusters of galaxies   316, 452
– Coma   453
– Virgo   453
supergiant (star)   182
supergranulation, Sun   219, 220
superluminal velocity, apparent   267, 434, 437, 438
supernova remnants (SNR)   347, 360

– Cas A   261, 348, 350, 351, 353, 363
– Crab nebula   261, 348, 349, 352
– Cygnus arc   352
– excitation of   347
– expansion of   348
– gamma radiation from   363
– optical   351
– Pup A   351
– radio emissions of   352
– RX J0852.0-4622.   353
– SN 1987A   353
– SN Kepler   351
– SN Tycho   351, 352
– types   352
– Vela   348, 349, 351, 352
– X-ray emission from   352
supernovae   259, 270, 289, 294, 360, 363, 403, 469, 483
– classification of   260
– energy of   291
– energy release of   261
– historical   260, 349
– in other galaxies   260, 261
– light curves of   260, 261, 291
– neutrino emission from   261, 262, 290
– origins of   261
– pre-   262, 294
– SN 1572 Tycho   243, 261
– SN 1604 Kepler   243, 259, 261
– SN 1987A in LMC   261, 262, 290, 353, 363
– spectra of   260
– theoretical model for   290, 291
– types   260
supernovae Ia   485
synchrotron electrons   429
– energy density of   428
– lifetime of   427, 433
– origin of   429
synchrotron radiation   122, 315, 349, 350, 352, 354, 355, 359, 396, 401, 425, 427, 431, 433, 436
– emission coefficient of   427
– theory of   426

T

T association   *see* stellar association
technetium, in giant stars   218

Subject Index

tectites 91
telescope
- achromatic 123
- focusing arrangements 126, 127
- for cosmic radiation 154
- for particles 154, 155
- Galilean 8, 123, 124
- gamma-ray 155, 156
- Gran Telescopio Canarias (GTC) 135
- Hubble Space (HST) 40, 142, 316, 322, 324, 335, 347, 348, 353, 373, 374, 404, 449, 454, 466
- infrared 151
- Keck 134
- Keplerian 123, 124
- large 132
- Large Binocular (LBT) 135
- magnifying power 124
- mm- and sub mm- 148
- mounting of 125, 126
- Multiple/Monolithic Mirror (MMT) 132
- New Technology (NTT) 133, 134
- optical 98, 123, 132, 142
- radio- 143, 145, 236
- Schmidt 128
- Spacewatch- 70, 75
- Subaru 134
- synthesis, radio 146
- tower (solar) 140, 141
- ultraviolet 141, 142, 160
- Very Large (VLT) 133–135, 316
- Wolter 158, 159, 229
- X-ray 158, 229
tensile strength, of satellites 49
thermodynamic equilibrium 109, 111, 198, 201, 489
- deviations from 199, 211, 229, 341, 510
- local (LTE) 111, 209
thermonuclear reactions 274, 275
3α process see helium fusion (burning)
3 kpc arm see spiral arms
3 K radiation see microwave background radiation
thymine 507
tidal friction 188
tides 36, 76
- acceleration 37
- friction of 37, 500
- interaction of galaxies 423, 425, 448, 460, 461
time 15, 16
- ephemeris (ET) 17
- mean solar 15
- sidereal 12, 15
- true solar 15
- universal (UT) 15, 17
Titan see Saturn, satellites of
torque 29, 30, 36
transition layer chromosphere/corona 231
transition, atomic 117
- bound–bound 117
- forbidden 337, 345
- free–bound 117
- free–free 117
Triton see Neptune, satellites and ring system of
tropic 13
tropopause 67
troposphere 67
T Tauri stars 251, 365, 370
Tully–Fisher relation 414
two-color diagram 185, 323, 324, 332, 434

U

UBV system 175, 331
U Geminorum stars 255
Universe
- age of 313, 484
- cosmological models of 491
- dark matter in 457, 458
- density fluctuations in 457
- density parameter of 473, 483–485, 491, 492
- density, critical 485
- development of (evolution) 486, 489, 490
- earliest stage 477
- early stages of 317
- Einstein–de Sitter 474, 475
- eras 489
- evolution of 317
- expansion of 313, 455, 468, 471, 485, 489
- inflationary 485, 491
- initial stages of 455
- matter 474, 475, 478
- matter density of 474, 483
- parameters of 491
- radiation 474, 479, 480
- scale factor 470, 471, 474, 476, 480
- standard model of 473, 485, 489
- steady state 492
- stellar radiation in 486
- structure of 316, 453, 455, 458, 459
uracil 507
Uranus 8, 23, 48, 80
- satellites and ring system of 82, 83
UV ceti stars see flare stars
UV excess 324, 393

V

variables (stars) 243, 244, 468
- cataclysmic 253
- eruptive 253
- Hubble-Sandage- 251
- luminous blue 251
- Mira (long-period) 180, 245, 247
- nomenclature 243
- pulsating 244
- semiregular 245
- spectrum see Ap stars
velocity dispersion
- in elliptical galaxies 414
- in galaxy clusters 449
velocity of light 26, 27, 101
Venus 21, 43, 44, 47, 52, 69
- atmosphere of 68
- interior of 53
- phases 21
- rotation of 60
- surface of 60–62
- surface temperature 51
virial theorem 31, 35, 278, 279
VLA (Very Large Array) 147
VLBI (Very-Long Baseline Interferometer) see interferometer
Voigt profile 203, 204

W

white dwarfs 165, 182, 184, 218, 253, 255, 288, 291, 296, 299, 374
- limiting mass 288, 290, 301

– mass–radius relation for   301
– structure of   301
Wien's displacement rule   110
Wien's Law   110
Wolf–Rayet stars   185, 253, 291, 293, 296, 363
worldview
– Copernican   37
– geocentric   7, 37
– heliocentric   6, 7, 20, 37
– Ptolemaic   37
W Virginis stars   244, 246, 312, 403

X

X-ray astronomy   315
X-ray bursters   258
X-ray emission   354, 415, 422, 431, 436, 442, 452
– energy source through accretion   256
X-ray pulsars   256

X-ray sources   253
– binary star systems   256, 257, 266
– galaxy clusters   450
– quasars   435
– stellar   251, 365
– supernova remnants   349
X-ray sources, individual
– burster 4U 1820-30   258
– Cen X-3   257
– Crab nebula   350
– Cyg X-1   258
– Cyg X-3   157, 258
– GRO J1655-40   267
– GRO J1744-28   259
– GRS 1915+105   267
– Her X-1 (HZ Her)   256–258
– LMC X-1   258
– LMC X-3   258
– M 87   415
– rapid burster MXB 1730-335   259
– SS 433   259

– Vel X-1 = HD 77 581   257
– X-ray nova Muscae   258
– 1E 1740.7-2942   268

Y

year   16
YY Ori stars   370

Z

Z Camelopardalis stars   255
Zeeman effect   221, 223, 355, 357
zenith   11, 12
zero age main sequence (ZAMS)
    see main sequence, zero age
zero geodesic   475
zodiac   13
zodiacal light   94
ZZ Ceti stars   245